16th International Conference on Calorimetry in High Energy Physics

(CALOR 2014)

Journal of Physics: Conference Series
Volume 587

Giessen, Germany
6 - 11 April 2014

ISBN: 978-1-5108-0077-9
ISSN: 1742-6588

Printed from e-media with permission by:

Curran Associates, Inc.
57 Morehouse Lane
Red Hook, NY 12571

Some format issues inherent in the e-media version may also appear in this print version.

Printed by Curran Associates, Inc. (2015)

For permission requests, please contact the Institute of Physics
at the address below.

Institute of Physics
Dirac House, Temple Back
Bristol BS1 6BE UK

Phone: 44 1 17 929 7481
Fax: 44 1 17 920 0979

techtracking@iop.org

Additional copies of this publication are available from:

Curran Associates, Inc.
57 Morehouse Lane
Red Hook, NY 12571 USA
Phone: 845-758-0400
Fax: 845-758-2634
Email: curran@proceedings.com
Web: www.proceedings.com

Table of contents

Volume 587

16th International Conference on Calorimetry in High Energy Physics (CALOR 2014)
6–11 April 2014, Giessen, Germany

Preface

Accepted papers received: 19 January 2015
Published online: 13 February 2015

Operating Calorimeters at LHC

Upgrade of Operating Calorimeters at LHC

Electronics and Readout

Calorimetry in Space and Neutrino Detection

Fields of Application

Preface

The XVIth International Conference on Calorimetry in High Energy Physics - *CALOR 2014* - was held in Giessen, Germany from 6-11 April 2014 at the Science Campus of the University. It was hosted by the *Justus-Liebig-University* and the *HIC for FAIR Helmholtz International Center*. The series of conferences on calorimetry were started in 1990 at Fermilab and are focusing primarily on operating and future calorimeter systems within the Hadron and High-Energy Physics community without neglecting the impact on other fields such as Astrophysics or Medical Imaging.
Confirmed by the impressive list of over 70 oral presentations, 5 posters and over 100 attendees, the field of calorimetry appears alive and attractive. The present volume contains the written contributions of almost all presentations which can be found at http://calor2014.de. Time slots of 15 or 30 minutes including discussion were allocated. The conference was accompanied by a small exhibition of several industrial companies related to the field.
The day before the opening of the scientific program, Richard Wigmans gave an excellent and vivid tutorial on basic aspects on calorimetry meant as an introduction for students and conference attendees new in the field. The opening ceremony was used to give an impression of the present and future status and the scientific program of the new FAIR facility nearby at Darmstadt presented by Klaus Peters from GSI. The conference program of the first day was dedicated to the performance and required future upgrade of the LHC experiments, dominated by ATLAS, CMS and LHCb. The program of the next day contained specific aspects on electronics and readout as well as calorimetry in outer space. Several contributions discussed in detail new concepts for hadron calorimeters within the CALICE collaboration completed by a session on sampling calorimeters.
The next sections were dedicated to operating and future calorimeters at various laboratories and covering a wide range of projectiles and beam energies, such as ELSA at Bonn, BELLE2, BESIII or future projects at JLab, FAIR or RHIC. The last group of contributions was focusing on new concepts including new detector materials or techniques taking into account the general demand on radiation hardness. Damage caused by a large fluence of hadrons appears to become the limiting factor for the detector performance in future collider experiments. The scientific program was completed by a single contribution on applications in medical imaging. Marcella Diemoz gave a personal conference summary underlining again the high quality of the presentations, lively discussions and the wide range of the physics program and the related detector systems.
The significant support by the *Justus-Liebig-University*, *HIC for FAIR* and sponsoring industrial partners allowed to keep the overall costs low and to support the participation of students and young scientists. I personally would like to thank the session chairs, the local organizing committee, the secretaries, technicians and students of the institute for the smooth operation of the conference. Finally, I would like to express my appreciation to all participants who made the conference successful by their high level contributions and discussion distributing their ideas and experiences.

Rainer W. Novotny

Chair of the Conference

International Advisory Committee:

Nural Akchurin, TTU
James E. Brau, Univ. of Oregon
Marcella Diemoz, INFN Roma I
Antonio Ereditato, Univ. of Bern
Tao Hu, IHEP Beijing
Michele Livan, Pavia Univ. & INFN
Steve Magill, ANL
Amelia Maio, LIPP Lisbon
Sven Menke, MPI Munich
Rainer W. Novotny, Univ. of Giessen
Youri Tikhonov, BINP Novosibirsk
Richard Wigmans, TTU
Ren-Yuan Thu, CALTECH

Local Organizing Committee:

Kai-Thomas Brinkmann, JLU Giessen
Peter Drexler, JLU Giessen
Eric Gutz, JLU Giessen
Claudia Höhne, JLU Giessen
Rainer W. Novotny, JLU Giessen
Klaus Peters, GSI Darmstadt & Univ. of Frankfurt
Hans-Georg Zaunick, JLU Giessen

List of Participants

Akchurin, Nural Texas Tech University
Albrecht , Malte Ruhr-Universität Bochum
Apresyan, Artur CALTECH
Auffray, Etiennette CERN
Bendel, Michael Technische Universität München
Biino, Cristina INFN
Bilki, Burak University of Iowa & Argonne National Laboratory
Bongi, Massimo University of Florence & INFN Florence
Bornheim, Adolf CALTECH
Bremer, Daniel JLU Giessen
Brinkmann, Kai-Thomas JLU Giessen
Boumediene, Djamel CNRS & IN2P3 & LPC
Capozza, Luigi University of Mainz

Journal of Physics: Conference Series **587** (2015) 011001 doi:10.1088/1742-6596/587/1/011001

Cario Argos, Fernando IFIC-UV Valencia
Chadeeva, Marina ITEP
Cowden, Christopher Texas Tech University
D´Alessandro, Raffaello Universita di Firenze & INFN Firenze
De Leo, Veronica University of Messina
del Re, Daniele Universita Sapienza di Roma & INFN Roma
Diehl, Stefan JLU Giessen
Diemoz, Marcella INFN Roma
Dormenev, Valera JLU Giessen
Drozdetskiy, Alexey University of Notre Dame
Dziewiecki, Michal WUT Warsaw
Eissner, Tobias JLU Giessen
Ereditato, Antonio University of Bern
Faltova, Jana Charles University Prague
Faraj, Sara Fibercryst
Feldbauer, Florian HMI Mainz
Fiascaris, Maria University of Chicago
Galluci, Giovanni INFN - Sezione di Pisa
Gatto, Corrado INFN - Sez. Napoli
Ghezzi, Alessio Universita di Milano Bicocca & INFN
Gras, Philippe CEA
Guz, Yury IHEP Protvino
Gutz, Eric JLU Giessen
Henriques, Ana CERN
Horn, Tanja Catholic University of America
Hu, Tao IHEP Beijing
Keshelashvili, Irakli University of Basel
Kistenev, Edouard Brookhaven National Laboratory
Korjik, Mikhail RINP Minsk
Krüger, Katja DESY
Künsken, Andreas RWTH Aachen University
Kuske, Till JLU Giessen
Laird, Edward Brown University
Lobanov, Artur DESY
Ma, Hong Brookhaven National Laboratory
Maio, Amelia LIPP
Matsubayashi, Eri Nagoya University
Menke, Sven MPI Munich
Metag, Volker JLU Giessen
Nanova, Mariana JLU Giessen
Neiser, Andreas University of Mainz
Nemecek, Stanislav FZU Praha
Nessi-Tedaldi, Francesca ETH Zürich

Nooren, Gert-Jan Universiteit Utrecht
Novgorodova, Olga TU Dresden & IKTP
Onel, Yasar University of Iowa
Pauwels, Kristof CERN
Perret, Pascal GR-SETAD University Ramon Llull
Pezzullo, Gianantonio University & INFN of Pisa
Pöschl, Roman CNRS & IN2P3 & LAL
Quagli, Tommaso JLU Giessen
Quittnat, Milena ETH Zurich
Rocco, Elena NIKHEF Utrecht University
Rodriguez Pineiro, David HIM Mainz
Rosenbaum, Christopher JLU Giessen
Rzayeva, Sevda Institute of Physics of Azerbaijan National Academy of Science
Sauvan, Jean-Baptiste Laboratoire Lepri nce-Ringuet CNRS & Ecole Polytechnique
Seabra, Luis LIPP
Schnell, Robert JLU Giessen
Schröder, Matthias DESY
Semenov, Pavel IHEP Protvino
Sicking, Eva CERN
Simard, Olivier CNRS
Solans Sanchez, Carlos CERN
Steen, Arnaud IPNL
Surin, Ilya Budker Institute of Nuclear Physics
Szalay, Marco MPI für Physik
Takeshita, Tohru Shinshu University
Thiel, Michaela University of Mainz
Tiemens, Marcel KVI
Tsai, Oleg University of California at L.A.
Urban, Martin HISKP Bonn
Valente, Roserio HIM Mainz
Vilasís-Cardona, Xavier La Salle - Universitat Ramon Llull
Wang, Zhigang IHEP Beijing
Wigmans, Richard Texas Tech University
Woody, Craig Brookhaven National Laboratory
Yang, Fan CALTECH
Zaunick, Hans-Georg JLU Giessen
Zhang, Liyuan CALTECH
Zhu, Ren-Yuan CALTECH
Zvolsky, Milan DESY

Peer review statement

All papers published in this volume of *Journal of Physics: Conference Series* have been peer reviewed through processes administered by the proceedings Editors. Reviews were conducted by expert referees to the professional and scientific standards expected of a proceedings journal published by IOP Publishing.

The CMS Electromagnetic Calorimeter: overview, lessons learned during Run 1 and future projections

Cristina Biino

Istituto Nazionale di Fisica Nucleare, sezione di Torino, v. P. Giuria 1, 10125 Torino, Italy

E-mail: `Cristina.Biino@to.infn.it`

Abstract. The Electromagnetic Calorimeter (ECAL) of the Compact Muon Solenoid (CMS) experiment at the LHC is a hermetic, fine grained, homogeneous calorimeter, containing 75,848 lead tungstate scintillating crystals. We highlight the key role of the ECAL in the discovery and elucidation of the Standard Model Higgs boson during LHC Run I. We discuss, with reference to specific examples from LHC Run I, the challenges of operating a crystal calorimeter at a hadron collider. Particular successes, chiefly in terms of achieving and maintaining the required detector energy resolution in the harsh radiation environment of the LHC, are described. The prospects for LHC Run II (starting in 2015) are discussed, building upon the experience gained from Run I. The high luminosity upgrade of the LHC (HL-LHC) is expected to be operational from about 2025 to 2035 and will provide instantaneous and integrated luminosities of around $5 \times 10^{34}/cm^2/s$ and 3000/fb respectively. We outline the challenges that ECAL will face and motivate the evolution of the detector that is thought to be necessary to maintain its performance throughout LHC and High-Luminosity LHC operation.

1. Introduction

The CMS detector [1] is a multipurpose particle physics experiment at the CERN Large Hadron collider (LHC) optimized to investigate electroweak symmetry breaking via the search of the standard model (SM) Higgs boson.

The performance of the high resolution CMS electromagnetic calorimeter is important because electrons and photons are essential ingredients in at least three of the Higgs boson decays channels: $H \rightarrow \gamma\gamma$, $H \rightarrow ZZ^{(*)} \rightarrow 4e^{\pm}$, $H \rightarrow WW^{(*)} \rightarrow e\nu e\nu$.

The key physics channel driving the design of the CMS electromagnetic calorimeter (ECAL) was $H \rightarrow \gamma\gamma$. This decay mode is the most sensitive for a low mass SM Higgs boson ($m_H < 150$ GeV). The branching ratio is very small, $\simeq$0.002 but the signature is clean: a narrow resonance of two high transverse energy (E_T) photons over a non resonant background of di-photon events [2]. The large irreducicle background originates from the QCD production of two photons while the reducible part comes from events in which at least one of the photons originates from misidentification of jet fragments.

The discovery potential increases with the instrumental invariant mass resolution and background rejection. This translates to a need for efficient photons and electrons identification, high energy and position resolution. The width of the diphoton resonance of a SM Higgs boson is totally dominated by the instrumental invariant mass resolution of the electromagnetic calorimeter. The ECAL design requirements were:

- Excellent energy and position/angle resolution up to $|\eta| <2.5$, to match the tracker coverage.

- Hermeticity, compactness and high granularity
- Fast response (~25 ns) and particle id, energy and isolation measurement at trigger level
- Large dynamic range (5 GeV to 5 TeV) and excellent linearity (at the per-mill level)
- Radiation tolerance (ECAL was designed for 14 TeV and $L = 10^{34} cm^{-2} s^{-1}$, and for a total luminosity of 500/fb)

In the following we discuss the challenges of operating the CMS electromagnetic crystal calorimeter at a hadron collider, in particular in achieving and maintaining the required energy resolution in the harsh radiation environment of the LHC. We summarise the role of ECAL in the discovery of the Higgs boson. We also present the prospect for the LHC Run II starting in 2015 and the challenges that ECAL will face with the High Luminosity (HL) upgrade of LHC, based on the experience gained during Run I.

2. The CMS electromagnetic calorimeter

The CMS electromagnetic calorimeter (ECAL)[3] (see Fig. 1) is a hermetic, homogeneous, fine grained lead tungstate ($PbWO_4$) crystal calorimeter. The choice of an homogeneous medium was made to obtain a better energy resolution by minimizing sampling fluctuations [4]. Very dense crystals offer the potential to achieve the required excellent performance and compactness. The CMS design enabled the electromagnetic calorimeter to fit within the volume of the CMS superconducting solenoid magnet.

The 75,848 crystals are arranged in a central barrel section (EB), with pseudorapidity coverage up to $|\eta| = 1.48$, closed by two endcaps (EE), extending coverage up to $|\eta|$=3.0. Crystals are projective and positioned slightly off-pointing ($\sim 3^0$) relative to the interaction point (IP) to avoid cracks aligned with particle trajectories. The calorimeter has no longitudinal segmentation, the measurement of the photon angle relies on the primary vertex reconstruction from the silicon tracker.

The crystal length in EB is 230 mm (220 mm in EE) corresponding to ~26 (25) radiation lengths. The transverse size of the crystals at the front face is 2.2×2.2 cm^2 in EB (2.86×2.86 cm^2 in EE). The total crystal volume is 11 m^3 and the weight is 92 t. The barrel calorimeter is organized into 36 supermodules each containing 1,700 crystals while the endcaps consist of two dees, with 3,662 crystals each.

A preshower detector (ES), based on lead absorber and silicon strips sensors (4,288 sensors, 137,216 strips, 1.90×61 mm^{2}with x-y view), placed in front of the endcaps at $1.65 < |\eta| < 2.6$, improves the photon-π^0 separation. The total thickness of the ES is ~3 radiation lengths.

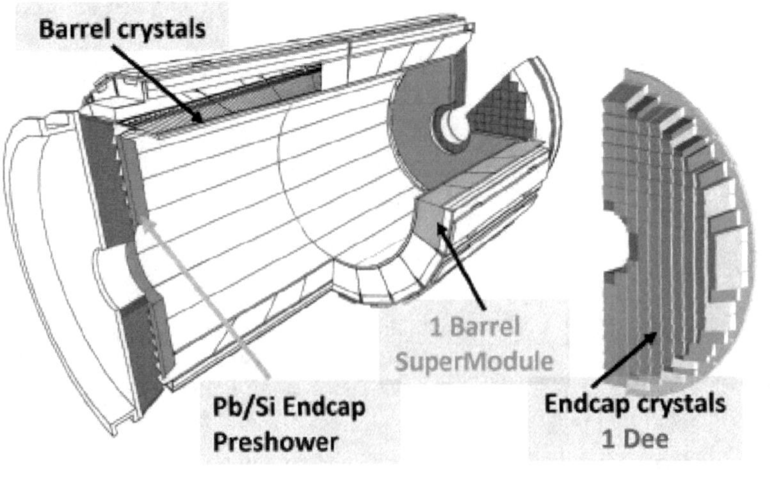

Figure 1. Schematic view of the CMS electromagnetic calorimeter.

2.1. Crystal properties

The main features of $PbWO_4$ scintillating crystals are high density (δ=8.28 g/ cm^3), extremely short radiation length and small Moliére radius (X_0=0.85 cm, R_M=2.19 cm), allowing the realization of a homogeneous compact calorimeter with high granularity. It produces fast signals, 80% of the light is emitted in 25 ns. This is important since the LHC collision rate is 40 MHz. The light emission peak is at ~420 nm. The crystals are transparent to their entire scintillation emission spectrum

The major drawbacks are: the reduced light yield (LY), only ~100 photons per MeV for a 23 cm long crystal, that requires the use of a photodetector readout system with internal gain; a strong light yield dependence on temperature ($\Delta LY/\Delta T$= -2%/^oC at ~18^oC) which imposes a requirement on temperature stability of ±0.05 ^oC in EB.

Most of the crystals were produced (~10,000 crystals/year) in Russia (Bogoroditsk Techno-Chemical plant) with a small contribution from China (Shanghai Institute of Ceramics).

The energy, position and time resolution of arrays of crystals have been throughly studied at beam tests with no magnetic field, no material upstream of the crystals, no radiation damage, and a negligible channel response variation. The obtained energy resolution, for central impact of electrons on a 3 × 3 crystal array, has a stochastic, a noise and a constant term [5]: σ_E/E = 2.8%/$\sqrt{E}$⊕ 0.128 GeV/E ⊕ 0.3% where E is measured in GeV. The constant term is dominated by the longitudinal non-uniformity of light collection. Material upstream of ECAL can result in photon conversion and electron Bremsstrahlung that can both affect all terms in the energy resolution. The CMS goal was to achieve a constant term below 1% [6].

The time resolution is also excellent (<100 ps for E>20 GeV) and has been measured at beam tests using the time difference between adjacent crystals belonging to the same electromagnetic shower. The ECAL time information can be exploited as an alternative method to determine the position of the primary vertex in events with low track multiplicity (see talk by Daniele Del Re at this conference).

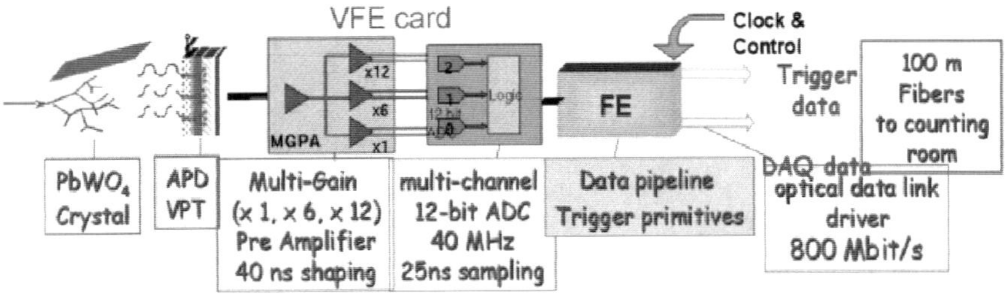

Figure 2. ECAL Front-End electronic chain.

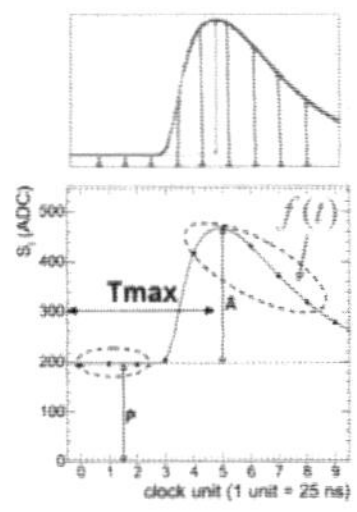

Figure 3. Pulse shape.

2.2. Photodetectors and electronics readout chain

For the purposes of light collection the crystals are equipped in the barrel with Hamamatsu avalanche photodiodes (APD, two for each crystal, 5×5 mm^2 each, 75% Q.E) read in parallel. The gain is set at ~50 and they are insensitive to the 4T magnetic field. In the endcap region the scintillation light from each crystal is readout by vacuum photo-triodes (VPT, 280 mm^2, 20% Q.E.). The gain is set at ~10 and they operate in a magnetic field almost parallel to their axis. The gain of the APDs is sensitive to temperature (-2.3 % ^oC). The temperature dependence of the VPT response is assumed to be negligible relative to the temperature sensitivity of the crystals. Accordingly a less stringent temperature stability requirement of 0.1^oC is assumed for the endcap dees.

To provide the desired resolution over the full energy range of signal events, the readout system measures energies over a wide dynamic range (between 50 MeV and 2 TeV); it is fast to minimize event pile-up, has low power consumption and uses radiation hard components. In order to minimize external noise contributions most of the readout chain is mounted directly on the detector. This has also the advantage of reducing the number of Gigabit optical links to the off-detector readout. The On-detector electronics (see Fig. 2) has been designed to read 5x5 crystals, forming a trigger-tower in the EB (a super-crystal in EE). In the VFE cards (Very Front End) the signals from photodetectors are pre-amplified and shaped by an ASIC Multi Gain Pre-Amplifier chip which consists of three parallel amplification stages with nominal gain 1, 6 and 12. Each of the three analog outputs are digitized in parallel by a multi channel 40 MHz, 12 bit ADC, with an integrated logic that selects the highest not saturated signal.

A time window of 10 samples is readout for every L1 Trigger. From the ten time samples we reconstruct (Fig. 3) the pedestal P, the signal amplitude A, and the time at the maximum T_{max} using a digital filtering technique, weights, fit and ratio methods. We subtract P on an event by event basis. The electronic noise is about 40 MeV/channel in EB.

Trigger primitives are generated from the summed amplitudes of 25 crystals in the FE (Front End) cards and sent to the Off-detector electronics. Electrons and photon candidates are formed at L1 by summing E_T in adjacent trigger towers.

All front-end ASICs were developed in 0.25μm technology, which is intrinsically radiation hard (see also talks by Jean-Baptiste Sauvan and by Philippe Gras at this conference).

3. ECAL operation and environmental stability

CMS is a complex experiment with $\sim$100M readout channels; nevertheless during collisions about 98% of the CMS total channels have been operational. The ECAL detector has been stably and efficiently running with very few single dead channels (99.1% active channels in EB, 98.4% in EE, 96.8% in ES), with little evolution in their number, and causing less than 1% downtime. This is a very good result for a detector that has no intrinsic redundancy. This has been possible due to the continuous work of a relatively small number of scientists, whose constant dedication has allowed ECAL and the CMS experiment to outperform any expectation. A number of bad channels with low voltage supply problems have been fixed in the current long LHC shutdown period.

Fluctuations in temperature affect the LY of the crystals and the APD gain. A cooling system maintains the environmental temperature to be stable within specifications: $\Delta T < 0.02^0 C$ in EB ($< 0.05^0 C$ in EE) [7]. Also the stability of the bias voltages has been within specifications. The contribution of environmental instabilities to the energy resolution constant term has been below 0.1%.

3.1. ECAL trigger

The single-photon L1 Trigger has been continuously operated unprescaled, with a low threshold (E_T=15 GeV) and with an efficiency $> 99\%$ for photons relevant for the $H \rightarrow \gamma\gamma$ analysis (leading photons with $E_T > 33$ GeV). This result has been possible following the implementation of the online rejection of spikes (anomalous isolated energy depositscaused by direct ionization of the APD sensitive volume [8]). If untreated, these would cause the eventual saturation of the bandwidth assigned to electron and photon triggers in CMS. Thanks to the flexibility of the front end electronics, we are able to suppress the spike contribution to the L1 trigger by a coarse analysis of the lateral energy deposit inside each trigger tower. The spike rejection obtained is 96% maintaining a trigger efficiency for electrons and photons of 99%.

The timing cut and a simple shower-shape algorithm is applied on the anomalous signals surviving the L1 Trigger at the next High Level Trigger (HLT).

3.2. ECAL response variation and corrections

The ECAL crystal response varies under irradiation due to the formation of colour centers that absorb the light and reduce the transparency of the $PbWO_4$ (we have no evidence yet of damage to the scintillation mechanism itself). This damage is partially recovered in a few hours at room temperature (thermal annealing of the colour centers). Damage and recovery during LHC cycles are monitored by laser light injected, during the LHC *orbit gaps*, into each crystal, through optical quartz fibres [9]. Blue light (447 nm) is used to while infrared and green light provide complementary measurements at other wavelengths. An optical switch directs light to one half-supermodule or one quarter-Dee in turn. A complete cycle takes ~45 min. The laser light is also injected into PN diodes to follow the laser pulse to pulse variations with an accuracy of 0.1%. ECAL signals are compared and normalized, event by event, to the reference PN diodes. The resulting transparency corrections (see section 4) are ready for prompt reconstruction in less than 48 hours.

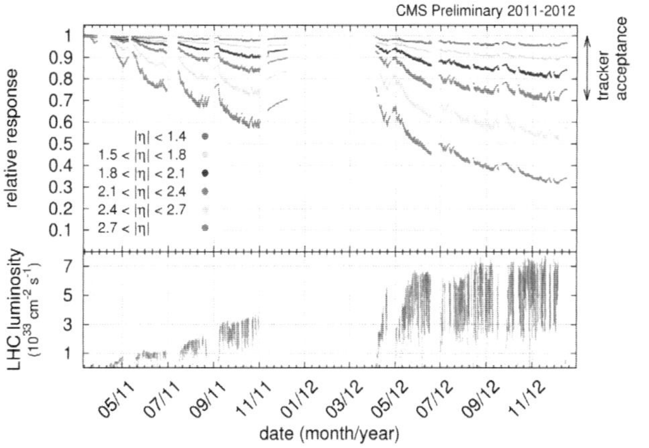

Figure 4. Top: ECAL response loss under irradiation: relative response variation as a function of time measured by the laser monitoring system and averaged in various η intervals, listed in the plot legend.
Bottom: LHC on/off cycles. Damage and recovery during LHC cycles is evident.

Fig. 4 shows the relative response variation measured by the laser monitoring system in 2011 and 2012 (Run I). The response is averaged in various pseudorapidity intervals. The LHC istantaneous luminosity varied from 10^{33}cm^{-2}s^{-1} in April 2011 to 7×10^{33}cm^{-2}s^{-1} at the end of 2012. Damage and recovery during LHC cycles is evident and the crystal transparency correction plays a crucial role in particular for the crystals in the endcaps. The observed response change is from 6% (in EB) to 70% (in EE). We observe steady recovery during the low luminosity Heavy Ions run (November 2012) and in periods without beam. During the present LHC shutdown we are following the transparency recovery taking data whenever possible.

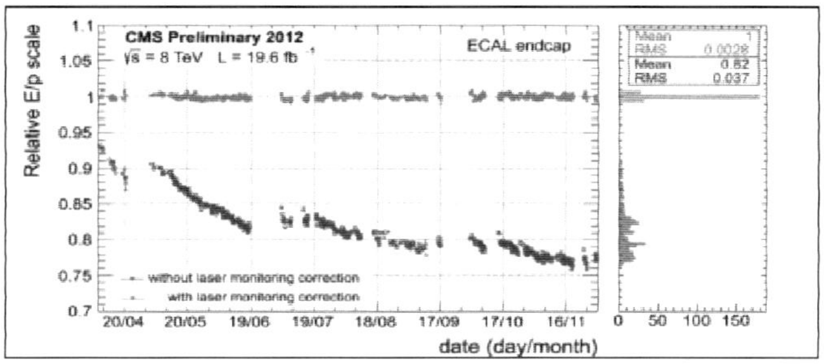

Figure 5. History plot for 2012 data of the ratio of isolated electron energy E, measured in the ECAL endcaps to the electron momentum p, measured in the tracker, before (red dots) and after (green dots) correction of the transparency loss, as measured by the laser monitoring system.

The corrections validity is checked regularly using i.e. electrons from W decays (Fig. 5).

Journal of Physics: Conference Series **587** (2015) 012001 doi:10.1088/1742-6596/587/1/012001

In 2012 we applied weekly transparency corrections at L1 and HLT to account for the response loss observed in the endcaps and to keep constant the trigger energy thresholds. See Fig. 6. In 2015 this will be extended also to the barrel because of the expected increase in LHC luminosity ($1\times10^{34}\text{cm}^{-2}\text{s}^{-1}$).

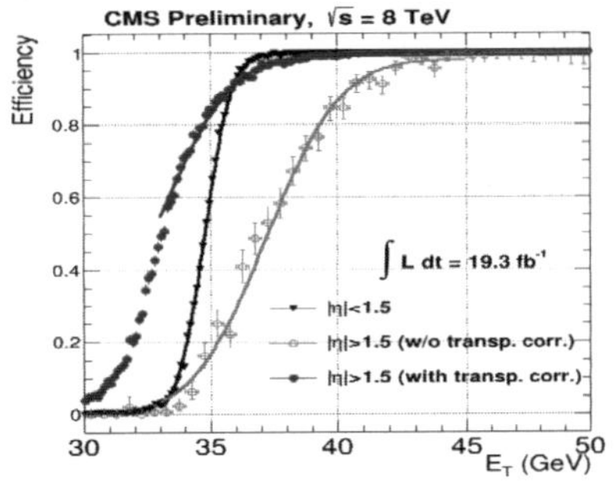

Figure 6. HLT trigger (E_T >33 GeV) efficiency in 2012: in EB in black; in the endcaps before weekly transparency corrections in red, and after in blue.

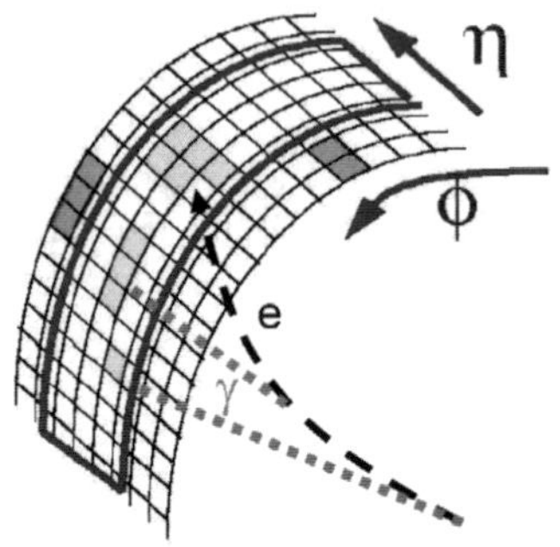

Figure 7. Superclusters of clusters along ϕ (magnetic bending direction). Dynamic clustering is necessary to recover energy radiated upstream of ECAL .

Secondary irradiation effects are: the slow increase of leakage current, and therefore noise, due to bulk damage in the ES sensors [10] and in the APDs and a small reduction of signal due to the conditioning of the VPT photocathodes.

4. Energy reconstruction and energy resolution

In the CMS calorimeter the crystal lateral dimensions are comparable to the Moliére radius. Therefore the electron and photon electromagnetic shower energy is deposited over several crystals especially if the photons/electrons interact in the upstream tracker material.

Clustering and containment corrections are fundamental to reconstruct the particle's original energy. Energy deposits are spread due to secondary emission (Bremsstrahlung and conversions) in the tracker material and distributed along ϕ by the intense CMS magnetic field. The clustering algorithm therefore uses a dynamic window in ϕ to form Superclusters [11] (see Fig. 7).

To obtain the most accurate estimate of electron and photon energy in a Supercluster of crystals we use the following formula:

$$E_{e/\gamma} = F_{e/\gamma}\ G\ \Sigma_i S_i(t)\ C_i\ A_i + E_{ES}$$

where the factors for the absolute energy calibration and related corrections are: G, the absolute energy scale factor (GeV/ADC count) and $F_{e/\gamma}$, the energy containment corrections (depending on the particle kind, geometry, clustering, upstream material). The factors related to the equalization of channel response are : A_i, the single channel amplitude (in ADC counts); C_i, the inter-crystal calibration coefficient; and S_i, the correction for time-dependent response variations (or crystal transparency correction).

After applying the *Supercluster algorithm*, the energy of the e/γ candidate is estimated from the sum over the index i of the signal amplitudes (A_i) of the individual crystals in the Supercluster, weighted with the channel dependent coefficients to correct for time response variation (S_i), to equalize the channel response (C_i), to calibrate the ADC to energy conversion (G) and to correct for imperfect clustering and geometric effects ($F_{e/\gamma}$). For EE clusters the preshower energy is also added (E_{ES}).

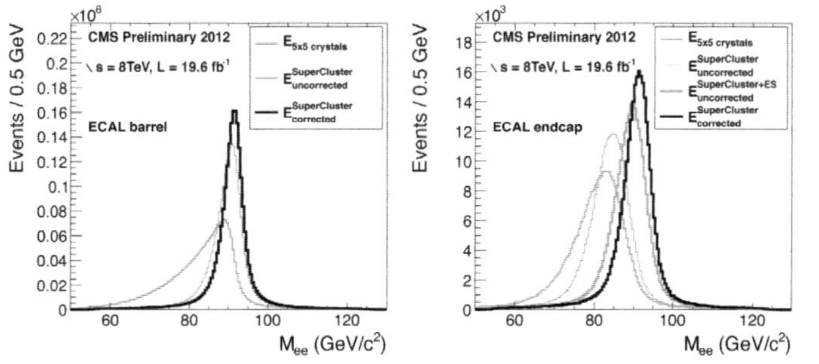

Figure 8. Effect of the dynamical clustering algorithm and energy corrections on the e^+e^- invariant mass resolution in EB (left) and EE (right) for Z^0 candidate events. For EE the effect of adding the preshower energy is also shown.

Fig. 8 shows the impact of the clusterisation process on the $Z \to e^+e^-$ energy scale and resolution from the incorporation of more sophisticated clustering and energy containment correction algorithms in EB and in EE: using fixed 5×5 crystals arrays, then the raw Supercluster energy and then including the F_e correction.

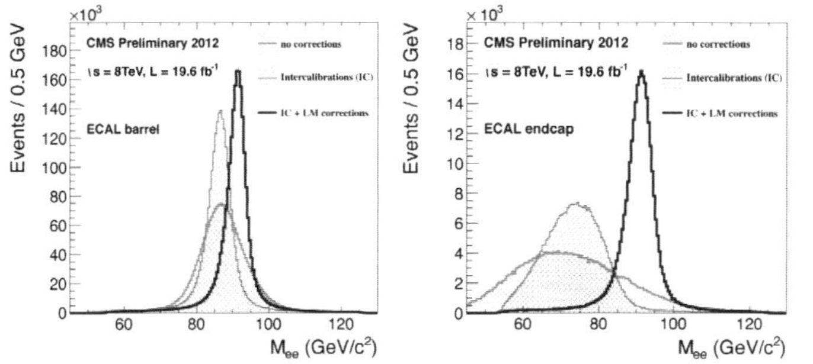

Figure 9. Effect of the intercalibration and transparency corrections on the e^+e^- invariant mass resolution in EB (left) and EE (right) for Z^0 candidate events.

The invariant mass of electron pairs from $Z \to e^+e^-$ at different levels of energy reconstruction is shown In Fig. 9: with both electrons in EB for the left plot (in EE, right plot) without any correction (violet), without time-dependent corrections S_i (red) and with all corrections (blue).

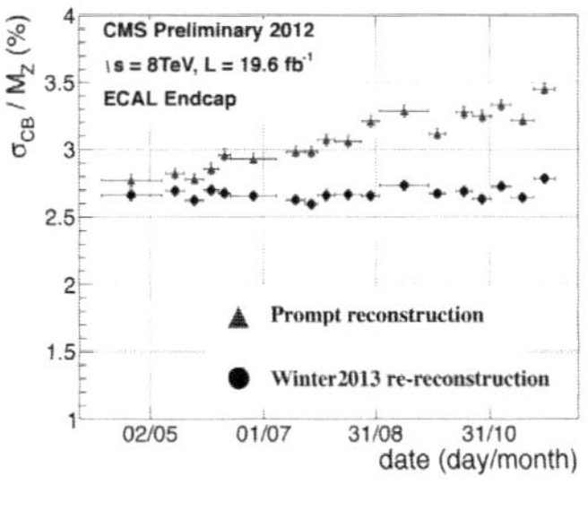

Figure 10. The $Z \to e^+e^-$ invariant mass resolution as a function of time for prompt (in blue) and re-reconstruction (in black). The width of the peak, fitted with a Crystal Ball (CB) function is plotted. Here we use the CB width as a measure of the mass resolution.

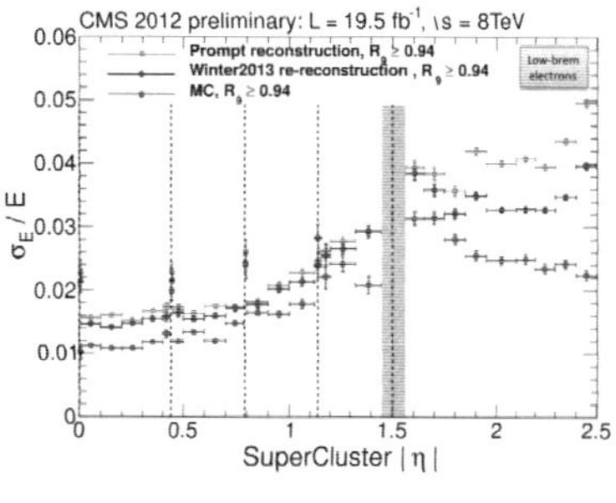

Figure 11. The relative electron resolution (derived from $Z \to e^+e^-$ events) plotted as a function of pseudorapidity. For prompt calibration constants (grey), with optimised calibration constants (blue) and expected from MC simulation (red).

Fig. 10 shows the stability of the resolution measured from $Z \to e^+e^-$ events in the endcaps as a function of time. The re-reconstruction of data, following a calibration based on the full data sample, improves the resolution stability and absolute resolution respect to the prompt (within 48 hours from data taking) reconstruction. The stability is at the per-mill level.

The variable $R_9 = E_{3\times3}/E_{SC}$ defined as the ratio of the energy in a simple 3×3 array of crystals to the energy in the Supercluster is used to identify electrons with little radiation in the tracker or unconverted photons. According to MC ~70% of the photons with $R_9 > 0.94$ are truly unconverted while all the photons with $R_9 < 0.94$ converted upstream of ECAL.

We derive electron energy resolution from the $Z \rightarrow e^+e^-$ peak width. The quantity σ_e/E is extracted, in η/R_9 bins, from an unbinned maximum likelihood fit to the invariant mass distribution of e^+e^- pairs. The result for electrons with R_9 >0.94 is given in Fig. 11. The energy resolution is better than 2% for $|\eta|$ <0.8 and between 2% and 5% elsewhere . The impact on the resolution from the material upstream of ECAL is particularly evident at η >1. Resolution is also degraded near detector cracks between ECAL modules (vertical lines in the plot). The differences between data (in blue) and MC (in red) come from various effects not yet perfectly simulated (such as imperfect description of the upstream material). The amount of material in front of ECAL is shown in Fig. 12. Fig. 13 shows the energy corrections as a function of η for different intervals of R_9.

The resolution of the MC used for the $H \rightarrow \gamma\gamma$ analysis has been tuned to match the data for different R_9 categories by adding an extra smearing term.

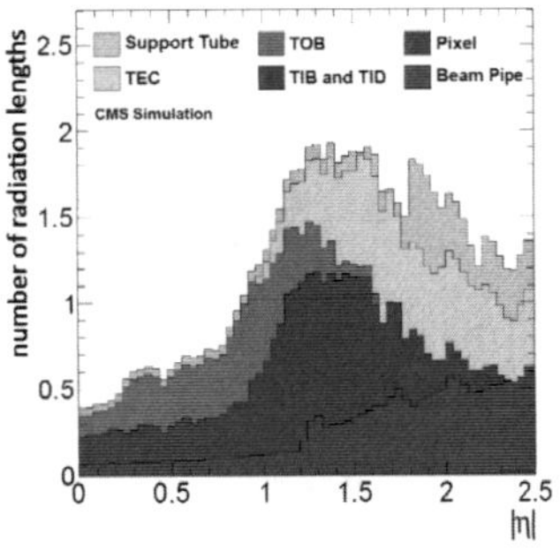

Figure 12. Material in front of ECAL as a function of η.

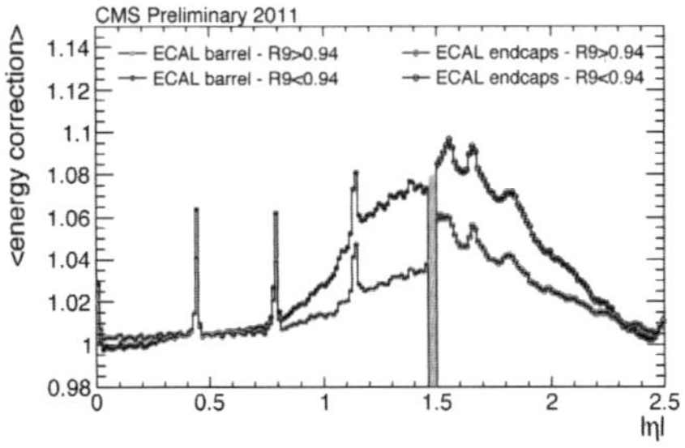

Figure 13. $W \rightarrow e\nu$ electron candidates. Particle energy corrections as a function of η in different intervals of R_9.

Calibrations procedure exploits different samples of events to intercalibrate, verify and tune monitoring and algorithmic corrections. For more details see talk by Alessio Ghezzi at this conference.

5. Role of ECAL in the Higgs discovery

On 4th July 2012 the CMS and the ATLAS collaboration announced the observation of a new particle, consistent with the SM Higgs boson, at a mass around 125 GeV [12] [13]. Five channels had been examined by CMS: $H \rightarrow \gamma\gamma$, $H \rightarrow ZZ^{(*)} \rightarrow 4\ leptons$, $H \rightarrow WW^{(*)}$, $H \rightarrow bb$, $H \rightarrow tt$, all critically dependent on ECAL.

The first two modes are the so called *golden modes* and have been used to extract the best and most recent measurement by CMS of the boson mass $M_{Higgs} = 125.7 \pm 0.3_{stat} \pm 0.3_{syst}$ GeV. The 68% confidence level contour plot of the Higgs mass estimation versus signal strength from the two golden modes and their combination is shown in Fig. 14. The decay rates are consistent with the SM predictions.

Journal of Physics: Conference Series **587** (2015) 012001 doi:10.1088/1742-6596/587/1/012001

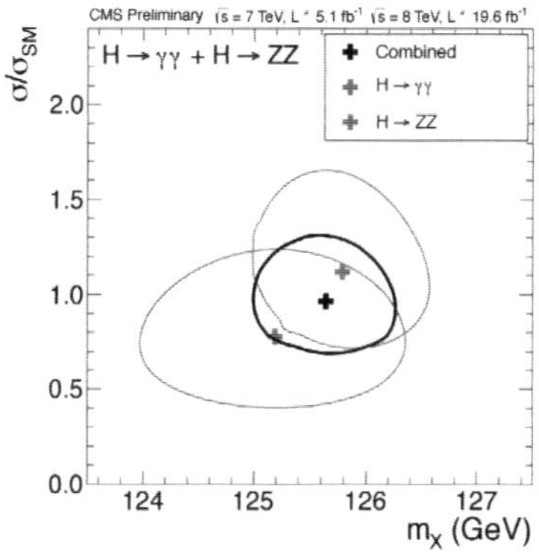

Figure 14. 68% confidence level contour plot of the Higgs mass (m_X) estimation versus signal strength (σ/σ_{SM}) from the $\gamma\gamma$ (green) and 4-leptons channels (red) and their combination (black). In the combination the signal strengths are constrained to the SM expectations for the Higgs boson.

5.1. The ECAL benchmark search channel: $H \to \gamma\gamma$

The experimental signature is a small narrow excess of events in the $M_{\gamma1\gamma2}$ invariant mass spectrum on a large falling background:

$$M_{\gamma1\gamma2} = \sqrt{2E_{\gamma1}E_{\gamma2}(1 - cos\theta_{\gamma1\gamma2}}$$

The search for the Higgs boson through its two photon decay channel depends on identifying efficiently the two photons, measuring accurately their energies and their relative opening angle.

In two photon events an important contribution to the invariant mass resolution comes from the resolution of the opening angle between the two photons. ECAL has no longitudinal segmentation and the photon direction is obtained from the shower position and the identification of the interaction vertex (IP).

The large length of the interaction region (~6 cm) would lead to a large uncertainty on the photon direction (see Fig. 15). But Higgs are produced in association with tracks from

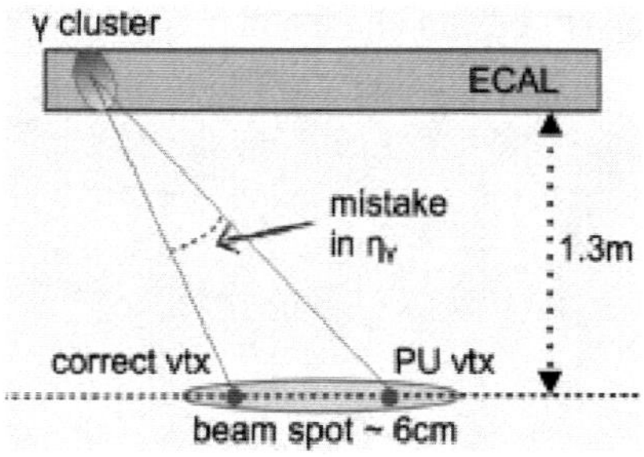

Figure 15. The photon direction is obtained from the position of the shower and of the interaction vertex. The large spread in the beam spot (~6 cm) transform in a large uncertainty on the photon direction. The diphoton vertex must be located to better than ~1 cm to make a negligible contribution to the mass resolution as compared to the ECAL energy resolution.

underlying events, initial state gluon radiation etc. The efficiency of correct assignment of the IP vertex, within 1 cm of the true vertex, has been estimated with simulation and data samples to be ~83% for $H \to \gamma\gamma$ events and close to 100% for events where the two photons have a combined transverse energy >100 GeV.

The combined shape of the expected SM Higgs invariant mass distribution for the $\gamma\gamma$ decay channel, estimated using the photon energy resolution is shown in Fig. 16 and Fig. 17 with an invariant mass resolution of 1.76 GeV for data taken at 7 TeV and 1.87 GeV for data taken at 8 TeV, for simulated M_{Higgs}=120 GeV and 125 GeV respectively.

Fig. 18 is an event display of a proton-proton collision producing two high energy photons, observed as large and isolated energy deposits in ECAL barrel. The invariant mass is around 125 GeV and the event is a candidate $H \to \gamma\gamma$ decay.

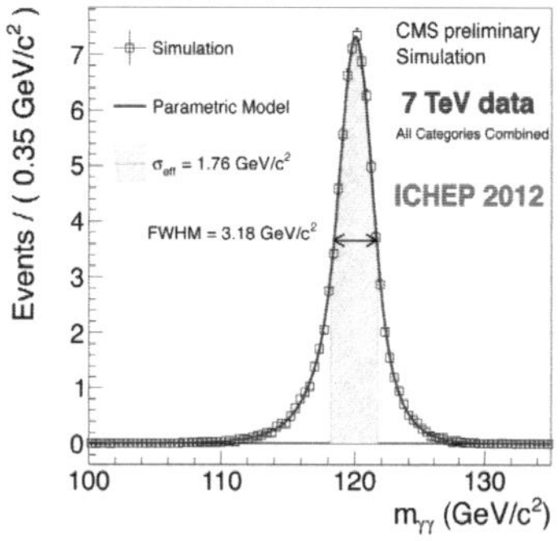

Figure 16. Expected preliminary SM Higgs invariant mass width, for the $\gamma\gamma$ channel, from simulation and 7 TeV data.

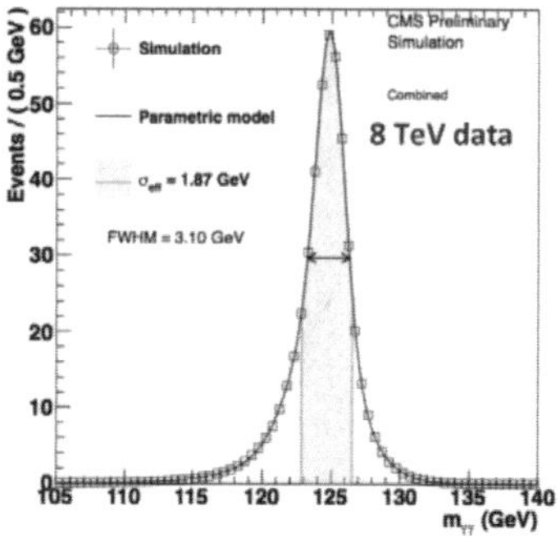

Figure 17. Expected preliminary SM Higgs invariant mass width, for the $\gamma\gamma$ channel, from simulation and 8 TeV data.

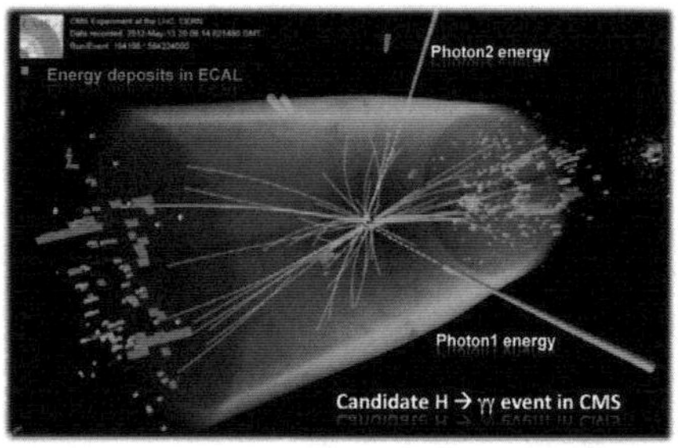

Figure 18. CMS event display of a candidate $H \to \gamma\gamma$ decay.

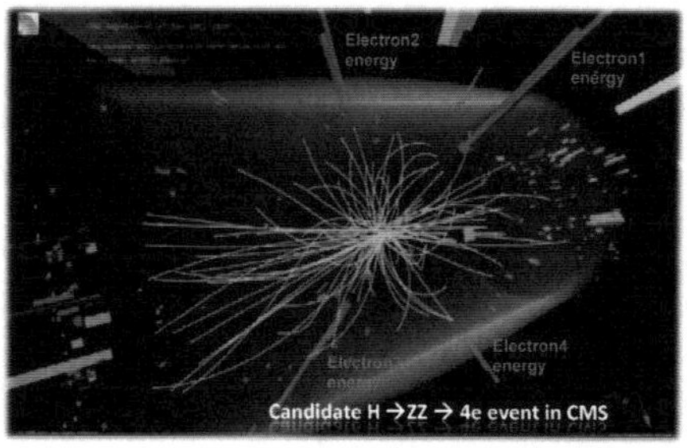

Figure 19. CMS event display of a candidate $H \to ZZ^{(*)} \to e^+e^-e^+e^-$ decay.

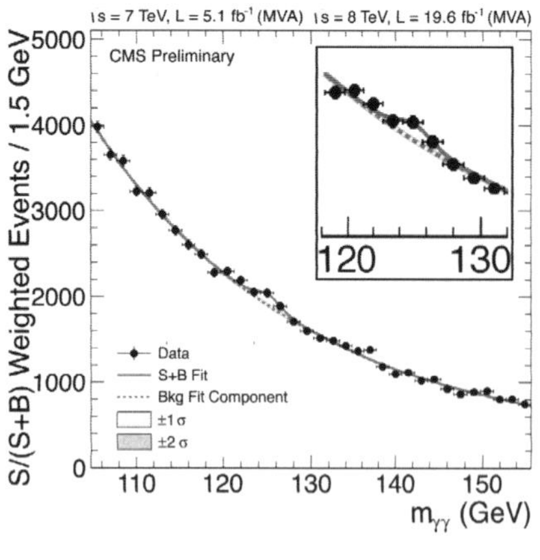

Figure 20. Excess of events in the diphoton channel at about 125GeV/c^2.

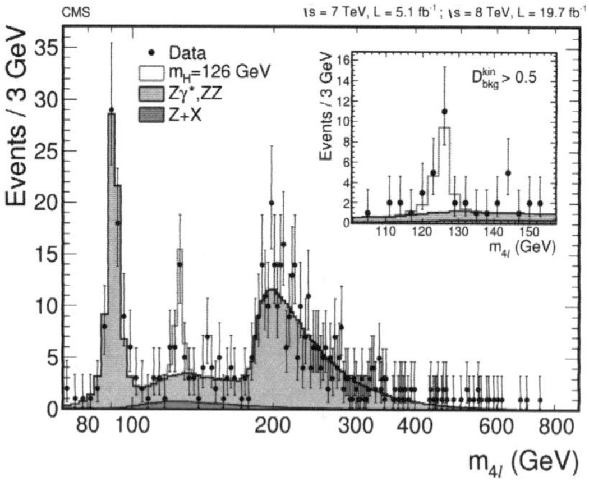

Figure 21. Excess of events seen in all 4 leptons channels at ~125 GeV/c^2.

An excess of events has been seen at a $M_{\gamma\gamma} = 125\text{GeV}/c^2$. Fig. 20 shows the diphoton mass spectrum (Moriond 2013). The signal significance is just over 3σ but better results available soon. This suggests the presence of a new boson with integer spin, but different from unity.

5.2. $H \to ZZ^{(*)} \to 4$ *leptons*

The search for the Higgs boson through its decay to two Z bosons subsequently decaying into 4 leptons (4 muons, 2 muons and 2 electrons, 4 electrons) has a low and well defined background.

The 4 electron final state is particularly challenging because the softest electron often has p_T <15 GeV. This is a difficult kinematic region due to the magnetic field and Bremsstrahlung. The tracker improves the ECAL electron energy measurement at very low p_T.

Fig. 19 is an event display of a p-p collision producing four high energy electrons, seen as large and isolated energy deposit in ECAL barrel. The invariant mass is around 125 GeV and the event is a candidate $H \to ZZ^{(*)} \to$4 leptons decay.

An excess of events has been seen in all 4 lepton channels at about 125GeV/c^2. Fig. 21 shows the four lepton mass spectrum. The signal significance is over 6σ.

6. Future perspectives

So far CMS has collected a total of about 30 fb^{-1} at 7 Tev in 2011 and at 8 TeV in 2012. Although until now the peak Lumi has been below foreseen maximum ($\sim 7 \times 10^{33} cm^{-2} s^{-1}$), the number of collisions per crossing has been higher due to LHC running at 20MHz. The experiments have been performing well and we have already experienced without problems ~80 reconstructed vertices in a special run in 2012. LHC Run II, that starts in 2015, will be at higher energy (≃13 TeV) and at an instantaneous luminosity up to $\sim 2 \times 10^{34} cm^{-2} s^{-1}$, but, from past experience in Run I, we expect that the ECAL will perform well under these conditions. We expect to collect ~300 fb^{-1} by the end of nominal LHC operation in 2023.

The High Luminosity LHC (HL-LHC or Phase 2, ~2025-2035) will provide unprecedented instantaneous ($\sim 5 \times 10^{34} cm^{-2} s^{-1}$) and integrated luminosity (~3000 fb^{-1}). The expected numbers of events per bunch crossing will be ~140 and radiation levels will be 6 times higher than for the nominal LHC design with a strong η dependence in the endcaps.

The lead tungstate crystals forming the EB will still perform well, even after the expected 3000fb^{-1} at the end of HL-LHC. Our main concern is the crystal transparency degradation with the integrated hadron fluence in the endcaps as indicated in Fig. 23. The simulation of the fractional response in EE has been tuned on data from highly irradiated crystals. In the fiducial region used for electron/photon physics (η <2.5), at about 500 fb^{-1}, the response drop to 10% of the original signal. The reduction of light output causes a progressive deterioration of energy resolution, shown in Fig. 24, and trigger efficiency, with a strong η dependence. We therefore plan to replace the ECAL endcaps before the start of HL-LHC.

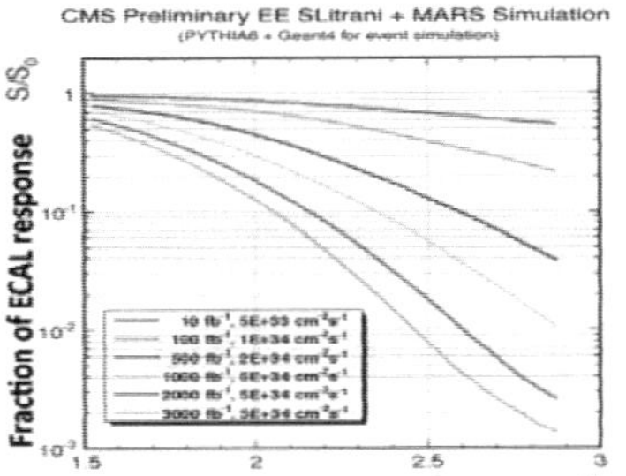

Figure 22. Simulation of fractional response from EE as a function of η, for different integrated luminosities.

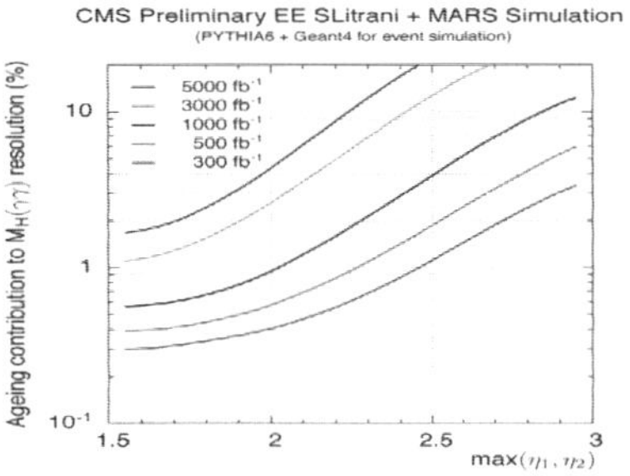

Figure 23. Deterioration of the energy resolution in EE as a function of η, for different integrated luminosities.

The APDs have recently been exposed to the levels of radiation expected at the end of HL-LHC. Although they will continue to be operational, there will be some increase in noise due to radiation-induced dark current that will require mitigation. Triggering on electromagnetic objects with ~140 pileup events necessitates a change of the front-end electronics. New developments in high-speed optical links will allow single-crystal readout at 40 MHz to upgraded Off-detector processors, allowing maximum flexibility and enhanced triggering possibilities. The VFE system will also be upgraded, to provide improved rejection of anomalous signals in the APDs as well as to mitigate the increase in APD noise. We are also considering lowering the EB operating temperature from 18 degrees C to $\sim 8^0$C, in order to increase the scintillation light output and reduce the APD dark current.

For the Endcap Calorimetry upgrade two conceptual designs are currently being considered: an e.m. calorimeter built with a Shashlik design and a finely segmented calorimeter (a la CALICE). Both approaches require significant R&D and extensive studies.

For details see talks by Burak Bilki, Milena Quittnat and Alexey Drozdetskiy at this conference.

7. Conclusions

The CMS electromagnetic calorimeter meets the high expectations of the CMS design. A new boson has been discovered in the decay channels $H \to \gamma\gamma$, $H \to ZZ^{(*)} \to 4$ *leptons* with significant contributions from ECAL. The excellent resolution achieved in the ECAL barrel drives the sensitivity for $H \to \gamma\gamma$ (1% di-photon mass resolution for unconverted photons in the center of the barrel).

Work is ongoing to further optimise the calibration and time-dependent response corrections, and we expect that the ECAL will continue to perform well during the upcoming LHC Run II. A better understanding of systematic effects, such as local containment fluctuations and the effects of upstream material, is leading to a closer agreement between data and simulation.

The HL-LHC will impose severe requirements on the performance and radiation tolerance of the detectors. The ECAL barrel will remain performant up to 3000 fb^{-1}, following planned modifications to the front-end electronics (FE electronics, increase granularity of L1 trigger processors, 40 MHz data stream readout, VFE with faster shaping time, cooling). The endcaps will need to be replaced prior to HL-LHC to enable the full exploitation of the HL-LHC physics potential. The ECAL barrel will perform up to 3000 fb^{-1} with some upgrade (FE electronics, increase granularity of L1 trigger processors, 40 MHz data stream, VFE with faster shaping time, cooling). The endcaps will work until 500 fb^{-1} but they need to be replaced to enable full exploitation of the HL-LHC potential.

References

[1] CMS Collaboration, *The CMS experiment at the CERN LHC*, J.Inst. 3 (2008) S08004, 2008.
[2] CMS Collaboration, *Letter of Intent*, CERN-LHCC-92-003, 1992.
[3] CMS Collaboration, *CMS Technical Proposal*, CERN-LHCC-94-38, 1994.
[4] CMS Collaboration *The e.m. calorimeter project: Technical Design Report*, CERN-LHCC-97-33, 1997.
[5] P.Adzic et al. *Energy resolution of the barrel of the CMS e.m. calorimeter*, J.Inst. 2 P04004, 2007.
[6] CMS Collaboration *Physics Technical Design Report*, Vol.1-2, CERN-LHCC-2006-001, 2006.
[7] CMS Collaboration *Performance and operation of the CMS e.m. calorimeter* J.Inst. 5 T03010, 2010.
[8] D.Petyt et al. *Anomalous APD signals in the CMS Electromagnetic Calorimeter* NIM A 695 293, 2011.
[9] M.Anfreville et al. *Laser monitoring system for the CMS $PbWO_4$ e.m. calorimeter* NIM A594 294-320, 2008.
[10] D.Barney et al. *Measurement of the bulk leakage current of silicon sensors of the CMS preshower detector after an integrated luminosity of 6.17 fb^{-1} at sqrt(s)=7 TeV* J.Inst. 8 P02004, 2013.
[11] CMS Collaboration *Energy, calibration and resolution of the CMS e.m. calorimeter* J.Inst. 8 P09009, 2013.
[12] CMS Collaboration *Observation of a new boson at a mass of 125 GeV with the CMS experiment at the LHC* Phys. Lett. B 716 30, 2012.
[13] ATLAS Collaboration *Observation of a new particle in the search for the Standard Model Higgs boson with the ATLAS detector at the LHC* Phys. Lett. B 716 1, 2012.

The CMS electromagnetic calorimeter calibration during Run I: progress achieved and expectations for Run II

Alessio Ghezzi, on behalf of the CMS Collaboration
University of Milano-Bicocca, Physics Department, P.zza Della Scienza 3, Milano
E-mail: alessio.ghezzi@mib.infn.it

Abstract. The CMS ECAL is a high-resolution, hermetic, and homogeneous electromagnetic calorimeter made of 75,848 scintillating lead tungstate crystals. It relies on precision calibration in order to achieve and maintain its design performance. A set of inter-calibration procedures is carried out to normalize the differences in crystal light yield and photodetector response between channels. Different physics channels such as low mass di-photon resonances, electrons from W and Z decays and the azimuthal symmetry of low energy deposits from minimum bias events are used. A laser monitoring system is used to measure and correct for response changes, which arise mainly from the harsh radiation environment at the LHC. The challenges of the different calibration techniques are discussed along with the performance evolution during Run I. The impact on physics performance is illustrated through the successful quest for the Higgs boson via its electromagnetic decays, and the subsequent mass measurement of the newly discovered particle. Conclusions are drawn for the performance to be expected from 2015 onwards, following the start of the LHC Run II.

1. Introduction

The Compact Muon Solenoid (CMS) experiment [1] is a general purpose experiment at the Large Hadron Collider (LHC) at CERN, designed to search for the standard model (SM) Higgs boson and for new physics beyond the SM. Many of these searches involve electrons or photons in the final state, and the electromagnetic calorimeter (ECAL) plays an essential role in their reconstruction and identification.

1.1. The CMS electromagnetic calorimeter

The CMS ECAL [2] has been designed to achieve an excellent energy resolution, which is a key requirement in the searches of the Higgs boson in final states involving electromagnetic particles and in particular in the $H \rightarrow \gamma\gamma$ channel, and to guarantee a good hermeticity, allowing good performances on the measurement of the missing energy.

It is a homogeneous and hermetic calorimeter containing 61200 lead tungstate ($PbWO_4$) scintillating crystals mounted in the barrel (EB), closed at each end by endcaps (EE) each containing 7324 crystals. The choice of the $PbW0_4$ with a radiation length X_0=0.85 cm and a Moliere radius R_0=2.19 cm ensures the compactness of the detector and the radiation hardness necessary to cope with the harsh environment of the collisions at LHC. The scintillation light is detected by avalanche photodiodes (APDs) and vacuum phototriodes (VPTs) in the EB and

EE respectively, yielding an average of 4.5 photoelectrons per MeV deposited in the crystals. In the EB, which covers the pseudorapidity region $|\eta|$ <1.48, the crystals have a truncated pyramidal shape (2.2×2.2 cm^2 on the frontal face, and a length of 23 cm) and they are organized in 36 supermodules, 18 on each side of the beam interaction point, each with 20 channels along ϕ and 85 along η, divided in four modules along η. The EE extends the coverage to $|\eta|$=3.0, with the crystals (2.86×2.86 cm^2 on the frontal face, and a length of 22 cm) arranged in an x-y grid.
A preshower detector (ES), based on lead absorbers equipped with silicon strip sensors, is installed in front of EE covering a fiducial region $1.65 < |\eta| < 2.6$.
The ECAL is installed inside the CMS superconductive coil and operates in the strong magnetic field of 3.8 T, used to reconstruct the momentum of charged particle in the CMS silicon tracker which covers the region up to $|\eta|$=2.5.

1.2. Energy reconstruction for electrons and photons

Electrons and photons deposit their energy over many ECAL channels. In fact the electromagnetic shower is not completely contained in a single crystal and, more importantly, the presence of material in front of ECAL causes radiation of energy (conversion of photons and bremsstrahlung from electrons) which is spread along ϕ by the strong magnetic field. Clustering algorithms are used to collect the particle energy, be it directly deposited in ECAL or radiated in the tracker material to then reach the calorimeter.
The electron or photon energy is then estimated as:

$$E_{e,\gamma} = F_{e,\gamma}[G \cdot \Sigma_i(C_i \cdot S_i(t) \cdot A_i) + E_{ES}] \quad (1)$$

where, the sum is performed over all the channels, labelled by i, which are part of the cluster. A_i is the digital amplitude measured in the channel, $S_i(t)$ is a time dependent correction that account for time variation of a channel response mainly due to changes in the crystal transparency, C_i is a relative calibration constant which takes into account differences in the crystal light yields and photodetector responses and G is a scale factor converting the digital scale into GeV. For clusters in the endcap region the corresponding energy in the preshower (E_{ES}) is added. Finally $F_{e,\gamma}$ is particle dependent correction applied to the clustered energy. It accounts for effects related to the geometry of the detector, the upstream material, and the clustering of energy emitted by bremsstrahlung or photon conversion.
This factorization of the various contributions to the electromagnetic energy determination allows for the channels' stability and intercalibration to be studied separately from material and geometry effects.

Measurements with electron beams [3] demonstrated that the ECAL crystals meet the requirement for an excellent energy resolution, with a stochastic term of 2.8% GeV$^{1/2}$ and a noise term of 12% GeV. The measured constant terms of 0.3% demonstrates an excellent shower containment and longitudinal uniformity of the light collection. Compared to the measurements with the electron beams, the energy resolution for electrons and photons *in situ* at CMS receives additional contributions from the presence of pile-up, from the material in front of ECAL, and from detector instabilities and channel-to-channel response spread. In order to profit on the excellent intrinsic energy resolution the two latter contributions must be kept to within 0.4%.

In the following the calibration of the ECAL response and the resulting performance in the data collected in the 2012 are illustrated. For detailed description of the ECAL performance in 2011 the reader is referred to [4].

2. Corrections for the change of the response in time

Radiation can create color centers in the crystals reducing their transparency and therefore reducing the measured response to the deposited energy. The color centers partially anneal with thermal energy so that the loss in transparency depends on the dose rate, which for ECAL varies along η. In absence of radiation a partial recovery of the transparency is observed.

While an intense R&D program led to the production of radiation resistant $PbWO_4$ crystals, a residual radiation damage remains and needs to be corrected for. Thus, the changes in transparency are measured by a dedicated laser monitoring system[5] (LM) which injects every 40 minutes a laser pulse (λ =440 nm, close to the peak of the scintillation light spectrum for $PbWO_4$) into each single crystal. The relative change in transparency (R/R_0) does not directly measures the relative change in response for the scintillation light (S/S_0) since the two have different spectra and the optical photons travel different paths to reach the photodetectors, but they can be related by a power law:

$$\frac{S}{S_0} = \left(\frac{R}{R_0}\right)^{\alpha} \tag{2}$$

The measured loss in transparency at the end of the LHC Run I in 2012 is less than 6% in EB and less than 30% in the EE region within the tracker acceptance ($|\eta|$ <2.5), while it reaches 70% in the most forward EE region. The corrections for transparency loss (LC) can be validated with collisions data, by monitoring the stability of the reconstructed invariant mass of π^0 and of the distribution of E/p for isolated electrons, where E is the energy measured in the calorimeter and p is the momentum measured in the tracker.

Applying the LC the stability is better than 0.1% in the barrel and 0.4% in the endcaps. As an example the stability during 2012 measured with the E/p distribution is reported in figure 1.

Measurements performed with electron beams on a small ensemble of crystals show a dispersion

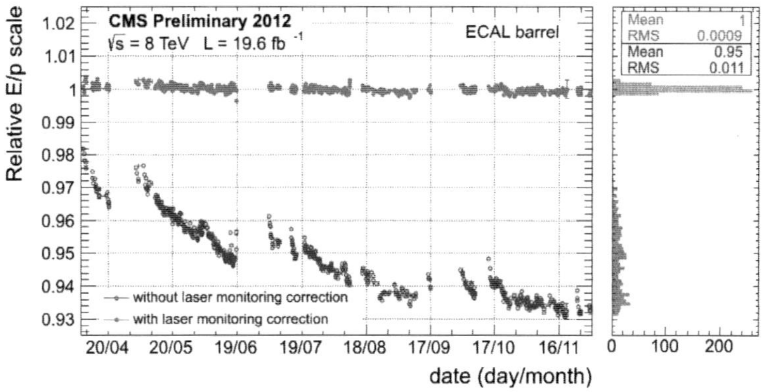

Figure 1. Stability of the E/p scale during 2012, before (red) and after (green) the LC, for the ECAL barrel. Each point is obtained with a statistic of 20000 electrons. The projections along the Y axis are shown in the histograms on the right side.

in the values of α of 10% [6]. In deriving the LC according to equation 2 a common value (one for EB and one for EE) is used. The resulting residual imperfection of the LC due to the dispersion in the values of α can be recovered by allowing a time dependence of the calibrations constants as explained in the next section.

3. Relative calibration and energy scale

The relative calibration ("intercalibration" IC) is obtained *in situ* employing collision data, after applying the LC, with different methods employing, the azimuthal symmetry of the energy flow ("ϕ-symmetry"), photons pairs from the π^0 decay, and isolated electrons (E/p). Hereafter these methods and their performances are briefly described, for a detailed description the reader is referred to [4] and references therein.

The **ϕ-symmetry** method is based on the equalization of the average energy measured in the different channels placed at the same η. It employs a dedicated data stream with a very reduced event information allowing to store a high rate of events ($\sim$ 1.5 kHz), and it reaches for each LHC fill cycle a statistical accuracy better than 0.2% (0.4%) in the barrel (endcap). The accuracy of the method is indeed limited to few percents by the systematic uncertainty in the knowledge of the material in front of ECAL. Since the systematic does not vary with time the method can be used to track possible time dependencies of the IC values. Indeed in 2012 a drift of the ICs was observed ascribable to imperfections of the laser corrections [7].

The **π^0 mass** method is based on the reconstruction of the peak in the spectrum of the invariant mass of unconverted photon pairs due to π^0 and η decays. The photons are reconstructed as 3×3 matrices of calorimeter channels, and an iterative method is used to determine the IC value of each single channel. It employs a dedicated data streams at high rate ($\sim$ 7 kHz) and it reaches a 0.5% precision in the central barrel, dominated by systematics when using a data sample corresponding to about 45 days of data taking. With the high pile-up of the LHC running in 2012 the reconstruction of the peak of π^0 in the region at $|\eta| > 2$ is challenging and only the peak from $\eta \to \gamma\gamma$ is used.

The **E/p** method is based on the comparison of the energy measured in the calorimeter (eq 1) to the momentum measured by the tracker for isolated electron, and an iterative procedure is used to extract the IC value for each channel. The method requires the full 2012 dataset and in the central barrel its precision reaches the systematic limit of 0.5%, while for $|\eta| > 1$ the statistical contribution is still significant.

These methods may be affected by η dependent systematics due for instance to the effect of pile-up or of the material in front of ECAL, therefore they are employed to obtain a relative calibration of the channels within a ring at constant η ("η ring"), and their results are combined accordingly to their precision. The precision of the different methods and of their combination is reported in figure 2. In the region $|\eta|$ >2.5, above the tracker acceptance, the E/p method can not be used and the high pile-up prevents the reconstruction of the invariant mass peak from π^0 or η, therefore only the ICs from the ϕ-symmetry are available.
To compensate for the imperfection in the LC, the time dependence observed in the ICs from the ϕ-symmetry method is propagated to the final calibration, in time intervals of the order of one month.

The relative calibration of the η rings (η scale) is obtained from the Z boson peak in the distribution of the invariant mass of electron pairs, selecting a sample of electrons with low emission of bremsstrahlung. The η scale is set to match the expectation from a detailed Monte Carlo simulation of the detector response to $pp \to Z\ X \to e^+e^-\ X$.
Finally the overall energy scale G is set, separately for EB and EE, such that the reconstructed Z peak in data matches the one in the MonteCarlo.

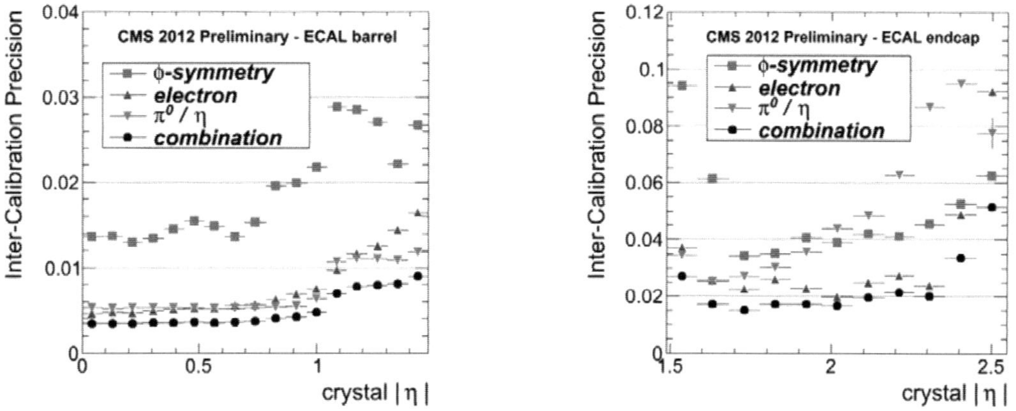

Figure 2. Precision of the calibration coefficients as a function of $|\eta|$, for the barrel (left) and the endcaps (right). The precisions of the different methods and of their combination are reported.

4. Performance of the detector

The energy resolution for electrons and photons plays a crucial role in the search for the Higgs with electromagnetic final states, and needs to be precisely estimated, in particular for the search in the Higgs decays channel $H \to \gamma\gamma$, since it affects the modelling of the expected signal. The accuracy of the calibration directly contributes to the energy resolution with a dilution factor ~0.7 due to the sharing of the energy among different channels. Indeed the actual energy resolution for electrons can be measured from the shape of the invariant mass distribution of electron pairs in the region dominated by $Z \to e^+e^-$ events. Figure 3 reports the measured energy resolution as a function of $|\eta|$. The calibration available at the beginning of 2012 has been used for a prompt reconstruction of the data. The performance of the prompt reconstruction (grey) can be compared with the one from a calibration based on the full 2012 dataset (blue) and an improvement can be seen in particular in the endcaps. The figure reports also the expected resolution from a Monte Carlo simulation. The discrepancy between data and Monte Carlo measured for electron is propagated to the Monte Carlo simulation of processes involving photons.

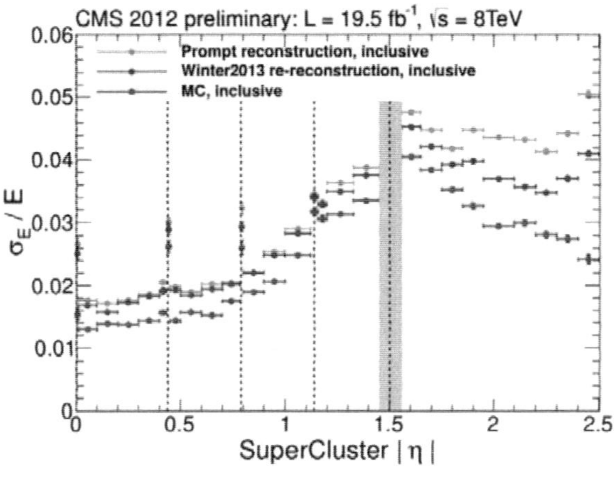

Figure 3. Energy resolution for electrons as a function of $|\eta|$. The performance with an initial (grey) and after the final (blue) calibration are reported, as well as the expected values (red) from a Monte Carlo simulation of the detector.

5. Plans for the LHC Run II

The LHC Run II in 2015 will be characterized by and increased center of mass energy ($\sqrt{s}$= 13 TeV) and by an increase of about a factor two in the instantaneous luminosity and therefore by an increase of the pile-up.
While the overall strategy for the calibration and monitoring of the ECAL performance is planned to follow the run1 strategy, some optimizations and development are foreseen. In particular the calibration data-streams for the π^0 and η methods are being optimized to cope with the increased pile-up and to improve the accuracy in the endcaps, and a dedicated stream with a reduced event information for the calibration with E/p is being developed to fully profit of the increase in the rate of isolated electron from $W \rightarrow e\nu$.
Additional methods have been developed to calibrate the endcap region employing an invariant mass constraint on $Z \rightarrow e^+e^-$ events. Preliminary studies indicates that this can improve the accuracy of the calibration in the endcap region covered by the tracker and it can also be extended to the region at $|\eta|$ >2.5 by reconstructing one of the two electrons employing only the calorimetric information. In fact, even if in this case the effectiveness of the electron identification is reduced, it has been shown that the level of background in the invariant mass region close to the Z mass remains marginal.
The calibration procedure with this method is being finalized and it is expected that it will be part of the calibration routine in 2015.

During the LHC Run I the imperfection of the LCs due to the spread of the value of α (equation 2) was effectively recovered by time-dependent ICs with a negligible impact on the energy resolution. With the increased dose rate expected in Run II it might become useful to determine the value of α per channel, in particular in the region at high η. Indeed during 2012 different methods have been developed to measure *in situ* the value of α, by comparing the variation of the response to a reference signal with the change in transparency measured by the laser monitoring system (R/R_0 in equation 2). As reference signal we have investigated the E/p distribution for isolated electrons, the π^0 peak in the invariant mass distribution of photon pairs and the energy flow in each channel. The different methods can determine the value of α with different granularity: in groups of 100 and 25 channels with respectively the E/p and π^0, and for each single channel with the energy flow. The study of the optimal combination of the different measurements is ongoing and different α values for the LC are expected to be available to be used in 2015, where relevant.

References

[1] CMS collaboration,*"The CMS experiment at the CERN LHC"*,2008 JINST 3 S08004
[2] CMS collaboration, *"The Electromagnetic Calorimeter Project: Technical Design Report"*, Technical Report CERN-LHCC-97-33, CMS-TDR-4 (1997)
[3] P. Adzic et al., *"Energy resolution of the barrel of the CMS electromagnetic calorimeter"*, 2007 JINST 2 P04004
[4] CMS Collaboration, *"Energy calibration and resolution of the CMS electromagnetic calorimeter in pp collisions at sqrts = 7 TeV"*, 2013 JINST 8 P09009
[5] M. Anfreville et al., *"Laser monitoring system for the CMS lead tungstate crystal calorimeter"*, Nucl. Instrum. Meth. A 594 (2008) 292
[6] A. Ghezzi et al., *"Analysis of the response evolution of the CMS electromagnetic calorimeter under electron and pion irradiation"*, CMS Note CMS-NOTE-2006-038, (2006).
[7] M Obertino, *"The challenges involving the calibration of the CMS Electromagnetic Calorimeter at the LHC"*, Proceedings of CHEF 2013

Journal of Physics: Conference Series **587** (2015) 012003 doi:10.1088/1742-6596/587/1/012003

Timing performance of the CMS ECAL and prospects for the future

Daniele del Re

Università Sapienza di Roma & INFN Roma

E-mail: daniele.delre@roma1.infn.it

Abstract. The CMS electromagnetic calorimeter (ECAL) is made of about 75000 scintillating lead tungstate crystals arranged in a barrel and two endcaps. The scintillation light is read out by avalanche photodiodes in the barrel and vacuum phototriodes in the endcaps, at which point the scintillation pulse is amplified and sampled at 40 MHz by the on-detector electronics. The fast signal from the crystal scintillation enables energy as well as time measurements from the data collected in proton-proton collisions with high energy electrons and photons. The stability of the time measurement required to maintain the energy resolution is on the order of 1 ns. The single-channel time resolution of ECAL measured at beam tests for high energy showers is better than 100 ps. The time resolution achieved with the data collected in proton-proton collisions at the LHC is presented. The time precision achieved is used in important physics measurements and also allows the study of subtle calorimetric effects, such as the time response of different crystals belonging to the same electromagnetic shower. In addition, we present prospects for the high luminosity phase of the LHC, where we expect an average of 140 concurrent interactions per bunch crossing (pile-up). It is currently being studied how precision time could be exploited for pileup mitigation and for the assignment of the collision vertex for photons. In this respect, a detailed understanding of the time performance and of the limiting factors in time resolution will be important.

1. Introduction

The primary goal of the Compact Muon Solenoid (CMS) experiment [1] is to explore particle physics at the TeV energy scale, exploiting the proton-proton collisions delivered by the Large Hadron Collider (LHC) at CERN [2]. The electromagnetic calorimeter (ECAL), which measures the energy of electrons and photons produced in LHC collisions, is located inside a solenoid magnet. It is a hermetic homogeneous calorimeter made of 75848 lead tungstate ($PbWO_4$) scintillating crystals: 61200 in the barrel (EB) and 7324 in each endcap (EE). The scintillation decay time of the crystals is comparable to the LHC bunch crossing interval of 25 ns, and about 80% of the light is emitted in 25 ns. For the light detection, the crystals are equipped with avalanche photodiodes in the barrel and vacuum phototriodes in the endcaps. The main purpose of the ECAL is the precise energy measurement, needed for many physics analyses. The stability of the time measurement required to maintain the energy resolution is on the order of 1 ns. In addition to the energy measurement, the combination of the scintillation timescale of $PbWO_4$, the electronic pulse shaping, and the sampling rate allow excellent time resolution to be obtained with the ECAL. This is important in CMS in many respects. The better the precision of time measurement and synchronization, the larger the rejection of backgrounds with a broad time distribution. Such backgrounds are out-of-time proton-proton interactions, cosmic rays, beam halo muons, and electronic noise. Precise time measurement also makes it possible to put constraints on the presence of particles predicted by models beyond the Standard Model, identifying photons from the decay of long-lived new particles, which reach the calorimeter out-of-time with respect to particles travelling at the speed of light from the interaction point [3]. To achieve these goals the time measurement performance both at low energy (1 GeV or less) and high energy (several tens of GeV for showering photons) becomes relevant.

The performance of the calorimeter has been measured before the collisions using test beam electrons, cosmics and beam splash events [4]. The resolution for large energy deposits (E>50 GeV) was estimated to be better that 20 ps and the linearity of the time response has been also verified. During collisions, there are many additional effects which could worsen the performance, like run-by-run variations, inter-calibration effects, energy-dependent systematics, and crystal damage due to radiation. A first measurement of the timing performance and intercalibration of the CMS ECAL during collisions has been reported in [5], where the time resolution was about 190 ps in the barrel and 280 ps in the endcap.

The purpose of the studies described in this document is to re-evaluate the time performance and check if the calorimeter is stable over time.

Detailed studies on time resolution are also important for the upgrade of the detector foreseen in 2024, when the LHC is expected to reach to peak luminosities of $10^{35} cm^{-2}s^{-1}$ and it will be necessary to replace some components of the current detectors, especially in the forward region. The increased luminosity will introduce a multiplicity of simultaneous events (pile-up) of about 140 interactions per beam crossing. The pile-up of energy deposits coming from different interactions deteriorates the calorimeter performances in terms of energy measurement and particle identification, as individual particles are less isolated. A possible strategy to reduce pile-up effects consists in complementing the calorimeters with an extremely high time resolution, which would enable the energy deposits coming from different interaction vertices to be resolved in time.

2. Measurement of time resolution at p-p collisions

The resolution is extracted by comparing the time measured in ECAL hits which are, or can be made, synchronous. Two methods are used:

- The first one is based on the comparisons of the time of the two electrons of a Z decay. The time of the electron corresponds to the one of the most energetic hit of the deposit (cluster). The time is corrected by the time-of-flight of the electrons, which is properly determined from the primary vertex position, obtained from the electron tracks. The two clusters are required to pass loose criteria on the shape of the deposit and the resulting invariant mass has to be consistent with the Z boson mass (60 GeV<m(ee)<150 GeV). The energy of each of the two hits must fall in the range 10 GeV<E<120 GeV. The resolution is extracted as the σ parameter of a gaussian fit to the core of the time difference distribution. The resulting resolution is plotted as a function of the effective amplitude, which depends on the amplitudes measured in the two crystals as $A_{eff} = A_1 A_2 \sqrt{A_1^2 + A_2^2}$. The amplitude is in units of ADC counts, with a dynamic range of 4000 and the electonic noise corresponds to about 1 count. This is shown in Fig. 1. The noise term is very similar to the one obtained prior to collisions [4] and the constant term turns out to be about 150 ps. The value obtained for the constant term is remarkable but it is still far from the test beam one, which was about 20 ps.
- the second method uses the two most energetic neighboring crystals of a photon cluster, with a similar amount of energy. The second requirement is to minimize the shower propagation effects. Additional selection criteria are applied, based on the shape of the ECAL cluster

and on the energy of the ECAL hit (E<120 GeV). The resulting resolution, as a function of the effective amplitude, is shown in Fig. 2. Here the constant term is smaller, about 70 ps, closer to the test beam results.

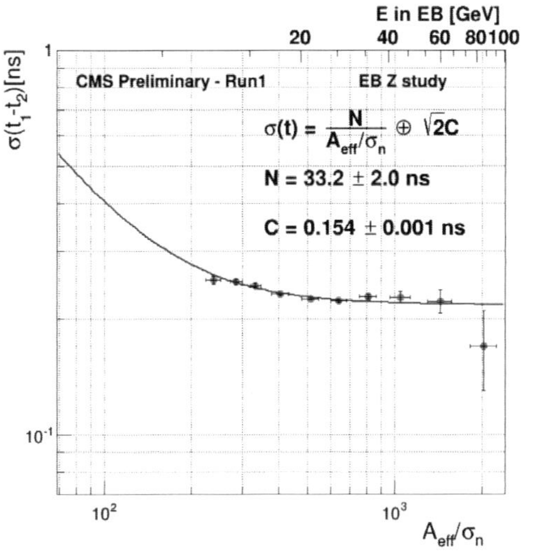

Figure 1. Barrel, electrons from Z decays: spread of the time difference as a function of the effective amplitude. A fit to the distribution, using the quadrature of a noise and a constant contribution, is superimposed.

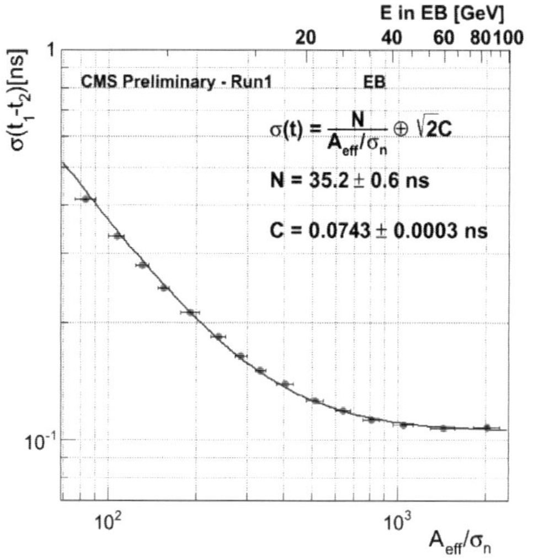

Figure 2. Study of neighboring crystals of a photon energy deposit: spread of the time difference as a function of the effective amplitude. A fit to the distribution, using the quadrature of a noise and a constant contribution, is superimposed.

A further study, comparing crystals belonging to either the same or different readout units, is also performed. The redaout unit corresponds to a trigger tower which is a 5x5 matrix of crystals. This is to see if the origin of this residual 70 ps constant term is due to residual timing jitter in the electronics or the clock distribution to the individual readout units. In Figures 3 and 4 the result for crystals in the same and different readout units are shown, respectively. The different constant term (67 ps vs 130 ps) does not explain the difference with respect to the test beam results but indicates that there are effects related to the electronics which need to be further investigated.

The stability of the time measurement as a function of time and pseudorapidity has been also checked. Both can reveal a dependence of the time measurement performance on the luminosity and radiation. The analysis of neighboring crystals in the same readout unit has been repeated in different run number ranges, Fig. 5 (this is equivalent to a study as a function of time), and in bins of pseudorapidity, Fig. 6. The spread of the time difference is extracted in an energy range where the constant term contribution is expected to dominate, i.e. for A_{eff}>30 GeV. The time resolution resulted to be very stable.

3. Physics potential of precise timing at HL-LHC

The high luminosity phase of LHC (HL-LHC) is complicated by the increased multiplicity of simultaneous events (pile-up): about 140 interactions per beam crossing, with a spread of approximately 6 cm and 300 ps along the beam axis, are anticipated. The pile-up of energy deposits coming from different interactions deteriorates the calorimeter performances in terms of

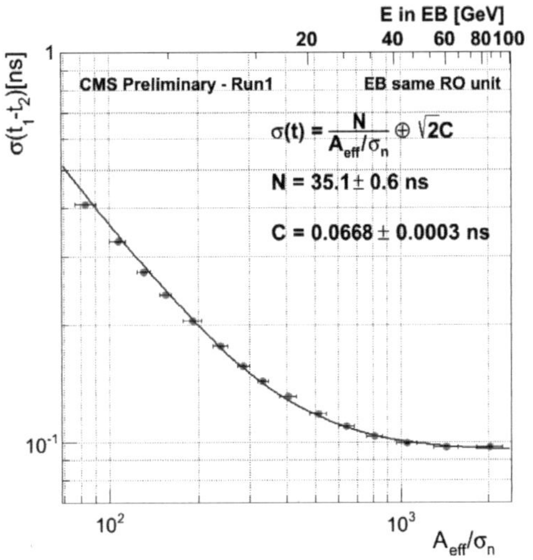

Figure 3. Study of neighboring crystals in the same readout unit of a photon energy deposit: spread of the time difference as a function of the effective amplitude. A fit to the distribution, using the quadrature of a noise and a constant contribution, is superimposed.

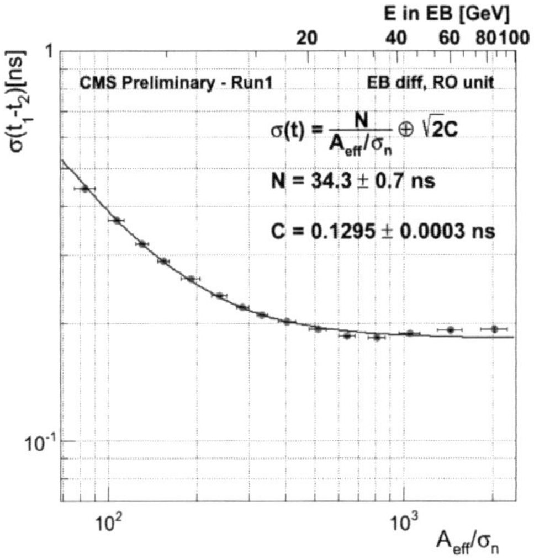

Figure 4. Study of neighboring crystals in different readout units of a photon energy deposit: spread of the time difference as a function of the effective amplitude. A fit to the distribution, using the quadrature of a noise and a constant contribution, is superimposed.

energy measurement and particle identification, as individual particles are less isolated. In order not to compromise the sensitivity to exclusive channels and the reconstruction of the missing transverse energy (MET), a crucial observable in searches for processes beyond the standard model, innovative solutions should be identified to mitigate pile-up effects. A possible strategy consists in complementing the transversal segmentation of the calorimeters with an extremely high time resolution, which would enable the energy deposits coming from different interaction vertices to be resolved in time. To set the scale, a time resolution of 30 ps corresponds to about 1 cm in flight length. This has to be compared with the size of the beam spot, which corresponds to about 6 cm. A TOF-based selection could then be able to achieve a pile-up rejection factor of about 6.

There are several possible uses of this time information: 1) It can be used to determine the primary vertex when there is a small track activity in the event. This is particularly important for the Higgs to two photons analysis, where there is a limited number of extra jets in the event and the vertex determination is crucial for the Higgs mass resolution; 2) The time information can be included in the pile-up jet rejection algorithms. Presently, they are based on tracks and on the jet shape but they are not very effective outside the acceptance of the tracking detector. The time would make these algorithms more performant in the forward region. 3) ECAL hits not consistent with the primary p-p collision can be filtered out with the timing. This reduces the ECAL occupancy, which is particularly relevant for reconstruction and trigger issues at 140 pile-up conditions. 4) The time information can be integrated with the CMS Particle Flow algorithm. Particle Flow is a reconstruction algorithm that combines all inputs from the different CMS detectors to improve the resolution of the physics objects, in particular jets and missing transverse energy. The Particle Flow algorithm currently applies timing cuts on the ECAL hits consistent with the resolution measured in the data. The gain in performance with an improved

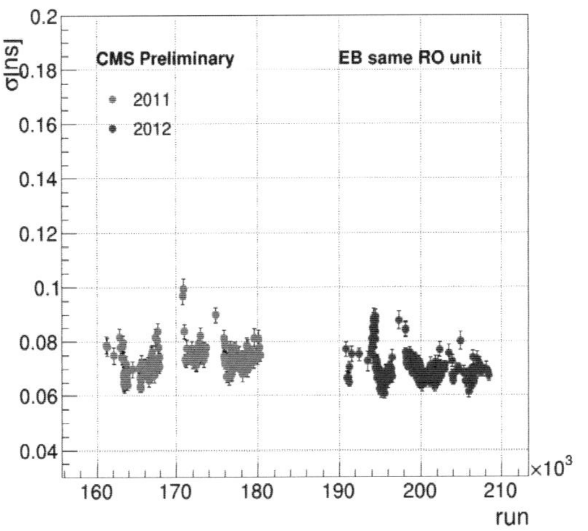

Figure 5. Stability of the time resolution as a function of the run number, estimated with the neighboring crystals method (same readout unit). The resolution corresponds to the σ of a gaussian fit to the core of the time difference distribution for A_{eff}>30 GeV.

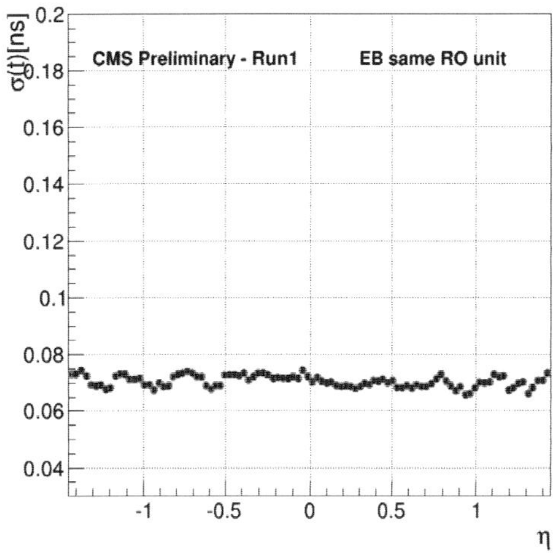

Figure 6. Stability of the time resolution as a function of the pseudorapidity, estimated with the neighboring crystals method (same readout unit). The resolution corresponds to the σ of a gaussian fit to the core of the time difference distribution for A_{eff}>30 GeV.

time resolution is currently being investigated in detail.

Here we present studies based on simulations, assuming that a detector capable of very precise time measurement can be integrated in the electromagnetic calorimeter in the forward region. We used samples of H $\rightarrow \gamma\gamma$ events with an average number of pile-up interactions of about 20. In Fig. 7 the performance in terms of vertex determination is shown as a function of the time resolution of the detector. The vertex of the two photons from the Higgs decay is reconstructed via a χ^2 fit which uses the measured time of the two photons and their expected time-of-flight, under the assumption they come from the same interaction point. The time fo the photon corresponds to the time of the most energetic crystal of the cluster. As shown, it is possible to reach a resolution in the vertexing better than 1 cm if the time resolution is about 30 ps. In Fig. 8 the impact of a precise time detector on the total energy reconstructed in the electromagnetic calorimeter is illustrated. Here ΣE_T^{ECAL} corresponds to the sum of the transverse energy of all ECAL cells above the electronic noise threshold. It can be noticed that the original distribution without pile-up can be restored by imposing a requirement on timing of $\pm$ 30 ps.

4. Conclusions

The resolution of the time measurement of the CMS electromagnetic calorimeter has been investigated with samples of electrons and photons taken during p-p collisions. In the barrel, the relative time resolution of adjacent channels in an electromagnetic cluster was found to be 70 ps for large energies if the channels belong to the same electronic readout unit. This resolution is 130 ps for channels belonging to different readout units. The timing performance is stable over time and at different pseudorapidities.

A possible use of a precise timing detector for the LHC Phase II upgrade is currently being studied. We show that, in events with two photons, the time information can be used to determine the primary vertex position with a resolution smaller than 1 cm, even without any

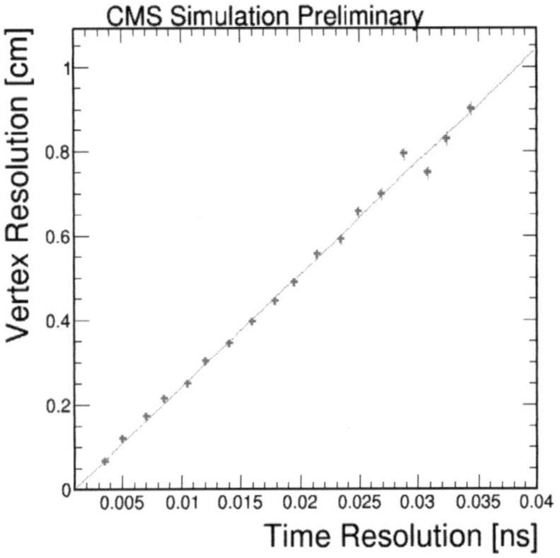

Figure 7. Resolution of the primary vertex determination using the time of two photons coming from the Higgs decay, as a function of the resolution of the timing detector.

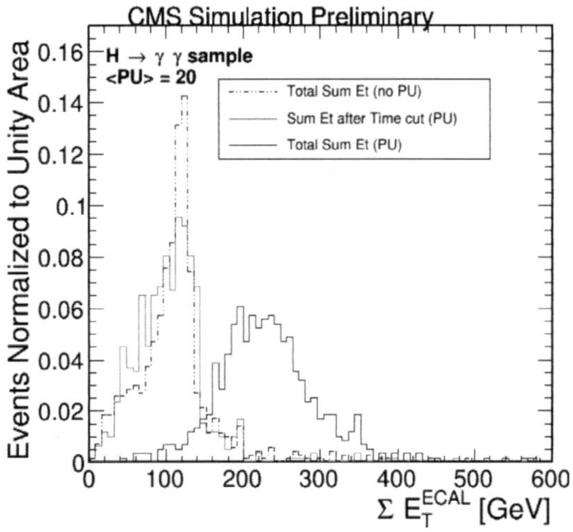

Figure 8. Impact of a selection based on time on the calorimeter-based ΣE_T variable.

tracking information and assuming a better than 30 ps resolution in time. A selection based on time also allows for a large reduction of the electromagnetic hit multiplicity, thus improving the energy measurements, resolution and trigger occupancy.

References

[1] S. Chatrchyan *et al.* [CMS Collaboration], "The CMS experiment at the CERN LHC," JINST **3**, S08004 (2008).

[2] L. Evans and P. Bryant, "LHC Machine," JINST **3**, S08001 (2008).

[3] S. Chatrchyan *et al.* [CMS Collaboration], "Search for new physics with long-lived particles decaying to photons and missing energy in *pp* collisions at $\sqrt{s} = 7$ TeV," JHEP **1211**, 172 (2012).

[4] S. Chatrchyan *et al.* [CMS Collaboration], "Time Reconstruction and Performance of the CMS Electromagnetic Calorimeter," JINST **5**, T03011 (2010).

[5] S. Chatrchyan *et al.* [CMS Collaboration], "Energy calibration and resolution of the CMS electromagnetic calorimeter in pp collisions at s = 7 TeV," JINST **8**, P09009 (2013).

Performance of jets at CMS

Matthias Schröder, on behalf of the CMS collaboration
Deutsches Elektronen-Synchrotron (DESY), Notkestraße 85, 22607 Hamburg, Germany
E-mail: matthias.schroeder@desy.de

Abstract. The calibration and reconstruction of jets critically relies on the performance of the calorimeters. Extending out to large pseudorapidities, the measurements depend on the interplay between forward calorimeters, central calorimeters, and the tracking system. The high number of additional pile-up interactions poses further complications. In CMS, these difficulties are overcome using the 'particle-flow' approach, which aims at reconstructing individually each particle in the event prior to the jet clustering. Measurements of the jet energy scale and the procedure for jet energy calibration in CMS are reviewed, which are performed with dijet, photon + jet, and Z + jet data collected in proton-proton collisions at a centre-of-mass energy of 8 TeV, corresponding to an integrated luminosity of 19.6 fb^{-1}. The effect of pile-up interactions and the state of the art mitigation techniques used in CMS as well as the main sources of uncertainty of the jet energy calibration are also presented.

1. Introduction

Quarks and gluons produced in high-energy processes, such as the proton-proton collisions at the CERN Large Hadron Collider (LHC), manifest as jets. Hence, jets provide experimental access to the underlying partonic process, and the precise understanding of their properties is of crucial importance in many physics analyses.

Here, an overview of the performance of jets at the Compact Muon Solenoid (CMS) experiment [1] at the LHC is given. The central feature of the CMS detector is a 6 m-diameter, superconducting solenoid providing a 3.8 T magnetic field. Within the magnet, closest to the interaction point, a silicon tracking detector is installed that measures the transverse momenta (p_{T}) of charged particles with a relative resolution of $\approx 0.7\,\%$ at 10 GeV up to pseudorapidities of $|\eta| < 2.5$. Surrounding the tracking volume but still within the solenoid, the calorimetric system covers the region up to $|\eta| < 3$. It consists of a lead-tungstate crystal electromagnetic calorimeter (ECAL) with an energy resolution of $\approx 0.6\,\%$ for 50 GeV electrons and a brass-scintillator hadronic calorimeter (HCAL) with a resolution of 18 % for 50 GeV pions. Steel and quartz-fibre hadron calorimeters extend the coverage to $|\eta| < 5$. Muons are identified in gas-ionisation detectors embedded in the steel flux-return yoke of the magnet.

In the following, the current status of the jet reconstruction and calibration at CMS is discussed, and pile-up mitigation techniques as well as tools to discriminate quark from gluon jets are reviewed. The presented results have been obtained with data collected during 2012 at a centre-of-mass energy of 8 TeV, corresponding to an integrated luminosity of up to 19.6 fb^{-1}.

2. Jet reconstruction at CMS

In most CMS analyses, events are reconstructed using a particle-flow (PF) algorithm, which aims at identifying and reconstructing the individual particles in the event, combining the information from all subdetectors [2]. For example, the energy of photons is directly obtained from the ECAL measurements; conversions in the dense tracker material are accounted for by adding tracking information. The energy of charged hadrons is determined from a combination of the track momenta with associated ECAL and HCAL energy deposits, and the energy of neutral hadrons is determined from the remaining calorimeter energy deposits. In both cases, the measurements are corrected for the non-linear response of the calorimeters to hadrons.

Jets are then clustered from the four-momenta of the reconstructed particles. Commonly, the anti-k_T jet clustering-algorithm [3] with a radius parameter of $R = 0.5$ is used in CMS, but other jet algorithms and reconstruction methods are also supported.

3. Properties of particle-flow jets

The PF-jet performance greatly benefits from two effects: Firstly, approximately 90% of the measured jet energy are reconstructed as charged hadrons and photons, which are measured dominantly with the high-resolution tracker and ECAL, while only the remaining 10 % are reconstructed as neutral hadrons, which are measured with the HCAL alone that has a poorer resolution [4]. Secondly, since individual particles are reconstructed, a dedicated calibration can be applied. As a consequence, the impact of the non-compensating calorimeters is mitigated, leading to a good overall jet response with flavour differences less than 4 % [4]. Furthermore, the jet energy resolution is greatly improved compared to jets reconstructed from calorimeter-information only [5].

Since tracking and thus vertex information is available, the impact of pile-up (PU) collisions that occur in addition to the hard interaction in the event can be mitigated already at the jet-clustering stage. As shown in Fig. 1, the additional $p_{T,\text{offset}}$ per jet due to PU particles amounts to approximately 700 MeV per primary vertex, leading to an average of ≈ 10 GeV being added to the p_T of each jet[1]. The average $p_{T,\text{offset}}$ per primary vertex is shown in Fig. 2 as function of η, split up into the contributions from different PF-particle types. In the central detector region with tracker coverage, approximately 50 % of the $p_{T,\text{offset}}$ is deposited by charged particles that can be unambiguously associated to PU vertices. These particles can be subtracted from the event prior to the jet clustering, thus effectively reducing the impact of PU.

4. Jet energy corrections

The jet energy corrections (JEC) relate the energy of the reconstructed jets on average to the energy at particle-level, which is independent of the detector response. CMS has adopted a factorised approach to JEC [5]. Given that the jet properties are well described by the simulation, the primary JEC factors are derived from simulated events relative to the generator jets, and only the remaining, small differences between the response in data and simulation are corrected for using data-driven methods. Since the latter corrections are derived as data-to-simulation ratios, potential biases of the methods cancel to first approximation.

The first factor in the JEC chain subtracts the $p_{T,\text{offset}}$ due to (the remaining) PU contributions. It is an average correction derived from the global per-event $p_{T,\text{offset}}$ density ρ and the jet area [5, 6]. The calibration of ρ is determined as a function of η and p_T from two samples of simulated events with identical hard interactions, where one contains in addition PU interactions. Residual differences to the ρ calibration in data are measured as a function of η with the random-cone method in zero-bias events [5] and corrected for.

[1] Taking into account an average of 21 PU collisions during the 2012 LHC run and the vertex-reconstruction efficiency of approximately 70 %.

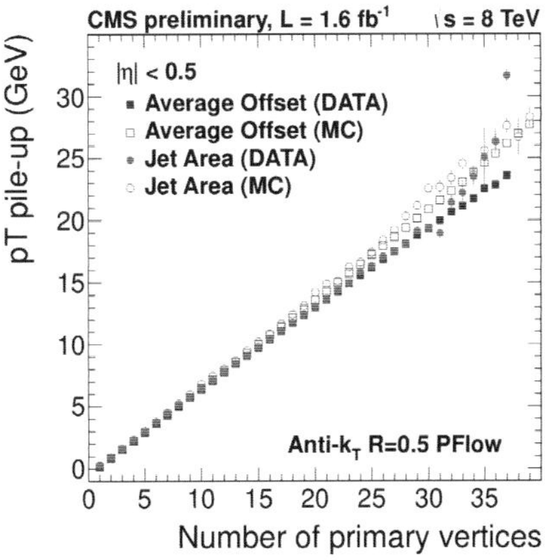

Figure 1. Average jet $p_{\mathrm{T,offset}}$ due to PU vs. the number of primary vertices in data and in simulation measured with the random-cone ('Average Offset') and the jet-area method. [4]

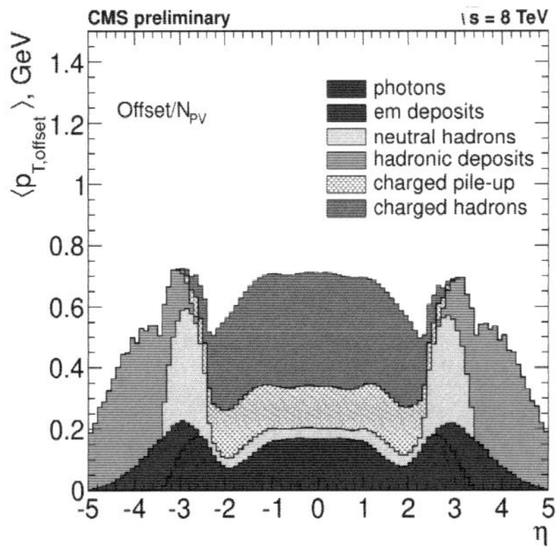

Figure 2. Average jet $p_{\mathrm{T,offset}}$ due to PU per primary vertex and PF-particle type vs. η. The solid red area marks contributions from charged particles associated to PU vertices. [4]

The second, primary JEC factor corrects for the p_{T} and η dependence of the average jet response due to the calorimeter non-linearities and p_{T} thresholds, the different detector properties along η, and geometric effects. It is obtained from simulated QCD-multijet events generated with PYTHIA6 [7] with tune Z2* processed through the full GEANT4-based [8] CMS detector simulation.

After these corrections, the jet energy scale (JES), i.e. the average response, in simulated events is at 1 within 1 % precision [4]. The data JES is validated using the measurements described in the following, and the residual differences to the simulation are converted into the final JEC factor, which is only applied to data.

4.1. Absolute scale from Z($\mu\bar{\mu}$) + jet events

The JES can be measured in events where the jet is balanced against a well-measured reference object. CMS uses Z + jet events with $Z \to \mu\bar{\mu}$, where the p_{T} of the Z boson ($p_{\mathrm{T}}^{\mathrm{Z}}$) can be measured with high precision (at the 0.1 % level) compared to the jet due to the excellent muon-p_{T} resolution of the CMS detector. Z(e$\bar{\mathrm{e}}$) + jet and photon + jet events are used as a cross check [5].

The jet is required to be in the central detector region $|\eta| < 1.3$ and back-to-back in azimuth with the Z boson. Two different estimators of the jet response are defined,

$$R_{\mathrm{bal}} = \frac{p_{\mathrm{T}}^{\mathrm{jet}}}{p_{\mathrm{T}}^{\mathrm{Z}}}\,, \qquad (1) \qquad\qquad R_{\mathrm{MPF}} = 1 + \frac{\vec{\not{E}}_{\mathrm{T}} \cdot \vec{p}_{\mathrm{T}}^{\,\mathrm{Z}}}{(p_{\mathrm{T}}^{\mathrm{Z}})^2}\,. \qquad (2)$$

For the p_{T}-balance method (1), the jet is assumed to fully capture the fragmentation products of the initial parton that balances reference object. For the *missing transverse energy projection fraction* (MPF) method (2), it is assumed that the missing transverse energy ($\vec{\not{E}}_{\mathrm{T}}$) in the event solely arises from mismeasurements of the hadronic recoil to the reference object, such that the jet response is given by the projection of $\vec{\not{E}}_{\mathrm{T}}$ on the reference object's $\vec{p}_{\mathrm{T}}$.

The absolute JES is defined as the mean of the R_{bal} and R_{MPF} distributions, respectively. It depends on the amount of additional hadronic activity in the event, which affects the p_{T} balance between the jet and the reference object. This is clearly visible when measuring the JES as a function of $\alpha = p_{\mathrm{T}}^{\mathrm{jet2}}/p_{\mathrm{T}}^{\mathrm{Z}}$, the fractional p_{T} of the second-leading jet, cf. Fig. 3 (left). The R_{MPF}

is less affected than R_{bal} since it depends less on the association of the particles to the leading jet.

The residual JEC factor is determined as the data-to-simulation response ratio[2]. It is measured for different α and extrapolated linearly to $\alpha = 0$ to suppress the influence of the additional hadronic activity on the result. The correction is at the level of 1 % in the central detector region, demonstrating the good description of the response by the simulation. A validation of the procedure in Fig. 3 shows excellent agreement of the JES in data and simulation after all corrections.

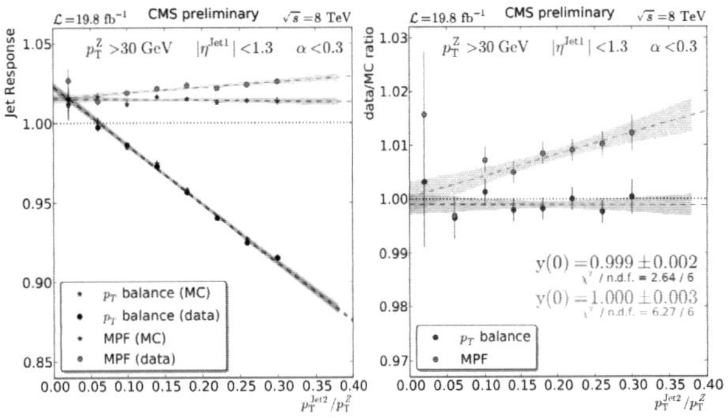

Figure 3. JES in data and simulation (left) and their ratio (right) after all corrections vs. $p_T^{\mathrm{jet2}}/p_T^Z$ measured in Z + jet events using the MPF and the p_T-balance response estimators. The JES is > 1 due to the different flavour composition compared to QCD-multijet events. [9]

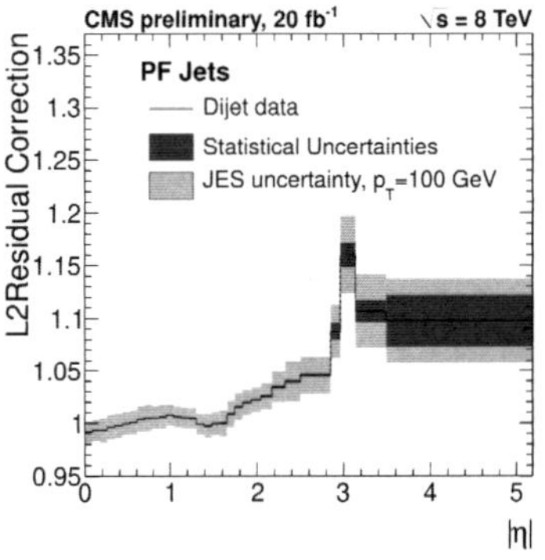

Figure 4. Relative residual JEC vs. η determined from dijet events using the MPF response-estimator. [9]

4.2. Relative scale from dijet events

Due to their relatively small production cross-section, Z + jet events are only used to measure the absolute JES in the central detector region. The JES in the forward regions is obtained from QCD-dijet events, which occur at a much higher rate [5]. They are selected requiring the two leading jets in the event to be back-to-back in azimuth, with at least one jet being in the central region $|\eta| < 1.3$. Even though there is no well-measured reference object, the two jets are expected to be balanced at parton level, and therefore, the scale relative to the JES of the central jet can be determined using slightly modified versions of the response estimators (1) and (2). The observed JES differences are converted into a relative residual JEC, which is below 5 % up to $|\eta| < 2.5$, cf. Fig. 4.

4.3. Jet energy correction uncertainty

After application of the described corrections, the JES at CMS is known to percent-level precision even in the difficult forward regions and better than 1 % for jets above 100 GeV in the central detector region, cf. Fig. 5. At low p_T, the uncertainties are dominated by the PU-correction uncertainty. At higher p_T, the dominant uncertainties are due to the fact that the simulation is used to extrapolate the Z + jet-derived JES to different flavour mixtures and p_T regions. In the forward regions with $|\eta| > 2.5$, the largest contribution stems from the uncertainty of the relative residual JEC, which is mostly due to the limited number of available dijet events.

[2] The MPF-based value, which is less affected by additional hadronic activity, is used for the JEC; it agrees with the p_T-balance result.

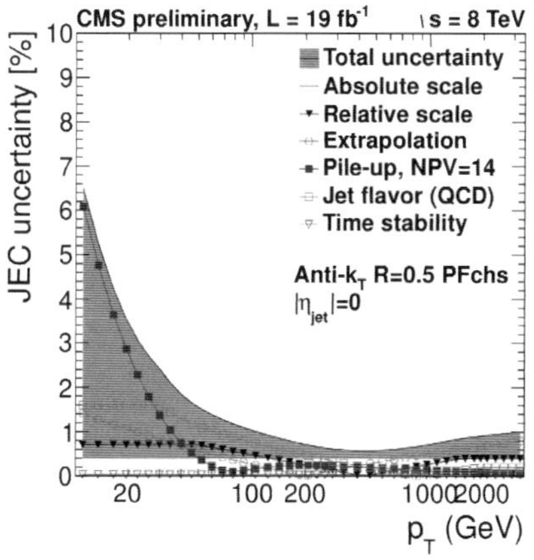

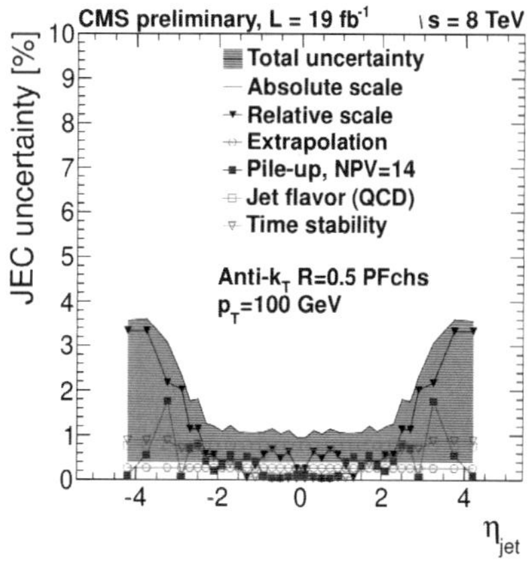

Figure 5. Relative JEC uncertainty and its components vs. p_{T} and η. [9]

5. Pile-up jet identification

In addition to contributing extra energy to the jets from the hard interaction, particles from PU collisions may also cluster and give rise to jets, referred to as PU jets. Even though these are typically very soft, they can overlap and form jets with sizable p_{T}, which occurs at a rate that grows quadratically with the number of PU collisions [10]. PU jets constitute a significant fraction of all low-p_{T} jets in the data collected during the 2012 LHC run, e.g. approximately 50 % of all 30 GeV jets in Z+jets events are PU jets, cf. Fig. 6, and this contribution will become even larger for the upcoming LHC runs with a much increased instantaneous luminosity.

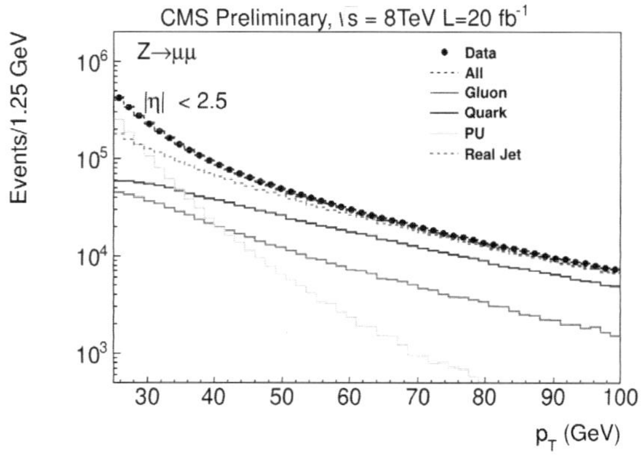

Figure 6. Jet p_{T} spectrum in Z + jets events in data and simulation, for which the contributions from PU and non-PU jets are indicated. [10]

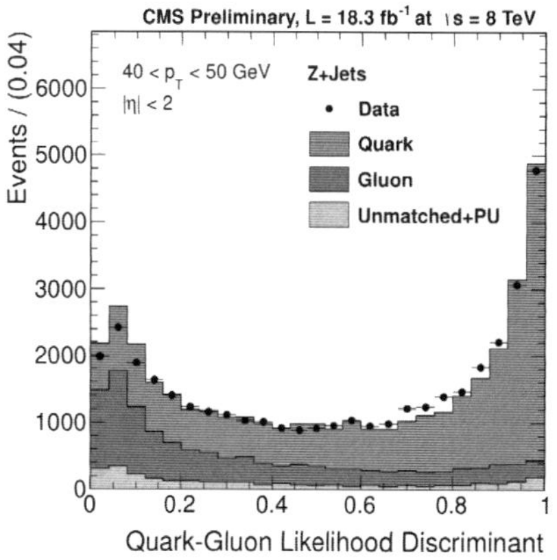

Figure 7. Quark-gluon jet discriminant in Z+jets events in data and simulation, for which the different flavour contributions are shown. [11]

CMS has developed a likelihood-based method to identify and reject PU jets [10]. The likelihood discriminator combines twelve jet-shape and vertex-information related variables that are sensitive to the characteristic substructure of PU jets, which are typically somewhat diffuse and contain many particles from PU vertices. The performance of the discriminator has been evaluated with simulated Z + jets events. A high PU-jet rejection efficiency is achieved of up to

90 – 95 % for central jets with $|\eta| < 2.5$ while retaining 99 % of the non-PU jets from the hard interaction; the performance degrades somewhat in the forward regions.

6. Quark-gluon jet discrimination

CMS has developed a tool to distinguish between jets initiated by light quarks and by gluons, which makes use of the PF event-reconstruction and exploits the different fragmentation properties of quarks and gluons [11]. In general, gluon jets consist of more and thus softer particles and are less collimated than quark jets. These differences are exploited in a likelihood discriminator, which combines three sensitive variables that are defined at the level of the PF jet-constituents: their multiplicity, the minor axis of the elliptical jet shape spanned by the second moments of the p_T-weighted constituents' directions, and $p_T D = \sqrt{\sum_i p_{T,i}^2} / \sum_i p_{T,i}$, which measures how the jet p_T is distributed among the constituents i. The impact of PU is mitigated by omitting charged particles that cannot be associated to the hard-interaction vertex.

The likelihood is built from jets in simulated QCD dijet events and parametrised as a function of p_T, η, and the average $p_{T,\text{offset}}$ density ρ for jets with $p_T > 30$ GeV and $|\eta| < 4.7$. A high quark-jet identification efficiency of up to 80 % at a gluon-jet rejection efficiency of 60 % is achieved. The discriminator output has been validated in Z + jets events, where good agreement is found between data and simulation as demonstrated in Fig. 7.

7. Summary

CMS commonly uses the particle-flow approach for jet reconstruction in order to optimally benefit from the high-resolution tracking system and electromagnetic calorimeter. Pile-up energy contributions to the jets can be reduced by up to 50 % already prior to the jet clustering by omitting charged particles originating from pile-up vertices.

Jet energy corrections are determined in a factorised approach. The dominant factors, which correct for the remaining impact of pile-up and the η and p_T dependence of the detector response, are derived from the simulation, which describes the data well. The obtained jet energy scale is validated in data, and small remaining differences to the simulation are converted into a residual correction factor. As a result, the overall jet energy scale at CMS is known to percent-level precision and better than 1 % for jets above 100 GeV in the central detector region.

Furthermore, CMS has developed advanced tools to reject pile-up jets and to discriminate between quark- and gluon-induced jets. The procedures exploit likelihood discriminators sensitive to the characteristic substructures of the different jet types.

In summary, jets at CMS are well-understood objects allowing high-precision measurements at the energy frontier, whereby many important final states become accessible for physics analysis. Hence, jets contribute in an essential way to the exploitation of the LHC physics-potential.

References

[1] CMS Collaboration 2008 *JINST* **3** S08004
[2] CMS Collaboration 2009 CMS-PAS-PFT-09-001
[3] Cacciari M, Salam G P and Soyez G 2008 *J. High Energy Phys.* **08** 063
[4] CMS Collaboration 2012 CMS-DP-12-012
[5] CMS Collaboration 2011 *JINST* **6** P11002
[6] Cacciari M and Salam G P 2008 *Phys. Lett. B* **659** 119 – 126
[7] Sjöstrand T, Mrenna S and Skands P Z 2006 *J. High Energy Phys.* **05** 026
[8] Agostinelli S *et al.* 2003 *Nucl. Instrum. Meth. A* **506** 250–303
[9] CMS Collaboration 2013 CMS-DP-13-033
[10] CMS Collaboration 2013 CMS-PAS-JME-13-005
[11] CMS Collaboration 2013 CMS-PAS-JME-13-002

The CMS Outer HCAL SiPM Upgrade

Artur Lobanov for the CMS Collaboration
Deutsches Elektronen-Synchrotron, Notkestrasse 85, D-22607 Hamburg
E-mail: artur.lobanov@desy.de

Abstract. The CMS Outer Hadron Calorimeter (HO) is the first large scale hadron collider detector to use SiPMs. By late January 2014 the installation of 1656 of 2376 channels was completed. The HO readout system provides for active temperature stabilization of the SiPMs to less than 0.1°C using Peltier coolers, temperature measurement, and software feedback. Each channel has independently controlled bias voltage with a resolution of 25 mV. Each SiPM is read out by 40 MHz QIE ADCs. We report on the system design, schedule and progress. The next phase for the detector is commissioning during 2014 before the 2015 LHC run. We report on the status of commissioning and plans for operation. We discuss the calibration strategy with local cosmic ray runs using the HO's self trigger ability. We discuss the plans for a global CMS operations run in November 2014.

1. The CMS Hadron Outer Calorimeter

The CMS central hadron calorimeter is a sampling calorimeter with brass absorber and plastic scintillator tiles with wavelength shifting fibers. The barrel hadron (HB) calorimeter is complemented with an outer calorimeter (HO) located just outside the cryostat and inside the magnetic flux return yoke [1]. It provides an additional calorimeter coverage of about 3λ thickness and ensures containment of highly energetic hadron showers, thus it works as a tail-catcher to improve the energy measurement of jets and missing transverse energy [2].

The HO is located in all 5 barrel rings of CMS and is split into 30 sections (ieta) along the Z-axis (beam-pipe). In the transverse plane, the HO consists of 12 sectors, each with 6 trays, totalling 72 sections (iphi), each 5°. The HO tiles contain two layers of scintillator in the central ring of the detector, and one layer everywhere else (see Fig. 1). The scintillation light is collected with wavelength shifting fibers and transmitted over clear fibers to front-end electronics placed close to the layers. Each phi-eta position is read out by a separate channel, summing to 2160 physical channels. A schematic layout of the calorimeters in CMS and the location of the HO scintillators is shown in Fig. 2. The tile size is chosen to match projective towers in the HB.

In addition, some readout channels are not connected to any scintillator and can be used for noise measurements and calibration. The total number of active electronics channels is 2376.

In the original design the wavelength shifted scintillator light was read out by hybrid photo diodes (HPDs). However, during operation in CMS they proved to be sub-optimal in the run conditions for several reasons:

- Discharge caused by the fringe field of the CMS magnet.
- Low gain and photo detection efficiency.
- Ageing.

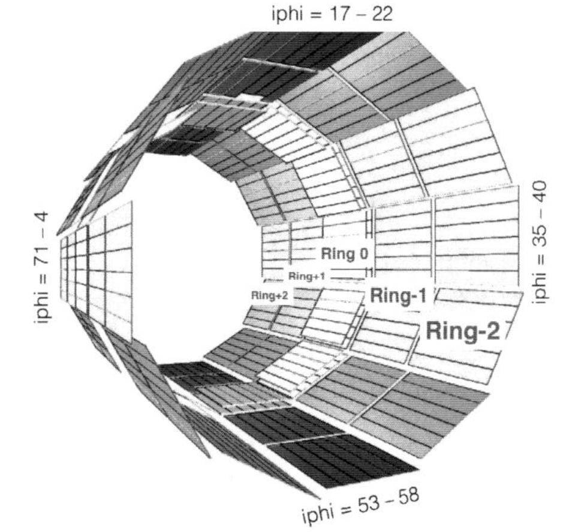

Figure 1. HO layout

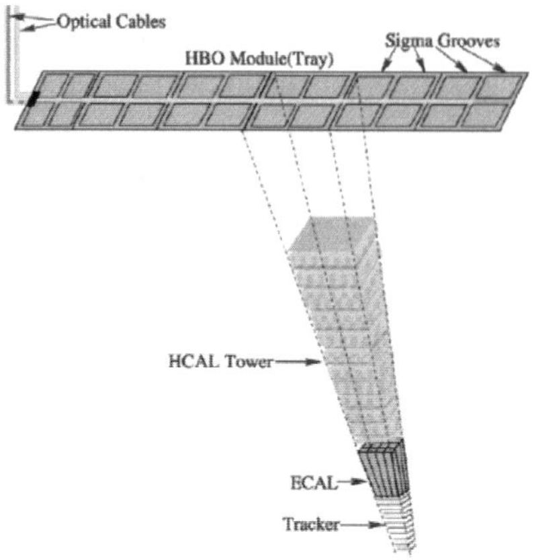

Figure 2. HO tiles corresponding to HB towers

The CMS HCAL group developed a drop-in replacement for the current front-end based on silicon photo multipliers (SiPMs) as photo-sensors [3]. The main advantages of SiPMs are the magnetic field insensitivity, relatively high photon-detection efficiency, and high gain. Also SiPM boards could easily fit into the limited space of the existing readout modules. In this configuration each eta-phi segment is readout by a single SiPM corresponding to a single electronics channel.

This upgrade design has been validated using laboratory measurements, test-beam data, and on the detector itself. The upgrade took place during the first LHC long shutdown in 2013-2014 - first, the existing readout modules (RM) were extracted from the detector, then, they were refurbished with SiPMs, and burned-in in a test stand. After verifying that the RMs were working properly, they were installed in CMS. Additionally, during the installation the commissioning of the upgraded parts began.

2. Commissioning

Primary reasons for commissioning during the installation are to identify problems with the new and existing hardware, to validate the installation, and, in case of malfunctions, to make repairs. This is of critical importance, as the barrel part of CMS will be closed until the next extensive upgrade.

The first commissioning step is a communication test with the readout system and the verification of slow control operation and channel response. Then, the measurement and optimization of SiPM operational variables is done as follows:

2.1. Temperature

The SiPM gain depends linearly on the temperature with a relative dependence of 8% gain shift per °C at a foreseen operating point of 1.5 V over-voltage [4]. This temperature dependence requires active control of the temperature of the SiPM with better than 0.1°C stability. Therefore, the SiPM temperature is stabilized by a Peltier element mounted on the back of the SiPM mounting board.

Instead of operating all SiPMs at the same temperature, it was chosen to fix the Peltier voltage at around 0.3 V. This minimizes the power consumption while providing a large range of cooling options.

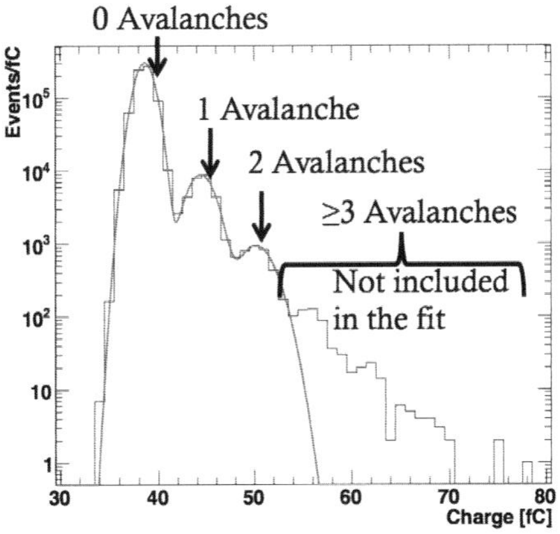

Figure 3. Pedestal spectrum of a SiPM with the gain fit

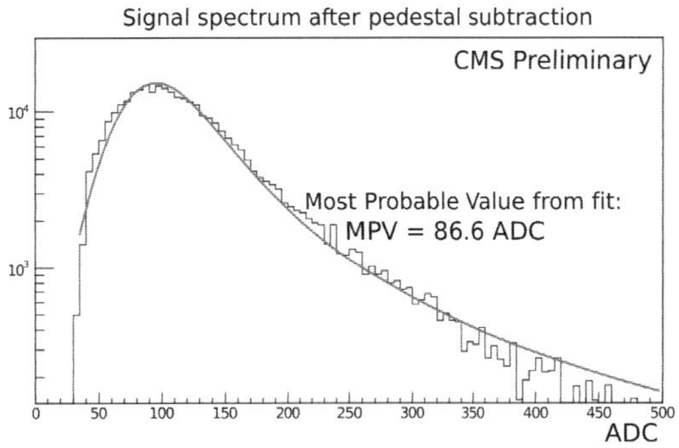

Figure 4. Single channel pedestal subtracted signal spectrum with coincidence trigger

2.2. Gain
The SiPM gain is adjusted by the operating voltage and can be measured from the single photo electron (SPE) spectrum as in Fig. 3.The distance between the SPE peaks (avalanches) is gain. All channels are set up to have gain around 6 ADC counts/photo-electron to prevent the ADC from saturation, but still have a good signal-to-noise ratio.

2.3. Pedestal
The SiPM signal is digitized corresponding to the LHC clock in 25 ns time-slices (TS), so this is the basic signal element from the ADC. To have a uniform response from all channels, the pedestal is set to 9 ADC counts per 1 TS. For a 4 TS sum, which is used in HO, this results in $\approx$ 36 ADC counts (0 avalanches in Fig. 3).

2.4. LED
An LED system is used to monitor the stability of the SiPMs outside of collisions. To adjust it to the new photo sensor type, an LED light calibration is performed. Because of different SiPM-fiber light coupling in Ring 0 and Rings 1-2, the LED amplitude set set to different values.

After the successful completion of these commissioning steps, the new readout hardware is ready for data taking.

3. Cosmic analysis

Cosmic runs with HO data is the verification of installation and operation of the new readout electronics. Additionally, an initial muon calibration can be obtained by extracting the MIP signal values of all channels.

Despite the location of the CMS detector, approximately 100 m below the surface, in certain parts of the underground cavern a 20 m wide access shaft allows a higher muon flux. During the long shutdown, some of the the CMS rings are placed in different positions including under the shaft. This biases the occupancy of the different scintillator tiles, but gives a high rate and yielding a higher statistics.

In contrast to global CMS runs which include data from its muon detector systems, the events for this analysis are taken locally, meaning only HO data is read out with the HO trigger. The HO trigger is setup as follows: each wheel with 12 sectors is split into 4 quadrants of 3 sectors each; a technical trigger processor constantly monitors for firing channels over a threshold of

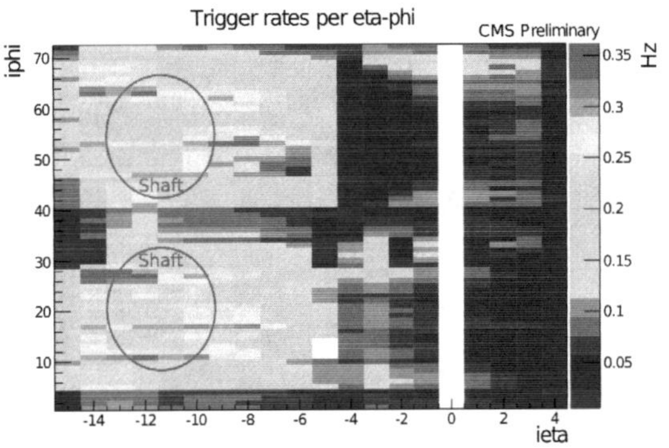

Figure 5. Trigger rate of each individual channel.

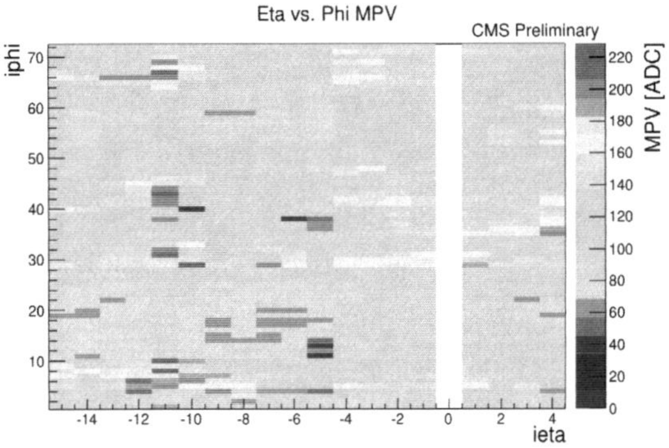

Figure 6. Landau fit most probable values (MPV) of individual channels

40 fC in 1 TS; if two channels from different quadrants are above threshold the trigger fires and the event is written to a disk. A time delay of 1 TS between the top and bottom quadrants is required to account for the time of flight through a diameter of approximately 8 m.

Trigger rates of all individual HO channels, obtained with this coincidence trigger configuration, are shown in figure 5. At the time of data taking, only 3 rings covering ieta -15 to +4 had been upgraded with SiPMs, but the ieta = 0 white stripe is due to HCAL coordinate system and not a physical gap in the detector. The empty spot at ieta -5 corresponds to the magnet chimney, where HO is not instrumented.

The highest rates are observed in the horizontal tiles (iphi = 15-25, 55-65) in the wheels under the shaft (YB-2: ieta -15,-11; YB-1: ieta -10,-5). The gap between the moving wheels and the central YB0 (ieta -4,4) is more than 5 m, so the drop in the rate around the edge at ieta -5 is considerably high. The fine structure of outliers along iphi lines is due to the different scintillator tiles sizes, which also leads to different light collection efficiencies. So ieta -15 shows lower rates because these tiles are half as big as the neighbours.

3.1. Muon calibration

For the muon calibration, the MIP value for each channel had to extracted. Theoretically, a MIP would give a Landau signal distribution for a thin layer, but because of the thickness of the scintillator the signal is smeared. Each channels' pedestal subtracted signal spectrum (where the signal is summed over 4 TS = 100 ns) is fitted with a Landau$\otimes$Gauss convoluted function and the most probable value (MPV) is taken as the MIP value. In case of noise contamination in the signal region, the tail is fitted with an exponential to improve the signal fit.

Figure 6 shows the MPV values for all individual HO channels. The MPV values are not uniform due to several factors:

- Ring 0 has 2 scintillator layers, while rings 1 and 2 have 2 layers.
- Altered light coupling of the fibers to the SiPMs in different rings.
- Different sizes of the scintillators lead to varying light detection efficiency.
- Fibers connecting SiPMs and tiles vary in length depending on the distance between scintillator and readout module. For YB-1,2 the readout is located between these rings (between ieta -11,-10) and for YB0 it is on both edges of the ring (ieta $\pm$4).
- Variation in iphi because of sector alignment and cosmic ray muon track path-length in the scintillator (track vs. tile angle): the lowest MPVs are measured at top and bottom tiles.

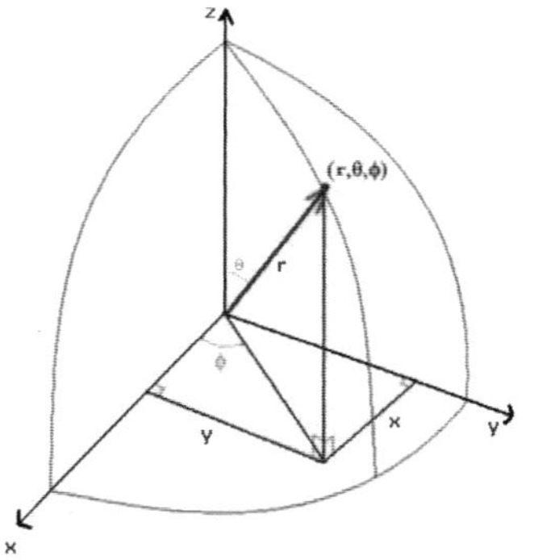

Figure 7. x and y are coordinates of the muon track vector projection on the horizontal plane. x,y = 0 corresponds to a vertical muon

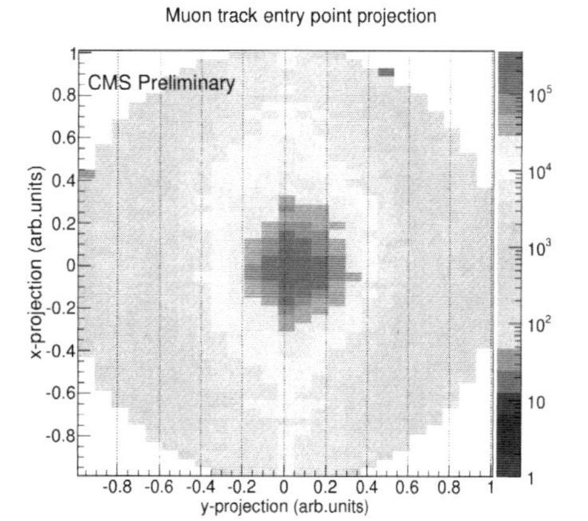

Figure 8. HO Muon entry point (as in Fig. 7) shifted because of shaft

The last variation is not hardware related, and causes a bias in the MIP value estimation as the calibration is performed for muons passing the tile perpendicular to it. The signal distribution is smeared and shifted towards higher values and, as each muon has a unique incident angle, the path-length correction has to be applied for each event. Therefore a signal correction algorithm has been developed. For each event the muon track is built using only two HO hits, and the angles between the track and the two scintillator planes are calculated. Then the recorded signal is multiplied by the cosine of this angle to get the value a normal muon would give in the tile.

The spatial distribution of the HO muon tracks in Fig. 8 shows a good reconstruction performance and agrees with the cavern layout.

Figure 9(a) shows the signal spectra with and without path-length correction for a side tile of HO, which has a large angle to the abundantly vertical muons coming from the sky. As expected, the uncorrected distribution is shifted to higher values and broader, than the corrected one. In

a)

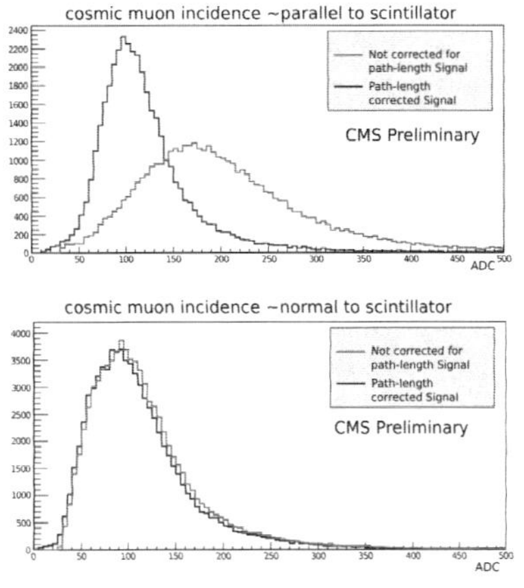

b)

Figure 9. Signal spectra for side (a) and top (b) tiles with and without muon path-length correction

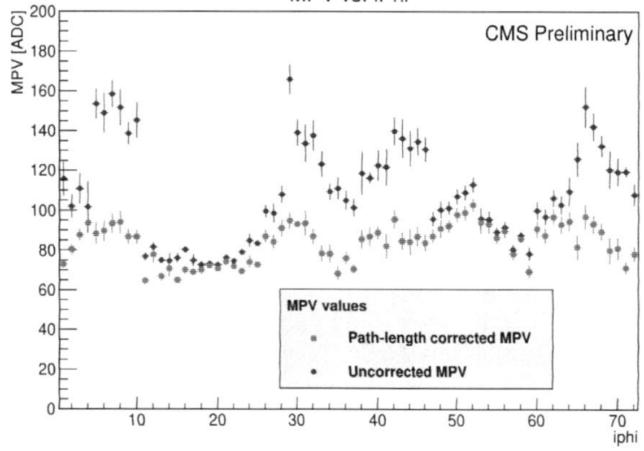

Figure 10. Averaged over YB-1,2 MPV profile along iphi

contrast, for a horizontally aligned tile, which is mostly hit normally by muons, the path-length correction doesn't introduce any visible shift as seen in Fig. 9b.

After the correction, the MPV values of different channels become more equal (compare blue histograms in Fig. 9 (a) and (b) with the remaining variation coming from hardware differences. Figure 10 shows an averaged profile of the MPV values before and after muon path-length correction. The average is taken for the rings -1 and -2 as these have the highest statistics due to the position under the access shaft. In the blue profile a modulation is seen which corresponds to the alignment of the HO sectors forming a circle. The correction flattens out the general modulation, but leaves local variations, which are related to the hardware.

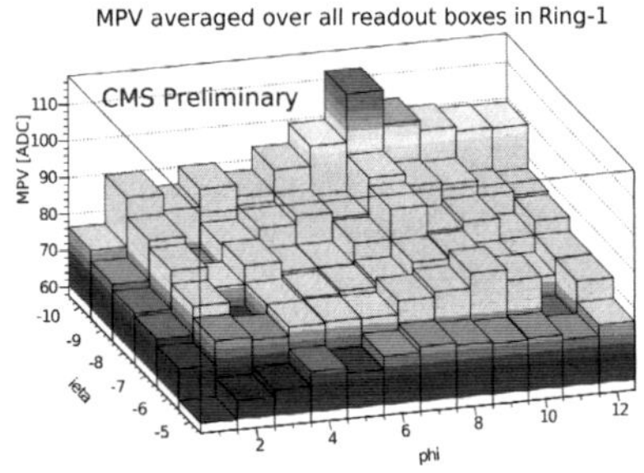

Figure 11. HO trigger layout

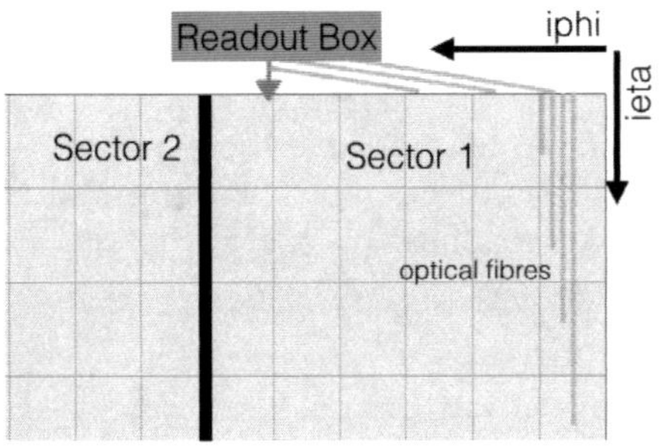

Figure 12. HO detector layout

A source of this remaining signal variation is the fiber light attenuation. From the cabling point of view, all sectors within a wheel are symmetric and should show the same behaviour of MPV variation with respect to the scintillator-SiPM distance. In rings -1 and -2 each readout box is connected to 12 iphi trays. Figure 11 shows the average MPV values over all readout boxes of ring -1 corresponding to ieta -10,-5. One can clearly see the location of the readout modules with SiPMs (where the distance to the tiles is the minimal) and the decrease to the sides.

The decrease along ieta is due to the cable length inside the scintillator tray, and along phi the MPV decreases with the length of the fibers connecting the trays with the readout boxes (see Fig. 12).

4. Summary

The CMS outer hadronic calorimeter is being upgraded with SiPMs during the first long shutdown of the LHC. The installed new readout is successfully commissioned and optimized for operation. A validation of the installation and calibration with cosmic muons is shown to be possible with an HO only trigger setup. Given the detector configuration, the obtained muon calibration is in agreement with the expectations from test-beam data.

As of April 2014 the whole HO detector is upgraded and commissioned. A global run with the CMS muons systems is planned for November 2014, and after the relaunch of the LHC in 2015, calibration with pp-collision data will become possible. The performance of the upgraded HO will allow the CMS RPC muon system to use HO data for a combined muon trigger.

References

[1] CMS Hadron Calorimeter Technical Design Report,, CERN/LHCC 97-31, CMS TDR 2, June 20th, 1997
[2] Abdullin S et al. (CMS Collaboration) 2008, *Eur. Phys. J.* **C57** 653-663
[3] Lutz B et al. (CMS Collaboration) 2012, *J. Phys.: Conf. Ser.* **404** 012018
[4] Anderson J et al. (CMS Collaboration) 2012, *Phys. Procedia* **37** 72 -78

Performance of MET reconstruction in CMS

Artur Apresyan

California Institute of Technology, Pasadena, CA, USA
ON BEHALF OF THE CMS COLLABORATION

E-mail: Artur.Apresyan@cern.ch

Abstract. The performance of missing transverse momentum reconstruction algorithms is presented using 8 TeV pp collision data collected with the CMS detector, corresponding to an integrated luminosity up to 19.7 fb^{-1}. The resolution of missing transverse momentum, including the effects of pileup interactions and advanced reconstruction algorithms, are measured using events with an identified Z boson or isolated photon.

1. Introduction

The Compact Muon Solenoid (CMS) detector [1] can detect most species of stable particles produced in proton-proton collisions delivered by the Large Hadron Collider (LHC). Notable exceptions are neutrinos and hypothetical neutral weakly interacting particles. Although these particles do not leave a signal in the detector, their presence can be inferred from the momentum imbalance in the plane perpendicular to the beam direction, known as missing transverse momentum and denoted by $\vec{E}\!\!\!/_{\mathrm{T}}$. Its magnitude is denoted by $E\!\!\!/_{\mathrm{T}}$ and is referred to as missing transverse energy (MET).

Missing transverse momentum plays a critical role in many physics analyses at the LHC. It is a key variable in many searches for physics beyond the standard model, such as supersymmetry and extra dimensions as well as for collider dark matter searches. It also played an important role in studies contributing to the discovery of a new boson at a mass of around 125 GeV, in particular in channels with the WW and $\tau\tau$ final states [2]. In addition, the precise measurement of $\vec{E}\!\!\!/_{\mathrm{T}}$ is critical for measurements of standard model physics involving W bosons and top quarks.

The $\vec{E}\!\!\!/_{\mathrm{T}}$ reconstruction is sensitive to detector malfunctions and various reconstruction effects resulting in particle momentum mismeasurements and particle misidentification. Precise calibration of all reconstructed physics objects is crucial for the $\vec{E}\!\!\!/_{\mathrm{T}}$ performance, and $\vec{E}\!\!\!/_{\mathrm{T}}$ is particularly sensitive to multiple proton-proton interactions in the same, earlier, and later bunch crossings (pileup interactions). Thus, it is essential to study $\vec{E}\!\!\!/_{\mathrm{T}}$ reconstruction in detail with data.

The central feature of the CMS apparatus is a superconducting solenoid, of 6 m internal diameter, providing a field of 3.8 T. Within the field volume are the silicon pixel and strip tracker, the crystal electromagnetic calorimeter (ECAL), and the brass/scintillator hadron calorimeter (HCAL). Muons are measured in gas-ionization detectors embedded in the steel return yoke. In addition to the barrel and endcap detectors, CMS has extensive forward calorimetry. A more detailed description of the CMS detector can be found in Ref. [1].

2. Data samples, particle reconstruction and event selection

The data samples used for studies presented in this paper were collected from February through December 2012 in proton-proton collisions at a centre-of-mass energy $\sqrt{s} = 8$ TeV, corresponding to an integrated luminosity of 19.7 ± 0.5 fb^{-1}.

CMS uses global event reconstruction, also called particle-flow (PF) event reconstruction [3], which consists of reconstructing and identifying each particle with an optimized combination of all subdetector information. For each event, hadronic jets are clustered from these reconstructed particles with the infrared and collinear safe anti-k_t algorithm, operated with a size parameter R of 0.5. The collision data are compared to samples of simulated events that are generated either using PYTHIA 6 [4], or MADGRAPH 5 [5] simulation programs. The generated events are passed through the GEANT4-based CMS detector simulation. The simulated events are weighted such that the distribution of the simulated pileup interaction multiplicity matches the expected distribution, as based on measurements of the instantaneous luminosities in data.

The $Z \to \ell^+\ell^-$ events, where ℓ is either a muon or an electron, are used in the $\not\!\!E_T$ resolution studies. In order to discriminate between prompt leptons and leptons which were produced inside a jet through decay of a hadron we quantify the isolation of the candidate particle by combining the transverse momenta of all other reconstructed particles inside a cone of $\Delta R < 0.3$ around the particles trajectory. Events with an invariant mass of the dimuon or dielectron system outside of the Z mass window $60 < m(ll) < 120$ GeV are rejected.

The $W \to e\nu$ are required to contain one well-identified, isolated electron satisfying $p_T > 30$ GeV and $|\eta| < 2.5$. Events with an additional electron passing the loose working point are vetoed if the additional electron also satisfies $p_T > 20$ GeV and $|\eta| < 2.5$.

In the semi-leptonic $t\bar{t}$ channel, we select single muon and single electron events from the full 2012 CMS dataset. Each event is required to pass either an e+jet or μ+jet trigger. We require at least 2 b-tagged jets with $p_T > 45$ GeV, at least 3 jets with $p_T > 45$ GeV, and at least 4 jets with $p_T > 20$ GeV. Exactly one identified and isolated lepton is required.

3. Reconstruction of MET

We define $\vec{\not\!\!E}_T$ as the imbalance in the transverse momentum of all particles in the final state of proton-proton collisions that interact via the electromagnetic or strong forces. Because of momentum conservation, $\vec{\not\!\!E}_T$ is the transverse momentum that is carried by weakly interacting or non-interacting particles, such as neutrinos. CMS has developed several distinct and complementary algorithms to reconstruct $\vec{\not\!\!E}_T$, already presented in [6]. The $\vec{\not\!\!E}_T$ reconstructed using a particle-flow technique (PF $\vec{\not\!\!E}_T$) is used in the majority of CMS analyses. It is defined as the negative vectorial sum over all PF particles transverse momenta.

In the following sections, we present a comparison of the performance of PF $\vec{\not\!\!E}_T$. In addition, two advanced $\vec{\not\!\!E}_T$ reconstruction algorithms specifically developed to mitigate effects from large numbers of pileup interactions are discussed. In all comparisons the $\vec{\not\!\!E}_T$ is corrected by correcting the p_T of the jets to the particle level p_T using the jet energy corrections, and recalculating the $\vec{\not\!\!E}_T$. In addition to the corrections discussed above, additional corrections to improve the performance of the $\vec{\not\!\!E}_T$ reconstruction are used in the analyses, such as corrections for the residual ϕ-asymmetry observed in $\vec{\not\!\!E}_T$.

4. Measurement of the MET resolution

The performance of the $\vec{\not\!\!E}_T$ reconstruction is assessed using events where an identified Z boson or isolated photon is present. While there is no genuine $\vec{\not\!\!E}_T$ in such events, we can induce it by removing the vector boson from the event reconstruction. Using the well-measured and well-understood vector boson momentum as a reference, we are able to measure the scale and

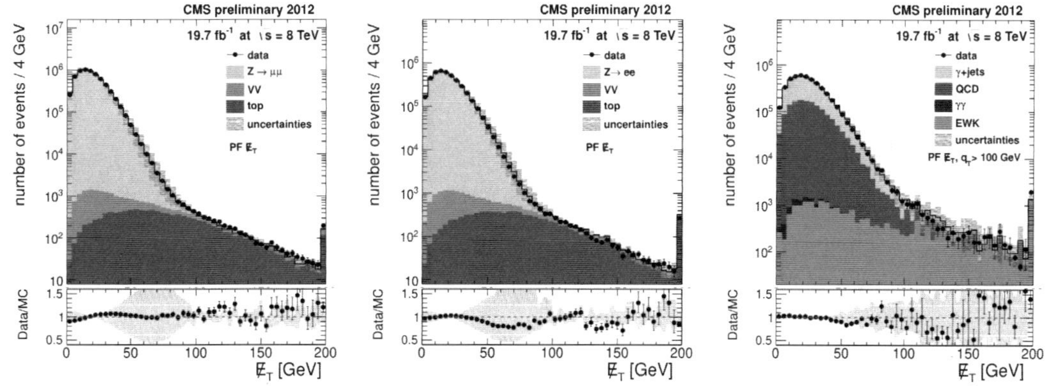

Figure 1. PF $E_{\mathrm{T}}^{\mathrm{miss}}$ distribution in events with $Z \to \mu^+\mu^-$ (left), $Z \to e^+e^-$ (middle), and photon events (right). The points of the lower section of the plots contain the statistical uncertainties of data and simulation, and the grey error band is the systematic uncertainty on the simulation.

resolution of $\not\!\!E_{\mathrm{T}}$ in an event sample with a global hadronic system that is kinematically similar to important Standard Model backgrounds such as $t\bar{t}$+jets and W+jets.

The PF $E_{\mathrm{T}}^{\mathrm{miss}}$ distributions in $Z \to \mu^+\mu^-$, $Z \to e^+e^-$, and photon events are presented in Fig. 1. Good agreement of data and simulation is observed in all distributions. Momenta of muons/electrons (photons) originating from Z decays (photons) are reconstructed with resolutions of $\sigma_{p_{\mathrm{T}}}/p_{\mathrm{T}}$ ~1–4 (1–3)%, while the energy of jets is reconstructed with resolutions of typically σ_E/E ~10–15%. Thus the $\vec{\not\!\!E}_{\mathrm{T}}$ resolution in Z or $\gamma+$jets events is dominated by the resolution with which the hadronic activity in the event is reconstructed.

Denoting the vector boson momentum in the transverse plane by $\vec{q}_{\mathrm{T}}$, and the hadronic recoil, defined as the vectorial sum of the transverse momenta of all particles except the vector boson (or its decay products, in the case of Z bosons), is $\vec{u}_{\mathrm{T}}$. Momentum conservation in the transverse plane requires $\vec{q}_{\mathrm{T}} + \vec{u}_{\mathrm{T}} + \vec{\not\!\!E}_{\mathrm{T}} = 0$.

The presence of a well measured Z boson or photon provides both a momentum scale, $q_{\mathrm{T}} \equiv |\vec{q}_{\mathrm{T}}|$, and a unique event axis, along the unit vector $\hat{q}_{\mathrm{T}}$. The hadronic recoil can be projected onto this axis, yielding two signed components, parallel ($u_{\|}$) and perpendicular ($u_{\perp}$) to the event axis. Since $u_{\|} \equiv \vec{u}_{\mathrm{T}} \cdot \hat{q}_{\mathrm{T}}$, and the observed hadronic system is usually in the opposite hemisphere from the boson, $u_{\|}$ is typically negative. The scalar quantity $-\langle u_{\|}\rangle/q_{\mathrm{T}}$ is refered as the $\vec{\not\!\!E}_{\mathrm{T}}$ response, and the distribution of $-\langle u_{\|}\rangle/q_{\mathrm{T}}$ versus q_{T} as a response curve.

The $\vec{\not\!\!E}_{\mathrm{T}}$ energy resolution is assessed with a parametrization of the $u_{\|} + q_{\mathrm{T}}$ and $u_{\perp}$ distributions by a Voigtian function, defined by the convolution of a Breit-Wigner (BW) distribution and a Gaussian (G) distribution:

The resolution curves, $\sigma(u_{\|})$ and $\sigma(u_{\perp})$ versus q_{T}, are shown in Fig. 2. The resolution increases with increasing q_{T}, and the data and simulation curves are in reasonable agreement for each channel. Figure 3 shows the resolution curves $\sigma(u_{\|})$ and $\sigma(u_{\perp})$ versus the number of primary vertices N_{vtx}, for both Z channels and the $\gamma+$jets channel. The offset of the curve is related to the resolution in Z or $\gamma+$jets events without pileup and the dependence with respect to N_{vtx} indicates how much the pileup degrades the $\vec{\not\!\!E}_{\mathrm{T}}$ resolution. For each additional pileup interaction, the PF $\vec{\not\!\!E}_{\mathrm{T}}$ resolution is degraded by around 3.4 GeV in quadrature.

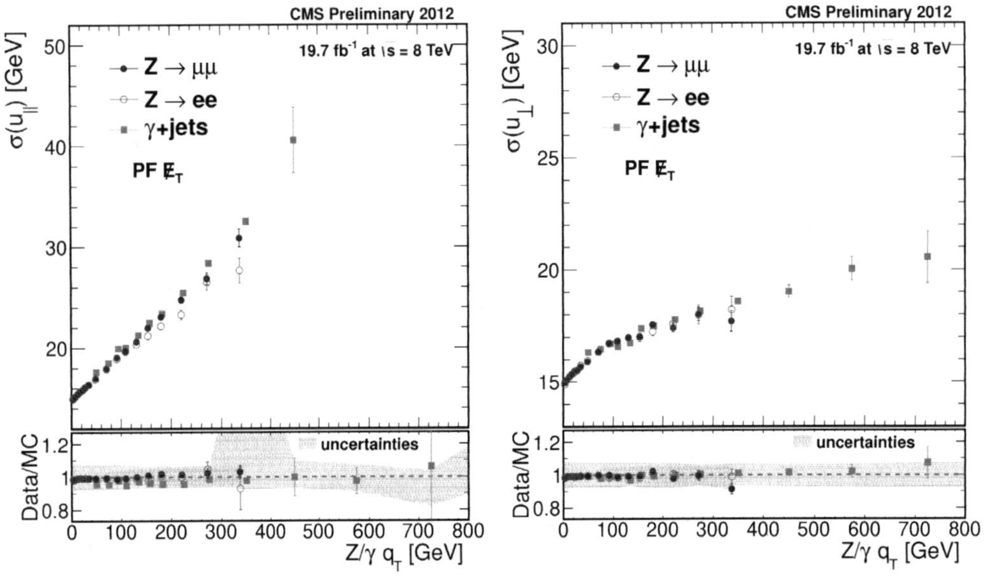

Figure 2. Resolution curves of the parallel recoil component (left) and perpendicular recoil component (right) versus Z/γ q_{T} for PF $\vec{E\!\!\!/}_{\mathrm{T}}$ for events with Z and γ. Results are shown for $Z \to \mu^+\mu^-$ events (full blue circles), $Z \to e^+e^-$ events (open red circles), and photon events (full green squares). The upper frame of each figure shows the response in data; the lower frame shows the ratio of data to simulation. The q_{T} value for each point is determined based on the average q_{T} value in data contributing to each point.

5. Pileup-mitigated MET

Since the vast majority of pileup interactions do not have significant $\vec{E\!\!\!/}_{\mathrm{T}}$ and the average value of $\vec{E\!\!\!/}_{\mathrm{T}}$ projected in any direction is zero, the effect on $\vec{E\!\!\!/}_{\mathrm{T}}$ response from pileup interactions is small. However, as shown in the previous section, pileup interactions have a considerable effect on the $\vec{E\!\!\!/}_{\mathrm{T}}$ resolution. In events where the recoil p_{T} is small and the instantaneous luminosity is around the mean value of 2012 data taking, which corresponds to approximately 20 pileup interactions, the contribution to the $\vec{E\!\!\!/}_{\mathrm{T}}$ resolution from pileup interactions is larger than the contribution from the hadronic recoil.

CMS Collaboration developed algorithms to reduce the effect of pileup interactions on $\vec{E\!\!\!/}_{\mathrm{T}}$ reconstruction, hereafter referred to as the "No-PU PF $E_{\mathrm{T}}^{\mathrm{miss}}$" and "MVA PF $E_{\mathrm{T}}^{\mathrm{miss}}$" algorithms. These algorithms are based on dividing each event into two components: particles that are likely to originate from the primary hard-scattering pp interaction (HS particles) and particles that are likely to originate from pileup interactions (PU particles). More details on the details of the algorithms are given in Ref. [7].

The resolution distributions as a function of N_{vtx} are shown in Fig. 3 and includes also as a reference the standard PF $\vec{E\!\!\!/}_{\mathrm{T}}$ algorithm. Significantly reduced dependence of the resolution on pileup interactions in both data and MC simulation. This reduced pileup dependence can significantly increase the sensitivity of searches for new physics. As an example, the sensitivity of the search for the Higgs boson decaying into tau pairs is improved by $\sim 20\%$ thanks to its use of the MVA PF $E_{\mathrm{T}}^{\mathrm{miss}}$ [8].

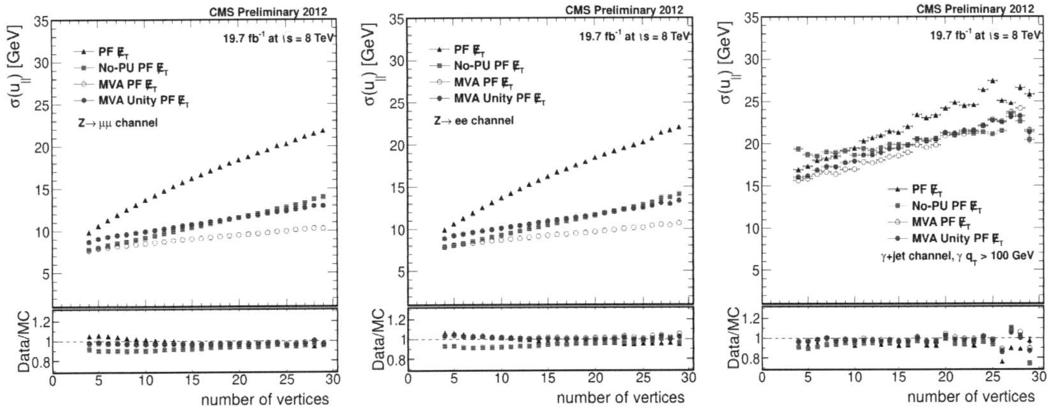

Figure 3. Parallel resolution as a function of the number of reconstructed vertices for PF $\vec{\not\!\!E}_{\mathrm{T}}$ (black triangles), No-PU PF $\vec{\not\!\!E}_{\mathrm{T}}$ (red squares), MVA PF $\vec{\not\!\!E}_{\mathrm{T}}$ (blue open circles),and MVA Unity PF $\vec{\not\!\!E}_{\mathrm{T}}$ (violet full circles) in $Z \to \mu^+\mu^-$ (left), $Z \to e^+e^-$ (middle) and $\gamma +$jets events (right).

6. MET Significance

The ability to distinguish between events with spurious $\vec{\not\!\!E}_{\mathrm{T}}$ and those with genuine $\vec{\not\!\!E}_{\mathrm{T}}$ is important for analyses using missing transverse energy variables. Fake $\vec{\not\!\!E}_{\mathrm{T}}$ may arise from object misreconstruction, finite detector resolution, or detector noise. To help identify such events, we have developed a missing transverse energy significance variable, which we will denote by "$\not\!\!E_{\mathrm{T}}$ Significance", or simply $\mathcal{S}$. On an event-by-event basis, the Significance evaluates the probability that the observed $\vec{\not\!\!E}_{\mathrm{T}}$ is inconsistent with a null hypothesis $\vec{\not\!\!E}_{\mathrm{T}} = 0$, given the full event composition and resolution functions for each object in the event. A high value of $\mathcal{S}$ is an indication that the $\vec{\not\!\!E}_{\mathrm{T}}$ observed in the event is not well explained by resolution smearing alone, suggesting that the event may contain unseen objects such as neutrinos or more exotic weakly interacting particles. A detailed description and definition of the Significance is given in [6].

The distributions of the $\vec{\not\!\!E}_{T}$ Significance are shown in Figs. 4 in data and simulation for both the $W \to e\nu$ and semi-leptonic $t\bar{t}$ channels. Some interesting features are apparent in the composition of simulation events in the Significance spectra. In the $W \to e\nu$ channel, events arising from zero true $\not\!\!E_{\mathrm{T}}$ physics channels, such as QCD and Drell-Yan events, are mostly found at low values of Significance compared to the broad distribution of nonzero-$\not\!\!E_{\mathrm{T}}$ events. The semi-leptonic $t\bar{t}$ channel has a significant nonzero-$\not\!\!E_{\mathrm{T}}$ background stemming from dileptonic $t\bar{t}$ decays. The dileptonic $t\bar{t}$ spectrum falls more slowly than the semi-leptonic $t\bar{t}$ signal as we approach the tail region of $\mathcal{S}$.

In the $W \to e\nu$ channel, there is a performance benefit in using $\vec{\not\!\!E}_{T}$ Significance, when compared to simpler background-discrimination variables such as $\not\!\!E_{\mathrm{T}}$ alone or the approximate Significance variable $\not\!\!E_{\mathrm{T}}/\sqrt{\sum E_{\mathrm{T}}}$. For example, choosing a working point with 50% signal efficiency yields a background efficiency of 8.2% using $\not\!\!E_{\mathrm{T}}$, 5.1% using $\not\!\!E_{\mathrm{T}}/\sqrt{\sum E_{\mathrm{T}}}$, and 4.0% using the Significance as a cut variable. In the semi-leptonic $t\bar{t}$ channel, $\mathcal{S}$ provides discrimination that is comparable to $\not\!\!E_{\mathrm{T}}$ and $\not\!\!E_{\mathrm{T}}/\sqrt{\sum E_{\mathrm{T}}}$. This reflects the fact that $\mathcal{S}$ is optimized for discriminating events which satisfy the null hypothesis ($\vec{\epsilon} = 0$) from those that do not. In the case of semi-leptonic $t\bar{t}$, the dominant background contribution comes from dileptonic $t\bar{t}$ decays with large, real, $\not\!\!E_{\mathrm{T}}$.

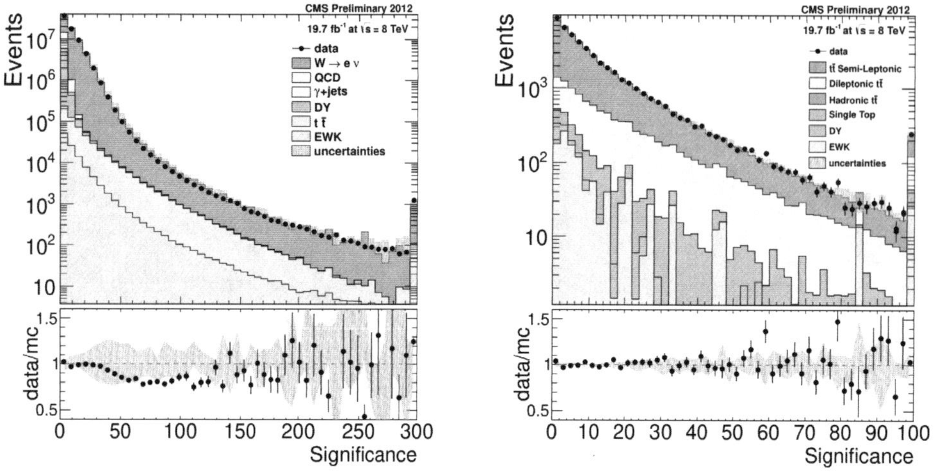

Figure 4. Distribution of $\vec{E\!\!\!/}_T$ Significance in the (Left) $W \to e\nu$ and (Right) $t\bar{t}$ channels.

7. Summary

The performance of $\vec{E\!\!\!/}_{\mathrm{T}}$ reconstruction algorithms has been studied using data collected in 8 TeV pp collisions with the CMS detector at the LHC. The performance of the $\vec{E\!\!\!/}_{\mathrm{T}}$ reconstruction algorithms is found to be well reproduced by the MC simulation. Pileup interactions are found to degrade the PF $\vec{E\!\!\!/}_{\mathrm{T}}$ resolution by 3.4 GeV in quadrature per additional pileup interaction. The performance of advanced $\vec{E\!\!\!/}_{\mathrm{T}}$ reconstruction algorithms developed to cope with large numbers of pileup interactions was also been studied, and shows a significantly reduced dependence on pileup interactions.

References

[1] CMS Collaboration, "The CMS experiment at the CERN LHC", JINST 3 (2008) S08004, doi:10.1088/1748-0221/3/08/S08004.

[2] CMS Collaboration, "Observation of a new boson at a mass of 125 GeV with the CMS experiment at the LHC", Phys. Lett. B716 (2012) 3061, doi:10.1016/j.physletb.2012.08.021, arXiv:1207.7235.

[3] CMS Collaboration, "Commissioning of the Particle-flow Event Reconstruction with the first LHC collisions recorded in the CMS detector", CMS Physics Analysis Summary CMS-PAS-PFT-10-001, (2010).

[4] T. Sjostrand, S. Mrenna and P. Z. Skands, "PYTHIA 6.4 Physics and Manual," JHEP **0605**, 026 (2006) [hep-ph/0603175].

[5] J. Alwall, P. Demin, S. de Visscher, R. Frederix, M. Herquet, F. Maltoni, T. Plehn and D. L. Rainwater *et al.*, "MadGraph/MadEvent v4: The New Web Generation," JHEP **0709**, 028 (2007)

[6] CMS Collaboration, "Missing transverse energy performance of the CMS detector", JINST 6 (2011) P09001, doi:10.1088/1748-0221/6/09/P09001, arXiv:1106.5048.

[7] CMS Collaboration, "Event Reconstruction in CMS and Performance for Jets, Taus, and ETmiss", CMS Physics Analysis Summary CMS-PAS-JME-12-002, (2013).

[8] CMS Collaboration, "Evidence for the 125 GeV Higgs boson decaying to a pair of τ leptons", CMS-HIG-13-004

Journal of Physics: Conference Series **587** (2015) 012007 doi:10.1088/1742-6596/587/1/012007

CMS Hadron Forward Calorimeter Phase I Upgrade Status

Yasar Onel[1]
On behalf of the CMS Collaboration

The University of Iowa, 203 Van Allen Hall, Iowa City, IA, 52242, USA

E-mail: yasar-onel@uiowa.edu

Abstract: The CMS Hadronic Forward Calorimeter has undergone upgrade maintenance during the LHC Long Shutdown 1. The Hamamatsu R7525 PMTs have been replaced with Hamamatsu R7600U-200-M4 PMTs, which have thinner window glass that reduces window-hit events. The R7600 PMTs also have multi-anode readout feature to further enable discrimination of window-hits while also allowing the recovery of true signal energy. Higher quantum efficiency of the R7600 PMTs improves calorimeter resolution. The new PMTs were tested and calibrated; new PMT baseboards were designed and tested, and can be configured to readout 1, 2, or 4 anodes of the R7600. New radiation hard (100Gy) QIE front-end electronics were designed for reading out the new PMTs and include a TDC with < 800ps resolution. New back-end electronics based on the microTCA industrial standard have been tested.

1. Introduction

The Compact Muon Solenoid (CMS) [1] is a general-purpose detector designed to run at the highest luminosity provided by the CERN Large Hadron Collider (LHC). Coverage between pseudorapidities (η) of 3 and 5 is provided by the steel/quartz fiber Hadron Forward (HF) calorimeter. The front face is located at 11.2 m from the interaction point and the depth of the absorber is 1.65 m. The signal originates from Čerenkov light emitted in the quartz fibers, which is then channeled by the fibers to photomultipliers. Machining 1 mm square grooves into steel plates, which are then diffusion welded together, creates the absorber structure. The diameter of the quartz fibers is 0.6 mm and they are placed 5 mm apart in a square grid. The quartz fibers, which run parallel to the beam line, have two different lengths (1.43 m and 1.65 m), which are inserted into grooves, creating two effective longitudinal samplings. The so-called "short fibers" start 22 cm inside the absorber, hence are mostly sensitive to hadron interactions. There are 13 towers in η, all with a size given by $\Delta\eta \approx 0.175$, except for the lowest-η tower with $\Delta\eta \approx 0.1$ and the highest-η tower with $\Delta\eta \approx 0.3$. The ϕ segmentation of all towers is 10°, except for the highest-η which has $\Delta\phi = 20°$. This leads to 900 towers and 1800 channels in the two HF calorimeter modules [2]. Details of the HF calorimeter design, together with test beam results and calibration methods, can be found in [3].

The photomultiplier tubes (PMTs) of the CMS HF calorimeter generate a large, fake signal when the PMT window is traversed by a relativistic charged particle due to Čerenkov light production at the

[1] To whom any correspondence should be addressed.

PMT window. These PMTs have circular windows with 2.54 cm diameter and 2 mm thickness at the center that gets thicker towards the rim (Hamamatsu R7525 [4]). This already-known problem was observed in the 2010 and 2011 CMS data to degrade data quality and to constitute a potential to interfere with rare physics events. An upgrade plan was therefore formulated for CMS HF, summarized in Table 1. Figure 1 describes the timeline of LHC operations in the next ~ 10 years.

Table 1. HF Upgrade Schedule

Upgrade	**Date**
PMT Replacement	LS1 - 2014
PMT Support Cabling	LS1 - 2014
Front-End Electronics	Before 2016 [a]
Back-End Electronics	Before 2016 [a]

[a] Although scheduled for LS2, likely will be installed prior to LS2.

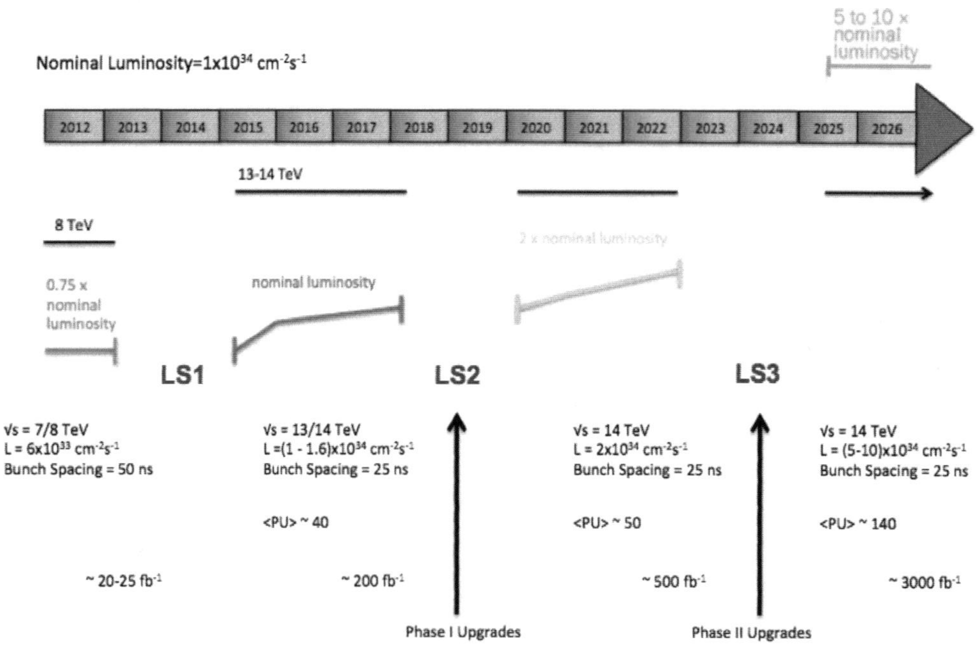

Figure 1. The timeline of LHC operations in the next ~ 10 years.

In the framework of the CMS HF upgrade program, several types of PMTs were tested and the four-anode R7600U-200-M4 by Hamamatsu [4] was selected as the replacement PMT for the upgrade [5]. The upgrade PMT has a square window of size 1.8 cm × 1.8 cm and thickness less than 1 mm indicating a significant reduction in the amount of glass seen by the traversing relativistic particles. The new PMT not only reduces the intrinsic level of background, but it also enables tagging of background events and recovering the underlying signal event (if any) by using the multi-anode features.

Journal of Physics: Conference Series **587** (2015) 012007 doi:10.1088/1742-6596/587/1/012007

A readout box (RBX) prototype was built to enable the tests of different readout options for the new four anode PMTs. The new readout boards provide the flexibility to switch between four-channel, two-channel and single-channel readout of the four anode PMTs where the four-channel readout option enables the full multi-anode functionality. Both the internal and external cabling of the RBX were also specific designs and selections, therefore form an integral part of the prototype.

The prototype RBX was tested in the CERN H2 beam line [6] in Summer - Fall 2011 with electron and muon beams to mimic calorimeter and background responses respectively.

HF front-end and back-end electronics will be upgraded in order to support multi-channel readout of the R7600 PMT. Testing has been completed and commissioning is ongoing, with a goal to install before Long Shutdown 2.

The front-end electronics, named QIE10 (Charge Integrator and Encoder), are radiation tolerant (up to 100 Gy) and include the addition of a TDC circuit. The TDC has a rising edge resolution of 390 to 780 ps (5 or 6 bits). This next generation QIE card has deadtimeless integration and digitization of charge in 25ns buckets; large dynamic range (3 fC to 330 pC) stored in 17 bits, and has digitization error less than the calorimeter resolution (2-3%). This increased data will require a new pipeline, and the SLHC beam conditions will require a radiation hard solution. The GigaBit Transceiver (GBTx) technology will serialize data from the QIE10 cards and transmit the data to the back-end electronics in the counting room. 288 new GBT data-link fibers will allow for the addition of the TDC bits, including the presence of multiple rising or falling edges. Upgraded HF back-end electronics will replace the current VME based system and will be installed into an industrial-standard microTCA crate. This is the first step of the Phase I electronics upgrade for HF.

Here we describe the necessity of the upgrade, the individual upgrade components and the status of progression in detail.

2. Upgrade components

The Phase I Upgrade of the CMS hadron Forward calorimeters include the replacement of the R7525 PMTs with the R7600 PMTs, the design of the new base and adapter boards to include the flexibility of switching between 1, 2 and 4-channel readout options, associated light guides, sleeves and cabling. Table 2 summarizes the upgrade components. In addition to the components listed, the fron-end and back-end electronics will also be upgraded.

Table 2. HF Phase I Upgrade Components

Upgrade	**Image**
Baseboard	

ROBOX Light Guides

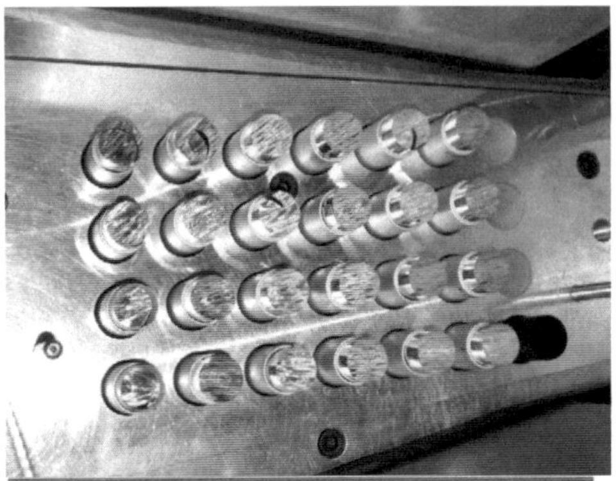

Light Sleeve

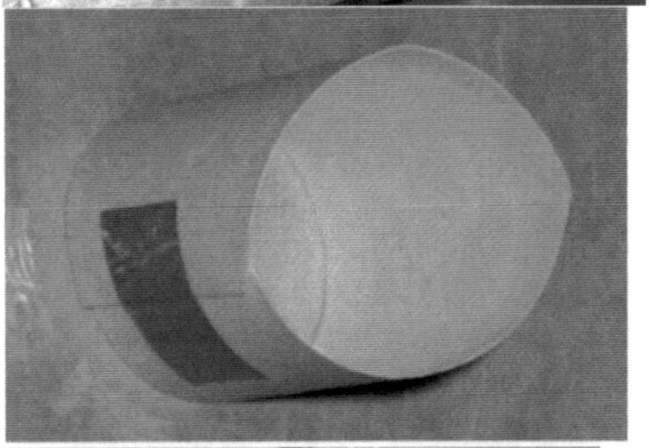

ROBOX Cabling

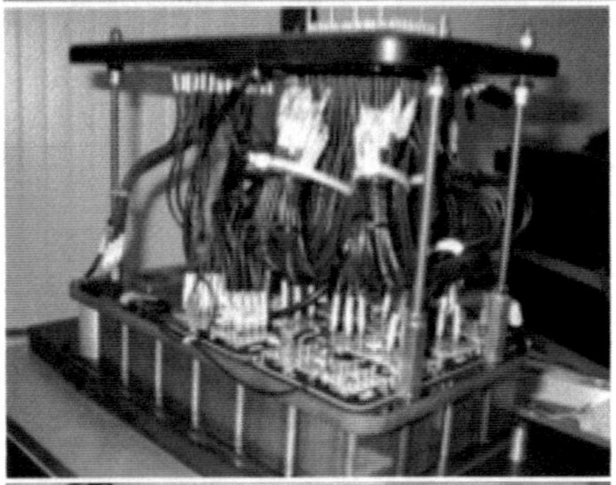

PMTs and Baseboard

Winchester Cable

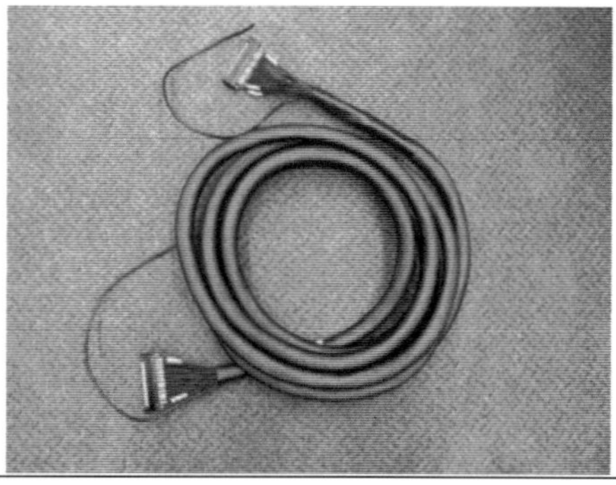

2.1. Upgrade PMTs

After it was discovered that muons traversing the CMS detector were interacting with the windows and bodies of the HF PMTs and were creating large anomalous signals in the detector, a program was developed to decide how to mitigate this signal contamination. Studies of the anomalous signal timing showed that muons from the interaction point arrived in the PMTs earlier than signals that traveled through the HF fibers. This resulted in the addition of a Time-to-Digital Converter (TDC) circuit to flag signals that arrived out of time with other HF signals. Studies of HF PMTs and multi-anode PMTs placed directly in beam in front of wire-chambers motivated the usefulness of a multi-anode PMT, as an interaction with one anode of the PMT window would be able to be discriminated by comparing the individual signals of the multiple anodes. The HF mixing modules should present light of roughly uniform intensity onto the face of the multi-anode PMTs, so a variation in light intensity between anodes larger than what would be expected statistically could be used to flag events. This upgrade has also motivated an updated readout box and accompanying front-end and back-end electronics.

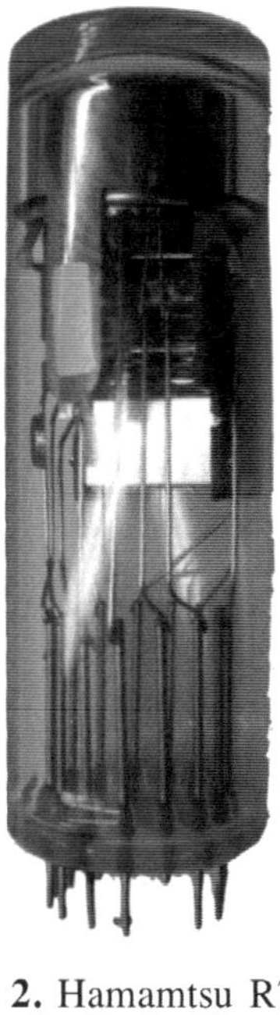

Figure 2. Hamamtsu R7525 – The original HF PMT

Figure 3. Hamamatsu R7600U-200-M4 – The 4-anode HF PMT replacement.

The R7600 PMTs have a thinner window that reduces the likelihood and intensity of a direct interaction with a particle. We have shown that the rate of interaction as well as the response of the interaction is greatly reduced regardless of angle of incidence [5]. The response of a MIP is reduced more than a factor of 5. The R7525 PMTs had glass bodies as well as glass windows; the R7600s contain glass only in its thin window, and the body is contained within a metal envelope.

A goal of the multi-anode PMT replacement was to design algorithms to tag and correct MIP interactions by comparing the relative levels of signal in each anode. The quad-anode design is most attractive for this purpose, though a two-channel readout design has been adopted for the current operation. The readout bases allow for configuring 1, 2, and 4 channel readout modes, enabling future operation of 4-channel readout. We have tested algorithms in test beam [8] and have shown that both readout modes can accurately flag MIP events, and have high efficiencies, (~98% 4-channel and ~90% 2-channel readout).

The actual signal within the MIP flagged events can be recovered whereas such events would be omitted with the old HF PMT readout. Since a light-guide mixing-module delivers roughly uniform light from the HF optical fibers to the face of the PMT, all 4 anodes should in principle receive almost equal amount of light. If a MIP hits the surface of the R7600, its Čerenkov radiation will be deposited into the quadrant it hits. One quadrant will produce a higher signal than the others, and different algorithms can be used to recover the signal that would have been recorded had the MIP not interacted with the PMT and contaminated the signal. This case is illustrated in Figure 4.

Figure 4. MIP interacting simultaneously with true HF Signal

The specifications of the R7600 PMTs exceed the optical demands of the HF detector. They have a higher quantum efficiency than the R7525, which will improve HF resolution. The mesh structure of the dynodes is less susceptible to interference from the magnetic field in CMS. Overall this is a straightforward replacement to the existing system, with no dramatic changes and significant retooling necessary.

2.2. Readout boards

HF requires 72 readout boxes, 216 baseboards with 1728 PMTs, and the ability for each 4-anode PMT to be read out in 1, 2, or 4 anode operations. This design challenge was met with a single PMT board, shown in Figure 5.

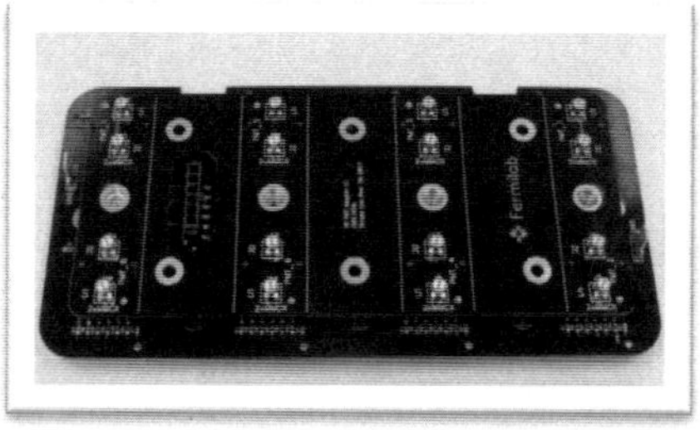

Figure 5. HF R7600 PMT Baseboard with adapter boards attached.

The R7600 baseboard has 8 power supplies (1 per PMT) with 3 sets of independent voltages (Vpc, Vd9, Vd10). The design incorporates conventional resistive bases and provides reference signals for the QIE readout card, and supplies ground connections to the PMT shields.

In order to comply with the different phases of the upgrade program, the adapter board is designed to provide convenient switching between different readout options. The adapter board uses through hole mounted MMCX (micro miniature coaxial) signal connectors. There are screw/standoff retainers for attachment to the base board.

2.3. Front-end electronics

An updated QIE, called QIE10 has been designed to digitize the PMT signal and the signal's arrival time information. It features deadtimeless integration and digitization of charge in 25 ns 'buckets'; the same clock as the LHC.

On board is a rising edge TDC with a resolution of less than 800 ps, which outputs a discrimination bit. It has a large dynamic range of 3 fC to 330 pC and holds this information in 17 bits. Its digitization error is less than the calorimeter resolution and is on the order of 2-3%, held in 6 bits. It digitizes in 4 ranges, stored in 6-bit mantissa and 2-bit exponent. It matches the input impedance of the new PMTs.

A major design consideration was the intense radiation environment of the High Luminosity LHC. The QIE10 features a radiation tolerant FPGA (ProASIC3E from Microsemi) manufactured using the AMS 0.35-micrometer SiGe BiCMOS process. This FPGA is responsible for synchronizing and formatting the data from several QIEs. It also determines the pulse width from the time discriminator output. The chip is tolerant to 100 Gy, though it is expected to receive a total ionizing dose of 1.5 Gy.

One feature of the MIPs that interact with the window of the PMTs is the difference in arrival time from true HF signal. The upgraded HF front-end electronics feature a TDC that determines the rising edge from the QIE10, in either 5 or 6 bit (390 or 780 ps) resolution. The falling edge is determined by the FPGA. The TDC data is transmitted in 4 bit codes corresponding to 3.125 ns intervals. A special code indicates multiple pulses.

A new data link was developed by CERN to carry this information to the counting room. GBTx is a 4.8 Gbps data link that serializes the QIE10 data for transmission to the back-end electronics. 88 bits of user data can be carried every bunch crossing, and the total use of this bandwidth depends on the number of additional fibers installed into HF. If each fiber is required to handle data from 6 channels, only the rising-edge TDC value provided by the QIE10 will be available. If only 50% more fibers

Journal of Physics: Conference Series **587** (2015) 012007 doi:10.1088/1742-6596/587/1/012007

(288) are installed, a falling-edge TDC can be provided along with the extra bits for determining multiple pulses.

2.4. Back-end electronics

The upgraded electronics will take advantage of a new industrial standard known as microTCA. This crate standard will replace the existing VME crate system. This is the first step of the Phase I electronics upgrade.

The new electronics improve the noise rejection and the granularity of the HF readout and trigger. A prototype microHTR data acquisition board performed very well during the heavy ion collision run shortly before LS1, and figures 6 and 7 show the improvement of the microTCA over the VME system.

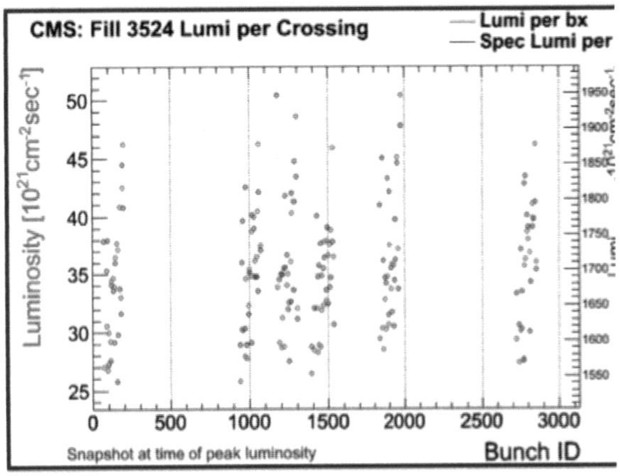

Figure 6. Lumi data from the existing VME system.

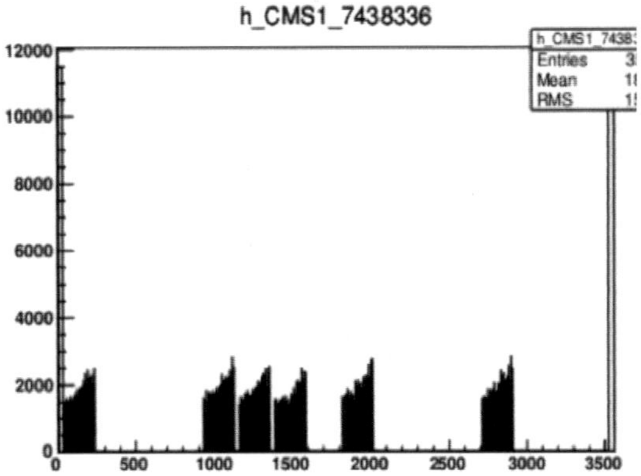

Figure 7. Lumi data from the new microTCA system.

The HF microHTR components were purchased in March 2013, and the boards are being produced in India.

3. Upgrade status

Nearly 1800 upgrade PMTs (R7600s) were tested at the University of Iowa. The system involved: 3 dark boxes, a VME and CAMAC data acquisition system, an XY scanner for surface linearity testing, 3 picoammeters, 3 digital oscilloscopes, UV and visible light power meters, UV and Blue LED light sources, 2 nitrogen lasers, DC tungsten light bulbs, optical tables for mounting optical elements in the dark boxes, picosecond LED pulsers and 1 double pulse generator. Figure 8 shows the Iowa test station.

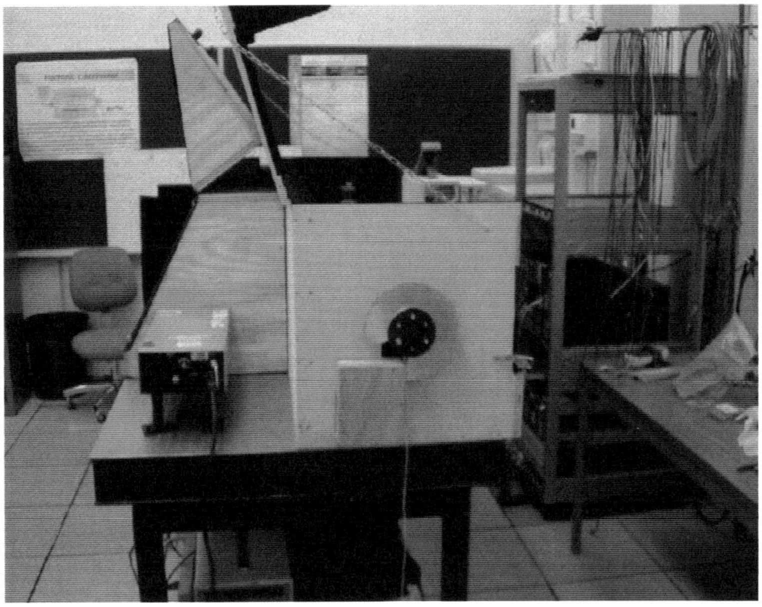

Figure 8. The University of Iowa PMT test station.

Reusing as much existing readout box infrastructure as possible was a major design goal of the project, while still meeting all criteria for proper PMT operation. Another huge consideration was the flexibility to change the number of channels read out per PMT. This capability became part of the baseboard design. These design considerations were prototyped and tested in CMS in the fall of 2010 during collisions with great success.

A slice-test readout box was designed and prototyped to directly compare HF PMT and R7600 readout in a real collision environment. Figures 9 and 10 show the hybrid HF Readout Box.

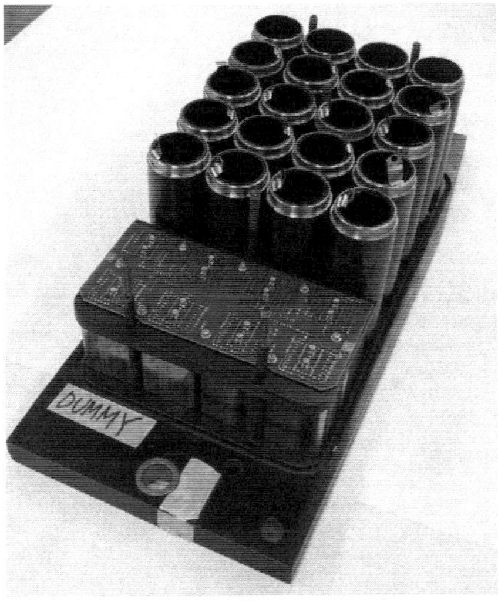

Figure 9. "Plate 2"; part of readout box with R7525s and R7600s for CMS HF.

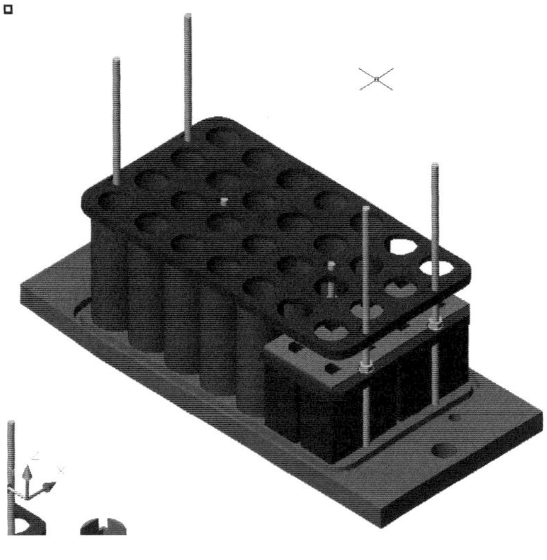

Figure 10. CAD model of readout box panel with HF PMT and R7600 for CMS HF.

Journal of Physics: Conference Series **587** (2015) 012007 doi:10.1088/1742-6596/587/1/012007

The slice test mixed 20 R7525s with 8 4-anode R7600 PMTs. This ROBOX was placed in HF at iPhi 67 and covered towers iEta 29 through iEta 32 in HF +. The R7600 placed at coordinate (iPhi67, iEta 29) was blacked out so only Cherenkov light generated by MIP interactions would be recorded. Analysis of the collision data was consistent with test beam data, namely the mean energy of the MIP event reduced from 120 GeV to ~35 GeV, and the rate was significantly lower. The energy spectra from the blackened R7525 and R7600 are shown in Figures 11 and 12 respectively.

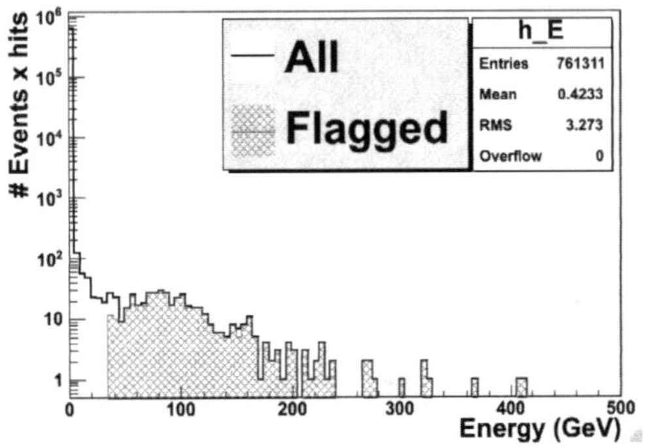

Figure 11. Energy spectrum from blackened R7525. Peak at ~120 GeV.

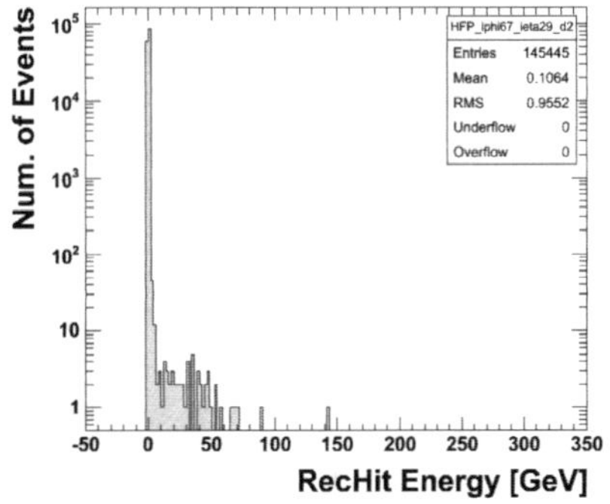

Figure 12. Energy Spectrum from R7600. Peak at ~35 GeV.

New shields (Figure 13) for the R7600 PMTs had to be manufactured, and the success of the pre-production prototype demonstrated that the design considerations are well understood.

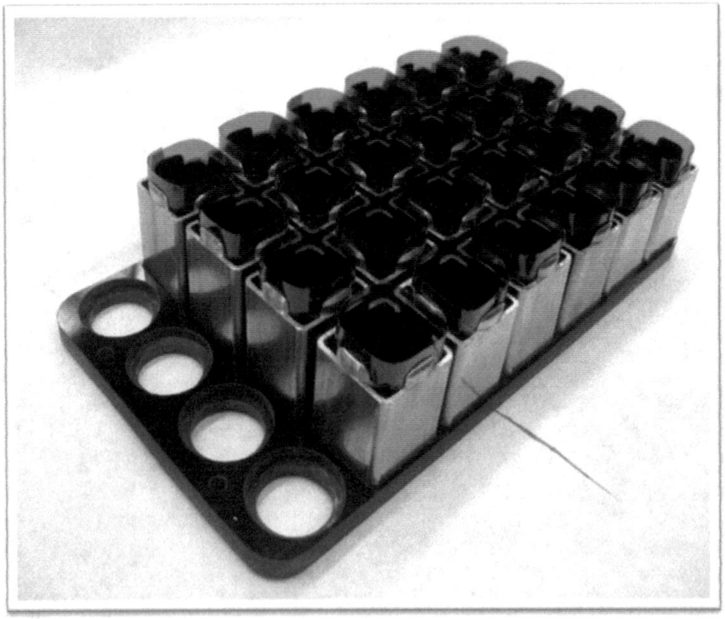

Figure 13. HF Pre Production Prototype Readout Box for R7600.

The upgrade readout box was tested at the CERN H2 test beam facility [8]. The operation was very successful, with no abnormal signals and no significant issues. The results show that we have a good understanding of the performance characteristics, and the R7600 PMTs have a uniform response under different readout conditions (1, 2, or 4 anode readout). A major success was the negligible response of MIPs with the tagging algorithm. The efficiencies of the algorithms were quantified with ~98% for 4-ch readout and ~90% for 2-ch readout. The 4-channel readout option has more controlling power on background elimination and is an unprecedented readout segmentation in an operational calorimeter.

By the time this note is written, the upgrade readout boxes are being assembled and tested at CERN. Following the installation and commissioning, the upgrade HF will be operational in 2015.

4. Physics implications of upgrade

The PMT upgrade stands to significantly reduce the occurrence and impact of anomalous signals, and at a time when HF will be very important to exotic physics searches as well as Higgs boson measurements.

Previous studies have shown the importance of HF to the measurement of Higgs production via vector boson fusion, whose signal includes very forward jets. These forward jets are not present in many background signals. Two-thirds of this signal would be lost without HF coverage, showing the importance of having a calorimeter at high-eta.

MIP window interactions can be tagged falsely as high pT jets and can bury real signal in this background. Previous studies show that the pT distributions of leading QCD jets in HF have an increased tail with PMT MIP events. Figure 14 shows that jets with transverse momentum, pT > 250 GeV, have a false increase of 68% significantly effecting rare physics signals. These extra HF false jets will lead to an increased background for the VBF Higgs search.

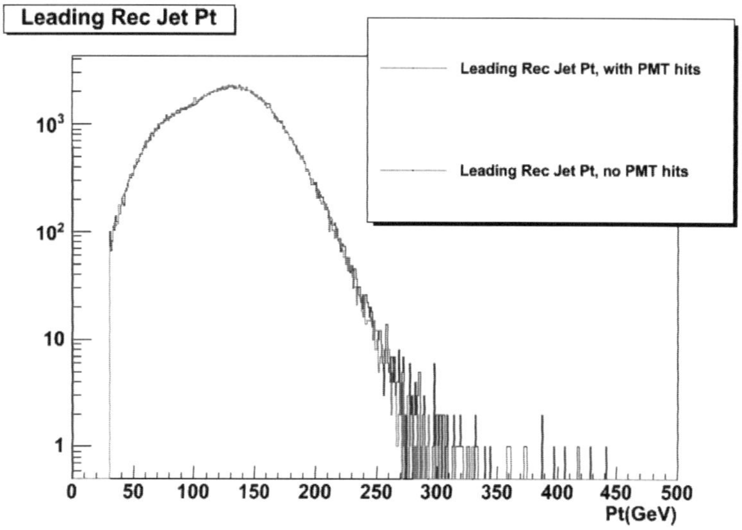

Figure 14. QCD Leading Jet with and without PMT window hits included.

Physics models with signatures of missing ET require HF at high-eta to improve the MET resolution of the CMS detector. SUSY, other exotic searches, which depend on high MET resolution, are also effected by the quality of the HF signal. HF itself reduces the MET tail (MET > 100 GeV) by a factor of 10 as compared to no HF hence is crucial for MET calculations.

5. Conclusions

The HF Phase I upgrade is underway with the construction of the upgrade readout boxes with complete upgrade components and readout options as of the day this note is written.

The multi-anode upgrade PMTs do not only reduce the rate of the background as a result of their physical specifications, but also enable the implementation of signal recovery algorithms pertaining to the multi-anode features. The design of the upgrade readout and adapter boards enables simple operation to switch between 1, 2 and 4-channel readout options. The new cabling significantly improves the space issues.

The front- and back-end electronics are developed to enable precision TDC for detailed signal shape information. The electronics is also associated with fast data links.

All the upgrade components are selected to provide the Hadron Forward calorimeters operations in the upgrade era with the desired performance characteristics. The components are extensively tested and the upgrade concept is validated with several beam tests. Progression of the upgrade is on schedule towards commissioning and operation planned in 2015.

6. References

[1] Chatrchyan S et al. (CMS collabration) 2008 *The CMS experiment at the CERN LHC* JINST 3 S08004

[2] Bayatian G L et al. (CMS collabration) 2006 *CMS physics: Technical design report*, CERN/LHCC-2006-001

[3] Abdullin S et al. (CMS HCAL collabration) 2008 *Design, performance, and calibration of CMS forward calorimeter wedges Eur. Phys. J.* C 53 139

[4] http://www.hamamatsu.com/

[5] Chatrchyan S et al. (CMS HCAL collaboration) 2010 *Study of various photomultiplier tubes with muon beams and Cerenkov light produced in electron showers* JINST 5 P06002

[6] http://nahandbook.web.cern.ch/nahandbook/default/h2/1 General.htm

[7] Zimmerman T and Sarraj M 1996 *A second generation charge integrator and encoder ASIC*, *IEEE Trans. Nucl. Sci.* 43 1683; Zimmerman T and Hoff J R 2004 *The design of a charge-integrating modified floating-point ADC chip* IEEE J. Solid State Circ. 39 895

[8] Chatrchyan S et al. (CMS HCAL collaboration) 2012 *Test of CMS hadron forward calorimeter upgrade readout box prototype* JINST 7 P10015

Journal of Physics: Conference Series **587** (2015) 012008 doi:10.1088/1742-6596/587/1/012008

The monitoring and data quality assessment of the ATLAS liquid argon calorimeter

Olivier Simard, on behalf of the ATLAS Liquid Argon Calorimeter Group
Centre National de la Recherche Scientifique (CNRS), LAPP-Annecy, 9 Chemin de Bellevue - BP 110, 74941 Annecy-le-Vieux CEDEX, France
E-mail: olivier.simard@lapp.in2p3.fr

Abstract. The ATLAS experiment is designed to study the proton-proton (pp) collisions produced at the Large Hadron Collider (LHC) at CERN. Liquid argon (LAr) sampling calorimeters are used for all electromagnetic calorimetry in the pseudo-rapidity region $|\eta| < 3.2$, as well as for hadronic calorimetry in the range $1.5 < |\eta| < 4.9$. The electromagnetic calorimeters use lead as passive material and are characterized by an accordion geometry that allows a fast and uniform response without azimuthal gaps. Copper and tungsten were chosen as passive material for the hadronic calorimetry; while a classic parallel-plate geometry was adopted at large polar angles, an innovative design based on cylindrical electrodes with thin liquid argon gaps is employed at low angles, where the particle flux is higher. All detectors are housed in three cryostats maintained at about 88.5 K. The 182,468 cells are read out via front-end boards housed in on-detector crates that also contain monitoring, calibration, trigger and timing boards. In the first three years of LHC operation, approximately 27 fb^{-1} of pp collision data were collected at centre-of-mass energies of 7-8 TeV. Throughout this period, the calorimeter consistently operated with performances very close to specifications, with high data-taking efficiency. This is in large part due to a sophisticated data monitoring procedure designed to quickly identify issues that would degrade the detector performance, to ensure that only the best quality data are used for physics analysis. After a description of the detector design, main characteristics and operation principles, this paper details the data quality assessment procedures developed during the 2011 and 2012 LHC data-taking periods, when more than 98% of the luminosity recorded by ATLAS had high quality LAr calorimeter data suitable for physics analysis.

1. Introduction
ATLAS [1] is a large general-purpose experiment at the Large Hadron Collider (LHC) at CERN. ATLAS started recording data from proton-proton collisions at a centre-of-mass energy of 7 TeV in 2010, at the beginning of what is referred to as Run-I. The integrated luminosity reached 45 pb^{-1} at the end of 2010. In 2011 and 2012, the LHC instantaneous luminosity gradually increased; the total integrated delivered luminosity reached 5.25 and 21.7 fb^{-1}, respectively, while the centre-of-mass energy was pushed to 8 TeV in 2012.

1.1. The ATLAS LAr Calorimeter
All calorimeters operating in ATLAS are sampling calorimeters and are essential to the experiment. Liquid argon (LAr) was chosen as the active medium for the electromagnetic calorimeter in the barrel and end-cap regions as well as for hadron calorimetry in the end-cap and forward regions. The barrel hadron calorimeter ($|\eta| < 1.7$), made of iron and plastic tiles

scintillator, referred to as the Tile Calorimeter, is discussed elsewhere in these proceedings. Within the LAr calorimeter system, three sub-detectors can be distinguished based on their purpose, location, and technology:

- The Electromagnetic Calorimeter (EM) is made of lead absorbers, interleaved with liquid argon, bent into a unique accordion shape allowing for full azimuthal coverage. The barrel (EMB) and end-caps (EMEC) account for the majority of the readout cells of the LAr system and cover the pseudo-rapidity η range up to 3.2. The four EM layers are referred to as the pre-sampler, strip, medium, and back layers.
- The Hadronic End-Cap Calorimeter (HEC) consists of parallel copper plates and liquid argon as active material. It is placed just behind the EMEC, in the same cryostat and covers η from 1.5 to 3.2.
- The Forward Calorimeter (FCal) is located in the HEC borehole, close to the beam pipe, extending the LAr coverage in η between 3.1 and 4.9. The FCal provides hadronic as well as electromagnetic calorimetry with three layers; an EM module using copper as absorber completed by two hadronic layers using tungsten. To cope with the higher rate in the forward region, the active gaps were made narrower compared to the other LAr calorimeters.

In total, there are 182,468 LAr readout cells. A negligible fraction of cells, 0.06%, were dead as of the end of Run-I. Calorimeter data are required to identify and measure electrons, photons, jets, and missing transverse energy. Furthermore, the EM calorimeter is the only sub-system of ATLAS that could measure the energy of unconverted photons. With its fine segmentation, the direction and thus the angle between two photons could be determined with high accuracy, leading to a good angular resolution, a key quantity in particle and primary vertex identification.

1.2. Operational Principle, Readout and Calibration

The charged particles of a shower development cross and ionize the argon in the thin 2-mm gaps. Electrons and ions are separated according to their charge with an electric field produced by applying high-voltage at the electrodes, where charges are collected. The electric signals produced by drifting electrons from adjacent gaps are summed to form cells, that are read out in groups of 128 by the front-end electronics. The ionization signal is triangular with a duration of roughly 450 ns in the EM calorimeter. It is thus shaped by a bipolar shaper to allow sampling at 40 MHz and meet the LHC specifications. The shaping occurs simultaneously in three gains to cover a wide dynamic range in energy. A configurable number n of digitized samples are sent out by the front-end boards for each event, which was set nominally to $n = 5$ for Run-I.

The amplitude A of the shaped pulse is proportional to the energy deposited in the cell. It is reconstructed online using an optimal filtering technique [2]. The amplitude A, the time offset τ, and pulse quality-factor Q were computed with the formulae

$$A = \sum_{i=1}^{n=5} a_i(s_i - p) \qquad A\tau = \sum_{i=1}^{n=5} b_i(s_i - p) \qquad Q \sim \sum_{i=1}^{n=5} (s_i - r_i)^2$$

where a_i and b_i are the optimal filtering coefficients, s_i are the digitized samples and p is the pedestal. The pulse quality-factor, also referred to as the Q-factor, is essentially a χ^2 figure of merit that measured the level of agreement between the observed and reference pulse shapes r_i. The optimal filtering coefficients were derived from the pulse shape and the noise autocorrelation. The electronics pedestals and gains were measured with the calibration system [3], which was used to inject known pulses close to the point where the electronics chain begins. Calibration runs were performed on a regular basis to monitor the stability of the constants and update the calibration when needed.

1.3. Experimental Conditions at the LHC

The LAr calorimeter performed very close to specifications, with high data-taking efficiency during Run-I, where the LHC conditions evolved quite rapidly, spanning from commissioning to nearly nominal design conditions. Despite the changes in operating conditions in Run-I, the ATLAS recording and data-quality efficiencies reached 93.5% and 95.8%, respectively. A summary and detailed accounting of efficiencies per data-taking period can be found at [4]. The success of the LAr calorimeter operations and data-quality teams was in large part due to a sophisticated data-monitoring procedure designed to quickly identify issues that would degrade the detector performance.

The following is a short overview of the data quality assessment procedures [5] developed during Run-I, when more than 98% of the luminosity recorded by ATLAS had high quality LAr calorimeter data suitable for physics analysis. The excellent performance of LAr calorimeter played a crucial role in the discovery of the Higgs boson that was announced in 2012 [6], providing the invariant mass resolution necessary to observe the small signal excess over the large background in the $H \to \gamma\gamma$ analysis.

2. Detector Conditions

All detector basic conditions, as the temperature and purity of the liquid argon within the three cryostats, were carefully monitored. Conditions were stored along with exact times or luminosity block number, which corresponded to one minute of data at most, in dedicated databases.

Among the important conditions to monitor was the distribution of High-Voltage (HV) to the 3250 sectors of the detector. The HV supply system comes with redundancy by design; for instance most sectors of the EMB were fed by two HV lines, providing 2 kV potential between electrodes. HV trips occurred during physics data-taking, with observed trends of increase after short stoppage periods or sudden luminosity steps taken by the LHC. While a correction to the signal is possible in the offline data, the loss of efficiency caused by non-nominal and varying HV conditions could be mitigated largely at the hardware level. An auto-recovery procedure initiated in the supplying module consisted in a slow HV ramp back up to the nominal value.

The full HV system behavior was also monitored continuously by experts during data-taking. Further HV module parameter adjustments were necessary, in rare cases, in regions where too many trips had occurred. A new generation of HV modules that are more robust were successfully tested and deployed at the very end of Run-I in 2013.

Figure 1 shows the behavior of a HV module around a trip and the following recovery in terms of voltages and currents. The luminosity blocks associated with the trip are not usable for physics. However, as shown in Figure 2, the missing transverse energy in the ramp-up luminosity blocks were shown to be similar to those found under nominal conditions. The physics data being usable during ramp-up, the data loss due to HV in 2012 was kept to a very low level. Only 0.46% of the total data set were not suitable for physics due to HV trips.

3. Data Integrity and Online Processing

The next step beyond having acceptable detector conditions was making sure meaningful data were propagated to online systems, trigger decisions, and event recording. Data integrity checks were continuously performed for all events by the LAr online software: front-end board data corruption, check for coverage or large detector regions loss, test for null or saturated digitized samples, ensure external database conditions, etc. In general, such rare problems could be corrected very rapidly by the relevant experts. In 2012, these problems could be dealt with almost automatically by the ATLAS online re-synchronization utility, further reducing the data-acquisition inefficiency.

Beyond their basic content, the data reduction and software trigger capabilities provided by LAr back-end electronics were also monitored to check their reliability. The online calculations

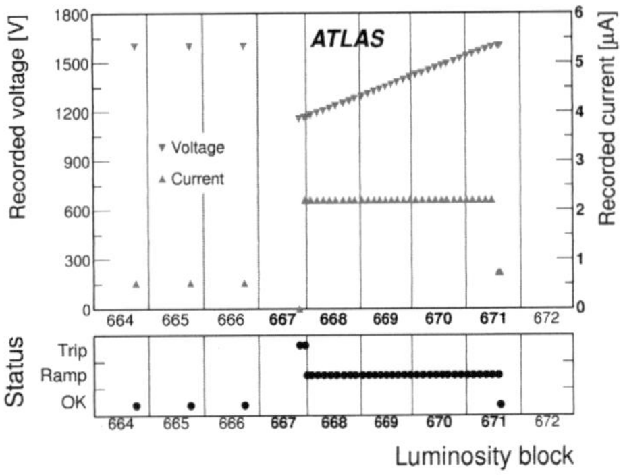

Figure 1. Typical HV trip of a line supplying one HEC sector, showing the evolution of voltage, current and status.

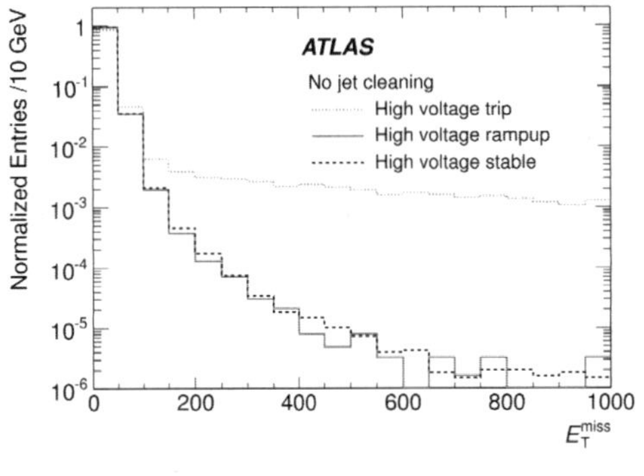

Figure 2. Missing transverse energy, in GeV, measured in 2011 data with various HV conditions.

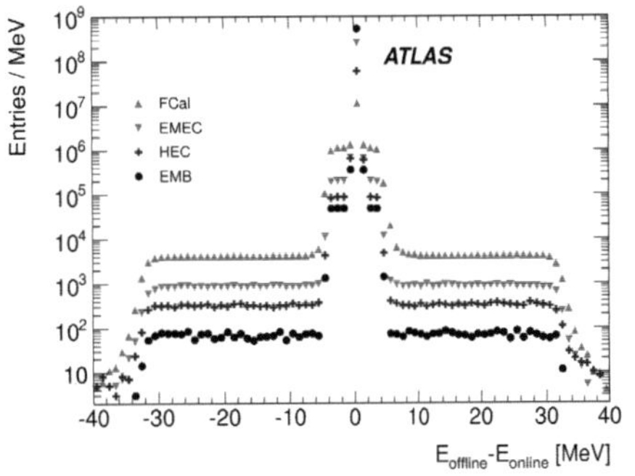

Figure 3. Difference between the energies computed online by the DSP and those recomputed offline in a ATLAS run.

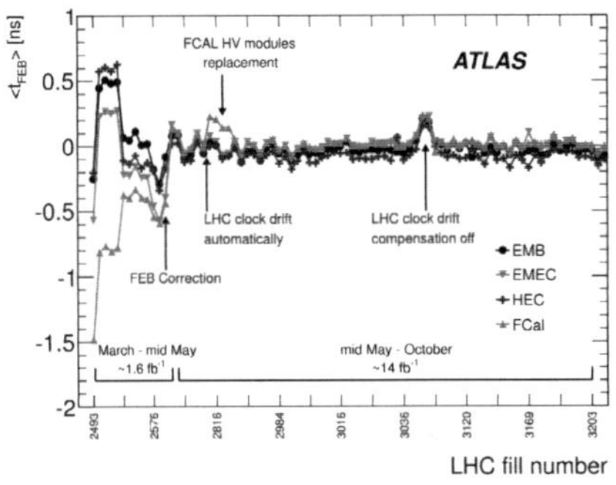

Figure 4. Evolution of the average FEB time per sub-detector as a function of LHC fill number in 2012.

of energy, time, and Q-factor, were compared with the full offline reconstruction for the 1-2% of the cells for which digitized samples were recorded. Figure 3 shows such a comparison for cell energies; the accuracy measured by the absolute difference was quite high and within the intrinsic digital signal processor (DSP) precision of 1-512 MeV [7]. Overall the various system diagnostics and monitoring tools ensured the excellent data integrity level of 99.98%.

4. Calorimeter Synchronisation

Bunch-crossing identification and precise timing were crucial for searches for exotic or very massive particles with long lifetime. Furthermore, timing could improve the energy resolution of the calorimeter, and the measurements of out-of-time signals and beam-induced backgrounds. Timing was therefore monitored using small samples of physics objects for every run. Global and fine adjustments were often made for Front-End Board (FEB) or group of 128 channels. Figure 4 diplays the average time per FEB, averaged over all FEBs. The dispersions were in the range of 0.10-0.17 ns. This stability level of 100 ps was comparable to estimations of the timing resolution obtained from offline analyses of electrons from W and Z bosons.

5. Treatment of Calorimeter Noise

Calorimeter noise, beyond electronics noise measured in routine calibration runs, was monitored and identified very carefully to minimize the rate of fake calorimeter clusters in event reconstruction. Useful quantities for identification were the observation of persistent deviations with respect to standard gaussian noise (3σ or more) and pulse quality with the Q-factor. These were gathered mostly in bunch-crossings without collisions, in self-triggering mode, to avoid the obvious effect of legitimate physics signals. Two main classes of noise were the large scale coherent noise and per-channel noise, described below.

5.1. Large scale coherent noise (bursts)

The large scale coherent noise affected whole events through sudden noise seen in several thousands of cells, leading to spectacular detector topology. Such an event is shown in Figure 5. In addition to the very large (fake) energy, patterns in groups of channels deviating from gaussian noise and with bad Q-factor were used to categorize them.

Occurrences of such phenomena lasted on average less than 1 μs, spanning over a group of events during collision data-taking. To avoid impacting various physics analyses, the trigger capabilities were used to identify, store, and analyze these events online and offline. Precise times were stored for each affected event, and a time window veto was imposed around them. The procedure was very successful in minimizing the fraction of data that should not be used for physics, and lead to data losses of the order of 0.25%.

An observed increase of large scale coherent noise rate with luminosity in 2012 further justifies to continue precise and dedicated monitoring in the future.

5.2. Per-channel noise

The event veto procedure clearly cannot be repeatedly applied for noise arising from small detector regions such as individual cells. Run-by-run analysis of deviating channels was such that two possible cases were taken into account by the data-quality, monitoring, and reconstruction teams. Conditionally-masked cells were identified on the basis of an unusual high proportion of events with large (bad) Q-factor. This indicated the reconstruction code to ignore the cell only in events where the Q-factor was large. Unconditionally-masked cells however were systematically recurring and skipped permanently in reconstruction, after confirmation by a group of experts. In all cases, when a cell was masked, the information from the neighboring cells was used instead to patch the cluster reconstruction.

The pre-sampler layer of the EMB had a significant number of such isolated channels requiring masking in 2011 and beginning of 2012. After adjusting the HV in some sensitive regions, the layer recovered while preserving a reasonable signal-to-noise ratio. Figure 6 shows how the HV reduction brought the proportion of cells flagged in the database from 10% to below 1%.

The total data loss in 2012 due to mishandling of per-channel noise was around 0.05%.

6. Achieved Performance, Summary and Outlook

The ATLAS collaboration has put in place an extensive and flexible system for flagging the problems relevant to data quality. All known problems were therefore identified and reported in a dedicated detector defect database. Browsing through that database, one finds that the overall LAr data-quality efficiency was quite high, near 99% at end of Run-I, despite the variety of features described in the above. Figures 7 and 8 show the physics data rejection by data-taking periods, for the 2011 and 2012 data sets, respectively.

Thus, the ATLAS LAr Calorimeter has achieved excellent performance and stability during the first three years of LHC operations, without significant hardware or software problems. The LAr working group aims at constantly improving the hardware, monitoring and data quality

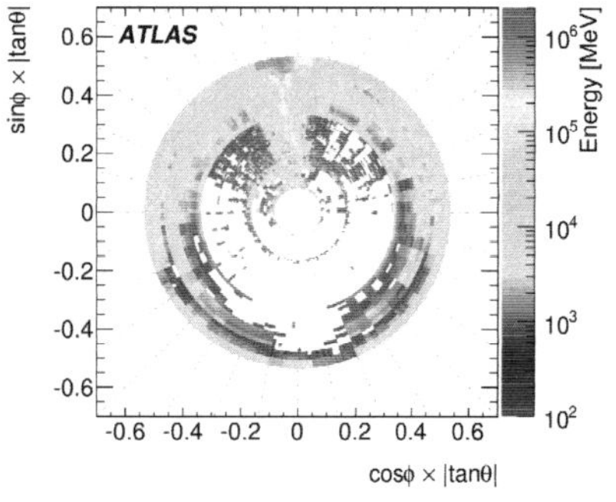

Figure 5. Typical coherent noise event in one end-cap. The energies of the layers have been summed in each bin of (θ,ϕ).

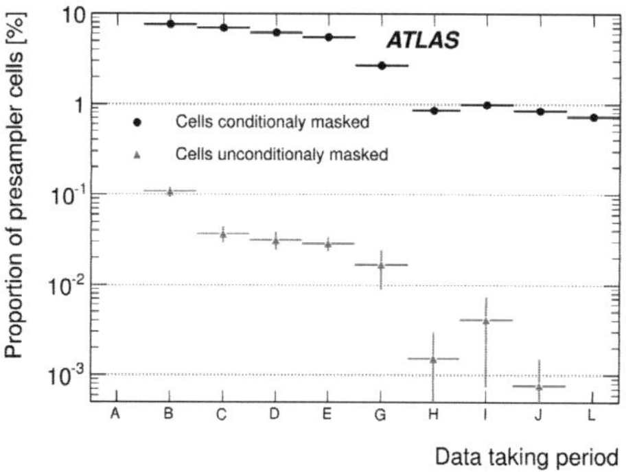

Figure 6. Proportion of pre-sampler cells masked as a function of time in 2012.

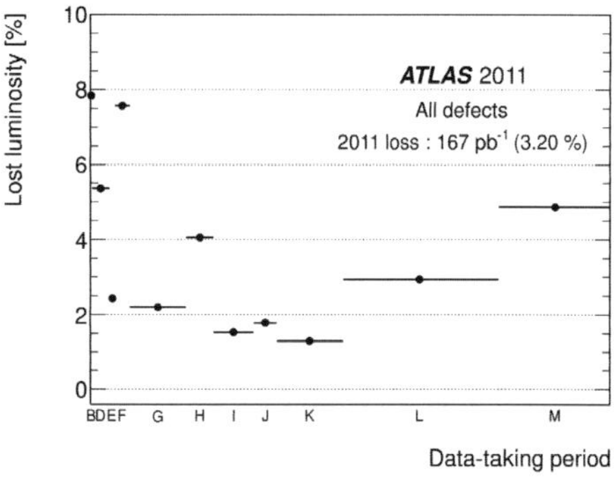

Figure 7. Evolution of the LAr data rejection by defect assignment as a function of data periods in 2011.

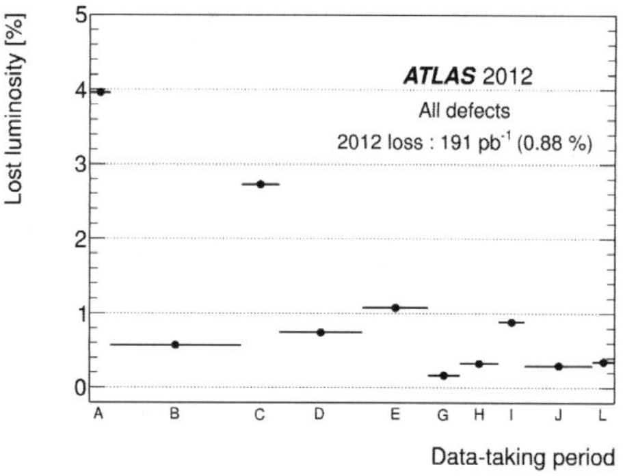

Figure 8. Evolution of the LAr data rejection by defect assignment as a function of data periods in 2012.

procedures, and looks forward to similar or better performance for the upcoming LHC operations (Run-II, 2015-2018) at higher collision energies and luminosities.

References

[1] ATLAS Collaboration, The ATLAS Experiment at the CERN Large Hadron Collider, JINST, 3:S08003, 2008.
[2] W.E. Cleland, E.G. Stern, Signal Processing considerations for liquid ionization calorimeters in a high rate environment, Nuclear Instruments and Methods in Physics Research A 338 (1994) 467-497.
[3] The ATLAS Liquid Argon Calorimeter Group, Performance of the electronic readout of the ATLAS liquid argon calorimeters, JINST, 5:P09003, 2010.
[4] ATLAS Experiment Public Results, Luminosity: `https://twiki.cern.ch/twiki/bin/view/AtlasPublic/LuminosityPublicResults`.
[5] ATLAS Collaboration, Monitoring and data quality assessment of the ATLAS liquid argon calorimeter, submitted to JINST, `http://arxiv.org/abs/1405.3768`.
[6] ATLAS Collaboration, Observation of a new particle in the search for the Standard Model Higgs boson with the ATLAS detector at the LHC, Phys. Lett. B 716 (2012) 1-29.
[7] A. Bazan et al., ATLAS liquid argon calorimeter back end electronics, JINST 2:P06002, 2007.

Calibration of the Tile Hadronic Calorimeter of ATLAS at LHC

Djamel BOUMEDIENE, on behalf of the ATLAS collaboration
CNRS/IN2P3/LPC Université Blaise Pascal, 24 avenue des landais, 63171 AUBIERE
E-mail: djamel.boumediene@cern.ch

Abstract.
The Tile Calorimeter (TileCal) is the central section of the hadronic calorimeter of the ATLAS experiment. The TileCal provides important information for reconstruction of hadrons, jets, hadronic decays of tau leptons and missing transverse energy. This sampling calorimeter uses iron plates as absorber and scintillating tiles as active medium. The light produced by the passage of charged particles is transmitted by means of wavelength shifting fibers to photomultiplier tubes (PMTs). The TileCal readout is segmented into about 5000 cells (longitudinally and transversally), each of them being read by two PMTs. A brief description of the individual calibration systems (Cs radioactive source, laser, charge injection, minimum bias) is provided. Their combination allows to calibrate each part of the data acquisition chain (optical part, photomultiplier, readout electronics) and to monitor its stability to better than 1%. The procedure for setting and preserving the electromagnetic energy scale during Run 1 data taking is discussed. The issues of linearity and stability of the response, as well as the timing adjustment are also shown.

1. Introduction

The TileCal Calorimeter of ATLAS [1] has three key elements: the scintillators, the photomultiplier tubes (PMTs) and the read-out electronics [2, 3]. The light is produced in scintillating tiles and converted into electric currents by the PMTs.Their signal is shaped and amplified with two gains. The sampling and the digitization is realised by ADCs. During collisions, if an event is selected by the trigger system, the digitized signals are collected and processed by a Read-Out Driver. In parallel to this, integrators measure the integrated current from the PMTs.

The reconstructed energy of each channel, $E(\mathrm{GeV})$, is derived from the raw response, $A(\mathrm{ADC})$, as follows:

$$E(\mathrm{GeV}) = A(\mathrm{ADC}) \cdot C_{ADC\to \mathrm{pC}} \cdot C_{\mathrm{pC}\to \mathrm{GeV}} \cdot C_{Cesium} \cdot C_{Laser}$$

This relation is established for each TileCal channel. The factors can evolve in time because of unstability of PMTs high-voltage, PMTs stress induced by high light flux or optics aging. The calibration systems are used to monitor the stability of these elements and provide per channel calibration.

While $C_{\mathrm{pC}\to \mathrm{GeV}}$ was fixed during dedicated test beam campaigns, the remaining calibration constants are provided by individual systems:

- spaced by weeks or months, calibrations of Tile optic components with movable Cesium radioactive gamma source;
- frequent calibrations of phototube gains with custom Laser calibration system;
- daily to weekly calibrations of digital gains and linearities with charge injection system (CIS) integrated on module front-ends;
- monitoring of beam conditions and Tile optics with the so-called integrator system (minimum bias).

The calibration tools follow different and partially overlapping read-out paths allowing for easier identification of potential failure and for crosschecks.

2. Calibration Systems of TileCal

2.1. The Cesium Calibration

The Cesium calibration system [4] is based on three movable radioactive sources using hydraulic control. The ^{137}Cs γ-sources move inside the calorimeter body. The channel response to the energy deposits is used to equalize the response of all the cells and maintain global response of the calorimeter at the electromagnetic scale.

Deviation of measured Cesium signals from expected values, corrected for the Cesium decay curve (-2.3%/year), are interpreted as gain variations and translated into calibration constants (C_{Cesium}).

The Cesium calibration is performed taking into account the presence of magnetic field produced by the ATLAS Toroid and Solenoid. It tests the optical chain and the photomultipliers, with a precision better then 3 per 1000, including the scintillators aging effect.

During this calibration, the Integrator is used to record the PMT responses making the Cesium measurement insensitive to what can effect the readout used for collision data. It has to be completed by a dedicated system, the CIS. On another hand, the frequency of the Cesium calibration can be insufficient to track fast drifts of the PMT responses. For prompt calibration, the Laser system is used between two Cesium scans.

2.2. The Laser Calibration

The gain of each PMT is measured using a Laser calibration system [5] that sends a controlled amount of light in the photocathode of each PMT in the absence of collisions. Both high and low gain regimes of TileCal read-out are monitored using two different light amplitudes. Deviations of any channel response with respect to its nominal response (at the time of the latest Cesium calibration) lead to a calibration constant: C_{Laser}.

The typical precision on the gain variation is better than 0.5% per channel when 10,000 Laser pulses are used. The Laser measurements are used for fast monitoring of TileCal and to correct the gain of some specific channels.

During 2011–2012, a few hundred of channels required a Laser calibration, most of these gain drifts were induced by luminosity effects.

Laser pulses are also sent during empty bunch crossings of the LHC, with a frequency of 1 Hz, and the timing of TileCal response is measured and used for channel per channel time calibration.

2.3. The Charge Injection System (CIS)

The CIS simulates physics signals in TileCal channels by injecting a known charge into the ADC and measuring the electronic response, as described in [6]. It provides then a quantitative relationship between the analogic physical signals from the Tile Calorimeter photomultiplier tubes and the electronic response of TileCal read-out channels and corrects for non-linearity in the analog signal processing. A set of CIS calibration constants ($C_{\mathrm{ADC}\rightarrow\mathrm{pC}}$) are regularly produced and applied to TileCal data.

The precision on the CIS calibration constant is about 0.7% for each channel.

2.4. The Minimum Bias integration

High energy proton-proton collisions are dominated by soft parton interactions, so-called Minimum Bias (MB) events. The response of the TileCal to signals induced by the MB interactions scale with instantaneous luminosity [7]. It can be used to measure the luminosity delivred to ATLAS and to monitor the stability of the channels and to provide an independent cross-check of the Cesium calibration.

The PMT currents are integrated over a time window of 10 ms. The MB data are continuously recorded during proton-proton collisions.

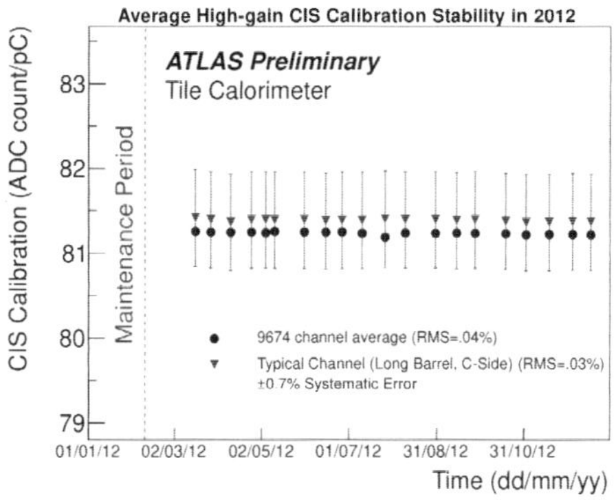

Figure 1. Time stability of the average high-gain readout calibration constants from February 2012 to December 2012 for 19421 ADC channels. The calibrations are measured in-situ with the Charge Injection System (CIS). The RMS values printed in the legend represent a measure of the fluctuations present in calibrations.

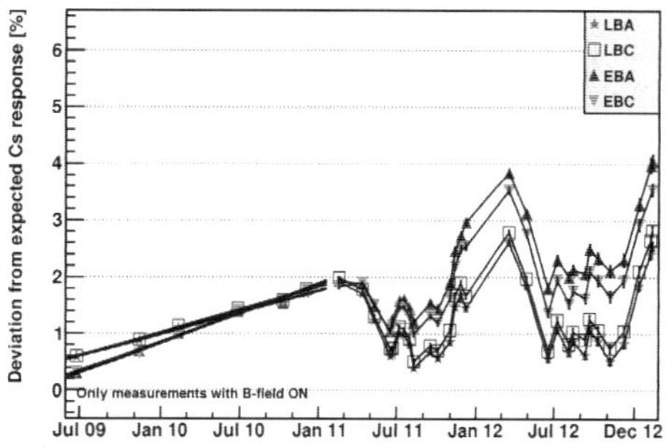

Figure 2. Deviation of TileCal response to Cesium sources per partition, in the inner radial layer (A layer), from 2009 to 2012.

3. Calibration and gain stability of TileCal

The four TileCal calibration systems were operated during Run 1 of LHC. Their performances, measurements and diagnostics are presented in the following sections. In addition to this, the High Voltage and the temperature are monitored in order to target possible unstable channels in the calibration process.

3.1. High Voltage and Temperature monitoring

High Voltage is monitored channel by channel while temperature is recorded in a representative sample of 2.5% of the total PMT number. Stability monitoring of the PMT high voltage with respect to its reference value, averaging over all PMTs for two running periods of nine months in 2011 and three months in 2012 shows a typical variation of 0.4V leading to a gain variation lower than 5 per 1000. The identification of channels with unstable High Voltages is used to target them in the calibration process. The stability of temperature averaging over all modules

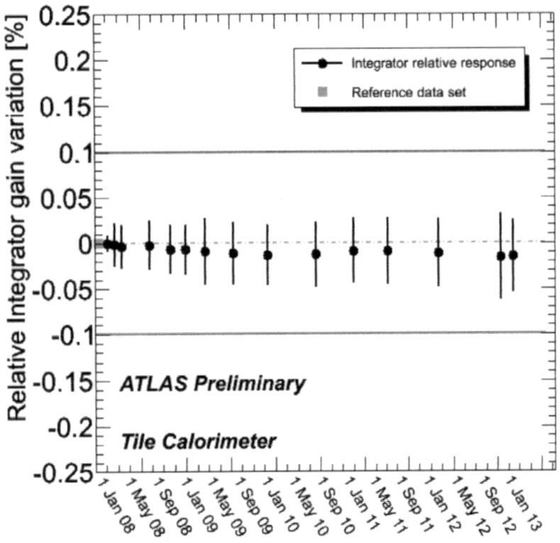

Figure 3. Comparison of channel responses, from 2008 to 2013, with respect to a reference sample taken in 2008. The error bars represent the standard deviation from the mean of the relative variation (precision per channel).

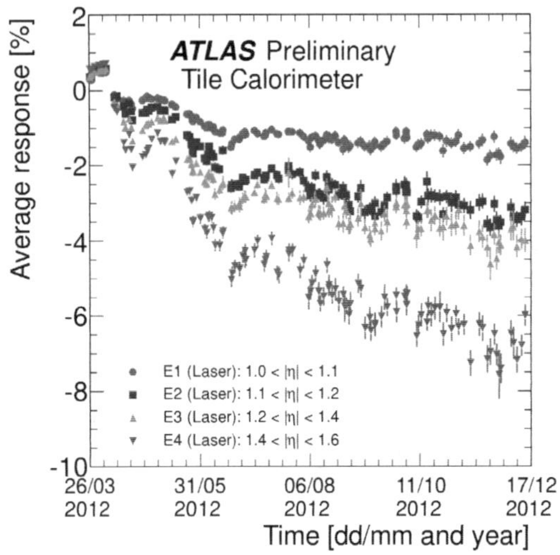

Figure 4. Gain variation measured by the Laser for the special cells E1 to E4. The measurement is shown as a function of time in 2012. A mean over modules is performed. Error bars are the uncertainty on the mean value.

during the same periods as for the HV monitoring analysis shows a typical variation of 0.2 °C. This small variation allows for no correction in the PMT gain since the gain dependence on the temperature is -0.2%/°C.

3.2. Readout stability and calibration

The mean of CIS calibration data for February 2012 was compared to the mean of CIS calibration data for December 2012. The RMS variation is approximately 0.1%. Channels with variation greater than 1.0% were targeted for recalibration. This in-situ calibration was applied to all collision data collected in LHC Run 1. The average read-out calibration constants, as provided by Charge Injection System over all LHC Run 1 and for all ADC channels, show a typical stability in time of 0.02% as illustrated on Figure 1 [8].

3.3. Global gain stability and calibration

An average up-drift of about 0.8% per year was measured by the Cesium calibration in 2009–2010. This up-drift effect is compensated by an opposit effect in 2011 and 2012. Since LHC reached a high luminosity regime, a sizable down-drift is seen when beam is on while the response recovers slowly when beam is off, *i.e.* during LHC technical stops, and during heavy ion collisions which have a low luminosity. The biggest down-drift is in the innermost part of the calorimeter - in A cells where up to -8%/year was observed in August 2011, while in the D cells of the barrel partition, down-drift is negligible. The deviation in the A TileCal layer is shown on Figures 2 [8].

Using the integrator system, the relative variation of the gains is obtained by comparing the response of 9389 channels (95% of all TileCal channels) measured at different dates, with the ones measured in January 2008. These measurements are shown on Figure 3 [8]. The stability of individual channels is better than 0.05% while the average stability is better than 0.01%.

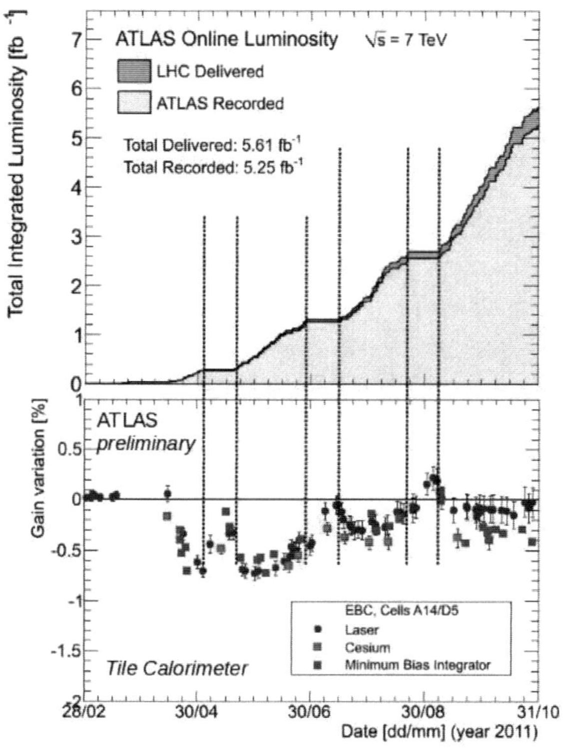

Figure 5. Gain variation measured by the Cesium, Laser and Integrator systems, averaged over one cell type (A14) in 2011, compared to LHC integrated luminosity (upper plot). Periods with no collisions are delimited by vertical dotted lines.

One particular case in TileCal are the Gap and Crack cells. They are located between long and extended barrel and named E1, E2, E3 and E4. They are characterized by a larger exposure to the irradiation and by the large amount of light they produce and that is thus read by the associated PMTs. E3 and E4 being located outside TileCal modules, at smaller radius, they are not monitored by the Cesium system. Their gains are equalized once per year using muons from collisions and by the Laser system for fast drifts. Figure 4 [8] shows the measured gain variation. It reaches -8% in the E4 cells after one year of collisions in 2012.

As shown on Figure 5 [8], the down-drift periods coincide with the periods of data taken with high instantaneous luminosity, while the up-drifts coincide with the technical stops (no collisions). The maximum variation is below 1% over all 2011 data taking period with an integrated luminosity of $5.6\ \text{fb}^{-1}$.

Such a luminosity induced drift is confirmed by the Laser system. It mostly affects cells at inner radius, that are the cells with higher current, as shown on Figure 6 [8] on a full TileCal map where the drift is clearly observed a few weeks after the first collisions of 2012. A comparison of the drifts observed by TileCal monitoring systems is also given on Figure 5 [8] for one typical cell. Their compatibility indicates that the gain variation is dominated by the PMTs gain variation.

4. Conclusion

The calibration systems of Tile Calorimeter were successfully operated during Run 1 of the LHC. Their performances were established. It was shown that these systems can be used in conjunction to monitor and correct fine unstabilities affecting the channels gain like PMT drifts induced by high instantaneous luminosity or to identify the small fraction of pathological channels.

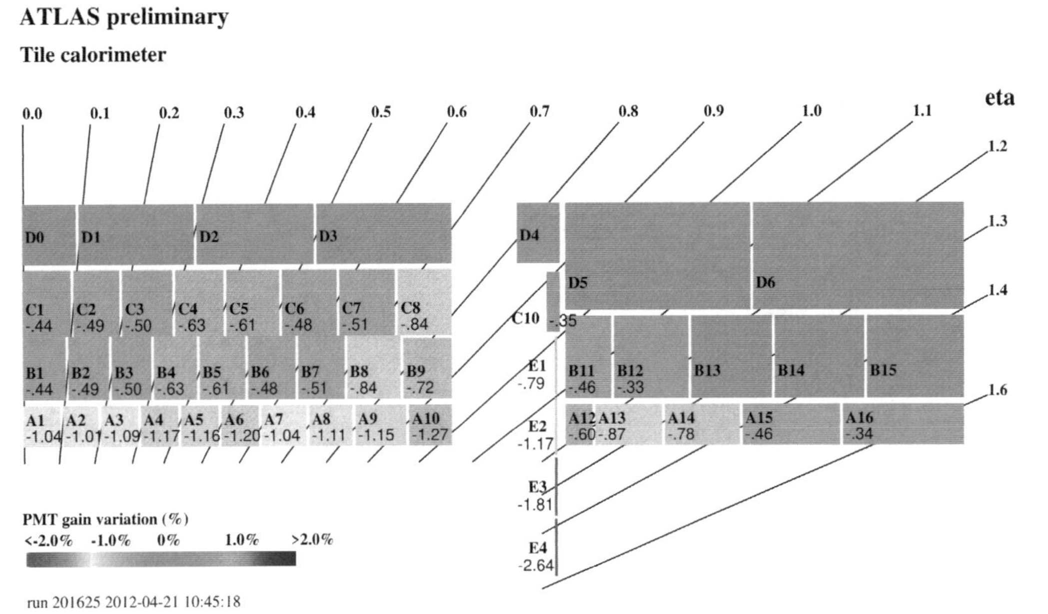

Figure 6. The mean gain variation of the 10000 channels is computed cell by cell as a function of η and radius, between the 19 March 2012 and the 21 April 2012 with the Laser system. The mean values are averaged over total azimuthal coverage of the detector.

References

[1] The ATLAS Collaboration, The ATLAS experiment at CERN Large Hadron Collider, JINST 3 (2008) S08003.

[2] The ATLAS Collaboration, Readiness of the ATLAS Tile Calorimeter for LHC collisions, Eur. Phys. J. C 70 (2010) 1193.

[3] ATLAS/Tile Calorimeter Collaboration, Tile Calorimeter Technical Design Report, CERN/LHCC 96-42 (1996). http://cdsweb.cern.ch/record/331062.

[4] E.A. Starchenko et *al.*, Cesium monitoring system for ATLAS Tile Hadron Calorimeter, Nucl. Instrum. Meth. A494 (2002) 381.

[5] S. Viret, et *al.*, The TileCal Laser monitoring, Nucl. Instrum. Meth. A617 (2010) 120–122.

[6] K.J. Anderson, et *al.*, Design of the Front-end Analog Electronics for the ATLAS Tile Calorimeter, Nucl. Instrum. Meth. A551 (2005) 469.

[7] The ATLAS Collaboration, Improved luminosity determination in *pp* collisions at $\sqrt{s} = 7$ TeV using the ATLAS detector at the LHC, Eur. Phys. J. C 73 (2013) 2518.

[8] TileCal collaboration public results. https://twiki.cern.ch/twiki/bin/view/AtlasPublic/ApprovedPlotsTileCalibration.

Journal of Physics: Conference Series **587** (2015) 012010 doi:10.1088/1742-6596/587/1/012010

Performance of the ATLAS Tile Calorimeter in *pp* collisions at the LHC

Maria Fiascaris, on behalf of the ATLAS Collaboration
CERN, CH-1211 Geneva 23, Switzerland
E-mail: maria.fiascaris@cern.ch

Abstract. The Tile Calorimeter is the central section of the ATLAS hadronic calorimeter at the Large Hadron Collider. This detector is instrumented for the measurements of hadrons, jets, tau leptons and missing transverse energy. Scintillation light produced in the tiles is transmitted by wavelength shifting fibers to photomultiplier tubes (PMTs). The resulting electronic signals from approximately 10000 PMTs are measured and digitized before being transferred to off-detector data-acquisition systems. After an initial setting of the absolute energy scale in test beams with particles of well-defined momentum, the calibrated scale is transferred to the rest of the detector via the response to radioactive sources. The calibrated scale is validated in situ with muons and single hadrons whereas the timing performance is checked with muons and jets. The data quality procedures used during the LHC data-taking and the evolution of the detector status during the LHC Run 1 are presented. The energy and the time reconstruction performance of the digitized signals is summarized and the calorimeter response to hadrons is investigated with collision data.

1. Introduction

The ATLAS Tile Calorimeter (TileCal) is the barrel hadronic calorimeter of the ATLAS experiment at the Large Hadron Collider (see figure 1). Covering the pseudorapidity[1] region -1.7 $< \eta <$ 1.7, together with the electromagnetic calorimeter TileCal provides precise measurement of hadrons, jets, taus and missing transverse energy. The performance requirements driven by the ATLAS physics programme are a jet energy resolution of $\sigma/E = 50\%/\sqrt{E(GeV)} \oplus 3\%$ and a response linearity within 2% up to 4 TeV.

The TileCal is a sampling calorimeter using plastic scintillator as active material and low-carbon steel as the absorber. It is divided into a long barrel (LB) in the central region (-1.0 $< \eta <$ 1.0) and two extended barrels (EB) spanning $0.8 < |\eta| < 1.7$. The barrel and extended barrel cylinders are segmented into 64 wedges (modules) in ϕ and in three longitudinal layers (A, BC, D) for a total thickness of approximately 7λ. Light emitted by the scintillating tiles is transported via wavelength shifting fibers to photomultiplier tubes (PMTs). Each TileCal three dimensional cell is read out by two PMTs on either ϕ side, providing redundancy and improved uniformity. The calorimeter granularity is $\Delta\eta$ x $\Delta\phi$ = 0.1 x 0.1 for A and BC cells and 0.2 x 0.1 for D cells. In the electronics readout, the signals from the PMTs are shaped and amplified

[1] Pseudorapidity η is defined as $\eta = -\ln\tan(\theta/2)$, where θ is the polar angle measured from the beam axis. The azimuthal angle ϕ is measured around the beam axis, with positive (negative) values corresponding to the top (bottom) part of the detector.

in two gains with relative ratio 1:64. Each signal is digitized at a rate of 40 MHz and stored in the front-end pipeline memory. Upon arrival of a first level trigger accept, seven samples from the appropriate gain are sent to the back-end electronics for energy and time reconstruction.

A detailed description of ATLAS and TileCal is provided in [1] and [2], respectively.

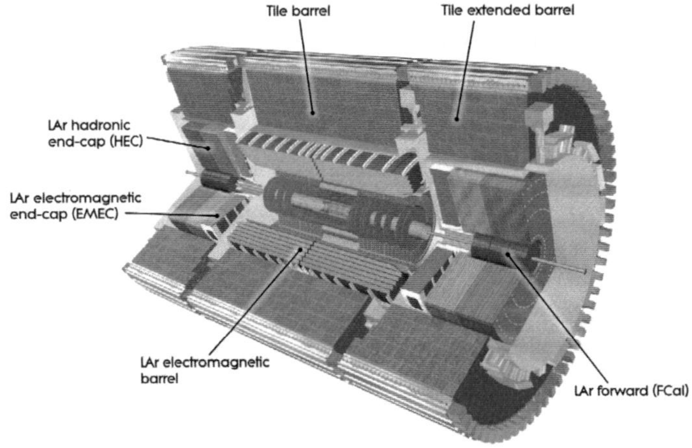

Figure 1. A cut-away view of the ATLAS calorimeters. The Tile Calorimeter consists of one central barrel and two extended barrels.

2. Detector operations

2.1. Data quality

In 2012 ATLAS recorded 21.3 fb^{-1} of pp collision data. Out of these, 95.5% were declared good for physics. The data quality efficiency for TileCal alone was 99.6%, with the main sources of inefficiency being problems with timing and with the readout system.

TileCal achieved this high data quality efficiency thanks to an effective monitoring system. A Data Quality Monitoring Framework (DQMF) collects information about the quality of the data in real-time and performs automatic checks. When problems are detected, a flag is raised in a graphical user interface to alert the shifter who can take immediate action. In addition, automatic recovery procedures are implemented in the Data Acquisition (DAQ) and Detector Control Systems (DCS) to minimize the need for manual intervention and the recovery time.

2.2. Low Voltage Power Supplies

An example of the importance of automatic recovery procedures is given by trips of the Low Voltage Power Supplies (LVPS). The LVPS power the front-end electronics of each module and therefore LVPS trips result in a full module being off. During the LHC Run 1, LVPS trips occurred at a rate of about 0.6 trips/pb^{-1} and the recovery time to power on LVPS and resume data-taking was about two minutes.

To prevent LVPS trips in Run 2, a new production of upgraded power supplies has been installed in the detector during the long LHC shutdown in 2013. The upgraded LVPS have already been tested under the LHC running conditions: 40 of them were installed in the detector in December 2011 and only one trip was observed in the whole of 2012 (compared to the about a hundred trips typically observed for old LVPS in the same area of the detector). The new LVPS also benefit from reduced electronic noise.

2.3. Detector status during Run 1

In TileCal, cells that deliver corrupted data are masked. The evolution of masked cells during 2011 and 2012 is shown in figure 2. The fraction of masked cells changed over time, increasing during the periods when the LHC was operational and decreasing to almost 0% during the maintenance period when the front-end electronics could be accessed and repaired. At the end of the LHC Run 1, in February 2013, about 3% of TileCal cells were masked. The majority of these were due to six modules being powered off because of failures in the LVPS. During the current LHC shutdown of 2013-2014, TileCal is undergoing major maintenance interventions to repair and consolidate all 256 modules and replace all LVPS with the upgraded ones. These activities aim to ensure high performance, high quality and robust operations during Run 2.

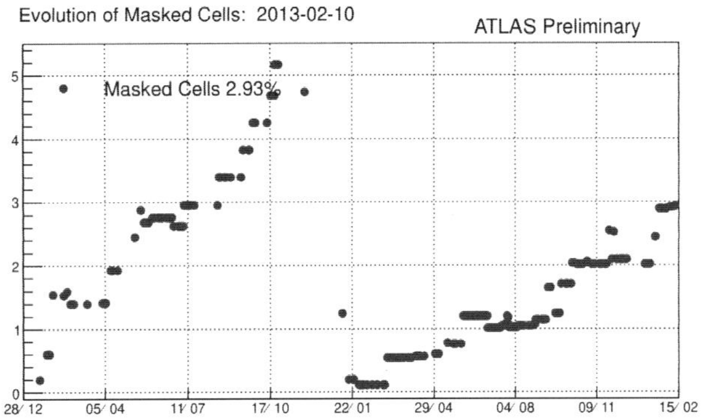

Figure 2. Percentage of masked TileCal cells as a function of time, from December 2010 to February 2013. A sharp increase in the fraction of masked cells corresponds to the loss of an entire module, for example due to failures in the LVPS. The sudden decrease between December 2011 and January 2012 corresponds to the maintenance period, when the front-end electronics could be accessed and repaired.

3. Performance

In TileCal, the Optimal Filter (OF) algorithm is used to reconstruct the pulse amplitude and time for each channel. The reconstructed channel energy is derived from the signal amplitude after appropriate calibration factors. The OF requires an initial knowledge of the time phase between the pulse peak and the LHC clock signal. The performance of the energy reconstruction is checked by comparing two different methods applied online and offline, see figure 3(a). The bias due to the phase variation can be reduced applying a correction using the time of the pulse. After the correction, online and offline energy reconstruction agree within 1% in the time range [-10 ns, 10 ns], as shown by the blue points.

Precise timing is crucial for accurate jet energy reconstruction. The 10000 TileCal channels were synchronized using the laser calibration system, cosmic ray events, beam splashes and collision events, as described in [2]. The cell time resolution is measured in data using muons and jets and found to be better than 1 ns for energies above 4 GeV, as shown in figure 3(b).

The inclusive cell energy response can be validated at the electromagnetic scale using collision events. Figure 4(a) shows the average cell energy as a function of ϕ. The response is uniform in ϕ and the data is well described by the simulation. Furthermore, an in-situ method is used to probe the calorimeter response to single isolated hadrons that shower in TileCal. The calorimeter response is characterized by the ratio E/p, where p is the charged particle momentum measured

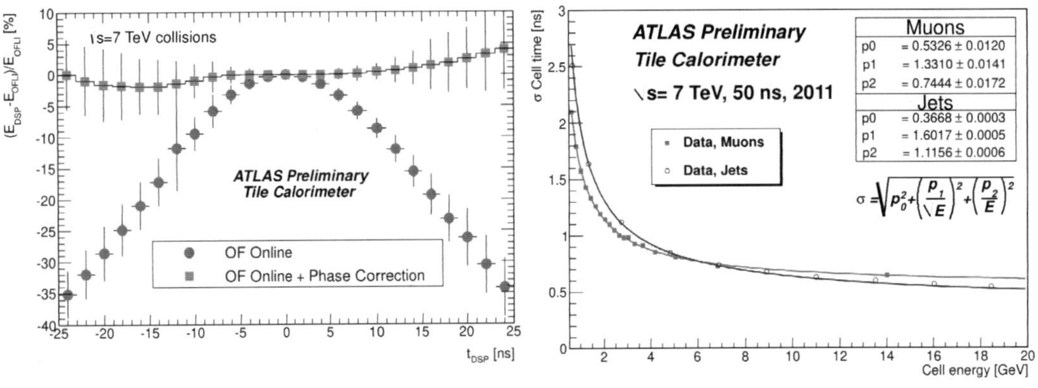

(a) (b)

Figure 3. (a) Relative difference between the online and the offline cell energy reconstruction as a function of the cell time. (b) Cell time resolution measured in collision data on jets and muons as a function of the cell energy.

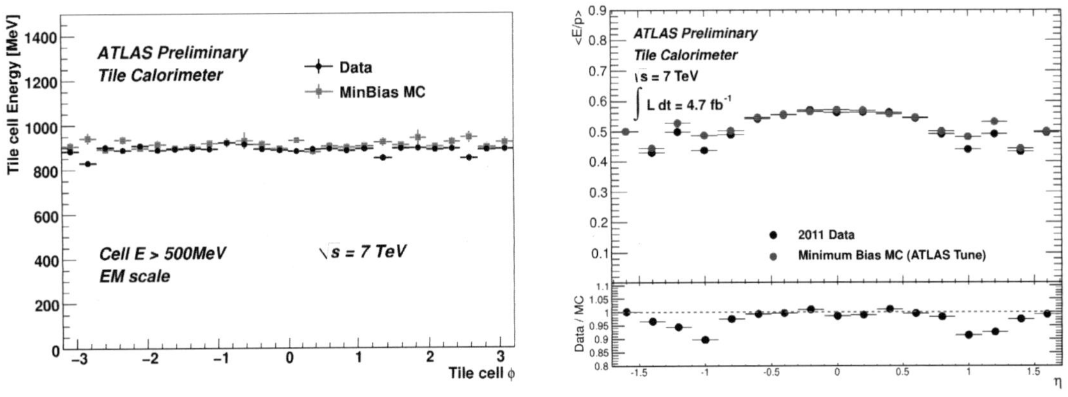

(a) (b)

Figure 4. (a) Average cell energy as a function of ϕ in collision events at $\sqrt{s}$ =7 TeV. Only cell energies above 500 MeV at the electromagnetic scale are considered. Non-diffractive minimum bias Monte Carlo events with the same energy cut are superimposed. (b) Mean value of E/p as a function of η for data at $\sqrt{s}$ =7 TeV and minimum bias Monte Carlo simulation. E refers to the energy deposited in TileCal by isolated charge particles with momentum p. The bottom plot shows the ratio data/Monte Carlo.

with high accuracy in the inner detector and E is the energy of three-dimensional clusters around the track extrapolated to the calorimeter. The E/p ratio measured in 2011 data at $\sqrt{s}$ =7 TeV is shown in figure 4(b) as function of η. The data is well described by the simulation, except in the region around $|\eta| = 1$ where a discrepancy of 10% is observed.

Journal of Physics: Conference Series **587** (2015) 012010 doi:10.1088/1742-6596/587/1/012010

4. Conclusion

The ATLAS Tile Calorimeter has been performing very well during the first LHC run, achieving a high data quality efficiency. By the end of the LHC Run 1, 97% of the detector was operational. At present, major interventions are ongoing in the ATLAS cavern to replace the problematic LVPS and ensure that the detector is in the best conditions for the beginning of Run 2. The performance of TileCal has been studied with collision data and shown to be good.

References

[1] ATLAS Collaboration 2008 *JINST* **3** S08003
[2] ATLAS Collaboration 2010 *Eur. Phys. J.* C **70** 1193

Journal of Physics: Conference Series **587** (2015) 012011 doi:10.1088/1742-6596/587/1/012011

ATLAS Tile Calorimeter: simulation and validation of the response

Jana Faltova, on behalf of the ATLAS Collaboration

Charles University in Prague, Institute of Particle and Nuclear Physics, Czech Republic

E-mail: jana.faltova@cern.ch

Abstract. The Tile Calorimeter (TileCal) is the central section of the ATLAS hadronic calorimeter at the Large Hadron Collider. Scintillation light produced in the tiles is readout by wavelength shifting fibers and transmitted to photomultiplier tubes (PMTs). The resulting electronic signals from approximately 10000 PMTs are measured and digitized before being further transferred to off-detector data-acquisition systems. Detailed simulations are described in this contribution, ranging from the implementation of the geometrical elements to the realistic description of the electronics readout pulses, including specific noise treatment and the signal reconstruction. Special attention is given to the improved optical signal propagation and the validation with the real particle data.

1. Introduction

The Tile Calorimer [1] is a central hadronic calorimeter in the ATLAS detector [2] at the LHC. It consists of a barrel (in the pseudorapidity region $|\eta| < 1.0$) and two extended barrels ($0.8 < |\eta| < 1.7$). It covers the area from 2280 to 3865 mm in the radial direction. The TileCal is a sampling calorimeter with alternating scintillating tiles (active medium) and steel (absorber). The calorimeter is segmented into three-dimensional cells. There are three radial layers (denoted as A, BC, D) and fine segmentation in $\Delta\eta \times \Delta\phi$ (0.1×0.1 in A, BC cells and 0.2×0.1 in D cells). Light emitted in the scintillator is transmitted by the wavelength shifting fibers to the photomultipliers. Each calorimeter cell is read out by two photomultipliers on both ϕ-sides of the cell.

Simulations of the Tile Calorimeter can be divided into three parts — simulation, digitization and reconstruction. The Geant4 toolkit [3, 4] is used in the simulation step. The geometry of the detector is thoroughly described and the passage of particles through matter is simulated. Collection of hits (energy deposits characterized by its position, energy and time) is created and used as the input for the digitization. Next, hits from the signal sample and hits from the minimum bias background events (in-time and out-of-time pile-up) are merged together in one collection. Seven samples spaced by 25 ns are derived corresponding to the situation in real data. Electronic noise is added to the individual samples. Thereafter, the read-out samples are processed in the reconstruction which is identical to the procedure applied in data. The energy is reconstructed by means of the Optimal Filtering Method and the cell energy is calibrated to the electromagnetic scale.

2. Noise description

Two main sources of disturbance might degrade the physics signal — electronic noise and a contribution from multiple minimum bias collisions overlaid with the hard scattering process (the pile-up). Good description of the noise is important for the creation of so-called topoclusters (clusters of cells based on noise significance) that are used for building the physics objects, especially hadronic jets.

The electronic noise was measured in data and found to have significant non-Gaussian tails. The cell noise is approximated with a double Gaussian function to a good precision. The double Gaussian model is used also in the Monte Carlo simulations where the noise is added to the individual samples. The constants used in the simulations are derived from the measurements in data cell by cell. A good agreement between data and Monte Carlo simulations is achieved as can be seen in figure 1. The measurement was performed using 2010 data where the pile-up term can be neglected. The noise contribution can be compared with data collected using a random trigger. The electronic noise comes mainly from the Low Voltage Power Supply (LVPS) which provides power to the front-end electronics. A new version of the LVPS is being installed and will be used for the next data-taking period (Run 2). The new LVPS leads to smaller electronic noise with more Gaussian behavior.

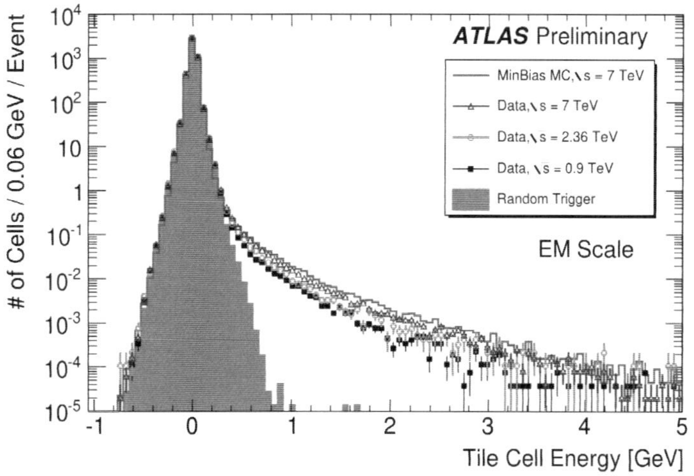

Figure 1. Cell energy spectrum measured in 2010 data. The pile-up term might be neglected.

Contrary to the electronic noise, the pile-up noise depends strongly on the radial layer. The largest contribution affects the A-cells (closest to the beam pipe) while the smallest pile-up appears in D-layer cells. Moreover, strong dependence on the pseudorapidity was found as shown in figure 2, where the noise level for all three radial layers (A, BC, D) and the gap/crack scintillators is plotted. The observed agreement between data and simulations is within $\pm 20\%$ in the region where pile-up is dominant. Currently, a simple model evaluates the pile-up term using the standard deviation of the cell energy distribution, the shape is not considered. An improved model, which takes into account the pile-up energy spectrum shape, is being prepared to be used in Run 2.

3. Optical signal propagation

The response of the photomultipliers is not flat in the azimuthal angle difference between the energy deposition point and the center of the cell ($\Delta\phi$), but it shows a non-negligible dependence

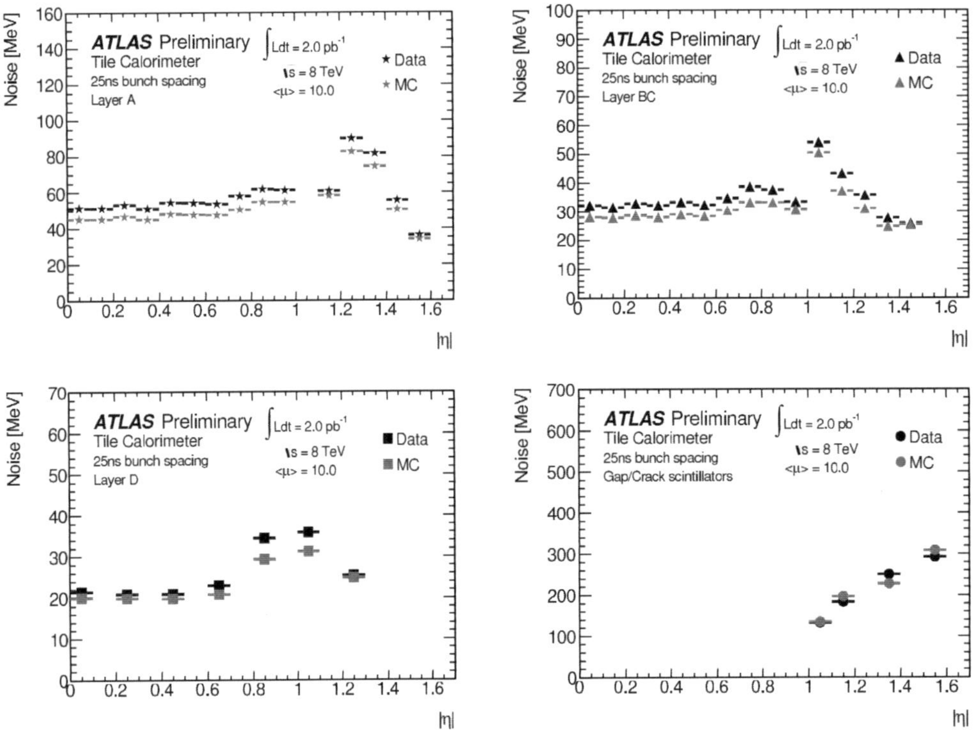

Figure 2. Cell noise distribution in the presence of pile-up measured with data collected at a centre-of-mass energy of 8 TeV with a bunch spacing of 50 ns and an average number of interactions 15.7 per bunch crossing.

referred to as U-shape. The dependence of the response on $\Delta\phi$ was measured using $W \rightarrow \mu\nu$ events in the 2012 collisions data. The measurement is performed separately in the barrel and in the extended barrels in each radial layer. The typical shape is shown in figure 3 (a) for barrel and 3 (b) for extended barrel. A steep dip at the center of the cell ($\Delta\phi = 0$) corresponds to the position of the apertures in the scintillating tiles [1]. The decrease at the edges of the cells ($\Delta\phi > 0.03$) is due to the geometrical effects.

The U-shape option has been recently added in the Monte Carlo simulations. Look-up tables are prepared based on the measurements in data and used in the simulations. The tables are normalized as follows: First, average value for one PMT is set to 0.5. Second, the look-up tables are rescaled to have the same sampling fraction for 100 GeV electrons at $\eta = 0.35$ in the center of the cell as the simulations without the U-shape. The comparison between simulations with and without U-shape can be seen in figures 4 (a) to (f). The layers A, BC and D in the central, resp. extended barrel are shown in figures (a), (b) and (c), resp. (d), (e) and (f).

4. Comparison between data and Monte Carlo simulations

4.1. Validation of the electromagnetic scale

The electromagnetic scale (EM) in TileCal is calibrated using electron beam data at the angle of 20 degrees. The validation of the EM scale was done using muons and pions in test beams and

[1] Each scintillating tile has two apertures in its phi-center. While a large rod runs through one hole and physically fixes the tile in the iron structure, a pipe used by Cesium calibration system is running though the second hole.

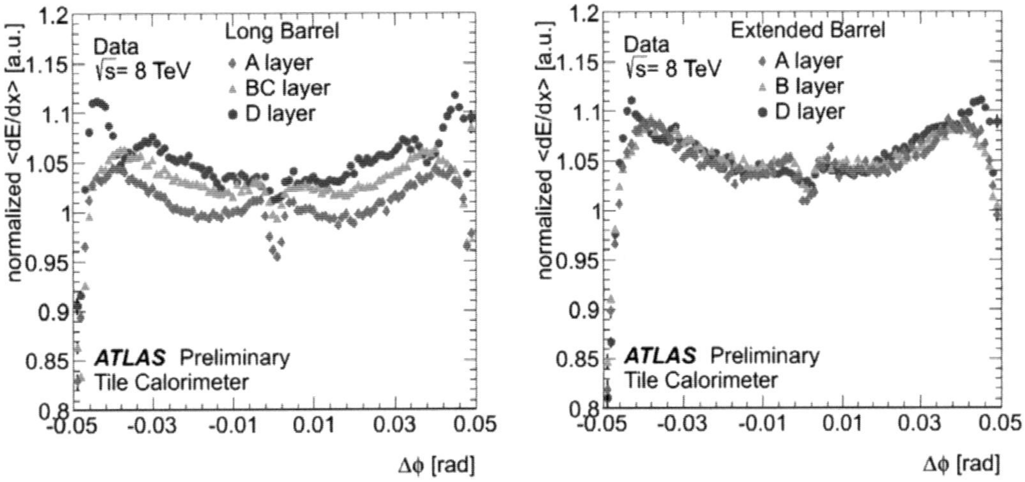

(a) (b)

Figure 3. Response dependence on the azimuthal angle difference between the muon track impact point and the center of the cell ($\Delta\phi$) measured in 2012 collisions data using $W \rightarrow \mu\nu$ events.

also cosmic muons, details can be found in [1, 5]. As an example, an analysis with cosmic muons is discussed here. A very good separation between signal and noise is achieved with S/N ratio of 29 for the sum of the three radial layers as shown in figure 5. The cell response uniformity, in terms of dE/dx, was measured with the muon tracks and was found to be better than 5% in all three layers. The EM scale measurement was compared with test beam results and it is consistent with each other within an uncertainty of 4%. The ratio of data and Monte Carlo predictions is close to one, as required (see figure 6).

4.2. Measurements of E/p ratio

The calorimeter response can be characterised by the ratio of energy and momentum (E/p) for isolated tracks. The deposited energy E is measured in the calorimeter while momentum p in the inner tracker. The energy reconstructed in the electromagnetic calorimeter is required to be smaller than 1 GeV in order to find hadrons with most of their energy deposited in TileCal. The results using 2011 collision data are shown in figure 7. A good agreement between data and Monte Carlo predictions is observed. The largest difference of 10% occurs in the gap region ($0.9 < |\eta| < 1.1$) where a precise description of the detector material is very difficult.

5. Conclusions

Data collected in TileCal have been compared with Monte Carlo simulations in a wide range of measurements — test beams, cosmics, collisions data. A good level of agreement was found at low signals (corresponding to noise contribution) and also at high energy signals. The light propagation in the Monte Carlo simulations has been improved by introducing the U-shape. Furthermore, more precise pile-up description will be used in the next data-taking period.

References

[1] Atlas Collaboration 2010, *Eur.Phys. J.* C **70** 1193
[2] Atlas Collaboration 2008, *JINST* **3** S08003
[3] Agostinelli, S. et al. 2003, *Nuclear Instruments and Methods in Physics Research* A **506** 250-303

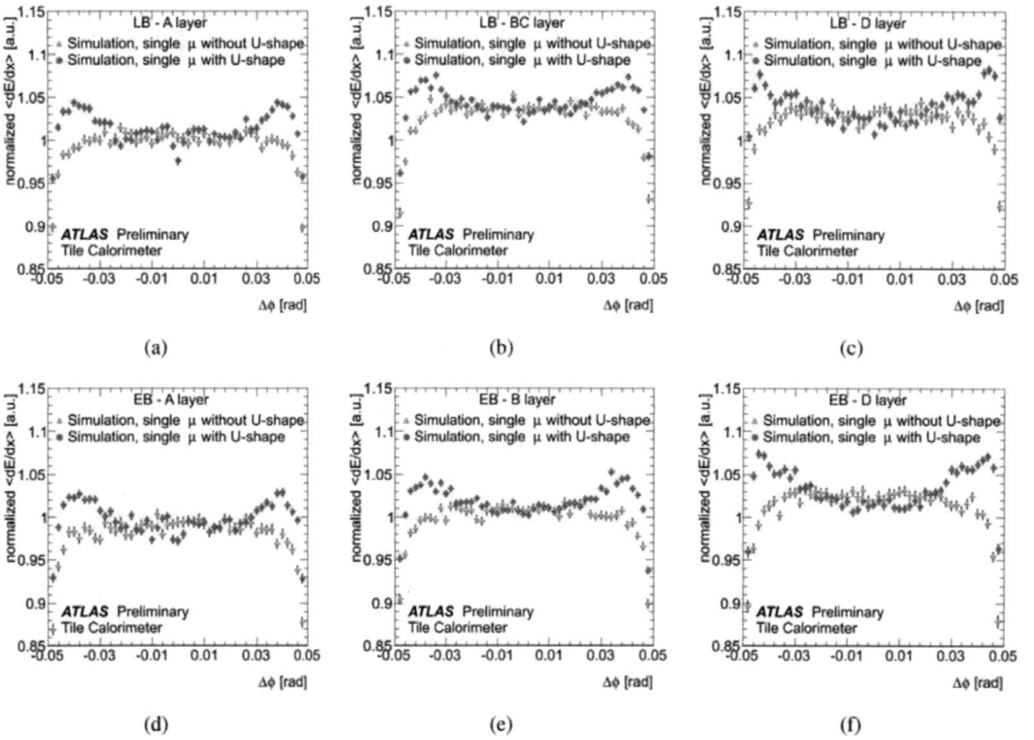

Figure 4. Response dependence on the azimuthal angle difference between the muon track impact point and the center of the cell ($\Delta\phi$) for Monte Carlo simulations with (red color) and without (green color) the U-shape.

[4] Allison, J. et al. 2006, *IEEE Transactions on Nuclear Science* 53 **No. 1** 270-278
[5] Atlas Collaboration 2009, *Nuclear Instruments and Methods in Physics Research* A **606** 362-394

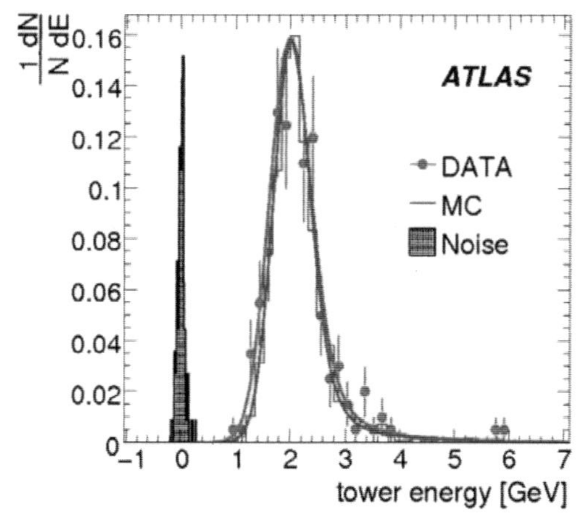

Figure 5. Example of the muon signal and noise for projective cosmic muons at $0.3 < |\eta| < 0.4$.

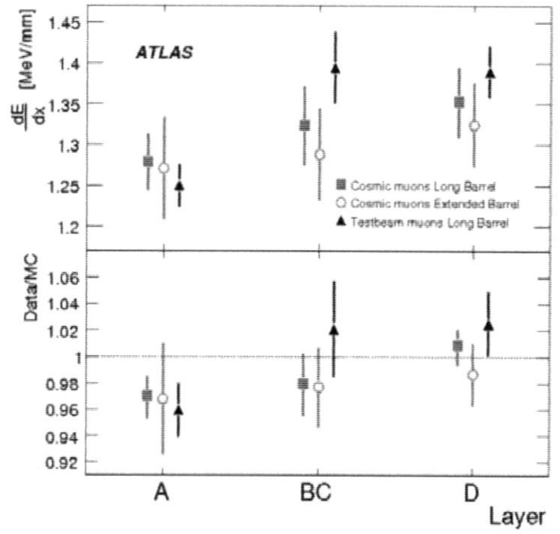

Figure 6. The truncated mean of the dE/dx for cosmic and test beam muons per radial layer. Comparison with Monte Carlo simulations is shown in the bottom part of the plot.

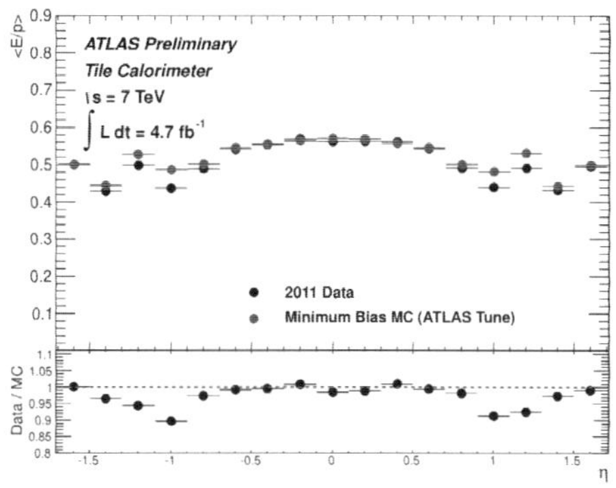

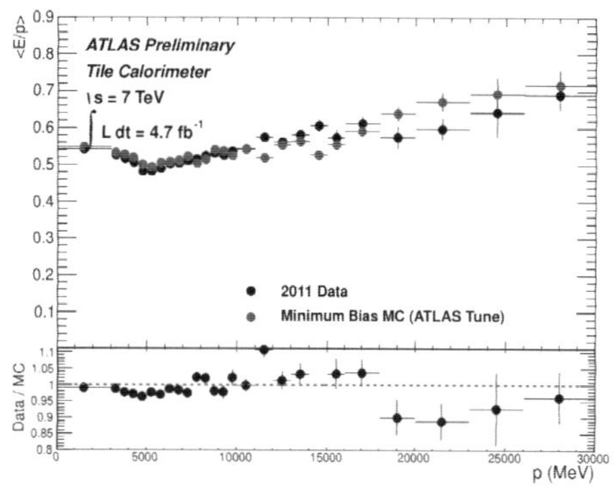

Figure 7. Energy over momentum (E/p) for isolated tracks using proton-proton collision data from 2011. The dependence on the pseudorapidity (momentum) is shown in the left (right) plot.

16th International Conference on Calorimetry in High Energy Physics (CALOR 2014) IOP Publishing
Journal of Physics: Conference Series **587** (2015) 012012 doi:10.1088/1742-6596/587/1/012012

Performance of the LHCb calorimeters during the period 2010-2012

Pascal Perret[1] and Xavier Vilasís-Cardona[2], on behalf of the LHCb Collaboration.

[1] Clermont Université, Université Blaise Pascal, CNRS/IN2P3, LPC, Clermont-Ferrand, France
[2] GR-SETAD, La Salle, Universitat Ramon Llull, Barcelona, Spain

E-mail: perret@in2p3.fr, xvilasis@salleurl.edu

Abstract. The calorimeter system of LHCb is subdivided into four sub-detectors which provide its longitudinal segmentation: a Scintillator Pad Detector, followed by a Preshower and, then, an electromagnetic, an hadronic calorimeter. After a description of these detectors, procedures developed to calibrate their response are discussed together with the degradation of response observed due to the harsh conditions (high radiation and current levels). First results on the calorimeter performance over the two most productive years of physics data taking, 2011 and 2012 are given.

1. Introduction

The LHCb experiment [1] is one of the four major experiments at the Large Hadron Collider (LHC) of CERN. The detector is a single-arm forward spectrometer covering the pseudorapidity range $2 < \eta < 5$, designed for the study of particles containing b or c quarks. It includes a high-precision tracking system consisting of a silicon-strip vertex detector surrounding the pp interaction, a large-area silicon-strip detector located upstream of a dipole magnet with a bending power of about 4 Tm, and three stations of silicon-strip detectors and straw drift tubes placed downstream. The combined tracking system has a momentum resolution $\Delta p/p$ that varies from 0.4% at 5 GeV/c to 0.6% at 100 GeV/c, and an impact parameter (IP) resolution of 20 μm for tracks with high transverse momentum (p_{T}). Charged hadrons are identified using two ring-imaging Cherenkov detectors (RICH). Photon, electron and hadron candidates are identified by a calorimeter system consisting of scintillating-pad (SPD) and preshower (PS) detectors, an electromagnetic calorimeter (ECAL) and a hadronic calorimeter (HCAL) which is described in details in the next section. Muons are identified by a system composed of alternating layers of iron and multiwire proportional chambers. The trigger consists of a hardware stage, based on information from the calorimeter and muon systems, followed by a software stage which applies a full event reconstruction.

The LHC delivered instantaneous luminosity at the LHCb interaction point of up to 4×10^{32} cm^{-2}s^{-1} for proton energies up to 8 TeV, which was equivalent to the average number of interactions per beam crossing 1.8, while the design value for the LHCb luminosity was 2×10^{32} cm^{-2}s^{-1} resulting in an average number of interactions per beam crossing of 0.4, much smaller than what the LHC is capable to provide. In 2011 a luminosity leveling procedure was introduced. By adjusting the transverse overlap of the beams at LHCb the instantaneous luminosity could be kept stable to within about 5% during a fill. The two most productive years were 2011 and 2012, when 1.107 fb^{-1} and 2.082 fb^{-1} of integrated luminosity have been recorded at beam energies 3.5 TeV and 4 TeV correspondingly.

The $b\bar{b}$ cross-section measured by LHCb at 7 TeV constitutes 79 μb [2] in LHCb acceptance, implying ~100 000 $b\bar{b}$ pairs produced every second. In each p-p collision about 1500 charged particles are produced resulting in sizeable radiation levels on detectors and readout electronics. In general, the apparatus has shown robust performance even in such a challenging environment.

The performance of the calorimeters relies both on their time alignment [3] and calibration [4] which require different approaches for each sub-detector and are described in section 3. Section 4 describes the detector ageing. Some performances in terms of particle identification are illustrated in section 5. Additional results are available in Ref. [4] and references therein.

2. Description of the LHCb Calorimeters

The LHCb calorimeter [5] has two main goals: the first one is to provide information to the hardware trigger level and the second is to measure the neutral particle energy and to take part to particle identification for offline analysis. To reach them it is subdivided into four systems. From the interaction point particles encounter at 12,3 m the SPD, which is designed to identify charged particles, followed by a wall of lead and the PS, then the ECAL and the HCAL. This ensures the characterisation of the longitudinal profile of the showers, along the LHC beam axis. All calorimeters use the same energy detection principle: scintillation light is transmitted to a photomultiplier tube (PMT) by wavelength-shifting (WLS) fibres. The single fibres for the SPD/PS cells are read out using multianode 64 channel photomultiplier tubes (MAPMT), while the fibre bunches in the ECAL and HCAL modules require individual phototubes. Optimal energy resolution requires the full containment of the showers from high energy photons. For this reason, the thickness of ECAL was chosen to be 25 radiation lengths (X_0). The PS is 10% as thick as the ECAL. The SPD/PS, ECAL and HCAL adopt a variable lateral segmentation into square cells since the hit density varies by two orders of magnitude over the calorimeter surface. A segmentation into three different sections has been chosen for the ECAL and projectively for the SPD/PS. Given the dimensions of the hadronic showers, the HCAL is segmented into two zones with larger cell sizes. The outer dimensions of the ECAL match projectively those of the tracking system, $\theta_x < 300$ mrad and $\theta_y < 250$ mrad while the square hole around the beampipe limits the inner acceptance approximately to projective polar angles $\theta_{x,y} > 25$ mrad. The main parameters of the sub-detectors are summarised in Table 1.

Table 1. Main parameters of the LHCb calorimeter sub-detectors

sub-detector	SPD/PS	ECAL	HCAL
number of channels	2×6016	6016	1488
overall lateral dimension in x, y	6.2 m × 7.6 m	6.3 m × 7.8 m	6.8 m × 8.4 m
cell size (mm) Inner	39.66 (SPD), 39,84 (PS)	40.4	131.3
Middle	59.5 (SPD), 59.76 (PS)	60.6	
Outer	119 (SPD), 119.5 (PS)	121.2	262.6
depth in z	180 mm, 2 X_0, 0.1 λ_{int}	835 mm, 25 X_0, 1.1 λ_{int}	1655 mm, 5.6 λ_{int}
light yield	~20 p.e./MIP	~3000 p.e./GeV	~105 p.e./GeV
dynamic range	0 - 100 MIP 10 bits (PS), 1 bit (SPD)	0 - 10 GeV E_T 12 bits	0 - 20 GeV E_T 12 bits

The SPD/PS detector consists of a 15 mm lead converter 2.5 X_0 thick, that is sandwiched between two almost identical planes of rectangular scintillator pads of high granularity with a total of 12032 detection channels.

The ECAL uses the shashlik calorimeter technology, i.e. a sampling scintillator/lead structure readout by plastic WLS fibres. The energy resolution of the ECAL modules was determined at the test beam, $\sigma_E/E = a/\sqrt{E} \oplus b \oplus c/E$ (E in GeV), where a, b and c stand for the stochastic, constant and noise terms respectively. Depending on the type of module and test beam conditions the stochastic and constant terms were measured to be 8.5% < a < 9.5% and b ∼0.8%. The noise term corresponds to 1.2 ADC count. The pedestal subtraction scheme accounts for an underestimation of the pedestal over a 12bit -10 GeV range and corresponds to $c = 0.003$ GeV.

The hadron calorimeter (HCAL) is a sampling device made from iron and scintillating tiles, as absorber and active material respectively. Its thickness is limited to 5.6 nuclear interaction lengths (λ_{int}) due to space limitations and the limited performance required. The special feature of this sampling structure is the orientation of the scintillating tiles that run parallel to the beam axis. A resolution of $\sigma_E/E = (69 \pm 5)\%/\sqrt{E} \oplus (9 \pm 2)\%$ (E in GeV) was determined.

3. Monitoring and calibration

The SPD calibration is performed by adjusting the discriminator threshold for each channel to yield a pre-defined efficiency to identify a charged track. The efficiency is computed from the ratio between the number of hits with a track pointing to a given cell over the total number of tracks pointing to it. The result of the efficiency as a function of the threshold can be compared with the theoretical expectation given by the integral of the convolution of a Landau and a Poisson distribution. The Landau distribution describes energy lost by charged particle traversing a thin layer of material while the Poisson distribution describes the fluctuations in the number of photoelectrons generated at the photocathode. The resolution in the electronics to set the threshold is 5% of E_{MIP} which directly sets the objective resolution at the 5% level.

The charge response of the 6016 PS channels is equalised according to the most probable energy deposition of minimum ionising particles (MIPs). The goal is to set the MIPs peak value to 10 ADC counts in the Front-End electronics (FE). It results in a 100 MIPs full dynamic range. Discrimination of e and γ's at the hardware trigger is then achieved with a typical threshold of 5 - 10 MIP. By tuning the high voltage (HV) of the MAPMT's one can adjust the gain of a group of 64 PS channels simultaneously. The FE numeric gain with tunable values in the range [1; 2] further allows a fine adjustment of the 64 channels response within a given board. This procedure results in a calibration of the PS better than 5%.

Both ECAL and HCAL are equipped with monitoring systems on the basis of light emitting diodes (LEDs). Each photomultiplier is illuminated by light flashes of fixed intensity, so the value of average PMT response can be used to follow up the behaviour of each readout channel (mostly defined by the stability of the PMT gain). In order to eliminate the effects, coming from possible variations of the intensity of LED flash, LEDs themselves are monitored by means of very stable silicon PIN photodiodes. The monitoring of ECAL and HCAL responses is performed online in parallel with data taking. LEDs flash with the frequency of about 11 kHz synchronously with one of accelerator service time slot not filled with protons and a subset at about 50 Hz is sent to a monitoring farm for analysis. The LED system allow to follow fastly the ageing of the detector. It is also used to detect malfunctioning cells and bad cable connections.

The fine calibration of ECAL uses fits to the invariant mass distribution of $\pi^0 \to \gamma\gamma$ decays. Photons are defined from 3×3 neutral clusters consistent with single photon signals. The cell with the highest energy deposit is called the seed. Each π^0 candidate mass is assigned to two cells, one for each photon seed. When enough statistics is collected (for the most peripheral cells it amounts up to few hundred millions of interactions), the π^0 mass distribution is fitted for each cell, and the energy-scale of the cell is corrected to center the mass peak onto the nominal π^0 mass (135 MeV/c^2). In addition, samples of pure electrons selected from conversions are used to follow fast ageing of ECAL (Section 4) by studying the energy to momentum ratio (E/p) distributions. This last method requires precise information from the tracking system. The final precision of the calibration is estimated to be better than 2%.

In addition to use of the LED system, more precise HCAL calibration is performed on the regular basis with a cesium radioactive source calibration system. Two 10 mCi ^{137}Cs sources (one per each detector half) are moved by a hydraulic system through each scintillator tile. The PMT response is measured with a dedicated system of current integrators. Relation between anode current and the measured particle energy is known from test beam studies, which showed also that this method facilitates a cell-to-cell intercalibration level 2-3%. Such precision is more than enough to fulfill all HCAL objectives in LHCb.

4. Detector ageing

Over the first years of almost continuous LHCb operation, signs of the degradation of detector response was observed in all four calorimeter sub-detectors. So far, two major reasons for it have been identified. The first is the radiation damage of scintillator tiles and WLS fibers, which manifests itself in the areas with largest hit densities, especially in the innermost regions around the beam pipe. This effect is proportional to the particle flux (charged and neutral particles). In the ECAL inner most region, up to ~0.25 Mrad/year are observed. For instance it was found that the precision of the ECAL monitoring based on LED suffers from the radiation degradation of transparency of polystyrene clear fibers used to transport the LED light to the ECAL photomultipliers. In order to improve the performance of the monitoring system, and especially in view of significant increase of LHCb working luminosity foreseen after 2018, the present plastic fibers were replaced by radiation hard quartz fibers [6].

For now, the most extensive studies of its effects was done for HCAL, where light yields can be controlled directly with the ^{137}Cs sources of the calibration system. The average relative decrease in the light yield for 44 most central HCAL cells is shown in Fig. 1 (left) as a function of the luminosity, delivered by LHC over 2011 and 2012, separately for each layer of scintillator tiles. As expected, the most affected by radiation layer is the frontal one, which corresponds to the shower maximum position, thanks to the additional 1.1 λ_{int} of ECAL. The second factor is

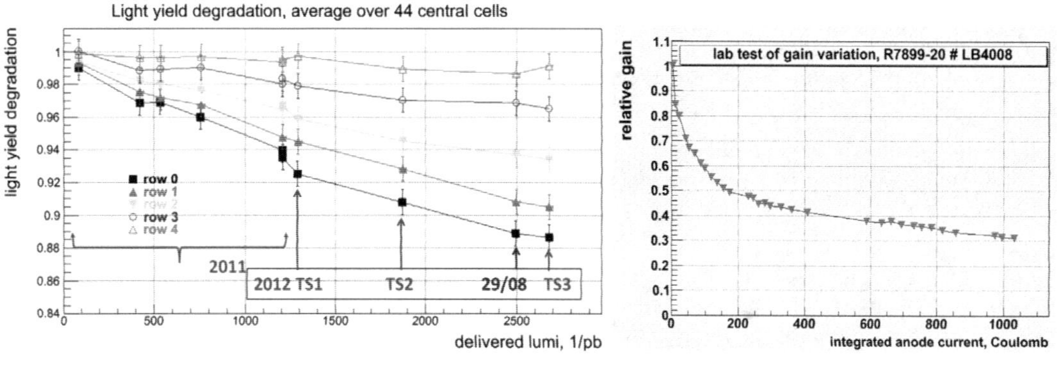

Figure 1. Left: relative decrease of HCAL light yield for different layers of scintillator tiles, averaged over the 44 innermost cells, as a function of the delivered luminosity. Right: PMT gain degradation as a function of the integrated anode current measured at lab.

PMT sensitivity losses. The effect was studied in laboratory, the tested photomultiplier being kept under continuous illumination by short light pulses to emulate the LHCb environment. Figure 1 (right) shows the relative change of PMT response with the integrated charge, which has passed through the device since the beginning of the test. The initial sharp decrease of the gain noticeably slows down after 150 coulombs, though it should be kept in mind, that this test did not reproduce exactly the real operational conditions with periods of inactivity between fills, allowing some recuperation of PMT sensitivities. This effect depends upon cell size and location. The actual amounts of integrated charge, obtained over 2011 and 2012, is estimated with current integrators, which readings are continuously logged. In case of HCAL, the total charge, which passed through the innermost photomultipliers during 2011, was on the level of 100 C. Prior to the beginning of 2012 data taking, the gains of all HCAL PMTs were decreased by a factor of two in order to reduce the sensitivity losses and also to keep the PMT anode currents within the recommended limits in the face of the increase of luminosity-setting. For ECAL, the maximal integrated charge was on the level of 20 C for 2011 and 37 C for 2012.

These are well known unavoidable effects which affect the trigger rate based on calorimeter information as well as detector performances. They required regular calibrations using the methods described in Sect. 3 to restore uniform detector responses. In addition the cumulative change in the response of the calorimeter sub-detectors could be estimated with cell occupancies in minimum bias events, thanks to the luminosity leveling system providing relatively stable conditions over long periods of data taking. This method is very promising and preliminary studies have shown that a precision of $\sim$ 1% could be achieved after few hours of data taking. It will be implemented for an automatic and online calibration for the next data taking period (Run 2) to adjust PMT high voltage after each LHC fill of data taking to obtain a stable trigger.

5. Performance of the LHCb calorimeters

5.1. Electron identification

The electron identification (ID) in the calorimeter system uses the background and signal likelihood distributions constructed for the ECAL, the PS and the HCAL sub-detector [4]. The performance is evaluated using the data recorded in 2011. The available statistics allows the electron ID efficiency to be measured using a tag-and-probe method. The method is applied to $B^{\pm} \to J/\psi K^{\pm}$ candidates having $J/\psi \to e^{+}e^{-}$, where one of the electrons is required to be identified by electron ID (e_{tag}) while the second electron is selected without using any information from the calorimeter system (e_{probe}). This second electron is used to estimate the efficiency of the electron ID and the performance of the electron hypothesis test. The efficiency and misidentification rate are obtained as a function of the e_{probe} momentum. By way of illustration of the typical performance, for electrons from $J/\psi \to e^{+}e^{-}$ decay in $B^{\pm} \to J/\psi K^{\pm}$ events the average identification efficiency in the complete calorimeter system is (91.9$\pm$1.3)% for a misidentification rate of (4.54$\pm$0.02)%. In addition information from the RICH detectors can be used to improved these performances allowing to reach an average identification efficiency up to 97% for a misidentification rate lower than 2%.

5.2. Isolated photon and neutral pion reconstruction

Energy deposits in ECAL cells are clusterised applying a 3x3 cell pattern around local maxima of energy deposition. Consequently the centers of the reconstructed clusters are always separated at least by one cell. If one cell is shared between several reconstructed clusters, the energy of the cell is redistributed between the clusters proportionally to the total cluster energy. The process is iterative and it converges rapidly due to relatively small ratio of the Moliere radius (3.5 cm) to the cell size. After the redistribution of energy of shared cells, the energy-weighted cluster moments up to the order 2 are evaluated to provide the (hypotheses-independent) cluster parameters, namely, the total energy, the energy barycenter position and the two-dimensional energy spread matrix. Neutral clusters are identified as those clusters that do not match to charged tracks.

The photon energy, E_c, is evaluated from the total cluster energy and the reconstructed energy deposit in the PS.

The performance of high energy photon reconstruction is illustrated by the reconstructed $B^0 \to K^{*0}\gamma$ and $B^0 \to \phi\gamma$ mass distribution [7] shown in Fig. 2. The mass resolution obtained for this radiative decay, $\sim$92 MeV/c^2, is dominated by the ECAL energy resolution. A comparison of the data to simulated samples shows that this corresponds to an intrinsic resolution in the photon energy of around 2% [4].

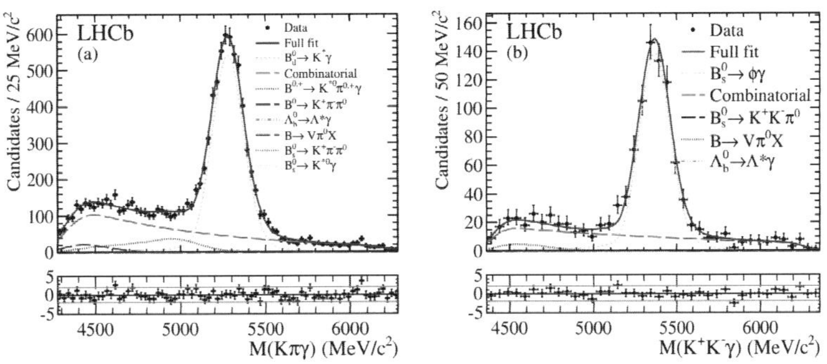

Figure 2. Mass distribution of the reconstructed $B^0 \to K^{*0}(K^+\pi^-)\gamma$ candidates (a) and $B_s^0 \to \phi(K^+K^-)\gamma$ candidates (b) obtained with an integrated luminosity of $\mathcal{L}$=1.0 fb^{-1} [7]. The black points represent the data and the fit result is represented as a solid blue line. The signal (green dotted line) and the various background contaminations are shown.

Neutral pions with low transverse energy are mostly reconstructed as a resolved pair of well separated photons. A mass resolution of 8 MeV/c^2 is obtained for such neutral pions. However, due to the finite ECAL granularity, photon pairs from the decay of sufficiently high momentum π^0 cannot be resolved as individual clusters. This is essentially the case for all π^0 decays with transverse momentum above 2 GeV/c. To reconstruct such π^0 candidates a procedure has been design to identify merged clusters. The algorithm consists of splitting each single ECAL cluster into two 3 $\times$ 3 subclusters built around the two highest deposits of the original cluster. The energy of the common cells is then distributed over the two assumed subclusters according to an iterative procedure based on the expected transversal shape of photon showers. The division of the energy depends on the barycenter position of each subcluster. Since the latter is a function of the energy distribution, this procedure requires a few iterations, but it is found to converge quickly. A $\pi^0 \to \gamma\gamma$ invariant mass resolution of $\sim$30 MeV/c^2 is obtained for these 'merged' π^0 candidates while it is close to $\sim$20 MeV/c^2 for 'resolved' π^0.

Acknowledgements

Work supported by CNRS/IN2P3, Region d'Auvergne (France); MICINN and GENCAT (Spain).

6. References

[1] LHCb Collaboration, Jr. A. A. Alves et al., *The LHCb detector at the LHC*, JINST, 3:S08005, 2008.
[2] LHCb Collaboration, R. Aaij et al., Phys. Lett., **B694** (2010) 209.
[3] C.Abellan Beteta et al., JINST, 7:P08020, 2012.
[4] C.Abellan Beteta et al., *Performance of the LHCb Calorimeters*, CERN-LHCb-DP-2013-004, 2013.
[5] S. Amato et al., *LHCb calorimeters: Technical Design Report*, CERN, Geneva, LHCb-TDR-2, 2000.
[6] Y. Guz and LHCb collaboration, CALOR 2014 conference proceedings, LHCb-PROC-2014-015.
[7] LHCb Collaboration, R. Aaij et al., Nucl. Phys. **B867** (2013) 1.

Journal of Physics: Conference Series **587** (2015) 012013 doi:10.1088/1742-6596/587/1/012013

Upgrade of the LHCb ECAL monitoring system

Yury Guz, on behalf of the LHCb Collaboration
IHEP, Protvino, Russia & CERN, Geneva, Switzerland
E-mail: iouri.gouz@cern.ch

Abstract. The LHCb ECAL is a shashlik calorimeter of 6016 cells, covering 7.68×6.24 m^2 area. To monitor the readout chain of each ECAL cell, the LHCb ECAL is equipped with a LED based monitoring system. During the LHC Run I (2009-2012) it was found that the precision of the monitoring suffers from the radiation degradation of transparency of polystyrene clear fibers used to transport the LED light to the ECAL photomultipliers. In order to improve the performance of the monitoring system, and especially in view of significant increase of LHCb working luminosity foreseen after 2018, the present plastic fibers have been replaced by radiation hard quartz fibers. The performance of the old LHCb ECAL monitoring system during LHC Run I and the design of the upgraded system are discussed here.

1. Introduction

The LHCb ECAL design is described in [1]. It is a shashlik calorimeter of 25 X_0 thickness, covering 7.686.24 m^2 area. It is subdivided into 6016 cells. The light produced in each cell is read out by a PMT; the PMT anode signal is digitized by an ADC.

To monitor the functionality of the readout chain of each ECAL cell and the stability of its characteristics, the LHCb ECAL is equipped with the LED based monitoring system. The value of average photomultiplier response to light flashes of fixed intensity is used to follow up the behavior of each readout channel, mostly determined by the stability of the PMT gain. The details of the monitoring system design can be found in [1].

The monitoring system is organized such that each LED illuminates a group of 9 or 16 ECAL cells. The 6016 cells of LHCb ECAL are illuminated by 456 LEDs. Red LEDs of type Ligitek LUR3333H, with emission peak at 630 nm, are used. The stability of LED flash magnitude is monitored by PIN photodiodes (HAMAMATSU S1223-01). Each PIN photodiode serves a group of 2 or 4 LEDs; the LED firing sequence ensures that two LEDs within same group never flash simultaneously.

The LEDs, PIN photodiodes and corresponding electronic boards are installed below and above the calorimeter, outside the high radiation area. The light is distributed to each PMT by means of 2–8 m long light guide made of polystyrene-based KURARAY Clear-PCM fiber of 1 mm diameter.

The LHCb ECAL monitoring system comprises 6016 light guides grouped into 456 bundles, 456 LEDs and 124 PIN photodiodes. The total length of clear fibers is 28.6 km.

2. Performance and use of the ECAL monitoring system

The ECAL monitoring system played an important role at the commissioning and initial calibration phase [2]. During the LHCb data taking, it is also kept continuously running. The

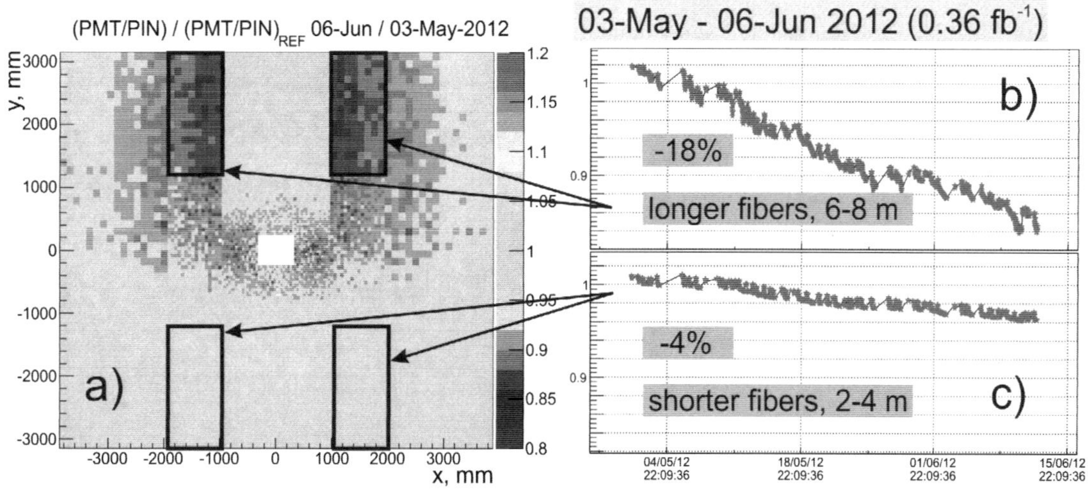

Figure 1. . Evidence of clear fiber degradation in LHCb ECAL: **a** the ratio of LED amplitudes on June 06, 2012 to those on May 02, 2012, for each ECAL cell; **b** the evolution of average LED signal for the two upper rectangular areas shown in **a**; **c** the evolution of average LED signal for the two lower areas.

LED pulses are sent to PMTs at approximately 50 Hz rate, at so-called empty LHC bunch crossings, in order to not interfere with physics events. The LED data, in form of amplitude spectra for every calorimeter cell, are stored to disk every 10–15 minutes.

As ECAL data participate in the LHCb L0 trigger decision [1], any instability of PMT gains affects the L0 rate. The gain of PMTs in the ECAL centre, in particular, degrades because of dynode system ageing at high particle rate [3]. The LED data could be used for monitoring the PMT gains and their adjustment when necessary. In order to keep the gain stable within 1-2%, the gain adjustments should be performed every several days.

However it was found that the precision of PMT gain monitoring with LED system is affected by radiation damage of the long light guides transporting LED light to calorimeter cells. It is illustrated in Figure 1. The particle rate in the four symmetrically placed areas, and therefore the average PMT gain degradation rate, is expected to be the same; however we see a 14% difference between upper and lower areas. It can be attributed to the difference in the transparency loss of the light guides, which is determined by their layout. For all these areas the LEDs are installed under ECAL. Therefore for the two upper areas the light guides are significantly longer than for the lower ones (6–8 m *vs* 2–4 m), and pass through the $y = 0$ plane, where the particle rate (dose rate) is significantly higher than in the selected areas themselves. According to the simulation studies [4] and recent measurements of the radiation doses in LHCb, the 14% difference shown in Figure 1 corresponds to a dose of ~10 krad accumulated mainly by ~1 m long parts of light guides at $-0.5 < y < 0.5$ m.

As the required ECAL calibration precision is ~1%, the LED monitoring data were not used for PMT gain following in LHCb Run I.

In order to enable the precise PMT gain following in ECAL, it was decided to replace in 2014 the present light guides to new ones, made of radiation tolerant quartz fibers.

We are planning to use the following ECAL calibration procedure. In order to stabilize the PMT gains, LED-based HV corrections will be performed every few days[1]. However the LED

[1] The method of stabilizing raw occupancies in the ECAL cells for minimum bias events, which is now under development, can also be used for the HV corrections.

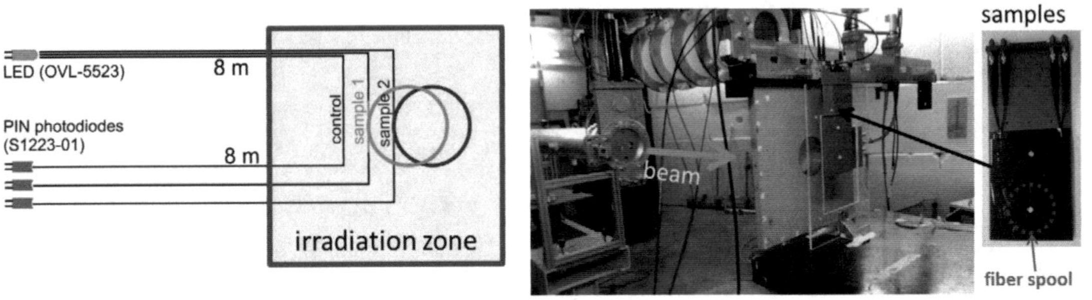

Figure 2. . The irradiation setup.

system monitors only PMT gain variations, and does not follow the degradation of light yield of ECAL cells arising from radiation damage of scintillator and WLS fibers of shashlik structure. The radiation degradation of ECAL cells proceeds slower than PMT gain variations; to take the former into account, precise offline calibration based on the π^0 peak position [5] will be performed every 2–3 month. The results of such calibration will be used to calculate the starting values of PMT HV for the next period.

The radiation degradation of the light guide transparency is determined, to a first approximation, by the dose integral over the fiber length. For LHC Run II, when LHCb will run at the luminosity of $4 \cdot 10^{32}\text{cm}^{-2}\text{s}^{-1}$, it does not exceed 100 krad·m per year (2 fb^{-1}).

After 2018, LHCb will be upgraded for work at 5 times higher luminosity, $2 \cdot 10^{33}\text{cm}^{-2}\text{s}^{-1}$ [6]. However, the dose rate at the ECAL front face, where tha light guides are placed, will increase by less than factor of 2, because the Preshower detector in front of ECAL will be removed. The precise calibrations at high luminosity running will be performed more frequently, according to the increase of the degradation rate of the ECAL light yield.

In any of these two modes of operation, the dose integral over the length of any light guide accumulated during a period between two precise calibrations will not exceed 100 krad·m. Therefore the radiation damage of light guides will not significantly affect the calibration if the degradation of transparency of the quartz fiber is less than 1% per metre for 100 krad. This can be considered as a requirement for the radiation tolerance of quartz fibers.

3. Selection of components

The radiation damage of quartz fibers for visible light was extensively studied for CMS HF [7, 8, 9]. It was found that the best radiation hardness for visible light can be obtained with quartz fibers with pure silica – high OH^- core and F-doped silica cladding. Their radiation hardness is at least two order of magnitude higher than that of Clear-PSM fibers.

The radiation loss of transparency of pure silica – high OH^- quartz fibers depends on the wavelength (see e.g. Figure 1 in [8]). Within the range of PMT photocathode sensitivity, the minimum radiation damage is observed between 400 and 500 nm. Obviously, the red LEDs of present monitoring system have to be replaced to blue ones. We have chosen Multicomp OVL-5523 with emission spectrum peaking at 460 nm. The choice criteria were good timing characteristics and highest signal amplitude obtained with existing ECAL LED drivers and PMTs, compared to other available LED types.

The LHCb ECAL HV setting scheme is such that the PMT gain is roughly proportional to the distance between the cell and the beam axis [1]. The PMTs in the very central cells work therefore at rather low gains, order of 10^3. In order to obtain LED signal in the middle of ADC range, the LED flash should produce 1–2·10^5 photoelectrons. The maximum amount of light

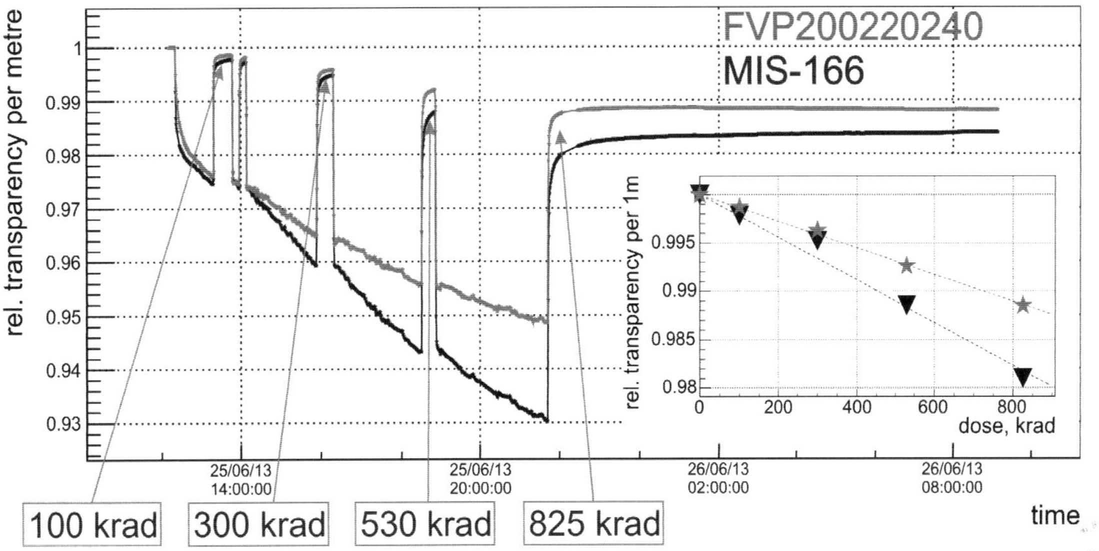

Figure 3. . Results of the irradiation test of quartz fibers: relative transparency degradation per 1 m of fiber during the test. At the insert: dose dependency of the long-term part of the transparency degradation, measured after 20 minutes annealing. Red stars: fiber FVP200220240; black triangles: MIS-166.

delivered to PMT is determined by the fiber diameter. With the chosen LED type and existing LED driver capabilities, the optimal fiber core diameter was found to be 200 μm.

The 200 μm diameter quartz fibers with pure silica / high OH^- core and F-doped silica cladding are available from several producers. All of them are expected to satisfy the requirements of the ECAL monitoring system. Our choice, made on the lowest price basis, was FVP200220240 fiber from Polymicro Technologies, Inc., USA.

4. Irradiation studies

As it was already mentioned above, a large array of data on radiation degradation of quartz fibers is available. Most of them are taken for multi-Mrad doses and sizeable transparency losses. Because of non-linear dependence of the degradation on dose, and reported dependence on dose rate, the extrapolation to relatively low doses and few percent degradation is not obvious. For this reason, it was decided to perform dedicated measurements[2].

The test was performed in June 2013 at the cyclotron of UCL (Louvain-la-Neuve, Belgium). Two types of fiber were tested: FVP200220240 (Polymicro) and MIS-166 (Fryazino, Russia), both with 200 μm diameter core. Two 5 m long samples wound on a 6 cm diameter spool were irradiated with a 62 MeV protons. The important feature of this facility is a broad beam with rather uniform profile: $\pm 10\%$ variations within 8 cm diameter core. The beam flux was set to $2{\cdot}10^8$ $p/\mathrm{cm}^2/\mathrm{s}$, which produced the dose rate in quartz of $\approx$ 28 rad/s.

The transparency of the two samples was continuously monitored during the irradiation and 12 h afterwards. The experimental setup is shown in Figure 2. The samples were connected to the light source and to the measurement equipment placed in the control room via 8m long light guides made of FVP200220240 fiber. To monitor possible degradation of these light guides, a dedicated control fibers of the same kind were used (see Figure 2). We studied only the spectrum

[2] The research leading to these results has received funding from the European Commission under the FP7 Research Infrastructures project AIDA, grant agreement no. 262025

Figure 4. . The production, installation and commissioning of the light guides.

range of interest: the blue LED Multicomp OVL-5523 selected for ECAL was used as a light source. The output light was measured by PIN photodiodes HAMAMATSU S1223-01.

The beam induced dark current in the samples was measured and found negligible, as expected: the 62 MeV protons do not produce Čerenkov light in quartz.

The result, expressed in terms of transparency of 1 m of fiber as a function of time, divided by the transparency before irradiation, is shown in Figure 3.

The samples were irradiated during 10 hours, with three 20-30 min periods without beam (and one 9 min period because of accelerator problems). These periods are clearly visible in the Figure, due to fast annealing of transparency loss. The radiation doses corresponding to the moments of beam stops were calculated on the basis of the theoretical dose rate and beam time, and have a common systematic uncertainty of ~10%. The total dose, 825 krad, roughly corresponds to the maximum dose accumulated by the clear fibers in ECAL after 5–10 years of work.

In Figure 3 one can see the fast degradation and fast annealing cycles corresponding to stopping and resuming of irradiation. The magnitude of transparency variation in these cycles depends on the previously accumulated dose; of course it should depend also on the dose rate. The annealing lasts ~10–20 minutes; there is no significant change of transparency during 12 hours after the test. The transparency value after annealing does not return to the initial value, there is a non-zero residual damage. Its dose dependence is shown at the insert in Figure 3, it is roughly linear.

For the LHCb ECAL application, the expected maximum dose rate for fibers will not exceed 0.01 rad/s, which is more than three order of magnitude less than at the irradiation test. Therefore the relevant value of fiber transparency degradation is expected to be very close to the residual one. Both types of fiber therefore satisfy our requirement of degradation not exceeding 1% per metre for 100 krad.

The direct measurement of the dose rate dependence of the fiber transparency variation is planned for July 2014 in UCL using our fiber samples which were irradiated there during the test in 2013.

5. Status of the upgrade work

The work on the ECAL monitoring system upgrade started in October 2013 with production of the new bundles of light guides made of quartz fiber. This work was performed at CERN. The main difficulty was that it was not possible to use standard fiber optic connectors, we had to develop and produce custom connectors to fit the optical inputs on existing ECAL modules. The bundle production has finished in February 2014.

The installation of bundles to ECAL took 3 months, February–April 2014. The commissioning phase for the moment (May 2014) is close to the end: all the cells are operational, the last thing left to do is to tune the amplitude and timing of LED flashes.

Few highlights from the bundle production, installation and commissioning are shown in Figure 4.

6. Conclusion

The accuracy of the present LED monitoring system of the LHCb ECAL appears to suffer from radiation damage of the light guides transporting LED light to PMTs and made of plastic clear fibers (KURARAY Clear-PSM). The upgrade consisting in replacement of the light guides to radiation tolerant ones made of quartz fiber will allow for the PMT gain following within few % accuracy. The upgrade work was finished in May 2014.

References

[1] Alves A *et al* 2008, The LHCb detector at LHC, *JINST* B **716**
[2] Machikhiliyan I and Lindahl V 2013, Calibration of photomultipliers and initial adjustment of the LHCb Electromagnetic Calorimeter for pilot LHC runs, LHCb-PUB-2013-010
[3] Guz Yu 2013, LHCb Calorimeter Upgrade, LHCb-CONF-2013-051
[4] Corti G and Shekhtman L 2003, Radiation background in the LHCb experiment, LHCb-2003-083
[5] Belyaev I M, Golubkov D Yu, Egorychev V Yu and Savrina D V 2014, *Instrum. Exp. Tech.* **57** 1 33
[6] LHCb Collaboration 2012, Framework TDR for the LHCb Upgrade, CERN/LHCC 2012-007
[7] Gurzhiev A N, Turchanovich L K and Vasilchenko V G 1996, *Instrum. Exp. Tech.* **39** 684
[8] Dumanoglu I *et al* 2002, *Nucl. Instr. Meth.* A **490** 444-455.
[9] Cankocak K *et al* 2008, *Nucl. Instrum. Meth.* A **585** 20

CMS Forward Calorimeters Phase II Upgrade

B Bilki
On behalf of the CMS Collaboration

University of Iowa, Iowa City, IA 52242, USA
Argonne National Laboratory, Argonne, IL 60439, USA

E-mail: burak-bilki@uiowa.edu

Abstract. The Phase II Upgrade of the CMS forward calorimeters (electromagnetic and hadronic) originates from the fact that these calorimeters will not be sufficiently performant with the expected HL-LHC (High Luminosity LHC) conditions. The major challenge is to preserve/improve the high performance of the current forward detectors with new devices that can withstand the unprecedented radiation levels and disentangle the very large event pileup. Here, we present an overview of the various upgrade options being considered by CMS, explaining the detector concepts and current/future beam test activities.

1. Introduction

The Compact Muon Solenoid (CMS) [1] is a general-purpose detector designed to run at the highest luminosity provided by the CERN Large Hadron Collider (LHC). The CMS detector calorimeter has been designed to detect the diverse signatures of new physics through the measurement of jets with moderate precision and by measuring missing transverse energy flow. The CMS experiment has a 4 T superconducting solenoidal magnet of length 13 m and inner diameter 5.9 m. The magnet determines many of the features of the CMS calorimeters because the barrel and end-cap calorimeters are located inside this magnet.

The CMS ECAL is a homogeneous crystal calorimeter and consists of a barrel (EB) section and two endcaps (EE) made in total of ~ 76k $PbWO_4$ crystals, with a 3 X_0 lead - silicon strip preshower (ES) detector in the endcap section. The scintillation light is read-out by a pair of avalanche photodiodes (APDs) for each EB crystal and a vacuum phototriode for each EE crystal. The small Moliere radius (~ 2 cm) in combination with the large number of crystals results in a fine granularity for the lateral shower shape. In the forward region, the granularity is further improved by the preshower detector (ES) $1.65 < |\eta| < 2.6$ which consists of 2 orthogonal planes of silicon strips. Further information about the ECAL can be found in [2-6].

The HCAL is used to measure the timing and energy of hadronic showers and their angle and position, needed for the generation of level-1 trigger primitives, the high-level trigger, and offline reconstruction of jets and missing transverse energy. The cylindrically symmetric hadron barrel (HB) calorimeter consisting of alternating layers of brass and plastic scintillator plates surrounds the lead tungstate electromagnetic calorimeter (EB). The HB design maximizes the number of interaction lengths inside the coil, which requires minimizing the amount of space devoted to the active medium. The scintillator tiles are read out with embedded wavelength shifting (WLS) fibers. Brass was chosen as the absorber material because it is non-magnetic. This design made construction relatively simple, lending itself to projective tower geometry, and eliminated uninstrumented gaps. The CMS HCAL contains 9072 readout channels organized into four subsystems: barrel (HB, 2592 channels), endcap (HE, 2592 channels), outer (HO, 2160 channels) and forward (HF, 1728 channels). The performance of the HB, HE, HO, and HF were also extensively investigated and are reported in [2, 6-9].

Detailed discussions of the status and future plans about the CMS calorimeters can be found in refs. 10-24.

Between 2010 and 2013, the LHC was operated with protons at beam energies of 3.5 and 4 TeV. The proton beams consisted of single bunches and trains of 150 ns (2010), 75 ns (2011) and 50 ns (2011 and 2012) bunch spacing. Performances well beyond what had been expected initially have been achieved with 50 ns beams, culminating in the discovery of a 125 GeV/c^2 Higgs boson by the ATLAS and CMS experiments [25-26]. The original design bunch spacing of 25 ns was only used for electron-cloud scrubbing runs at injection and for collision tests in view of future operation. The cycle structure evolved over the years, and the operational β^* for ATLAS and CMS was lowered in steps from 3.5 m (2010) to 0.6 m (2012). An overview of LHC operation between 2010 and 2013, so called Run 1, is provided in ref. 27.

The projected timeline of the LHC operations is given in Figure 1. Currently, the LHC is in the Long Shutdown 1 (LS1) period and the CMS along with the LHC is going through various upgrade processes (see e.g. [17]). The start-up in 2015, Run 2, is aimed at collision energy of 13 TeV and an instantaneous luminosity at the nominal value i.e. $1\text{x}10^{34}$ $cm^{-2}s^{-1}$. The performance reach is specified in terms of levelled luminosity. This implies that the peak luminosity that is virtually required to reach such a leveled luminosity is much higher. More information about the levelling approaches can be found in [28].

At the end of Run 2, LHC is planned to reach the collision energy of 14 TeV and instantaneous luminosity of $1.6\text{x}10^{34}$ $cm^{-2}s^{-1}$. Nominal bunch spacing of 25 ns will be used for this run period on. The expected average pileup will be around 40. The target integrated luminosity before Long Shutdown 2 (LS2) is ~ 200 fb^{-1}.

During Run 3, which will start after LS2 in 2020, LHC will reach 2x nominal luminosity, again in a levelled approach. The collision energy and the bunch spacing parameters will be at their nominal values, 14 TeV and 25 ns respectively. The projected average pileup will be ~ 50. Run 3 is aimed at an integrated luminosity of 500 fb^{-1}.

Long Shutdown 3 (LS3) between 2023-2025 is the planned time interval for the significant detector upgrades necessary for the operations beyond 2025. Following LS3, in Run 4, the LHC will operate in the High-Luminosity LHC (HL-LHC) regime. This will be realized by instantaneous luminosities between 5 to 10 times the nominal luminosity at the nominal collision energy and bunch spacing. The target integrated luminosity in the HL-LHC era is ~ 3000 fb^{-1}.

This article reports on the necessity of CMS forward detector upgrades for the HL-LHC era and describes the possible upgrade options considered by CMS. Here, the forward region is considered to be $1.4 < |\eta| < 5$, where η is the pseudorapidity. The upgrade detectors described are planned to be installed during LS3.

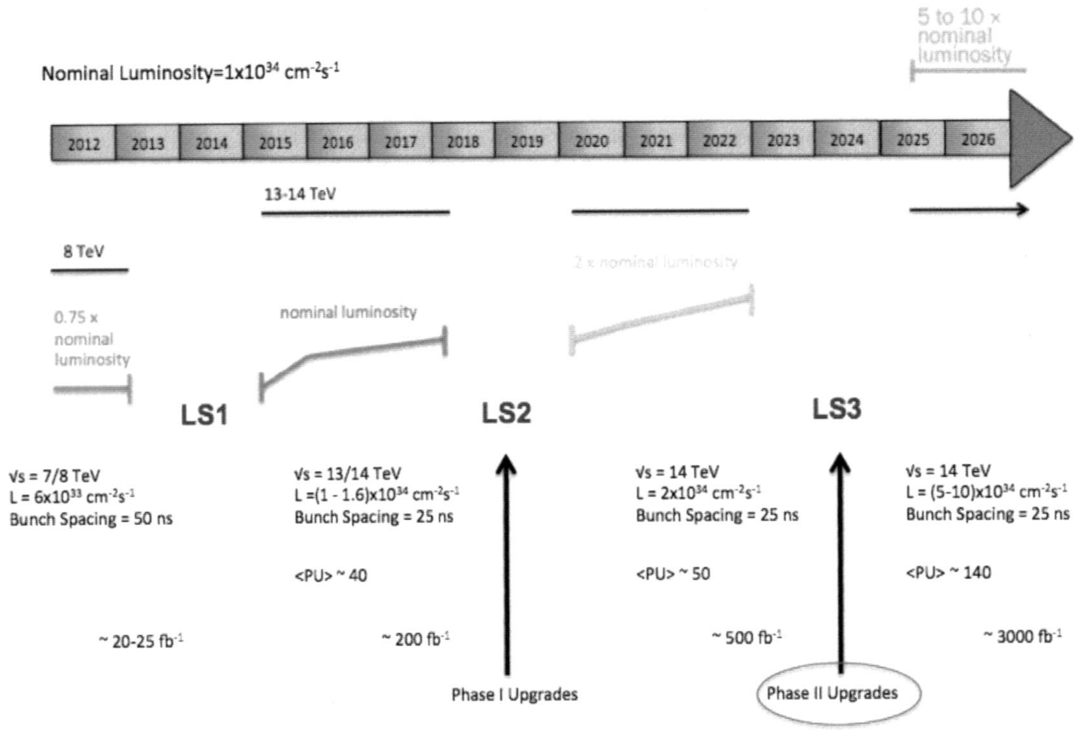

Figure 1. The timeline of LHC operations towards HL-LHC.

2. Performance projections of the CMS forward calorimeters

The ECAL Endcap detectors, covering $1.4 < |\eta| < 3$, will experience progressive deterioration of energy resolution and trigger efficiency. Figure 2 shows the fraction of the ECAL response at various operating conditions to the nominal response of the ECAL as a function of pseudorapidity. The performance degradations show strong dependence on the pseudorapidity. The EE detectors should be replaced for optimal operation in the HL-LHC era.

Figure 3 shows the response degradation of the Hadron Endcap calorimeters for different conditions and as a function of the pseudorapidity. HE covers $1.3 < |\eta| < 3$ and L1 and L7 represent the longitudinal samplings at 1 and 7 layer depths. The green solid and dashed lines are based on the data acquired between 2010 and 2013, 500 fb^{-1} and 3000 fb^{-1} lines (blue and red, solid and dashed) are predictions based on the measurements. The performance degradation dictates that the HE detectors will need to be upgraded/replaced for the HL-LHC running conditions.

The coverage between $3 < |\eta| < 5$ is provided by the Hadron Forward (HF) calorimeters. Figure 4 shows the fractional performance for three pseudorapidity sections as a function of integrated luminosity. The points are based on the laser calibration data acquired in 2010-2011 and the solid lines are the predictions. The predictions are not corrected for the expected recovery of the HF fibers during non-exposure periods. The HF calorimeters will be sufficiently performant through HL-LHC and there is no upgrade planned by CMS.

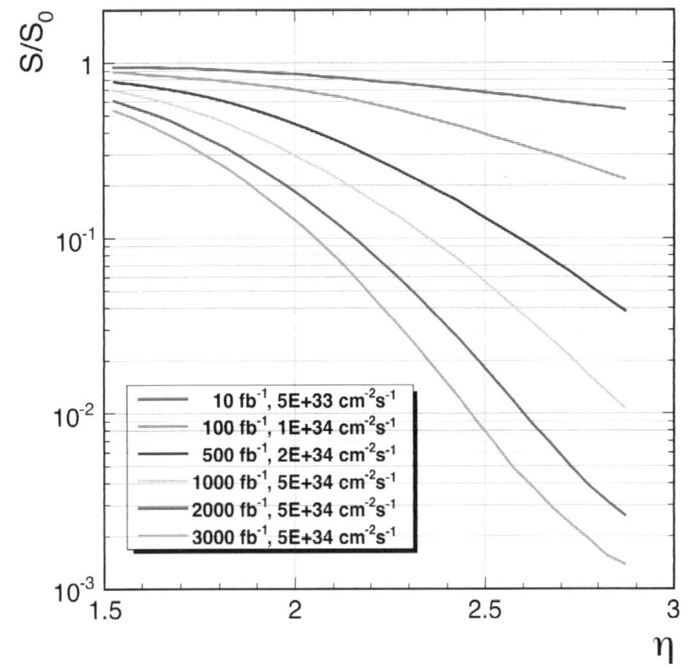

Figure 2. Simulated fraction of ECAL response to 50 GeV electrons under different operating conditions as a function of pseudorapidity.

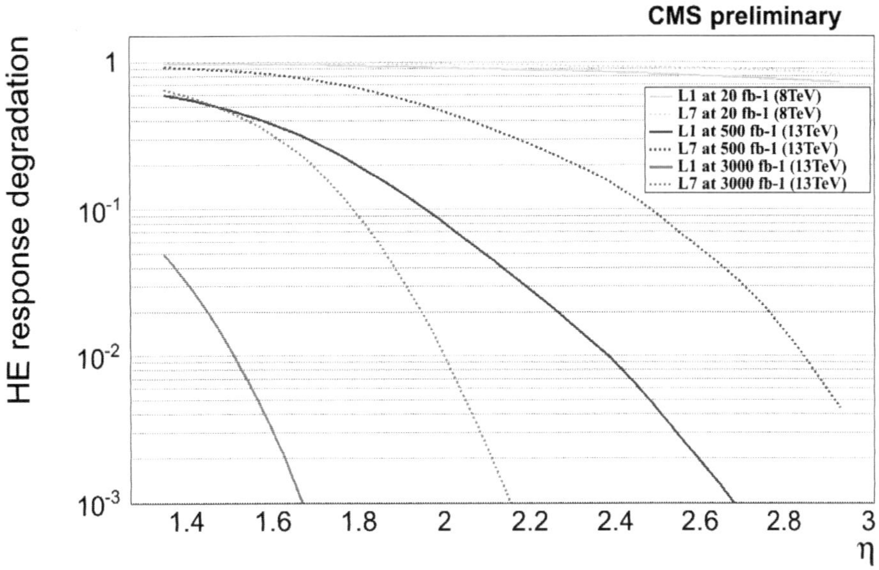

Figure 3. Response degradation of the Hadron Endcap calorimeters at different operating points for two different longitudinal segmentations in the calorimeter and as a function of pseudorapidity.

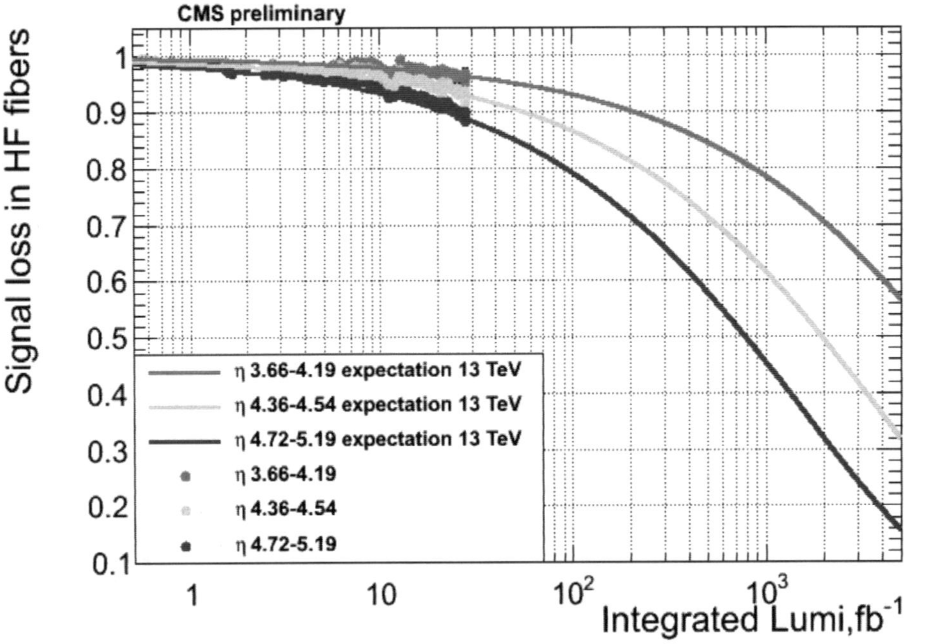

Figure 4. Response degredation of the Hadron Forward calorimeters at three pseudorapidity sections as a function of integrated luminosity.

3. Concepts considered for Phase II Upgrade

There are two options considered by CMS for the Phase II Detector Upgrade: Shashlik electromagnetic calorimeter + HE rebuild; High Granularity Calorimeter.

3.1. Shashlik EE + HE rebuild

This option is based on replacing the EE + ES with the shashlik design ECAL and replacing the active media of the HE and extending the segmentation of the calorimeter.

The expected performance of various EE replacement detectors has been studied for long and of these, an architecture based upon W absorber layers and LYSO sampling layers read out with quartz capillaries with waveshifter cores and GaInP photosensors has been found to potentially provide the best electromagnetic energy resolution ~ $10\%/\sqrt{E}$ in the absence of radiation damage and pileup.

Figure 5 shows the design of the basic shashlik module. A single module consists of 28 2.5 mm thick W plates and 29 1.5 mm thick LYSO crystals both having 14 mm lateral sizes. The length of a single module is 114 mm. The readout is provided by 4 wavelength shifting quartz capillaries, which are coupled to 1 or 2 GaInP photodetectors. The module also consists of a calibration fiber running through the center.

Table 1 shows the comparison of the parameters of shashlik EE and $PbWO_4$ EE. The shashlik modules have shorter length and smaller transverse size resulting in a larger number of modules. The modules also have smaller Moliere radius and radiation length. There are also substantial differences in terms of the light yield and temperature dependence.

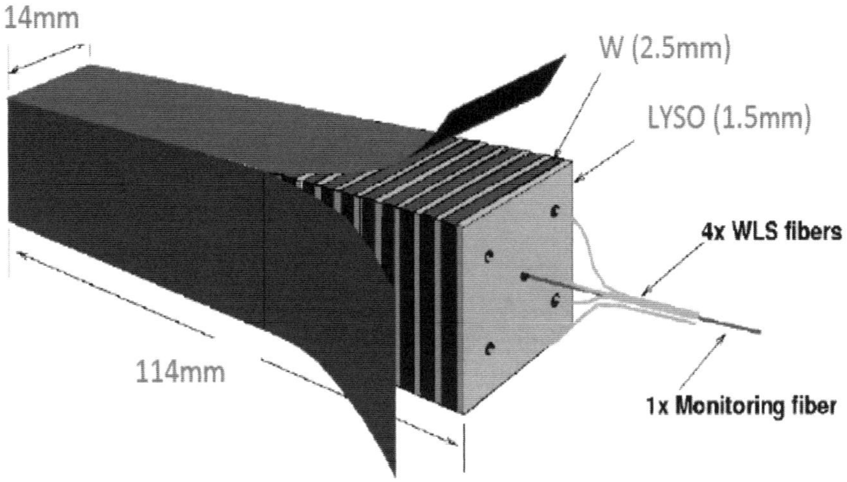

Figure 5. Design of the basic shashlik module.

Table 1. Parametric comparison of the shashlik EE and $PbWO_4$ EE.

	W/LYSO(Ce)	$PbWO_4$
Length (mm)	114	220
Transverse size (mm)	14	28.6
# modules for 2 endcaps	60,800	14,648
Average Moliere Radius (mm)	13.7	21
Average Radiation Length Xo (mm)	5.1	8.9
Light Yield (relative to NaI)	85	0.3
Emission Wavelength	420	420
Decay time (ns)	40	25
Light Output (p.e./MeV)	6	4.5
Temp Dependence (%/C)	-0.2	-2.2

The major advantages of LYSO are brightness and density. Stability of the scintillation mechanism and material costs are the outstanding R&D issues. Other crystals, in particular CeF_3, are being investigated. The merits of CeF_3 include its lighter components that lead to less hadronic radiation damage at a given dose and recovery from hadronic radiation damage (unlike LYSO in which hadronic radiation damage is cumulative). In addition, the scintillation time constant is shorter (below 30 ns versus 40 ns for LYSO); the peak emission is in the ultraviolet (310-340 nm versus 430 nm for LYSO) and so will require a different wavelength shifter that emits in the 400 to 500 nm range. Finally, CeF_3 is relatively more insensitive to temperature compared to LYSO.

Figure 6 shows the thick quartz wall (left) and quartz rod (right) options as the wavelength shifting capillaries. The thick quartz wall capillary has an outer diameter of 1.2 mm (D) and an inner diameter of 0.4 mm (d). The core is a liquid wavelength shifter with the ends plugged. Photodetector is coupled to the quartz annulus and the far end is mirrored. The quartz rod has 1 mm diameter with powdered wavelength shifter on the outer surface as a thin layer. The wavelength shifter for both cases is envisaged as J2 (Y11) or DSB1 with other alternatives possible.

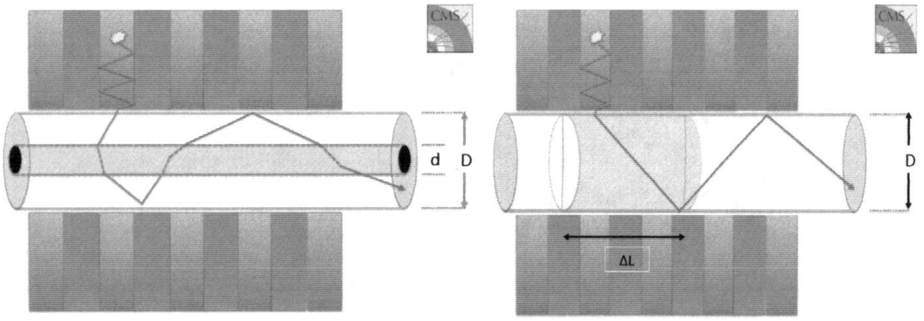

Figure 6. Possible designs of wavelength shifting capillaries: Thick quartz wall (left) and quartz rod (right).

The major advantage of the quartz capillaries is radiation hardness of both quartz and the wavelength shifter. The R&D requirements are the irradiation tests and stability studies of various options and test beam measurements. Cerium-doped quartz fibers are also considered among alternatives.

It has been shown that photodetectors with similar performance to SiPMs with Ga compounds, specifically GaAs, can be fabricated. The development of photosensors from GaInP that have the potential of being 200,000 times as radiation hard as Si based photosensors is underway. The fabrication of SPADs (single photon avalanche devices) made from GaInP is completed and the SPADs achieved the expected low dark currents of 1 pA. Figure 7 shows the single photon response for a single 50 μm SPAD using a 50Ω probe and no amplifier. It is worth pointing out that GaInP is not considered to be an "exotic" material in the semiconductor industry and indeed, is used along with GaAs in many applications. It now appears that this will be the leading contender for the shashlik EE.

Figure 8 shows the proposed readout scheme for the shashlik EE. The major point in the design is the capability to build upon existing strategies i.e. this can be made compatible with the readout scheme foreseen for the ECAL barrel at Phase II or with the QIE10 electronics aimed for HE at Phase I.

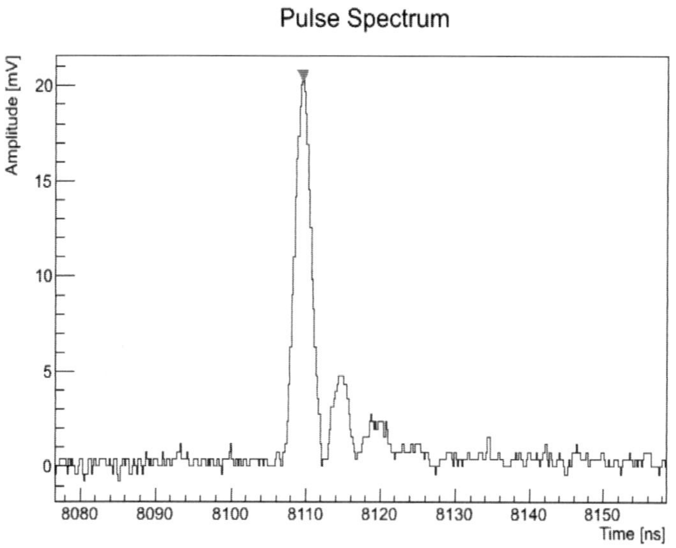

Figure 7. Single photon response for a single 50 μm SPAD using a 50Ω probe and no amplifier

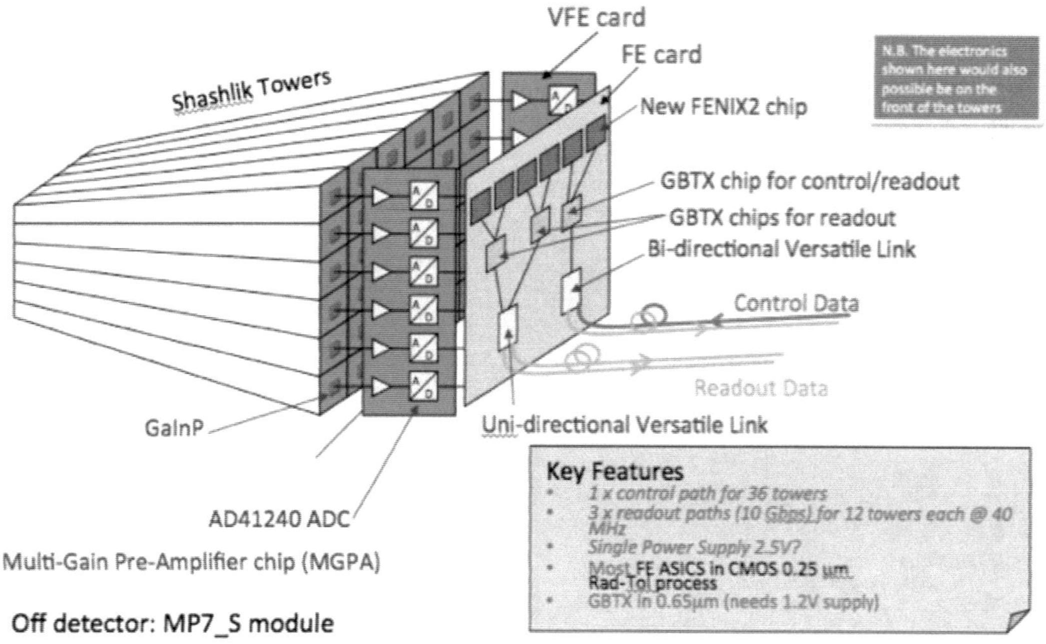

Figure 8. Proposed readout scheme for the shashlik EE.

The two potential roles of HE rebuild in the Phase II upgrade are the fully functioning hadron calorimeter behind shashlik EE and the backing hadron calorimeter of the HGC (high granularity calorimeter). In this context and in the view of the HL-LHC conditions, the proposed R&D paths for the HE rebuild can be summarized as follows:

- Study of radiation damage to the CMS HE and of possible alternative solutions
  - Radiation study of present readout
  - Development and radiation damage study of finger tiles, liquid scintillator tiles and green emitting scintillators
- Finger scintillator option for HE upgrade beyond Phase I
  - Understanding of radiation damage of present tiles at 30 fb^{-1}
  - Radiation damage of present tiles up to 7 Mrad (700 fb^{-1})
  - Radiation damage of finger tiles up to 25 Mrad (3000 fb^{-1})
- Crystal fiber R&D
  - Layer of LuAG crystal fibers coupled to quartz capillaries.
- Development of radiation-hard WLS Films, Tiles, and Fibers
  - Quartz plate R&D
  - Development of UV-absorbing WLS fibers

While following these R&D paths, the mechanical design will also be studied to enable larger longitudinal and lateral coverage.

3.2. High Granularity Calorimeter (HGC)

A high granularity calorimeter with a detailed sampling in both the hadronic and the electromagnetic sections with excellent pointing capability could have significant performance benefits. The approach that has been developed is a sampling calorimeter with layers of silicon detectors that feature very high longitudinal and lateral granularities in the electromagnetic and the front hadronic calorimeter sections, and a coarser segmentation backing hadronic calorimeter section which is under the development framework of the HE rebuild option.

The major components of the HGC are summarized below:

- *Electromagnetic Calorimeter:*
  - 30 samplings of lead/copper total of 25 X_0
    - 10 layers of 0.5 X_0 / 10 layers of 0.8 X_0 / 10 planes of 1.2 X_0.
    - Pad size 0.9 cm^2 for first 20 layers, 1.8 cm^2 for the last 10 layers.
  - 420 m^2 of silicon pad detectors.
  - 3.7M channels.
- *Front Hadronic Calorimeter*
  - 4 interaction lengths.
  - 12 layers of brass/silicon each 0.33 interaction lengths.
  - Pad size is 1.8 cm^2
  - 1.4M channels.
- *Backing calorimeter*
  - Five interaction lengths (e.g. sampling of 0.5 interaction lengths).
  - Radiation levels are lower, so plastic scintillators can be used.

Table 2 shows the parameters of the electromagnetic and front hadron calorimeter sections of the HGC. The detector is envisaged to have an unprecedented granularity resulting in 5.1M readout channels utilizing 670 m^2 of Silicon as the active medium.

Table 2. Parameters of the electromagnetic (EE) and front hadron (FH) calorimeter sections of HGC.

	EE	FH	Total
Area of silicon (m^2)	420	250	670
Channels	3.7M	1.4M	5.1M
Detector Modules	19,000	11,000	30,000
Weight One Endcap (tonnes)	16	63	79
Number of plates	30	12	42
Front end power (kW)	70 – 80	20 – 30	90 - 110

All Silicon sensors will be made with the standard 320 µm thick wafers. For regions of low radiation, 300 µm depletion depth is foreseen. The medium and high radiation regions will be equipped with 200 µm and 100 µm depletion length sensors respectively.

The front-end electronics design for HGC is shown in Figure 9. It utilizes an input capacitor of 50 - 100 pF, a time constant of the order of 15-20 ns and a single MIP response larger than 15k e^- before damage. On-chip digitization for each channel will be at 40 MHz. The sum of 4 (2) adjacent pads will serve as the Level 1 Trigger primitives with 8-10 bit resolution. The target power consumption per channel is 15 mW.

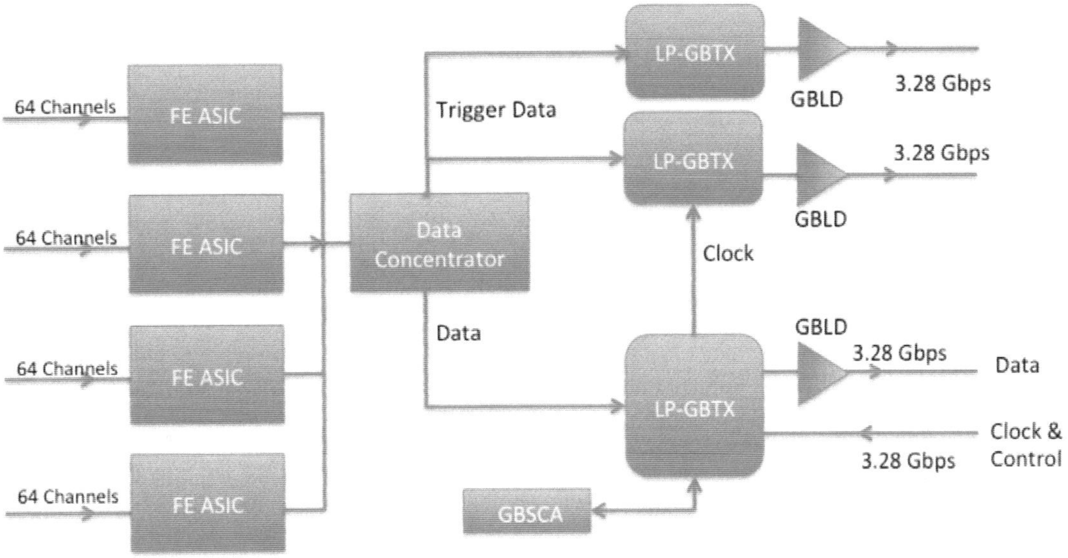

Figure 9. Front-end electronics design of HGC.

4. Beam tests

The CMS Forward Calorimetry Upgrade groups have already started the test beam effort in December 2013 at Fermilab Test Beam Facility (FTBF) [29]. The groups perform their tests under the collaboration of Fermilab T-1041 "CMS Forward Calorimetry R&D" Experiment [30]. All groups are planning on future beam tests at Fermilab and CERN as well as irradiation tests at CERN and other non-HEP facilities.

As an example, Figure 10 shows the picture of the shashlik calorimeter prototype test setup.

Figure 10. Shashlik calorimeter prototype at FTBF test setup.

5. Conclusions

Due to high radiation environment exposed by the High Luminosity LHC conditions, CMS endcap electromagnetic and hadron calorimeters will experience significant performance degradation and will need to be replaced/upgraded. In this context, CMS considers two upgrade scenarios:

- Shashlik EE + HE Rebuild: The endcap electromagnetic calorimeter as well as the endcap preshower detector will be replaced by the shashlik type electromagnetic calorimeter, and the hadron endcap calorimeter will be improved in terms of active media and lateral and longitudinal coverage.
- High Granularity Calorimeter: A high granularity calorimeter based on Silicon as the active medium will be constructed in two sections with varying segmentation - electromagnetic and front hadronic. The backing hadron calorimeter with a coarser segmentation will be built with the HE rebuild options.

Both scenarios have rigorous R&D plans on active media (e.g. scintillators, silicon sensors, crystals), readout components (e.g. fibers and photodetectors) and readout electronics (e.g. on-detector electronics).

6. References

[1] Chatrchyan S et al. (CMS Collaboration) 2008 JINST 3 S08004
[2] Bayatian G L et al. (CMS Collaboration) 2006 CERN/LHCC 2006-001
[3] Adzic P et al. (CMS ECAL Collaboration) 2006 Eur. Phys. J. C 44 1–10
[4] Adzic P et al. (CMS ECAL Collaboration) 2007 JINST 2 P04004
[5] Chatrchyan S et al. (CMS Collaboration) 2013 JINST 8 S09009
[6] Abdullin S et al. (CMS ECAL and HCAL Collaborations) 2009 Eur. Phys. J. C 60 359–373
[7] Abdullin S et al. (CMS HCAL Collaboration) 2008 Eur. Phys. J. C 55 159–171
[8] Abdullin S et al. (CMS HCAL Collaboration) 2008 Eur. Phys. J. C 53 139–166
[9] Bayatian G L et al. (CMS Collaboration) 2008 CMS-NOTE-2008/010
[10] Biino C 2014 (CMS Collaboration) *The CMS Electromagnetic Calorimeter: lessons learned during run 1, overview and future projections* talk given at this conference
[11] Ghezzi A 2014 (CMS Collaboration) *The CMS electromagnetic calorimeter calibration during Run I: progress achieved and expectations for Run II* talk given at this conference
[12] Del Re D 2014 (CMS Collaboration) *Timing performance of the CMS electromagnetic calorimeter and prospects for the future* talk given at this conference
[13] Ghezzi A 2014 (CMS Collaboration) *The CMS electromagnetic calorimeter calibration during Run I: progress achieved and expectations for Run II* talk given at this conference
[14] Schroder M 2014 (CMS Collaboration) *Performance of jets in CMS* talk given at this conference
[15] Lobanov A 2014 (CMS Collaboration) *The CMS HO SIPM Upgrade* talk given at this conference
[16] Apresyan A 2014 (CMS Collaboration) *Performance of MET in CMS* talk given at this conference
[17] Onel Y 2014 (CMS Collaboration) *CMS Hadron Forward Calorimeter Phase I Upgrade Status* talk given at this conference
[18] Akchurin N 2014 (CMS Collaboration) *Combined Forward Calorimetry (CFC) for Phase II CMS Upgrade* talk given at this conference
[19] Gras P 2014 (CMS Collaboration) *The CMS electromagnetic calorimeter barrel upgrade for High- Luminosity LHC* talk given at this conference
[20] Drozdetskiy A 2014 (CMS Collaboration) *Evolution studies of the CMS ECAL endcap response and upgrade design options for High-Luminosity LHC* talk given at this conference
[21] Laird E 2014 (CMS Collaboration) *The CMS central hadron calorimeter upgrade* talk given at this conference
[22] Cowden C S 2014 (CMS Collaboration) *Object Reconstruction with Non-pointing Combined Forward Calorimeter (CFC)* talk given at this conference
[23] Sauvan J B 2014 (CMS Collaboration) *Performance and upgrade of the CMS electromagnetic calorimeter trigger for Run II* talk given at this conference
[24] Kunsken A 2014 (CMS Collaboration) *MPPC Photon Sensor Operational Experience in CMS* talk given at this conference
[25] Aad G et al. (ATLAS Collaboration) 2012 Phys. Lett. B 716 1–29
[26] Chatrchyan S et al. (CMS Collaboration) 2012 Phys. Lett. B 716 30–61
[27] Alemany-Fernandez R et al. 2013 CERN-ACC-NOTE-2013-0041
[28] Giovannozzi M 2013 AIP Conference Proceedings 1560 686
[29] http://www-ppd.fnal.gov/FTBF/
[30] http://www-ppd.fnal.gov/FTBF/TSW/PDF/T1041_tsw_signed.pdf

16th International Conference on Calorimetry in High Energy Physics (CALOR 2014) IOP Publishing
Journal of Physics: Conference Series **587** (2015) 012015 doi:10.1088/1742-6596/587/1/012015

Combined Forward Calorimetry Option for Phase II CMS Endcap Upgrade

Nural Akchurin [1]

Texas Tech University, Department of Physics, Lubbock, Texas, USA

E-mail: `Nural.Akchurin@ttu.edu`

Abstract. Traditionally, EM and HAD compartments are thought to be separate and are often optimized individually. However, it is possible to optimize a robust and economical combined calorimeter system for myriad physics objectives. By employing event-by-event compensation afforded by the dual-readout technique, we have shown that excellent jet performance can be attained with a longitudinally un-segmented calorimeter that is calibrated only with electrons. In addition, the linear hadronic energy scale renders complex off-line correction schemes unnecessary. The proposed replacement of the CMS EE and HE calorimeters with a single Combined Forward Calorimeter (CFC) shows excellent jet performance complemented by good EM object detection. In this paper, we give brief snapshots on basic design criteria, timing characteristics of Cherenkov and scintillation pulses, trigger generation criteria and performance under high radiation fields. Although CMS has recently chosen different technologies for its endcap calorimetry in Phase II, the concepts developed here are likely to remain valuable for some time to come.

1. Introduction

The Combined Forward Calorimeter (CFC) can be thought of as two independent calorimeters occupying the same volume (the clear and scintillator fibers as two active media in a single absorber) that provide complementary information. The Cherenkov (Q) part originating in the clear fibers provides information on the EM content as well as the fast timing signal, whereas the scintillation (S) part predominantly gives information on ionization (dE/dx) losses. The ratio of the two signals enables us to measure the EM fraction (f_{em}) of a shower event-by-event and to remove the source of one of the larger fluctuations in energy measurement. The proposed calorimeter gives excellent energy measurement and it is likely to help mitigate pile-up effects because of its fast (Cherenkov) response and fine transverse granularity.

It is possible to optimize a robust and economical combined calorimeter (ECAL+HCAL) system for myriad physics objectives. We have shown that excellent performance can be attained with a longitudinally un-segmented calorimeter that is calibrated only with electrons in beam tests [1, 2, 3, 4] and in standalone simulations [5]. In addition, the linear hadronic energy scale renders complex off-line correction schemes unnecessary. Our studies targeted optimization of crucial design aspects: the absorber (tungsten *vs* brass), tower orientation

[1] The CFC Collaboration: Baylor University, CERN, Fermilab, Kansas State University, University of Notre Dame, IHEP-Portvino, Rutgers University, Saha Institute of Nuclear Physics - Kolkata, Texas A&M University, Texas Tech University, University of Virginia.

(quasi-pointing *vs* non-pointing), fiber spacing (packing fraction), cerium-doped fused-silica (SiO2:Ce3+) fiber structures (LY, decay time, core/clad/buffer sizes) for rad-hard scintillation light, detailed photon propagation in fibers, pulse formation using variety of photodetectors, effective longitudinal segmentation using timing information and pulse shape.

Fine transverse segmentation is one of the key features that contributes strongly to efficient particle ID and background suppression at the high luminosity LHC. The measurement of f_{em} on an event-by-event basis is also a powerful discriminant: $f_{em} \approx 1$ for electromagnetically interacting particles, and < 1 for hadronically interacting ones. Precise pulse shape information on the other hand, reveals space-time structure of the shower profile inside the calorimeter. Finely sampled pulse shape opens up new possibilities in object reconstruction because appropriate treatment of the pulse shape effectively segments the calorimeter into many sections in depth [5].

2. Physical Design

The baseline absorber material is cartridge brass (70% Cu and 30% Zn). The angular coverage extends from $\sim 25^o$ to $\sim 2^o$ corresponding to $1.5 \leq |\eta| \leq 4.0$. The span of 23° can be either covered by a non-pointing or by a quasi-pointing geometry (Figure 1).

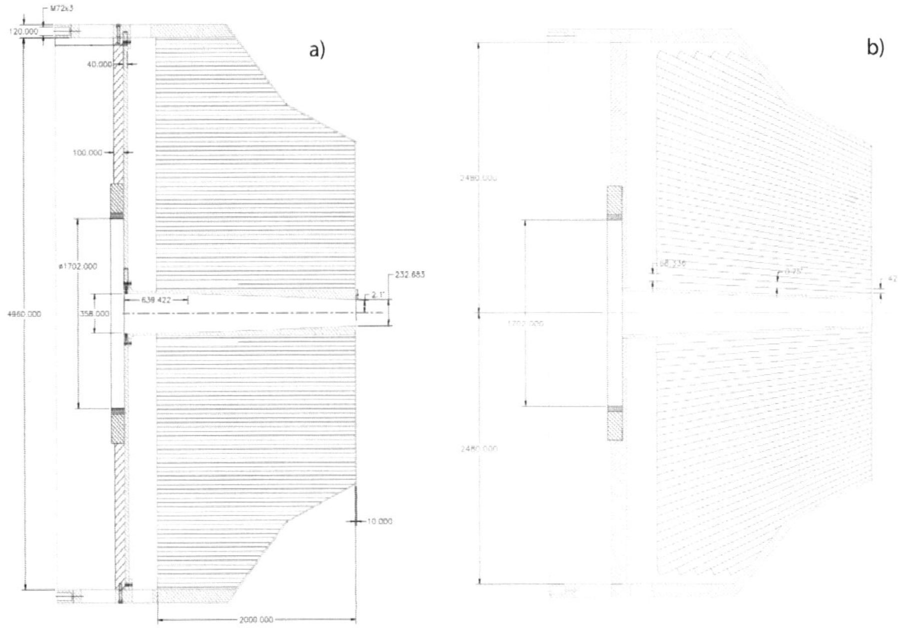

Figure 1. Two options are considered for all brass absorber: a) non-pointing tower geometry where the fibers are positioned parallel to the beam axis (Option I), and b) quasi-pointing geometry where the geometric axis of the towers nearly points toward the interaction point (Option II).

Variations on the two options are possible and depend on the overall goals of the experiment, such as the upgrades of muon systems, forward calorimeters, forward pixel detectors, and the beam pipe. For example, the inner annulus ($\sim 2.4 \leq |\eta| \leq 4$) can be constructed out of (replaceable) tungsten in order to accommodate an additional muon chamber in the back (Option III). This approach will potentially provide better performance in resolving overlapping jets (tungsten) and flexibility (replaceable). As this replaceable insert is compact, the impact on cost is relatively small. The fourth possibility is to entirely replace brass with a tungsten alloy (Option IV). The clear advantage in this case is the shorter overall length in z-direction ($\sim$120 cm in tungsten *vs.* $\sim$200 cm in brass). Unless explicitly noted, all results given here are based on Option I.

Two different kinds of fibers scintillating (S) and clear (Q) are embedded in the absorber and they are nominally spaced 2.0 mm apart, alternating in a square-grid pattern. The fiber dimensions are assumed to be identical for both types: 600 μm diameter core, 50 μm thick cladding, and 50 μm thick buffer. In the case of clear (Cherenkov) fiber, the core is high-OH$^-$ fused-silica, the clad is hard-polymer and the buffer is made out of acrylate. The numerical aperture is 0.33. The scintillating fibers are made of rad-hard SiO2:Ce3+ core and are surrounded by similar cladding and buffer materials. The packing fraction of active materials by volume is 3.5% of each type. The clad plus buffer materials and air take up 5.4% and 3.3% of the calorimeter volume, respectively. The effective Molière radius and radiation length are 2.0 and 1.82 cm, respectively.

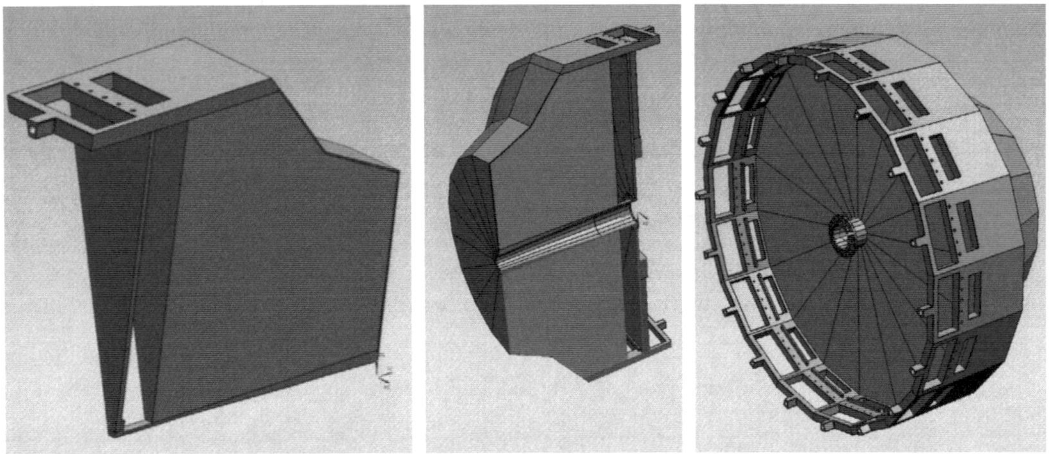

Figure 2. The CFC extends the coverage to $|\eta| \approx 4$. Behind the 10.2 λ_I^{eff} absorber, space for photodetectors, electronics, and services is accessed through the side ports. The entire calorimeter consists of 36 individual 20-degree wedges.

The brass absorber is 200 cm long (10.2 λ_I^{eff}) in the beam direction. In the back of the absorber, 23 cm is reserved for the readout and services. Steel plates provide mechanical strength and match the existing support brackets in CMS, thereby minimizing integration issues (Figure 2). The transverse tower size is 2×2 cm^2 and 4×4 cm^2 at larger radius ($r \geq 180$ cm). The transverse segmentation is the same throughout the calorimeter, *i.e.* HAD tower size is the same as the EM tower size. The smaller towers contain 50 fibers per channel (either S or Q) and read out separately. The total number of channels is 65,196 for S and 65,196 for Q for the entire calorimeter (see Table 1).

The absorber fine structure (grooves or holes) can be made in several different ways:

- As in CMS forward calorimeters multiple grooved (cold rolled) plates are stacked and (diffusion) welded together to form large blocks of the absorber [6]. The grooved plates may also be produced by stamping, etching or machining.
- Thin precision punched (perforated) plates can be stacked together to form large absorber blocks.

Table 1. The nominal parameters of the brass absorber option.

Parameter	Value
Absorber	Brass (70% Cu + 30% Zn)
Total Weight (kg)	235,376
Nominal Fiber-to-fiber Spacing (mm)	2.0
Fiber Core dia. (mm)	0.6
Fiber Core+Clad dia. (mm)	0.7
Fiber Total dia. (mm)	0.8
Hole dia. (mm)	0.9
Nuclear Interaction Length, λ_I (cm)	19.7
Radiation Length, X_o (mm)	18.2
Moliere Radius, ρ_M, (mm)	20.0
Transverse Segmentation ($r < 180$ cm)	2×2 cm^2
Transverse Segmentation ($r \geq 180$ cm)	4×4 cm^2
Longitudinal Segmentation	by timing on Q-channel (200 ps)
Number of channels	130,392
Total Fiber Length (Q or S) (km)	9,955

- Emerging 3D metal printing technologies are now capable of "printing" stainless steel, brass and tungsten. The bond jetting technique, for example, is able to print relatively large blocks albeit with reduced density (60% of metal). The selective laser sintering technique produces higher density blocks with good precision.

3. Readout Geometry, Signal Formation, and Electronics

Figure 3 shows the transverse granularity of the CFC as implemented in simulation. The fiber bundles are formed using Optical Guiding Unit (OGU) which guides the fibers to either scintillating or clear fiber bundle. The diameter of the small bundles is ~6.5 mm and the larger ones measure ~13 mm.

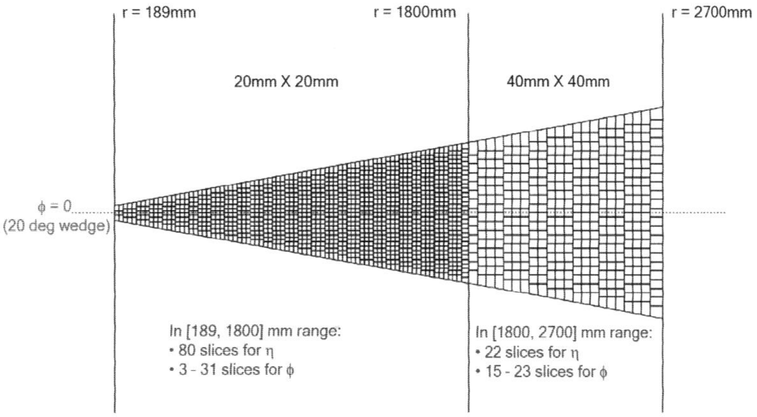

Figure 3. The transverse segmentation for a 20-degree wedge results in 1368 channels of small (2×2 cm^2) and 443 channels of larger (4×4 cm^2) towers. In a wedge, total number of towers is 1811.

A typical SiPM is taken as a benchmark photodetector, however others such as multi-channel VPT and GaInP sensors need also to be considered. Figure 4 shows the pulse shape for 1 GeV electrons impinging brass absorber with fiber-to-fiber distance set at 2.4 mm, as an example. The pulse is sampled at 5 GHz and the amplitude is shown in units of fC/200 ps. The light yield is ~30 photons/GeV resulting in 10 pe/GeV. The SiPM gain is assumed to be 250,000 (~15 fC/pe). The QIE noise is ~2 fC. The timing features of the pulse include SiPM and the "3 ns preamp/electronics" time constants. The response is 150 fC/GeV.

The Cherenkov photon light yield smoothly varies as a function of the impact point: the EM showers generate 30 photons/GeV at $\eta = 1.5$ and 25 photons at $\eta = 3.5$. The scintillating fiber model consists of fused-silica core with parametrized scintillation properties. The photon yield per unit energy is constant within 2% (~1820 photons/GeV for EM showers).

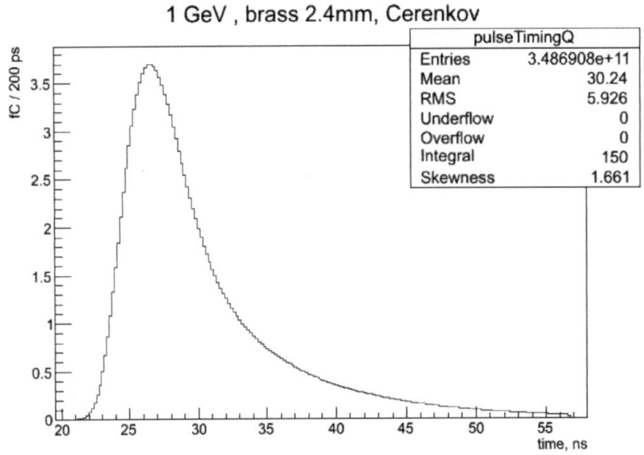

Figure 4. The simulated SiPM pulse shape for the Cherenkov pulse for 1-GeV electrons.

Figure 5 shows the conceptual CFC readout. The signal from the Cherenkov photodetector (Q-channel) is split into two paths. The first path is a conventional chain as in the CMS HCAL Phase I upgrade. The signal is digitized on detector and sent to the counting room for trigger generation. The digital pipeline holds data for the HLT up to 20 μs. The second path is for a detailed analysis of the Cherenkov signal using a multi-GHz pulse-shape sampler. The analog pipeline holds the signal up to 20 μs. When the CFC receives an L1A, the analog signal is extracted from the pipeline and digitized in a narrow time window of interest (5-10 ns). The digitized signal is then sent to the HLT.

The first path is very similar to the CMS HCAL Phase I electronics. The QIE can be upgraded from non-linear 8 bits to non-linear 10 bits for a larger dynamic range (15-16 bits equivalent) with a resolution better than one percent. The sampling frequency is also to be increased from 40 to 80 MHz to allow a crude pulse shape analysis for L1.

A 5 GHz pulse shape sampler is already commercially available: the Domino Ring Sampler (DRS4) series is one example. The DREAM/RD52 collaboration used DRS4 in beam tests and successfully measured the longitudinal shower shape utilizing timing information [7]. The next generation sampler, DRS5, is under development at PSI and will be ready for production in 2015. It provides up to 10 GHz sampling, up to 3 μs latency, and dead-time-free readout.

We considered the 40 MHz QIE10 as a benchmark system and assumed 10 fC/pe and 30% PDE for the SiPMs. QIE10 consists of 8 bits (6 mantissa and 2 exponent) and is non-linear, extending from 3 fC to 330 pC (10^5 dynamic range). The LSB is 3.2 fC, and at high energies, the QIE10 digitization error is 1%. The CFC EM energy resolution at high energies is below

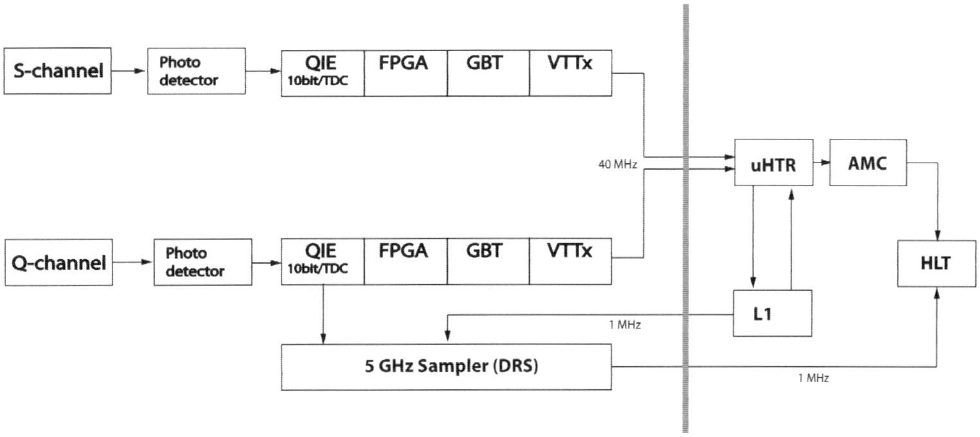

Figure 5. While a part of the CFC readout system is based on existing (CMS HCAL) readout technology, a multi-GHz sampler on the Cherenkov side brings additional capabilities such as effective longitudinal segmentation and precision timing for the front-end. The back-end is on the right.

2%. There is insufficient headroom, a situation that can be remedied with an additional bit in exponent (see Figure 6).

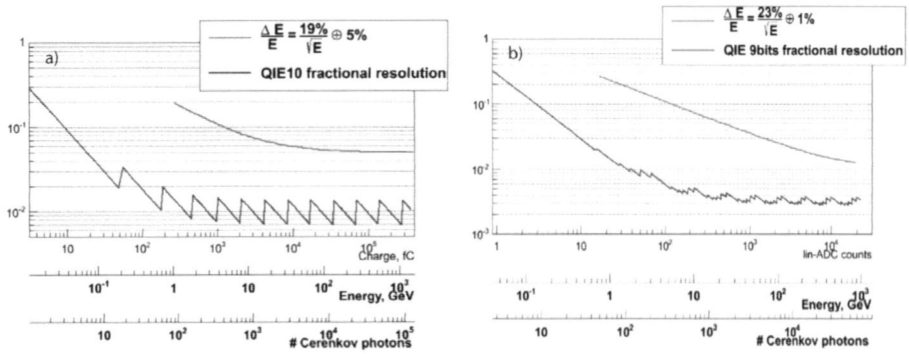

Figure 6. A case study: 2 bits exponent and 6 bits mantissa in a QIE10 system (a). When an additional bit is added to the exponent (3 bits or 8 ranges), the digitization precision improves (b). The EM energy resolution curves are for comparison purposes only.

The baseline approach for effective longitudinal segmentation is accomplished by the analyses of the signal shapes. Figure 7.(a) shows the level of timing precision that was achieved in a test beam using a DREAM module: the distribution of the time difference between the PMT signals from 80 GeV electrons measured in the calorimeter and the trigger counters upstream. This difference was obtained by measuring the start of the calorimeter signals with the DRS chip (a simple threshold), where the domino wave was started by the trigger signal. We see that a time resolution of 0.55 ns is achieved in this way. The same measurement for 180 GeV pions yielded a very different time distribution, Figure 7.(b). The slope of 0.8 ns reflects the fact that the light is produced at a depth z which varies like $\exp(-z/\lambda_{\rm int})$. Since the interaction length is ~25 cm in this detector, the time resolution of 0.55 ns means that the depth at which the light is produced can be determined with a precision better than 20 cm. The starting time of the DRS signals indeed measures the depth at which the light is produced in the fibers, by comparing the DRS information with the depth measured in two other ways, namely from the longitudinal profile measured in the leakage counters (Figure 8.(a)) and from the lateral displacement of the center-of-gravity of the calorimeter signals with respect to the coordinates measured in the upstream wire chambers (Figure 8.(b)) [8]. Experimental data clearly show that timing information provides sufficiently precise determination of "center of light" inside the calorimeter.

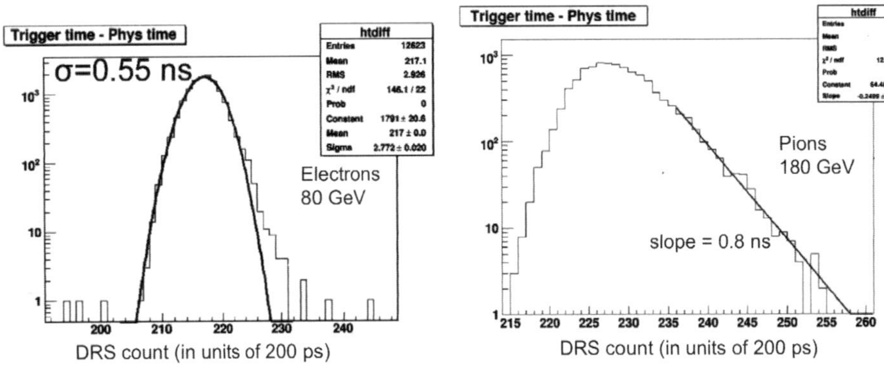

Figure 7. The starting time of signals from 80 GeV electron (left) and 180 GeV pion (right) showers in a DREAM module, measured with the DRS readout. The readout was triggered by the passage of a beam particle through the upstream trigger counters.

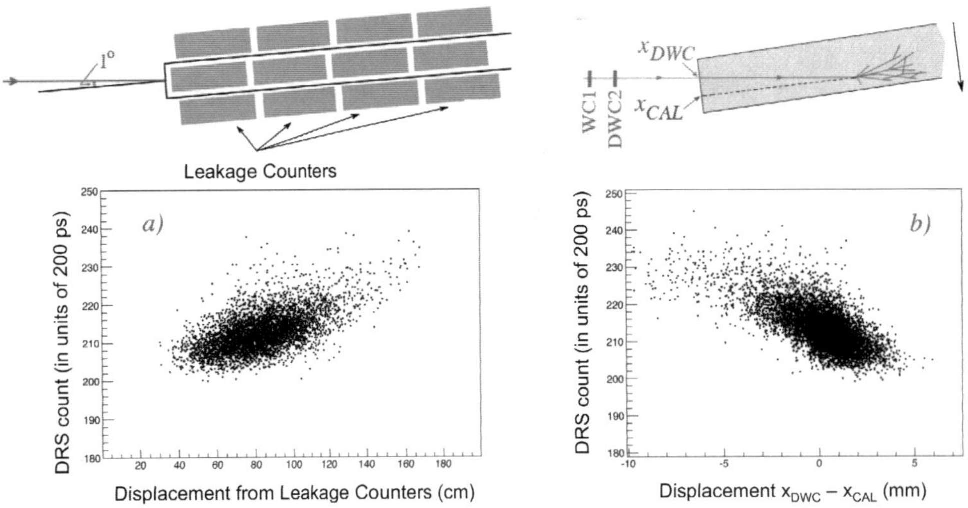

Figure 8. The depth of the light production in the unsegmented fiber module versus the depth measured from the leakage profile (a) or from the displacement of the center-of-gravity with respect to the coordinates measured in the upstream wire chambers (b).

The detailed time structure of each event makes it possible to obtain crucial information about the depth at which the light is produced. By using the fact that light travels at a speed of c/n in the fibers, while the particles producing the light travel at c, the starting time of the signals makes it possible to measure the depth at which the light is produced, as illustrated in Figure 9.

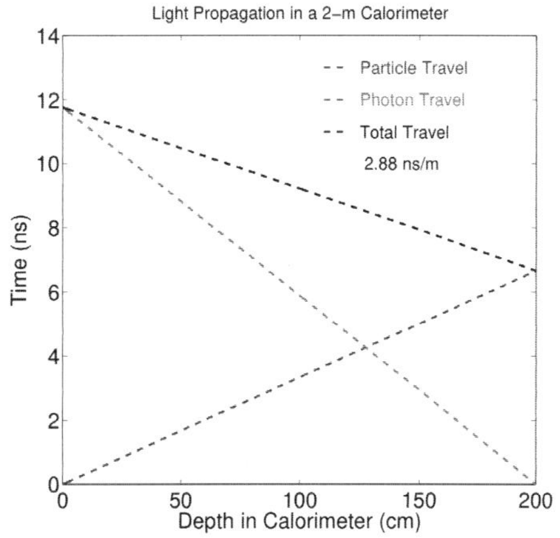

Figure 9. The dependence of the starting time of the signals on the depth (z) inside the calorimeter at which the light is produced. Also shown are the time traveled by the beam particles from the front face of a 2-m long brass calorimeter to this depth z and the dependence of the time traveled by the light through the fibers from z to the photodetector.

There is a significant (measured) difference between the distribution of the starting time of pion and electron induced showers as shown in Figure 7. Since the depth distribution pattern of the light produced by the latter is very similar for all showers, the of the starting time distribution of the electron showers is a good measure for the precision in the depth of the light production that can be obtained for individual showers. Note that the timing precision that is derived from the EM showers was 0.55 ns and was not optimized to achieve maximal longitudinal segmentation. In addition to the effective longitudinal segmentation, this timing information also makes it possible to measure, and correct for, the attenuation (intrinsic and radiation induced) characteristics of the fibers.

4. Performance

The CFC calorimeter builds on the experience gained working with several operating calorimeters and the R&D prototypes in the last decades. More specifically, the "spaghetti" calorimeters, such as SPACAL, HF, and DREAM/RD52, have given us increasingly wider and deeper perspectives in calorimetry. We studied various tungsten alloys in detail but only the results for the brass absorber are given here. Although the dual-readout technique provides hadronic response linearity with excellent energy resolution, it might be possible to construct CFC naturally compensating as some results presented here may suggest. In order to make realistic progress on this issue, appropriate prototypes need to be tested in high energy beams to verify the simulation results.

Figure 10 shows the EM energy resolutions for the scintillating part as a function of the fiber spacing. The resolution is computed from signal distributions (*rms* divided by the mean) at five different energies. If we are to take 2-mm fiber-to-fiber spacing as a reference, the stochastic term is better than 20% while the constant term is below 2%. The same set of plots are shown in Figure 11 for clear (Cherenkov) fibers. In the case of the Cherenkov part, the dominating factor is the photo-statistics whereas in the case of the scintillator part, the sampling fluctuations dominate over photo-statistics.

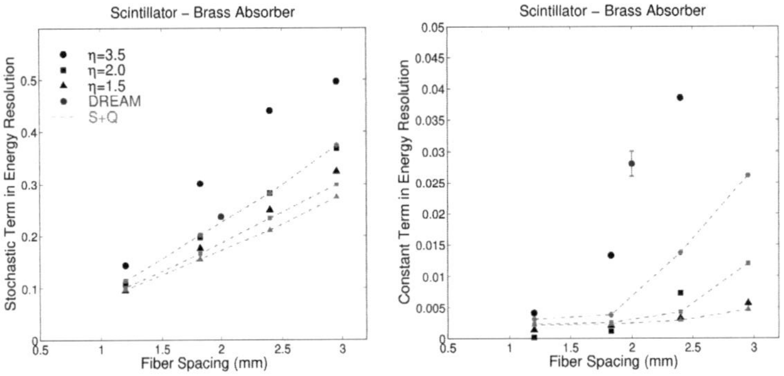

Figure 10. The EM energy resolutions, stochastic (left) and constant terms (right), for scintillating fibers are shown for brass absorber at three pseudorapitidy regions as a function of fiber-to-fiber spacing. The red markers indicate the resolutions achieved if scintillation and Cherenkov signals are summed. The DREAM data point (blue) is included as a reference point.

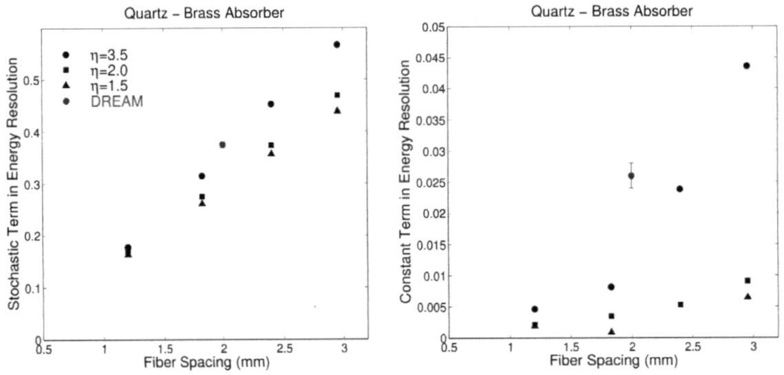

Figure 11. The EM energy resolutions, stochastic (left) and constant terms (right), for clear Cherenkov fibers are shown for brass absorber at three pseudorapitidy regions as a function of fiber-to-fiber spacing. The DREAM data point (blue) is included as a reference point.

The scintillator part gives excellent (raw) hadronic energy resolution, better than $40\%/\sqrt{E}$ for 2-mm fiber-to-fiber spacing (Figure 12). As expected, the energy resolutions for jets are not significantly different than that of the single hadrons. The response to hadrons (signal/GeV) is easily linearized and the resolution is further improved using the dual-readout technique as shown in Figure 13.

Figure 14 displays the energy-weighted Cherenkov photon arrival times for electrons and pions as well as the *rms* spread of these distributions as a function of particle energies. It is evident that signals from hadron showers arrive $\sim$0.5 ns earlier compared to that of electron initiated showers for all configurations. The *rms* distributions are also distinctly different between the two types of showers: EM showers are tightly distributed in time ($\sim$ 50 ps) whereas due to stochastic nature of neutral pion production in hadronic showers, the *rms* distributions are wider ($\sim$ 400 ps) for pion initiated showers. Figure 15 illustrates how the timing information from a non-pointing geometry contributes in sharpening the E_{T} definition. Using a simple correction scheme, we are able to improve the E_{T}/E distributions to narrow gaussian distributions. For more details, see

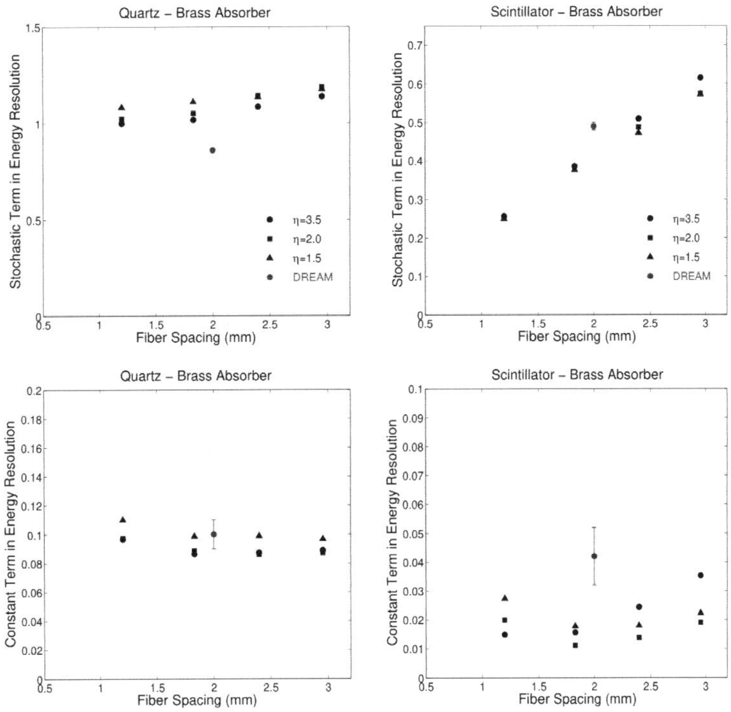

Figure 12. The HAD energy resolutions for both quartz (left) and scintillating fibers (right) are shown at different pseudorapitidy regions as a function of fiber-to-fiber spacing. The DREAM data point is shown in blue as a reference.

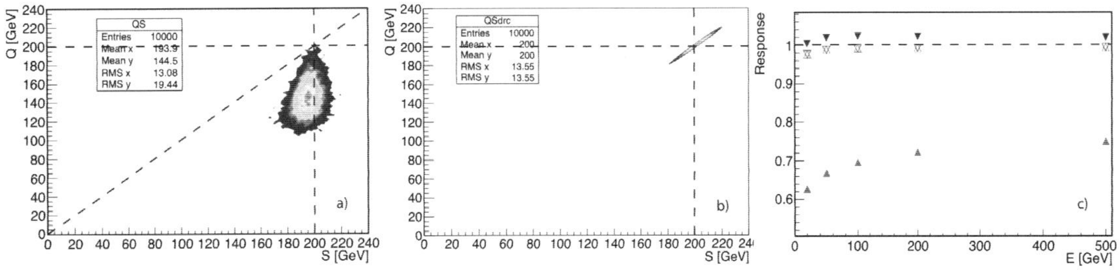

Figure 13. Before (a) and after (b) dual-readout corrections are applied to the scintillation and Cherenkov signals for the case of fiber-to-fiber 2.4 mm in brass for 200 GeV charged pions at $\eta = 2.0$.

[5] in these proceedings.

5. Trigger Design

The CMS L1 trigger is able to perform identification of muons, electrons, photons, jets and missing transverse energy and the CFC will contribute to the identification of photons/electrons and jets. At L1, we consider the transverse shower shape, $f_{\rm em}$ and potentially the timing information. An example of energy distributions in X-Y plane for electrons and pions are shown in Figure 16. The fine granularity is used to construct a variable sensitive to the transverse shower size, utilizing a ratio of energy deposition in two rectangular areas with different sizes E(X×Y)/E(X'×Y'), where E(X×Y) is the energy deposition in area of X×Y cells.

We studied a simplified L1 electron/photon particle identification. We used two variables:

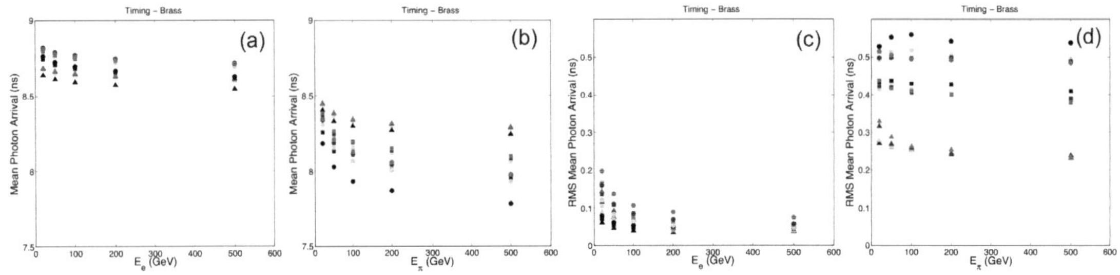

Figure 14. The Cherenkov photon timing characteristics for electron and pion initiated showers in brass are shown: mean arrival times for EM showers (a) and for HAD showers (b). The *rms* spread of these arrival times are displayed in (c) for electrons and (d) for hadrons. The markers represent the η location: circle $|\eta|$=3.5, square $|\eta|$=2.0, and triangle $|\eta|$=1.5, and the colors symbolize the fiber-to-fiber spacing: black 1.2 mm, green 1.8 mm, blue 2.4 mm and red 3.0 mm.

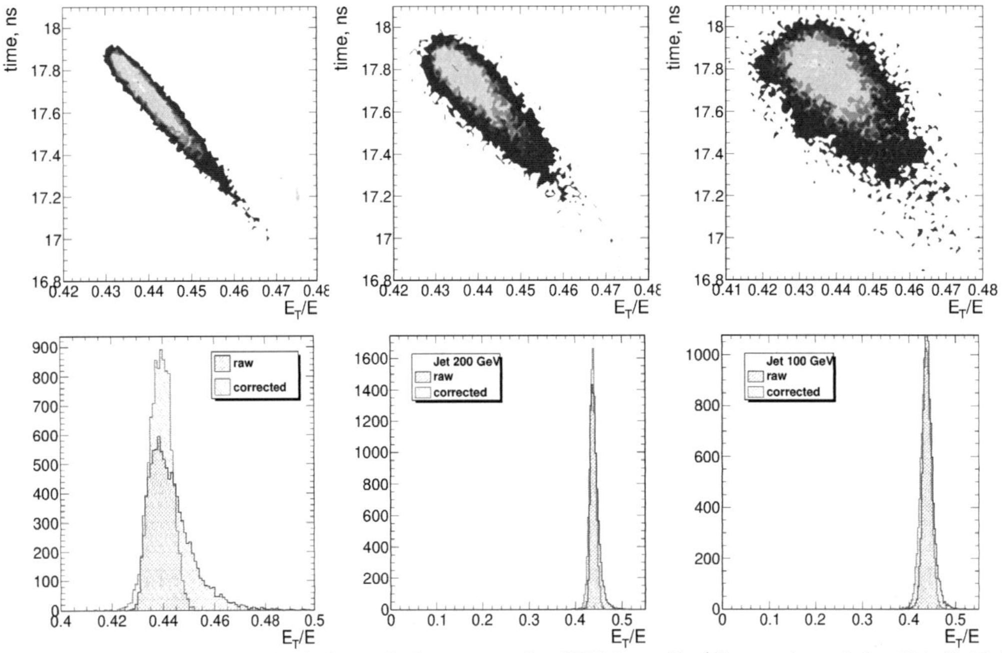

Figure 15. The average arrival time of photons to the SiPMs *vs* E_{T}/E are plotted for 500 GeV (left column), 200 GeV (middle column) and 100 GeV (right column) jets at $|\eta|$ = 1.5 for Option I. In the bottom row, a simple correction scheme is applied to sharpen the E_{T}/E ratio in order to suppress possible tails.

the transverse shape and f_{em}. The transverse shape is defined as E(2×7)/E(X×Y), where X×Y is η dependent:

- η=1.5 : E(2×7)/E(5×31)
- η=2.0 : E(2×7)/E(5×21)
- η=3.5 : E(2×7)/E(5×9)

The EM fraction is evaluated by the ratio of the quartz and scintillator fiber signals collected from the same area used in the denominator of the transverse shape variable. The performance

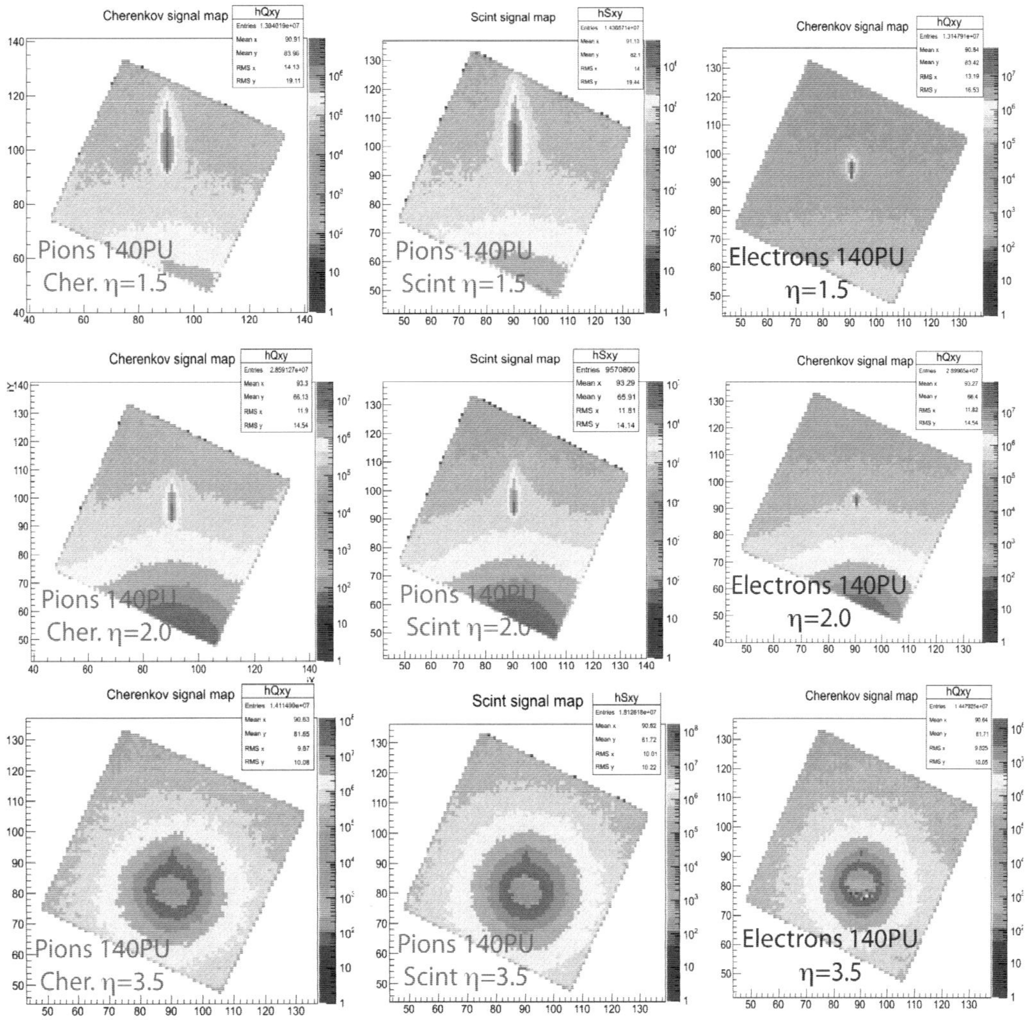

Figure 16. The energy distributions in X-Y plane for pions and electrons in the presence of 140PU at $|\eta| = 1.5, 2.0$ and 3.5. Left column is for Cherenkov and the middle column is for scintillation for pions. The right column is for electrons. A pixel corresponds to a physical tower in Option I.

of the particle identification is evaluated on samples with single electrons(signal) and pions (background) in the energy range of 20-500 GeV. We choose 95% signal efficiency as a reference point and evaluate the background rejection for the cases of 0 and 140 PU. The result is shown in Figure 17 as a function of the transverse energy (E_{T}) and for η = 1.5, 2.0 and 3.5.

6. Radiation Tolerance

The effects of radiation damage to clear fused-silica fibers have been well studied and it has been shown that the induced absorption depends on the wavelength and accumulated dose by the following equation:

$$\mu_{\mathrm{IND}}(\lambda) = \frac{1}{4.343}\alpha(\lambda)\left(\frac{D}{100}\right)^{\beta(\lambda)} \tag{1}$$

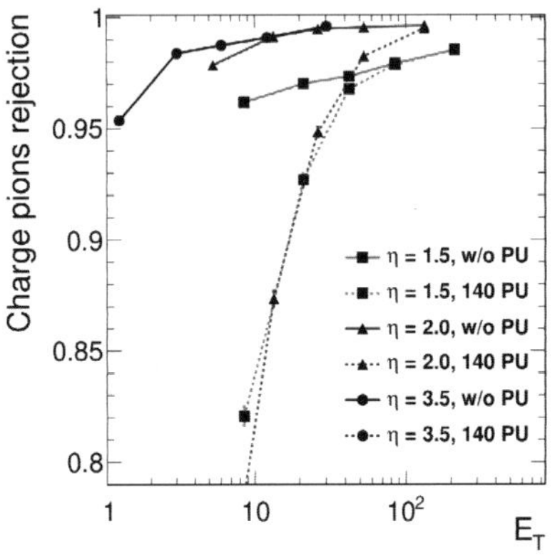

Figure 17. The electron identification performance at the L1: charge pion rejection is shown for a fixed electron selection efficiency of 95% for 0 and 140 PU.

where D is the accumulated dose in MRad and α and β are functions of the wavelength λ. The numerical values α and β for fused-silica fibers are taken from [9, 10].

The FLUKA simulation tool was used to estimate the expected dose and neutron fluences up to after 3000 fb^{-1}. The typical maximum dose inside the absorber ranges from $\sim$ 15 kGy to $\sim$ 16 MGy. From the dose maps available, the non-uniform μ_{IND} of the fibers is calculated along their length according to Eq. 1. The non-uniformity of radiation damage to the fibers results in the addition of a constant term to the energy resolution. Figure 18 show this constant term as a function of η for two wavelengths. For 450 nm light at $\eta = 4$, the additional constant term is of the order of a few percent.

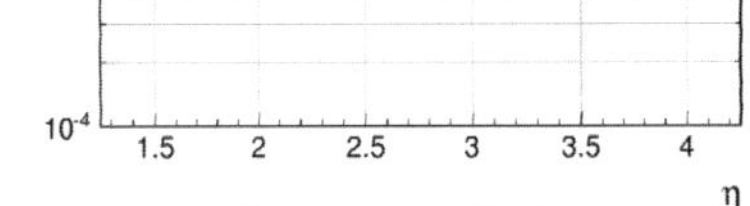

Figure 18. The additional constant term for two particular wavelengths is shown here as a function of η after 3000 fb^{-1}. This shows at $\eta = 4$ and additional $\sim 1-2\%$ constant term for 450 nm light.

The development of rad-hard cerium-doped scintillating fibers to a level at which they can be deployed in calorimeters in Phase II started recently and some promising results have been already presented in this conference [11, 12, 13]. Based on progress in rad-hard inorganic scintillating materials [14, 15], on the CMS HF experience [9, 10], and on developments in fiber production technologies, the R&D on rad-hard scintillating fibers is likely to succeed in the coming years. Aside from calorimetry, other types of scintillating fiber detectors will also benefit from these advances.

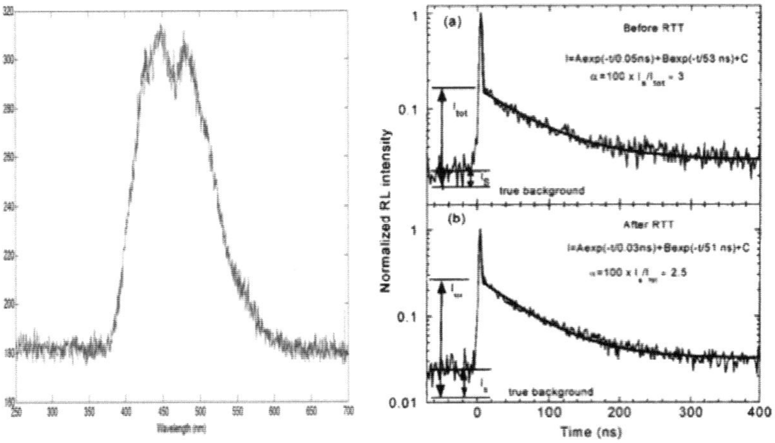

Figure 19. The emission spectrum of SiO2:Ce3+ fiber peaks around 470 nm (left) when it is excited at 337 nm by a pulsed N_2 laser. The decay time is about ~50 ns (right) with small contributions from faster and slower components [14].

Figure 19.(left) shows the emission spectrum of a SiO2:Ce3+ cladless fiber produced in Milan-Bicocca and characterized at TTU. The light emission peaks at around 470 nm, and the light yield (although not shown in this figure) is expected to be approximately twice that of BGO. The dominant decay time is ~50 ns (Figure 19.(right)). Shorter decay times are desirable, and synthesis strategies to shorten the decay times down to ~5 ns could involve tuning activator concentration along with the use of co-dopants. Another dopant, praseodymium, for example, would give ~10 ns in shorter wavelengths.

References

[1] N. Akchurin *et al.*, for EM performance, see Nucl. Instr. and Meth. A 536 (2005) 29; Nucl. Instr. and Meth. A 548 (2005) 336; Nucl. Instr. and Meth. A 686 (2012) 125; Nucl. Instr. and Meth. A 735 (2014) 130.
[2] N. Akchurin *et al.*, for muon detection, see Nucl. Instr. and Meth. A 533 (2005) 305.
[3] N. Akchurin *et al.*, for hadron and jet detection, see Nucl. Instr. and Meth. A 537 (2005) 537; Nucl. Instr. and Meth. A 584 (2008) 304.
[4] N. Akchurin *et al.*, for particle ID using dual-reaout technique, see Nucl. Instr. and Meth. A 735 (2014) 120.
[5] C. S. Cowden, in these Proceedings.
[6] S. Abdulline *et al.*, Eur. Phys. J. C 53 (2008) 139166.
[7] F. Scuri *et al.*, IEEE Nucl. Sci. Symp. Conf. Rec., 2010, 1711-1717.
[8] Report to SPSC 2013, DREAM/RD52 Collaboration.
[9] K. Cankocak *et al.*, Nucl. Instr. and Meth A 585 (2008) 20-27.
[10] N. Akchurin, C. Cowden, P. Dudero, CMS DN-2013-018.
[11] E. Auffray, in these Proceedings.
[12] M. Korjik, in these Proceedings.
[13] V. Dormenev, in these Proceedings.
[14] S. Braccini *et al.*, 2012 JINST 7 T02001 doi:10.1088/1748-0221/7/02/T02001; M. Neff *et al.* , Optical Materials 33 (2010) 1-3; N. Chiodini *et al.*, Applied Physics Letters, Vol 81, No 23 (2002) 4374; M. Neff *et al.*, Optical Material 31 (2008) 247-251; A. Vedda *et al.*, Appl. Phys. Lett. 85 (2004) 6356; A. Fasoli *et al.*, Journal of Non-crystalline Solids 355 (2009) 1140; and N. Chiodini *et al.*, Chemistry of Materials 24 (2012) 677.
[15] D. Sangla *et al.*, J. Cryst. Growth, 312 (2009) 125; C. Dujardin *et al.*, J. Appl. Phys. 108 (2010) 013510; M. Sugiyama *et al.*, J. Cryst. Growth, 33 (2011) 905.

The CMS electromagnetic calorimeter barrel upgrade for High-Luminosity LHC

Philippe Gras
on behalf of the CMS collaboration
CEA/IRFU Saclay, France
E-mail: philippe.gras@cern.ch

Abstract. The High Luminosity LHC (HL-LHC) will provide unprecedented instantaneous and integrated luminosity. The lead tungstate crystals forming the barrel part of the CMS Electromagnetic Calorimeter (ECAL) will still perform well, even after the expected 3000 fb^{-1} at the end of HL-LHC. The scintillation light from the crystals is measured with avalanche photodiodes (APDs). Although the APDs will continue to be operational, there will be some increase in noise due to radiation-induced dark-currents. Triggering on electromagnetic objects with ~140 pileup events necessitates a change of the front-end electronics. New developments in high-speed optical links will allow single-crystal readout at 40 MHz to upgraded off-detector processors, allowing maximum flexibility and enhanced triggering possibilities. The very-front-end system will also be upgraded, to provide improved rejection of anomalous signals in the APDs as well as to mitigate the increase in APD noise. We are also considering lowering the ECAL barrel operating temperature from 18°C to about 8 ~10°C, in order to increase the scintillation light output and reduce the APD dark current.

1. Introduction

The CMS experiment [1] is located at CERN at the Large Hadron Collider (LHC). At the end of year 2022 the LHC will have accumulated 300 fb^{-1}. After this first operation phase, the LHC will be upgraded to raise its peak luminosity to $5 \cdot 10^{34}\,\mathrm{cm}^{-2}\mathrm{s}^{-1}$. At that luminosity, 300 fb^{-1} will be integrated each year to reach a total amount of data of 3000 fb^{-1} at the end of year 2035 [2]. The LHC experiments, including CMS, will be also upgraded to meet the challenges brought by this unprecedented luminosity level and optimise the detectors for the rich physics program foreseen for HL-LHC.

2. The central electromagnetic calorimeter of the CMS experiment

The current central electromagnetic calorimeter of CMS [3] (barrel part of the calorimeter, or *ECAL barrel*) is a homogeneous calorimeter made of 61,200 lead tungstate, $PbWO_4$, crystals. It is highly granular, $\Delta\eta \times \Delta\phi = 0.087 \times 0.087$, $\Delta \ln(\tan\theta/2) = 0.087$ in the longitudinal plan with θ the angle with respect to the the beam axis and 0.087 radian in the transverse plan. The crystals are 26 radiation lengths long and their square front size is ~ 10 Molière radii ×10 Molière radii. The ECAL barrel is made of 36 supermodules, which were assembled on the surface, before being shipped down into the cavern where the CMS experiment is located and inserted into the CMS detector. The scintillation light is measured with avalanche photodiodes (APDs). To maximize the collection surface two APDs are glued on each crystal, the sum of

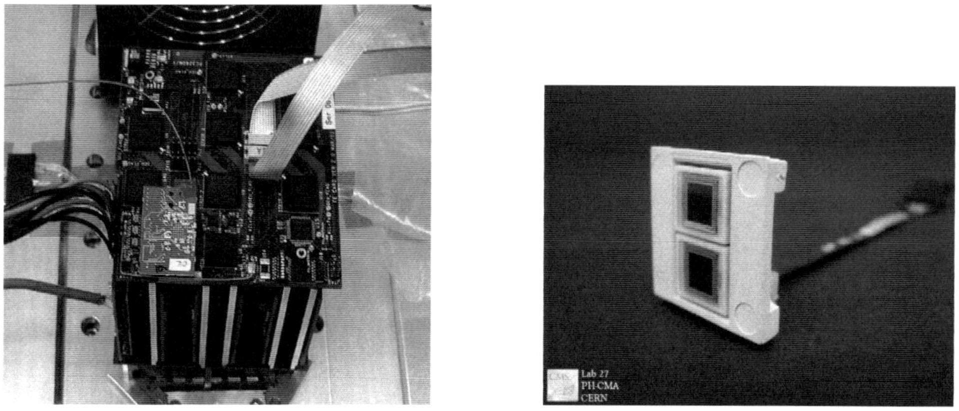

Figure 1. CMS ECAL barrel readout. (Left) Front-end readout electronics, covering the readout of 5×5 crystal channels. The APDs are connected on the bottom board. A vertical board handles five crystals, one additional vertical board (on left side in the picture) is dedicated to low voltage supply. (Right) Picture of an APD capsule. To extend the collection surface two APDs are used per crystal andtheir signals are merged.

the two signals being read out. A picture of the APD pair assembly can be found in Fig. 1. The crystals are grouped for the readout electronics in entities of 25 crystals, 5×5 in η-ϕ. The data stream is split in two on the detector: one stream of events selected by the first trigger level with the full granularity offered by the calorimeter, an average of about 100,000 events per second and one stream to the trigger system read at 40 MHz with a reduced granularity. For the trigger system stream, each readout unit, which covers 25 crystals, delivers one signal which is the sum of signals over these crystals. A picture of the electronics of one readout unit is shown in Fig. 1.

The irradiation sustained by the crystal induces a reduction of their transparency. The crystals have the capacity to recover the transparency loss in absence of collisions, e. g. during the LHC machine fill, the transparency increases. Fig. 2 shows channel response changes during 2011 and 2012 operations of the electromagnetic calorimeter. The response is measured with a laser, at the rate of one measurement point every 40 min. The channel response includes the electronics chain. For the barrel this response changes is dominated by the crystal transparency variations. The barrel parts corresponds to the two red and green curves at the top of the figure, while the other curves correspond to the forward part of the ECAL constituted by the *endcaps*. The loss is limited to 10% in the most forward barrel pseudorapidity ranges represented in this plot. Loss will be larger for HL-LHC, reaching values currently reached in the endcaps. The laser monitoring system will be used as it is currently done for the barrel and endcaps to correct for the response variations.

The irradiation also effects the APDs, it induces damage in the silicon bulk, creating a leakage current called *dark current*. The net effect is an electronic noise proportional to the square root of this dark current, hence to the square root of the integrated luminosity. Measurement of the dark current during 2011 and 2012 is shown in Fig. 3. A particle can also deposit energy directly into the silicon bulk generating a signal. There are two handles to discriminate this anomalous signal from the signal from an electromagnetic shower: the signal shape, the anomalous signal is faster, narrower and arrives earlier, and correlation between channels, since an electromagnetic shower will systematically induce a signal on a group of neighbouring crystals, as opposed to anomalous signal which appears in one crystal. Fig. 4 illustrates this discrimination. The two-

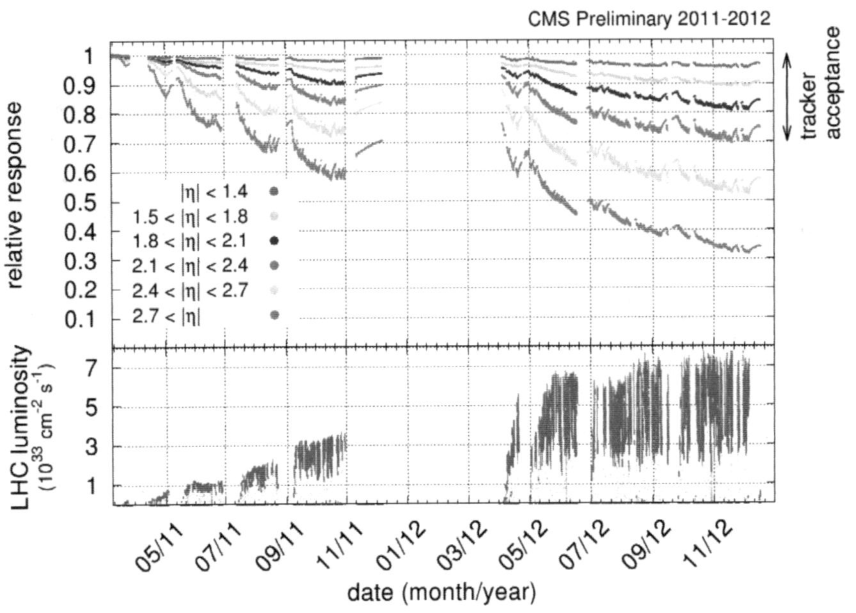

Figure 2. Relative response to laser light (440 nm) measured by the ECAL laser monitoring system, averaged over all crystals in bins of pseudorapidity, for the 2011 and 2012 data taking periods. The two red and green top curves correspond to the barrel part, the curves below correspond to the endcap part of the calorimeter.

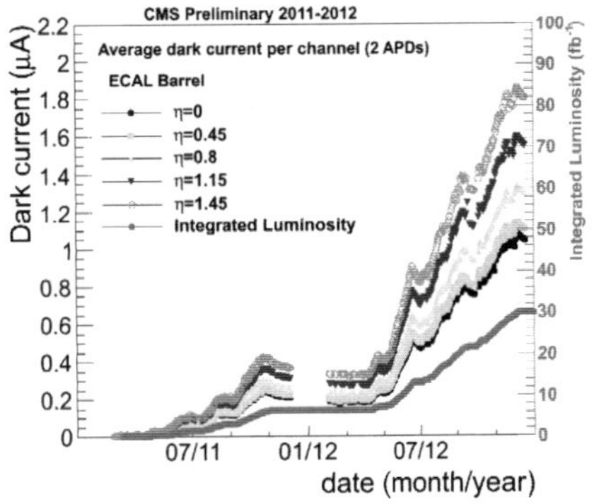

Figure 3. APD dark current measured during the 2011 and 2012 years of operation. The dark current increase is proportional to the integrated luminosity, apart from the limited recovery during the period without beam.

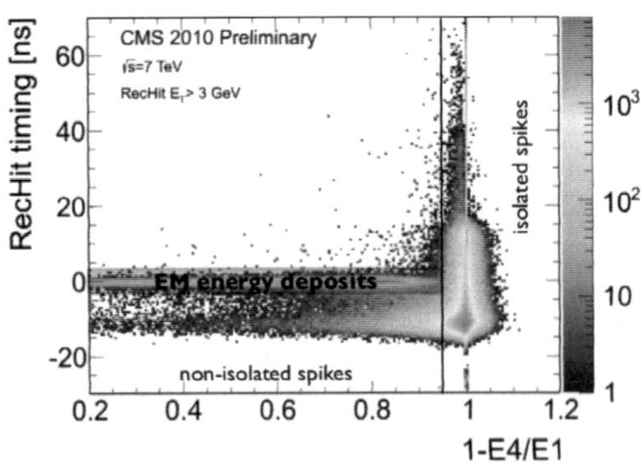

Figure 4. Anomalous signal discrimination. The plot represents the correlation between two discriminating variables of anomalous and scintillation signals, the timing of the signal maximum and a ratio of signal measured in adjacent crystal to the considered crystal. Anomalous signals can be rejected highly efficiently with the use of these two variables.

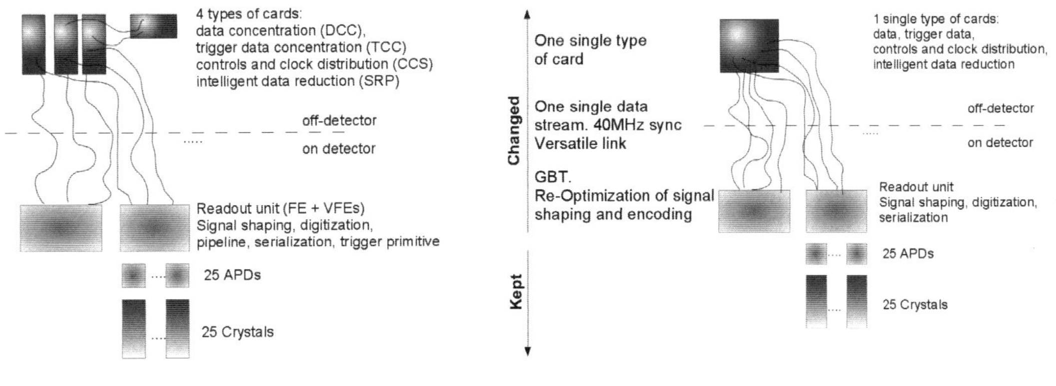

Figure 5. Architecture of the (Left) current and (Right) future readout of the CMS ECAL. Current readout includes two data streams from the front-end electronics (electronics located on the detector) to the off-detector electronics. Data collection and controls of the front-end electronics is handled by four types of electronics boards. The *FE* and *VFE* referred in the schema are respectively the top horizontal board and the vertical boards in left picture of Fig. 1. In the new architecture a single data stream is used.

dimensional histogram shows the correlation of the signal timing with a variable sensitive to the localisation of the signal, defined for a crystal hit as one minus the ratio of the total transverse energy measured in the four closest crystals to the transverse energy measured in this crystal. In this histogram, a minimum transverse energy of 3 GeV is required on the crystal hits. The rate of anomalous signals is proportional to the instantaneous luminosity, so a factor 20 increase is expected for HL-LHC for the same transverse energy threshold, but the rate sharply decreases when the transverse energy threshold is increased.

3. Upgrade of the central electromagnetic calorimeter of the CMS experiment

Extrapolation of dark current evolution from 2011-2012 data to HL-LHC provides an estimation of 400 MeV noise at the end of HL-LHC. Complementary irradiation tests are being performed in order to verify this extrapolation. This noise level is considered adequate to reach required precision. Operation at a lower temperature, 8°C to 10°C, which would mitigate the increase of dark current is under study. The upgrade for the HL-LHC will concern the readout electronics from the output of the APDs up to the data acquisition (DAQ) and the trigger systems. Part of the laser monitoring system will also be upgraded.

The CMS trigger system for HL-LHC will keep the current concept of two levels of triggers: a Level-1 trigger implemented in dedicated electronics and a High Level Trigger implement in software and run in a cluster of computers. The Level-1 trigger rate will be increased by a factor 10 to reach 1 MHz and the tracker information will be added for the Level-1 trigger decision. While current ECAL electronics support a trigger latency up to 6.3 μs, 20 μs will be needed with the new trigger system. Fig. 5 illustrates the current architecture of the ECAL barrel readout. A preprocessing is performed for the trigger in the electronics located on the detector and two data streams go from the on-detector to the off-detector electronics, one to feed the Level-1 trigger system and one for the data of the selected events. The functionality of the off-detector electronics is split in four types of boards: detector control cards (DCC) to collect triggered event data and send them to the CMS DAQ system, trigger control cards (TCC) to collect the data for the trigger decision and send them to the CMS calorimeter trigger system, the clock and control system cards (CCS) to control, configure the on-detector electronics, provide it the clock

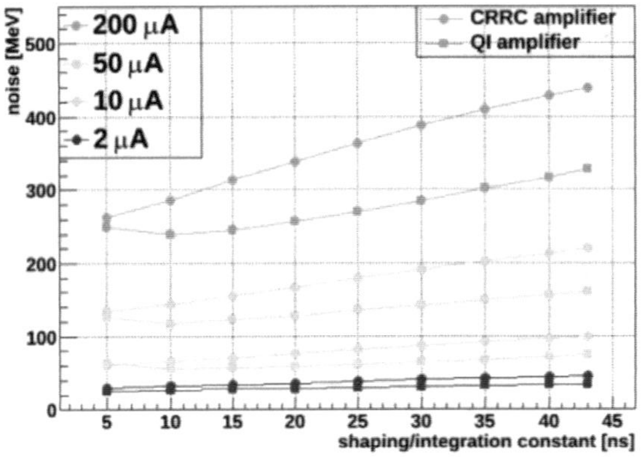

Figure 6. Level of the electronics noise induced by the dark current in function of the signal preamplifier parameters and type. Two options are considered a CR-RC shaping and a charge integration in a fixed time window.

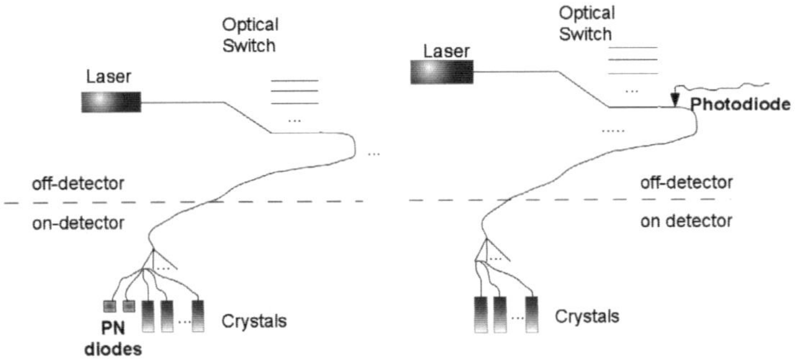

Figure 7. Laser monitoring system (Left) before and (Right) after the upgrade. The crystals are illuminated by groups of 800 or 900. The laser is located outside the detector in the service cavern. A remotely controlled optical switch is used to direct the light in the different groups of crystals. Fibres are split inside the detector supermodules to feed each crystal. In the upgraded system, the diodes used to measure the light injected into the crystals will be moved to outside the detector.

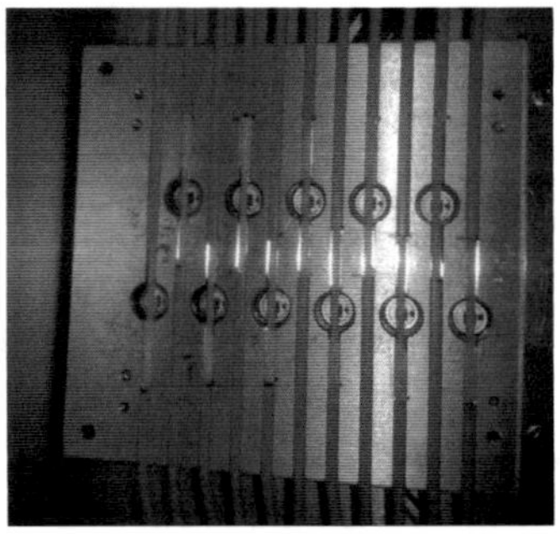

Figure 8. Picture of the prototype of laser light intensity measurement setup. Shield is removed on a small part of the fibre, the fibre is unpolished and a photodiode is placed in front of the unpolished part.

and transmit it the Level-1 trigger accept signal received from the CMS trigger system. The upgrade of the electronics will exploit the developments in data-link and reduction of cost-to-bandwidth ratio, and the improvement of FPGA integration to use a single data stream. After the upgrade, the channels will be read at 40 MHz without on-detector trigger pre-processing, and the full off-detector functionality will be implemented in a single card type. In the current architecture the on-detector electronics includes a data pipeline to retain the event data the time the Level-1 trigger decision is taken. With the new architecture, this pipeline is moved to the off-detector electronics. The data link from on-detector to off-detector electronics will be done using the versatile link technology and companion GBT chips which were developed at CERN [4]. For each readout unit (25 crystals) three upstream links will be used for the detector data and one downstream for the controls of the electronics and the transmission of the clock.

We will take the advantage of the electronics change that is mandatory for the trigger system to optimise the shaping and the encoding of the signal. Figure 6 shows the electronics noise with an CR-RC amplifier, which is used in current electronics, for different APD dark current levels. This plot shows that for an increased dark current level it is useful to shorten the shaping time in order to minimise the effective induced noise. Reduction of shaping time constant will also help in mitigating the pile-up of events from nearby bunch crossings (since there will be an average of 140 hard interactions in a bunch crossing with the HL-LHC beam luminosity). As shown on this figure further improvements can be obtained by using a different approach for the signal readout. The curve with box markers shows the noise level if the signal is integrated in a fixed-time window. Use of the concept of the QIE (Charge Integrator and Encoder) developed for the CMS hadronic calorimeter [5] is very promising and extension of the design to the needs of the electronic calorimeter is under study. Finally, including a precise time encoding in the on-detector electronics to ship the signal time together with its amplitude is under studies. It will provide a powerful handle to mitigate the effects of the pile-up and to reject anomalous signals. The encoding of the signal amplitude is also being revised. The current system provides a quantisation step of about 35 MeV, while a ~ 100 MeV quantification should be sufficient to keep the quantisation noise small with respect to the noise induced by the APD dark current.

Finally, the laser monitoring system measures the light injected into the crystals with PN diodes located on the detector inside the supermodules, as represented in the sketch of Fig. 7. The readout of these PN diodes uses the same link technology than the readout of the APDs and is similar to the other readout units from the point of view of the DCCs. A first option will be to upgrade the electronics to support the new links. An alternative option under study is to measure the injected light outside of the detector on the main fibre, upstream the fibre splitting, as represented in Fig. 7. The measurement will be performed with a photodiode measuring light diffused from the fibre surface locally unpolished for this purpose, like in the picture of Fig. 8. This option has the advantage to limit interventions on the supermodules and to be cost effective. A prototype is being put in place for the LHC run 2, starting in 2015, and will be operated in parallel of the current system. That will allow an evaluation of this option and especially of the measurement precision which can be reached.

4. Summary

The HL-LHC will enable the acquisition of 10 times more data that will have been acquired with the LHC. The extremely large luminosity, resulting in a very large number of events per bunch crossing which get piled-up and in stringent irradiation conditions, brings new requirements for the experiment detectors. The plan for the upgrade of the barrel part of the electromagnetic calorimeter of CMS has been presented. The detector itself, a homogeneous lead tungstate calorimeter read out by avalanche photodiodes will be kept as such. The full electronic chain from the output of the APDs down to the CMS data acquisition system and the CMS trigger system will be changed. Part of the laser monitoring system which measures the detector response variation, mainly due to the irradiation effect on the lead tungstate crystal transparencies, will also be upgraded.

References

[1] S. Chatrchyan, et al., "The CMS experiment at the CERN LHC", JINST **3** (2008) S08004
[2] M. Lamont, "LHC, HL-LHC and Beyond", PoS EPS - **HEP2013**, 149 (2014).
[3] CMS collaboration, "The CMS electromagnetic calorimeter project. Technical Design Report", CERN-LHCC-97-033
[4] P. Moreira et al. "The GBT-SerDes ASIC prototype", JINST **5** (2010) C11022.
[5] A. Baumbaugh et al., "QIE10: a new front-end custom integrated circuit for high-rate experiments", JINST **9** (2014) C01062.

Journal of Physics: Conference Series **587** (2015) 012017 doi:10.1088/1742-6596/587/1/012017

Object Reconstruction in Non-Pointing Geometry

C S Cowden [1]
Texas Tech University, Department of Physics, Lubbock, Texas, USA
E-mail: christopher.cowden@ttu.edu

Abstract. The Combined Forward Calorimeter (CFC) option for the Phase II upgrades covers the endcap region ($1.5 < |\eta| < 4$) and its design calls for embedded clear and rad-hard scintillation fibers in an absorber matrix. The object construction using complementary features of Cherenkov and scintillating signals coupled with fast timing information results in some unique capabilities for high efficiency particle identification, precision energy measurement, background suppression and triggering. These topics along with possible pile-up mitigation techniques at the HL-LHC will be discussed in detail.

1. Introduction

The Combined Forward Calorimeter (CFC) was proposed as a replacement for the CMS end-cap calorimeters for Phase II upgrades expected to operate in the HL-LHC. The CFC would replace both hadronic and electromagnetic calorimeters into a combined system. This device collects signals in two types of fibers: (1) a scintillating Cerium doped quartz fiber and (2) a clear un-doped quartz fiber. The Cherenkov signal (Q) produced in the un-doped quartz and the dE/dX sensitivity in the scintillator (S) provide the tools needed to remove fluctuations from the electromagnetic fraction on an event-by-event basis using the dual readout technique [1, 2, 3, 4].

It would also be desirable to extend the current CMS end-cap coverage to be consistent with the proposed tracker coverage. Therefore, the CFC proposal is to cover the range $1.5 < |\eta| < 4$ in order to avoid a transition in calorimeter systems at $\eta = 3.5$ to the HF detector which is currently in place. This design is described in more detail in another contribution in these proceedings [5].

The following discussion assumes a configuration in which all towers and fibers run parallel to the beam. Fibers are placed 2 mm apart alternating between quartz and scintillating. The calorimeter segments are 2 cm^2 in the transverse plane, and this segmentation is constant throughout the depth of the calorimeter. A depth segmentation is obtained by digitizing the quartz channel with a 5 GHz sampler.

2. Particle Identification

Some conservative assumptions regarding the capabilities of processing timing information in the level 1 (L1) trigger decision were made. Particle identification performance is based on the

[1] The CFC Collaboration: Baylor University, CERN, Fermilab, Kansas State University, University of Notre Dame, IHEP-Portvino, Rutgers University, Saha Institute of Nuclear Physics - Kolkata, Texas A&M University, Texas Tech University, University of Virginia.

assumption that the L1 trigger level digitizes signals at 40 MHz. The longitudinal segmentation based on time slicing is included later in the offline particle identification.

Based on the transverse shower shape and Q/S (quartz and scintillator, respectively) ratio, a reasonable e/γ background rejection can be achieved. Figure 1 and Figure 2 show distributions of these variables at $\eta = 1.5$ and $\eta = 2.0$, respectively, both with and without in-time pileup. The transverse shower shape is defined here as the ratio of energy in a 7×2 ($\eta \times \phi$) cluster of towers to a 10×42 cluster of towers – $\frac{E(7\times2)}{E(10\times42)}$.

Q/S provides a measure of the electromagnetic fraction (f_{em}) since f_{em} only depends upon Q/S and the detector compensation (e/h). A ratio near 1 indicates the shower has a large f_{em} fraction.

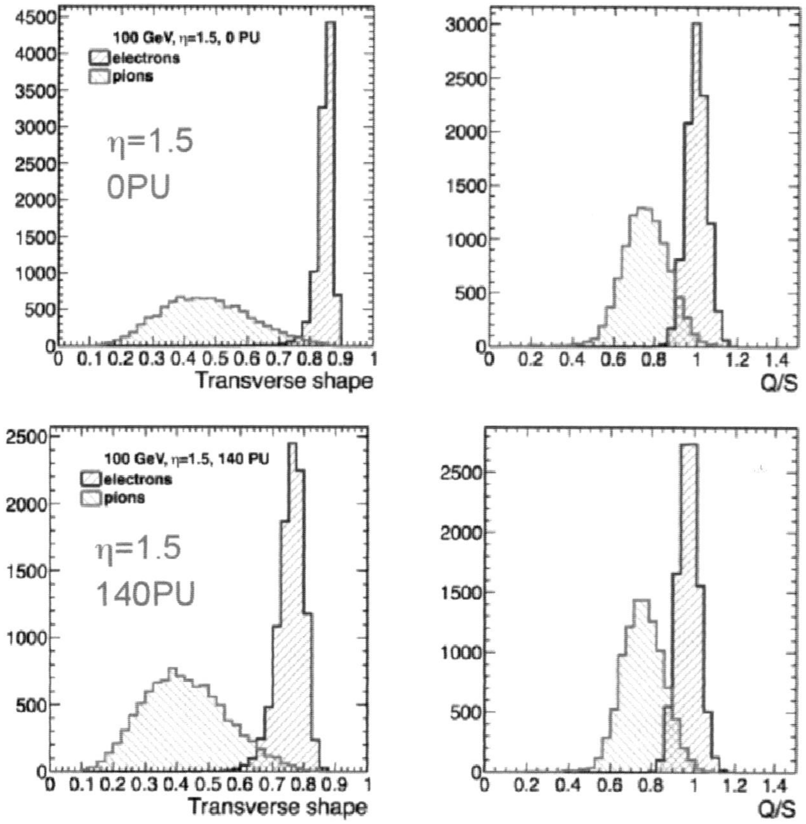

Figure 1. Distributions of variables used in particle identification with and without in-time pileup at $\eta = 1.5$.

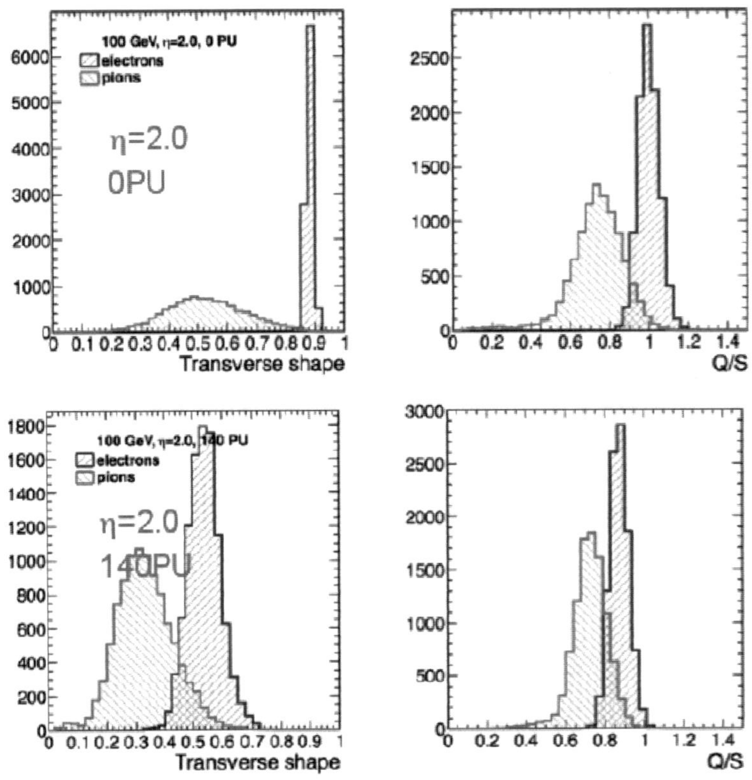

Figure 2. Distributions of variables used in particle identification with and without in-time pileup at $\eta = 2.0$.

The amount of pileup present in clusters increases at higher pseudo-rapidities. However, transverse shower shapes remain a good observable for background rejection. Some improvements to the front-end electronics have been suggested which could introduce some timing aspects into the L1 decision. For example, a TDC or 80 MHz digitization rate could provide a coarse depth segmentation to help identify hadrons.

Figure 3 shows a study comparing electron/pion separation in the non-pointing geometry to a quasi-pointing geometry. Although the cluster sizes and shapes were not fully optimized, the two options show similar performance.

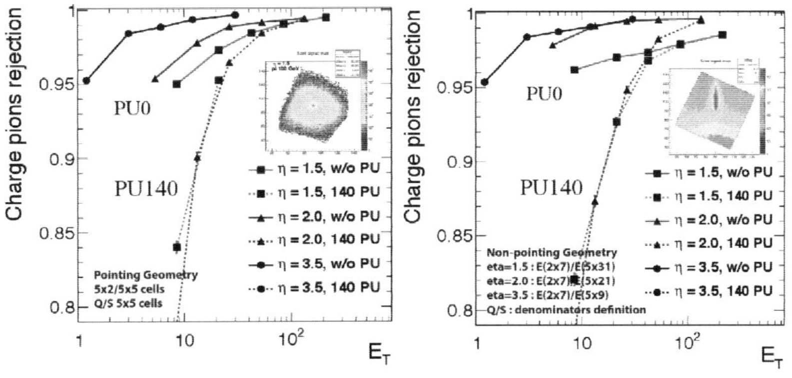

Figure 3. Charged pion rejection rates with the online algorithm are compared for a signal (electron) identification efficiency of 80% with and without in-time pileup, at various rapidities. Also shown is this same rejection rate comparison in a quasi-pointing geometry.

The offline particle identification employs the timing information obtained by the 5 GHz digitizer on the Q channel to discriminate between hadronic and electromagnetic showers. Figure 4 shows this discriminating power at various pseudo-rapidities for a fixed signal (electron) efficiency of 95%. In this case, the charged pion rejection is $\sim$ 99%.

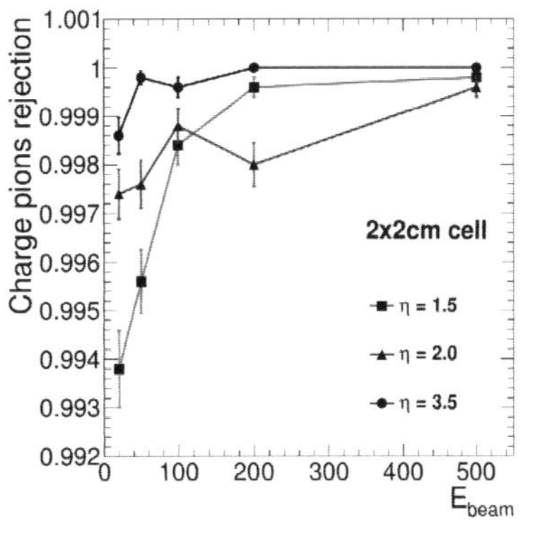

Figure 4. Charged pion rejection rates with the offline algorithm are compared for a signal (electron) identification efficiency of 95% at various rapidities without in-time pileup.

3. Timing Features

The position of a shower in $\eta - \phi$ space is determined to first order by the location of the shower at the face of the detector. In this approximation, showers which develop deeper are incorrectly found at a smaller η. These showers contribute to a small tail in the E_T resolution resulting from this incorrect position. This tail can be corrected by taking the time into account. Figure 5 shows the anti-correlation of time to E_T/E. Also shown is the E_T/E resolution before and after the time correction to E_T.

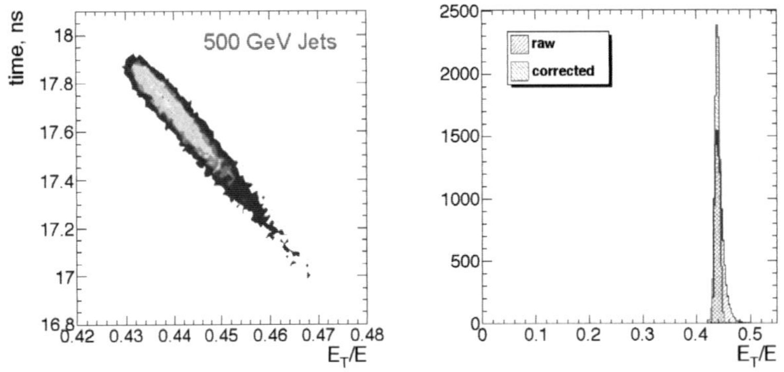

Figure 5. The figure on the left shows the distribution of mean shower time from the fast 5 GHz sampler against E_T/E.

As has been stated previously, with a fast digitizer, it is possible to slice pulses in the time domain to provide a measure of longitudinal segmentation. Figure 6 shows that both in and out of the presence of $E[n] = 140$ vertices, showers originating from pions arrive before showers originating from electrons. Since hadronic showers typically penetrate the absorber deeper than do electromagnetic showers, the time of arrival of signal at the back of the detector is earlier for hadrons. The correlation of deeper position and earlier times results from particles traveling faster than the speed of light in the fiber.

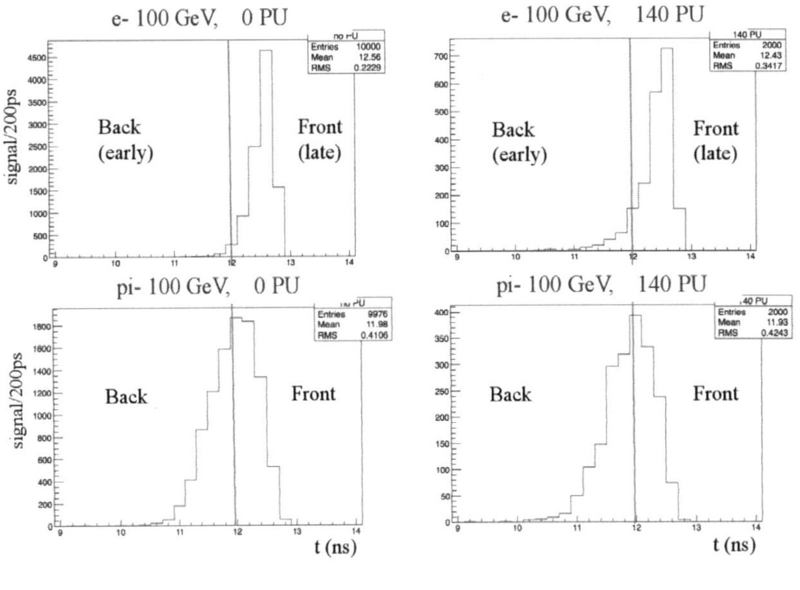

Figure 6. Shower profiles are compared between electron and pion showers in the case of pure signal, or in the presence of $E[n] = 140$ minimum bias vertices. A coarse depth segmentation is implied by the vertical line.

The intrinsic time resolution of this geometry has a dependence upon the angle made by the incoming particle. In the high η region, where this angle is more shallow, the time resolution is broadest and, conversely, the intrinsic resolution is best at lower η. Figure 7 shows the time resolution of electromagnetic showers as a function of $1/\sqrt{E}$ for three angles.

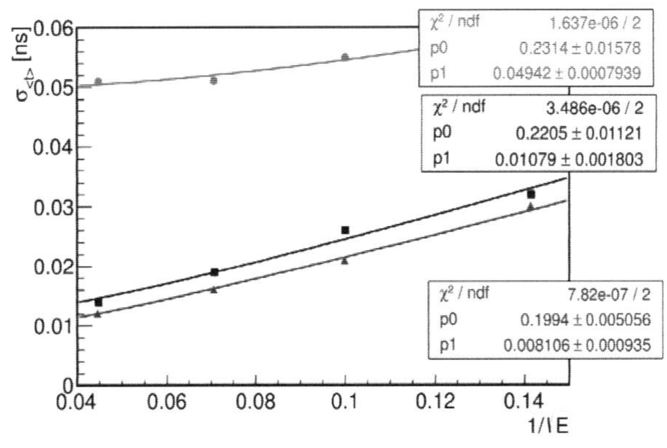

Figure 7. The EM time resolution as a function of $1/\sqrt{E}$ for three angles the incoming particle makes w.r.t to the fiber axis.

Signal photons populate various helical modes. These modes lead to a dispersion in the time of arrival for photons produced by a single charged particle. These modes can be characterized by the effective speed along the fiber axis or, alternatively, as the effective index of refraction, n_{eff}. The distribution of n_{eff} is shown in Figure 8 for a pion shower.

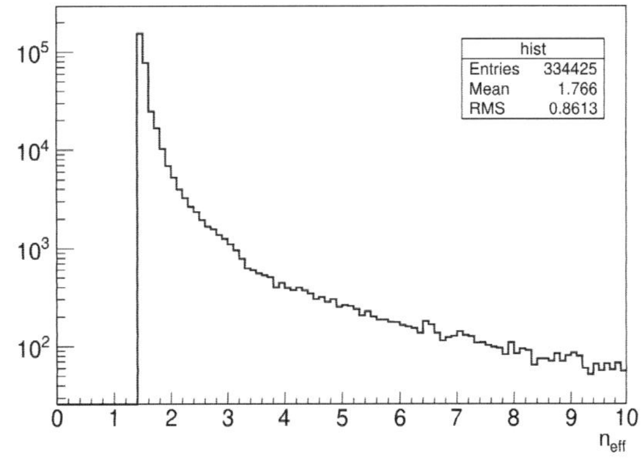

Figure 8. The effective index of refraction for pion induced showers.

4. Conclusions

The CFC would have a fine transverse segmentation for both EM and hadronic showers which help provide good object identification and background rejection in the HL-LHC environment. The inclusion of pulse shape analysis in offline reconstruction adds an interesting perspective on 3D energy reconstruction from a device that does not suffer from the inter-calibration issues of longitudinally segmented calorimeters.

5. References

[1] N. Akchurin et al., Nucl. Instr. and Meth. A 536 (2005) 29; Nucl. Instr. and Meth. A 548 (2005) 336; Nucl. Instr. and Meth. A 686 (2012) 125; Nucl. Instr. and Meth. A 735 (2014) 130.
[2] N. Akchurin et al., Nucl. Instr. and Meth. A 533 (2005) 305.
[3] N. Akchurin et al., Nucl. Instr. and Meth. A 537 (2005) 537; Nucl. Instr. and Meth. A 584 (2008) 304.
[4] N. Akchurin et al., Nucl. Instr. and Meth. A 735 (2014) 120.
[5] N. Akchurin, in these proceedings.

The ATLAS liquid argon calorimeter: upgrade plans for the HL-LHC

Olga Novgorodova, on behalf of the ATLAS Liquid Argon Calorimeter Group.
TU Dresden, IKTP, D-01063 Dresden
E-mail: Olga.Novgorodova@cern.ch

Abstract. The ATLAS detector was designed and built to study proton-proton collisions produced at the LHC at centre-of-mass energies up to 14 TeV and instantaneous luminosities up to 10^{34} cm^{-2}s^{-1}. Liquid argon (LAr) sampling calorimeters are employed for electromagnetic and hadronic calorimetry. The luminosity for the proposed High Luminosity LHC phase (HL-LHC) will increase up to 5×10^{34} cm^{-2}s^{-1}, with the goal of accumulating an integrated luminosity of 3000 fb^{-1}. This is well beyond the values for which the detectors were designed. The electromagnetic and hadronic calorimeters will be able to tolerate the increased particle flux, but the performance of the forward calorimeter (FCal) will be affected. Two possible solutions for keeping the current performance are being discussed. The readout electronics will also need to withstand larger radiation environment. In the hadronic endcap calorimeter (HEC) cold GaAs preamplifiers are located inside the endcap cryostats. The properties of these devices have been investigated in recent proton and neutron irradiation tests to determine whether they must be replaced. In addition, the entire front-end readout system is not expected to survive the integrated luminosity at the HL-LHC and will be replaced. The description of the new readout system is presented.

1. Introduction

The ATLAS detector [1] at the LHC was designed for measurements of proton-proton collisions at centre-of-mass energies up to 14 TeV and instantaneous luminosities up to 10^{34} cm^{-2}s^{-1}.

The ATLAS LAr calorimeter, shown in Figure 1, is a sampling calorimeter with liquid argon as an active medium. It consists of barrel and endcap parts enclosed in corresponding cryostats. The barrel cryostat contains only the electromagnetic calorimeter (EMB), covering a pseudorapidity range of $|\eta| < 1.475$. The endcap cryostat contains the electromagnetic endcap (EMEC), covering a range of $1.375 < |\eta| < 3.2$, the hadronic endcap (HEC), covering the region of $1.5 < |\eta| < 3.2$ and a forward calorimeter (FCal) covering the region $3.1 < |\eta| < 4.9$.

The EMB and EMEC were constructed using an accordion geometry for full φ coverage, with copper and kapton electrodes and lead as the absorbing material. The electrodes are positioned between the absorber plates by honeycomb spacers. The calorimeter is segmented into three longitudinal layers for $|\eta| < 2.5$ and two coarser-granularity layers for $|\eta| > 2.5$.

The HEC is segmented into four layers constructed in parallel-plate geometry. It has copper absorbers and copper and kapton electrodes. The FCal consists of three modules constructed of electrode rods parallel to the beam pipe, sitting in a metal matrix absorber. The LAr gap between the rods and matrix are smaller than in the rest of the LAr detector in order to endure the high particle fluxes in the forward regions. The first module of the FCal is for electromagnetic measurements and consists of a copper matrix and 269 μm liquid argon gaps. The other two modules are for hadronic measurements and consist of tungsten matrices with 376 and 508 μm liquid argon gaps.

Although the nominal LHC experimental program is still in progress, plans for a High Luminosity LHC (HL-LHC) are already being developed for operation of the collider and associated detectors at luminosities of up to 5×10^{34} cm^{-2}s^{-1}, with the goal of accumulating an integrated luminosity of 3000 fb^{-1}.

The LHC baseline plan for 2013-2030 [2] includes long shutdowns (LS) in which upgrades will be made for the next Phase. In the next long shutdown (LS2) during 2018-2019, an upgrade for Phase-I, it is planned to upgrade the injector system and secure the ultimate LHC luminosity. And the long shutdown (LS3) over 2023-2025, upgrade for Phase-II or HL-LHC phase, will increase an instantaneous luminosity up to 5×10^{34} cm^{-2}s^{-1}.

The proposed instantaneous and integrated luminosities are both well beyond the values for which the detectors were designed. The electromagnetic and hadronic calorimeters will be able to tolerate the increased particle flux, but the performance of the forward calorimeter (FCal) will be affected. Two solutions for mitigation of the response degradation are under study: one option is to build and install a replacement FCal with smaller LAr gaps. A second approach involves the installation of a small Mini-FCal calorimeter in front of the FCal. These options are discussed in section 4. The currently preferred Mini-FCal option is a LAr/Cu sampling calorimeter for which detailed design and performance studies are underway.

The electronics readout, originally qualified for an integrated luminosity of only 700 fb^{-1}, will also potentially be affected. This concerns especially the hadronic endcap calorimeter (HEC) that employs cold GaAs preamplifiers located inside the endcap cryostats. Individual ASIC positions are located at the outer circumference of the HEC with r = 204 cm and 430 cm $<$ z $<$ 600 cm. The properties of these devices have been further investigated in recent proton and neutron irradiation tests to determine whether they must be replaced. The measured gain and linearity degradation is used to determine the performance expected for calorimetric measurements in HL-LHC conditions, discussed in section 3.

Beside the specific HEC radiation concerns, the entire front-end readout system is not expected to survive the integrated luminosity at the HL-LHC. Furthermore, the requirements of long-latency data buffering and multi-level hardware triggering by the future ATLAS trigger system are not compatible with the current front-end electronics. As these electronics are much more accessible than the HEC preamplifiers, a decision has been taken to replace the complete

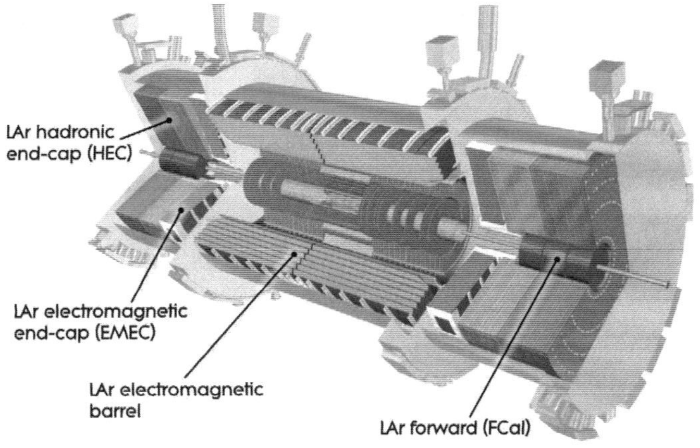

Figure 1. The ATLAS detector.

readout chain. The new readout system must be able to process about 180000 channels at 40 MHz in real time, with 16 bit dynamic range and a total data rate of about 140 Tbit/s. To meet these challenging environmental and performance parameters, dedicated ASICs for low-latency, low-power analog-to-digital conversion and for fast, low-latency optical transmission are being developed, along with FPGA-based high-bandwidth on-line data processing modules, discussed in section 2.

2. Readout Upgrade

The front-end crates (FEC) are installed at the two ends of the barrel cryostat at the outer ends of each endcap vessel. The crates house the front-end electronics which amplify, process and digitize the calorimeter signals. There are three points that can be addressed by an upgrade of the FEC: the radiation damage of the on-detector electronics, natural ageing effects of electronic components and the possibility to improve the ATLAS trigger system. The FEC electronics for HL-LHC phase will be upgraded in steps. First, additional new trigger boards will be installed in LS2. Then there will be the replacement of the primary readout and calibration boards in LS3. The FEC will contain four types of boards, implementing four different functionalities: the upgraded front-end boards (FEB2) reading out signals from 128 calorimeters cells, the LAr trigger digitizer boards (LTDB and LTDDB), the calibration boards and the control boards for the clock and configuration signals handling [3]. Figure 2 shows the front-end architecture planned for the HL-LHC phase. The calibration boards, which are not used during collisions, are not shown. The control board functionality may or may not be integrated directly into the new generation front-end boards. The upgrade benefits from the high granularity of the calorimeter to provide more information for the triggering system. This also requires development of high-speed data transmission elements.

The new readout, described above, will have an optical transmission of all data to the back-end. Also front-end components are planned to be more radiation-hard and with trigger buffers located off-detector. The upgraded trigger readout foreseen in the LS2 will be used as a low-latency Level-0 trigger in the HL-LHC Phase. This also requires upgrading the back-end electronics.

2.1. The Upgraded Front-End Board

The new front-end boards are designed to improve radiation tolerance and to increased the size of the trigger pipelines which will allow more advanced trigger algorithms to be applied. The main functional blocks are: analog preamplification and shaping, production of (summed) analog signals for the LTDB, analog-to-digital conversion of all signals at a rate of 40 MHz, multiplexing and serialization of digital data, and transmission over high-speed optical links. The analog preamplification and shaping stages will be integrated in a single ASIC. The Liquid Argon Preamp and Shaper (LAPAS) test-chip, fabricated in 2009 in IBMs 8WL SiGe process, validated the design approach of implementing a wide dynamic range single ended preamp followed by low power differential shaping stages with multiple gains to achieve the required 16 bit resolution. Less expensive SiGe process alternatives are currently being explored: IHP SG25H3P and IBMs 7WL. These studies will evolve into LS3 applicable ADCs. As an alternative, the qualification of commercial 12-bit ADCs for LS2/LS3 upgrade applications was done. The best performance was shown by a Texas Instruments chip (ADS5272) which tolerated radiation up to 1 kGy, has a low power consumption (113 mW) and a low latency (163 ns).

Data of 8 ADC channels are encoded, multiplexed, serialized and send out by the Link On Chip (LOC). A development in Silicon-on-Sapphire (SoS) 0.25 μm technology allows 8 Gbps

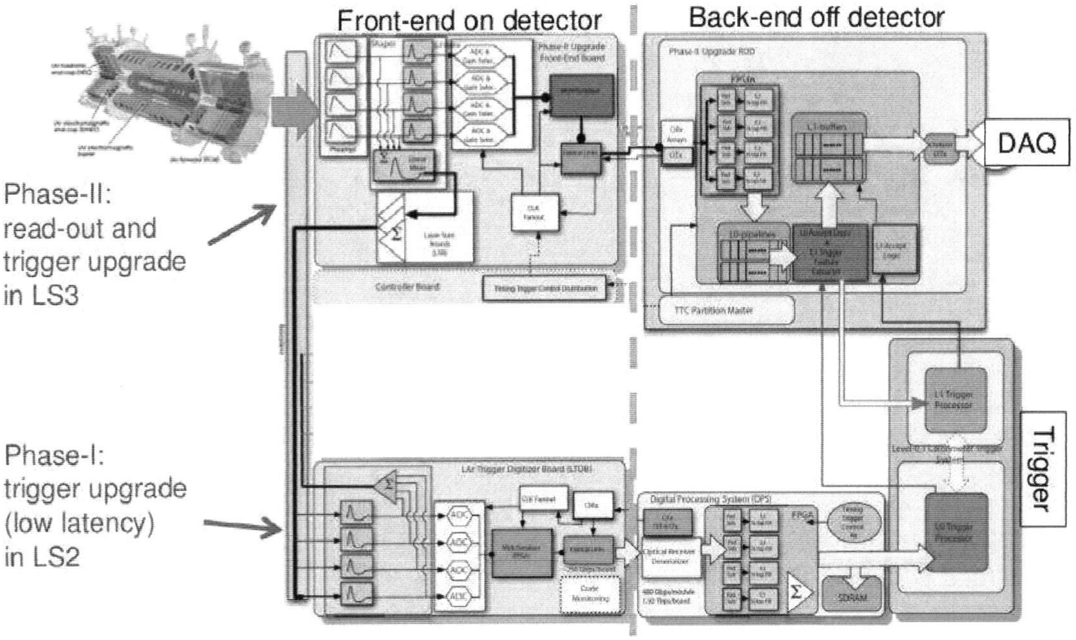

Figure 2. LS3 front-end readout architecture for the ATLAS liquid argon calorimeters. The calibration boards, which are not used during regular data-taking, are not shown. The LTDB will have been installed in LS2, superseding the original Tower Builder Board.

and will be applied at 5 Gbps in the LS2 upgrade. In addition, there are planned a custom interface and serializer VCSEL pairs from the CERN Versatile Link [4]. All together for LS3, a 12×10 Gbps CERN VCSEL arrays and an improved SoS process or a next-generation CERN GBTx [5] need to be developed.

The data processing will be done on the basis of LS2 upgrade R&D. There are 14 FEB2s connected to one Pre-processor board covering 32 towers of size $\Delta\eta \times \Delta\varphi = 0.1 \times 0.1$. The incoming data rates per board are estimated to be 14×90 Gbps of digitized input data. At the output, low-latency data transmission of about 250 Gbps and 5 to 10 Gbps is expected to the Level-0 and Level-1 trigger systems, respectively. This requires high-bandwidth ATCA Pre-Processor boards implemented on high performance FPGAs and treating 1.2 Tbps input from the FEB2s and delivery of summed signals to the trigger system. The new FEB2 should fulfill the required tasks: energy reconstruction, bunch-crossing identification, data pipeline and delivery of summed signals to the trigger system.

3. HEC Low Voltage Power Supplies and HEC Cold Electronics

The radiation doses in ATLAS including at the positions of the HEC ASICs have been studied originally for the design luminosity of the LHC [3]. To extract the Si NIEL fluences for the HL-LHC from these simulations the following assumptions have been made: 14 TeV proton-proton collisions with $\sigma_{inelastic} = 79.3$ mb and an integrated luminosity of 3000 fb^{-1} after 10 years of running the HL-LHC. A safety factor of 2 was used to account for uncertainties in the simulation. This was the result of dedicated measurements in critical areas of the detector in order to validate the simulated particle fluxes. Figure 3 (left) shows the expected ATLAS Si-NIEL fluences induced in the ASICs as a function of the location of the readout cell in the HEC which will be impacted by the corresponding ASIC.

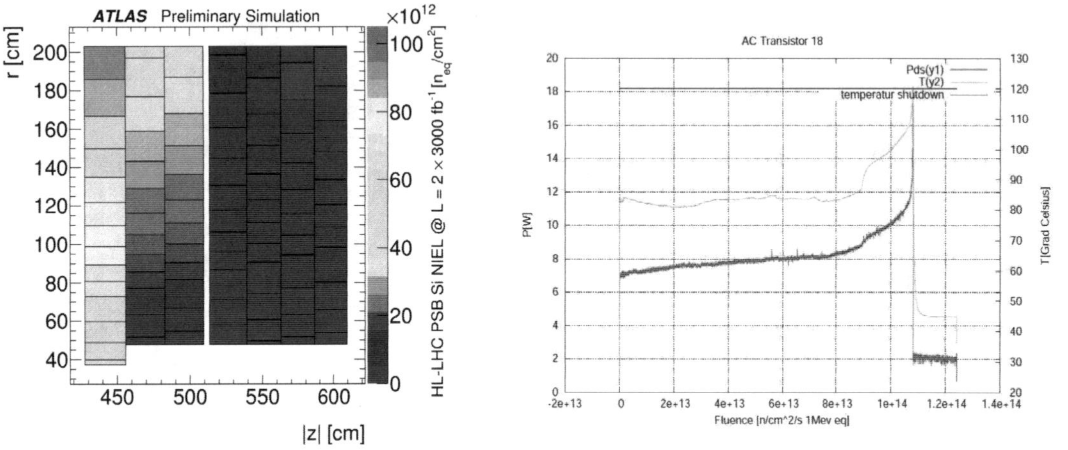

Figure 3.
Left : Si-NIEL fluence in ATLAS under HL-LHC conditions after 3000 fb[1] and with an applied safety factor of 2 to account for simulation uncertainties. The color coded fluences in the ASICs are shown at the r - z locations of the corresponding readout regions of the HEC. Right: Example measurements with a IXFH12N90P power MOSFET by IXYS showing the power loss (red) and temperature (green) of the device under test as a function of fluence up to 1.4×10^{14} n/cm^2.

Total ionizing dose (TID), single event effects (SEE) and neutron backgrounds measured in the 2012 run agree with background simulations using FLUKA and GEANT4. The electrical power required by the FE electronics is delivered by a Low Voltage Power Supply (LVPS). The expected neutron dose (from simulation) for HL-LHC for HEC LVPS exceeds 1.8×10^{12} n_{eq}/cm^2 including a safety factor two. For the HEC cold electronics it reaches up to 10^{14} n_{eq}/cm^2. Due to the limited radiation tolerance of the current electronics, the 8 HEC LVPS should be replaced. Radiation qualification of power MOSFETs and logic devices has thus been started.

All together, 16 types of power MOSFETs were evaluated by measuring power loss as function of fluence (up to 1.4×10^{14} n/cm^2), an example is shown in Figure 3 (right). All tested transistors qualify for DC/DC converter application with the new NIEL requirement. The updated radiation requirements may allow use of radiation-tolerant FPGAs in the digital part. Detector control electronics (Embedded Local Monitor Board - ELMB) also need be developed in radiation tolerant technology [6].

The readout electronics of the HEC will have to withstand larger radiation environment at the HL-LHC compared to their design values. The preamplifier and summing boards (PSBs) are equipped with GaAs ASICs. These ASICs were irradiated with neutrons and protons with fluences surpassing several times ten years of operation of the HL-LHC. The setup and measurements are described in [7].

The three different transistor technologies that were tested are Si CMOS FET in SGB25V 250 nm technology from IHP, SiGe Bipolar HBT (IHP SGB25V 250 nm and IBM 8WL- BiCMOS 130 nm), and the GaAs FET currently used in ATLAS, either the Triquint CFH800 250 nm transistors themselves or integrated into the HEC BB96 Preamplifiers (PAs) and Systems [8]. Later, the irradiated boards were re-tested in liquid nitrogen to simulate the conditions inside the HEC cryostats.

The investigation of the GaAs ASICs performance was studied as a function of neutron fluence with two parameters: the forward transmission coefficient of the Systems and the linearity of the PAs. The linearity of the PAs was characterized by a power-law fit to their response to triangular input pulses of varying amplitudes.

Figure 4 shows the result of the warm and cold investigations for the currently installed BB96 ASICs. The warm results show only a small degradation of the HEC electronics up to HL-LHC conditions, whereas the cold measurements indicate a much more severe performance

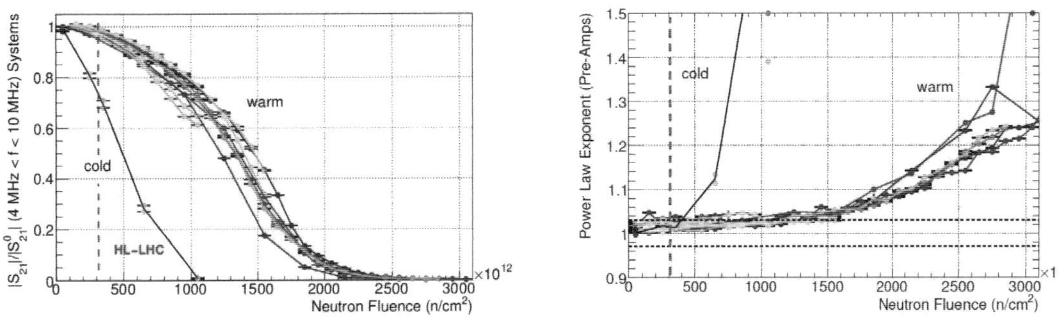

Figure 4. Forward transmission coefficient, |S21|, of the BB96 Systems normalized to the value before irradiation (left), and linearity of the BB96 PAs in terms of a power-law exponent (right). Dense data points connected by colored lines correspond to the in situ measurements at room temperature st the Nuclear Physics Institute (NPI) in *Řež*, Czech Republic, whereas the black line represents the cold results obtained at Max Planck Institute for Physics (MPP). The red dashed line indicates the HL-LHC limit (Si) (including the safety factor of 5, recently lowered to two), taking into account that 4.1×10^{14} n_{eq}/cm^2 NIEL in ATLAS corresponds (Si) to 3.0×10^{14} n_{eq}/cm^2 NIEL at NPI (for GaAs devices) [9].

degradation. The red line corresponds to HL-LHC conditions with safety factor 5. This has recently been reduced to 2, after further validation of the background simulations. With this reduced SF, results show that the GaAs ASICs are good for HL-LHC operation.

In addition to the problems associated with the increased radiation levels, the current electronics has to survive until the end of the HL-LHC phase and work for approximately 30 years. Therefore there is a need to estimate the longevity of the pre-amps [3]. Activities to measure mechanical and electrical stability of the chips and system are ongoing.

4. Forward Calorimeters

The Forward Calorimeter (FCal) was designed with very narrow LAr gaps in order to avoid problems due to ion build-up that would distort the electric field. At HL-LHC luminosities and at high η region, the gaps in the first FCal sections are no longer sufficiently narrow. In this area the pulse shapes are expected to degrade due to Ar^+ build-up, which leads to field and signal distortion, and due to increased current draws over the 1 or 2 MΩ protection resistors used to deliver the high voltage currents, which leads to a voltage drop. Due to high expected particle flux at 5×10^{34} cm^{-2} s^{-1}, a temperature increase of a liquid argon may lead to LAr bubble formation and has to be studied. Figure 5 shows the FCal cylindrical electrodes, parallel to the beamline, located within an absorber matrix. In the right figure the schematic structure of a unit cell, including the absorber matrix is shown. It comprises a cylindrical copper tube into which a metal absorber rod of slightly smaller diameter has been inserted. Each electrode consists of a rod at high voltage, a tube at ground, and a PEEK fibre that maintains the gap between the rod and tube. In the Figure 6 (left) the simulated combined effect of Ar^+ build-up and reduced HV on the nominally triangular physics pulses for the LAr cells is illustrated for cells at $\mid \eta \mid = 4.7$, for a range of instantaneous luminosities. The amplitude is reduced with increased luminosities. The main effect is coming from the HV drop due the higher currents flowing across the 1 or 2 MW HV protection resistors, which are also located within the cold volume of the endcap cryostats.

At a luminosity of 1×10^{34} cm^{-2} s^{-1} about 30 Watts of heat is expected to be deposited in each FCal and especially 18 W in first FCal module (FCal1), as shown in Figure 7 (left) the most left calorimeter. This heat flows radially to the cooling coils on the inner cylindrical surface of the

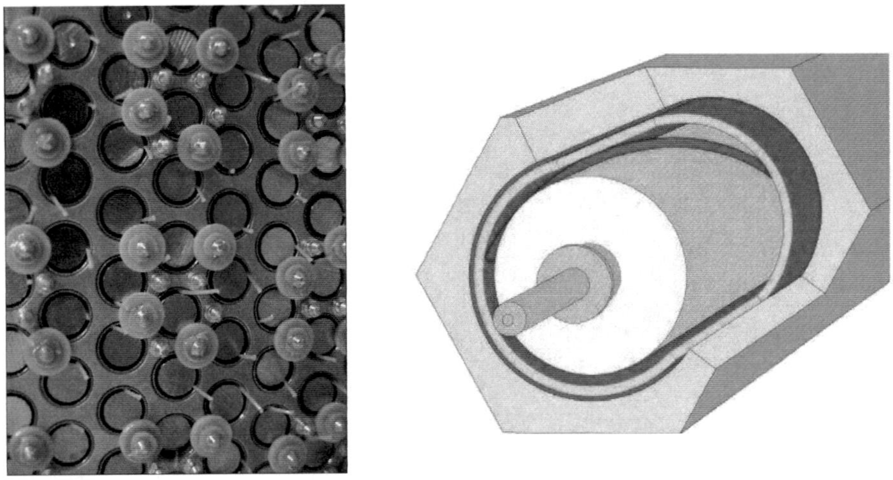

Figure 5. The ROD FCal matrix (left) and structure of a LAr unit cell in the FCal1 modules (right).

endcap cryostat cold vessel. The greatest impedance to this heat flow is at the 12 mm LAr-filled gap between the FCal support tube and the inner bore of the HEC. Heat simulations for FCal for 2012 luminosities agree well with temperature probe data. The measured temperature increase is 0.1 K. The simulation for 6×10^{34} cm^{-2} s^{-1} shows however a relative temperature increase of about 2.1 K, as shown in Figure 7 (right), which may lead to a bubble formation. To study the heat transport in the region between FCal and HEC a mock-up module is being prepared. The test module is displayed in Figure 6 (right). At the left is a heater plane which heats an aluminum plane representing the wall of the FCal support tube. Then the heat flow will go through an changeable liquid argon gap to an irregular copper surface constructed to resemble the inner bore of the HEC. The HEC will be cooled from the right side with liquid argon, since the copper structure of the HEC serves as main cooling source. In between temperature probes will be installed to measure temperature differences which will be used to determine the effective thermal conductivity of LAr gap. Finally the mock-up module can be rotated in the cryostat to simulate the different slices of the HEC inner ring. The results of this test will be used to validate the simulation used to predict the module heating.

To solve the problems of pulse degradation and bubble formation two options are considered as discussed below.

4.1. Upgrade Scenarios for the ATLAS Forward Region

The ATLAS Hilum Collaboration [10] has experimentally investigated the performance of the three endcap LAr calorimeters in a high-intensity test beam in Protvino. The LHC luminosity of 10^{34} cm^{-2} s^{-1} minimum bias events energies were translated to Protvino energies and obtained a corresponding beam intensity of 6.7×10^{8} p/s for the FCal(269), which is affected by large systematic errors. Their results indicate that while the present FCal1 performance would degrade significantly at high luminosity, a similar structure with an argon gap of 119 μm shows no signal degradation up to intensities an order of magnitude greater than those anticipated for the HL-LHC. This is shown in Figure 8 (right). There is no critical intensity observed within the intensity range studied in contrast to other FCal module.

High particle flux tests showed that a liquid argon calorimeter with this electrode structure can function at the rates expected in a Mini-FCal at the HL-LHC [10]. These narrow gap FCal electrode can therefore be employed in the sFCal and Mini-FCal designs.

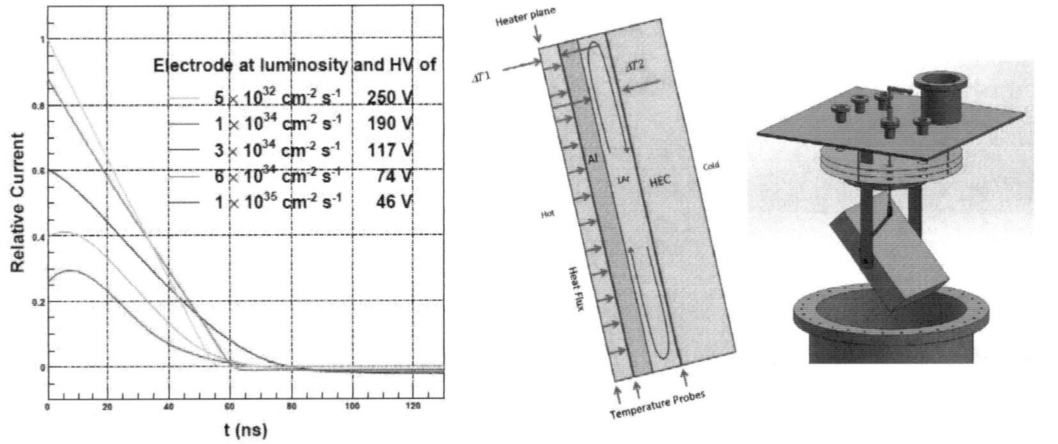

Figure 6. Left: Simulated FCal1 pulse shapes for different instantaneous luminosities, at $|\ \eta\ | = 4.7$,where the ionization rate is highest. The effects of the ion-buildup and HV sag are included. Middle: The mock-up module schematics. Right: The simulated module in the box to be installed in the cryogenic module.

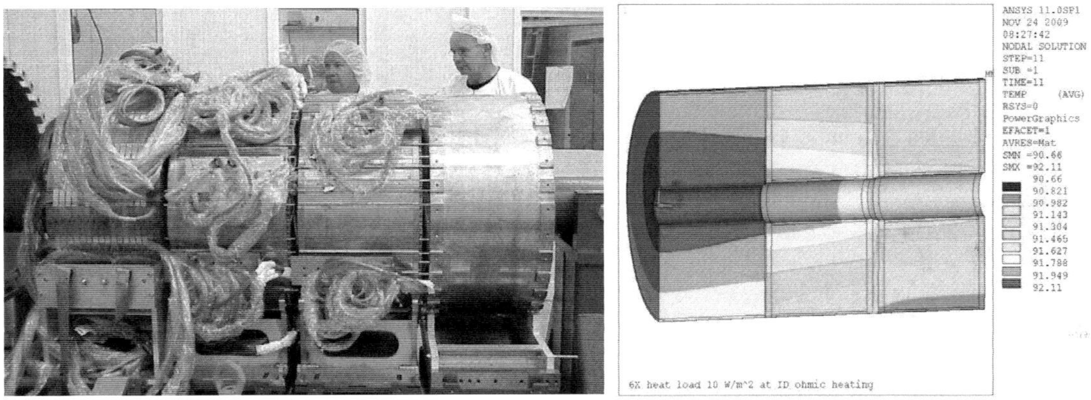

Figure 7. Left: The FCal modules from left to right: FCal1, FCal2, FCal3 and a plug (left). Right: The temperature distribution in the FCal modules (FCal1 at the left) for a luminosity of $6{\times}10^{34}$ cm^{-2} s^{-1}.

4.1.1. The sFCal Option Replacement of the current FCal with the sFCal is one of the options. This option is easier due to known technology to design and optimize the sFCal. It is basically a modified version of FCal with smaller LAr gaps ($\sim$ 100 μm in the FCal1), new cold electronics boards (which hold the HV resistors), and cooling loops. This option is described in [3]. But the installation of the sFCal requires the opening of the cold volume of the endcap cryostats which are welded shut, so protection of the detector is also a major issue. This procedure is complicated and can be done only in a long shutdown. In addition, radiation protection issues are to be taken into account.

4.1.2. The Mini-FCal Option The Mini-FCal option was designed to handle the problems expected at the HL-LHC without the need to open the endcap cryostats and thus reducing the potential risk to the detector. The idea is to install a small calorimeter just in front of the current FCal. This would result in a shorter and simpler installation procedure than opening cryostats. Reduced time on working on the detector would result in a lower radiation exposure

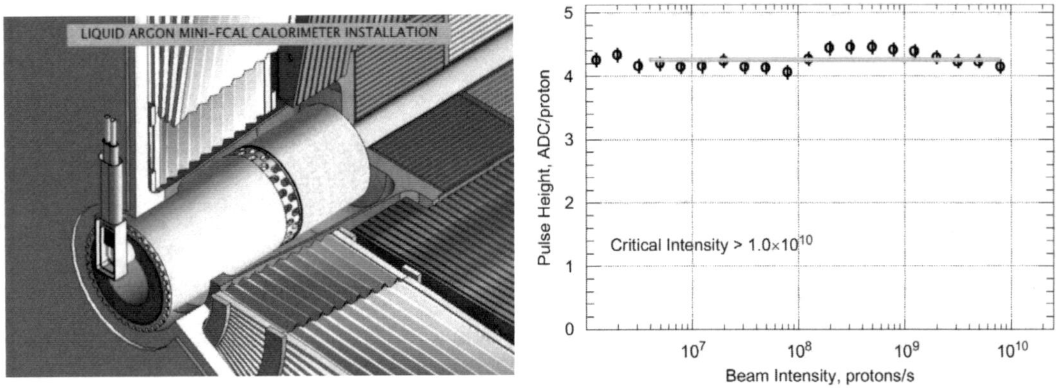

Figure 8. Left: A schematic view of the LAr Mini-FCal option (light blue-green) located in the vacuum between a new replacement inner warm tube and the cold wall of the cryostat. The cold vessel of the existing cryostat consists of the cone that surrounds the cryostat, the bulkhead between the cryostat and FCal1 and the small cold tube that runs inside of the FCals. Right: Dependence of the FCal signal on Protvino beam intensity (119 μm argon gap). The solid line shows a fit to the data.

to installers. Simulations of this option are ongoing to show the reduction both the heat load and the instantaneous ionization rate in the FCal1 to acceptable levels. A schematic view of one Mini-FCal design, located between an inner warm tube and the cold wall of the cryostat, is shown in Figure 8 (left). The Mini-FCal detector must be extremely radiation hard and several technologies have been considered. Currently preferred is an FCal1-like LAr calorimeter with copper absorbers and the same 100 μm gaps proposed for the sFCal.

The disadvantage of this option is that it will require additional material upstream of the detector due to the necessity of an additional cryostat for Mini-FCal and by the associated services. This has the potential to degrade the overall detector performance. Design optimization and performance simulations for the LAr Mini-FCal option are ongoing.

5. Conclusions

In order to survive the expected luminosity at the HL-LHC, the ATLAS experiment will be upgraded to meet the experimental challenges: increased data rates, occupancy and radiation fluence. The goal of HL-LHC upgrade is to meet the challenging pile-up and radiation requirements, to ensure longevity of the system and to maintain or improve the existing detector performance. The LAr front-end and back-end electronics will be replaced. The current R&D for LS2 trigger readout developments is ongoing and will naturally continue. There are many activities ongoing to clarify the longevity and performance of critical components in the LAr endcap area, such as, HEC cold electronics, FCal detector and its possible upgrade with new sFCal or Mini-FCal detectors. A decision on which upgrade path will be selected for the forward region and for the endcap calorimetry is planned to be completed by 2015 in order to be prepared for a construction phase well ahead of the LS3 and the beginning of the HL-LHC phase.

References

[1] G. Aad et al., The ATLAS experiment at the CERN Large Hadron Collider, Journal of Instrumentation, vol. 3, no. 08, p. S08003, 2008.

[2] L. Rossi and O. Bruning, "High Luminosity Large Hadron Collider: A description for the European Strategy Preparatory Group", CERN-ATS-2012-236.

[3] ATLAS Collaboration, "Letter of Intent for the Phase-II Upgrade of the ATLAS Experiment", CERN, Geneva, CERN-LHCC-2012-022. LHCC-I-023, Dec. 2012.

[4] L. Amaral et al., The Versatile Link, "A Common Project For Super-LHC", 2009 JINST 4 P12003.

[5] P. Moreira et al., "The GBT project", proceedings of the Topical Workshop on Electronics for Particle Physics (TWEPP 2009), Paris, France, CERN-2009-006.

[6] B. Hallgren, H. Boterenbrood, H.J. Burckhart, H. Kvedalen, "The embedded local monitor board (ELMB) in the LHC front-end I/O control system", CERN and NIKHEF, Amsterdam, Sep 2001, Prepared for Conference: C01-09-10.12, p. 325-330 Proceeding

[7] F Adamov et al., "Upgrade plans for the Forward and Hadronic-Endcap Calorimeter of ATLAS for the high luminosity stage of the LHC", ATL-COM-LARG-2013-002

[8] J. Ban, H. Brettel, W. Cwienk, J. Fent, L. Kurchaninov, et al., "Cold electronics for the liquid argon hadronic endcap calorimeter of ATLAS", Nucl.Instrum.Meth. A556 (2006) 158168. doi:10.1016/j.nima.2005.10.114.

[9] S. Menke, "Proton and Neutron Irradiation Tests Readout Electronics of the ATLAS Hadronic Endcap Calorimeter", Proceedings of the IEEE Nuclear Science Symposium, Anaheim, October 2012.

[10] A. Glatte et al., "Liquid argon calorimeter performance at high rates", *Nucl. Instrum. Meth.* **A669**, 47-65, 2012.

16th International Conference on Calorimetry in High Energy Physics (CALOR 2014) IOP Publishing
Journal of Physics: Conference Series **587** (2015) 012019 doi:10.1088/1742-6596/587/1/012019

Upgraded Trigger Readout Electronics for the ATLAS LAr Calorimeters for Future LHC Running

Hong Ma, On behalf of the ATLAS Liquid Argon Calorimeter Group
Physics Department, Brookhaven National Laboratory, Upton, NY 11973, USA
E-mail: hma@bnl.gov

Abstract.
The ATLAS Liquid Argon (LAr) calorimeters produce almost 200K signals that are digitized and processed by the front-end and back-end electronics for every triggered event. Additionally, the front-end electronics sums analog signals to provide coarse-grained energy sums to the first-level (L1) trigger system. The current design was optimized for the nominal LHC luminosity of $10^{34}\text{cm}^{-2}\text{s}^{-1}$. In order to retain the capability to trigger on low energy electrons and photons when the LHC is upgraded to higher luminosity, an improved LAr calorimeter trigger readout is proposed and being constructed. The new trigger readout system makes available the fine segmentation of the calorimeter at the L1 trigger with high precision in order to reduce the QCD jet background in electron, photon and tau triggers, and to improve jet and missing E_{T} trigger performance. The new LAr Trigger Digitizer Board is designed to receive the higher granularity signals, digitize them on-detector and send them via fast optical links to a new Digital Processing System. The reconstructed energies of trigger readout channels after digital filtering are transmitted to the L1 system, allowing the extraction of improved trigger signatures. This contribution presents the motivation for the upgrade, the concept for the new trigger readout and the expected performance of the new trigger, and describes the components being developed for the new system.

1. Introduction

The successful operation of the ATLAS detector [1] at the Large Hadron Collider (LHC) during the first data taking period (Run 1, 2009-2013) has demonstrated its extraordinary capability for proton-proton and heavy ion physics. During the next period (Run 2, 2015-2018), the LHC is expected to reach the design luminosity and energy. This will be followed by a series of upgrades over the next decade. The first phase of the upgrade (Phase-I) is expected to be completed in 2018-2019, bringing the instantaneous luminosity to $2-3\times10^{34}\text{cm}^{-2}\text{s}^{-1}$, exceeding the LHC nominal design luminosity. ATLAS is constrained by readout dead time to a 100 kHz event rate at Level-1 (L1) trigger for most of the front-end electronics. The requirement for studying physics processes that depend on low p_{T} lepton triggers dictates that the L1 trigger needs to be significantly improved to handle the increased background rates to lepton triggers when the detector is operated in the high pile-up environment with an average number of collisions per crossing $\langle\mu\rangle$ of 55 at $\mathcal{L}=2\times10^{34}\text{cm}^{-2}\text{s}^{-1}$. The upgrade of the Liquid Argon (LAr) calorimeter trigger readout aims at retaining the capability of triggering on low transverse energy (E_{T}) electromagnetic (EM) objects when the detector is operated during the third period (Run 3, 2019-2022) after the Phase-I upgrade.

After describing the current LAr calorimeter readout and trigger, the expected performance of the upgraded triggers for electrons and jets, and the technical implementation of the new trigger readout are summarized. The details of the design are described in the Technical Design Report [2] for ATLAS LAr Calorimeter Phase-I Upgrade, which was presented to and endorsed by CERN LHCC in Dec 2013, and approved by the CERN Research Board in March 2014.

2. The current LAr readout and calorimeter trigger system

The ATLAS LAr calorimeters produce almost 200K signals, which are digitized and processed by the front-end and back-end electronics for every triggered event. In addition, the front-end electronics sums analog signals of the readout channels to provide coarse-grained energy sums to the L1 calorimeter trigger system. The current design was optimized for the nominal LHC luminosity of $10^{34}\text{cm}^{-2}\text{s}^{-1}$. The trigger readout consists of electromagnetic and hadronic trigger towers, each with granularity of $\Delta\eta \times \Delta\phi = 0.1 \times 0.1$ for most parts of the calorimeter, with a quantization scale of 1 GeV. EM objects are identified as clusters of energy with fixed size in the grid of trigger towers, with possible isolation energy requirements around the objects and a hadronic energy veto. Jets are identified on a grid of even coarser granularity of $\Delta\eta \times \Delta\phi = 0.2 \times 0.2$.

The trigger rate for a single 25 GeV EM object is estimated to be 120 kHz at $3 \times 10^{34}\text{cm}^{-2}\text{s}^{-1}$ with the L1 trigger of Run 1. A general trigger menu that enables a broad physics program implies that no more than 20 kHz can be allocated to single EM object triggers at the L1. This would require an E_{T} threshold of 50 GeV (or 35 GeV with isolation requirements) and even higher at higher luminosity. It would significantly reduce the acceptance for a variety of important physics measurements and searches based on EM objects, such as Higgs production in association with a W or Z boson, which are tagged by the leptonic decays of the W's and Z's.

The ability to reject the background to EM objects in offline reconstruction takes advantage of the high granularity of the calorimeter readout. The shower shapes of EM objects differ significantly from the background from jets. Discriminating variables, such as the ratio of the energy in 3×7 cells over the energy in 7×7 cells centered at the cluster position and lateral shower width within a window of 3×5 cells have been successfully used in the offline analysis.

The design of the LAr trigger readout upgrade is based on improving the granularity of the information and the precision of the energy measurements at the L1 within the constraint of the current front-end electronics, such that shower shape variables are calculated within the latency to reject jet backgrounds for EM triggers.

3. Configuration of the new trigger readout and its expected performance

The ATLAS LAr EM calorimeter is highly segmented, as illustrated in Fig. 1. The calorimeter cells in each sampling layer have different size by construction. The trigger tower signals are created through several stages of on-detector analog electronics: in the shaper ASICs, in the Layer Sum Boards on the Front-End Boards, and in the Trigger Tower Boards installed on-detector. Typically, a total of 64 read-out channels are summed to form an EM trigger tower.

The new finer granularity scheme is based on "Super-Cells ", which provide information in each calorimeter layer with finer segmentation ($\Delta\eta \times \Delta\phi = 0.025 \times 0.1$ in the front and middle layers of the EM calorimeter for $|\eta| \leq 2.5$). This scheme is detailed for the EM barrel calorimeter in Table 1 and is illustrated in Fig. 2. For the strip and middle layers, where the bulk of the EM shower energy is deposited, this represents a factor of 4 higher transverse granularity.

The digitization precision of the Super-Cell signals is improved by at least a factor of 4 compared to the existing L1 system: the quantization scale and the dynamic range of the digitizers are optimized in each η-region and for each layer of the calorimeter to achieve sensitivities at the level of the Super-Cell electronic noise or better. The transverse energy deposited in the Super-Cells is calculated at each bunch crossing through optimized

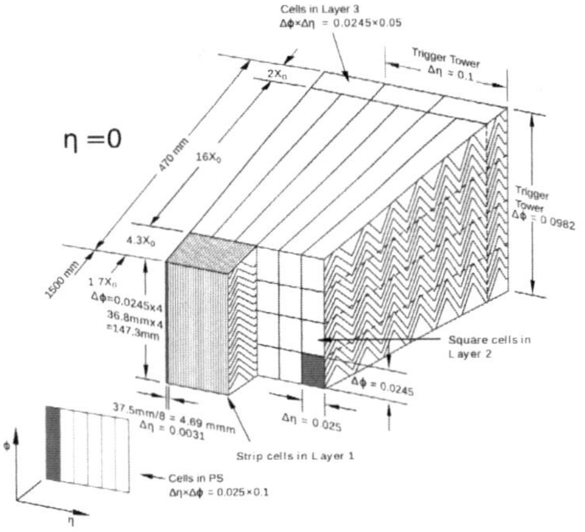

Figure 1. The segmentation of the readout cells in the LAr EM Barrel calorimeter.

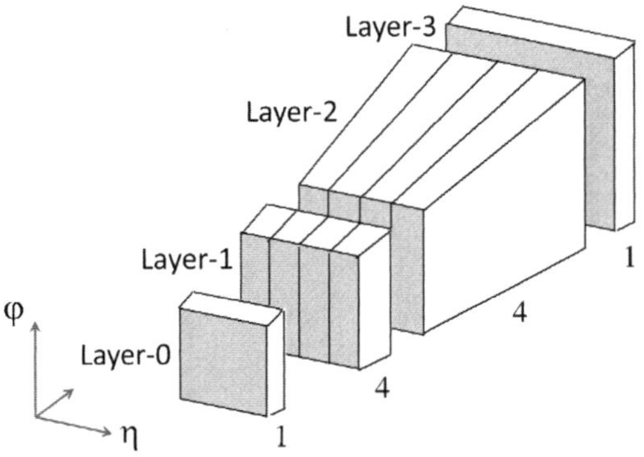

Figure 2. The segmentation of the Super-Cells of the EM calorimeter.

Table 1. Comparison of the current Trigger Tower granularity vs. the proposed Super-Cell granularity in the LAr EM barrel calorimeter, in terms of both elementary cells and $\Delta\eta$ and $\Delta\phi$. The number of elementary cells grouped for the trigger readout in η and ϕ are indicated by n_η and n_ϕ, respectively.

	Layer	Elementary Cell	Trigger Tower		Super-Cell	
		$\Delta\eta \times \Delta\phi$	$n_\eta \times n_\phi$	$\Delta\eta \times \Delta\phi$	$n_\eta \times n_\phi$	$\Delta\eta \times \Delta\phi$
0	Presampler	0.025 × 0.1	4 × 1	0.1 × 0.1	4 × 1	0.1 × 0.1
1	Front	0.003125 × 0.1	32 × 1		8 × 1	0.025 × 0.1
2	Middle	0.025 × 0.025	4 × 4		1 × 4	0.025 × 0.1
3	Back	0.05 × 0.025	2 × 4		2 × 4	0.1 × 0.1

algorithms. Techniques similar to the optimal filtering currently implemented in the LAr online signal processing will provide results close to the energy resolution obtained from the offline reconstruction.

Preliminary studies, shown in Fig. 3, demonstrate the significant improvement in the energy resolution using the layer-weighted 3 × 2 Super-Cell cluster E_T^{SC} over the existing L1 E_T^{L1}, when the measurements are compared to the offline reconstructed cluster energy, E_T^{offline}. This resolution improvement translates into an improved efficiency for EM objects in the trigger for a given offline E_T, sharpening the turn-on curve and thus making it possible to lower the offline E_T threshold by several GeV for the same trigger rate.

Similarly, the improved energy resolution of the LAr calorimeter upgraded trigger electronics allows for an improvement in the energy resolution of jets reconstructed in the L1 trigger system. Figure 4 compares the trigger efficiency as a function of the highest p_T jet in QCD dijet events between the current and the upgraded systems for a online L1 jet threshold of 20 GeV.

The finer granularity of the Super-Cells enables a more sophisticated analysis to reject jet backgrounds to EM triggers than in the current system through the use of shower shape variables. A few discriminating variables have been investigated:

$\boldsymbol{R_\eta}$ Given a 3×2 group of Super-Cells in $\eta \times \phi$ centered on the highest-energy Super-Cell in the middle layer, R_η is defined as the transverse energy measured in the 3×2 group divided by

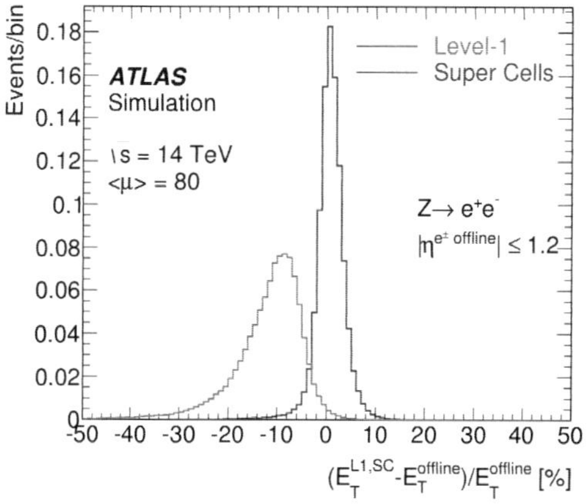

Figure 3. Difference in percent between $E_{\mathrm{T}}^{\mathrm{L1}}$ and $E_{\mathrm{T}}^{\mathrm{offline}}$ (red) compared to that between $E_{\mathrm{T}}^{\mathrm{SC}}$ and $E_{\mathrm{T}}^{\mathrm{offline}}$ (blue) for simulated $Z \rightarrow e^+e^-$ events in the LAr EM barrel calorimeter. The electron candidates are required to be within $|\eta| \leq 1.2$.

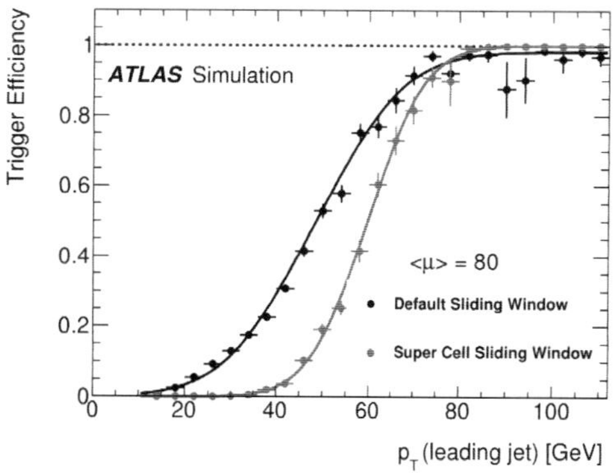

Figure 4. The trigger efficiency as a function of the highest p_{T} offline jet for $\langle\mu\rangle = 80$ in simulated QCD dijet events for the default sliding window algorithm (black points) and the sliding window algorithm based on Super-Cells (red points) for jets within $|\eta| < 2.5$.

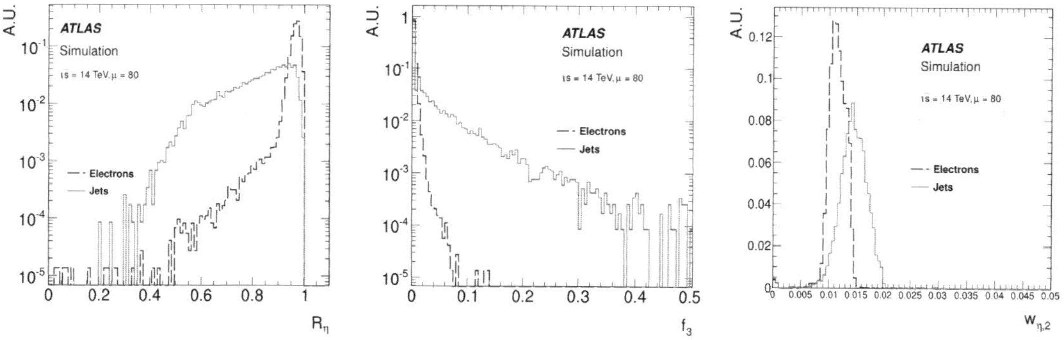

Figure 5. Distributions of variables allowing to distinguish between electrons (black) and jets (red) with $p_{\mathrm{T}} > 20$ GeV and with the upgraded L1 trigger: R_η (left), f_3 (middle) and $w_{\eta,2}$(right). Each distribution is normalized to unit area.

the transverse energy measured in a 7×2 group;

$\boldsymbol{f_3}$ The ratio of the transverse energy measured in the back EM layer in an area of size $\Delta\eta \times \Delta\phi = 0.2 \times 0.2$ to that deposited in all three layers for an EM cluster; the energies in the front and middle EM layers are reconstructed in the area $\Delta\eta \times \Delta\phi = 0.075 \times 0.2$;

$\boldsymbol{w_{\eta,2}}$ The spread of the shower in the middle EM layer in a 3×2 Super-Cell region, defined as an energy weighted root-mean-squared.

In addition, hadronic isolation ($E_{\mathrm{had}}^{\mathrm{core}} \leq 1$ GeV) can be required for each electron candidate, where $E_{\mathrm{had}}^{\mathrm{core}}$ is defined as the transverse energy deposited in a $\Delta\eta \times \Delta\phi = 0.2 \times 0.2$ region of the hadronic calorimeter behind the EM shower. A multi-dimensional optimization is performed to minimize the trigger rate while keeping a high signal efficiency.

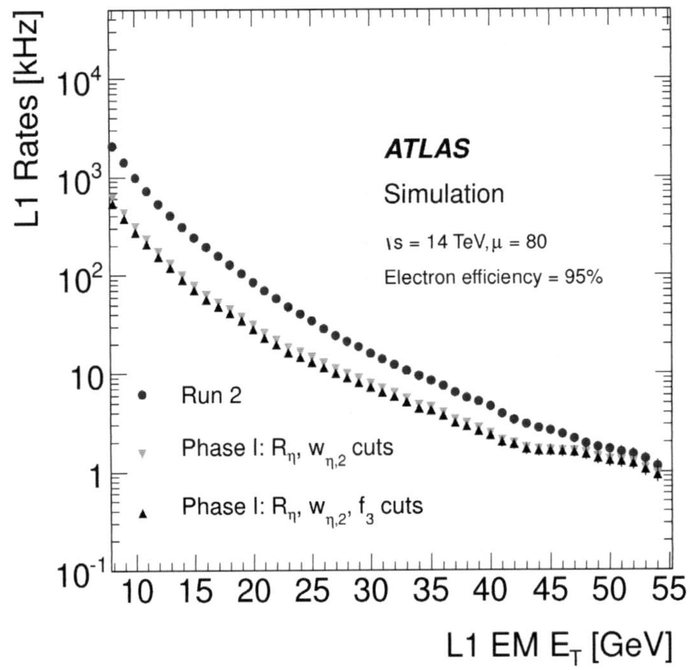

Figure 6. L1 trigger rates for a 95% electron efficiency as a function of the EM E_T threshold assuming Run 2 conditions (blue points) and for Run 3 conditions, at $\mathcal{L} = 3 \times 10^{34}$ cm^{-2} s^{-1}, for two sets of cuts (green and black triangles), as measured from a sample of simulated minimum bias events with $\langle\mu\rangle = 80$.

Figure 6 provides a comparison of the Run 2 L1 trigger performance to that of the upgraded L1 trigger for a 95 % electron efficiency in both cases. It shows that for a given bandwidth of e.g., 20 kHz, the L1 threshold, which would need to be 28.5 GeV assuming Run 2 conditions, could be lowered by 7 GeV to 21.5 GeV with the proposed upgrade.

4. Design of front-end readout and signal processing

The schematic of the LAr calorimeter readout with the upgraded trigger readout is shown in Fig. 7. The current readout components will remain mostly intact. The new components are indicated by the red outlines and arrows on the diagram. The analog signal sums for the trigger will be modified such that the sums for Super-Cells will be formed first, which are then further summed into trigger towers so that the current L1 trigger will continue to function. The new components include new layer sum boards on the existing front-end boards to produce finer granularity Super-Cell signals for each layer, and new base planes in front-end crates for analog signal transfer. New LAr Trigger Digitizer Boards (LTDB) receive and digitize analog Super-Cell signals and transmit digital Super-Cell signals to the backend, where a LAr Digital Processing System (LDPS) extracts transverse energy values for each Super-Cell at every bunch crossing and transmits data to the L1 calorimeter trigger system.

Each LTDB will process up to 320 Super-Cell signals. The signals will be digitized at 40 million samples per second by 12-bit ADCs on the LTDB after one additional stage of shaping. In the current design, the least significant bit is 32 MeV for the front layer, and 125 MeV for

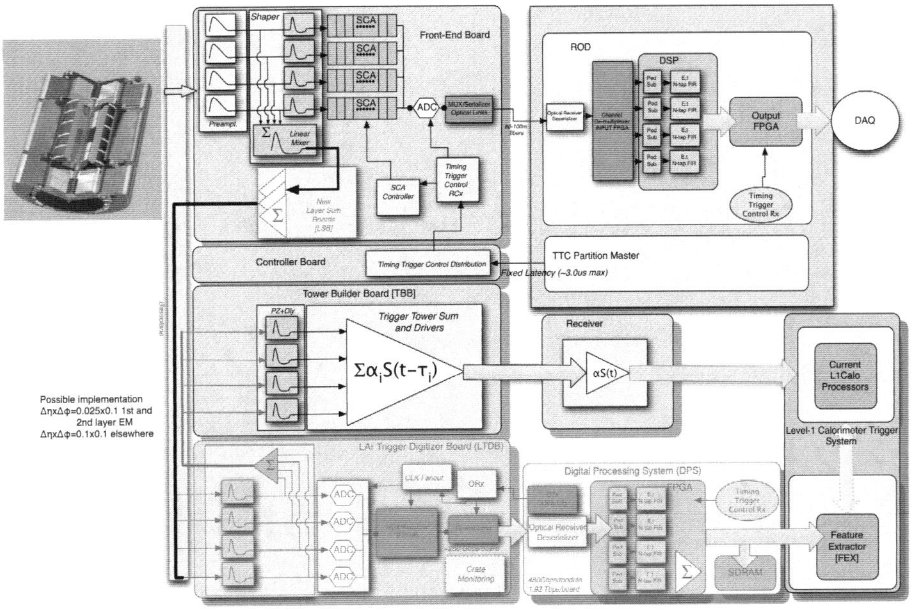

Figure 7. Schematic of Liquid Argon Calorimeter readout with the upgraded trigger readout.

the middle layer, in E_{T}. Both custom designed ADC and commercial off-the-shelf option are being considered. Data of 8 ADC channels will be serialized and transmitted at 5.12 Gbps to the backend. Each LTDB will typically have 40 fibers. A total of 124 LTDBs will be needed, delivering 25 Tbps digital information to the LDPS.

The LDPS is made up of 31 LAr Digital Processing Blades (LDPBs) housed in three ATCA shelves. Each LDPB consists of one carrier board equipped with four Advanced Mezzanine Cards (AMCs), for a total of 124 AMCs. The AMC is designed around a powerful FPGA with high-speed transceivers that will process the data of up to 320 Super-Cells within the latency budget (17 bunch crossings for LDPS). In addition to receiving $\sim$ 25 Tbps for 34,000 Super-Cells, the LDPS transmits $\sim$ 41 Tbps over optical fibers to the L1 calorimeter trigger system, within which the physics objects such as electrons, photons, taus and jets are identified [3].

5. Summary

ATLAS has designed a new LAr calorimeter trigger readout in preparation for its operation after LHC Phase-I upgrade. The new system makes available the fine segmentation of the calorimeter at the trigger level with high precision in order to reduce the QCD jet background in electron, photon and tau triggers, and to improve jet and missing E_{T} trigger performance. It will enable ATLAS to retain low trigger thresholds when it is operated at luminosity of $3 \times 10^{34}\mathrm{cm}^{-2}\mathrm{s}^{-1}$. The upgraded trigger readout is planned to be used as the low-latency Level-0 trigger after the Phase-II upgrade, during which a fully digital LAr front-end readout will be put in place [4].

References

[1] ATLAS Collaboration, The ATLAS Experiment at the CERN Large Hadron Collider, 2008 JINST 3 S08003
[2] ATLAS Collaboration, ATLAS Liquid Argon Calorimeter Phase-I Upgrade Technical Design Report, CERN-LHCC-2013-017, `https://cds.cern.ch/record/1602230`
[3] ATLAS Collaboration, Technical Design Report for the Phase-I Upgrade of the ATLAS TDAQ System, CERN-LHCC-2013-018, `https://cds.cern.ch/record/1602235`
[4] O. Novgorodova, The ATLAS liquid argon calorimeter: upgrade plans for the HL-LHC, these proceedings

Upgrade of the ATLAS Tile Calorimeter Electronics

F Carrió [1], on behalf of the ATLAS Tile Calorimeter System

[1]Instituto de Física Corpuscular (CSIC-UV)

E-mail: fernando.carrio@cern.ch

Abstract. The Tile Calorimeter (TileCal) is the hadronic calorimeter covering the central region of the ATLAS experiment at LHC. The TileCal readout consists of about 10000 channels. The bulk of its upgrade will occur for the High Luminosity LHC phase (Phase-II) where the peak luminosity will increase 5 times compared to the design luminosity (10^{34} $cm^{-2}s^{-1}$) but with maintained energy (i.e. 7+7 TeV). An additional increase of the average luminosity with a factor of 2 can be achieved by luminosity levelling. This upgrade is expected to happen around 2024. The TileCal upgrade aims at replacing the majority of the on- and off-detector electronics to the extent that all calorimeter signals will be digitized and sent to the off-detector electronics in the counting room. To achieve the required reliability, redundancy has been introduced at different levels. Three different options are presently being investigated for the front-end electronic upgrade. Extensive test beam studies will determine which option will be selected. 10 Gbps optical links are used to read out all digitized data to the counting room while 5 Gbps down-links are used for synchronization, configuration and detector control. For the off-detector electronics a pre-processor (sROD) is being developed, which takes care of the initial trigger processing while temporarily storing the main data flow in pipeline and derandomizer memories. One demonstrator prototype module with the new calorimeter module electronics, but still compatible with the present system, is planned to be inserted in ATLAS this year.

1. Introduction

The Tile Calorimeter [1] (TileCal) is a sampling hadronic calorimeter which covers the most central region of the ATLAS [2] experiment at CERN. TileCal detector is made out of steel absorber plates and plastic scintillator and it is divided in three cylindrical parts along the beam axis as figure 1a shows: a central Long Barrel (subdivided in Long Barrel A and Long Barrel C) and two Extended Barrels (Extended Barrel A, Extended Barrel C). Each barrel is composed of 64 modules which are divided in cells. The PhotoMultiplier Tubes (PMTs) and on-detector electronics used to read out the cells are hosted in the outermost part of the modules, called drawer, as figure 1b shows. TileCal contains a total of 256 drawers comprising ~10000 readout channels.

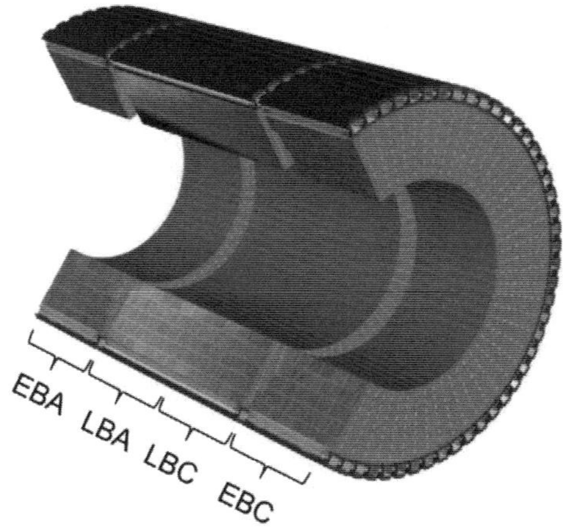

Figure 1a. Tile Calorimeter detector

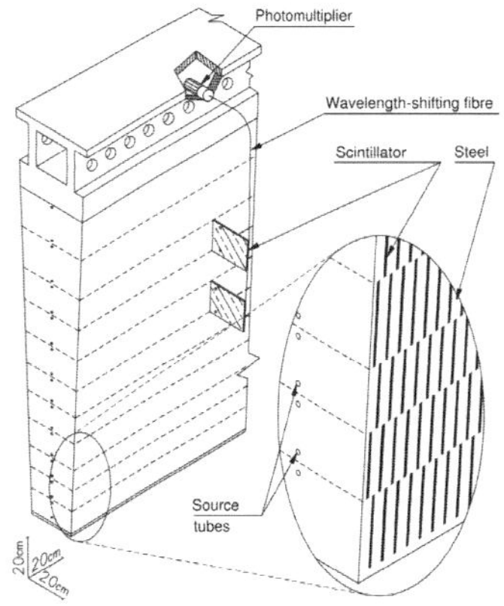

Figure 1b. TileCal module

Light generated by charged particles crossing the plastic scintillating tiles are collected by wavelength shifting fibres and carried to the PMTs. The analog signals produced by PMTs are conditioned and digitized by the on-detector electronics every 25 ns, while analog tower sums are transmitted to the Level-1 trigger system for trigger decision. Once the on-detector electronics have received the Level-1 trigger accept signal, the digitized data for the selected event is transmitted to the off-detector electronics.

In the off-detector electronics, the Read-Out Drivers (ROD) [3] receive the digitized data coming from the on-detector electronics at a maximum Level-1 trigger rate of 100 kHz. RODs are in charge of performing data processing tasks in real time and sending the processed data to the Read-Out Buffers (ROB) in the Level-2 trigger. A total of 32 ROD modules are needed to read out the ~10000 channels.

Figure 2 shows the dataflow of the current readout architecture.

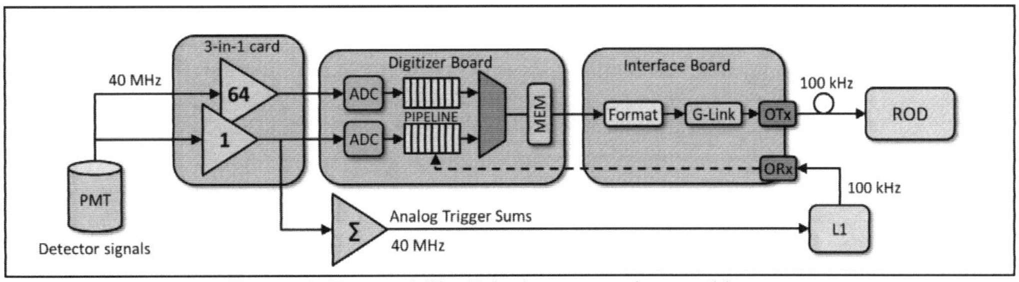

Figure 2. Present Tile Calorimeter readout architecture

2. Tile Calorimeter Upgrade plans for HL-LHC

The Phase-II Upgrade [4] of the LHC, called High Luminosity LHC (HL-LHC), is scheduled around 2024 and plans to increase the luminosity to a value of around $5\text{-}7\cdot10^{34}\text{cm}^{-2}\text{s}^{-1}$ with a pileup close to 200 events in order to reach a total integrated luminosity of 3000 fb^{-1}.

A full redesign and replacement of the readout electronics of TileCal is mandatory due to the increase of radiation level and the implementation of a new readout architecture with a full-digital trigger system raising the total bandwidth from 165 Gbps up to 80 Tbps. With the purpose of operating with high radiation doses, the new readout electronics system will include redundant optical fibers between on- and off-detector electronics, radiation-tolerant electronics, redundant low voltage power supplies and will use reliable protocols with error correction for data transmission.

The readout architecture for Phase-II points to a full digital readout where the on-detector electronics will transmit digital data from all the channels to the off-detector electronics for every bunch crossing. Then, the off-detector electronics will pre-process and transmit the data to the first level of trigger with improved precision and granularity.

Figure 3 shows a diagram of the upgraded TileCal readout architecture for the HL-LHC.

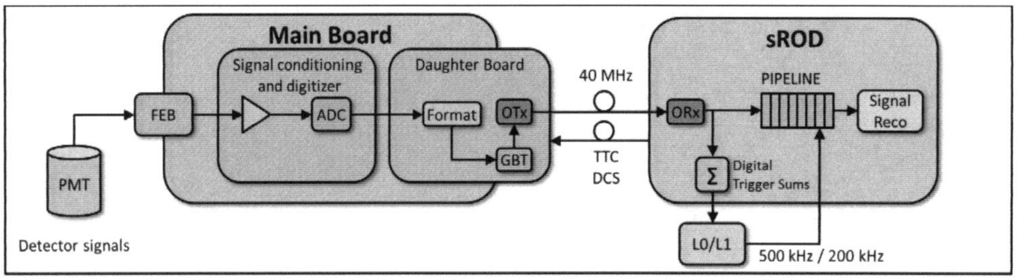

Figure 3. Upgraded Tile Calorimeter readout architecture for HL-LHC

2.1. The Tile Calorimeter Demonstrator Project

The Tile Calorimeter Demonstrator Project activities are intended to evaluate and qualify the new electronics before the complete replacement of the on- and off-detector electronics for Phase-II Upgrade. These activities not only comprise the development and test of electronics which fulfil the future conditions, but novel trigger algorithms, monitoring and control software and new mechanics.

Furthermore, the Demonstrator Project plans to replace one of the TileCal drawers with the new electronics and install it into the detector during the Long Shutdown 1 (LS1) scheduled for 2013-14. Since the interface of the current Level-1 trigger system is analog, the Demonstrator drawer will provide analog trigger signals keeping backward compatibility with the present system. Due to this functionality, this upgraded drawer is so-called hybrid drawer.

3. Tile Calorimeter electronics for HL-LHC

3.1. On-detector Electronics

As figure 3 shows, Front-End Boards (FEB) are connected directly to the output of the PMTs and provide conditioned signals to the readout electronics in the Main Board, integrated data for minimum bias monitoring and electronics for the calibration of PMT gains. The Daughter Board is responsible for gathering and transmitting the digitized data to the off-detector electronics.

3.1.1. Front-End Boards

Three FEB options are being evaluated for the HL-LHC. The first FEB option is an improved version of the present 3-in1 card [5] designed by the Enrico Fermi Institute of the University of Chicago. This FEB has been designed using Commercial Off-The-Shelf components and provides analog outputs with two different gains, as well as calibration capabilities for PMTs. The modified 3-in-1 card shows better linearity and lower noise than the current version and has passed successfully the radiation tests.

The second option is a custom ASIC called Charge Integrator and Encoder (QIE)[6], developed in joint collaboration between the Argonne National Laboratory (ANL) and Fermilab. The QIE ASIC divides the signal coming from the PMTs with a current splitter followed by a gated integrator providing four different ranges (16/23, 4/23, 2/23, 1/23). A 6-bit flash ADC digitizes the selected range covering a dynamic range of 17 bits. The QIE version 10.5 also includes a Time-to-Digital Converter (TDC) with sub-nanosecond resolution for out-of-time pulse identification.

The third option for the FEB is an ASIC called Front-end for ATLAS TileCal Integrated Circuit (FATALIC) [7] developed by The Laboratoire de Physique Corpusculaire de Clermont-Ferrand (LPC). FATALIC includes a 3-gain current conveyor associated with shaping stages which covers the full dynamic range of the PMT signal. Three 12-bit ADCs, designed by LPC and called Twelve bits ADC for s-ATLAS TileCal Integrated Circuit (TACTIC), read out the conditioned signals from the current conveyor stage. Both chips have been designed using the IBM 130 nm CMOS technology. Final version of FATALIC will combine both ASICs functionalities in a single chip.

The modified 3-in-1 card has been selected as FEB for the hybrid drawer, since it is the unique FEB which provides analog outputs required for the current trigger system.

3.1.2. Main Board

The Main Board is responsible for the control, monitoring and readout of the FEBs and low voltage power supplies, and provides a digitized data to the Daughter Board. Three different types of Main Board will be designed in order to evaluate the three FEB options.

The Enrico Fermi Institute has designed the Main Board that will be used in the Demonstrator Project which is able to host and manage up to 12 modified 3-in-1 cards, so one drawer will include four Main Boards. This Main Board includes 4 Altera Cyclone IV FPGAs for control and configuration purposes, and 12-bit Linear Technology LTC 2264 ADCs for the digitization of the analog signals. The Main Board is functionally divided into halves each of them hosting 6 FEBs. Each half has a separated low voltage power supply and is diode-ORed with the other half, ensuring the completely functionality if the power supply for one half stops working. Figure 4 shows a picture of the first Main Board prototype for the Demonstrator.

Figure 4. First prototype of the 3-in-1 Main Board

3.1.3. Daughter Board

The Daughter Board [8] is in charge of the high speed communication between the on- and off-detector electronics in the counting room. The Daughter Board collects, formats and transmits the data to the off-detector electronics using redundant high speed links at data rates of 10 Gbps. It also implements slow control functionalities as the distribution of Detector Control System (DCS) commands needed for the control and monitoring of the Main Board and the high voltage power supplies.

The Daughter Board has been developed by the Stockholm University. The third generation of the Daughter Board includes two Xilinx Kintex 7 FPGAs, two Quad Small Form-factor Pluggable (QSFP) optical modules, two GBTx chips and an FPGA Mezzanine Connector (FMC) for the communication with the Main Board. Figure 5 shows a picture of a third generation Daughter Board.

Figure 5. Third generation of the Daughter Board

3.2. Off-detector Electronics

The super Read-Out Driver (sROD) [9] is the key component of the off-detector electronics. This system is responsible for receiving and processing the data coming from the detector, sending pre-processed data to the trigger system, interfacing the DCS and the on-detector electronics, as well as the reception and distribution of Trigger, Timing and Control (TTC) information towards the detector.

The sROD demonstrator module, developed by the IFIC-Valencia group, is a prototype for the Demonstrator Project able to read out a complete hybrid drawer. The sROD demonstrator board has been designed in a double mid-size Advanced Mezzanine Card (AMC) form factor which can be plugged in an Advanced Telecommunications Computing Architecture (ATCA) carrier or in a Micro Telecommunications Computing Architecture (μTCA).

This module contains two FPGAs for data processing: one Xilinx Virtex 7 and one Kintex 7 FPGAs with embedded GTX transceivers capable of data rates above 10 Gbps. Figure 6 shows the connectivity between the two FPGAs and the optical modules and peripherals. The Virtex 7 FPGA is connected to 4 QSFP modules providing high speed communication with the on-detector electronics at a maximum data rate of 160 Gbps in each direction. This FPGA performs signal reconstruction algorithms and handles the communication between the DCS and the on-detector electronics.

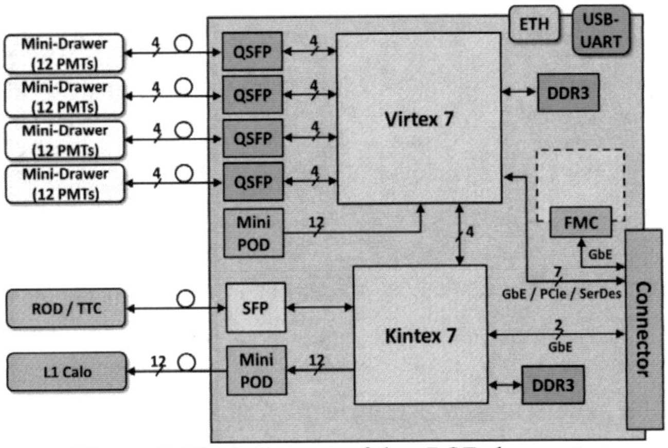

Figure 6. Block diagram of the sROD demonstrator

A Small Form-factor Pluggable (SFP) module connected to the Kintex 7 interfaces the hybrid drawer with the current system, sending data from the hybrid drawer to the present RODs and receiving TTC information which is decoded in the Kintex 7. The Avago MiniPOD transmitter is used to transmit the pre-processed data to the trigger system from the Kintex 7 FPGA. Figure 7 shows a picture of the PCB of the sROD demonstrator.

Figure 7. sROD demonstrator module

3.3. Optical Links

QSFP modulators based on Mach-Zehnder interferometers from Luxtera will be used as data links between the on- and off-detector electronics at data rates of 40 Gbps per module [10].

Several radiation tests carried by ANL have shown that while electro-optical chip exceeds the required radiation levels, the microcontroller used to configure and monitor the QSFP module does not fulfil the requirements to operate into the detector. For that reason, flash-based FPGAs are being evaluated to replace the original radiation soft microcontroller.

4. Conclusions

The increased luminosity for the HL-LHC implies the complete redesign of all the readout electronics in TileCal. The new readout architecture will include a full digital readout and trigger fitting the new requirements in the future HL-LHC.

The Demonstrator Project aims to gain experience and evaluate the new electronics before its full replacement around 2024. This project plans to install a hybrid drawer with the new readout electronics into the detector during the LS1 and based on the results more upgraded drawers could be installed during the next Long Shutdown 2 scheduled for 2018-2019.

References

[1] The ATLAS Collaboration 2010 Readiness of the ATLAS Tile Calorimeter for LHC collisions *Eur. Phys. J.* **C70**

[2] The ATLAS Collaboration 2008 The ATLAS Experiment at the CERN Large Hadron Collider *JINST* **3** S08003

[3] Valero A et al. 2007 ATLAS TileCal Read-Out Driver production *JINST* **2** P05003

[4] The ATLAS Collaboration 2012 Letter of Intent for the Phase-II Upgrade of the ATLAS Experiment *CERN-LHCC-2012-022*

[5] Tang F et al. 2013 Upgrade Analog Readout and Digitizing System for ATLAS TileCal Demonstrator *Proc. Nuclear Science Symposium and Medical Imaging Conference (Seoul)*

[6] Baumbaugh A et al. 2013 QIE10: a new front-end custom integrated circuit for high-rate experiments *JINST* **9** C01062

[7] Pillet N et al. 2011 FATALIC, a wide dynamic range integrated circuit for the TileCal VFE ATLAS Upgrade *Topical Workshop on Electronics for Particle Physics (Vienna)*

[8] Muschter S et al. 2014 Development of a digital readout board for the ATLAS Tile Calorimeter upgrade demonstrator *JINST* **9** C01001

[9] Carrió F et al. 2014 The sROD module for the ATLAS Tile Calorimeter Phase-II Upgrade Demonstrator *JINST* **9** C02019

[10] Drake G et al. 2013 Modulator based high bandwidth optical readout for HEP detectors *JINST* **8** C02023

Performance and upgrade of the CMS electron and photon trigger for Run 2

Jean-Baptiste Sauvan
On behalf of the CMS Collaboration

Laboratoire Leprince-Ringuet, CNRS/Ecole Polytechnique, Palaiseau, France

E-mail: `sauvan@llr.in2p3.fr`

Abstract. The CMS experiment implements a sophisticated two-level online trigger selection system that achieves a rejection factor of nearly 10^5. The first level (L1) trigger is based on coarse information coming from the calorimeters and the muon detectors while the high-level trigger combines fine-grain information from all sub-detectors. In the near future the LHC will increase its centre of mass energy to 13 TeV and progressively reach an instantaneous luminosity of $2\times10^{34}\,\mathrm{cm}^{-2}\mathrm{s}^{-1}$. In order to guarantee a successful and ambitious physics program under this challenging environment, the CMS trigger and data acquisition system must be consolidated. In particular the calorimeter L1 trigger hardware and architecture will be changed. The aim is to maintain the current thresholds and improve performance. This programme will be achieved using the μTCA architecture with fast optical links and latest generation FPGAs. Sophisticated object reconstruction algorithms, as well as online pile-up corrections, are being developed that will make use of these new hardware capabilities. For electron and photon reconstruction and identification, the first version of the new algorithms has been tested against the current algorithms. It shows a reduction of the trigger rate by a factor of two for isolated objects, an improved energy resolution of about 30%, and a position resolution reduced by more than four.

Introduction

The Compact Muon Solenoid (CMS) detector [1] has been designed to study proton–proton and heavy ion collisions produced by the LHC. Only a few hundreds of events per second can be recorded offline and a trigger system is used to select events of interest [2]. The trigger is organised in two stages: the Level-1 (L1) hardware-based trigger reduces the rate to 100 kHz, and the High-Level Trigger (HLT), based on a farm of computers, reduces the rate to 300 to 600 Hz. This system performed extremely well during the Run 1 of the LHC from 2010 to 2012, even with the significant pile-up conditions at the end of the run. After the first long shutdown the LHC will restart in 2015 with an instantaneous luminosity greater than $10^{34}\,\mathrm{cm}^{-2}\,\mathrm{s}^{-1}$ and an average number of pile-up events of 45 or higher. To avoid a significant increase in energy thresholds, which would be detrimental for physics, an upgrade of the L1 trigger system is required for Run 2 [3].

In the following, the current electron and photon L1 trigger and the new architecture that is being installed for the Run 2 are presented. Future algorithms for the reconstruction and identification of electrons and photons based on this new architecture are described and their expected performance are shown.

1. Current electron and photon level-1 trigger

The CMS ECAL is composed of a barrel (EB) and two endcaps (EE) and is made of lead tungstate ($PbWO_4$) scintillating crystals. In the EB, matrices of 5×5 crystals (in the pseudorapidity (η) and azimuthal angle (ϕ) directions) are combined into trigger towers. In the EE, crystals are arranged in a more complicated way due to the X-Y geometry of the detector, but the resulting trigger towers mimic an η/ϕ layout similar as in the EB. These trigger towers form the basic inputs to the electron and photon L1 trigger. The transverse energies E_T measured by the crystals in a given trigger tower are summed by the front-end electronics into a trigger primitive that is sent to Trigger Concentrator Cards (TCC) situated off-detector.

The TCCs transmit groups of trigger primitives to the Regional Calorimeter Trigger (RCT), which builds and identifies electron and photon candidates (denoted in the following as e/γ candidates) formed by pairs of trigger towers. The identification of the candidates is based on:

- the lateral extension of the energy in the most energetic tower (which is compact for electromagnetic showers) that is encoded into a Fine Grain veto bit;
- the associated E_T deposited in the hadronic calorimeter (HCAL) that is required to be small (typically $< 5\%$ of the E_T deposited in the ECAL);
- the E_T deposited in the four corners around the candidate, used to separate isolated from non-isolated candidates (isolated candidates must have at least one of the four corners with $E_T < 3.5\,\text{GeV}$).

Successfully identified candidates are passed to the Global Calorimeter Trigger (GCT) with a regional granularity of 4×4 trigger towers (an RCT region of interest). The GCT builds L1 jets as well as global event quantities (E_T sum and missing E_T), sorts the e/γ candidates, and sends the different calorimetric objects to the Global Trigger (GT). The GT also uses the L1 muon information and decides whether a given event is accepted or not.

2. Architecture of the future calorimeter level-1 trigger for Run 2

A new architecture is needed for Run-2 data taking because of the increased instantaneous luminosity and the greater pile-up. Without improved architecture and reconstruction algorithms, a substantial increase of trigger thresholds would be required by the end of the Run 2. Typically a double-electron trigger, with thresholds of 13 GeV on one leg and 7 GeV on the other leg, was giving a rate of about 5 kHz in 2012 in Run 1. This would increase to about 50 kHz for the expected Run-2 conditions. This is not manageable given the total bandwidth of 100 kHz. Refined algorithms, with improved identification criteria and pile-up corrections, are needed in order to reduce the rates in such conditions while maintaining high signal efficiency and low thresholds. Such algorithms need to view the full calorimeter at the level of trigger towers and require large computing power. These requirements can be fulfilled with higher bandwidth optical links as well as faster and larger Field-Programmable Gate Arrays (FPGAs) compared to the current hardware.

The new calorimeter L1 system [3] is being installed. It is based on μTCA crates, replacing the current VME crates and offering more bandwidth through external high-speed serial links. The system is composed of two processing layers instrumented with Virtex-7 FPGAs from Xilinx. The first layer is designed for data formatting and pre-processing. The second layer will be used for object reconstruction and identification. Two types of cards form the two layers and are based on the Advanced Mezzanine Cards (AMC) specifications. The cards are optimised for different tasks. The so-called CTP7 layer-1 cards, displayed in Fig. 1 (left), have a regional view of the calorimeter, that can potentially be extended through dedicated connections to the μTCA backplane allowing data sharing across different calorimeter regions. They are used for pre-processing data that only needs a reduced view of the detector. The so-called MP7 layer-2 cards, displayed in Fig. 1 (right), on the other hand, have access to the full calorimeter at trigger tower level on single FPGAs via a large number of optical inputs. These cards are suitable for running algorithms that need a large view of the detector.

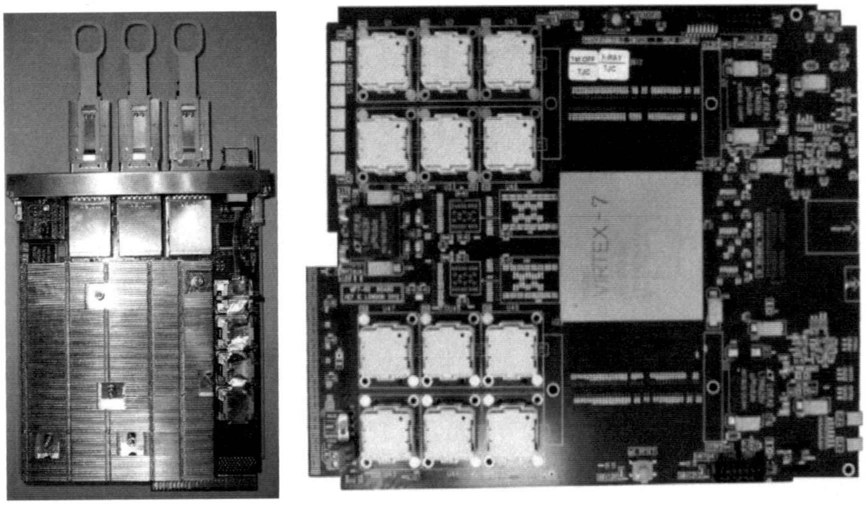

Figure 1. The CTP7 (left) and MP7 (right) cards developed for the first and second layers of the upgraded calorimeter L1 system

The current system will nevertheless be kept, with small changes, in order to be able to run both the current and upgrade systems in parallel for commissioning the upgrade system. The current Serial Link Boards (SLB) used to send trigger primitives from the TCCs to the RCT will be replaced by a new optical version (oSLB) featuring a different configuration with 4.8 Gb/s on a single link (compared to the 4×1.2 Gb/s for the current copper links). Duplicated optical outputs will be sent to the RCT and to the first layer of the upgraded system. In order to accommodate the new optical links, the Receiver Mezzanine (RM) of the RCT will be replaced by an optical version (oRM).

3. Future algorithms for electrons and photons at level-1

Sophisticated algorithms have been developed based on the capabilities of the new architecture. Improvements with respect to the current electron and photon reconstruction are:

- improved energy reconstruction with a dynamic clustering of trigger towers;
- better position reconstruction, with the tower granularity instead of the granularity of an entire RCT region;
- additional discrimination of electrons (or photons) and jets, based on the shape of the reconstructed clusters;
- more precise isolation criteria that account for the level of pile-up, without boundaries.

A better containment of the electron and photon energy (in particular for showering electrons and converted photons) is obtained with dynamic clustering. The maximum size of the clusters is limited (at most 8 trigger towers can be clustered) in order to minimize the impact of pile-up energy deposits while including most of the electron or photon energy. An extended region in the ϕ-direction (at most 5 trigger towers) and a narrow region in the η-direction (at most 2 trigger towers) have been chosen (see Fig. 2 (left)) since electron and photon showers spread mostly along the ϕ-direction due to the magnetic field.

The clusters of towers are built from a seed tower (the red tower on Fig. 2 (left)). The first and second neighbour towers (the orange and yellow towers on Fig. 2 (left)) are dynamically clustered if connected to the seed. The sum of the ECAL E_T of the seed and clustered towers is taken as the raw E_T of the cluster. A calibration is applied to this raw energy with factors depending on the η-position of the seed tower. The factors are obtained from $Z \to ee$ events and are defined as the ratio of the offline fully reconstructed electron E_T to the associated raw L1 cluster E_T.

The position of the reconstructed cluster is first taken from the centre of the seed tower. It is then refined, within the seed tower, based on the distribution of energy in the cluster.

This dynamic clustering leads to a large variety of cluster shapes that depend on the distribution of energy around the seed tower. Some cluster shapes are very unlikely to come from an electron or a photon. This is the case for large clusters involving many trigger towers, which probably come from jets. Small clusters (containing typically one to four trigger towers) are more compatible with electron or photon showers. This information, which is not available for a fixed size clustering, brings additional discrimination power between electrons (or photons) and jets.

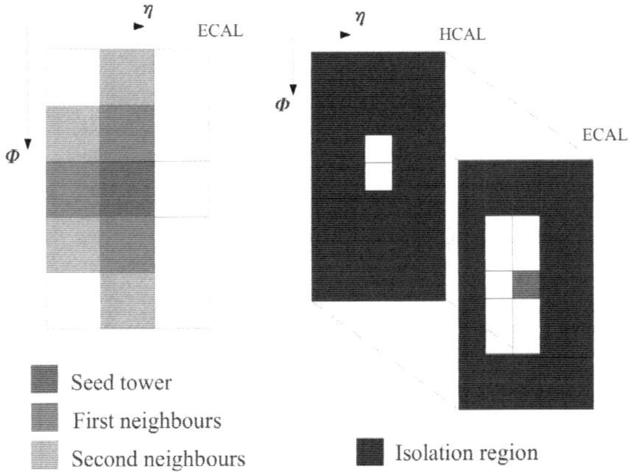

Figure 2. The L1 e/γ clustering and isolation. A candidate is formed by clustering neighbour towers (orange and yellow) if they are linked to the seed tower (red). A candidate is considered as isolated if the E_T in the isolation region (blue) is smaller than a given value.

The full view of the detector in the second layer of the upgrade L1 calorimeter trigger system presents trigger tower information without boundaries. It is possible to implement more precise isolation criteria that account for the level of pile-up event-by-event. The E_T deposited in an isolation region as depicted in Fig. 2 (right) is computed. This isolation region has a size of 5×9 trigger towers excluding the footprint of the e/γ candidate in ECAL (a 2×5 region) and HCAL (a 1×2 region). If this E_T is smaller than a given threshold the candidate is tagged as isolated, and as non-isolated otherwise. The threshold is a function of η and depends on an estimator for the number of pile-up interactions in the event. This pile-up estimator is the number of trigger towers in the event with E_T greater than a given value. The threshold values are tuned to have a 90% efficiency, constant with pile-up and with η.

These algorithms are currently being implemented in firmware. A first version is being tested in the layer-2 cards.

4. Expected performance

The upgrade algorithms are being tested during their development and compared to the current algorithms. The expected performance obtained with an initial version are presented here.

The trigger efficiency is measured with electrons from $Z \to ee$ events, using a tag and probe method. The dataset is from the $6\,\mathrm{fb}^{-1}$ of data collected in the last period of 2012. Both the tag and the probe electrons are required to pass tight identification criteria in order to reduce background contamination to a negligible level. The invariant mass of the dielectron system must be consistent with the Z boson mass ($60\,\mathrm{GeV} < M_{ee} < 120\,\mathrm{GeV}$). The tag electron must trigger the event at L1 while the probe electron is used for efficiency studies. The trigger efficiency is given by the fraction of probes with an associated L1 cluster above a given trigger threshold, as a function of the offline reconstructed probe E_{T}. The efficiency curves for a trigger threshold of 20 GeV (L1_SingleEG20) obtained with the current and upgrade algorithms are shown in Fig. 3, for probe electrons in the EB and in the EE.

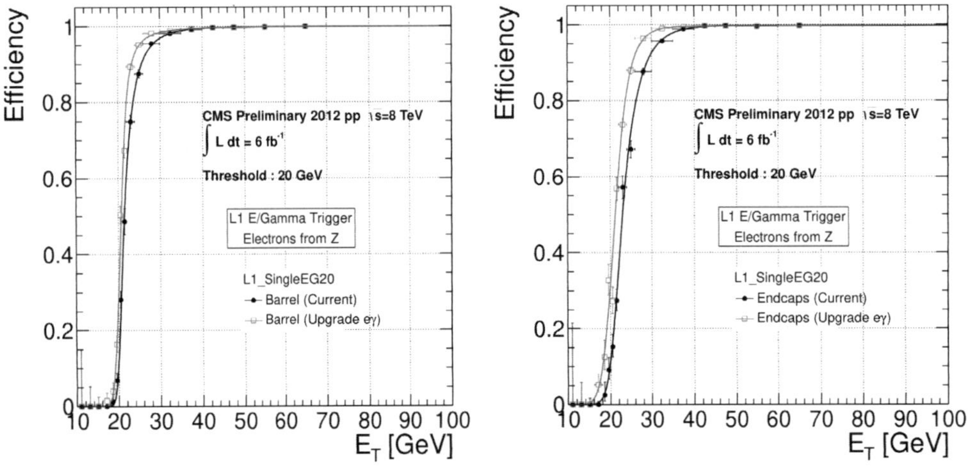

Figure 3. Electron trigger efficiency at L1 as a function of the offline reconstructed E_{T}, in the EB (left) and EE (right), for a threshold of 20 GeV. In both cases the efficiencies obtained with the upgrade and current algorithms are shown, without isolation criteria.

The curves corresponding to the upgrade algorithms are steeper, which is a direct consequence of the dynamic clustering. The energy deposited by electrons is better clustered and leads to a better energy measurement of about 30%. In the endcaps, the upgrade improvement comes from a more precise calibration of the upgrade clusters. The EB and EE turn-on curves are now closer to each other.

The better granularity available in the layer-2 of the upgrade L1 system (granularity of trigger towers), compared to the granularity of the current system (granularity of RCT regions), leads to improved spatial location of the L1 e/γ candidates. The position measurement, based on the energy distribution in the cluster, is improved by more than a factor of four, as can be seen in Fig. 4 (left).

The new capability of the clustering algorithm to provide shape information, to discriminate between electron (or photon) clusters and jet clusters, reduces the rate of fake electrons and photons with negligible loss of signal efficiency. These rates can be reduced further, with a signal efficiency loss of about 10%, by requiring the candidates to be isolated. In this case a reduction of the rate by a factor of about two is obtained, with respect to the current algorithms, for similar signal efficiencies. Figure 4 (right) shows the relative rates of triggered events as a function of the offline threshold at which the trigger reaches 95% of the plateau efficiency. The results for the current and upgrade algorithms are shown with and without isolation criteria.

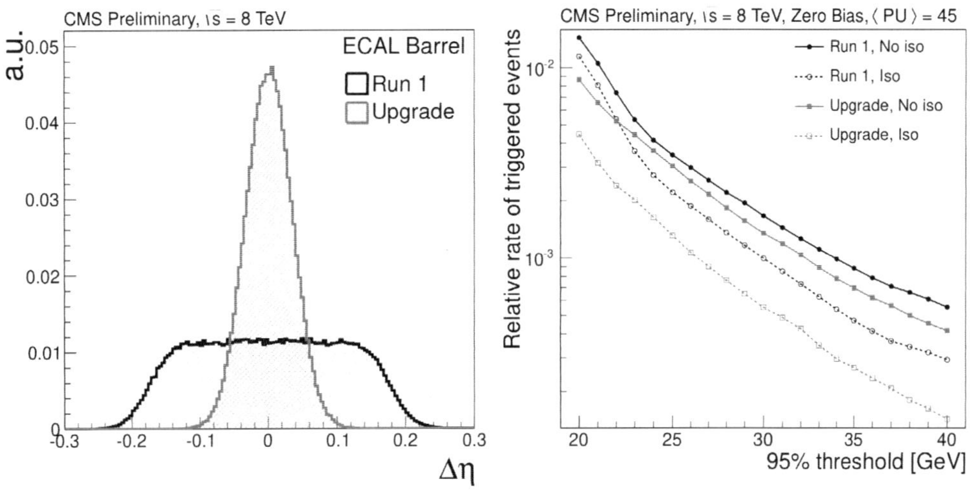

Figure 4. (Left) The difference in pseudorapidity for L1 electrons with respect to the offline reconstructed pseudorapidity, for the current and upgrade algorithms. (Right) Relative rate of triggered events from 8 TeV minimum bias data for an average pile-up of 45, obtained with the current and the upgrade algorithms, both with and without their respective isolation requirements.

Conclusion

An upgrade calorimeter L1 trigger system is being developed for CMS. It will be commissioned in parallel with the current system at the beginning of the Run 2 in 2015, and will be used for physics starting from 2016. This new system is composed of cards based on latest generation FPGAs with high bandwidth optical links. Sophisticated algorithms are being implemented for the reconstruction and identification of electrons and photons at L1 that make use of the improved capabilities of the new system. An initial version of these algorithms brings substantial improvement. The position resolution is improved by more than a factor of four, and the energy resolution is improved by about 30%. A reduction of the trigger rate by a factor of two is obtained for isolated candidates, while maintaining a similar signal efficiency to that for the current isolated candidates.

Acknowledgment

The work of the author of this paper has been partly funded by the P2IO LabEx (ANR-10-LABX-0038) in the framework "Investissements d'Avenir" (ANR-11-IDEX-0003-01) managed by the French National Research Agency (ANR).

References

[1] CMS Collaboration, *The CMS experiment at the CERN LHC*, JINST 3 (2008) S08004
[2] CMS Collaboration, *CMS TriDAS project: technical design report, vol. 1: the trigger systems*, CERN-LHCC-2000-038, CMS-TDR-6-1 (2000)
[3] CMS Collaboration, *CMS technical design report for the level-1 trigger upgrade*, CERN-LHCC-2013-011, CMS-TDR-12 (2013)

Journal of Physics: Conference Series **587** (2015) 012022 doi:10.1088/1742-6596/587/1/012022

MPPC photon sensor operational experience in CMS

Andreas Künsken for the CMS Collaboration

III. Physikalisches Institut B, RWTH Aachen University, Otto-Blumenthal-Straße, D-52074 Aachen, Germany

E-mail: kuensken@physik.rwth-aachen.de

Abstract. During the first long shutdown of the LHC the Outer Hadron Calorimeter of CMS has been equipped with silicon photomultipliers that replace the existing hybrid photodiodes while maintaining most of the readout chain. The new photon sensors are Hamamatsu Multi-Pixel Photon Counters with an active area of 3 mm × 3 mm and a cell pitch of 50 µm. Different means of determining the device's breakdown voltage and gain are presented. The stability of the overall system temperature as well as the stability of the gain and the breakdown voltage of the silicon photomultipliers are studied. The correlation between the different methods of determining the gain and breakdown voltage is tested to find whether they yield comparable results.

1. Introduction

The hadronic calorimeter (HCAL) of the compact muon solenoid (CMS) detector at the Large Hadron Collider (LHC) consists of different detector subsystems, namely the barrel calorimeter (HB), the endcap calorimeters, the very forward calorimeters and the outer hadronic calorimeter (HO) [1, 2]. The HO is placed directly outside the superconducting coil and was designed as a tail catcher to improve the energy measurement of jets leaking through the solenoid [3]. In the barrel region CMS consists of five rings around the solenoid with ring 0 placed in the middle. HO is built up of two layers of scintillator in ring 0 and one layer of scintillator in rings ± 1 and ± 2 each. The geometry of the scintillating tiles of HO matches the projected tower geometry of HB as is shown in Fig. 1. To achieve a better light collection and to guide the light, the scintillators have grooves in which wavelength shifting (WLS) fibers are embedded. Outside the trays in which the scintillators are placed, the light from the WLS fibers is coupled to clear fibers that guide the light to the photon sensors. During the first long shutdown of the LHC (2013 - 2015) the previously used hybrid photodiodes (HPDs) were replaced by silicon photomultipliers (SiPMs) due to insufficient performance. The new readout was designed as a "drop-in" replacement in order to maintain the rest of the readout chain [4].

2. The new photon sensor readout design

The previously used HPDs had 18 individual pixels that were read out separately. To match this setup, the new printed circuit board (PCB) was designed with the SiPMs placed at the pixel positions giving the same number of readout channels as before. The arrangement of the SiPMs on the PCB together with the position of the HPD window marked by two white circles can be seen in Fig. 2. As the SiPMs are solid state devices, they are sensitive to changes of the ambient temperature. In order to stabilize the temperature environment, a Peltier element is

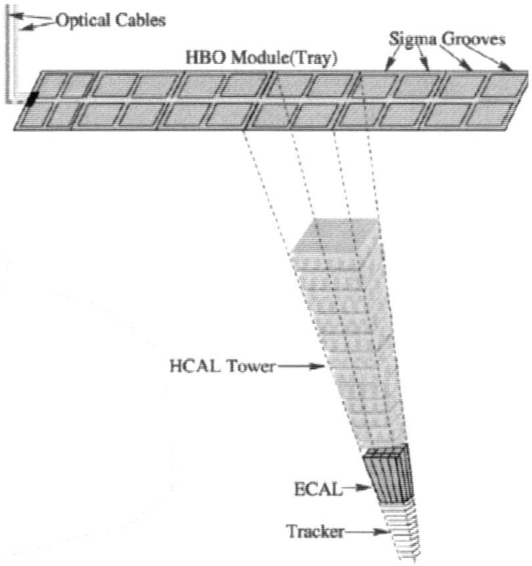

Figure 1. Geometry of the HO tiles matching the towers of the barrel calorimeter. Adapted from [3].

Figure 2. PCB carrying the SiPMs that are placed inside the white circles which mark the position of the old HPD. Adapted from [4].

placed on the backside of the PCB. An optical decoder unit routes the four fibers coming from one scintillator tile onto one SiPM so one channel corresponds to one scintillator tile. As ring 0 has two layers of scintillator and thus the doubled number of fibers arriving at the SiPM, the area of the SiPM is not large enough to cover all fibers. Therefore, a light mixer with a specular reflecting surface is placed between the fiber end and the SiPM surface in order to distribute the light over the whole SiPM surface. With the light mixer, the loss of light is independent of the fibre but homogeneously distributed.

2.1. The installed SiPM model

The SiPM chosen for HO is a Hamamatsu Multi-Pixel Photon Counter (MPPC) in a surface-mounted device (SMD) housing. It has a cell pitch of 50 µm and an active area of 3 mm × 3 mm. The device is identical to the Hamamatsu model S10931-050P. The required operating voltage is of the order of 70 V with a gain that is typically around 6×10^5 when operated 1 V above breakdown voltage. One important quantity of the SiPM is the change of the gain with the temperature. As the gain depends linearly on the over-voltage, which is the bias voltage subtracted by the breakdown voltage, the change of the breakdown voltage with temperature can be transformed into a change of gain with temperature which yields a value of $-8\,\%\,\mathrm{K}^{-1}$ at a foreseen operating point of 1.5 V over-voltage [5].

3. Analysis of the operating parameters

To get the first impression of the performance and the stability of the SiPMs different operating parameters are examined. These are the temperature, the gain and the breakdown voltage. While the temperature is only studied regarding the stability over time, the gain and the breakdown voltage are tested against both the stability over time and the comparability of different methods.

3.1. The temperature

Due to the fact that the breakdown voltage depends on the temperature of the SiPM, also the gain depends on the temperature if the bias voltage stays constant. This implies that the ambient temperature around the SiPM needs to be monitored to see whether possible deviations in the gain are caused by a temperature change in the device. This is depicted in Fig. 3. For clarity reasons only some selected runs are shown. As can be seen, the temperature stays well within a variation of below 0.1 °C once the set temperature at the PCB is reached. This is shown over a period in 2014 from the end of January to the beginning of March which emphasizes that any changes in gain or breakdown voltage, if observed, are not caused by changes in the temperature environment.

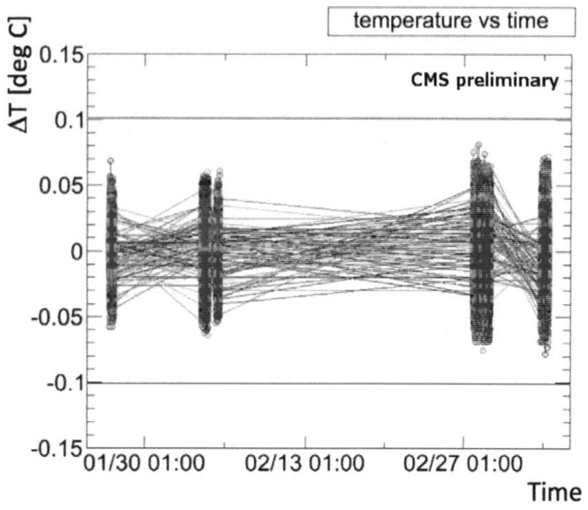

Figure 3. Temperature difference to the target temperature of the SiPMs over time for all installed SiPMs.

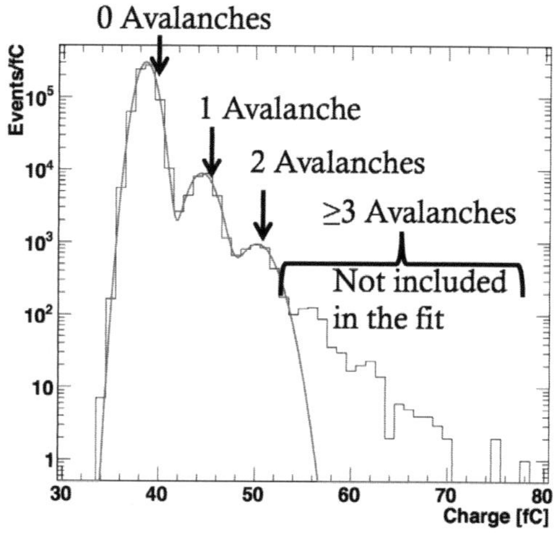

Figure 4. Pedestal spectrum of an SiPM with a fit applied to determine the gain.

3.2. The gain

There are two methods used to determine the gain of an SiPM in HO. The first uses the fact that an SiPM generates a small signal even if there is no light hitting the detector. This pedestal spectrum is exemplified in Fig. 4. The different peaks in the spectrum develop due to different numbers of cells giving a signal at the same time in one SiPM. While in general the distance between the peaks of any order can be used to determine the gain, in HO only the pedestal peak and the peaks corresponding to one and two cells giving a signal are used for measuring the gain. Alternatively, one can generate short light pulses with an LED[1] onto the SiPM. With the intensity of the light pulse being not too large and assuming N photons reaching the SiPM the signal should be equal to N×gain. Furthermore, using Poisson statistics the sigma of the photon number is $\sqrt{N}$ so the sigma of the measured signal equals $\sqrt{N}\times$gain. Dividing the variance of the signal by the mean of the signal one gets sigma^2/mean = gain. The relative variation of the gain versus time using the pedestal method for a single SiPM mounting board is depicted in Fig. 5. For a comparison with the LED method refer to section 4.1. As can be seen, the gain for the tested SiPMs is stable over a time from the middle of February to the beginning of March in 2014 and the relative variation of the gain lies within 2 %. Looking at the distribution

[1] **L**ight-**E**mitting **D**iode

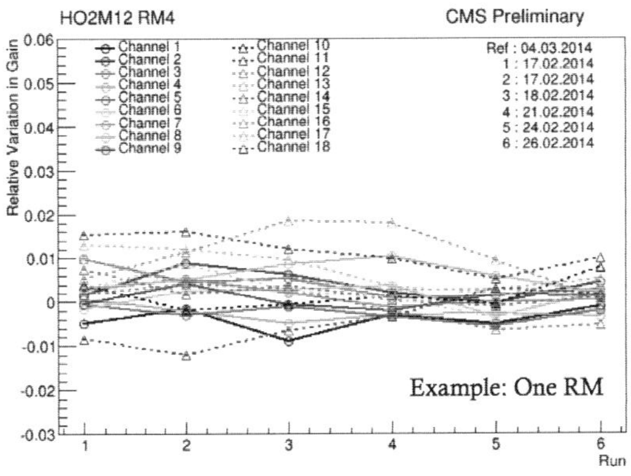

Figure 5. Relative variation of the gain over time determined using the pedestal method shown for one PCB (18 channels).

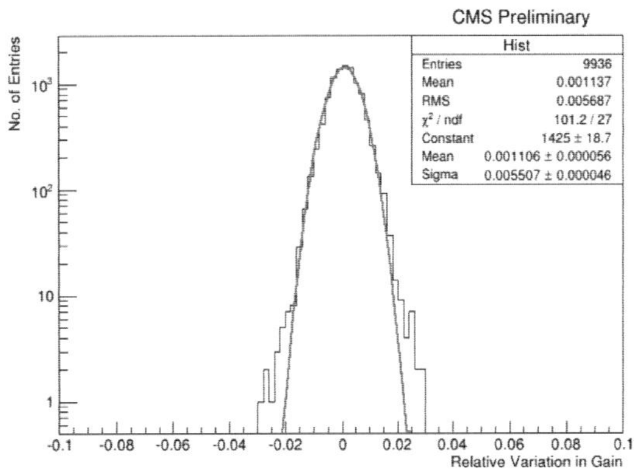

Figure 6. Distribution of the relative variations in gain for all installed SiPMs.

of the relative variation over time for all installed SiPMs, which is shown in Fig. 6, it can be found that the distribution has a width of only 0.5 % and that all variations lie within 3 %. This demonstrates that the gain determination works as expected and the operation of the SiPMs with a stable gain is possible.

3.3. The breakdown voltage

For the determination of the breakdown voltage, there is also the possibility to use either the pedestal spectrum of the SiPM or the signal of a short LED pulse. For the first method the gain is determined for different applied bias voltages. The gain depends linearly on the operating voltage so the resulting curve is a straight line which can be extrapolated to gain 0 as is shown

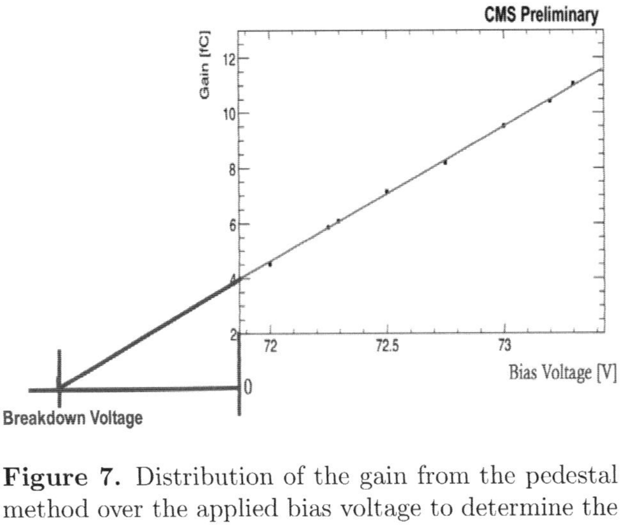

Figure 7. Distribution of the gain from the pedestal method over the applied bias voltage to determine the gain.

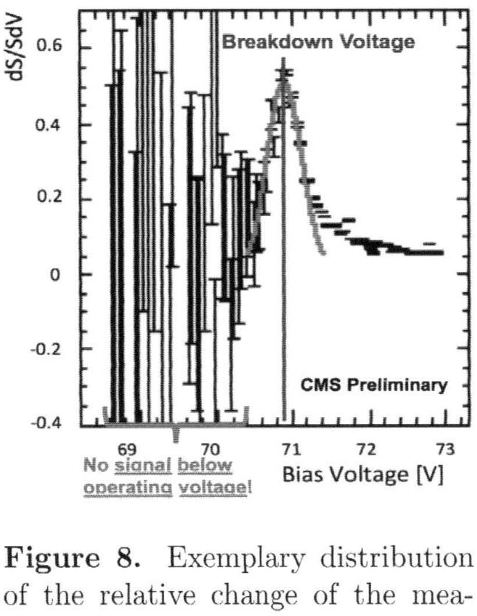

Figure 8. Exemplary distribution of the relative change of the measured LED signal with the bias voltage.

in Fig. 7. At the corresponding bias voltage the breakdown of a cell is just about to start which is the definition of the breakdown voltage. The second way is to use the relative change of the measured signal when pulsing an LED and varying the bias voltage. As shown in Fig. 8 the distribution of $\frac{\mathrm{dS}}{\mathrm{SdV}}$ peaks at that voltage, at which the change of the relative signal is at maximum which is then used as the breakdown voltage. The distribution of the variation in

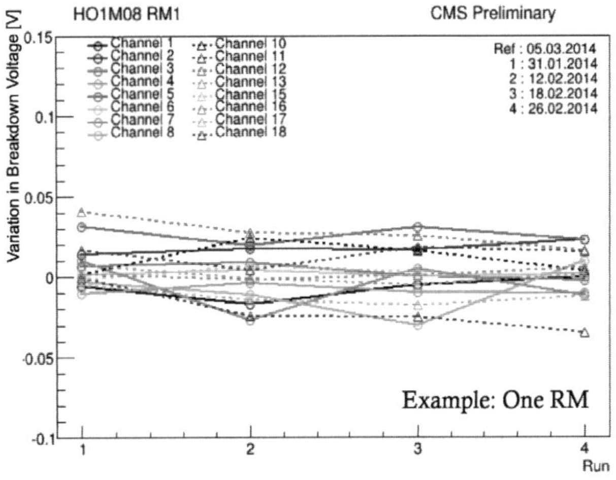

Figure 9. Distribution of the breakdown voltage over time for one PCB (18 channels).

breakdown voltage over time for one PCB using the LED method is shown in Fig. 9. The time period reaches from the end of January to the beginning of March in 2014. The breakdown voltage is found to be stable within 50 mV which demonstrates the reliability of the breakdown voltage determination. For comparison with the pedestal method refer to section 4.2.

4. Correlation of the employed methods

To estimate the comparability of the two methods for determining the gain and the breakdown voltage it is useful to plot the result of one method against the result of the other method for each channel. This helps spotting systematic differences between the procedures.

4.1. Gain correlation

The correlation plot for the gain measurements for all installed SiPMs for a specific run is shown in Fig. 10. All gain values are centralized around a certain value close to the design gain of 6 ADC counts per photoelectron and the distance in gain between most of the SiPMs is small. The difference in the center of the distribution on the y-axis compared to the center on the x-axis can be explained with the fact, that the noise of an SiPM is not Gaussian, which is assumed when calculating the gain with the pedestal method. This way, a systematic offset between the two methods is introduced which shows itself in the correlation plot. However, the plot shows that the different means of determining the gain yield comparable results.

4.2. Breakdown voltage correlation

The correlation plot for the breakdown voltage is presented in Fig. 11. Here the values follow a straight line which is expected as the different SiPMs do not all have the same breakdown voltage. The offset can be explained with the fact that the breakdown voltage is defined differently for the two methods. For the pedestal method the breakdown voltage is the point where amplification starts while for the LED method the breakdown voltage is defined as the point of maximum relative change of the signal with respect to the bias voltage, which lies above the breakdown

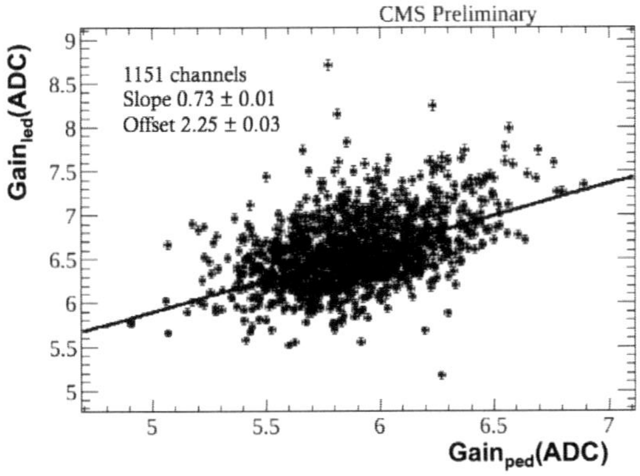

Figure 10. Correlation plot for the gain measurements.

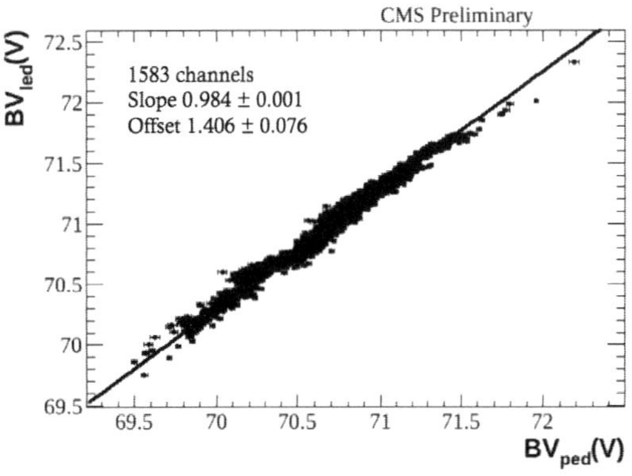

Figure 11. Correlation plot for the breakdown voltage measurements.

voltage from the pedestal method. It is possible to use the offset from the correlation plots to correct the breakdown voltage obtained with the LED method for this shift.

5. Summary

It was shown that the newly installed SiPMs in HO are operated in a stable temperature environment. With this precondition it was possible to perform measurements of the gain and the breakdown voltage of the devices. Two methods using the pedestal spectrum of an SiPM or light from an LED were introduced. It was found that both yield comparable results for the gain and for the breakdown voltage. Furthermore, it was demonstrated that in HO the SiPMs can be operated in a stable way over a long time.

Acknowledgments

I would like to thank Ketino Kaadze and Benjamin Lutz for their strong support with the plots and the preparation of the conference talk.

References

[1] CMS Collaboration 2008, The CMS experiment at the CERN LHC, *JINST* **3** *S08004*
[2] CMS Hadron Calorimeter Technical Design Report, CERN/LHCC 97-31, CMS TDR 2, June 20th, 1997
[3] Abdullin S *et al.* (CMS Collaboration) 2008, *Eur. Phys. J.* **C57** *653-663*
[4] Lutz B *et al.* (CMS Collaboration) 2012, *J. Phys.: Conf. Ser.* **404** *012018*
[5] Anderson J *et al.* (CMS HCAL Collaboration) 2012, *Phys. Procedia* **37** *72 -78*

16th International Conference on Calorimetry in High Energy Physics (CALOR 2014) IOP Publishing
Journal of Physics: Conference Series **587** (2015) 012023 doi:10.1088/1742-6596/587/1/012023

Correction for PMT temperature dependence of the LHCf calorimeters

Eri Matsubayashi for the LHCf collaboration

Solar Terrestrial nvironment Laborator ,Nago a niversit , Nago a, Japan

mail `matsubayashi@stelab.nagoya-u.ac.jp`

Abstract. L Cf is an e periment to measure the ver forward production of particles at L C in order to calibrate the hadronic interaction models used to simulate cosmic ra air showers. The results obtained from $\sqrt{s}$=7 TeV proton proton collisions have large s stematic errors resulting from the energ scale shift of the reconstructed π^0 mass. It was found that one of the major sources of this shift is the temperature dependence of the MT response used in the L Cf detectors. In order to correct the variation of the MT gain b temperature, two t pes of temperature variation were considered. As a result, s stematic errors resulting from the energ scale of Arm2 was improved b 3 %.

1. Introduction

L Cf e periment was motivated to calibrate the hadronic interaction models used to simulate cosmic ra air showers with the e perimental data obtained at L C. L Cf was speci call designed for measurements of the ver forward pseudo rapidit ; $\eta >$ .4 production spectra of neutral particles which are most relevant to air showers development.

The L Cf data ta ing at $\sqrt{s}$ 7 Te proton proton collisions has nished in the middle of Jul 2 1 and from these data gamma ra spectra and neutral pion spectra have been anal ed 1 2 . owever these results have large s stematic errors resulting from the energ scale shift. The energ scales of the L Cf calorimeters were determined b measuring the beams with nown energ at C RN S S in 2 7 3 . This energ scales were tested b the reconstructed invariant π^0 masses from the data obtained at L C. owever the reconstructed invariant mass pea deviated from the rest mass of π^0 135 Me b 2 sigma. This disagreement was ta en into account as a part of the s stematic errors of the energ scale.

This large s stematic error caused b energ scale is a big problem for L Cf detectors because this error becomes larger as the energ becomes higher, and the s stematic error contributes a half of total error above 3 Te . Therefore in order to proceed further anal sis, we need to reduce this error.

It was found that one of the ma or sources of this error is temperature dependence of MT response used in the L Cf detectors, while it was not corrected in the previous anal ses. There was correlation between the reconstructed mass and the temperature of the MT holders in the 2 1 data. In order to correct the variation of the MT gain b temperature, two t pes of temperature variation were considered.

In this paper, the detail of the data correction for MT temperature dependence is presented together with a brief introduction of the L Cf and its future prospect.

2. LHCf calorimeters

L Cf has two independent detectors, called Arm1 and Arm2 installed in the instrumentation slots of the neutral particle absorbers TANs located ±14 m from the interaction point 1 I 1 in L C. ecause of the dipole magnet installed in front of the detectors, charged particles are swept and onl neutral particles come to the detectors. ach detector has a pair of shower calorimeters composed of 44 radiation lengths of tungsten and 16 sampling la ers of plastic scintillator and 4 imaging la ers as shown in Figure.1 and Figure.2.

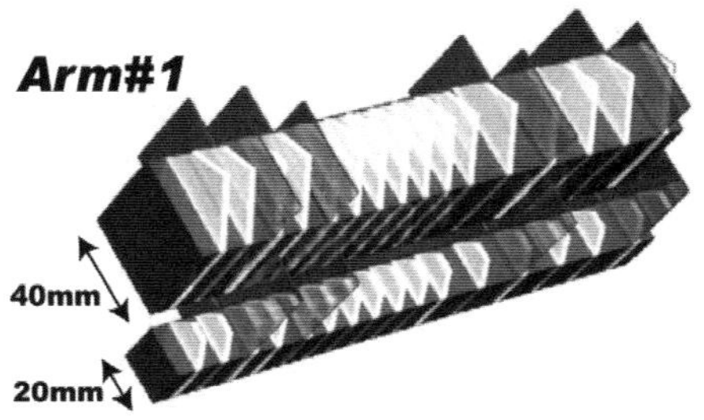

Figure 1. The schematic view of the Arm1 calorimeter.

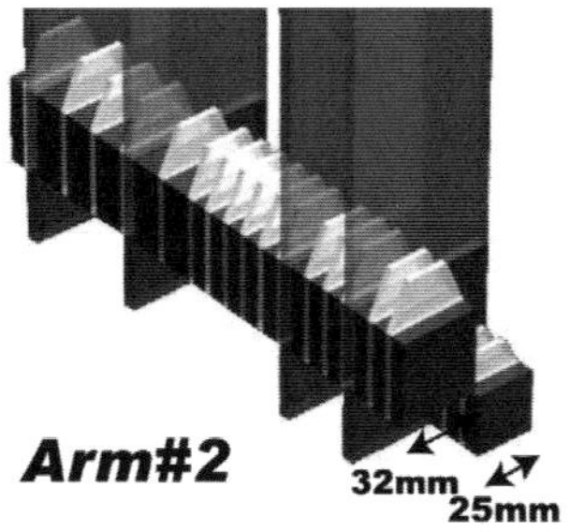

Figure 2. The schematic view of the Arm2 calorimeter.

3. Energy scale

The energ scale of each calorimeter was determined b the S S calibration in 2 7 b measuring 15 e electron beams. Charge Q_i measured in each MT was converted to deposited energ dE_i b a conversion factor A_i obtained from calibration for all the MTs i 1~32 ,

$$dE_i \quad Q_i \times A_i. \qquad 1$$

This energ scale can be tested from the data obtained at L C. Since the L Cf detector is composed of two calorimeters, the detectors are able to measure the energ of two photons individuall generated b the deca of a π^0 meson. The invariant mass $m_{\gamma\gamma}$ of the photon pair is reconstructed using the opening angle θ calculated from the incident positions and energies E_{g1} and E_{g2} of the photon pair.

$$m_{\gamma\gamma} \quad \sqrt{E_{g1} \cdot E_{g2} \cdot \theta^2}. \qquad 2$$

comparing $m_{\gamma\gamma}$ with π^0 rest mass the energ scale is veri ed.

A part of the e perimental data obtained on Ma 15th and 16th 2 1 L C Fill 11 4 during proton proton collisions at $\sqrt{s}$ 7 Te were used for anal sis, and we found the pea position of $m_{\gamma\gamma}$ distribution was slightl higher than the π^0 rest mass and the disagreement were .1 in Arm1 and 3. in Arm2. This disagreement is larger than errors of calibration ±3.5 .

4. PMT temperature dependence

In order to nd the cause of the energ scale problem, $m_{\gamma\gamma}$ were reconstructed in long term from 1 April to 25 Ma in 2 1 . Figure.3 presents the time variation of $m_{\gamma\gamma}$ as a function of the L Cf R N number. The blac points and red points represent $m_{\gamma\gamma}$ and temperature of the MT holder, respectivel . From Figure.3, it is clear that $m_{\gamma\gamma}$ correlates with the MT holder temperature. It was found that the MT AMAMATS R74 has .25 °C temperature dependence for its gain 4 . We suspected that the one of the ma or reasons of the energ scale shift could be in uence of the MT temperature.

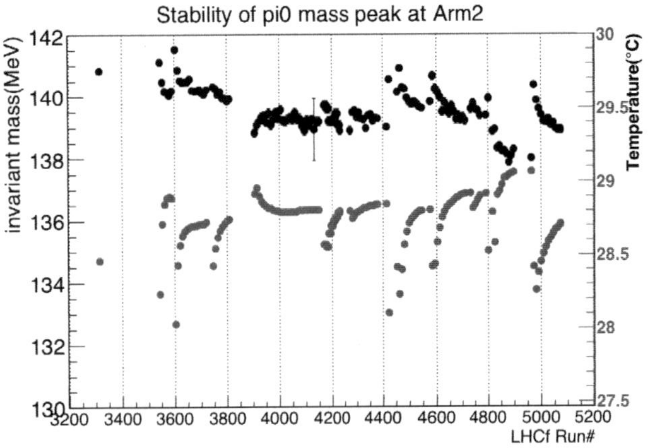

Figure 3. The time variation of $m_{\gamma\gamma}$ and the temperature of the MT holder as a function of the L Cf R N number. lac points and red points represent $m_{\gamma\gamma}$ and temperature of the MT holder, respectivel .

5. Correction for the temperature dependence

In order to correct the MT gain variation, correction terms G_i were introduced in e uation 1 ,

$$dE_i \quad G_i \times Q_i \times A_i. \qquad 3$$

owever, as the MT temperature T_{PMT} was not directl measured during the time of measurement at L C and calibration at S S, it was estimated b the elapsed time t since high voltage was applied and the high voltage value ,

$$T_{PMT} \ t, V \quad T_{ref} + \quad T \ V \ \ 1 - exp \ -t/\tau \ . \qquad 4$$

It was assumed that T_{PMT} rose b the e ponential function time constant is τ and T_{PMT} reached $T_{ref} +$ T where T_{ref} is an ambient temperature and $Delta$T is an as mptotic temperature increase.

stimated MT temperature was divided into two parts. One is the time depending component named correction A, and the other is the time independent component named correction . The correction A corrects for the temperature variation during the L C operation because of Joule heat from the MT breeders. On the other hand, the correction corrects for the temperature di erence between the measurement at L C and the calibration at S S because of di erence of applied high voltage and ambient temperature.

ta ing these corrections into account, G_i is e pressed as

$$G_i \quad \frac{1}{1 - a \cdot \{T_{ref}^{SPS} + \quad T_i^{SPS} - \ T_{ref}^{LHC} + \quad T_i^{LHC} \ + \quad T_i^{LHC} \ -t/\tau \ \}}. \qquad 5$$

In order to calculate G_i, T_i and τ of all MTs in Arm2 were measured using platinum thermometers attached on the MTs in the acr lic MT holders. We applied two values of high voltage called high gain around 6 and normal gain around 5 applied in the S S calibration and the L C operation, respectivel . The time constants τ were almost same for 32 MTs and the average of τ was 342 $\pm$ sec. On the other hand, T_i were di erent for each MT correlated with the location in the MT holders. T_i^{SPS} ranged from 5.2 °C to 12. °C and T_i^{LHC} ranged from 3.4 °C to .1 °C.

As for T_{ref}^{SPS} and T_{ref}^{LHC}, the temperature data were measured in situ b nearb sensors. T_{ref}^{SPS} values were about 3 °C higher than T_{ref}^{LHC}.

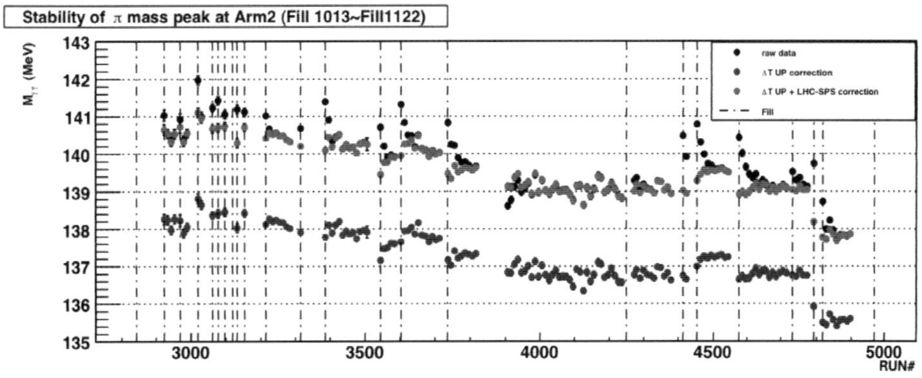

Figure 4. The improvements of the $m_{\gamma\gamma}s$ b corrections. The hori ontal a is shows R N number, the vertical a is shows the reconstructed invariant mass. The blac points show the results before corrections, the red points show the results of the correction A, the blue points show the results of the correction A and correction . The vertical dashed lines show the beginning of the lls.

6. Result of correction

After correction for the MT gain of Arm2 during the measurement at the $\sqrt{s}$ $7\,TeV$, $m_{\gamma\gamma}s$ were reconstructed. Figure.4 shows the variation of the $m_{\gamma\gamma}s$ after the correction A and the correction A+ . The hori ontal and the vertical a is are the same as Figure.3. In Figure.4 the blac , red and blue points show the results before correction, after the correction A and after the correction A and , respectivel . The vertical dashed lines show the beginning of the lls.

correction A, the amplitude of the short term variation of $m_{\gamma\gamma}s$ during a single ll were improved from 1.5 Me to .5 Me . adding the correction $m_{\gamma\gamma}s$ decreased b about 2 Me as whole periods. In the case of Fill 11 4, which was used for this anal sis, $m_{\gamma\gamma}$ decreased from 13 . Me to 137.4 Me , that means the disagreement decreased from 3. to 1. . ecause of this improvement, s stematic errors resulting from the energ scale of Arm2 was improved b about 3 .

owever $m_{\gamma\gamma}s$ were going down b about 3 Me step b step after R N 36 . The cause of these changes is probabl a radiation damage of the scintillators because there is correlation between the corrected $m_{\gamma\gamma}$ and dose measured at the bac of the Arm2 detector.

7. Conclusions

The L Cf results at $\sqrt{s}$ 7 Te proton proton collisions had large s stematic errors resulting from the energ scale shift. It was found that one of the ma or reasons of the energ shift is the temperature dependence of the MT gains. In order to correct this e ect, the temperature variation during the L C operation because of the Joule heat from the MT breeders and the temperature di erence between the time of calibration at S S and operation at L C were considered. these corrections, energ scale shift of Arm2 was improved from 3. to 1. , and s stematic error caused b the energ scale were improved b about 3 .

owever this correction is still preliminar e orts because a temperature coe cient of MT gain was ta en from the catalog values. It is necessar to ta e into account actual temperature coe cient of each MT b the measurement.

This correction method will be applied to the Arm1 data having a larger shift of $m_{\gamma\gamma}$ than the Arm2. In near future the spectra of photons and π^0 will be updated with smaller s stematic uncertaint . This stud is also fed bac to the measurement at $\sqrt{s}$ 13 Te proton proton collisions planed in 2 15.

8. Acknowledgement

We than the C RN sta and the ATLAS collaboration for their essential contributions to the successful operation of L Cf. This wor is partl supported b rant in Aids for Science Research b M XT of Japan, the Mitsubishi Foundation in Japan and Istituto Na ionale di Fisica Nucleare INFN in Ital . The receipt of JS S Research Fellowship M and TM , INFN fellowship for non Italian citi ens M and KN and the L S Leadership evelopment rogram of Nago a niversit 'QF ' from JS S and M XT of Japan M are also ac nowledged. A part of this wor was performed using the computer resource provided b the Institute for the Cosmic Ra Reseach ICRR , niversit of To o.

References

1 O.Adriani, h s.Lett.B 7 3 2 11 12 134
2 L Cf Collaboration, h s. Rev. 6, 2 1 2 12
3 T.Mase et al., Nuclear Instruments and Methods in h sics Research A 671 2 12 12 136
4 AMAMATS ,Technical information, Metal package photomultiplier tubes R74 series and subminiature photosensor modules

Journal of Physics: Conference Series **587** (2015) 012024 doi:10.1088/1742-6596/587/1/012024

Development of Low-Noise / Low-Power Preamplifier for the Readout of Inorganic Scintillators and their Mass Production Test System

I. Keshelashvili, W. Erni, F. Müller, M. Steinacher, B. Krusche

Department of Physics, University of Basel, Klingelbergstrasse 82, CH-4056 **Basel**, Switzerland

E-mail: i.keshelashvili@unibas.ch

Abstract. The development of the preamplifer and all test procedures done by our group during the past several years at the University of Basel will be discussed. In addition, we present the results from the testing of the first 1'500 channels of the preamplifiers for the Crystal Barrel experiment using a specially developed automatized test system.

1. Introduction

Photomultiplier tubes like vacuum-photo-tetrode (VPTT) and semiconductor photosensors like avalanche photodiode (APD) have the important advantage over multi-dynode photomultipliers (PMT's) that they can be operated in reasonably strong magnetic fields. They have, however, much lower gains than PMT's, which requires well adapted preamplifiers. We have developed, the so called Basel-LNP (Low-Noise / Low-Power) preamplifier, which is a very sensitive charge to voltage converter/amplifier. The basic concept is a discrete element design preamplifier (that can be easily adapted to different requirements) for the first stage of the readout of low amplification photosensors. Two version of the LNP have been optimized for the readout of the forward end-cap of the electromagentic calorimeter of the PANDA detector, which is under design for the planned FAIR facility. One will be sued for the inner sections equipped with VPTT's, the other for the outer sections equipped with APD's. For this detector a very high dynamic range ($\sim$10'000) and high count-rates (max. 1 MHz) are expected. In addition, crystals and preamplifiers have to be operated at temperatures of -25°C, which strickly limits the acceptable energy dissipation of the LNP. Another version of the LNP has been optimized for an upgrade of the Crystal Barrel detector at the ELSA accelerator (Bonn) which consists of CsI(Tl) crystals. The upgrade from the current readout of the scintillators by photodiodes to APD's will provide so far missing timing information from the scintillators (particularly important for triggering). A special feature of this system is the compensation of gain changes of the APD's due to temperature variations by a feed-back of the temperature information on their HV supplies.

1.1. The Crystal Barrel detector

The Crystal Barrel detector [1, 2] is a large acceptance electromagnetic calorimeter designed to measure charged and neutral particles up to a few GeV. The current setup counts 1'230 CsI(Tl)

crystals. Each crystal is 16.1 X_0 radiation length thick and covers $6^o \times 6^o$ degree in polar and azimuthal angle between 30°–150° and $12^o \times 6^o$ between 150°–156° (only 30 crystals). In the past, it was designed to operate inside a strong solenoid magnet at the LEAR (Low Energy Antiproton Ring) ring at CERN. Due to the presence of a strong magnetic field, the readout of the crystals used PIN diodes. This has the large disadvantage that no timing information is provided. Now, at the Electron Stretcher Accelerator ELSA at the University of Bonn the device is no longer operated in magnetic fields and mainly used for measurements of final states with high photon multiplicity. Ideally, one would like to trigger on the photons, but since the detector provides no timing information most triggers have to rely on signals from charged particles in a cylindrical scintillating fiber detector surrounding the target. This is in particular a problem for reactions off quasi-free neutrons, where no recoil proton can provide the trigger. Therefore it will be equipped with large area ($\sim$1cm^2) avalanche photodiodes [3]. Each crystal of the detector will be read out by two APD's of the type Hamamatsu S11048(X3). The dual channel, single board Basel-LNP preamplifier will be the first stage in the front-end electronics (FEE) followed by a signal shaper and the readout electronics.

1.2. The PANDA detector

The PANDA detector will be one of the major detectors of the FAIR facility at the GSI Darmstadt. It is planned as a high luminosity, internal experiment operating with intense antiproton beams on a fixed target at the HESR (Hight Energy Storage Ring). Basically all branches of the planned experimental program [4] require an electromagnetic calorimeter [5] with excellent energy, position, and time resolution. In order to meet this requirement, the PANDA EM calorimeter will have more than 15'000 PWO crystals of 22 radiation lengths X_0 and with a front face of $\sim$2×2 cm. In contrast to the PWO crystals used in the CMS experiment at CERN, the PANDA experiment will be equipped with second generation scintillators from the recently developed material which roughly doubles the light output. Furthermore, the whole PANDA EMC will be cooled down to -25° degree Celsius because at this temperature the number of photons emitted by PWO is three times higher than at room temperature. The PWO crystals in the electromagnetic calorimeter will be distributed in three main groups: forward end-cap having 3'856 crystals, central barrel having 11'360, and backward end-cap having 528 crystals. The PANDA EMC will be placed inside the 2 T solenoid magnet which excludes the use of conventional multistage PMT's. Therefore the crystals will be read out by specially developed Hamamatsu Large Area Avalanche Photodiodes (LAAPD, tow sensors per crystal). Only in the extreme forward direction close to the beam the crystals will be equipped with Vacuum-Photo-Tetrodes (VPTT, Hamamatsu R11375-01), in order to guarantee radiation hardness and to cope with the high count rates. The two types of preamplifiers designed in Basel for this end-cap will be the only active electronic devices located inside the cold area.

2. Preamplifier

The design of the preamplifiers for the Crystal Barrel detectors allows amplification of two separate APD signals (each scintillator is read out by two APD's) with two electrically identical channels placed onto one Printed Circuit Board (PCB). The electronic circuit of a single channel is shown in Fig.: 1. The Basel-LNP preamplifier is an operational amplifier with a field-effect transistor JFET at its input with the feed-back capacitance C12 (6.8pF). At the input, we have first the decoupling capacitor C1 (4.7nF) followed by low-noise protection diodes D1 and D2. The output rise time ($\sim$40 ns) and the noise level is mainly set by the JFET, while the R11+R12 direct feed-back resistors determine the signal fall-time constant ($\sim$60μs). This preamplifier is powered by +8 V and -2 V, dissipating 12 mA and 6 mA current respectively. The maximum current with 10 kHz count-rate and at the maximum amplitude is $\sim$18 mA (+8 V) and $\sim$9 mA (-2 V).

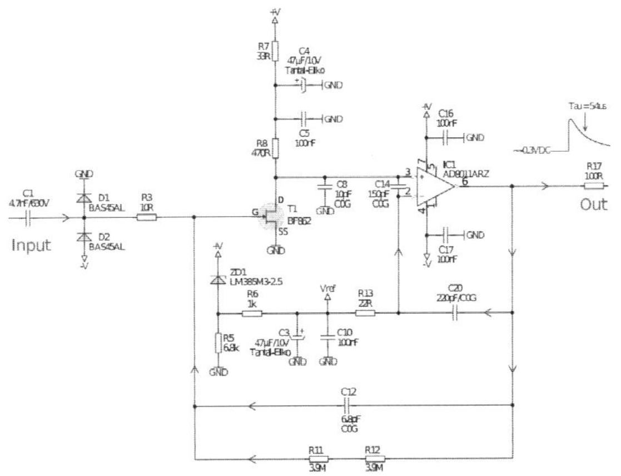

Figure 1. The circuit diagram of the Crystal Barrel preamplifier.

The preamplifier design concept for the PANDA APD and VPTT preamplifier are similar to this circuit. The main difference is the different gain, adapted to the two types of photosensors and the different energy range. The PCB design is also different due to the different geometry of the crystals. In case of the PANDA EMC, the two preamplifiers are powered symmetrically with ± 6 V. In Fig.: 2, the typical output signals (middle and zoomed), measured with a custom made test system are shown (described in the next section). The extensive laboratory tests of

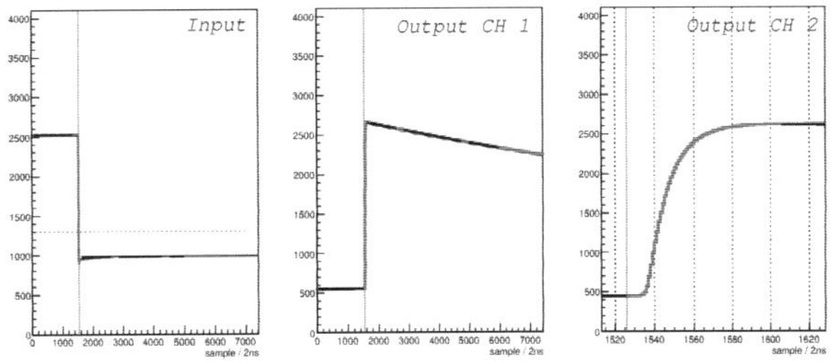

Figure 2. Typical signals form Crystal Barrel preamplifier. Left: input signal from the FPGE generator. Middle: output signal from the LNP preamplifier channel 1, about 14 μs sampling. Right: zoomed in output channel 2, showing $\sim$40 ns rising time.

the LNP preamplifiers have shown their excellent amplitude linearity (less than 1% deviation over the whole dynamic range), its immunity for pile-up effects (two signals within 1μs or 1 MHz have below 1% deviation up to 50% of maximum signal), and its radiation hardness. Irradiation with γ-rays (^{60}Co surface doses of 30kGy) showed only 2% increase of the gain, which could be explained by degradiation of the feed-back capacitance.

3. Test System

Access to the preamplifiers will be very restricted after assembly of the PANDA forward end-cap and the situation for the Crystal Barrel detector after the upgrade is similar. Therefore, all individual preamplifier modules have to undergo very detailed and careful pretesting and characterization procedures. The test system was designed for the test of all important parameters. The complete test protocol for one module runs for $\approx$ 5 minutes time. As a first step, the high voltage is set to the maximum value in order to test all preamplifier components for current leakage. The main focus in this test is to check the ceramic capacitors and resistors of the bias voltage for possible damages. The second test involves measurements with several different amplitudes and different hit rates. The raw information from a sampling ADC is saved into output files, which allows the analysis of the line shape of the signal, its time characteristics and amplitude linearity. It is also monitored that the noise level is within acceptable limits. The test procedure is generated by a Linux based computer system controlling the four main components, which are shown in Fig.: 3. The first device (starting from the left hand side), is

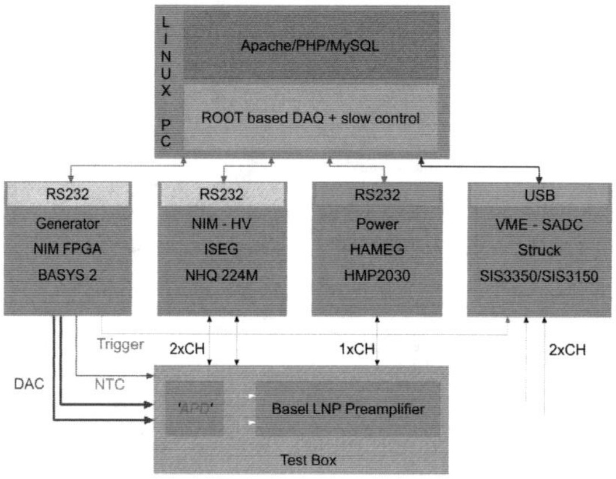

Figure 3. Setup of the test system with its main components.

a custom-built, single-width NIM module which combines a standard Digilent Basys 2 FPGA board with two standard and two custom made peripheral modules (Pmod) shown in Fig.: 4. The standard Pmods are a RS232 serial-converter interface and a 16 bit digital-to-analog (DAC) converter. They define the reference level for the output voltage generator, which uses Schmitt triggers (shown in Fig.: 4). The forth custom peripheral module generates the TTL trigger signal for the flash ADC. The high voltage test (second module from the left), uses a dual channel, high

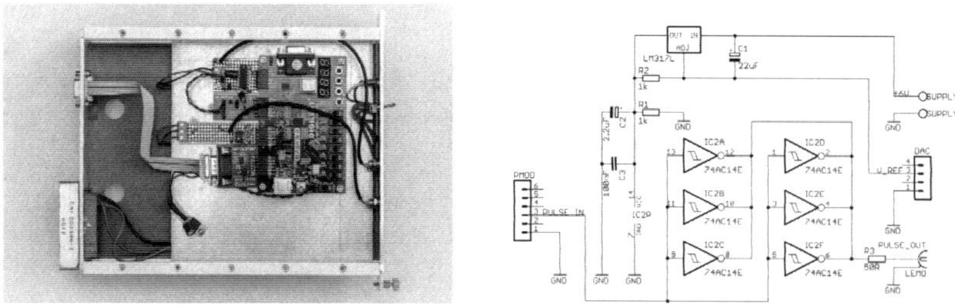

Figure 4. Left: Custom made NIM–FPGA based pulse generator module. Right: Schematic of the output driver.

precision ISEG NHQ 224M module (single width NIM standard), which is RS232 compatible and capable to measure voltage and current with 1 nA resolution [7]. For the monitoring of the power consumption during the test, the programmable power supply HAMEG HMP2030 has been used [8]. It is also RS232 compatible and is monitoring the voltage (+8 V and -2 V) and the corresponding current consumption. The third channel of the power supply has been used as a relay to switch from the high voltage test to the amplitude or signal testing mode.

It was especially electrically separated, in order to avoid any possible damage of the amplifier. The fourth and last module is the VME readout system capable to digitize all input and output signal shapes using a 500MS/s flash ADC from Struck [10]. The readout of the VME bus was done with a high performance USB to VME interface [9].

The signal generated by the FPGA puls generator (shown left on Fig.: 2) is passively split (R1, R2, R3 resistors) inside the test box (Fig.: 5) to simulate charge bursts (C1, C2 capacitors, generate charge out of voltage) of the photo sensor. The capacitors C3 and C4 on the right hand side of Fig.: 5 are equivalent to the capacitance of the APD under bias voltage. The

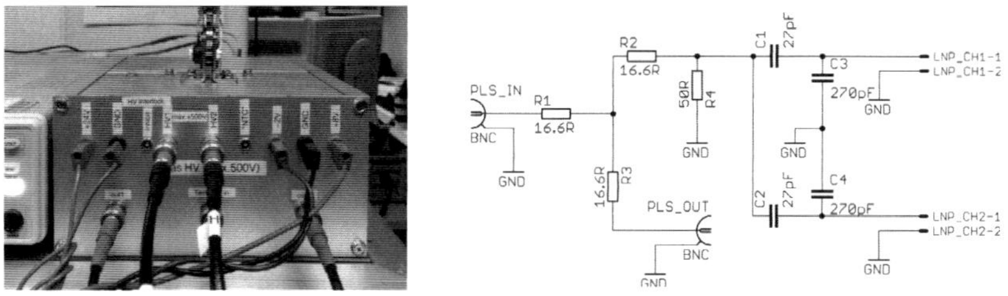

Figure 5. The test box for the Crystal Barrel preamplifiers.

software controlling the test system has a modular architecture and is based on the CERN ROOT package [11]. It has a graphical interface and can be easily customized to select the optimal test procedure.

4. Results

So far, we have tested 756 dual channel preamplifiers, which is effectively more than 1500 separate channels of the Crystal Barrel type preamplifier. Out of them, we have rejected only about 2.5% due to different failures. This results where much below of the expected rejection rate. Less than 1% of the preamplifiers where electrically damaged, probably due to malfunctioning of some electric components. The rest was rejected due to the so-called tombstone effect caused by a special soldering procedure which, in fact, created this problem for the surface mountable device (SMD) components, shown in Fig.: 6. As a first step, the components of the bias voltage

Figure 6. The rejected Crystal Barrel type preamplifier due to the tombstone effect (this is an SMB component that is partially or completely lifted off one end of the soldering pad).

are tested for their absolute maximum voltage of +500V (that will never be the case in the real experiment). In this test, we simply secure the normal operation of the resistors and the ceramic capacitors. The whole HV test takes about 30 sec with 100V/sec ramping time. During this test, the voltage stability test and the current leakage is monitored, see left hand side of

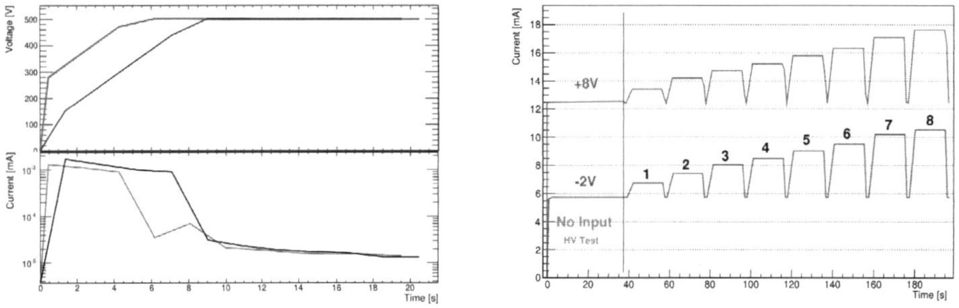

Figure 7. Left hand side: High voltage test, top the voltage and bottom the current vs time. Right hand side: Current consumption of the preamplifier during the test.

Fig.: 7. The right hand side of the same picture shows the current consumption during the whole test. The first ~40 seconds shows the minimal power consumption (without input signal) of the preamplifier. After the HV test, the power consumption of eight different amplitudes, evenly distributed over the whole dynamic range of the output, is measured with the maximum expected count-rate of 10 kHz. The gain of each channel is extracted by linear fitting of the eight different

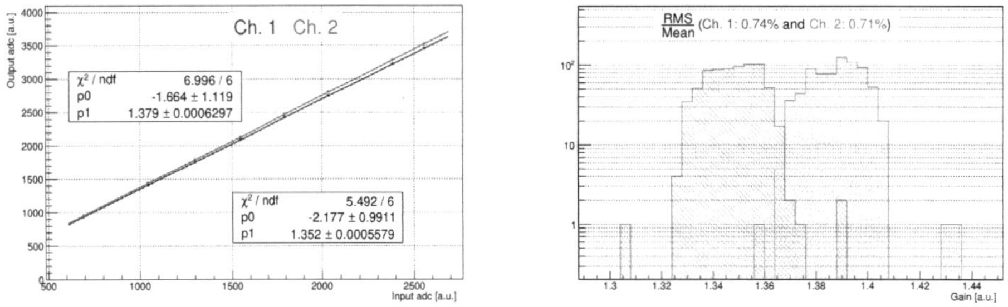

Figure 8. Left hand side: The graph of an output voltage dependence vs. input voltage and Right hand side: The gain distribution of the first (red) and second (green) channels is shown.

amplitudes. It is shown at the lefthand side of Fig.: 8. The slope (gain) and the amplitude linearity is extracted for each preamplifier channel and is stored into the database (together with other important parameters) for a quick overview and the selection of the best preamplifier modules for the experiment. The right hand side of Fig.: 8 shows ~ 0.7% (RMS/Mean) spread of the gain distribution, which is significantly lower than the light output differences between different crystals.

References

[1] E. Aker et al. 1992 NIM A321 69.
[2] G. Suft et al. 2005 NIM A538 416
[3] Christoph Wendel (for the CBELSA/TAPS Collaboration) 2009 J. Phys.: Conf. Ser. 160 012006
[4] Physics Book - PANDA `http://www-panda.gsi.de/archive/public/panda_pb.pdf`
[5] Technical Design Report - PANDA EMC `http://www-panda.gsi.de/archive/public/panda_tdr_EMC.pdf`
[6] Basys^{TM}2 Spartan-3E FPGA Board `http://www.digilentinc.com/BASYS2`
[7] ISEG NHQ 224M `http://www.iseg-hv.com/en/products/product-details/product/15/`
[8] HAMEG HMP2030 `http://www.hameg.com/0.723.0.html`
[9] SIS3150-USB USB2.0 to VME interface `http://www.struck.de/sis3150usb.htm`
[10] SIS3350 4 Channel 500 MS/s 12-bit ADC/Digitizer `http://www.struck.de/sis3350.htm`
[11] CERN ROOT library `http://root.cern.ch`

Journal of Physics: Conference Series **587** (2015) 012025 doi:10.1088/1742-6596/587/1/012025

A Triggerless readout system for the $\bar{P}ANDA$ electromagnetic calorimeter

M. Tiemens on behalf of the PANDA Collaboration

KVI - Center For Advanced Radiation Technology, University of Groningen, Zernikelaan 25, 9747 AA Groningen, The Netherlands

E-mail: m.tiemens@rug.nl

Abstract. One of the physics goals of the future $\bar{P}ANDA$ experiment at FAIR is to research newly discovered exotic states. Because the detector response created by these particles is very similar to the background channels, a new type of data readout had to be developed, called "triggerless" readout. In this concept, each detector subsystem preprocesses the signal, so that in a later stage, high-level phyiscs constraints can be applied to select events of interest. A dedicated clock source using a protocol called SODANET over optical fibers ensures proper synchronisation between the components. For this new type of readout, a new way of simulating the detector response also needed to be developed, taking into account the effects of pile-up caused by the 20 MHz interaction rate.

1. Introduction

A new experiment is being developed to perform precision measurements in the energy range between 1.7 and 5.5 GeV, where Quantum Chromodynamics becomes non-perturbative for charmonium systems[1]. The experiment is called $\bar{P}ANDA$ ($\bar{P}$ ANnihilations at DArmstadt), and one of the mysteries that it hopes to unravel is the generation of hadron masses, as for these composite particles, the sum of the parts does not add up to the whole (in fact, for e.g. protons, the sum of the valence quark masses only adds up to about 1% of the proton mass [1]). Furthermore, over the last years, several new charmonium-like hadrons were discovered, called XYZ states [2], possibly consisting of more than 3 quarks. $\bar{P}ANDA$ will be able to perform precision measurements on the properties of these and other predicted, as-of-yet unobserved (such as glueballs), exotic states of matter, which can be directly populated through the proton-antiproton interactions.

The $\bar{P}ANDA$ experiment will be build at the new Facility for Antiproton and Ion Research (FAIR), currently under construction in Darmstadt, Germany. FAIR's main accelerator will be used to create an antiproton beam, which will be stored in the High Energy Storage Ring (HESR). The $\bar{P}ANDA$ experiment will be placed inside the HESR. The antiproton beam impinges on a proton target, made of frozen hydrogen pallets [3]. In these so-called fixed-target experiments, particle production will be forward boosted, leading to an asymmetric detector design, see Figure 1. The $\bar{P}ANDA$ detector will follow the "traditional" onion-like design, where the detector subsystems are placed nested in each other to be able to measure all the properties

[1] Charmonium is the bound state of a charm and an anticharm quark

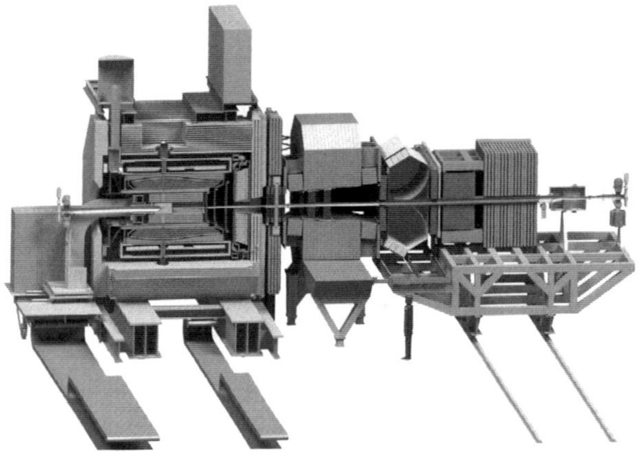

Figure 1. CATIA drawing of the $\bar{P}ANDA$ detector. *Courtesy:* A. Cebulla, FZ Juelich, 2014.

of the particles produced in the annihilations. This paper will mainly focus on one subsystem, the electromagnetic calorimeter (EMC). The scintillation crystals are made of second generation lead tungstate ($PbWO_4$, PWO), featuring an improved light yield over first generation PWO by a factor 2. The crystals are cooled down to -25 °C, improving the light yield by another factor 2. This will enable an energy resolution of 1% at 1 GeV. The crystals will be 20 cm long, which is estimated to be about 22 radiation lengths, with the front and back face about 2×2 cm. The complete EMC consists of three parts: a Barrel, containing 11,360 scintillation crystals and a Backward and Forward Endcap, containing 528 and 3856 crystals, respectively.

2. The readout of the detector

Traditionally, the data stream being produced is reduced using hardware triggers, based on the topology of the response in one of the detector subsystems. However, for $\bar{P}ANDA$, this approach is insufficient. The reason for this, is that there are many background channels with a similar final-state topology as the decays of interest. A typical example is the decay of a recently discovered exotic state candidate, the $Y(4260)$. A strong decay mode is through a J/ψ particle and two neutral pions. Prominent final-state products of these particles feature an electron-positron pair and 4 photons, respectively:

$$p\bar{p} \rightarrow Y(4260) \rightarrow J/\psi\pi^0\pi^0 \rightarrow e^+e^- + 4\gamma$$

However, there are many decay modes with a similar final-state topology, some examples of which are shown here:

$$p\bar{p} \rightarrow J/\psi\eta\eta \rightarrow e^+e^- + 4\gamma,$$
$$p\bar{p} \rightarrow J/\psi\eta\pi^0 \rightarrow e^+e^- + 4\gamma,$$
$$p\bar{p} \rightarrow J/\psi\pi^+\pi^-\pi^0\pi^0 \rightarrow e^+e^- + 4\gamma.$$

The presence of this large number of background channels with a similar final-state topology renders the triggered readout unusable, since the hardware triggers, which in this case will be based on final-state photon multiplicities, will allow events from these background channels to pass, so a sufficient reduction factor cannot be achieved when searching for $Y(4260)$ events.

2.1. Solution
The proposed solution for $\bar{P}ANDA$ is to reconstruct events in realtime during data taking, after which high-level physics-inspired constraints can be imposed to perform event selection before sending data to storage. In the previous example, taking the invariant mass of the final state particles into account allows to distinguish between the desired signal and the one generated by background events. To illustrate this, the concept of this "triggerless" readout of the detector will be explored below for the EMC.

3. The readout chain for the EMC
For the $\bar{P}ANDA$ EMC, the readout chain can be divided into *five parts*:

- The front-end electronics, consisting of *photosensors, preamplifiers,* and *digitisation modules.* These are placed inside the detector, and will therefore be exposed to high levels of radiation.
- An intelligent, self-triggering, dead-time-free readout, starting with the *digitisation modules*, and including *data concentrators* and *compute nodes.* The data concentrators and compute nodes are placed outside the detector. Data will be transported from the digitisation modules using high-speed (5 Gbps) optical links.

3.1. Photosensors
For the EMC, the readout starts with the photosensors. Two types will be used:

- Hamamatsu Silicon Avalanche Photo Diodes (APDs). These feature a high quantum efficiency in the order of 70-80%. These are large area APDs, with an active area of 7×14 mm, two of which will be fitted to each crystal to cover most of the back face. This increases the light collection, offers redundancy (see Section 3.4) and allows the detection of a nuclear counter event.
- Hamamatsu Vacuum Photo Tetrodes (VPTTs). The quantum efficiency for these is around 20%. VPTTs are radiation hard and can handle high hit rates, and will therefore be used in the most forward region of the EMC (i.e. at the centre of the Forward Endcap). One VPTT is needed for each crystal.

3.2. Preamplifiers
The preamplifiers have been discussed in detail by I. Keshelashvili, see [4]. See also [5] for the second type of preamplifier (the APFEL ASIC) that will be used in the $\bar{P}ANDA$ experiment.

3.3. Digitisers
After the preamplifiers, the signal is fed to a 64-channel, 14-bit Sampling Analogue-to-Digital Converter (SADC), which continuously digitises the input signal at a rate of 80 MHz. Since each APFEL preamplifier offers two outputs, and the LNP preamplifiers feature a dedicated splitter, addressing inputs with two different gains on the digitisers, every digitiser can read out 32 photosensors. The hardware design of this device was made by P. Marciniewski, and the firmware was developed at KVI-CART [6].

After the digitisation stage, two Xilinx Kintex-7 Field-Programmable Gate Arrays (FPGAs), a chip with fully programmable logic blocks for fast, task-specific calculations, are used to perform preprocessing of the signal. First, DC offsets in the baseline of the signal are corrected for, so all amplitudes are measured with respect to the same reference point. The next stage is pulse detection. If the amplitude of a pulse crosses a set threshold, a discrimination is made between single or so-called pile-up pulses. A pile-up of pulses occurs when a second particle hits

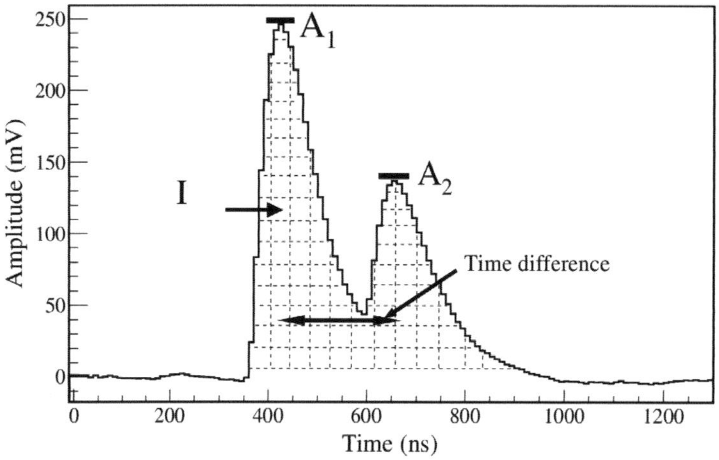

Figure 2. A typical pile-up pulse (after digitisation) [8]. The symbols are explained in the text.

the detector within the response time, superimposing the second pulse on the first (see Figure 2). For an interaction rate of 20 MHz, assuming a Poisson distribution of events, the probability for a pile-up to occur is about 9%, and for multiple pile-up, it is less than 0.5%. The digitiser discriminates by considering the ratio between the amplitude and the area [7, 8] of the peak, whose relation[2] is given by:

$$I = k(A_1 + A_2), \tag{1}$$

where I is the area of the waveform, k a constant to be determined by calibration, and A_1 and A_2 the amplitude of the first and the second peak, respectively. Rewriting this into a ratio (taking the ratio with the amplitude of the first pulse):

$$I/A_1 = k(1 + A_2/A_1), \tag{2}$$

if A_2 is zero (i.e., no pile-up pulse), $I/A_1 = k$, therefore by setting a threshold for the ratio, it is possible to discriminate between the two cases. In Figure 3, a measurement on this ratio performed by G. Tambave *et al* [7] is shown.

If a pile-up pulse is identified, the complete waveform is sent to the data concentrators. If a single pulse was detected, the amplitude of the pulse, being a measure for the energy deposited in the crystal, is extracted, and the associated timestamp is calculated using Constant Fraction Discrimination. Finally, only this information is sent on to the next stage. This process is called "Feature Extraction".

Some additional features of the SADCs include:

- A compact design (160×100 mm), which is needed due to space restrictions in the detector volume.
- Advanced debugging capabilities, including the possibility to read out the baseline and complete waveform of pulses.
- Fast rebooting (a few milliseconds) after radiation damage to the device's configuration register is detected.

[2] This relation only holds if the pulse shape is stable, which was found to be the case [7]

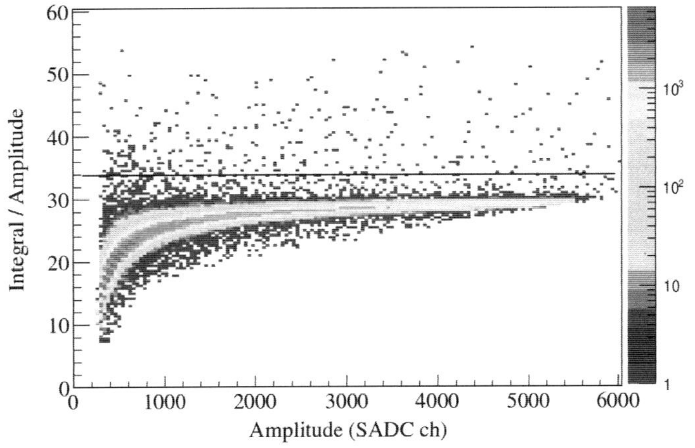

Figure 3. The offline calculated I/A ratio. The majority of points (coming from the more common single pulses) lie at a line equal to $I/A = k$, allowing to set the black horizontal line as a threshold for discriminating pile-up pulses (at 3σ significance above k).

3.4. Data Concentration

The next stage in the readout chain are the data concentrators, whose current hardware is based on TRBv3 developed at GSI [9]. For redundancy, each APD of a crystal will be read out by a separate digitiser, therefore, if one of the digitisers ceases to function, it is still possible to read out the crystals connected to this digitiser. By recombining this information, the amount of data can be reduced.

3.4.1. Pile-up Recovery The data concentrators feature a recovery algorithm for the pile-up pulses being sent from the digitisers[3] [8]. The algorithm is inspired by Eq. (2). From the calibration to single pulses, the constant k can be determined, and the total integral I and amplitude A_1 of the first pulse can be obtained from the waveform. A_1 can be determined unambiguously if the second pulse comes after the rise time of about 50 ns, as then, it won't affect the amplitude of the first pulse. The amplitude A_2 of the pile-up pulse then follows directly from Eq. (2), since all other parameters are determined. To test this procedure, an experiment was performed with a tagged photon beam in MAMI, Mainz [7]. The amplitude recovery was tested by taking the amplitude of the first pulse as A_1 and recovering the amplitude of the second (left picture in Figure 4), and by taking the amplitude of the second pulse as A_1 and recovering the first (right picture in Figure 4). From the FWHM of the recovered pulses, Figure 4 shows that in both cases it is possible to recover the amplitude of the other pulse, showing the viability of the method.

3.4.2. Synchronisation In experiments using a hardware trigger, the synchronisation between the different parts of the detector is also automatically provided by the trigger itself. The absence of a hardware trigger in the $\bar{P}ANDA$ experiment calls for a different means to ensure proper synchronisation. To this end, a dedicated clock source using a protocol called SODANET over optical fibers is being used, that provides synchronisation to the data concentrators, which then redistribute the clock signal to the digitisers. The SODANET clock has a jitter of about

[3] As can be seen from Figure 2, the recovery of pile-up pulses is limited. If the second pulse follows the first one within the rise time (about 50 ns), the two pulses will appear as one.

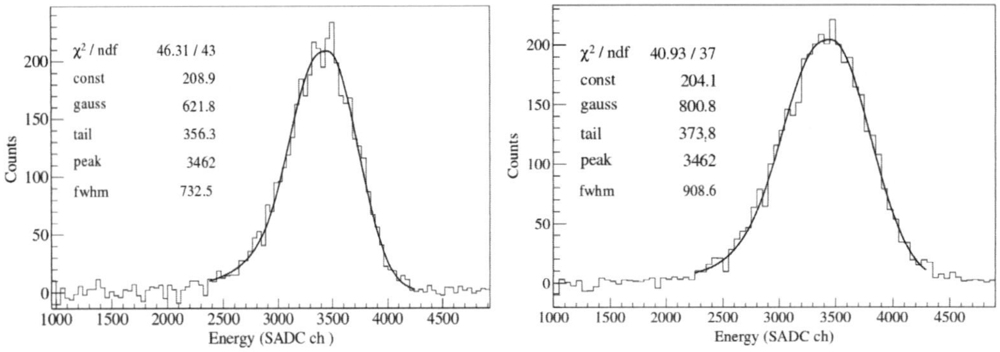

Figure 4. Recovered pile-up pulses, by taking the first peak as A_1 (left), and the second peak as A_1 (right). The figure is taken from [7].

20 ps and continuously monitors the propagation time of the clock signal, compensating for fluctuations when neccesary.

3.5. Compute Nodes
The final event selection takes place in the Compute Nodes. Before data is sent to them, it is sorted in the so-called Burst Building Network. This network collects information belonging to the same time period from the data concentrators of all subsystems, making complete event information for this period available to the Compute Nodes. This information is sent to one of the FPGAs on the Compute Nodes for event selection. In the example of section 2, requiring the invariant mass of combinations of two photons to lie at the pion mass eliminates any background containing η particles (as the mass of the η particle is much higher), and requiring the combined invariant mass of all final state particles to have a peak at 4260 MeV eliminates more background channels. Only events that satisfy these constraints will be sent to storage for offline processing and analyses. By using techniques such as these, a data reduction factor of about 10^3 is expected to be achieved (for all physics channels of interest), leaving 200 MB/s being sent to storage.

4. Status of hardware development
As of the moment of this writing, there are working prototypes for all parts of the readout chain, with the performance of these devices close to their required ones. The firmware, with almost complete functionality, is developed for the digitisers, data concentrators and clock-distribution network (SODANET). The hardware for the EMC digitisers is in the final stage of development, but for the data concentrators, a redesign is neccesary to increase the speed of the optical links.

Implementation and comprehensive testing of an online version of the pile-up recovery algorithm has to be done, as this has only been tested offline, and the topology of the Burst Building Network needs to be optimised. For the latter, a good understanding of the expected data flow is required.

5. Simulating the Data Flow
In order to determine the requirements for the detector components and evaluate the performance of the reconstruction algorithms, an accurate simulation of the expected data flow is needed. The following discussion focusses on the EMC, but most statements hold for other subsystems as well.

In a typical simulation, an event is generated, which then creates several energy depositions in the EMC crystals, called hits. After this, the next event is generated, along with its associated hits, after which this process is iterated. Each generated event is essentially independent of the previous one. The $\bar{P}ANDA$ experiment has to deal with very high interaction rates of 20 MHz, leading to individual crystal hit rates in excess of 500 kHz in the most forward region, creating pile-up of hits from different events in the detector. Furthermore, protons and antiprotons are composite particles, allowing a direct production of a large variety of secondary particles with different lifetimes. Furthermore, the different detector subsystems also have different response times. This will increase the probability for hits of one event overlapping with those from another further. Hence, it makes sense to adjust the simulation procedure accordingly, switching from this sequential (also referred to as "event-based") simulation type to one which takes pile-up into account. This is called a "time-based" simulation.

5.1. The Time-based Simulation

For the time-based simulation, first, a data set is generated as described above. Then, the timestamps of the hits are reassigned, creating pile-up between events; an impression of which is shown in Figure 5. The reconstruction algorithms will need to disentangle this data stream,

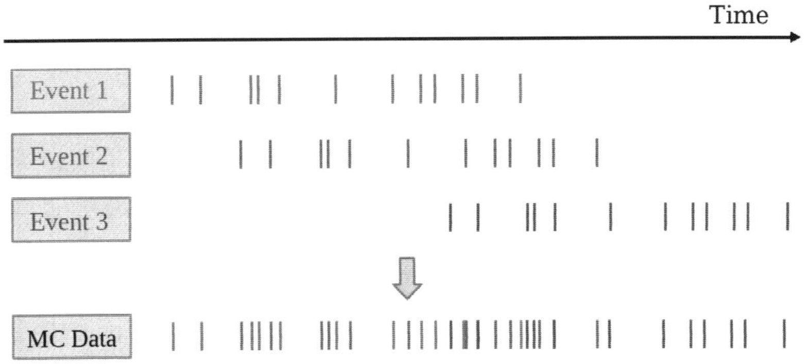

Figure 5. Artist's impression of the generation of the time-based MC data flow. *Courtesy:* T. Stockmanns, FZ Juelich, 2014.

taking pile-up and geometrical overlap into account. Figure 5 shows that the data stream can be cut into bunches in time. Apart from being forward-boosted, particle decays will essentially be isotropic, implying that it is very unlikely for hits from different events within such a time bunch to end up in the same part of the detector, allowing to separate events in space.

This presumption needs testing. The particle shower created by particles hitting the EMC crystals will in general spread out through several of them, creating groups of hits, called clusters. The hits in these clusters will need to be recombined in order to find the particle that spawned the shower. This is called cluster finding, or clustering, for short. An algorithm to perform this task is currently under development by the author, but a preliminary version already shows, using the MC thruth information, that only about 1% of these clusters contain hits from different events. A large number of low energy split-off 'clusters' (mostly consisting of a single hit) was observed, likely contributing to the "contamination" of the clusters. Such low energy split-offs occur when one of the particles in the particle shower scatters out of the crystal and into a neighbouring one. This is often a next-to-nearest neighbour, as low energy particles have a larger mean free path in the scintillation material.

The current tests were done using mono-energetic, 1 GeV photons. The next step would be to check to performance of the algorithm for a more realistic case, involving particles with multiple photons in the final state, e.g.

$$\bar{p}p \to h_c \to \eta_c\gamma \to \pi^0\pi^0\eta\gamma \to 7\gamma.$$

Once the simulation tests are successful, the algorithm will be implented into an FPGA for testing in an experiment with one of the EMC prototypes.

6. Summary

This paper discussed the new triggerless readout concept that is envisioned for the future $\bar{P}ANDA$ experiment. This new type of readout is needed because $\bar{P}ANDA$ will have a very high interaction rate, and because the topology of the final-state particles is very similar for the decay channels of interest and its associated background channels, rendering the conventional approach for data reduction unusable.

For the same reason, the simulation of the data flow also required adaptations. The time-based simulation takes into account the effects of pile-up, i.e. when hits of different events start to overlap. An algorithm for clustering, currently under development, is part of the solution to disentangling the resulting data flow.

References

[1] Beringer J *et al* (Particle Data Group) 2012 *Phys. Rev.* D **86** 010001
[2] Liu X 2014 An overview of XYZ new particles *ArXiv* 1312.7408
[3] PANDA Collaboration 2012 Targets technical design report *http://www-panda.gsi.de/html/reports.php*
[4] Keshelashvili I 2014 Development of Low Noise/Low Power Preamplifier for the readout of inorganic scintillators and its mass production test system *This proceedings series*
[5] Wieczorek P and Flemming H 2010 Low noise preamplifier ASIC for the PANDA EMC *IEEE Nucl. Sc. Symp. Conf. Rec. (NSS/MIC)* p1319-22
[6] Kavatsyuk M, Hevinga M, Konorov I, Lemmens P J J, Marciniewski P, Schakel P, Schreuder F, Speelman R, Tambave G *et al* 2011 Trigger-less readout electronics for the PANDA electromagnetic calorimeter *IEEE Nucl. Sc. Symp. Conf. Rec. (NSS/MIC)* p43-47
[7] Tambave G 2013 Electromagnetic calorimeter studies for charmonium physics (PhD thesis - University of Groningen) *http://irs.ub.rug.nl/ppn/356608891*
[8] Tambave G *et al* 2012 Pulse pile-up recovery for the front-end electronics of the PANDA electromagnetic calorimeter *JINST* **7** 11001
[9] Frohlich I, Kajetanowicz M, Korcyl K, Krzemien W, Palka M, Salabura P, Schrader C, Skott P, Strobele H *et al* 2007 A general purpose trigger and readout board for HADES and FAIR-experiments *15th IEEE-NPSS Real-Time Conference* p1-6

New EMC electronic channel for the SND detector at VEPP-2000 e^+e^- collider

I K Surin[1,*], M N Achasov[1,2], A G Bogdanchikov[1], V P Druzhinin[1,2], V B Golubev[1,2], S V Koshuba[1,2], D P Kovrizhin[1,2], S I Serednyakov[1,2], A I Tekut'ev[1] and Yu V Usov[1]
[1]Budker Institute of Nuclear Physics, 11 akademika Lavrentieva prospect, Novosibirsk, 630090, Russia
[2]Novosibirsk State University, 2 Pirogova street, Novosibirsk, 630090, Russia

* corresponding author e-mail: I.K.Surin@inp.nsk.su

Abstract. A new measuring electronics channel with FADC is proposed for the electromagnetic calorimeter of the SND detector, taking data at e^+e^- collider VEPP-2000. This system is intended to raise detection efficiency of the low-speed anti-neutrons, produced in $e^+e^-\rightarrow$n anti-n reaction. The proposed channel provides time resolution of about 1 ns.

1. Introduction

One of the important experiments with SND[1] at e^+e^- collider VEPP-2000[2] is $e^+e^-\rightarrow$ anti-n reaction cross measurement section near threshold[3]. Low speed of the anti-neutrons allows to use time measurements for event selection of this process. Time delay between anti-neutrons annihilation signal and signal from ultrarelativistic particles is about from 3 to 9 ns. Existing electronic channel can't provide time resolution better than 5 ns. Due to this, development new electronic was started. This new channel is also needed for EMC electronics operation rate increasing.

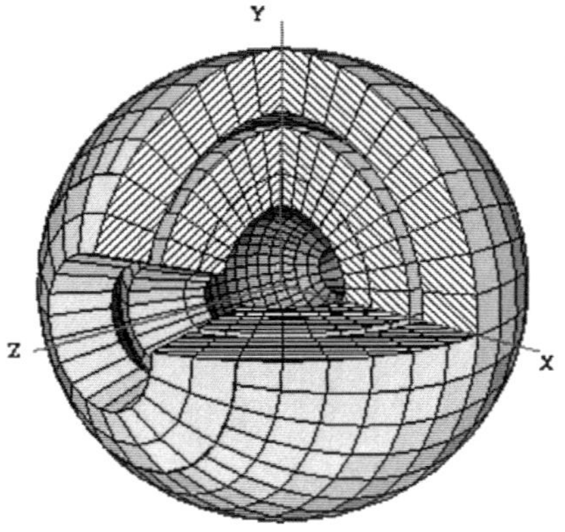

Figure 1. SND calorimeter general parameters:
Total weight of NaI – 3.5 tons
Total crystals – 1632
Thickness – 13.4 X_0NaI=(2.9+4.8+5.7) X_0 (34.7 cm)
VPT readout
$0.9\cdot 4\pi$ solid angle
Angular segmentation $\Delta\varphi=\Delta\theta=9°$
Energy resolution $4.2\% / \sqrt[4]{E(GeV)}$
Angular resolution $0.82° / \sqrt{E(GeV)} \oplus 0.63°$

2. Brief overview of the SND EMC

The three-layer spherical NaI(Tl) electromagnetic calorimeter (fig. 1) is the main part of SND. Each calorimeter layer is divided into separate counters in the following way: in the azimuthal direction it is divided into 40 equal parts by planes with 9 degrees angle between adjacent ones. The total solid angle covered by the calorimeter is equal to 0.9 of 4π (more parameters see below). As photosensitive devices for the calorimeter counters the vacuum phototriodes are used. The quantum efficiency of their photocathodes is about 15%, average gain is 10 and light collection efficiency is about 10%. Signals from phototriodes are amplified by charge-sensitive preamplifiers located directly on the counters. Output signals are transferred to shaping amplifiers via 20m-long twisted pairs. For trigger requests the calorimeter crystals are logically organized into "towers" (fig. 2). A tower consists of counters located within 18 degrees interval in the same polar and azimuthal directions in all three layers.

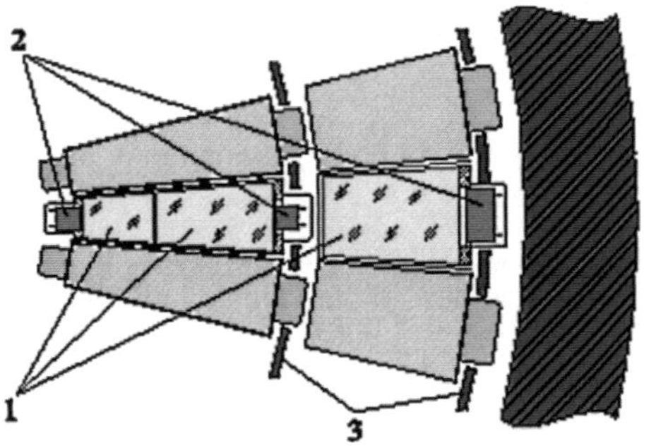

Figure 2. NaI(Tl) crystals layout inside the calorimeter: 1 — NaI(Tl) crystals, 2 — vacuum phototriodes, 3 — aluminum supporting hemispheres.

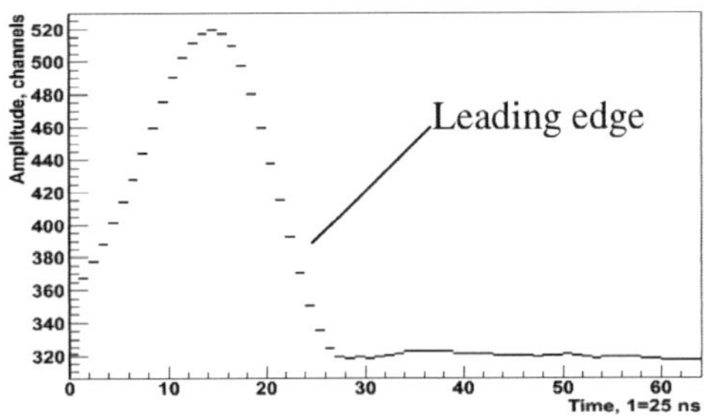

Figure 3. Signal oscillogram.

3. New EMC electronic channel

Figure 4 is illustrated the new EMC electronic channel. The signal from VPT (or calibration generator) comes to module containing computer-controlled attenuator and shaping amplifier and further goes to the new FADC module (24 channels, 12 bit, freq 40 MHz). FLT signal is used as "stop" for FADC oscillogram reading. T module provides time measurement between FLT signal and the pulse of 40 MHz clock generator (it's needed for time resolution measurement).

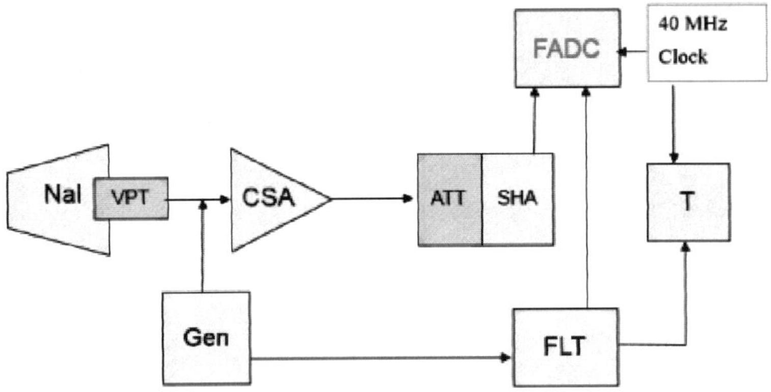

Figure 4. FADC – flash ADC (24 channels, 12 bit, freq 40 MHz), NaI – NaI(Tl) scintillator, VPT – vacuum phototriode, CSA – charge-sensitive preamplifier, Gen – calibration generator, SHA – shaping amplifier, ATT – computer-controlled attenuator, FLT – first-level trigger, T – time-to-digital converter

Obtained by FADC oscillograms are processed for measurement of time and amplitude resolution. Several methods of oscillogram shape fitting were investigated. As a result of this investigation, optimal method of fitting was found. The method is based no approximation signal oscillogram by cubic spline.

Time resolution is calculated as a difference between time measured by T module and fractional tick (a time less than 25 ns) obtained by oscillogram fitting.

4. Measurement results

Pulse calibration generator is very important for all these measurements, because its signal has the same shape as a signal obtained from VPT output. Using this generator gives us opportunity to find correct experimental devices settings for experiments with e^+e^- beams.

To obtain objective data about amplitude and time resolution of new electronic channel the measurements have to be carried out in real experiments with beams. In this way we can consider possible outside influences to these measurements.

Figures 4 and 5 illustrate dependences of time and amplitude resolutions from energy deposition inside single NaI(Tl) crystal. The results were obtained using the calibration generator.

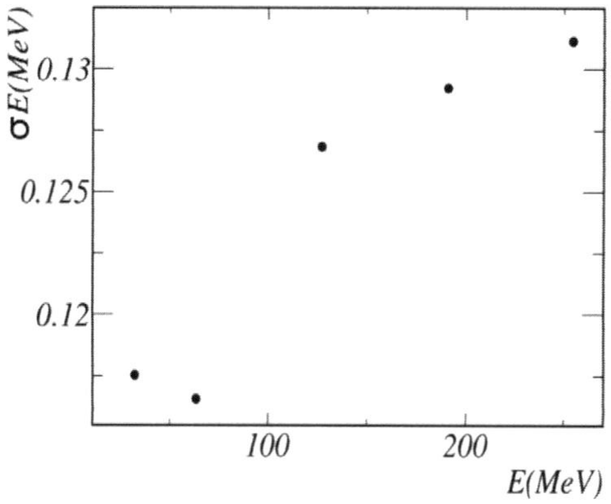

Figure 4. Amplitude resolution

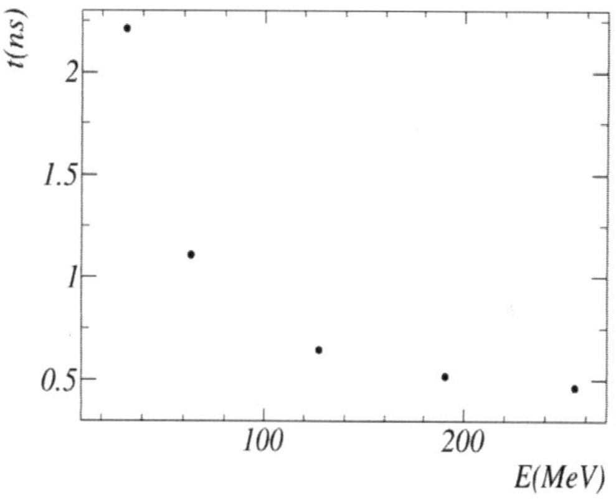

Figure 5. Time resolution

As you can see, the amplitude resolution has weak dependence on energy deposition. High amplitude resolution better than 0.15 MeV has been obtained.

Energy deposition dependence of time resolution is shown in fig. 5. Such kind of dependence indicates, that new electronic channel will provide high time resolution about 1 ns or better, if energy deposition in crystal is more than 100 MeV.

Obtained time resolution is determined by several factors. There are CSA noise, signal shape fluctuation caused by statistical fluctuation of photoelectron flux, discreteness of FADC (25 ns) and T module (0.15 ns).

Time resolution was also measured in real experiment SND at e^+e^- collider VEPP-2000 (see fig. 6). Beams energies were c.m.s. energy about 1 GeV and currents of electron and positron beams were about 30 mA. If energy deposition is more than 200 MeV, time resolution is being about 1.2 ns. This result already has corresponded requirements for new electronic channel.

5. Conclusion

New EMC electronic channel for the SND detector has been developed and tested. High amplitude resolution better than 0.15 MeV has been obtained. Time resolution ~1.2 ns has been obtained.

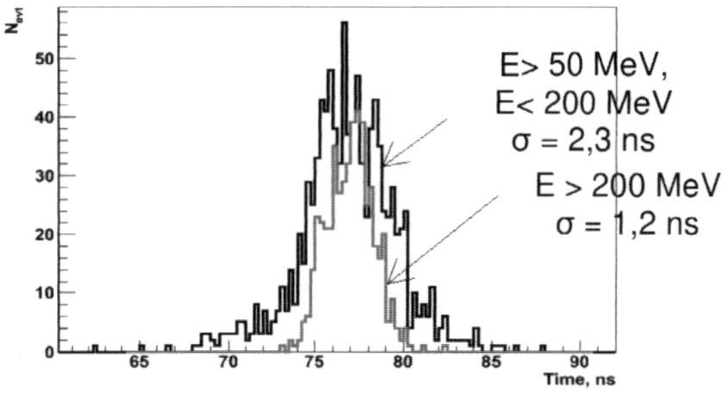

Figure 6. The time resolution, which was obtained on experimental Bhabha events.

Acknowledgments
This work was supported by the Ministry of Education and Science of the Russian Federation, RFFI grants 12-02-00065-a, 12-02-01250-a, 13-02-00375-a, 13-02-00418-a, 14-02-00129-a, 14-02-31375-mol_a, Russian Federation President's grant for supporting leading scientific Schools NSh 2479.2014.2, Russian Federation President's grant MK-4285.2014.2

References

1. M. N. Achasov et al, *Nuc. instr. Meth. A* vol 449, p 125 (2000)
2. D. Berkaev et al, *Nucl. Phys. Proc. Suppl* vol 225-227,p 303, (2012)
3. A.A.Botov, A.D.Bukin, D.A.Bukin, V.B.Golubev, V.P.Druzhinin, S.I.Serednyakov, K.Yu.Skovpen, *Nucl. Phys. B, Proc. Suppl.* vol 162, p 41 (2006).

Journal of Physics: Conference Series **587** (2015) 012027 doi:10.1088/1742-6596/587/1/012027

Progress of the HERD detector

Zhigang Wang , Ming Xu

Institute of High Energy of Physics, Chinese academy of science

E-mail: wangzhg@ihep.ac.cn

Abstract. The High Energy cosmic Radiation Detection facility, HERD, is proposed for the observation of high-energy electrons, gamma-rays and cosmic-rays onboard China's Space Station. The HERD has a capability to observe the electrons(without separation between e+ and e-) and gamma-rays in 10 GeV-100 TeV and the cosmic-rays in 100 GeV-1000 TeV with high energy resolution . The HERD is advantageous in terms of energy resolution, geometrical factor and energy detection range.

1. Introduction

Dark matter is one of the most important mysteries of physics today. Accounting for over a quarter of the universe's mass-energy balance, it can be observed indirectly through its interaction with the visible matter but has yes not to be detected. Cosmic ray discovered in 1912, their sources and propagation mechanisms are still subject of intense research. The all–particle energy spectrum of cosmic rays around PeV range the so-called knee is observed, but experimental data in the knee region contradict each other.

The HERD mission onboard China's Space Station is planned for operation starting around 2020. It is designed for observing very high energy electrons [1] [2] and gamma rays [3] as well as protons and heavy nuclei. The objective of the HERD mission is to explore a new frontier at the higher energies for the origin of cosmic-rays(CR), the propagation of CR and to search for dark matters [4] [5]. We will measure electrons and gammarays from 10 GeV to 100 TeV, with an excellent energy resolution. At the same time we will measure protons and heavy nuclei from 100 GeV to 1000 TeV.

2. HERD instrument

A schematic configuration of the HERD detector is presented in Fig. 1. The HERD detector consists of two main parts: the micro silicon strip tracker, five sides covered of the instrument, for incident particle trajectory and nucleon charge measurement; the LYSO cubic homogeneous electromagnetic calorimeter, for the incident gamma and electron energy measurement;

The micro silicon strip tracker consists of a top part and 4 side parts. The top tracker as presented in Fig. 2, is constituted of silicon strips and thin tungsten foils, the latter act as converter for gamma-rays. There are seven tracking plane of x-y layers of silicon strips, each with the dimension of 100 cm ×100 cm, and five converter layers of tungsten interspersed into layer 2-6, with thickness 3 × 1 mm + 2 ×2 mm = 2 X0. The first layer of silicon determines if the incident particle is a gamma or a charged particle. The top tracker provides charge identification, trajectory measurement, back scatter rejection and some early shower development of gamma and electron. The other four sides are covered with silicon strip detector the same as the top one, only with the dimension of 100 cm × 70 cm, each side with three layers without tungsten foils interspersed, for chemical composition of the primary cosmic ray and the trajectory measurement.

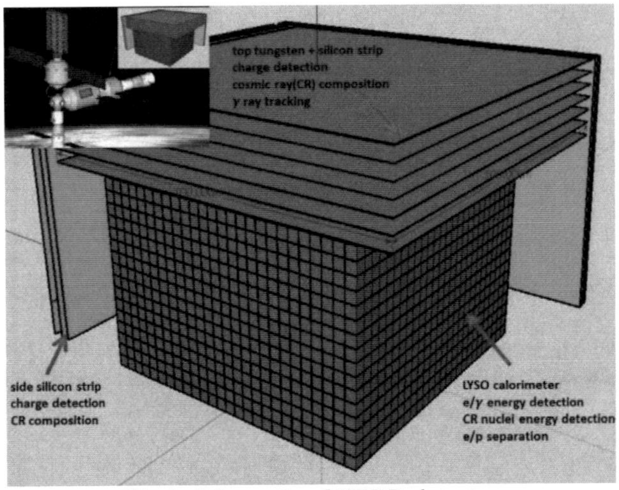

Fig.1. Schematic of HERD detector

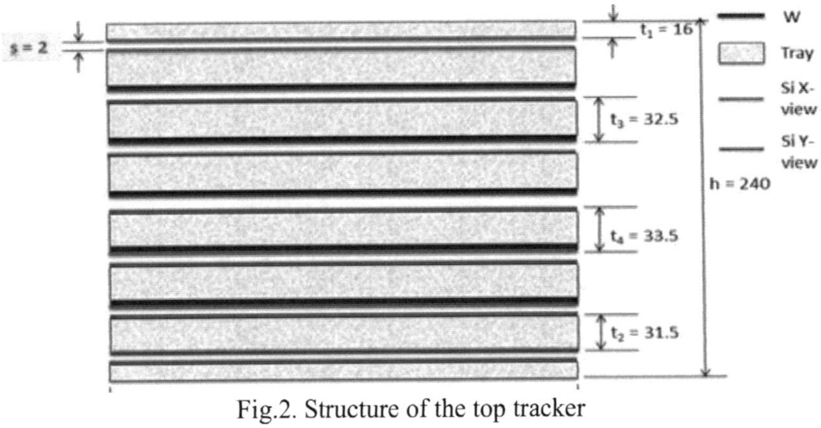

Fig.2. Structure of the top tracker

The calorimeter is made of 9261 LYSO scintillator, the side length of cubic LYSO cell is 3cm. It is a 3D imaging calorimeter, used to measure the development of particle shower. The homogeneous calorimeter detect not only the electromagnetic shower induced by electron and gamma-rays but also the hadronic shower induced by cosmic ray nuclei.

The scintillation light of crystals is absorbed by a 300 um diameter wavelength shifiting(WLS) fiber that is attached to the surface of the scintillator with a spiral structure. There are totally 20 thousand readout channels, all the fiber ends are bundled in one compact bunch, which has a size of only a couple of centimeters. The fibers at the end of the bundle can be glued together and polished making a "fiber optic plate"-like structure. The fiber bundle is coupled to the input-window of image intensifier. The shower development profile of the event in the detector is translated into the surface of the fiber optic plate (FOP). This image on the FOP can be photographed by using an externally triggered ICCD.

3. Simulation study on performance

The performance of the HERD is studied by simulations with the use of GEANT4 and FLUKA. As shown in Fig. 3., The energy resolution of electromagnetic shower with the consideration of only energy deposition in crystals is about 0.1% at 200GeV. If assume per minimum ionization particle produce 10 photoelectrons, the resolution increase to about 0.5%. Take into account other influence, the energy resolution for electron and gamma would be better than 1%. For cosmic ray nuclei, the energy resolution is around 20% between hundreds of GeV to PeV. Since the electromagnetic shower development is different from the hadronic shower in energy distribution, calorimeter can be used to do particle identification of the electron and proton. Several variables are used in TMVA to do the particle identification. As shown in Fig.4., the background efficiency is about 5×10-6 while maintain the electron identification around 90%. Because the calorimeter is a homogenous detector, it can accept particle from five sides, the effective geometrical factor for electron and proton are large than 3.7 m2sr@200GeV and 2.6 m2sr@100TeV, respectively. This is one of the foremost advantages of HERD compare with other missions.

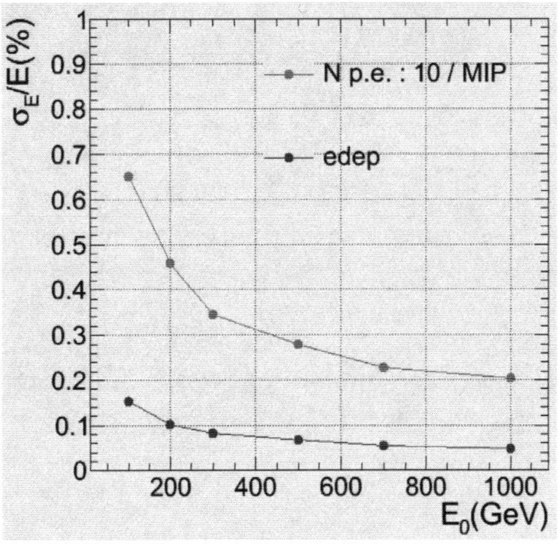

Fig.3. Energy resolution of calorimeter

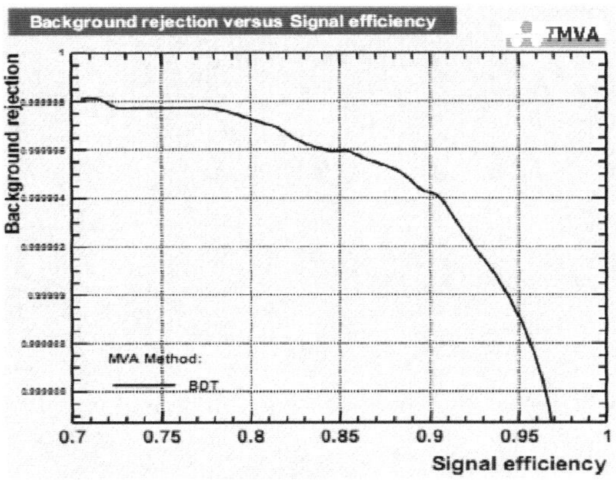

Fig.4. Particle identification of calorimeter

4. Cosmic ray test of prototype

A calorimeter prototype is constructed to verify the readout method of ICCD. Twenty four CsI(Na) crystals with the side length of 3 cm are used to constructed a 2×2×4 vertex. The 24 WLS fibers are attached to an image intensifier and then coupled to a commercial ICCD to readout the fiber signal. The ICCD is trigged by the coincidence of two plastic scintillators. The distribution of typical psssing through muon event is shown in Fig.5. When a cosmic muon pass through six crystals, the output of the fiber signal will be received by ICCD. The energy resolution of cosmic muons deposited in six crystals is 14% @84MeV, this is satisfied to the requirement of HERD.

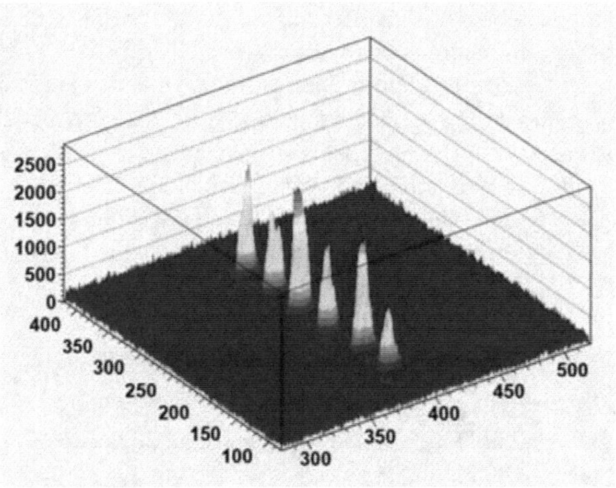

Fig.5. Distribution of a typical passing through muon event

5. Summary

The HERD mission will perform observations of very high energy electrons, gamma-rays and cosmic ray. Monte Carlo simulation results show it will be a powerful space detector in energy resolution, effective geometrical factor and particle identification. The cosmic ray test of prototype verify the feasibility of the ICCD readout.

Reference

[1] *J. Nishimura et al. Emulsion chamber observations of primary cosmic-ray electrons in the energy range 30-1000 GeV. Apj, 238:394–409, May 1980.*

[2] *C. S. Shen and C. Y. Mao. Anisotropy of High Energy Cosmic-Ray Electrons in the Discrete Source Model. Astrophysical Letters, 9:169, August 1971.*

[3] *M. Pohl et al. Electron Acceleration in Supernova Remnants and Diffuse Gamma Rays above 1 GeV. Apj, 507:327–338, November 1998.*

[4] *L. Bergström, J. Edsjö, and C. Gunnarsson. Neutralino gamma-ray signals from accreting halo dark matter. PRD, 63(8):083515, April 2001.*

[5] *H.-C. Cheng, J. L. Feng, and K. T. Matchev. Kaluza-Klein Dark Matter. PRL, 89(21):211301, October 2002.*

16th International Conference on Calorimetry in High Energy Physics (CALOR 2014) IOP Publishing
Journal of Physics: Conference Series **587** (2015) 012028 doi:10.1088/1742-6596/587/1/012028

Performance of the AMS-02 Electromagnetic Calorimeter in Space

G.Gallucci (for the AMS-02 ECAL group)
INFN - Sezione di Pisa, Largo Bruno Pontecorvo 3, 56127 Pisa, Italy
E-mail: giovanni.gallucci@pi.infn.it

Abstract. AMS-02(Alpha Magnetic Spectrometer) is an high energy particle detector developed to operate on the International Space Station. AMS-02 was installed on ISS on May 2011 and is expected to operate for 10-20 years collecting about 160-320 billions of events. The main goals of the experiment are the detection of primordial antimatter and dark matter by studying spectra and flux of different cosmic ray components (protons, electrons, nuclei, positrons, antiprotons, gamma rays, etc) in the high energy range (1-2000 GeV). Identification of electrons, positrons and photons is provided by the Electromagnetic Calorimeter (ECAL), a fine grained lead-scintillating fibers sampling calorimeter that allows for a precise three-dimensional imaging of the longitudinal and lateral shower development. It provides an excellent reconstruction of electromagnetic shower energy and a highly efficient rejection of the hadronic background. Thanks to the 3D shower reconstruction capability, ECAL allows a stand-alone determination of the incoming particle direction, with unprecedented angular resolution. As a result, ECAL is able to identify high energy photons coming from galactic or extragalactic sources.

1. The AMS-02 Experiment

The AMS-02 is a TeV precision multipurpose spectrometer designed to operate in space, outside the International Space Station (ISS). Its launch and subsequent installation on the ISS took place successfully on May 2011. The main goals of AMS are the precise measurements of cosmic ray composition and flux (protons, electrons, ions, strange matter, etc...), the indirect search of Dark Matter (positron fraction, anti-proton fraction, etc...) and search and study of primordial anti-matter (anti-helium, etc...). Moreover, it could study the diffused gamma rays in the GeV to TeV range.

2. The AMS-02 Detectors

The layout of AMS-02 detector consists of a Transition Radiation Detector (**TRD**), a Time of Flight (**TOF**), a Magnetic spectrometer, an Anti-Coincidence scintillator counter system (**ACC**), a Ring Image Cherenkov (**RICH**) and an electromagnetic calorimeter (**ECAL**) as shown in figure 1:

- The **TRD** is composed by 5248 proportional tubes of 6mm diameter and maximum length of 2 m filled with Xe and CO2 (90:10 mixture). The tubes are organized in 16-tube modules and mounted in 20 layers. Between two layers there is a 20 mm thick polypropylene/polyethylene radiator . The TRD is used to distinguish between electrons and protons up to few hundred GeVs and to independently identify nuclei using $\frac{dE}{dx}$ [1].

- The **TOF** consists of four planes of plastic scintillators. Each plane contains eight or ten paddles and each paddle is readout by two or three phototubes on each end. It is used for charged particle triggers, timing information, beta and dE/dx measurements[2].
- The Magnetic spectrometer is composed by a silicon tracker (**TRK**) and a permanent magnet of 0.14 Tesla. The TRK is made by 192 ladders, each containing double-sided silicon sensors, mounted on 9 layers. Seven layers are in inside the magnet volume and constitute the inner tracker. One layer is located on top of TRD and the last one is located between the RICH and the ECAL. The TRK provides the rigidity, the charge and the charge sign of particles [3].
- The **ACC** is made by 16 curved scintillator planes 0.8 m long surrounding the inner tracker inside the magnet. They are used as veto to reject events entering the detector through the sides and, in general, to detect unwanted particles that enter or leave the inner tracker volume transversely [4].
- The **RICH** consists of two dielectric radiators with different refractive index and different beta thresholds. The Cherenkov light is detected by 10880 photosensors of 8.5x8.5 mm2. It is used to measure velocity of cosmic rays [5].
- The **ECAL** is shown in the next section.

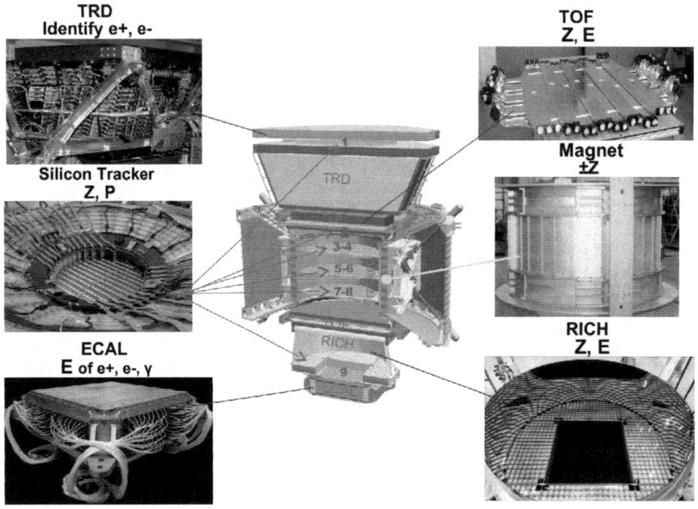

Figure 1. Sketch of the AMS-02 detector with the different sub detectors.

3. The Electromagnetic calorimeter: ECAL

The calorimeter has been designed to precisely reconstruct the longitudinal and the lateral profiles of electromagnetic showers and to measure the deposited energy. For these characteristics it is used to provides energy and axis direction of electromagnetic showers as well as reject the protons using the different shape of hadronic showers.

3.1. Structure and light collection

ECAL is a lead-scintillating fiber sandwich with an active area of 648x648 mm^2 and a thickness of 166mm for a weight of ~ 500 Kg [6]. The calorimeter consists of 9 modules (called *superlayers*) with a thickness of 18.5 mm (figure 2a). Each superlayer is composed by grooved lead foils alternate with 1mm scintillating fiber layer glued to the foils by mean of optical epoxy. The resulting ECAL volume is 58% of lead, 33% of fibers (~ 50000) and the remaining part of optical glue. The fibers run only in one direction so, in order to obtain a 3D image, the 9 modules are stacked with fibers alternatively parallel to x-axis (five superlayers) and y-axis (4 superlayers). Each superlayer is readout by 36 four-anode photomultipliers (PMTs), only at one edge. In order to avoid dead zones the PMTs are place alternatively on the two opposite ends. The entire calorimeter is readout by 324 PMTS, 1296 anodes. Each anode covers an active area equal to 9x9 mm^2 corresponding to 35 fibers called cell (figure 2b). Each cell correspond to ~1 radiation length and half Moliere radius. Thus the calorimeter consists of 18 longitudinal independent samplings and 72 lateral samplings. The entire structure corresponds to 17 radiation lengths but only 0.7 nuclear interaction lengths. To maximize light yield and reduce crosstalk between cells, scintillating light is collected by individual light guides wrapped in an aluminum foil with chromium and quartz coating (figure 3). Furthermore for each PMT, the light guides and the front end electronics are shielded from residual magnetic filed by a 1mm thick soft iron square tube(figure 4). In order to have a good linearity in a large energy range (from few MeVs up to about 60 GeVs), each anode signal is split into two different channels and digitized. The first channel (High Gain) is amplified and is sensitive to Minimum Ionizing Particles (MIP) that release around 7 MeV/cell. The second one (Low Gain) is not amplified and is used for high energy measurement . The gain ration between the two is about 33. Also the signal of the last dynode of each PMT is readout to ensure a redundant measurement of energy deposition and to provide input to the trigger logic.

3.2. ECAL trigger

ECAL is involved in the trigger for electromagnetic particles and provides a standalone trigger for photons. When a high energy-deposit is recorded in the innermost layers of the calorimeter, a fast signal is generated and processed by the trigger board which can enable the event acquisition. Also the incident particle direction is evaluated by taking, for x and y sides, the average position of the fired PMTs. A cut at 20 degrees is applied to select particles inside AMS-02 geometrical acceptance and reject charged particles entering by side. The efficiency for photons is about 20% at 1 GeV and reaches about 99The one-orbit average ECAL trigger rate is about 115 Hz (approximately 10% of total AMS-02 rate).

4. ECAL calibration and performances at Test Beam

ECAL was tested at CERN on H4 and H8 beam lines, alone and after the integration with rest of apparatus. It was exposed to electrons and positrons beams with energy in the range 6-250 GeV and to primary proton beam line of 400 GeV. Using test beam data, equalization and calibrations procedures have been created and verified and the calorimeter performance has been measured.

4.1. Light attenuation

In order to measure light attenuation along fibers we used the MIP peak (figure 5). The light attenuation is parametrized using two different attenuation constants, λ_s and λ_f fast and slow respectively:

$$A(x) = fe^{\frac{-x}{\lambda_f}} + (1-f)e^{\frac{-x}{\lambda_s}} \tag{1}$$

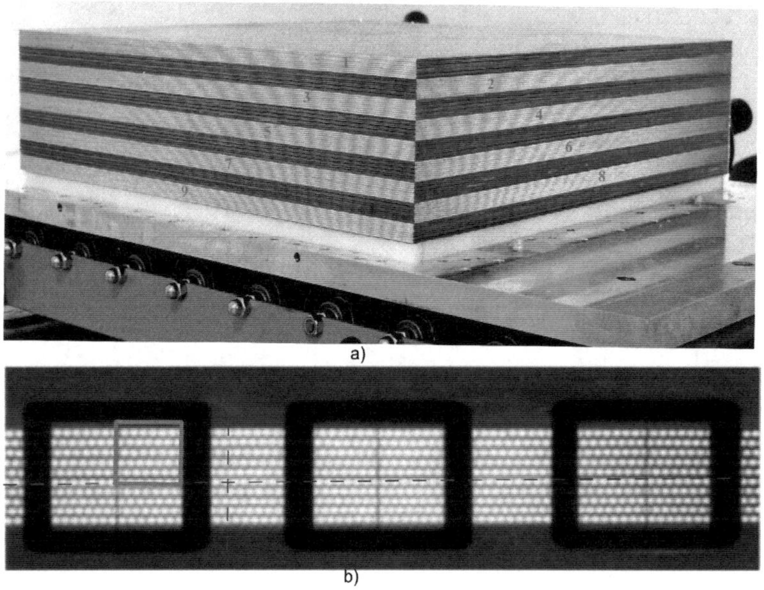

Figure 2. **a)** ECAL Structure. **b)** One of the 9 superlayers of the calorimeter with the footprint of the PMT and single cell superimposed.

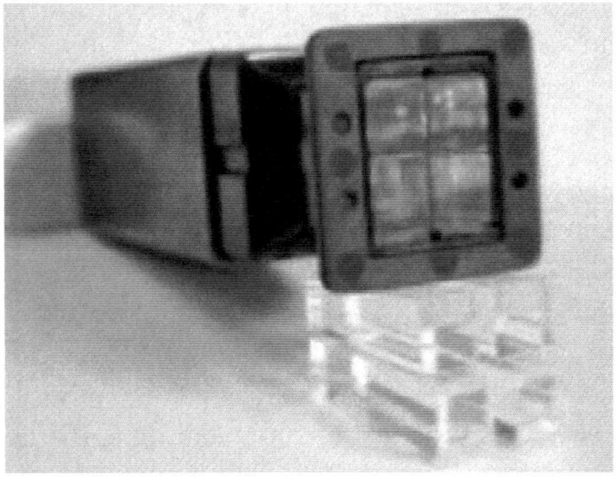

Figure 3. Scheme of the PMT light collection system.

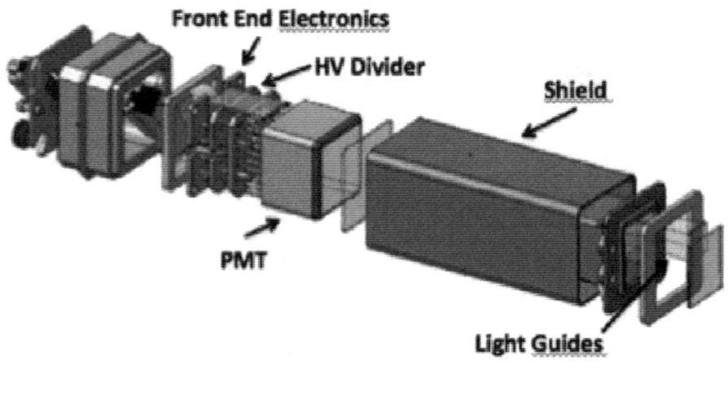

Figure 4. Light guides and polycarbonate support.

with A the attenuation factor and x the distance travelled by the light along the fiber. Typical value are $f = 0.17$,$\lambda_f = 110mm$ and $\lambda_s = 2605mm$ with a spread shown in figure 5.

4.2. Equalization

Equalization is done adjusting each High Voltage so that the cell response to MIP is the same. Because of intrinsic fluctuation of the four different anodes of each PMT and the fact that 84 PMT pairs are fed by the same HV lines, a offline correction is needed. Only HG channels are equalized but HG-LG ratio is constantly evaluated.

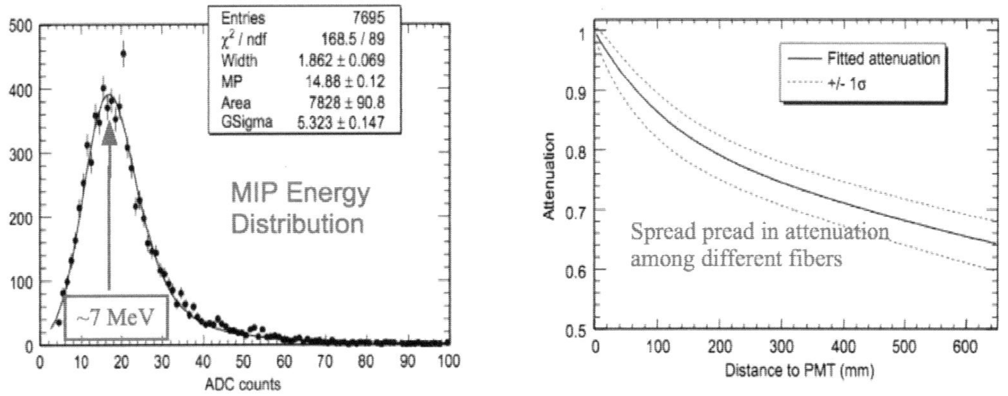

Figure 5. Left: Typical distribution of MIP signal in an ECAL cell. **Right:** Light attenuation along the fibers. The two dashed lines represent the typical spread in attenuation.

4.3. Radiation length

Using the different electron and positron beams, we measured the X_0 of calorimeter. Figure 6a shows the energy longitudinal distribution for 10 and 180 GeV. The distribution is well parametrized from a Rossi function []. Figure 6b shows the distribution of the max as function of the energy beam and we obtained:

$$X_0 = 0.98 \pm 0.02\ cm \tag{2}$$

as expected from the project.
The total thickness of ECAL, corresponding to $17.0X_0$, still contains ~75% of energy of 1 TeV electrons.

4.4. Anode Efficiency and rear leakage

We observed that anode efficiency in not uniform, in particular it decreases at the cell border with respect to the centre. Farther there is a small dead space at the edge of each cell. This effect appears to be linearly dependent on the ratio between the sum of the maximum energy deposited in each layer (S1) and the sum of (S1) plus the content of the two cells adjacent to the maximum in each layer (S3). This ratio, called S1/S3, tends to 0.5 for particles hitting ECAL between two cells, corresponding to maximal energy loss (~10%), and to 1 for particles hitting ECAL in the cell centre, corresponding to minimal energy loss. Figure 7 shows the ECAL energy response before and after correction.

A second correction is applied to take into account of the energy leakage after the 17 radiation lengths of ECAL. This loss is proportional to the energy deposited in the last part of the shower. We used the energy fraction of the last two layers of ECAL (~2 X_0) and computed the following formula:

$$\frac{E_{true}}{E_{dep}} = \alpha(E_{dep}) + \beta\frac{E_{LS}}{E_{dep}} + \gamma\frac{E_{LS}}{E_{dep}}^2 \tag{3}$$

with E_{dep} the deposited energy corrected for anode efficiency, E_{LS} the deposited energy in the last two layers. α, β and γ has been measured at test beam. α has a slightly energy dependence and it has been corrected using an iterative process.

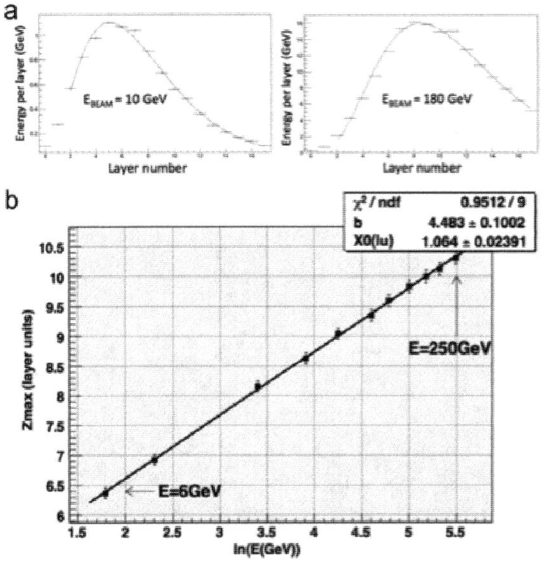

Figure 6. **a)**Typical longitudinal electromagnetic shower profiles for two different beam energies. **b)**Position of the peak of the Rossi function vs the logarithm of the beam energy. A straight fit provides the parameter X_0 in layer units.

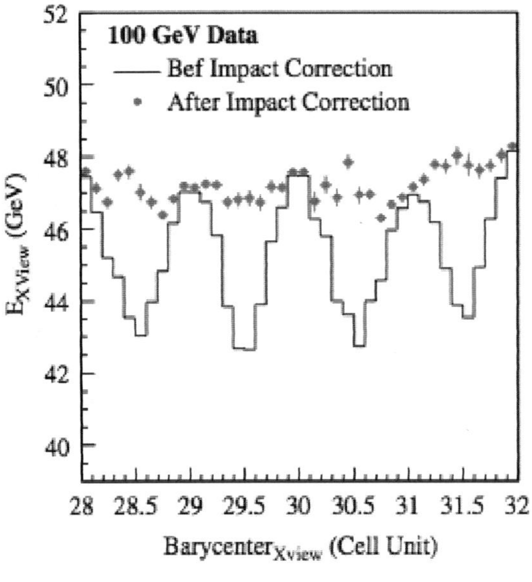

Figure 7. Energy deposited as a function of the impact point before (histogram) and after (red dots) impact correction.

4.5. Energy resolution and linearity

Using the different energy beams we measured the linearity and the energy resolution of ECAL. The linearity is within 1% in the range 6-250 GeV (figure 8), energy resolution is well parametrized by (figure 9):

$$\frac{\sigma(E)}{E} = \frac{(10.4 \pm 0.2)\%}{\sqrt{E(GeV)}} \oplus (1.4 \pm 0.1)\% \quad (4)$$

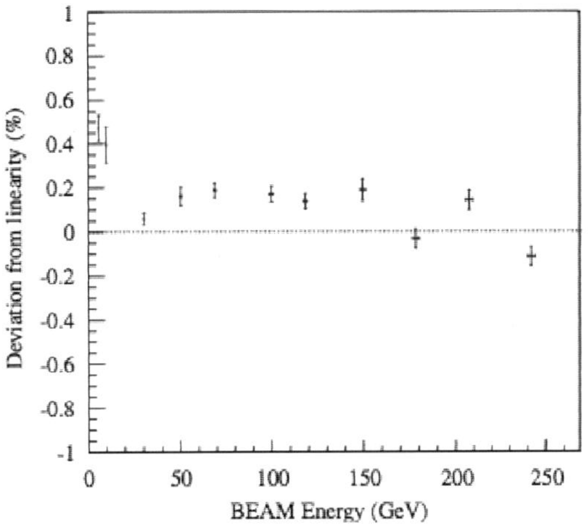

Figure 8. The energy linearity is held within 1%.

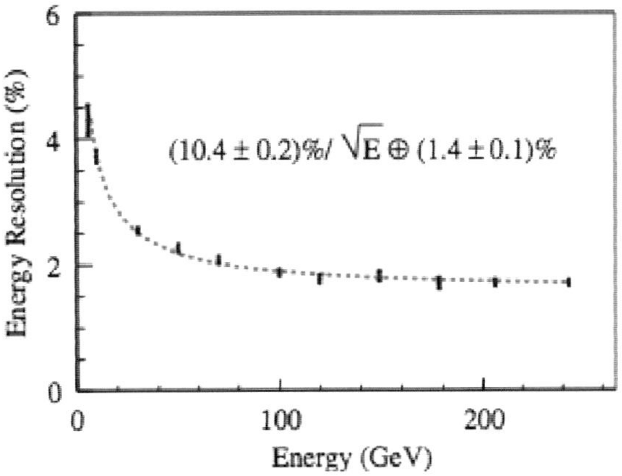

Figure 9. ECAL energy resolution from Test Beam.

4.6. Angular resolution
The high granularity of calorimeter allows a good reconstruction of shower axis and, consequently, a good angular resolution. Shower directions, in x and y side, are determined by fitting the axis positions in each layer. Three different methods has been developed to do it:

(i) the Centre Of Gravity (COG), evaluated from the energy weighted centres of the cells belonging to the shower:

$$y_{COG}(x_{COG}) = \frac{\sum y_i(x_i)E_i}{\sum E_i} \tag{5}$$

where E_i is the energy deposited in the i-th cell and $y_i(x_i)$ is its position;

(ii) the Neighbour Cells (NC) method, which uses the ratio of the energy deposited in the cells adjacent to the most energetic one as a function of the impinging position;

(iii) the Lateral Fit method, in which the energy distribution of the cells in the layer is fitted using the following parametrization:

$$f(x) = \frac{2(x - x_C)R^2}{((x - x_C)^2 + R^2)^2} \tag{6}$$

where the lateral width R and the axis position x_C are the output of the fit.

Figure 10 summarizes angular resolution for the three methods.

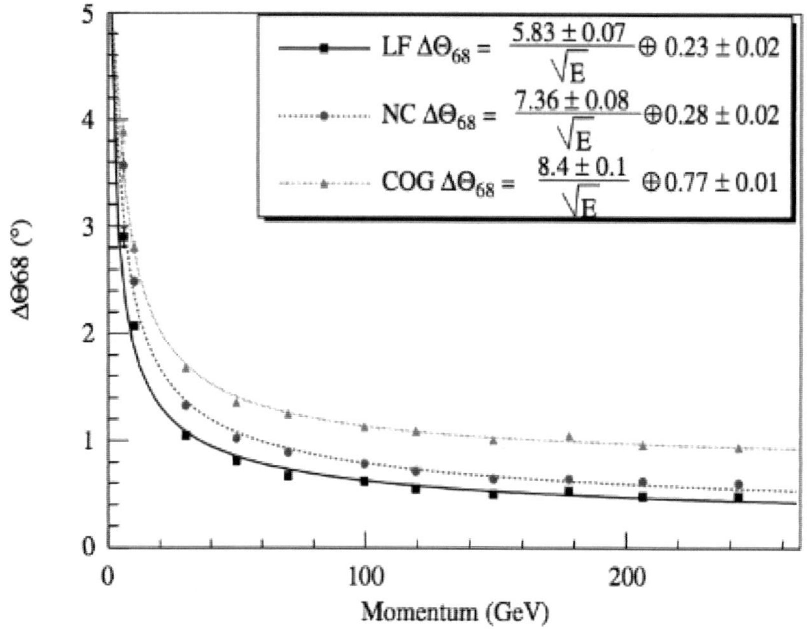

Figure 10. Angular resolution as a function of Test Beam momentum for the three analysis methods.

5. ECAL performances on flight

The launch was a very stressing test for apparatus. Otherwise the hard environment conditions in the space needed for a continuously monitoring of calorimeter response and performance. After three years of data taking no major problems have happened and only 1 PMT is dead and 1 PMT is noisy. No fiber aging has been observed.

Journal of Physics: Conference Series **587** (2015) 012028 doi:10.1088/1742-6596/587/1/012028

5.1. Equalization and light attenuation

Using cosmic ray MIPs it is possible to perform the same tests and measurement done on ground. Figure 11 shows the light attenuation measured in two different periods compared to the on ground one. The thermal environment in space is severe: Sun light as well as the position of the solar panels onboard the ISS can affect the thermal configuration of the instrument producing temperature changes of several degrees. A dependence between the PMT gain and temperature was observed in flight data. The gain is anti-correlated with temperature and change ~1% for 4 degrees (figure refMIPTemp). An offline correction has been developed and implemented.

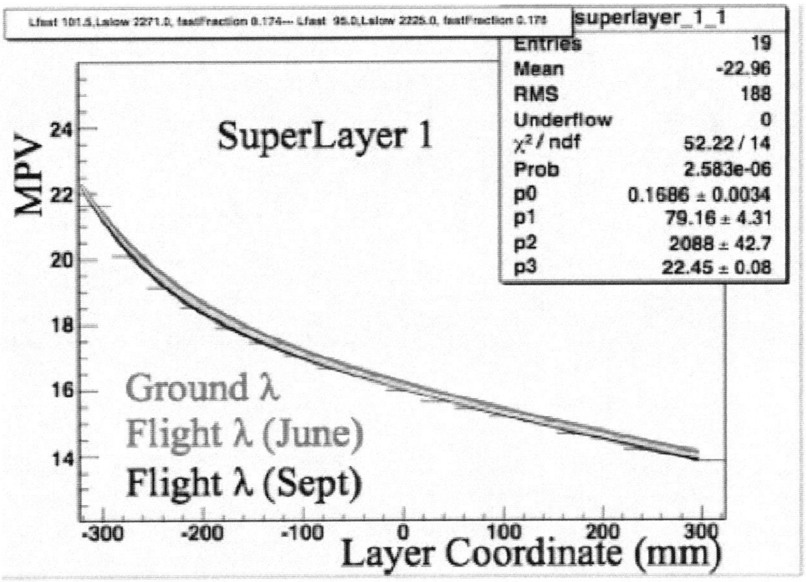

Figure 11. Attenuation length measured in two different data taking periods (June and September 2011) compared to the test beam one.

5.2. Energy scale

Energy scale has been checked using MIP peak as well as the ratio between the reconstructed energy and the pulse measured by the tracker. Figure 13 shows that energy scale is stable at 1%.

6. Proton rejection

As mentioned before, high granularity of calorimeter provides a good 3D reconstruction of the shower development. It allows the identification of electromagnetic particles with respect of protons using the different shapes of showers. Several approaches have been studied and developed to achieve the necessary separation between electrons and protons: cut based methods and multivariate analysis. The best method for ECAL has been a Boosted Decision Tree (BDT). To avoid any bias the training samples (background and signal) were selected from ISS data using the TRD and the tracker. Figure 14 shows the distribution of the BDT for electrons (and positrons) and hadrons. Adding the match between the reconstructed energy and the pulse measured by the tracker, we obtained a proton rejection more than 10^4 until 300 GeV and still 10^3 at 1 TeV (figure 15).

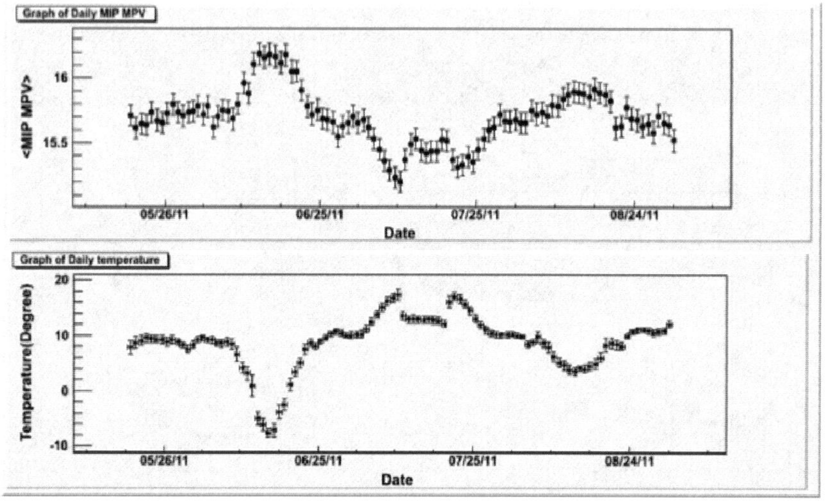

Figure 12. Up: MIP Peak as function of the time. **Down:** Temperature as function of the time in same period of the above picture.

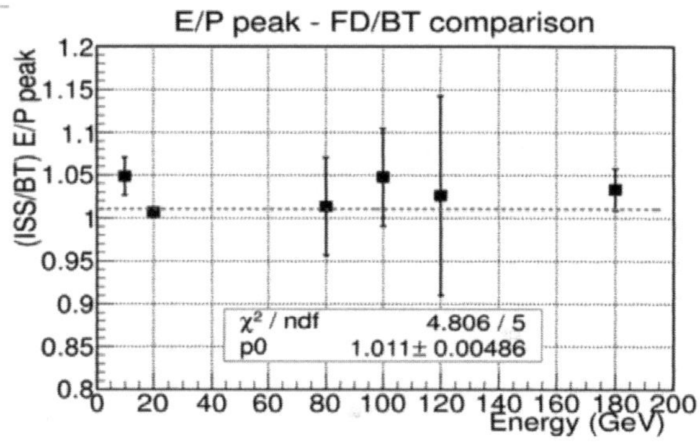

Figure 13. The energy linearity is stable at 1%.

7. Photon study

The material before ECAL is only 0.6 X_0 so we expected ~60% of photons interact directly in the calorimeter. Applying the selection cuts summarized in table 1 to the ECAL standalone trigger data we selected a photon sample. Backtracking electromagnetic shower axis and correcting for exposition time we obtained a picture of photon sky in which we could recognize known photon sources (figure 16). The two black holes in figure 16 are blind regions due to the ISS orbits. Reversing the procedure we used know positions of photon sources to measure the angular resolution of ECAL obtaining expected values.

8. Conclusions

The AMS-02 electromagnetic calorimeter design has been described, as well as its performance with Test Beam and flight data. After 3 years of data taking, ECAL is performing according to expectations. Energy resolution and electromagnetic shower identification already achieved

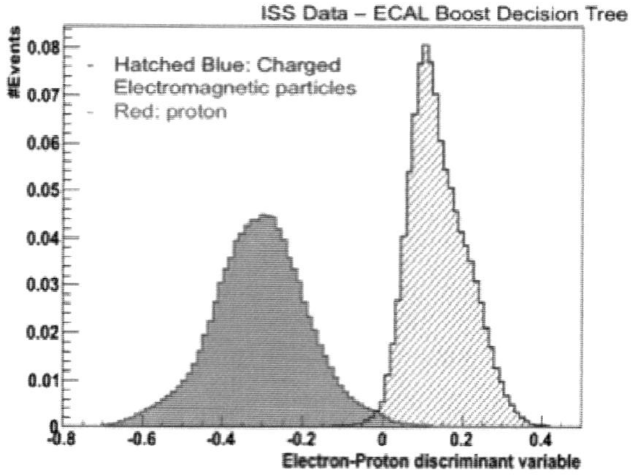

Figure 14. BDT variable distribution for electrons and positrons (blue distribution), as well as for protons (red distribution) for all energies.

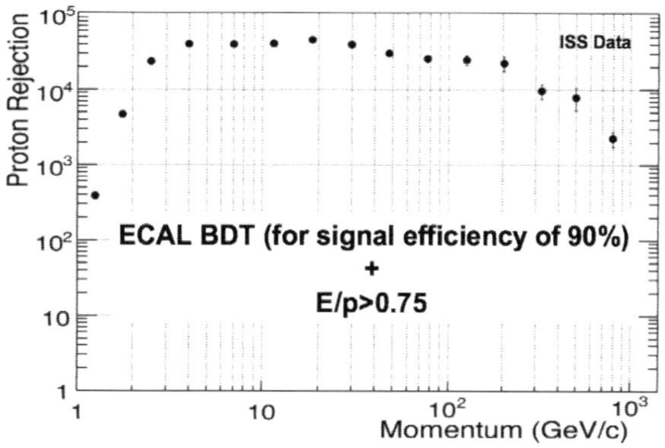

Figure 15. Proton rejection as function of proton momentum. ECAL BDT plus a 0.75 cut on $\frac{E}{P}$ ratio are applied.

Table 1. Selection cuts to obtain photon sample.

Detector	Cut
TRD	No tracks reconstructed
Tracker	No tracks reconstructed, no hits correlated with shower axis
TOF	Limits on hits numbers. Time difference between Upper and Lower TOF
ACC	Limits on the number of hits
RICH	Limits on the number of hits
ECAL	Shower axis within ECAL fiducial volume
	Extrapolation of shower inside active upper TOF area
	Electromagnetic shower shape

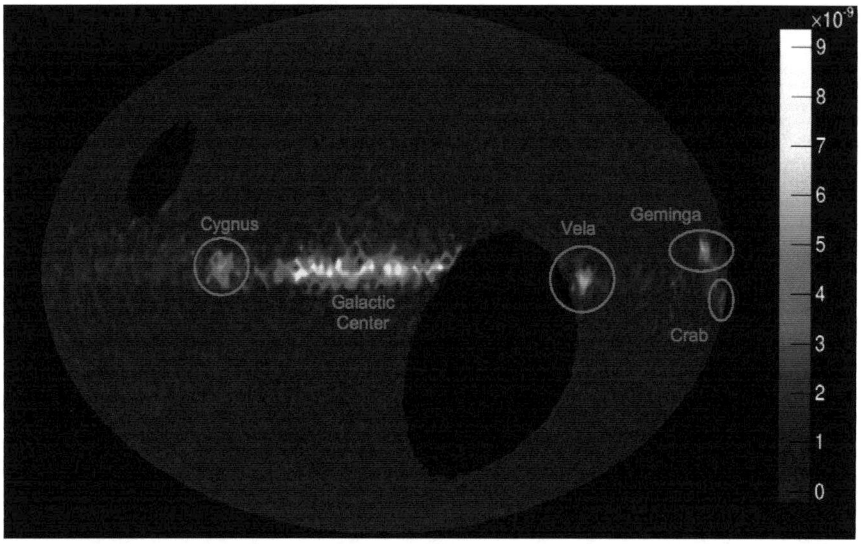

Figure 16. Photon sky obtained using ECAL reconstructed photons. Inside the red circles there are known photon sources (Vela, Galactic Center, etc...). Black holes are blind regions due to the ISS orbits.

relevant scientific results on electron/positron measurement [7]. Otherwise angular and energy resolution allow to search for substructures in the diffused gamma flux from center of galaxy or other possible dark matter candidate sources.

9. References

[1] Th. Kirn, Nucl. Instrum. Methods Phys. Res., Sect. A 706, 43 (2013); Ph. Doetinchem et al., Nucl. Instrum. Methods Phys. Res., Sect. A 558, 526 (2006); F. Hauler et al., IEEE Trans. Nucl. Sci. 51, 1365 (2004).

[2] A. Basili, V. Bindi, D. Casadei, G. Castellini, A. Contin,A. Kounine, M. Lolli, F. Palmonari, and L. Quadrani,Nucl. Instrum. Methods Phys. Res., Sect. A 707, 99 (2013); V. Bindi et al., Nucl. Instrum. Methods Phys. Res., Sect. A 623, 968 (2010).

[3] B. Alpat et al., Nucl. Instrum. Methods Phys. Res., Sect. A 613, 207 (2010);K. Luebelsmeyer et al., Nucl. Instrum. Methods Phys. Res., Sect. A 654, 639 (2011).

[4] Ph. von Doetinchem, W. Karpinski, Th. Kirn, K. Lu belsmeyer, St. Schael, and M. Wlochal, Nucl. Phys. B, Proc. Suppl. 197, 15 (2009).

[5] M. Aguilar-Benitez et al., Nucl. Instrum. Methods Phys. Res., Sect. A 614, 237 (2010); M. Aguilar-Benitez et al., in Proceedings of the 30th ICRC, 2007 (Universidad Nacional Auto noma de Me xico, Mexico City, 2008), Vol. 2, p. 461; P. Aguayo et al., Nucl. Instrum. Methods Phys. Res., Sect. A 560, 291 (2006); B. Baret et al., Nucl. Instrum. Methods Phys. Res., Sect. A 525, 126 (2004); J. Casaus, Nucl. Phys. B, Proc. Suppl. 113, 147 (2002).

[6] C. Adloff et al., Nucl. Instrum. Methods Phys. Res., Sect. A 714, 147-154 (2013) .

[7] AMS Collaboration (M. Aguilar et al.), Phys.Rev.Lett. 110 141102 (2013).

Journal of Physics: Conference Series **587** (2015) 012029 doi:10.1088/1742-6596/587/1/012029

CALOCUBE: an approach to high-granularity and homogenous calorimetry for space based detectors

M Bongi[1,2], O Adriani[1,2], S Albergo[3,4], L Auditore[5,6], M G Bagliesi[7,8], E Berti[1,2], G Bigongiari[7,8], M Boezio[9], L Bonechi[2], S Bonechi[7,8], V Bonvicini[9], S Bottai[2], P Brogi[7,8], G Carotenuto[10], A Cassese[2,11], G Castellini[11], P W Cattaneo[12], D Cauz[9,13], P Cumani[9,14], R D'Alessandro[1,2], S Detti[2], M Fasoli[15], A Gregorio[9,14], A Lamberto[5,6], P Lenzi[2], P Maestro[7,8], P S Marrocchesi[7,8], A Mezzasalma[5,6], M Miritello[16], N Mori[1,2], P Papini[2], G Pauletta[9,13], G F Rappazzo[4,5], A Rappoldi[12], S Ricciarini[2,11], P Spillantini[1,2], O Starodubtsev[1,2], A Sulaj[7,8], A Tiberio[1,2], A Trifirò[5,6], M Trimarchi[5,6], E Vannuccini[2], A Vedda[15], G Zampa[9], N Zampa[9] and B Zerbo[4]

[1] University of Florence, Department of Physics and Astronomy, via G. Sansone 1, I-50019 Sesto Fiorentino (Firenze), Italy
[2] INFN Firenze, via B. Rossi 1, I-50019 Sesto Fiorentino (Firenze), Italy
[3] University of Catania, of Physics and Astronomy, via S. Sofia 64, I-95123 Catania, Italy
[4] INFN Catania, via S. Sofia 64, I-95123 Catania, Italy
[5] University of Messina, Department of Physics, sal. Sperone 31, I-98166 Messina, Italy
[6] INFN Messina, sal. Sperone 31, I-98166 Messina, Italy
[7] University of Siena, Department of Physical Sciences, Earth and Environment, I-53100 Siena, Italy
[8] INFN Pisa, via F. Buonarroti 2, I-56127 Pisa, Italy
[9] INFN Trieste, via Valerio 2, I-34127 Trieste, Italy
[10] Institute for Polymer, Composites and Biomaterials (CNR), piazzale E. Fermi 1, I-80055 Portici (Napoli), Italy
[11] IFAC (CNR), via Madonna del Piano 10, I-50019 Sesto Fiorentino (Firenze), Italy
[12] INFN Pavia, via A. Bassi 6, I-27100 Pavia, Italy
[13] University of Udine, Department of Chemistry, Physics and Environment, via delle Scienze 206, I-33100 Udine, Italia
[14] University of Trieste, Department of Physics, via Valerio 2, I-34127 Trieste, Italy
[15] University of Milano-Bicocca, Department of Materials Science, via Cozzi 55, I-20125 Milano, Italy
[16] MATIS (CNR), via S. Sofia 64, I-95123 Catania, Italy

E-mail: massimo.bongi@fi.infn.it

Abstract. Future space experiments dedicated to the observation of high-energy gamma and cosmic rays will increasingly rely on a highly performing calorimetry apparatus, and their physics performance will be primarily determined by the geometrical dimensions and the energy resolution of the calorimeter deployed. Thus it is extremely important to optimize its geometrical acceptance, the granularity, and its absorption depth for the measurement of the particle energy with respect to the total mass of the apparatus which is the most important constraint for a space launch. The proposed design tries to satisfy these criteria while staying within a total mass budget of about 1.6 tons. Calocube is a homogeneous calorimeter instrumented with Cesium iodide (CsI) crystals, whose geometry is cubic and isotropic, so as to detect particles arriving from every direction in space, thus maximizing the acceptance; granularity is obtained by filling the cubic volume with small cubic CsI crystals. The total radiation length in any direction is more than adequate for optimal electromagnetic particle identification and energy measurement, whilst the interaction length is at least sufficient to allow a precise reconstruction of hadronic showers. Optimal values for the size of the crystals and spacing among them have been studied. The design forms the basis of a three-year R&D activity which has been approved and financed by INFN. An overall description of the system, as well as results from preliminary tests on particle beams will be described.

1. Introduction

Calocube is a project funded by the Italian National Institute for Nuclear Physics (INFN) for the study of a calorimeter with an innovative design, whose characteristics make it very appealing for next-generation space experiments dedicated to the investigation of high-energy cosmic rays.

Its scientific goal is the measurement of the energy spectrum of the nuclear component of cosmic rays in the region of the knee (E $\sim 10^{15}$ eV), where the spectral index shows a significant change, and of the electromagnetic component above the TeV region. The main issues for these kind of measurements are the steep fall of the spectrum with increasing energy, which makes large acceptance (> 3 m^{2}sr) a necessary requirement, and the dominant proton background, which requires a very good hadron/electron rejection power ($> 10^5$) for the identification of electromagnetic particles. In order to reveal possible structures in the spectrum, a further requirement is an excellent energy resolution both for the electromagnetic component ($\sim 1\% - 2\%$) and for the hadronic one ($< 30\%$).

The geometrical acceptance of conventional cosmic-ray telescopes is limited by their ability to detect only particles impinging on one face of the apparatus. The peculiarity of the design of Calocube is its capability of detecting particles coming from any direction, and not only those on its upper surface. To ensure that the quality of the measurement does not depend on the arrival direction of the particles, the calorimeter will be designed as homogeneous and isotropic as possible. In addition, to achieve a high discrimination power for hadrons and nuclei with respect to electrons, the sensitive elements of the calorimeter need to have a fine 3-D sampling capability which can be used to analyze the topological differences in the development of electromagnetic and hadronic showers. Finally, to improve the energy resolution for hadronic showers we plan to study advanced compensation techniques that rely both on the measurement of the fraction of electromagnetic energy of the shower through the detection of the Cherenkov light, and on the detection of neutrons which are produced in hadronic interactions within the shower. A further key point of the instrument is the capability of directly measure the charge Z of the incident particle, allowing a complete measurement of the cosmic ray spectra without any need of additional dedicated detectors, hence greatly simplifying the structure and reducing power consumption and weight. Most of the above requirements are shared by homogenous hadronic calorimeters for future collider experiments and the development of Calocube could contribute to the optimization of the advanced compensation and leakage-correction techniques which are of common interest.

2. The Calocube detector

As a starting point for the optimization of the design of the detector, a geometry consisting of a cubic matrix made up by cubes of sensitive material has been studied (see Fig. 1). The calorimeter is composed by 20 planes ($78 \times 78 \times 4$ cm^3), each consisting of 20×20 cubic Tallium-activated Cesium iodide (CsI(Tl)) scintillating crystals. The dimensions of the crystals ($3.6 \times 3.6 \times 3.6$ cm^3) are chosen to match the Molière radius of the material of the calorimeter, in order to optimize the e/h rejection power. CsI(Tl) has been selected because of the very

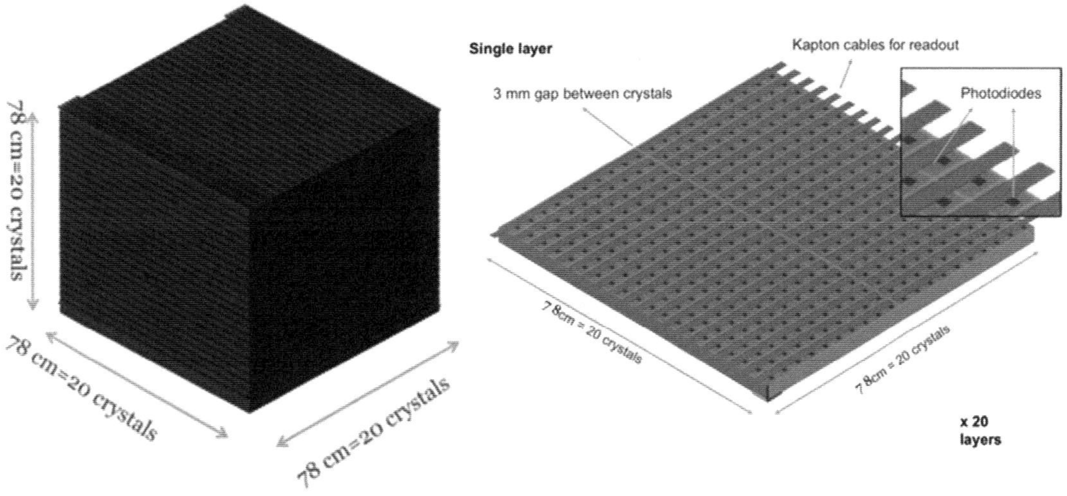

Figure 1. Conceptual design of the 3-D highly-segmented calorimeter. The left panel depicts the complete cubic detector, while in the right panel some details about a single plane are shown.

high light yield (54000 ph/MeV) and because of its intermediate density (4.5 g/cm^3), which allows to reach a very large geometric factor with an acceptable overall mass (1683 kg of active material in this configuration). The distance between the crystal has been tuned to a 3-mm gap in order to maximize the acceptance without any significant degradation of the performance of the calorimeter. The very large depth of the calorimeter (39 X_0, 1.8 λ_I for vertically incident particles) allows a complete containment of the electromagnetic showers up to the highest energies, and the measurement with high efficiency of the hadronic particles up the PeV region. A total of 8000 crystals will be used for the whole calorimeter assembly. This geometry has no privileged direction, thus achieving performances that are independent of the direction of incidence of the particle. The high granularity in every direction provides accurate information regarding the formation of the shower inside the device. This information could be exploited to distinguish electrons from hadrons and to help correcting for the lack of containment.

To cover the huge required dynamic range (from ≤ 1 MIP for non interacting protons, useful for calibration purposes, up to 10^7 MIPs in one single crystal for PeV interacting hadrons) the photosensors and the readout electronics must be carefully chosen and designed. In the basic design of the apparatus the scintillation light produced in one crystal is read out with two different photodiodes, one with larger area (9.2×9.2 mm^2 in the present design) for the low energy particles, and one with smaller area (0.5×0.5 mm^2) for the highest energy particles. Rows of flat kapton cables are used to extract signals from the calorimeter (see Fig. 1). To make optimal use of the excellent linearity of the photodiodes, a front-end chip with a large dynamic range is mandatory. In our proposal we have selected the CASIS chip [1], that has been specifically developed by INFN Trieste for calorimetric space applications. Every chip consists of 16 independent channels, with a charge sensitive amplifier and a correlated double sampling system, with automatic switching between two different gains, depending on the injected charge. The small power consumption of a single channel (2.8 mW), associated with the low noise (3000 e$^-$ for 100 pF input capacitance) and the large maximum input charge (53 pC) are perfectly suited for the Calocube design. By properly combining the dual gain of

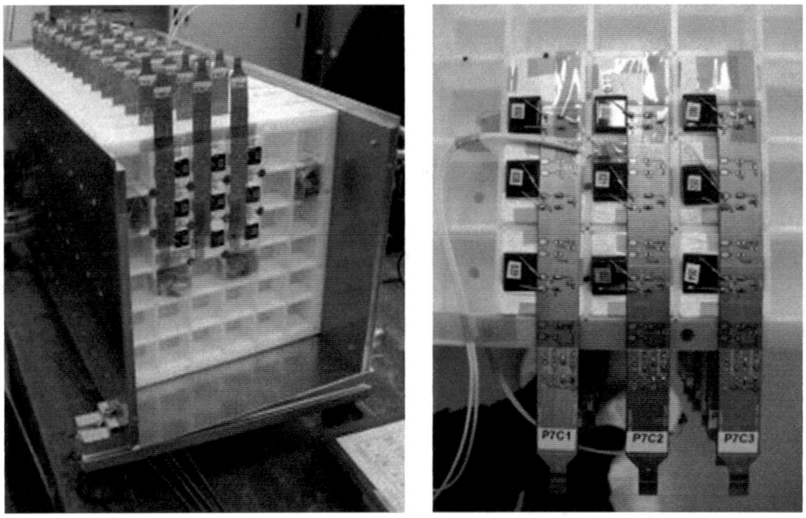

Figure 2. Left panel: picture of the 126-crystals Calocube prototype. Right panel: detail of a single layer, showing the 9 large area photodiodes placed on the crystals, that are wrapped in white teflon tape, and the kapton cables used to read-out the signals.

the chip and the pair of photodiodes connected with a single crystal, we can efficiently cover the whole required dynamic range.

The performance of the detector in this configuration have been estimated from a detailed Monte Carlo simulation based on the FLUKA package [2]. The resulting geometric factor when accepting particles from 5 sides is about (1.91 m^2sr $\times$ 5) = 9.55 m^2sr, while an electromagnetic energy resolution better than 2% for TeV particles and a hadronic energy resolution $\sim$ 35% for 1-100 TeV protons can be expected (using the scintillation light output only), with an effective geometric factor of about 3-4 m^2sr. Further details about the Monte Carlo studies can be found in ref. [3].

3. SPS test on the Calocube 126-crystals prototype

In parallel to simulation studies, prototype construction and testing activities have been carried out. A matrix of 126 CsI(Tl) crystals (14 layers, 3 $\times$ 3 crystals per layer) has been assembled in a modular mechanical structure (see Fig. 2). Each crystal was coupled with a large area 9.2×9.2 mm^2 Excelitas VTH2090 photodiode which ensures a linear response in the range from 0.3 to 10^4 MIP, and 9 CASIS chips have been used for the readout.

This prototype has been tested on high-energy particle beams at the CERN SPS accelerator facility in February 2013. Ions with A/Z=2 and 30 GeV/n and 12.8 GeV/n have been used to characterize the front-end electronics and to study the performance of the calorimeter. The excellent low-noise performance obtained can be seen from the plot in Fig. 3 that shows a very good separation of the ^{2}H and ^{4}He peaks observed in the first layer of the prototype, as well as a clear separation of the 1 MIP signal from the pedestal value. The non-interacting Deuterium peak can be resolved with a S/N $\sim$ 14. A preliminary study of the capability of this kind of calorimeter as a charge detector is reported in Fig. 4. The left plot shows the energy deposited in the central crystal of the first layer for a sample of events which have not suffered an inelastic interaction in the first two layers ($\sim$ 0.2 λ_I). These events have been selected by requiring that the signals in the two layers are consistent with the dE/dx of a fully stripped ion

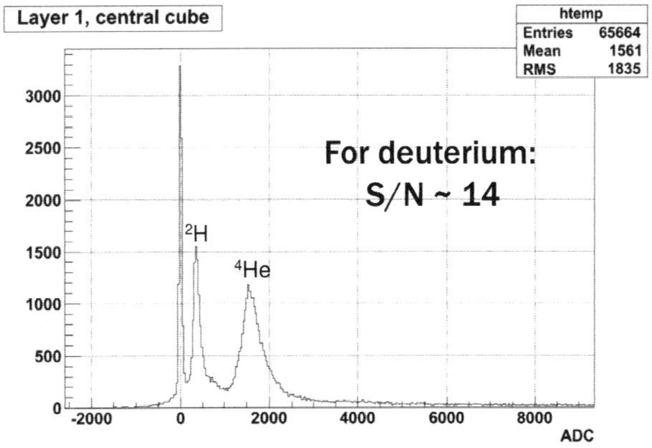

Figure 3. Signal (ADC units) of non-interacting 30 GeV/n ^{2}H and ^{4}He detected in the first layer of the calorimeter prototype.

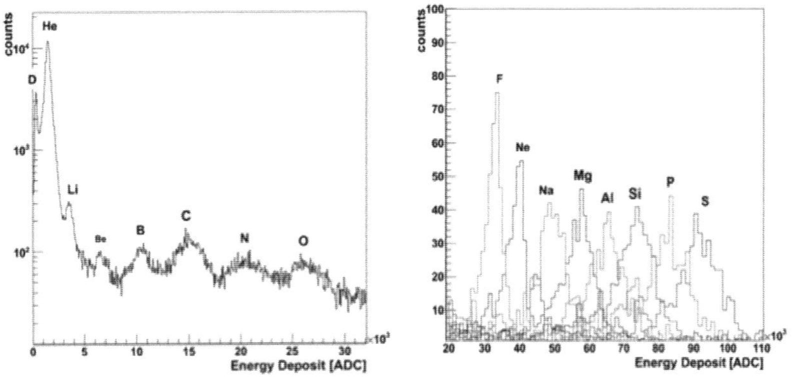

Figure 4. Preliminary results showing the charge identification capabilities of the calorimeter prototype. Left panel: energy deposited in the central crystal of the first layer for a sample of non-interacting ions. Right panel: sample of high-Z events whose charge has been selected by a precise low-thickness silicon tracker located upstream of the prototype, providing a high-purity charge tagging by means of 12 independent dE/dx measurements.

of charge Z. The pulse height distributions corresponding to nuclei from Deuterium to Oxygen are clearly visible. High-purity samples of heavier nuclei up to Sulphur were tagged by means of an array of silicon pixel and microstrip sensors positioned upstream the calorimeter, as shown in the right plot of Fig. 4. Even though the charge separation is quite poor (worse than 0.3e in the BCNO region) compared to the typical performance of a charge detector, these results are encouraging, considering that the instrument prototype was not optimized for this purpose and the identification capabilities can be improved with a different configuration of sensitive elements on the faces of the calorimeter.

A first glance at the performance in the large energy release region can be obtained from the left plot of Fig. 5, which shows a good linearity in the response of the prototype from the low-Z (^{2}H) up to the high-Z ions (Fe). The right panel of Fig. 5 shows a preliminary estimate

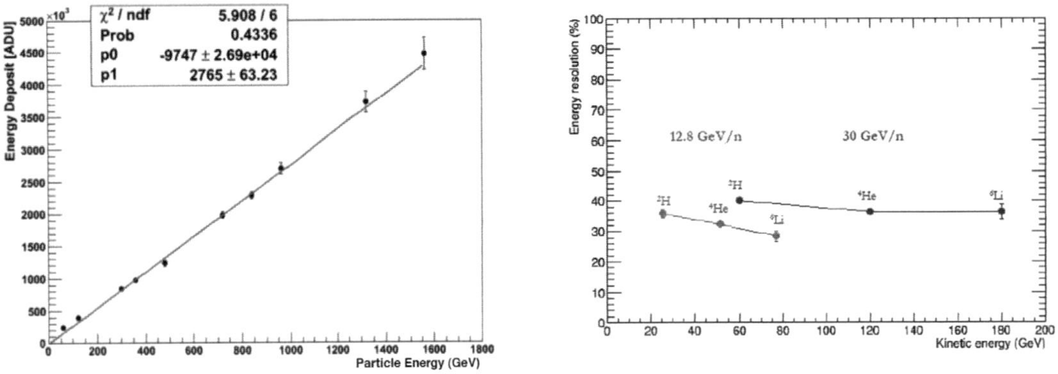

Figure 5. Left panel: preliminary result showing the linearity of the response of the prototype to the 30 GeV/n beam as a function of the energy of the various A/Z=2 ions. D, He, B, C, O, Mg, Si, S, Ti and Fe nuclei have been selected by the external silicon tracker. Right panel: preliminary estimate of the energy resolution of the prototype for showers starting on the second layer as a function of the energy of the incoming ion (the error bars refer to statistical errors only and further analysis work is in progress to assess the systematic uncertainties).

of the energy resolution of the prototype as a function of the energy of incoming interacting nuclei. Events with showers starting on the second layer of the calorimeter are selected in order to ensure a good longitudinal containment.

4. BTF tests on crystals for dual readout

The limitations imposed by the use of this kind of instrument in a space experiment drastically restrict its weight and size, making it impossible to build a total containment hadronic calorimeter. The energy resolution is therefore dominated by longitudinal losses which, because of the intrinsic fluctuations between the electromagnetic and the hadronic components during the development of the shower, cause large fluctuations in the amount of detected energy. To partially compensate for the effect of the fluctuations, a line of development of the Calocube project pursues the possibility to detect simultaneously the scintillation and the Cherenkov light, which carry information on the different components of the shower. In this context, an activity of characterization of the optical properties of CsI(Tl) crystals and other scintillating materials to optimize their application for a dual readout technique is currently underway: the possibility to improve transparency and emission efficiency by adjusting the concentration of the doping impurities is also being considered.

Some of the features of the Cherenkov light that can be used to discriminate it from the scintillation light are its prompt emission and its rapidly decreasing spectrum with increasing wavelength, which basically makes it detectable in the UV band. Accordingly, we are investigating different choices of light collection and detection systems. These include optical surface treatments of the scintillating material in order to better collect/convert the UV Cherenkov light, such as the deposition of thin wavelength shifter layers directly on the crystals, or the use of suitably engineered nanomaterials to act as dichroic filters.

Preliminary tests with single crystals have been done in September 2013 at the Beam Test Facility (BTF) of the DAFNE accelerator of the INFN Frascati National Laboratory. Several combinations of scintillating crystals (PWO, BSO, CsI(Tl)) and wrapping materials (white teflon, black teflon, mylar) have been exposed to a beam of electrons with $E = 500$ MeV. Crystals were mounted on a circular platform that could be rotated with respect to the beam

Figure 6. Setup of the test beam at BTF. The circular platform supporting the crystal could be rotated by ±30° with respect to the incoming beam line (indicated by an arrow). Two detectors (PT1 and PT2) mounted on opposite sides of the scintillator collected the scintillation and the Cherenkov light.

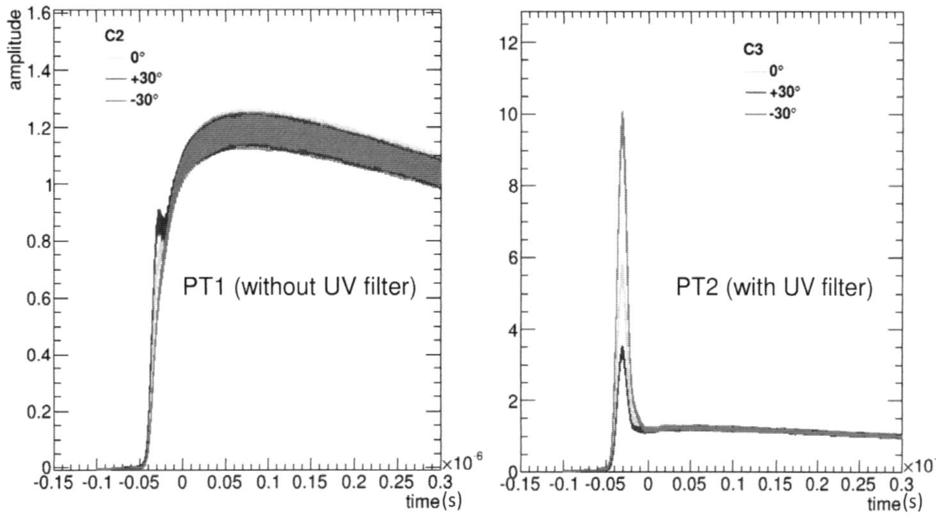

Figure 7. Preliminary results showing the possibility to extract the Cherenkov signal from the CsI(Tl) crystals. The plots show the time profile of the signals of the two phototubes for the three different incidence angles described in Fig. 6.

line, in order to study the angular dependance of the light output; signal were read out by a couple of sensors placed on opposite sides of the crystals (see Fig. 6). As an example, the plots in Fig. 7 shows preliminary results for the time profile of the signal, averaged over many events, obtained from a 3.6 cm CsI(Tl) cube wrapped in black teflon and read by two Hamamatsu R8900 series phototubes with a 23.5×23.5 mm^2 effective area for three different incidence angles. The phototube PT1 received directly the light produced in the crystal (left plot), while on PT2 a UG11 UV transmitting filter was applied (right plot). While the response of PT1 is nearly independent from the incidence angle, as expected for the scintillation component, the signal collected by PT2 shows clearly a fast peak whose intensity is larger when the phototube is placed downstream the incoming particle with respect to the crystal. The results seem to indicate that it is possible to use the directional and timing characteristics of emitted light to disentangle the Cherenkov signal from the scintillation in CsI(Tl).

5. Conclusions

An innovative calorimetric solution for direct measurements of high-energy cosmic rays in space is being developed by the Calocube collaboration. The basic idea is to build a large-acceptance, deep, homogeneous and isotropic detector: a cubic calorimeter composed of a 3-D lattice of small CsI(Tl) crystals able to detect particles from 5 sides has been optimized for this purpose. Monte Carlo simulations show that this design can reach the required acceptance and that its granularity, coupled with the large depth, will guarantee an excellent performance in the e/p separation, as well as in the energy resolution both for electromagnetic and hadronic particles. Preliminary results of a test beam with a prototype of the apparatus show good performances both in the energy measurement and in the charge discrimination capabilities. Studies of the applicability of dual redout techniques to enhance the hadronic performances are in progress and a test beam with single crystals indicate the possibility to successfully use the Cherenkov light emission.

References

[1] Bonvicini V et al. 2010 *IEEE Trans. Nucl. Sci.* **57** 2963–70
[2] Battistoni G at al. 2007 *AIP Conf. Proc.* **896** 31–49
Ferrari A at al. 2005 *CERN-2005-10, INFN/TC 05/11, SLAC-R-773*
[3] Mori N et al. 2013 *Nucl. Instr. Meth.* A **732** 311-5

Journal of Physics: Conference Series **587** (2015) 012030 doi:10.1088/1742-6596/587/1/012030

Search for Majorana neutrinos with the SNO+ detector at SNOLAB

A. Maio[abc] for the SNO+ collaboration
[a]Laboratório de Instrumentação e Física Experimental de Partículas, Av. Elias Garcia, 14, 1,1000-149 Lisboa, Portugal,
[b]Departamento de Física, Faculdade de Ciências da Universidade de Lisboa, Campo Grande, Edifício C8, 1749-016 Lisboa, Portugal,
[c]Centro de Física Nuclear da Universidade de Lisboa, Av. Prof. Gama Pinto, 2, 1649-003 Lisboa, Portugal,

E-mail: amelia@lip.pt

Abstract. The SNO+ experiment is adapting the Sudbury Neutrino Observatory (SNO) detector, in order to use isotope-loaded liquid scintillator as the active medium. SNO+ has multiple scientific goals, the main one being the search for neutrinoless double beta decay, the most promising signature for the possible Majorana character of neutrinos and for the absolute neutrino mass. Measurements of neutrinos from the Sun, the Earth, Supernovae and nuclear reactors are additional goals of the experiment. The detector consists of a 12m diameter spherical vessel, filled with 780 tonnes of Tellurium-loaded liquid scintillator, and surrounded by about 9500 PMTs. It is shielded by a large volume of ultra-pure water and the underground location at SNOLAB, Canada. This talk will review the Physics goals and current status of SNO+.

1. Introduction

The SNO+ experiment [1] at SNOLAB [2] in Canada aims to search for Majorana neutrinos through the observation of the neutrinoless Double Beta Decay (DBD). The neutrinoless-DBD (or $0\nu\beta\beta$-decay) signal is proportional to an effective mass, $<m_\nu>$, dependent on the neutrino mixing parameters U_{ek} and on the difference of squared masses measured by oscillations. The observation of $0\nu\beta\beta$-decay would contribute to the determination of the neutrino absolute mass scale, the mass hierarchy and, in addition, reveal the Majorana character of the neutrino [3].

Since the late 1940s, early 1950s, that the search for Double Beta Decay with the emission of two neutrinos ($2\nu\beta\beta$) has been the subject of several experiments. The DBD nuclear process consists on $(A, Z) \rightarrow (A, Z + 2) + 2e^- + 2\nu$ and it is possible for a few isotopes with extremely long half-lives, so very small event rates. This process has been observed for the first time in 1950 in the decay of ^{130}Te in geochemical experiments and only in 1987 it has been observed by direct counting by Michael Moe's group in the decay of ^{82}Se. Later on many more of these events were observed [4].

The quest for Majorana neutrinos through the observation of neutrinoless Double Beta Decay ($0\nu\beta\beta$), i.e., $((A,Z) \rightarrow (A,Z + 2) + 2e- + 0\nu$, is a key topic research in fundamental physics in view of the neutrino properties as well as the completion or extension of the Standard Model of Particle physics. While the signature of $0\nu\beta\beta$ – a monoenergetic peak - at the $Q_{\beta\beta}$ is clear when compared to the continuous spectrum of $2\nu\beta\beta$, the process has so far not been observed. From this non-observation, limits on the $0\nu\beta\beta$ half-life and, correspondingly, on the effective neutrino mass, have been placed by several experiments. The best current limits are around 10^{25} years and 200 meV, respectively [4]. Sensitivities of about 100 meV are expected for currents experiments or under preparation, using ultrapure materials with very low radioactivity, employing active or passive shielding and located in underground laboratories [5].

Since the design of SNO+ is driven by the detection of low energy electrons from $0\nu\beta\beta$-decay, measurements of low energy pep/CNO solar neutrinos, Supernovae neutrinos as well antineutrinos from nuclear reactors and geoneutrinos are additional goals of the experiment.

2. The SNO+ experiment

2.1. Detector overview

The SNO+ detector [1] is based on the SNO detector but with a liquid scintillator that replaces the heavy water used by SNO. The project and design of SNO+ is driven by the detection of electrons from the DBD of radioisotopes such as ^{150}Nd, and more recently ^{130}Te, the choice made for the first phase of the SNO+ experiment. The natural Tellurium is dissolved in a liquid scintillator based on the idea of Raghavan for the Borexino detector [6], using ^{136}Xe. So, the basics of detection in SNO+ is the full energy loss of ionizing particles with production of scintillation of the active material. The scintillation light is detected by a system of photomultipliers (PMTs) and the number of photo-electrons measured is proportional to the energy of the particle (electrons, alphas, muons, etc.). So, in the language of Experimental High Energy Physics the SNO+ detector can be considered a homogeneous calorimeter, and since the final result is the counting of electrons, SNO+ can be considered also a counting detector.

Figure 1. Artistic view of SNO+ in its cavity 2km deep.

Figure 2. SNO+ physicist cleaning the top part of the AV.

SNO+ is located 2 km underground at SNOLAB and consists of a 12 m diameter acrylic vessel (AV) containing 780 tonnes of liquid scintillator. The scintillator will be loaded with 0.3% of natural tellurium, equivalent to 2340 kg, or 800 kg of ^{130}Te, in the initial phase. The AV is located inside a cavity in the rock, and a geodesic structure around the AV supports the system of 9500 photomultipliers tubes with QE ~20% coupled to a set of reflectors to increase the geometric coverage to 55%. The acrylic vessel and the PMT structure are immersed in 7000 tonnes of ultrapure water (10^{-13} to 10^{-14} g of U/Th per g of water) to shield the liquid scintillator from radioactivity from the rock surrounding the underground cavity and the PMTs. The diffusion of radon from the rock is stopped by a thick liner of Urylon. The AV is sustained by a system of ropes to compensate the buoyancy of the scintillator with a density of 0.86g/cm^3. The ropes have been carefully chosen minimising the radioactivity impurities of the order of ppb for the U/Th isotopes and ppm for ^{40}K. An artistic view of the detector is shown in figure 1, and in figure 2 a SNO+ physicist finalising the cleaning of the top part of the detector is seen and the hold-down rope system is also visible.

A new readout system, 20 times faster, and new calibration systems were designed and are being implemented. A new optical system of 96 LEDs/plastic fibers for calibration and monitoring the PMT response, including time and charge information, and a system of 20 quartz fibers for transparency measurements, have been built and partially installed in the detector (see talk of L. Seabra in these proceedings [7]).

A mechanism for the deployment of radioactive sources is designed and under construction. Natural radioactive sources from the background, as well as new radioactive sources (AmBe for neutrons and γ-rays, ^{16}N, ^{24}Na, ^{48}Sc, ^{57,60}Co, ^{65}Zn for γ-rays and ^{90}Y for betas) are being developed for setting the absolute energy calibration. Natural sources part of the background, such as ^{14}C, ^{208}Tl, ^{214}Bi, ^{210}Po, etc... will also be extensively used.

2.2. Neutrinoless DBD search with liquid scintillators

Measuring events of extremely small rates and with energies in the range of 1 to 5 MeV needs large volume detectors such as SNO, and large mass of the DBD isotope, even at small loading and a very low background. To reach the low levels of Borexino [8] background SNO+ requires: (i) a very good intrinsic radiopurity of the scintillator, (ii) a fiducial volume to reject external backgrounds, (iii) the choice of materials with ultra-low levels of radioactivity, (iv) the purification of cosmogenically activated contaminants; (vi) the knowledge of irreducible background from solar neutrinos, (vii) and for $Q_{\beta\beta}$> 2 MeV, the identification of many backgrounds with delayed coincidence (alpha or beta) tagging .

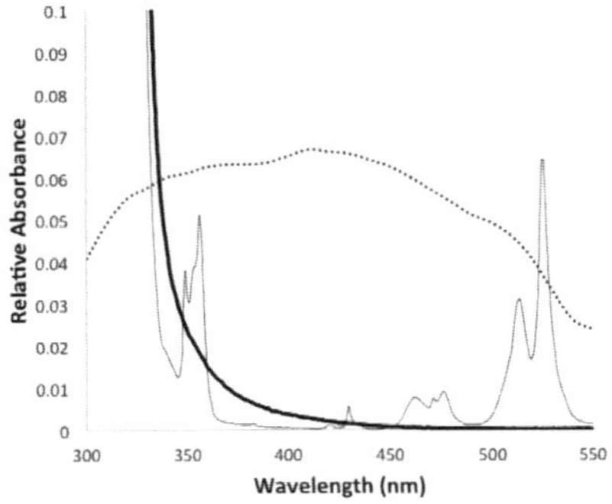

Figure 3. Absorbance vs λ (nm) for 0.3% Te (—) and 0.3% Nd (---) loaded scintillator without 2nd WLSs or fluors, and SNO+ PMTs QE(λ) (····).

Figure 4. The transparent liquid scintillator.

The liquid scintillator chosen by the SNO+ collaboration after a program of R&D was Linear AlkylBenzene (LAB) with 2g/l of the PPO fluor (2,5-diphenyloxazole) that optimises the match of the scintillating light with the PMT quantum efficiency spectra as shown in figure 3. The main requirements of the scintillator were: (i) chemical compatibility with the acrylic vessel (ii) high radiopurity (Borexino levels) (iii) high light yield, (iv) good optical transparency (figure 4) and low scattering, (attenuation lengths larger than 10 m were achieved, but probably will decrease after loading with ^{130}Te) [9] and absence of absorption peaks in the region of the PMT maximum QE (λ), fast response, of about 5 ns, (vi) safety, i.e. low toxicity and high flash point (150ºC). To minimise cosmogenic activation, a purification system for distillation and water extraction is presently being installed at the underground laboratory. A scintillation light yield of ~10000 photons/MeV corrected for the PMT QE of ~20%, geometric coverage of 55% and absorption in the scintillator, resulted on typical values of 200-300 detected photoelectrons/MeV [9]. As is typical of liquid scintillator experiments, SNO+ has intrinsically worse energy resolutions than solid state detectors, making it harder to distinguish the $0\nu\beta\beta$ decay peak from background γ-ray lines and from the two-neutrino DBD continuous spectrum. However, this is compensated by having a large isotope mass and very low backgrounds, and other ways of reducing them, as will be described in the next section.

2.3. Data analysis
The ionizing particles, from isotopes decay or cosmic rays reaching the scintillator, excite the scintillator producing photons that travel through the volume and are detected by the PMTs. The number of photoelectrons viewed by the PMTs is corrected by the PMT QE and by the geometric efficiency coverage and attenuation coefficient depending on distance. The electronics allows the measurement of the time of arrival and the charge of each PMT hit, within a 400ns window. The event position is reconstructed with the PMT timing information. A fiducial volume of about 20% of the liquid scintillator is defined, in order to reject the external backgrounds. The event energy is obtained from the number of hits corrected by its position. Delayed coincidences are used to identify some of the radioactive backgrounds [10].

The DBD analysis is done with the energy spectrum after removing or constraining the background and the candidate events to the $0\nu\beta\beta$ peak are counted. So, we can consider SNO+ a counting experiment using calorimeter techniques.

3. Neutrinoless DBD with SNO+

3.1. The source of DBD-decay for SNO+
With the aim of searching for Majorana neutrinos through DBD processes, several isotopes candidates have been considered. Some even-even nuclei are usually β-decay energetically forbidden but double decay is allowed. From the 35 naturally occurring isotopes energetically DBD allowed, among them ^{78}Ge, ^{130}Te, ^{136} Xe, ^{150}Nd, etc, only 10 of them were observed [4]. The main characteristics of those DBD isotopes are their extremely long half-life (>10^{18} years), a $Q_{\beta\beta}$ of the order 1 MeV to 5 MeV, and a natural abundance about 10% or less. Only one known DBD-isotope, ^{130}Te, presents a high natural abundance of about 34.1% although with a low $Q_{\beta\beta}$ of 2.53 MeV and a half-life 7.14×10^{20} years. ^{150}Nd has a higher $Q_{\beta\beta}$ of 3.4 MeV but its natural abundance is only 3.6 %. SNO+ has chosen ^{130}Te as first option.

3.2. DBD Background and signal
The sources of background for neutrinoless DBD searches with SNO+ detector are internal and external. Beta particles from $2\nu\beta\beta$ are an irreducible background although lower for ^{130}Te, but demanding a good energy resolution. Position resolution is also an important factor for the energy reconstruction, and can be important to avoid pile-up in case of high $2\nu\beta\beta$ rates. Solar ^{8}B neutrinos are

also an irreducible background but its rate and distribution is known from SNO and Super-Kamiokande.

External γ-rays from the U/Th chains and ^{40}K from the rock can be suppressed by the ultrapure water shielding surrounding the AV; cosmic ray muons are suppressed to negligible levels by the underground location. Internal background from the U/Th in the water used in the process of diluting the Tellurium in the scintillator (TeLS) is assumed to be of the order of ~10^{-14} to 10^{-15}g of U/Th per g of TeLS as in SNO. Alpha (α) or beta (β) emitters, with short half-lives, or close to them in the decay chain, such as ^{214}Bi, ^{212}Bi, ^{208}Tl, ^{210}Po can be rejected by delayed coincidences. The radiopurity of the mixture TeLS requires several steps of purification. Cosmogenic activation on the surface is responsible for the production of unstable isotopes through cosmic ray interaction on the bulk material requiring a final purification step to be done at the SNOLAB underground laboratory.

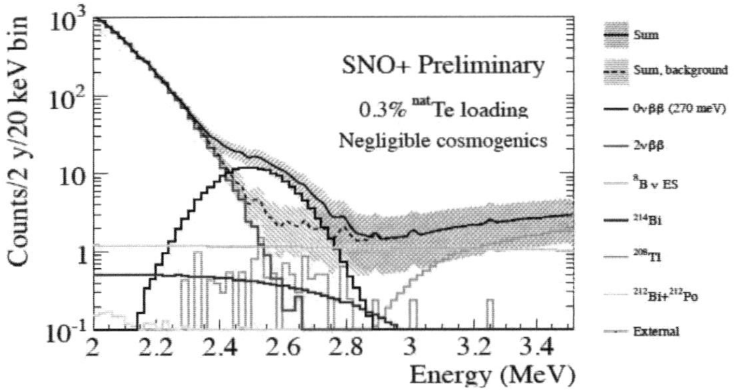

Figure 5. MC simulation of the SNO+ background and DBD (2vββ and 0vββ) signals.

In figure 5 is depicted a preliminary Monte Carlo simulation of the background in SNO+ and the neutrinoless double decay spectrum expected for a loading of 0.3% of natural Tellurium and for two years of SNO+ data taking. It is well visible the neutrinoless double beta decay peak superimposed to the end of the 2vββ-decay continuous spectrum. After five years of data taking the expected neutrino mass sensitivity is about 100meV, which could be improved by achieving higher loading of Tellurium in the scintillator.

4. Reactor and Earth Antineutrino searches with SNO+

Measurements of anti-neutrinos from nuclear reactors and the mapping of geoneutrinos reaching the SNO+ detector, contributing to the knowledge of the mantle composition and radiogenic heat production of the Earth, are additional goals of the SNO+ experiment.

The first observation of geoneutrinos, produced in the planet's crust and mantle (figure 6), are recent but are already being used to test Earth models; anti-neutrinos produced in nuclear reactors have been used in neutrino physics, namely for the study of neutrino oscillations [11]. In figure 7 it is shown a preliminary simulation of anti-neutrino oscillations seen at SNO+ where the peaks from neutrinos from different reactors in Canada are well visible.

Anti-neutrinos from reactors were detected for the first time by C. L. Cowan Jr., F. Reines et al (1956) through the interactions in liquid scintillator detectors like SNO+. The signature of the process, can be clearly identified by the delayed coincidence of the positron annihilation followed by a neutron capture in which the positron energy and the neutron initial direction follows the anti-neutrino kinematics. Anti neutrinos can then be identified with SNO+ even under the large event rate of DBD.

Also Supernovae explosions could be detectable with ^{130}Te loading, while low energy solar neutrinos will be searched with pure scintillator only.

Figure 6. The Earth: crust and mantle are visible.

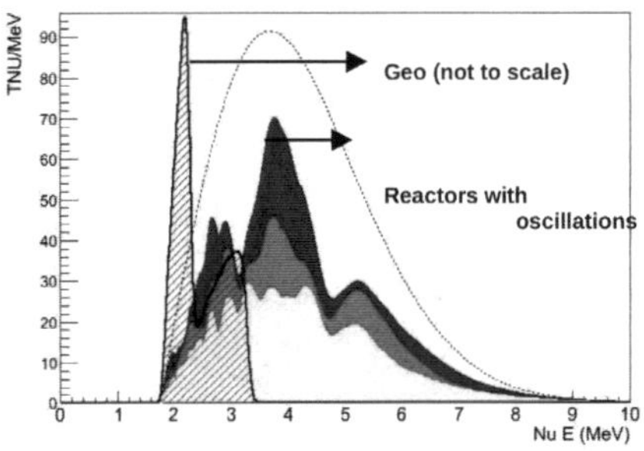

Figure 7. Simulation of anti-neutrino oscillations seen by SNO+.

5. Status and outlook

SNO +, a large volume liquid scintillator loaded with a large mass of ^{130}Te will probe the nature and magnitude of the effective neutrino mass with the sensitivity about 100 meV, and will measure anti-neutrinos from reactors and the Earth and also Solar and Supernova neutrinos. Water filling of the detector has started and SNO+ is almost ready for the water phase commissioning by the end of 2014. The new trigger and readout as well as the calibration systems are almost finished.

SNO+ scintillator phase is expected for 2015, initially with unloaded scintillator. First measurements of the background before loading it natural Tellurium are also planned. Neutrinoless DBD search will start with 0.3% Tellurium loading but could be later increased to 3%.

6. Acknowldgments

My thanks go to José Maneira, Sofia Andringa and to Luís Seabra, for helping me with my first public action in SNO+.

This research was supported in part by EU FEDER funds through the COMPETE program (Operational Program for Competitiveness Factors) and by national funds from Portugal through FCT (Fundação para a Ciência e a Tecnologia) within the project PTDC/FIS/115281/2009.

7. References

[1] Chen M 2005, Nucl. Phys. B (Proc. Suppl.) 154, 65-68 and http://snoplus.phy.queensu.ca
[2] Duncan F, Noble AJ, Sinclair D 2010, Ann. Rev. of Nuclear and Particle Science 60, 163-180 and www.snolab.ca
[3] Zuber K 2011 Neutrino Physics, 2nd edition, Taylor & Francis
[4] Barabash AS 2011, Phys. Atom. Nucl. 74 603-613, nucl-ex:1104.2714
[5] Weinheimer C., 2013, arXiv:1307.3518v2
[6] Raghavan RS 1994, Phys. Rev. Lett. 72 10
[7] Seabra L 2014, these proceedings
[8] Borexino Collaboration 1991, Proposal (Univ. of Milan, AT&T Bell Labs)
[9] Biller S 2014, to appear in Proc. of TAUP 2013, arXiv:1405.340v1 and Lozza V 2012, IOP J. Phys. Conf. Series 375 042050
[10] Maneira JM 2013, IOP J. Phys. Conf. Series 447 012065
[11] Andringa S 2014, to appear in Proc. of Nucl. Phys. in Astrophys. IOP J. Phys.: Conf. Series

16th International Conference on Calorimetry in High Energy Physics (CALOR 2014) IOP Publishing
Journal of Physics: Conference Series **587** (2015) 012031 doi:10.1088/1742-6596/587/1/012031

The LED and fiber based calibration system for the photomultiplier array of SNO+

L. Seabra[a], R. Alves[b], S. Andringa[a], S. Bradbury[c], J. Carvalho[b], K. Clark[d], I. Coulter[e], F. Descamps[f], L. Falk[d], L.Gurriana[a], C. Kraus[g], G. Lefeuvre[d], A. Maio[ahj], J. Maneira[ah], M. Mottram[d], S. Peeters[d], J. Rose[i], J. Sinclair[d], P. Skensved[k], J. Waterfield[d], R. White[d], J. Wilson[l], for the SNO+ collaboration

[a]Laboratório de Instrumentação e Física Experimental de Partículas (LIP), Av. Elias Garcia, 14, 1,1000-149 Lisboa, Portugal,
[b]Laboratório de Instrumentação e Física Experimental de Partículas and Departamento de Física, Universidade de Coimbra, 3004-516 Coimbra, Portugal,
[c]School of Physics and Astronomy, University of Leeds, Leeds LS2 9JT, United Kingdom,
[d]Dept. of Physics and Astronomy, University of Sussex, Falmer Campus, Brighton BN1 9QH, United Kingdom,
[e]Dept. of Physics, Oxford University, Denys Wilkinson Building, Keble Road, Oxford, OX1 3RH, United Kingdom,
[f]Nuclear Science Division, Lawrence Berkeley National Laboratory, Berkeley, California 94720, USA,
[g]Dept. of Physics and Astronomy, Laurentian University, Sudbury, Ontario P3E 2C6, Canada,
[h]Departamento de Física, Faculdade de Ciências da Universidade de Lisboa, Campo Grande, Edifício C8, 1749-016 Lisboa , Portugal,
[i]Dept. of Physics, University of Liverpool, Liverpool L69 7ZE, United Kingdom,
[j]Centro de Física Nuclear da Universidade de Lisboa, Av. Prof. Gama Pinto, 2, 1649-003 Lisboa, Portugal,
[k]Queen's University, Physics Dept., Kingston, Ontario K7L 3N6, Canada,
[l]School of Physics and Astronomy, Queen Mary, University of London, 327 Mile End Road, London, E1 4NS, United Kingdom.

E-mail: `lseabra@lip.pt`

Abstract.
A new external LED/fiber light injection calibration system was designed for the calibration and monitoring of the photomultiplier array of the SNO+ experiment at SNOLAB. The goal of the calibration system is to allow an accurate and regular measurement of the photomultiplier array's performance, while minimizing the risk of radioactivity ingress. The choice in SNO+ was to use a set of optical fiber cables to convey into the detector the light pulses produced by external LEDs. The quality control was carried out using a modified test bench that was used in QC of optical fibers for TileCal/ATLAS. The optical fibers were characterized for transmission, timing and angular dispersions. This article describes the setups used for the characterization and quality control of the system based on LEDs and optical fibers and their results.

1. Introduction

The SNO+ (Sudbury Neutrino Observatory) detector has many Physics goals including the search for neutrinoless double beta decay, the detection of pep and CNO solar neutrinos, the detection of geo-neutrinos and the study of reactor neutrino oscillations [1](see talk of A. Maio in these proceedings [2]).

The detector, installed in the Creighton mine near Sudbury, Canada at a depth of 2000 meters, was previously used with heavy water as active element (for the detection of solar neutrinos) and will now use liquid scintillator (LAB, PPO and ^{130}Te). The detector is composed by a 12 m diameter spherical acrylic vessel with 5 cm thickness that will be filled with 780 tons of active material. Around the vessel there is a geodesic structure with ~9000 PMTs. The cavern where the detector is installed will be filled with purified water and, because the density of the liquid scintillator is lower than the heavy water, the acrylic vessel will have a rope basket attached to the cavern floor. The detector depth serves as a barrier to cosmic rays but the detector is still not free from unwanted radiation. Radioactive decays of residual impurities present in the scintillator and detector materials constitute an important background. The radiopurity requirements are very high for the liquid scintillator ($1 \times 10^{-17} g/g$) and progressively relaxed for more external materials like acrylic, water and PMTs. The events originated by gammas from external materials concentrate in the most external layers of the scintillator volume and can be removed with a position selection of the events. The position measurement is based on time information given by PMTs and therefore the precision depends on the quality of the synchronization between PMTs.

In the calibration system (Figure 1), optical fibers, placed in the PMTs support structure, emit cones of light that will pass through the whole detector and will illuminate one group of PMTs on the opposite side of the detector. The principal requirements on this light cone are the time and angular dispersion. The time dispersion of light at the fiber exit should not be much higher than the resolution in time of the SNO+ components. The irreducible components are PMTs jitter (1.6 ns) and scintillator decay time (5 ns) so the desired time dispersion is of the order of a few ns [3].

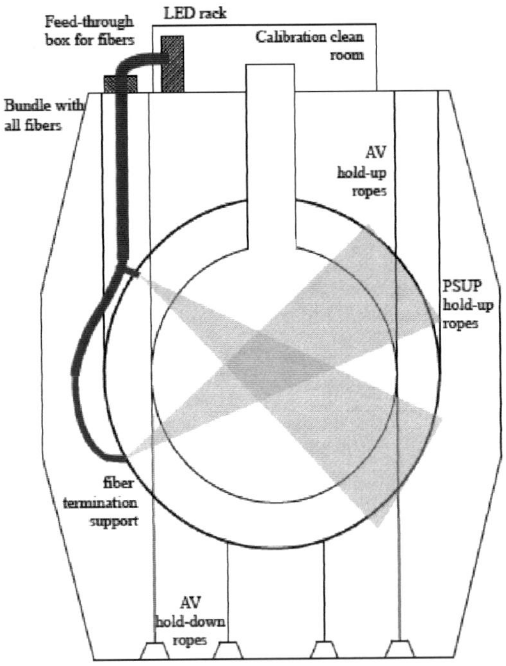

Figure 1. Schematics of the detector calibration system where the light produced by LEDs travel through optical fibers into several points in the PMT structure. Each optical fiber emits a cone of light that will pass through the whole detector and will illuminate one group of PMTs on the opposite side of the detector.

The angular dispersion of the optical fibers is an important parameter on the global system project because the total number of fibers is dependent of the cone aperture. For redundancy reasons, and to account for possible fiber failure, all the ~9000 PMTs must be illuminated, at least, by two light cones. Furthermore, this redundancy allows for cross-synchronization between the PMT regions illuminated by different fibers. For installation reasons the number of fibers should not be higher than strictly necessary. 96 double optical fibers plus spares were installed in the PMT structure, each of them connected to one LED placed outside the detector cavern.

2. System Components

2.1. Optical fibers

The optical fibers are 45 meters long with single cladding with a numerical aperture of 0.51 and a refractive index of 1.49. The core with 1 mm of diameter is composed by polymethyl methacrylate (PMMA) and the cladding by a fluorinated polymer. Each fiber cable has two optical fibers (one as spare) coated with black polymer. The wet-end connector is a duplex latching and the dry-end connector is a ST. Each optical fiber is connected to a custom made support plate that is attached to the side of a PMT in one of the geodesic nodes in the PMT structure. This choice ensures a uniform coverage of all PMTs.

2.2. LED

The selected LED has the highest possible peak wavelength in order to achieve the optimal direct transmission. The limiting factor is the quantum efficiency of the SNO photomultiplier tubes and the absorption probability in the scintillator mixture. The optimal peak wavelength was found to be 505.6 ± 2.6 nm. The chosen LED has a low differential resistance in the near linear region which allows for a pulse width of the same order as the Transit Time Spread of the SNO PMTs.

2.3. Driver

The driver circuit is capable of operating with a dynamic range of frequencies, between 10 Hz and 10 kHz, in order to perform monitoring at low frequencies and calibrations at high frequencies. It can generate pulses with a full width half maximum (FWHM) of the order of 4 ns.

3. Characterization of optical fibers and Quality Control

All the 110 optical fibers were subject to characterization and quality control measurements for transmission, timing and angular dispersions. For the time dispersion the measurements were made using a LED driver that gives a signal in the order of 3 ns of FHWM. The signal travels through the optical fiber into a PMT where it is read by an oscilloscope (Figure 2). From the measurements a low time dispersion was obtained with a FHWM of the order of 4 ns (Figure 4).

For angular dispersion a two-dimensional scan of one optical fiber was done using a PMT in single photo-electron mode with a very low LED intensity and a neutral density filter on the duplex connector. Quality control was carried out using a test bench that was used in QC of optical fibers for TileCal/ATLAS [4][5]. This base automated test-bench was adapted to check the light output at different angles in a transverse scan, in order to cross-check the variation of the angular opening between fibers. The peak value and the integration of the angular scan gave also the variability in terms of light transmission. Figure 3 shows the interior of the dark-box, adapted for this test. Three duplex fibers, being illuminated independently from outside the dark-box, could be checked in each single run. The edges of some of the angular distributions were cut for some of the fibers and showed some distortions, but the central part was consistent for all the six tested positions, with the variations in between the 220 fibers

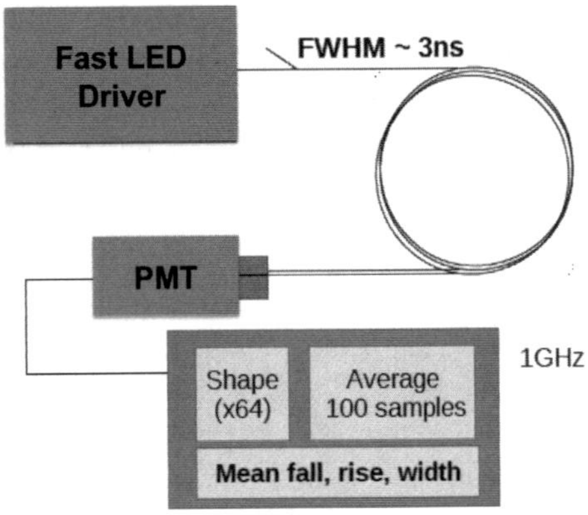

Figure 2. Set-up for the time dispersion measurements.

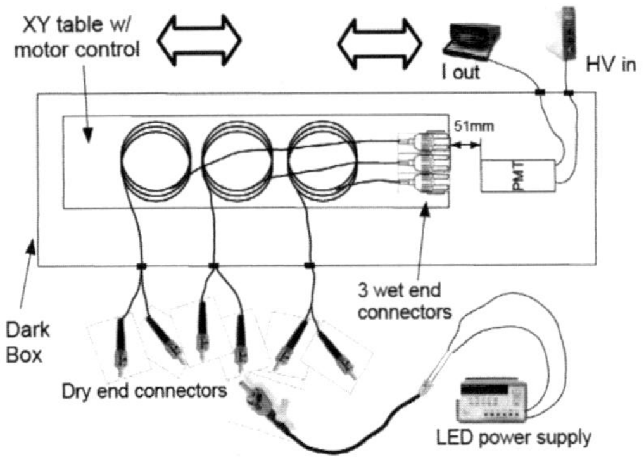

Figure 3. The interior of the dark-box, showing the mounting of three 45 m duplex fiber cables in front of the PMT.

shown in Figure 5, where they are compared with the higher resolution test of a single fibre in the two-dimensional scan given an aperture angle of 14.5 degrees. Taking in consideration the results in QC for transmission, timing and angular dispersions some optical fibers were selected as spares. Nevertheless theses results reveal a good uniformity between optical fibers.

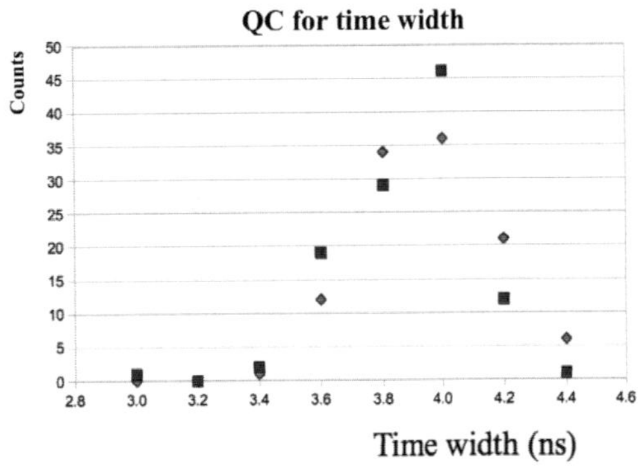

Figure 4. Time width dispersion of all 110 duplex fibers with an injected signal of 3 ns. The blue dots and the red are for different fibers from the duplex latching connector.

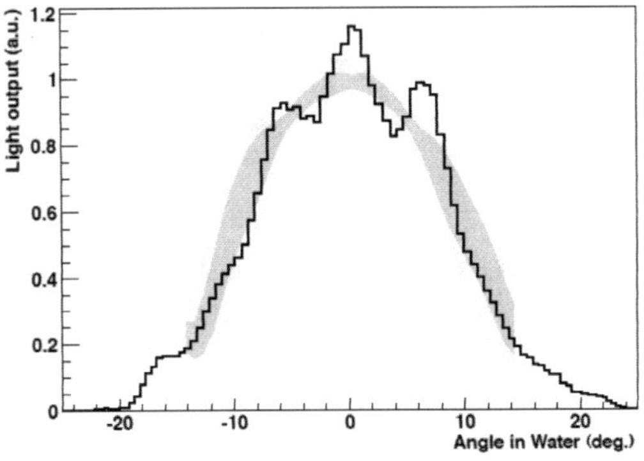

Figure 5. The black histogram is the profile of the detailed 2D measurement. The bands summarize the stability measurements of the 110 duplex cables: the yellow band shows the overall spread, the green band represents the measurement uncertainty.

4. Commissioning

The installation of the new calibration system is in progress: 36 of 96 light injections fibers are already fully installed and have been tested in commissioning runs of SNO+ electronics, with the full dark detector and the scintillator and water regions filled with air. On the 2D view of Figure 6 is possible to see a spot far from the main charge collection that corresponds to the light reflections in the acrylic vessel near the fiber injection. In order to validate the measured angular distribution, the commissioning results were compared to simulations that used the existing measurements as input. A good match was obtained for the distribution of intensity versus angle, confirming that the system will have enough coverage for all PMTs.

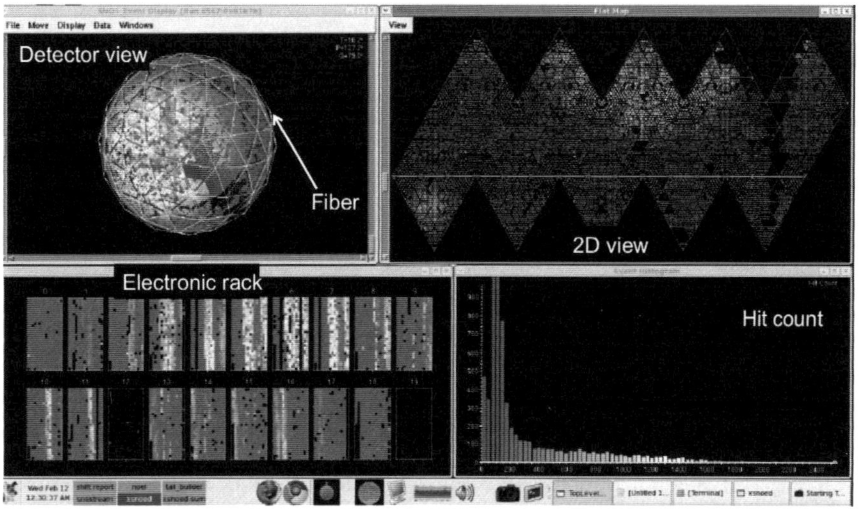

Figure 6. Display of a commissioning dry run with a single fiber injection where 11.2 Million triggers were obtained. Each dot corresponds to a PMT where the color indicates the collected charge.

5. Summary

The calibration of ~9000 PMTs can be done with the external LED/fibers system using only 96 optical fibers, with aperture angles 14.5 degrees in water, which guarantees a full detector coverage. The calibration signal created by the Driver+LED+fiber has a time width of $\sim 4\ ns$, sufficient for the synchronization and time versus charge calibration. The new calibration system is under installation and the first commissioning shows promising results.

6. Acknowledgments

Research supported by EU FEDER funds through the COMPETE program 674 (Operational Program for Competitiveness Factors) and by national funds from Portugal through 675 FCT (Fundação para a Ciência e a Tecnologia) within the project PTDC/FIS/115281/2009.

7. References

[1] Chen, M.C. "The SNO+ Experiment" 2008 *Proceedings of the 34th International Conference on High Energy Physics.*
[2] A. Maio "Search for Neutrino Double Beta Decay with SNO+" 2014 *Proceedings of the CALOR 2014.*
[3] B. Caccianiga, D. Franco, D. Giugni, P. Lombardi, S. Malvezzi, J. Maneira, G. Manusardi, L. Miramonti, G. Ranucci, O. Smirnov, "A multiplexed optical-fiber system for the PMT calibration of the Borexino experiment" 2003 *Nuclear Instruments and Methods in Physics Research A* **496** 353-36
[4] J.Santos, A. Maio, J.G.Saraiva, J. Silva and L. Gurriana 2007 *Characterization of plastic optical fibers with pulsed LEDs", in Nuclear Instruments and Methods in Physics Research A* **580** 306-309
[5] M. David et al 1998 *Tilecal-pub-2008-003*

R&D for a highly granular silicon tungsten electromagnetic calorimeter

Roman Pöschl, on behalf of the CALICE Collaboration

Laboratoire de l'Accélérateur Linéaire, Centre Scientifique d'Orsay, Bâtiment 200, 91898 Orsay, France

E-mail: poeschl@lal.in2p3.fr

Abstract. This article reports on first experience with the technological prototype of a highly-granular silicon-tungsten electromagnetic calorimeter as envisaged for the detectors at a future lepton collider. In the focus of the analysis is the performance of a highly integrated Application Specific Integrated Circuit designed to meet the requirements in terms of dynamic range, compactness and power consumption. The beam test results show that the circuit will allow a future detector with a signal over noise ratio of at least 10:1. To minimise the power dissipation the ASIC will be operated in a power pulsed mode. So far no conceptual problem was revealed but the studies show the way for further work. The prototype is read out by a DAQ system conceived to meet the needs of a trigger less system with a huge number of readout cells.

1. Introduction

The next major worldwide project in high energy particle physics will be a linear electron positron collider at the TeV scale to complement and extend the scientific results of the LHC. The most advanced proposal for this linear collider is the International Linear Collider (ILC). Here, electron and positrons will be collided at centre-of-mass energies between 0.2 and 1 TeV. The detectors which will be installed around the interaction point are required to achieve a jet energy resolution of $30\%/\sqrt{E}$, thus a factor two better than the energy resolution achieved for a typical detector at LEP. The reconstruction of the final state of the e^+e^- will be based on so-called particle flow algorithms (PFA) [1]. The goal is to reconstruct every single particle of the final state which in turn demands a perfect association of the signals in the tracking systems with those in the calorimeters. As a consequence this requires a perfect tracking of the particle trajectories even in the calorimeters. State of the art Particle Flow algorithms such as the Pandora PFA [2] or ARBOR [3] promise to reach jet energy resolutions between 3% and 4% over an jet energy range from 45 GeV up to the TeV region [4, 5]. At hadron colliders the technique of Particle Flow is applied successfully for the jet reconstruction by the CMS collaboration [6, 7].

To meet the requirements of a detector optimised for PFA, the detectors have to cover the whole solid angle and have to feature an unprecedented high granularity. The calorimeters have to be placed inside the coil of the super-conductive solenoid of the detectors. This puts tight constraints on the available space for the installation of the detectors. The design of the calorimeters have to take the following guidelines into account

- Optimisation of the number of calorimeter cells.

- Choice of the absorber material and the infra-structural components such as cooling, power supplies, readout cables and the very front end electronics.

For the electromagnetic calorimeter at lepton colliders these criteria has lead to the choice of tungsten with a radiation length of X_0=3.5 mm, a Molière of R_M=9 mm and an interaction length of λ_I=96 mm. In the years 2005 to 2011 the CALICE collaboration has performed beam campaigns at DESY, CERN and FNAL in order to demonstrate the principle of highly granular calorimeters and to confront the concept of particle flow with real data. The first results of the data analysis have been published in three articles [8, 9, 10, 11]. The work is now focused on a technological prototype that addresses the engineering challenges which come along with the realisation of highly granular calorimeters. The key parameters of the new prototype are

Figure 1 shows in its left part a schematic view of the technological prototype that can house up to 30 layers with a total depth of 24 X_0. The mechanical housing is realised by a tungsten-carbon reinforced epoxy (CRP) composite, which supports at the same time the absorption medium and ensures the mechanical integrity of the detector.

Figure 1. *Left: Schematic view of the technological prototype with dimension. Center: Mechanical structure of the technical prototype with a short slab in front and one inserted on the right. Right: 2.5 m alveolar end-cap layer.*

A large prototype structure has been produced in 2012, close in scale to a barrel module for the ILD detector concept (see Fig. 1 middle). It features 15 × 3 1.5 m-long alveola interleaved with Tungsten plates. As shown on the right part of Fig. 1, some long alveolar layer prototypes of 2.5 m in length have been moulded in order to validate the technological specific production process for mechanical structures as needed for ILD.

The left part of Fig. 2 shows a cross section through two calorimeter layers which form a slab. The sensitive parts will be mounted on both sides of tungsten board embedded in an "H" made of carbon fibre, equipped with a thermal drain, wrapped and inserted into the alveoli of the mechanical structure. The silicon wafers are composed of high resistive silicon. An example of a silicon wafer matrix is shown in the right part of Fig. 2. These silicon wafers are glued onto an interface card, called PCB hereafter which carries the ASICs for the readout of the silicon wafers.

2. The silicon sensors

The silicon sensors are the central component of the detector, and also the most expensive. Work is on-going on the sensor design, particularly on the sensor edge. Silicon sensors of various designs and producers have been purchased. These include large $9 \times 9\,\text{cm}^2$ sensors from Hamamatsu

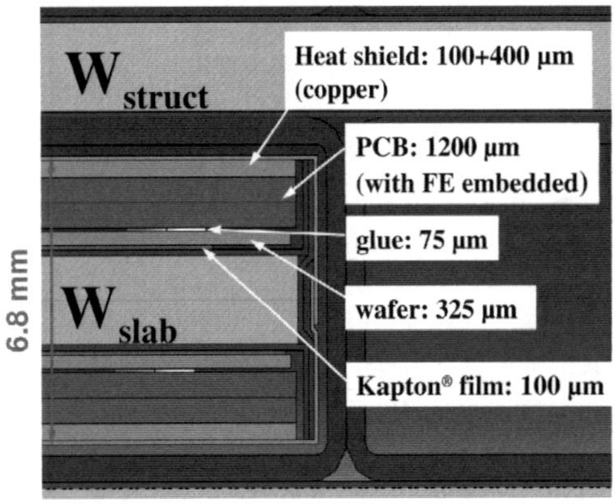

Figure 2. *Left: Cross section through one slab of the prototype with the thickness of the various components. Right: Hamamatsu Photonics silicon sensor: 324 pixels of* $5 \times 5\,\mathrm{mm}^2$.

Photonics (HPK) subdivided in 324 pixels (256 in more recent versions). The HPK sensors have excellent electrical characteristics, and a number of them have been successfully tested in beam tests. An undesirable feature was identified in beam tests with the physics prototype: a large energy deposition near the guard rings provokes a cross-talk between the (electrically floating) sensor guard rings and the pixels at the sensor edge. Dedicated numerical and experimental studies on "baby wafers" of the effect have shown that the effect can at least be attenuated by about a factor of 80 by segmenting the guard ring [12]. Studies with wafers without any guard rings show promising results as well.

Recent tests with an infrared laser shooting in the guard-ring reproduced the cross-talk effect, opening the way to a validation of the design on a test bench.

Simpler and/or "open" designs, which could then be manufactured by non-specialist companies and therefore reducing the sensor cost, have been developed.

3. The SKIROC ASIC

The main purpose of the beam tests carried out in the years 2012 and 2013 was to understand the performance of the highly integrated SKIROC2 circuit [13]. Each of the ASICs reads 64 calorimeter cells and realise the measuring of the analogue signal, the digitisation and the zero suppression such that only a limited number of channels are finally sent to the data acquisition. The ASIC is designed in Silicon Germanium 0.35μm technology. Its outer dimensions are $7.5 \times 8.7\,\mathrm{mm}$. The signal from the silicon pad is is amplified by a variable gain charge sensitive amplifier and the split into a trigger chain and a chain which actually shapes signal that it is readout by a sample and hold stage. The trigger chain provides an internal trigger fro signals as small as about 1/2 of a Minimal Ionising Particle (MIP). This corresponds to a charge of about 0.4 fC. The signal is stored in a 15 bin deep buffer. The buffer is emptied if either all of the 15 buffers of one of the 64 channels is full or if the clearing of the buffer is initiated by an external signal. The signals stored in the buffer are finally digitised by a 12-bit Wilkinson ADC internal to the ASIC.

The implementation of the 15-bin deep buffer is motivated by the beam structure of the International Linear Collider. The beam arrives in so called bunch trains containing up to 3000 bunches. Each bunch train is about 1 ms second long followed by a time gap of about 200 ms. This time structure of the ILC beam allows to operate the ASIC in power pulsed mode. Roughly speaking the front end electronics will only be fully enabled during the millisecond of a bunch

Figure 3. *Left: Picture of one layer of the SiW Ecal setup tested in 2012 and 2013. Right: Experimental setup at DESY..*

train of particles.and the bias currents will be shut down during the gap between the bunch trains.

4. Data Acquisition System

The DAQ is divided in four parts. The detector interface (DIF) is reading data directly from the chips and format their data stream in packets. These packets are aggregated by two levels of aggregation cards. The data concentrator card (DCC) aggregates data from the DIF to use the full capacity of the protocol (similar to FastEthernet) on HDMI cable. The Giga concentrator card (GDCC) converts this streams to Giga-Ethernet encoding. The acquisition clock and the detector-level commands are sent through a specific card named the Clock and Control Card (CCC) which convert the analogue signal into numeric words. This commands words are faned-out by both GDCC and DCC to the DIF. A software suite, named Calicoes, handles the data acquisition and control-command of the whole system. A data acquisition chain acquires Gigabit data flow. It is able to dispatch this data through several kind of media (files, shared memory and TCP socket). It verifies the integrity of the data flow and uncaps the data to produce physical events and feed an event builder. The software suite includes also a complete control and command system, allowing to configure all the cards from a single XML configuration file.

5. Beam test setup

Between March 2012 and July 2013 four beam test campaigns have been carried out at the DESY Testbeam facility of which a detailed description can be found elsewhere. Electrons with an energy between 1-6 GeV are delivered to the beam test area. For the beam tests simplified detector layers were fabricated, see left part of Fig. 3. The layers consisted of one silicon wafer glued onto a PCB. This PCB was equipped with four SKIROC ASICs packed in a regular PQFP package. This layers were operated in continuous powering mode in the 2012 running and in power pulsed mode during 2013. During the beam tests up to eight layers were available for tests. and assembled into the mechanical structure as shown in the right part of Fig. 3.

6. Data analysis

Altogether up to 2048 read out cells were available for the data taking. For details, in particular on the analysis of the 2012 data, the reader is referred to [14][1] and [15]. Here a brief summary

[1] This is an internal CALICE-Note that can be made available upon request. The work is about to get published soon in a refereed journal

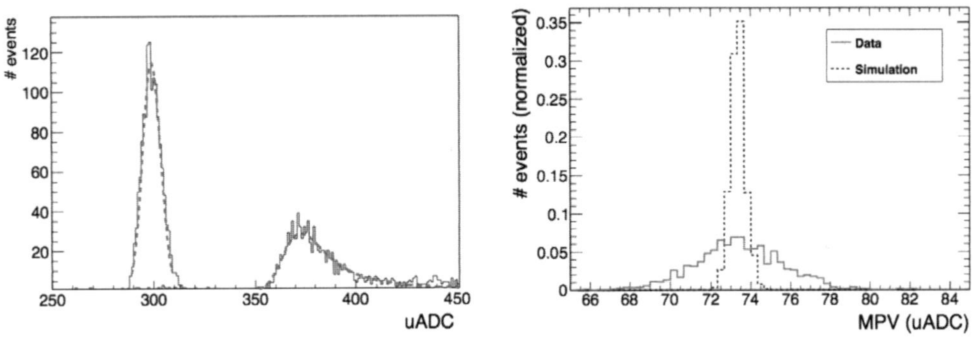

Figure 4. *Left: Pedestal and MIP distribution of one channel. Right: Most probable value of the Landau fit for all active channels in data compared with simulation.*

of the calibration procedure and the data analysis is given.

The first step of the analysis to clean the data from signals caused by hardware issues. The analysis allowed to disentangle effects due to the ASIC from effects rom peripheral devices.. Effects of the ASICs manifest themselves for example by events in which nearly all channels of an ASIC carry a signal. These kind of events are provoked by voltage dips on the ASIC due to high in-rush currents to the ADC in case of e.g. large signals. This shortcoming can e.g. be remedied by a bypass capacitor that stabilises the power supply of the pre-amplifier. Another observation is empty events. Empty events occur if the same signal triggers a readout in two consecutive bunch crossing due to an non-optimal timing of the ASICs w.r.t. the arrival of particles. Empty events won't occur once the timing of the ASICs are coupled to a machine clock. A filter to select undesired events has been developed.

6.1. Calibration of detector modules

The ASIC SKIROC2 [13] is operated in auto-trigger mode. Therefore the calibration starts with the settings of the trigger thresholds. Within the ASIC the signal is shaped and propagated to a sample and hold unit. The readout of the signal happens at a given position determined by the time of the trigger arrival and and internal delay. The goal is to read out when the shaped pulse is at it's maximum. Finally the analogue signal is digitised in Wlikinson-type 12-bit ADC.

The thresholds are determined in a so called threshold scan. For this the internal threshold of the trigger is raised until no or only a few hits of the detector noise are recorded. One threshold threshold is set for all 64 channels of the ASIC. In the next version of the ASIC the threshold will be set on a cell-by-cell basis.

In the hold scan the signal of minimal ionising particles (MIP) is sampled for different delay values. The optimal value hold value is obtained when the Most Probable Value, MPV, of the MIP signal is at maximum as a function of the delay time.

At the example of one channel the left part of Fig. 4 demonstrates the excellent separation of the MIP signal from the detector noise. In [14] the Signal-over-Noise ratio is found to be always above 10:1 for different settings of the SKIROC2 ASIC. The MIP signal is fitted with a Landau distribution and the MPV of that distribution is the calibration constant of a given channel. The right part of Figure 4 shows the distribution of the MIP signal for all active channels in 2012 data taking period. The data are compared with a simulation of the detector. While the position of the MIP is very well reproduced by the simulation the dispersion of the signal cannot be reproduced. The origin of this dispersion is under study.

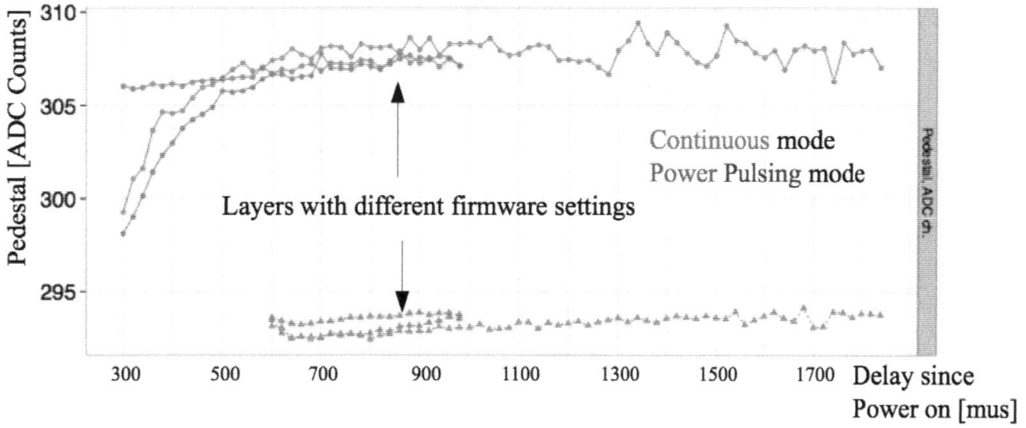

Figure 5. *Pedestal position as a function of time elapsed after the enabling of the bias currents for detector operation in power pulsed mode. The upper and lower lines correspond to two different detector settings. The red line shows the pedestal position for the continuous powering mode.*

6.2. Study of power pulsing

The data taking in 2013 was dedicated to the study the behaviour of the detector in power pulsed mode. Power pulsing implies a periodical shut-down and relaunch of the bias currents that polarise the various stages of the ASIC. After the relaunch a certain delay has to be taken into account until the ASIC is fully operational. The Fig. 5 shows the pedestal position as a function of the time of the signal arrival. In this study two different settings were tested one in which signals can arrive earliest after about $300\,\mu s$ and another with an allowed arrival time of at least $600\ \mu s$. It is clearly visible that the pedestal stabilises after about $600\,\mu s$. The time of stabilisation has been subject to a systematic study for all pixels across a layer. In the left part of Fig. 6 the time dependence of one one cell is approximated by the function $a \cdot (1 - \exp(-(x - 300\mu s)/t)) + c$ where x is the constant of the exponential increase of the pedestal before the stable plateau. As shown in the right part of Fig. 6 this constant is around $100\mu s$ for all cells of the studied layer.

In the following only signals are taken into account that arrive at least $600\,\mu s$ after the relaunch of the bias currents. In the left part of Fig. 7 the correlation of the pedestals between continuous and power-pulsing modes are shown. The right part gives the difference for MIP signals in both operation modes for four ASICs of a layer. The pedestal positions are very well correlated between the two operation modes. The right part shows that for two ASICs the difference in the MIP position is less than 1ADC count equivalent to about 1.5%. The ASICs revealing a bigger difference and therefore worse agreement are connected by relatively long lines to the Si Pads. In [14] it is shown that for these ASICs also the signal over noise ratio is worse than for ASICs connected with short lines to the Si Pads. In the third region the trigger thresholds were too high for various reasons. Details are left to the finalisation of the data analysis.

6.3. Detector simulation

For a deeper understanding of the detector response the detector has been implemented into a detailed GEANT4 simulation. Great care has been taken to the digitisation step of the simulation which allows to reproduce the hardware behaviour in terms of for example residual

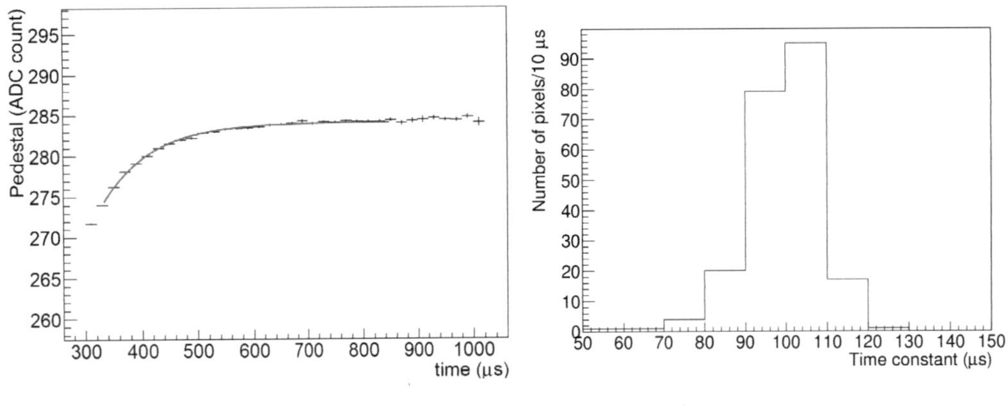

Figure 6. *Left: Time dependence of the pedestal position as time elapsed after the enabling of the bias currents for detector operation in power pulsed mode for one pixel. The functional dependence is approximated by* $a \cdot (1 - \exp(-(x - 300\mu s)/t)) + c$. *Right: Time constant* x *for all pixels across a layer.*

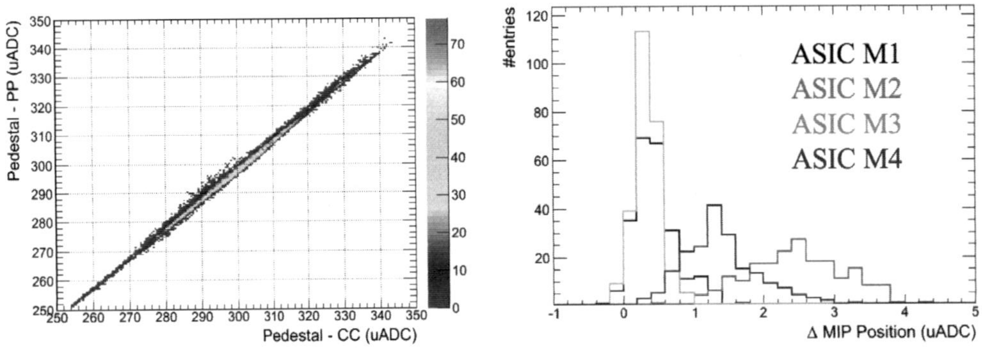

Figure 7. *Left: Correlation between pedestals recorded in continuous and power-pulsed operation of the SKIROC2 ASIC. Right: Difference between the MPV of the MIP signal recorded in continuous and power-pulsed operation of the SKIROC2 ASIC..*

detector noise or the influence of the trigger threshold In Fig. 8 the simulation are compared with data for runs in which the mechanical structure was equipped with tungsten plates. The simulation reproduces the data very well.

7. R&D plans and steps towards a real detector

The next few years will be used to progressively complete the technological prototype with up to 30 short ASUs and at least one long layer with up to 10 ASUs. This prototype is to be tested in beam test campaigns in the years 2015-2016.

The very next step is to produce ASUs for four wafers and read out by 16 ASICs. The final layout of the calorimeter for a linear collider detector depends significantly on the success of the PCB manufacturing. As for the sensors a close collaboration with industrial partners is needed. An option for maximum compactness is embedding the naked dies of the front end chips into the PCB. The application of thin BGA packaging is considered at least as a safe intermediate

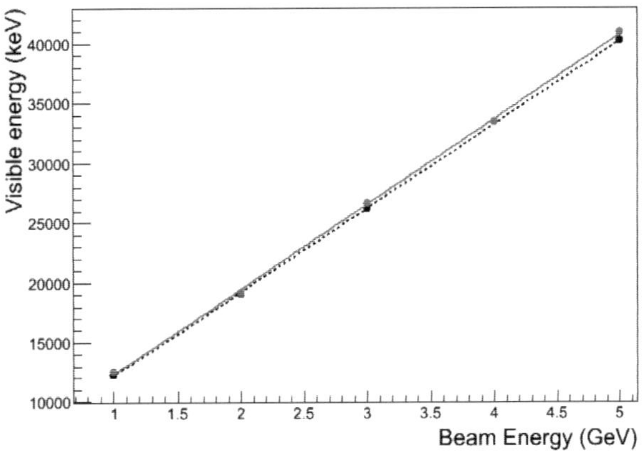

Figure 8. *Comparison between data and simulation of the detector response to electrons between 1 and 6 GeV for runs in which the detector layers were interleaved with tungsten plates.*

step and currently under study.

The final SKIROC ASIC will feature an on-chip zero suppression. As this is a major step it has been agreed to produce first an intermediate version, dubbed SKIROC2b, to assure that (minor) flaws observed with the current SKIROC2 ASIC can be completely remedied. A production run of SKIROC2b ASICs is planned for the end of 2014.

Further development of links with silicon sensor manufacturers is essential in order to better understand the eventual cost of such a detector, and to prepare for possible mass production.

The use of power pulsing in the front end electronics is a central tenet of the detector design. The results obtained in the 2013 beam test campaign are very promising in this respect but need to be confirmed as the system evolves. The validation need to incorporate an estimation of the final power dissipation of the detector.

Many steps of detector construction will be outsourced to industry, and one aim of the technological prototype is to choose techniques which are well adapted to an industrial process. This however requires still scrutinising the available assembly tools and techniques as e.g. the robotic gluing system. In parallel with this, a significant quality control control process will have to be developed, with tests of detector components being carried out at various stages of their integration into detector elements.

References

[1] J.-C. Brient and H. Videau, "The Calorimetry at the future e+ e- linear collider" *eConf* **C010630** (2001) E3047, `arXiv:hep-ex/0202004 [hep-ex]`.

[2] M. Thomson, "Particle Flow Calorimetry and the PandoraPFA Algorithm" *Nucl.Instrum.Meth.* **A611** (2009) 25–40, `arXiv:0907.3577 [physics.ins-det]`.

[3] M. Ruan, "Arbor: a new approach to the Particle Flow Algorithm" in *proceedings of Calorimetry for the High Energy Frontier - CHEF 2013*. 2013.

[4] T. Behnke, J. E. Brau, P. N. Burrows, J. Fuster, M. Peskin, *et al.*, "The International Linear Collider Technical Design Report - Volume 4: Detectors" `arXiv:1306.6329 [physics.ins-det]`.

[5] L. Linssen, A. Miyamoto, M. Stanitzki, H. Weerts, and D. J. Feldman, "Physics and Detectors at CLIC: CLIC Conceptual Design Report" `arXiv:1202.5940 [physics.ins-det]`.

[6] **CMS Collaboration**, "Jet Energy Corrections and Uncertainties. Detector Performance Plots for 2012.".

[7] M. Schröder, "Performance of Jets in CMS" *Proceedings of CALOR14* (2014) .

[8] **CALICE Collaboration**, J. Repond *et al.*, "Design and Electronics Commissioning of the Physics Prototype of a Si-W Electromagnetic Calorimeter for the International Linear Collider" *JINST* **3** (2008) P08001, arXiv:0805.4833 [physics.ins-det].

[9] **CALICE Collaboration**, C. Adloff *et al.*, "Response of the CALICE Si-W Electromagnetic Calorimeter Physics Prototype to Electrons" *J.Phys.Conf.Ser.* **160** (2009) 012065, arXiv:0811.2354 [physics.ins-det].

[10] **CALICE Collaboration**, C. Adloff, Y. Karyotakis, J. Repond, J. Yu, G. Eigen, *et al.*, "Study of the interactions of pions in the CALICE silicon-tungsten calorimeter prototype" *JINST* **5** (2010) P05007, arXiv:1004.4996 [physics.ins-det].

[11] **CALICE Collaboration**, C. Adloff *et al.*, "Effects of high-energy particle showers on the embedded front-end electronics of an electromagnetic calorimeter for a future lepton collider" *Nucl.Instrum.Meth.* **A654** (2011) 97–109, arXiv:1102.3454 [physics.ins-det].

[12] **CALICE Collaboration**, R. Cornat, "Semiconductor sensors for the CALICE SiW EMC and study of the cross-talk between guard rings and pixels in the CALICE SiW prototype" *J.Phys.Conf.Ser.* **160** (2009) 012067.

[13] S. Callier, F. Dulucq, C. de La Taille, G. Martin-Chassard, and N. Seguin-Moreau, "SKIROC2, front end chip designed to readout the Electromagnetic CALorimeter at the ILC" *JINST* **6** (2011) C12040.

[14] T. Frisson *et al.*, "Beam test performance of the SKIROC2 ASIC" *CALICE Internal Note submitted to Nucl.Instrum.Meth. A* (2014) . https://twiki.cern.ch/twiki/pub/CALICE/CaliceInternalNotes/CIN-023.pdf.

[15] J. Rouëné, "Construction and testing of a large scale prototype of a silicon tungsten electromagnetic calorimeter for a future lepton collider" *Nucl.Instrum.Meth.* **A732** (2013) 470–474.

Acknowledgments

The research leading to these results has received funding from the European Commission under the FP7 Research Infrastructures project AIDA, grant agreement no. 262025, the French ANR program (Project: CALIIMAX-HEP, 2010 BLANC 0429 01), the French IN2P3 Quarks and Leptons program and the Japanese JSPS program. The authors would like to thank the DESY test beam facility team for their support and kind hospitality. The author would like to thank the CALOR2014 team for the organisation of the conference and the opportunity to present the results.

Prototype tests for a highly granular scintillator-based hadron calorimeter

K Krüger (for the CALICE collaboration)
DESY Hamburg, Notkestrasse 85, 22607 Hamburg, Germany
E-mail: katja.krueger@desy.de

Abstract. Within the CALICE collaboration, several concepts for the hadronic calorimeter of a future linear collider detector are studied. After having demonstrated the capabilities of the measurement methods in "physics prototypes", the focus now lies on improving their implementation in "technological prototypes", that are scalable to the full linear collider detector. The Analog Hadron Calorimeter (AHCAL) concept is a sampling calorimeter of tungsten or steel absorber plates and plastic scintillator tiles read out by silicon photomultipliers as active material. In the AHCAL technological prototype, the front-end chips are integrated into the active layers of the calorimeter and are designed for minimal power consumption. The versatile electronics allows the prototype to be equipped with different types of scintillator tiles and SiPMs. The current status of the AHCAL engineering prototype is shown and recent beam test measurements as well as plans for future hadron beam tests with a larger prototype will be discussed.

1. Introduction

At a future linear collider, one of the physics goal is the distinction of the hadronic decays of W and Z bosons which requires an unprecedented jet energy resolution of $3-4\%$ in a wide range of jet energies from 40 to 500 GeV. These values can be reached by using particle flow algorithms (PFAs) which reconstruct each particle within a jet from the detector part with the best energy resolution. In this way only neutral particles are reconstructed in the calorimeter, while for charged particles the information from the tracker is used. A good performance of the PFA requires a fine detector granularity and a compact calorimeter layout contained within the magnet coil. The CALICE collaboration develops such highly granular options for the electromagnetic and the hadronic calorimeter for a future linear collider detector.

The Analog Hadron Calorimeter (AHCAL) is a sampling hadron calorimeter concept based on scintillator tiles coupled to silicon photomultipliers (SiPMs) as active component. Iron and tungsten have been used as absorber material in beam tests of a large "physics prototyp" which have demonstrated the capabilities of the AHCAL concept. Now the focus lies on improving the implementation in a "technological prototype", that is scalable to the full linear collider detector.

2. AHCAL physics prototype

The $\sim 1\,\text{m}^3$ large AHCAL physics prototype was operated between 2006 and 2012 in several beam test campaigns in electron, muon and hadron beams at DESY, CERN and Fermilab. It has been used to measure the AHCAL energy resolution for electrons [1] and hadrons [2]. Its high granularity and good energy resolution allow for detailed studies of hadronic showers [3] which are used also to validate the simulation of hadronic showers in the GEANT4 toolkit [4]. Recent results from the beam tests of the AHCAL physics prototype include the measurement of longitudinal and radial shower profiles of pions [5] (figure 1) and the validation of the performance of the PandoraPFA algorithm [6] by overlaying pion showers measured in beam tests [7] (figure 2). Overall, the studies show a good performance of the AHCAL concept and a reasonable to good agreement of the simulation with data if a modern physics list like FTFP_BERT or QGSP_BERT is used in GEANT4.

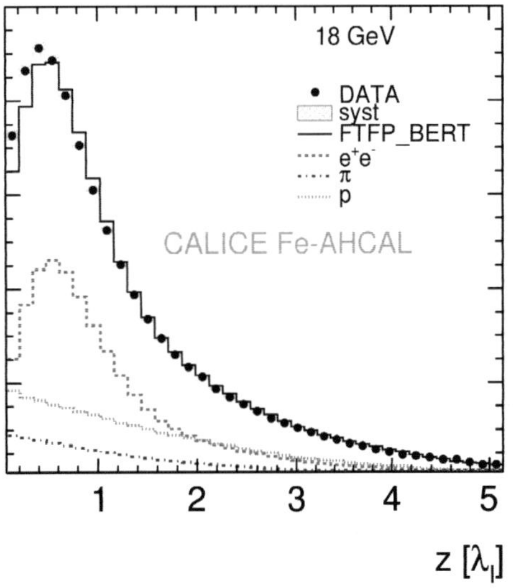

Figure 1. Mean longitudinal shower profiles from shower starting point for 18 GeV pions for data and for GEANT4 simulated data for the FTFP_BERT physics list.

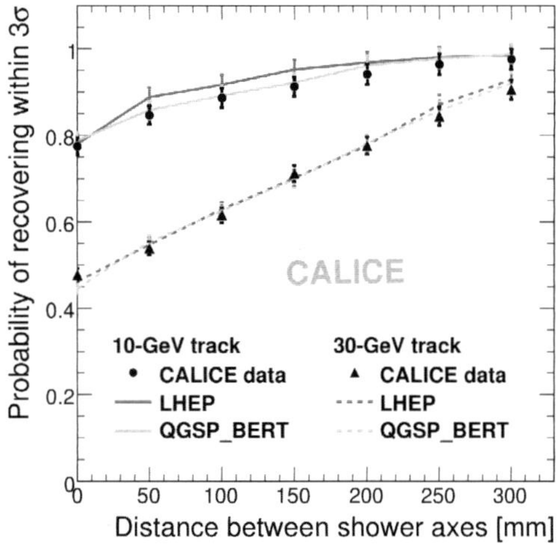

Figure 2. Probability of recovering a neutral 10 GeV hadrons energy within 3 standard deviations from its real energy as a function of the distance from charged 10 GeV and 30 GeV hadrons for beam data and for GEANT4 simulated data for two physics lists.

3. AHCAL technological prototype

The goal of the development of the AHCAL technological prototype is the scalability to a full linear collider detector. It is based on a fully integrated design with front-end readout electronics embedded in the active layers to cope with the high channel density without dead zones. In order to avoid cooling in the active layers, this imposes stringent requirements on the power consumption of the electronics that can only be realized by switching off parts that are not needed during some data taking phases (power-pulsing). For the calibration and monitoring of all individual channels an LED system is also integrated into the layers.

The AHCAL technological prototype is based on $3 * 3 * 0.3\,\mathrm{cm}^3$ tiles of plastic scintillator, resulting in ~ 8 mio. channels for the complete hadronic calorimeter. The availability of new SiPMs with reduced noise and very small device-to-device variations opens up new possibilities for the tile design. Several different tile options with and without wavelength shifting (WLS) fiber for the coupling of the tile to the SiPM are studied. One goal of the optimisation of the tile design is an easier and faster production and assembly of the detector. The design with a WLS fiber (figure 3) offers a better signal uniformity across the tile, while the design without

WLS fiber (figure 4, design based on [8]) is easier to produce. Changing the SiPM position from the side of the tile to the top would allow for the use of large scintillator plates [9] ("megatiles") which cover several SiPMs (figure 5). These different tile options also have implications for the assembly and calibration procedure which are currently investigated.

Figure 3. AHCAL scintillator tile with wavelength shifting fiber

Figure 4. AHCAL scintillator tile without wavelength shifting fiber

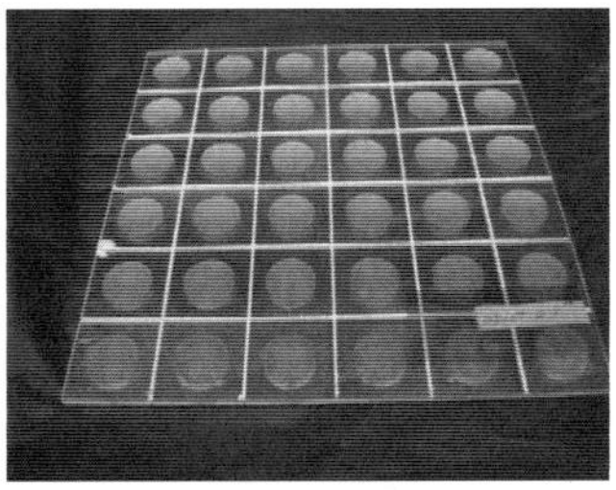

Figure 5. Large AHCAL scintillator tile ("megatile") for SiPMs mounted above the tile

The layout of the AHCAL for the International Linear collider Detector (ILD) [10] is planned to consist of two endcaps and a barrel, which is segmented into two rings along the beam direction. Each of the half-barrels is further segmented into 16 half-octants with a thickness of 40 to 48 layers (figure 6). The layers in turn are segmented into 18 HCAL Base Units (HBUs) arranged into 3 slabs with a length of 6 HBUs each (figure 7).

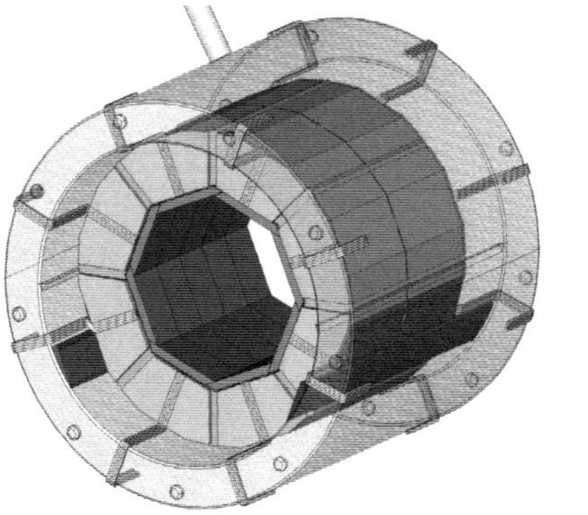

Figure 6. AHCAL barrel layout for ILD.

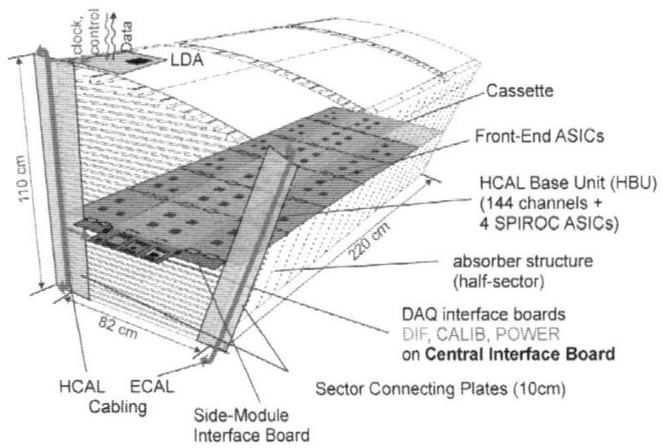

Figure 7. Layer structure of an AHCAL barrel half-octant.

Several configurations with small numbers of HBUs have been operated successfully in the lab and studied in beam tests. A slab with the full length of 6 HBUs (figure 8) was tested with and without power pulsing in the lab and showed good signal uniformity over the full lenght of the slab of 2.1 m. A layer of 2*2 HBUs was calibrated with electrons from the DESY test beam used as minimal-ionising particles (MIPs) and used to measure hit times in hadron showers in the SPS test beam at CERN. Most recent, a stack of 8 HBUs (figure 9) was calibrated with MIPs in the DESY test beam (figure 10) and then used to measure electron showers. The active layers as well as the steel absorber have the geometry planned for ILD. As next step we plan to partly equip an ILD sector such that a few layers can be used to identify the shower start and a few larger layers can be used to measure the correlation of the size and the time of hits in a hadronic shower. This configuration will be tested in muon and hadron beams at CERN. These tests are important steps toward the realisation of a large AHCAL technological prototype that demonstrates the feasability of building a hadronic calorimeter according to the AHCAL concept for a future linear collider detector.

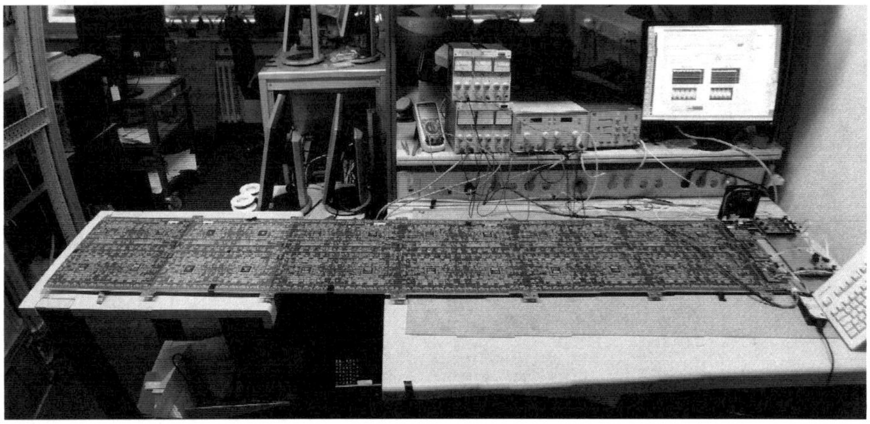

Figure 8. Slab of 6 HBUs tested in the lab.

4. Acknowledgments

The research leading to these results has received funding from the European Commission under the FP7 Research Infrastructures project AIDA, grant agreement no. 262025.

References

[1] CALICE collaboration (Adloff C *et al.*) 2011 Electromagnetic response of a highly granular hadronic calorimeter *JINST* **6** P04003 (*Preprint* arXiv:1012.4343 [physics.ins-det])

[2] CALICE collaboration (Adloff C *et al.*) 2012 Hadronic energy resolution of a highly granular scintillator-steel hadron calorimeter using software compensation techniques *JINST* **7** P09017 (*Preprint* arXiv:1207.4210 [physics.ins-det])

Figure 9. Steel stack with 8 layers of 1 HBU each.

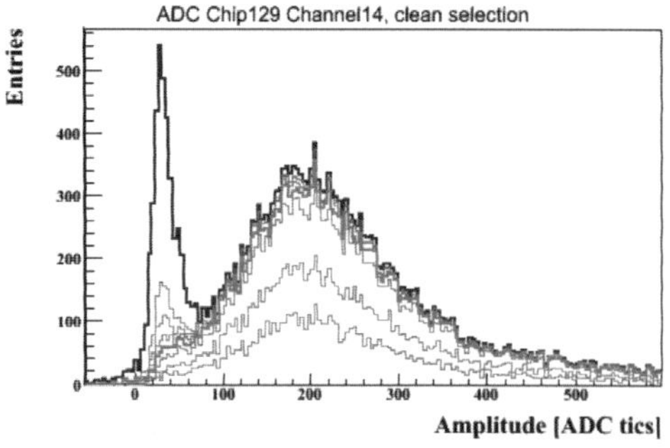

Figure 10. Energy deposition of a MIP traversing several HBUs before (blue) and after cuts on the hit times (red). The noise peak at very low amplitudes is reduced significantly.

[3] CALICE collaboration (Adloff C *et al.*) 2013 Track segments in hadronic showers in a highly granular scintillator-steel hadron calorimeter *JINST* **8** P09001 (*Preprint* arXiv:1305.7027 [physics.ins-det])
[4] Geant4 collaboration (Agostinelli S *et al.*) 2003 Geant4 - a simulation toolkit *NIM* **A506** 250–303
[5] CALICE collaboration (Adloff C *et al.*) 2013 Validation of GEANT4 Monte Carlo models with a highly granular scintillator-steel hadron calorimeter *JINST* **8** P07005 (*Preprint* arXiv:1306.3037 [physics.ins-det])
[6] Thomson M A 2009 Particle flow calorimetry and the PandoraPFA algorithm *NIM* **A611** 25–40 (*Preprint* arXiv:0907.3577 [physics.ins-det])
[7] CALICE collaboration (Adloff C *et al.*) 2011 Tests of a particle flow algorithm with CALICE test beam data *JINST* **6** P07005 (*Preprint* arXiv:1105.3417 [physics.ins-det])
[8] Simon F and Soldner C 2010 Uniformity studies of scintillator tiles directly coupled to SiPMs for imaging calorimetry *Nucl. Instrum. Meth.* A **620** 196–201 (*Preprint* arXiv:1001.4665 [physics.ins-det])
[9] Blazey G *et al.* 2009 Directly coupled tiles as elements of a scintillator calorimeter with MPPC readout *Nucl. Instrum. Meth.* A **605** 277–281
[10] Behnke T *et al.* 2013 The International Linear Collider Technical Design Report - Volume 4: Detectors *Preprint* arXiv:1306.6329 [physics.ins-det]

Journal of Physics: Conference Series **587** (2015) 012034 doi:10.1088/1742-6596/587/1/012034

Shower characteristics of particles with momenta up to 100 GeV in the CALICE scintillator-tungsten hadronic calorimeter

Eva Sicking for the CALICE and CLICdp collaborations
CERN, 1211 Geneva 23, Switzerland
E-mail: eva.sicking@cern.ch

Abstract. We present a study of showers initiated by 1–100 GeV positrons, pions, kaons, and protons in the highly granular CALICE analogue scintillator-tungsten hadronic calorimeter. The data were taken at the CERN PS and SPS. The analysis includes measurements of the calorimeter response to each particle type and studies of the longitudinal and radial shower development. The results are compared to several Geant4 simulation models.

1. Introduction

A hadron sampling calorimeter with highly granular readout, as required for Particle Flow Analysis [1], and with tungsten as absorber, is proposed as barrel hadronic calorimeter for experiments at future multi-TeV e^+e^- colliders such as the Compact Linear Collider (CLIC) [2]. The usage of the dense absorber material tungsten allows, on the one hand, for a deep calorimeter system which can cope with TeV-jets and on the other hand, it limits the diameter of the surrounding solenoid of the experiment.
Test beam experiments are used to characterise this detector concept. The results are compared to predictions of Monte Carlo simulations in order to estimate the accuracy of the simulation models used in detector development studies for CLIC.

2. Experimental setup

The CALICE collaboration has built a sandwich calorimeter prototype with alternating layers of tungsten (W) absorber plates [3] and scintillator readout planes of the CALICE Analogue Hadronic Calorimeter (AHCAL) [4].
Each tungsten absorber layer is 1 cm thick and is made of a tungsten alloy with an admixture of 5.25 % nickel and 1.76 % copper. The alloy has a nuclear interaction length of $\lambda_\mathrm{I} = 10.80\,\mathrm{cm}$ and a radiation length of $X_0 = 0.39\,\mathrm{cm}$.
Each AHCAL layer contains a 0.5 cm thick scintillator layer that is subdivided into tiles read out with wavelength shifting fibres and silicon photomultipliers (SiPM). In the central part of an AHCAL layer, the tiles have the highest granularity of $3 \times 3\,\mathrm{cm}^2$. Towards the outer part of the readout layer, the granularity decreases to $6 \times 6\,\mathrm{cm}^2$ and $12 \times 12\,\mathrm{cm}^2$.
The initial W-AHCAL prototype used in 2010 consisted of 30 layers. For the 2011 experiments at higher beam momenta, this was extended to 38 layers. The full dimensions of the W-AHCAL prototypes are $0.9 \times 0.9 \times 0.75\,\mathrm{m}^3$ and $0.9 \times 0.9 \times 1\,\mathrm{m}^3$. This corresponds to 3.9 (4.9) nuclear

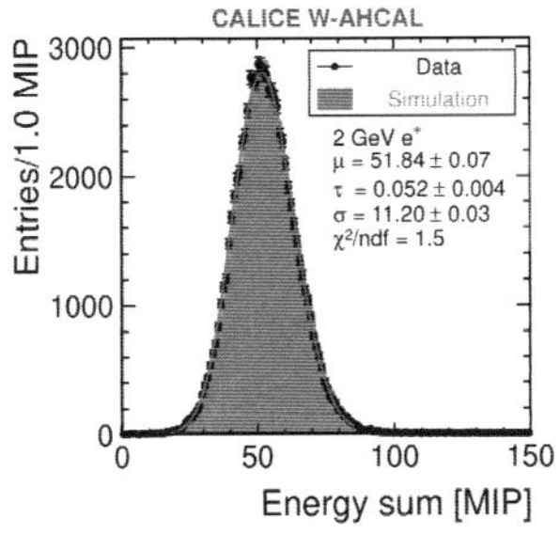

Figure 1. W-AHCAL energy sum distribution of 2 GeV e^+ [9].

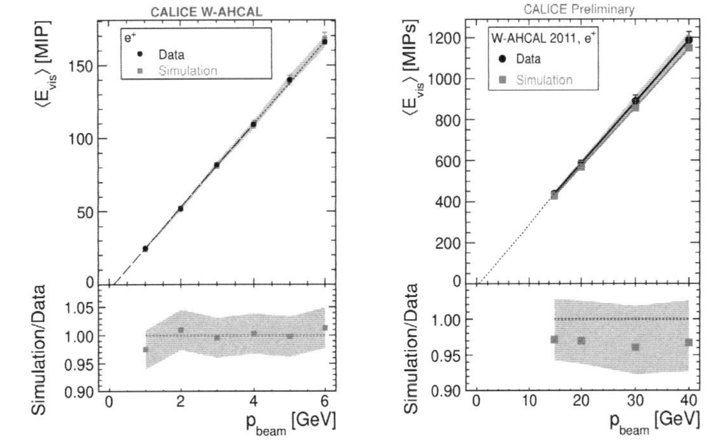

Figure 2. Mean visible e^+ energy as a function of the beam momentum at 1–6 GeV [9] (left panel) and at 15–40 GeV [10] (right panel).

interaction lengths, 85 (107) radiation lengths, and 6480 (7608) readout channels.

In the following, we present results from W-AHCAL test beam experiments performed in 2010 at the CERN PS with beam momenta from 1 GeV to 10 GeV and in 2011 at the CERN SPS at 10–100 GeV. Mixed beams of e, μ, π, K, and p were provided in both test beam campaigns. At the SPS, also dedicated e beams were provided. In order to select events based on particle type and trajectory, information from two in-beam Cherenkov counters, three wire chambers, and two trigger scintillators were used [3, 5]. For the rejection of muons from in-flight decays, information from the highly granular W-AHCAL itself could be exploited. The W-AHCAL response was calibrated using dedicated muon beams to create signals from minimum ionizing particles (MIP) [6].

Results of the W-AHCAL test beam measurements are compared to Geant4 simulations [7]. Here, physics lists with a data driven high precision (HP) model for neutron simulations are used as neutrons are expected to play an important role in the neutron-rich tungsten absorber.

3. Shower characteristics

3.1. Positrons

Due to the dense absorber material tungsten, electromagnetic showers in the studied momentum range deposit energy only in a few W-AHCAL cells. At low beam momenta, this results in a non-Gaussian energy sum distribution with tails at high energies as shown in figure 1. This behaviour is reproduced by simulations. The signal shape can be described well by the Novosibirsk fit function [8]

$$f(x) = A \cdot \exp\left[0.5 \cdot \left(\frac{\ln^2[1 + \Lambda \cdot \tau \cdot (z - \mu)]}{\tau^2} + \tau^2\right)\right] \quad \text{with} \quad \Lambda \equiv \frac{\sinh(\tau \cdot \sqrt{\ln 4})}{\sigma \cdot \tau \cdot \sqrt{\ln 4}} \tag{1}$$

in which μ is the peak position, σ the width, and τ the tail parameter. The fit is used in the following to extract the mean visible energy, $\langle E_{\text{vis}} \rangle = \mu_{\text{Novosibirsk}}$, and the width of the energy sum peak, $\sigma_E = \sigma_{\text{Novosibirsk}}$.

Figure 2 shows the positron mean visible energy as a function of the beam momentum for data and simulations for the two beam momentum ranges. The shaded error bands indicate the systematic uncertainties of the data points dominated by the uncertainty of the MIP scaling

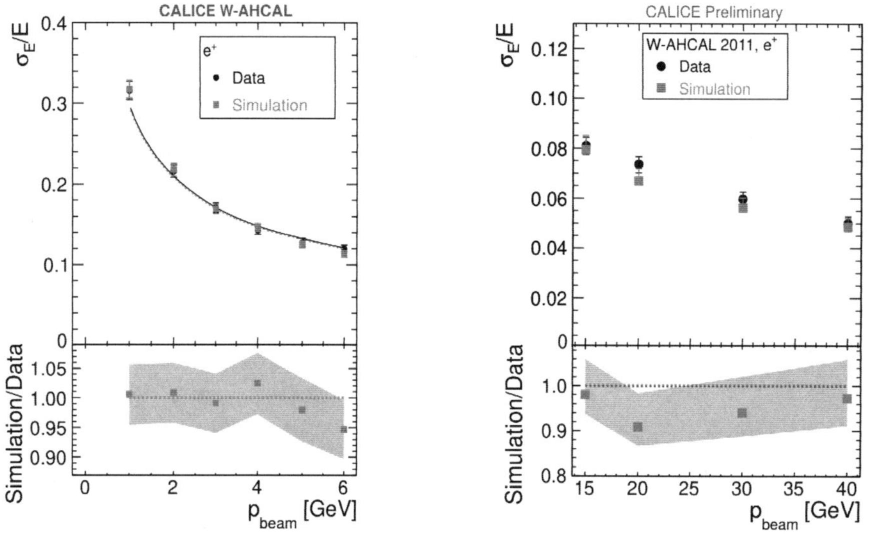

Figure 3. e^+ energy resolution as a function of the beam momentum at 1–6 GeV [9] (left panel) and at 15–40 GeV [10] (right panel).

factor, the stability of the detector response, and the uncertainty of the SiPM saturation scaling factors. The bottom panels of the figures display the ratio of simulation over data. Within the uncertainties, data and simulation agree with each other. The visible energy increases linearly with the beam momentum, as expected.
Figure 3 shows the positron energy resolution as a function of the beam momentum. The energy resolution in the beam momentum range 1–6 GeV (left panel) is fitted with the fit function $\frac{\sigma_E}{E} = \frac{a}{\sqrt{E[\mathrm{GeV}]}} \oplus b \oplus \frac{c}{E[\mathrm{GeV}]}$, where a is the stochastic term, b is the constant term, and c is the noise term. In the following, the noise term is fixed to the measured noise in the fiducial volume of the positron shower of $c_{e^+} \approx 36\,\mathrm{MeV}$. The results of the positron energy resolution fit are $a_{e^+} = (29.6 \pm 0.5)\,\%$ and $b_{e^+} = (0.0 \pm 2.1)\,\%$, which are in agreement with the values extracted from a fit of the simulated positron energy resolution of $a_{e^+} = (29.2 \pm 0.4)\,\%$ and $b_{e^+} = (0.0 \pm 1.5)\,\%$. The positron energy resolution for the high momentum data set is not fitted as the energy range 15–40 GeV is too small to obtain stable fit results.

3.2. Hadrons

At low beam momenta, the hadron energy sum distributions have a non-Gaussian shape with a high-energy tail (cf. figure 4). This tail is reproduced by Geant4 simulations and decreases with increasing beam momentum. The visible energy and the width of the energy sum distribution at low beam momenta are therefore estimated using the mean and the RMS of the energy sum histogram. At high beam momenta, $\langle E_{\mathrm{vis}} \rangle$ and σ_E are extracted using Gaussian fit functions of the 80 % most central entries of the energy sum distributions. Figure 5 shows the π^+ mean visible energy as a function of the available energy $E_{\mathrm{available}} = \sqrt{p^2_{\mathrm{beam}} + m^2_{\pi}}$ for the low momentum data and as a function of the beam momentum for the high momentum data. The Geant4 physics list QGSP_BERT_HP gives the best description in the low momentum range. For the high momentum data, QGSP_BERT_HP and FTFP_BERT_HP describe the data equally well.

Figure 6 shows the π energy resolution. For the low momentum range, the energy resolution fit results are $a_{\pi^+} = (61.8 \pm 2.5)\,\%$ and $b_{\pi^+} = (7.7 \pm 3.0)\,\%$, for positive pions and

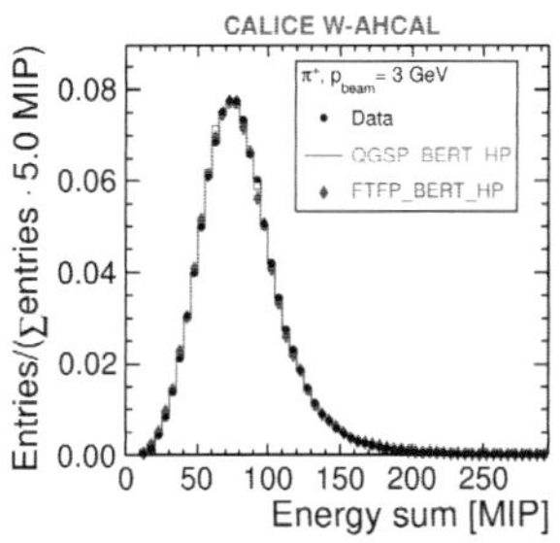

Figure 4. W-AHCAL energy sum distribution of 3 GeV π^+ [9].

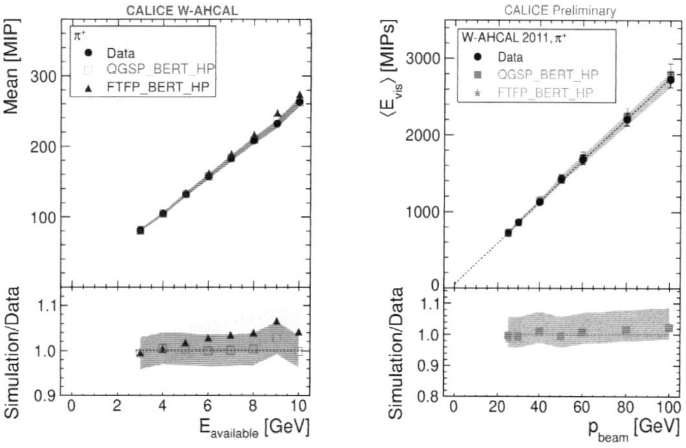

Figure 5. Mean visible π^+ energy as a function of the beam momentum at 3–10 GeV [9] (left panel) and at 25–100 GeV [10] (right panel).

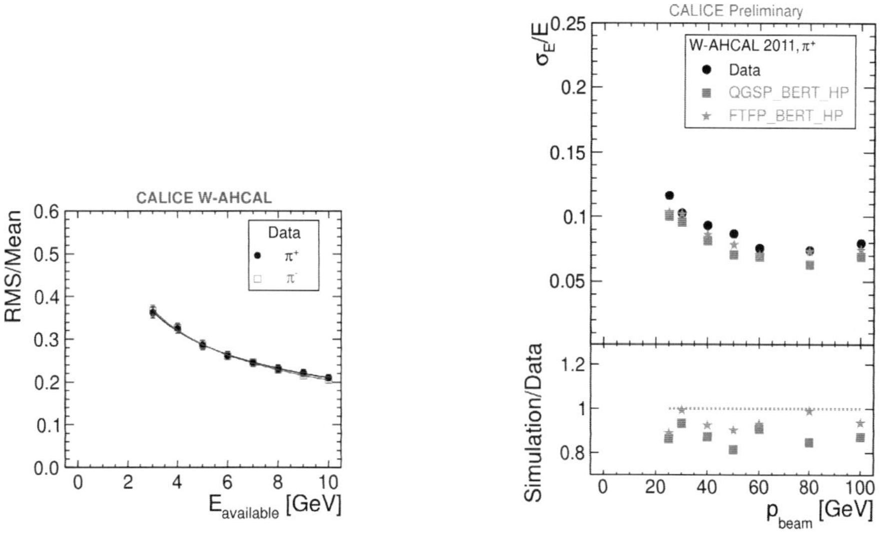

Figure 6. Energy resolution for $\pi^\pm$ as a function of the beam momentum at 3–10 GeV [9] (left panel) and for π^+ at 25–100 GeV [10] (right panel).

$a_{\pi^-} = (63.9 \pm 2.4)\,\%$ and $b_{\pi^-} = (3.2 \pm 6.9)\,\%$ for negative pions. The noise term is fixed to the measured noise of $c_\pi \approx 70\,\mathrm{MeV}$. In the right panel of figure 6, the π^+ energy resolution measured at 15–100 GeV is compared to Geant4 model predictions. QGSP_BERT_HP and FTFP_BERT_HP show slightly smaller energy resolutions. The high momentum pion energy resolution is not fitted as the limited beam momentum range does not allow constraining the fit parameters sufficiently.

Figure 7 shows the longitudinal shower development relative to the measured shower start for 100 GeV π^+. QGSP_BERT_HP overestimates the energy deposition in the first part of the shower but reproduces well the energy deposition after one λ_I. On the contrary, FTFP_BERT_HP underestimates the energy deposition in the first part of the shower and overestimates the

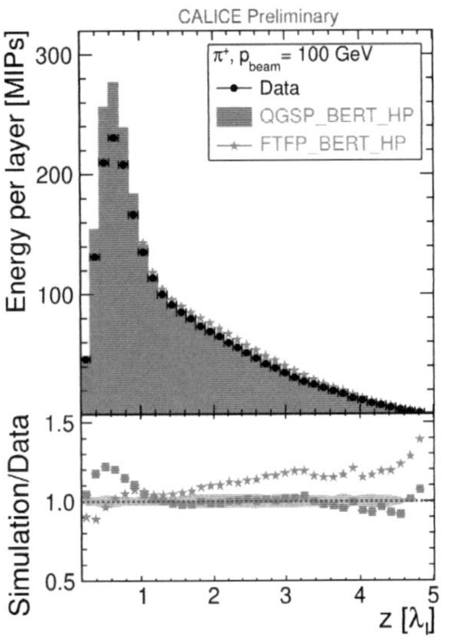

Figure 7. Longitudinal profile for 100 GeV π^+ [10].

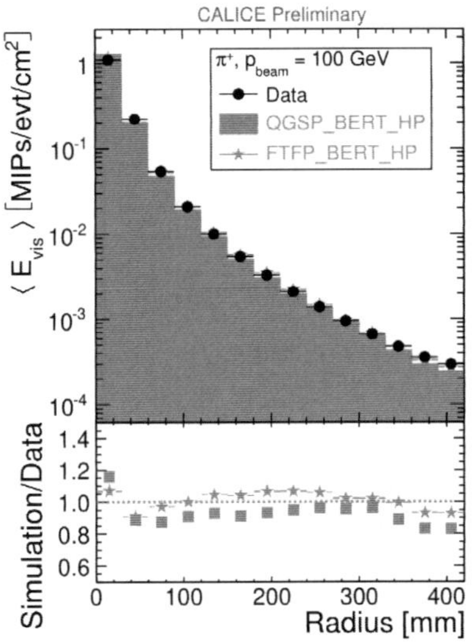

Figure 8. Radial profile for 100 GeV π^+ [10].

energy deposition after one λ_I.

The radial shower development for 100 GeV π^+ is shown in figure 8. Both models predict a slightly higher energy deposition in the core of the shower than observed in data. In the radial shower tail, QGSP_BERT_HP underestimates the energy deposition slightly while FTFP_BERT_HP shows a good agreement with data at this energy.

The calorimeter response of e^+, π^+, K^+ and p are compared to each other in figure 9. The response is similar for all hadrons in the studied momentum range while the electrons show a slightly steeper slope. This behaviour is reproduced by Geant4 simulations (figures not shown).

4. Summary

We have presented analysis results of test beam experiments with beam momenta up to 100 GeV recorded with a highly granular scintillator-tungsten HCAL prototype. The W-AHCAL response increases linearly with the available particle energy. The response is similar for e, π, K, and p for the studied particle energy range. The stochastic term of the W-AHCAL energy resolution is found to be $a_e \approx 30\,\%$ for positrons and $a_\pi \approx 62\,\%$ for pions. The pion energy resolution of the W-AHCAL is similar to that achieved by the steel AHCAL ($a_{\pi,\ \text{Fe-AHCAL},\ 10\text{–}100\,\text{GeV}} \approx 58\,\%$ [11]), while the W-AHCAL has a larger positron energy resolution than the Fe-AHCAL ($a_{e,\ \text{Fe-AHCAL},\ 10\text{–}50\,\text{GeV}} \approx 22\,\%$ [6]). These alterations in the energy resolution are expected due to the larger number of X_0 per layer in the tungsten AHCAL. The Geant4 W-AHCAL detector simulation is validated using electromagnetic showers. For a good agreement between data and simulation results of hadron showers in the W-AHCAL, the HP model extension for high precision neutron tracking is needed. The agreement between measured hadron shower characteristics and predictions of the QGSP_BERT_HP and FTFP_BERT_HP models are on the percent level.

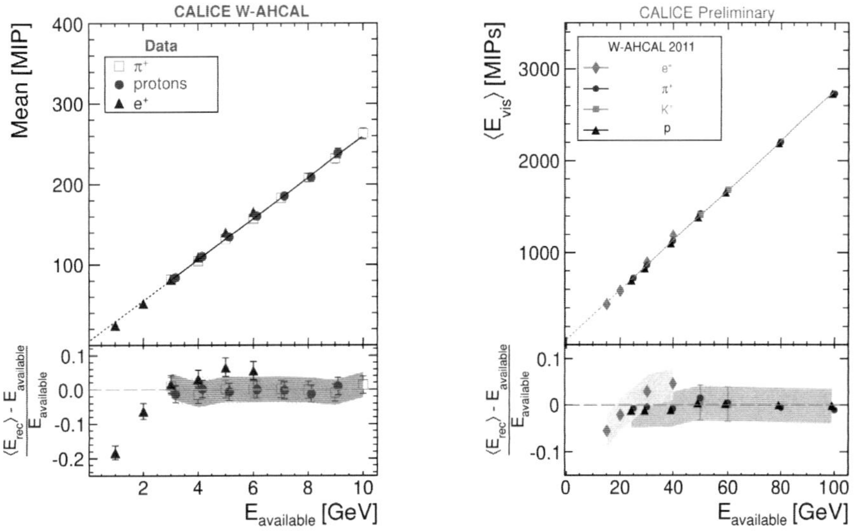

Figure 9. Mean visible energy as a function of the available energy for e^+, π^+, and p at 1–10 GeV [9] (left panel) and for e^+, π^+, K^+, and p at 10–100 GeV [10] (right panel).

References

[1] Thomson M A 2009 *Nucl. Instrum. Meth.* A **611** 25
Marshall J S, Münnich A and Thomson M A 2013 *Nucl. Instrum. Meth.* A **700** 153
[2] Linssen L, Miyamoto A, Stanitzki M and Weerts H (Eds.) 2012 Physics and Detectors at CLIC: CLIC Conceptual Design Report *Preprint* physics.ins-det/1202.5940 CERN-2012-003/ANL-HEP-TR-12-01/ DESY-12-008/KEK-Report-2011-7
[3] Dannheim D, Klempt W and van der Kraaij E 2012 Beam tests with the CALICE tungsten analog hadronic calorimeter prototype *Public note* LCD-Note-2012-002/AIDA-NOTE-2013-001
[4] Adloff C *et al.* [CALICE collaboration] 2010 *JINST* **5** P05004
[5] Dannheim D, Elsener K, Klempt W, Lucaci Timoce A and van der Kraaij E 2013 Particle Identification with Cherenkov detectors in the 2011 CALICE Tungsten Analog Hadronic Calorimeter Test Beam at the CERN SPS *Public note* LCD-Note-2013-006
[6] Adloff C *et al.* [CALICE collaboration] 2011 *JINST* **6** P04003
[7] Agostinelli S *et al.* [GEANT4 collaboration] 2003 *Nucl. Instrum. Meth.* A **506** 250
Allison J *et al.* [GEANT4 collaboration] 2006 *IEEE Trans. Nucl. Sci.* **53** 270
[8] Ikeda H *et al.* [Belle collaboration] 2000 *Nucl. Instrum. Meth.* A **441** 401
[9] Adloff C *et al.* [CALICE collaboration] 2014 *JINST* **9** 01004
[10] Adloff C *et al.* [CALICE collaboration] 2013 Shower development of particles with momenta from 10 to 100 GeV in the CALICE Scintillator-Tungsten HCAL *Public note* CAN-044
[11] Adloff C *et al.* [CALICE collaboration] 2012 *JINST* **7** P09017

Results of the CALICE SDHCAL technological prototype

Arnaud Steen, on behalf of the CALICE Collaboration

Institut de Physique Nuclaire de Lyon, 4 rue Enrico Fermi, Villeurbanne Cedex, 69622, France

E-mail: steen@ipnl.in2p3.fr

Abstract. The SDHCAL prototype was completed in 2012, and exposed to beams of pions, electrons of different energies at the SPS of CERN for a total time period of 5 weeks. The data are being analyzed within the CALICE Collaboration. Preliminary results indicate that a highly granular hadronic calorimeter conceived for PFA application is also a powerful tool to measure hadronic particle energy. In addition it was found to discriminate efficiently pions from electrons. The use of multi-threshold readout mode shows a clear improvement of the resolution at energies exceeding 30 GeV with respect to the binary readout mode. New ideas to improve on the energy resolution using the topology of hadronic showers such as the Hough Transform technique are studied.

1. Introduction

The CALICE Semi-Digital Hadronic CALorimeter (SDHCAL) prototype achieved in 2012, was built for several purposes and it is one of the hadronic calorimeter candidate for the International Large Detector (ILD) of the future International Linear Collider (ILC). The main goal of the SDHCAL prototype is to show that highly granular gaseous calorimeters can provide a powerful detector for the hadron energy measurement and for the application of the Particle Flow Algorithm (PFA). First studies using Hough Transform methods tends to confirm that the tracking capability of such detectors will be a useful tool for the PFA and for the estimation of the detector behaviour *in situ*. Another motivation for the SDHCAL was to build a detector compatible with the futur linear collider requirements in terms of efficiency, low power consumption and compactness. In this paper, a brief description of the SDHCAL prototype and the experimental set-up during test beams will be presented. Then, in section 2, the event building procedure will be introduced and detector behaviour presented. In section 3, the methods for the particle selection and the results for the energy resolution will be discussed. In the last section, a tracking method within hadronic showers using Hough Transform technique will be presented.

1.1. The SDHCAL prototype

The SDHCAL prototype is a sampling calorimeter composed by 48 Glass Resistive Plate Chambers (GRPC) as active layers [1]. The absorber layers are stainless-steel plates with a thickness of 2 cm. The total depth of the prototype represents $6\lambda_I$. Each GRPC has a transverse size of 1 m^2. The cathode and the anode are glass plates with a thickness of 1.1 and 0.7 mm respectively. The gas gap between these electrodes is 1.2 mm. The readout of each chamber

Journal of Physics: Conference Series **587** (2015) 012035 doi:10.1088/1742-6596/587/1/012035

is done by 9216 pads of 1 cm^2. This represents a total number of electronic channels of more than 440000 for the whole prototype. Readout signal from those pads is digitized and recorded in 2-bit format within HARDROC2 ASIC [2]. The aim of the thresholds is to have additional information of the number of particles which cross the pad and improve the energy resolution [3].

1.2. Test beam and set-up

The SDHCAL prototype has been exposed to muons, pions and electrons during test beam campaigns. In the following, only data collected in August 2012 at CERN H6 beam line of the SPS are considered. The gas mixture was 93% of TetraFluoroEthane(TFE), 5% of CO_2 and 2% of SF_6. The high voltage applied on the glass electrodes was 6.9 kV. The threshold values were set at 114 fC, 5 pC and 15 pC (0.1, 4, 12.5 MIP). The data acquisition was performed using the trigger-less mode. In this mode, all hits are recorded in the internal ASIC memory until one of them is full. Then a RamFull signal is sent and all the ASICs stop their acquisition to allow the readout of the data collected by all ASICs. After the data transfer, the acquisition restarts automatically. Because of the use of the trigger-less mode an event building procedure (see section 2.1) and an event selection (see section 3.1) are needed. Moreover no Cherenkov counter was used to discriminate particles. To reduce the noise and the energy consumption, the power-pulsed mode was used. In this mode the detector electronics are switched off during the time period separating two beam spills. The spill duration at SPS is about 9 seconds within a cycle of 45 seconds.

2. Event Building and detector behaviour

2.1. Event building

The event building procedure is based on a time clustering method. The time is recorded through the ASIC internal time counter with a step of 200 ns. To build physical events (cosmic and beam particles) a histogram of hit time is done (Fig. 1.a). A physical event candidate is found if the number of hits in one time slot exceeds 7. A time clusterisation is then applied by gathering all hits in the two adjacent time slots. Rejected bins are used to estimate the electronic noise per time slot. From the noise distribution, shown in Fig. 1.b, the contribution of the noise on the total number of hit per physical event should not be larger than one.

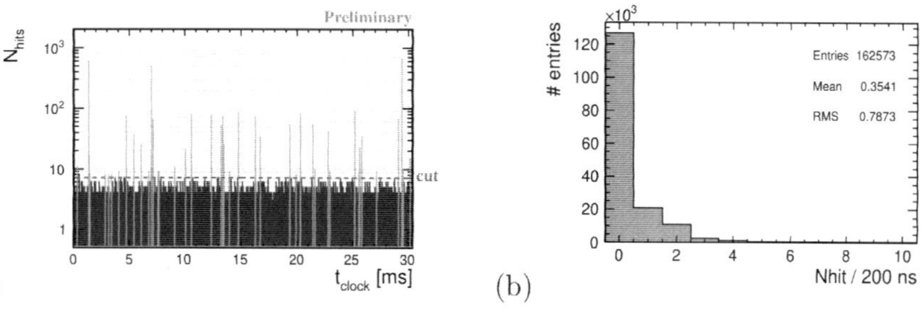

(a) (b)

Figure 1. (a): Hit time spectrum. Each bin corresponds to 200 ns. Bins in green ($\#_{hit} > 7$) are the candidates to build a physical events. (b): Distribution of the number of noise hits in a time slot of 200 ns for the whole detector.

2.2. Detector performance

The SDHCAL performance is studied using the detector reponse to muons. The efficiency and the pad multiplicity of one layer are estimated using reconstructed tracks from other layers. A muon selection is needed because of the trigger-less mode. To select those particles, hits of one layer are grouped into clusters if they share an edge. Then a linear fit is applied to determine the expected track impact. The layer is counted as efficent if one cluster is found at a transverse distance lower than 2.5 cm to the expected track impact. The multiplicity is the number of hits of the closest cluster if any. The effienciency and the multiplicity per layer are shown in Fig. 2.

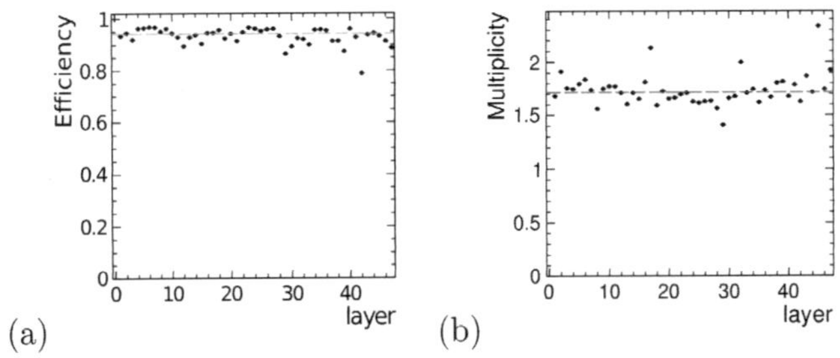

Figure 2. (a): Efficiency per layer. The red line represents the average efficiency ($\simeq$ 95%). (b): multiplicity per layer. The red line represents the average multiplicity ($\simeq$ 1.73).

3. SDHCAL prototype results

3.1. Particle identification

At SPS, the pion beam is contaminated by different kind of particles. There is a contamination by electrons, muons and protons [4]. Cosmic muons are also recorded in data samples. Because of the absence of Cherenkov counter, the pion selection is done using topological variables and in the following protons will be treated as pions. To eliminate the muons, it is requested that the ratio between the total number of hits and the number of fired layers is higher than 2.2. In addition, radiative muons are rejected by requesting that the ratio between the number of layers in which the hit dispersion is higher than 5 cm and the total number of fired layers is greater than 20%. The electron contamination could be quite high, especialy at low energy, even though the presence of a 4 mm thick lead absober is supposed to eliminate most of them. To remove those particles, the fact that electrons start their shower in the first plates is used (X_0=1.75 cm) by requesting that the shower is starting after the fifth layer. The first interaction plane (FIP) is defined as the first layer with at least 4 hits and with the three following layers containing at least 4 hits. To avoid the rejection of many hadronic showers, the previous criterion ($FIP \geq 5$) is applied only if the number of fired layers is lower than 30. Indeed, electromagnetic showers are well contained in 30 layers even at high energy as it can be seen in Fig. 3.a. With this selection, at low energy the part of rejected hadronic shower is not negligible but the effect is only statistical and the energy resolution is not affected. The power of this selection has been tested on electromagnetic showers and can be seen in Fig. 3.b. Additional criteria are used to eliminate neutral and multi-particles. Multi-particle events are rejected if the hit dispersion in the first five layers is greater than 5 cm. Neutral particle events are dropped by requesting that the number of hits in the first five layers is exceeding 4. Fig. 3.c shows the distribution of number of hits for a 50 GeV pion run before and after appling the selection criterions.

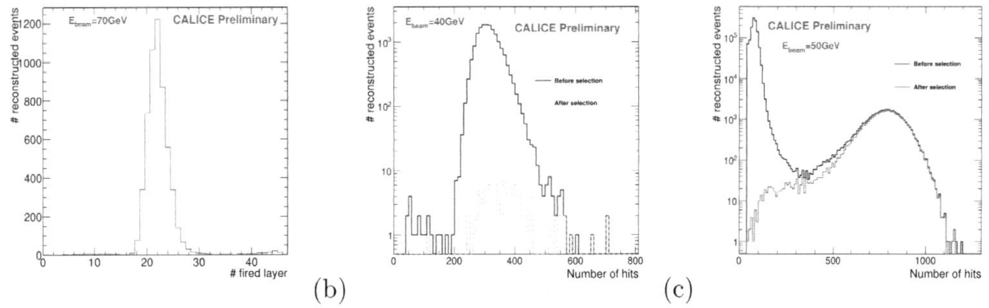

(a) (b) (c)

Figure 3. (a): Distribution of the number of fired layers for a 70 GeV electron run after muon rejection. (b): Distribution of number of hits for a 40 GeV electron run before (solid black line) and after (dotted green line) electron rejection. (c): Number of hit distribution for a 50 GeV pion runs before (black line) and after (red line) the whole selection

3.2. Energy resolution

The hadronic showers recontructed energy (Eq. 1) is calcuted after the previous selection using the number of hits per threshold (N_i).

$$E_{reco} = \alpha(N_{hit})N_1 + \beta(N_{hit})N_2 + \gamma(N_{hit})N_3 \tag{1}$$

The coefficients α, β and γ are quadratic function of total number of hits and they are determined using a χ^2 minimisation (Eq. 2). The choice of such a parametrization comes from a previous study [3] which shows that this parametrization is needed to restore linearity at high energy.

$$\chi^2 = \sum_{i=1}^{N} \frac{(E_{beam} - E_{reco})^2}{E_{beam}} \tag{2}$$

In Fig. 4.a the evolution of these parameters with the total number of hits is shown. These coefficients are then used to estimate the reconstructed energy of incoming particles. Reconstructed energy distributions are fitted with a Crystal Ball function defined in [3] to take into account the tail on the left part of those distributions (Fig. 4.b). A gaussian fit is also performed and used for estimation of systematic uncertainties. The mean reconstructed energy for pion showers and the relative deviation of the pion mean reconstructed energy with respect to the beam energy are shown in Fig. 5.a. The relative deviation of the pion mean reconstructed energy with respect to the beam energy is lower than 5% on a large energy scale ([7.5-80] GeV). Figure 5.b shows the hadronic energy resolution of the SDHCAL prototype which reaches 9% at 80 GeV. Since no gain correction was appplied, this results are very encouraging.

4. Tracking within hadronic showers using Hough Transform

Hadronic showers can contain many tracks generated by Minimum Ionising Particles (MIP) which can cross several layers before being stopped. Imaging calorimeters like the SDHCAL prototype can reconstruct the path of such tracks and use them to monitor the detector efficiency. Tracks reconstruction may be also helpful for the estimation of the reconstructed energy and for the application of the Particle Flow Algorithm. The Hough Transform technique is particulary interesting for the track detection within a dense environment like hadronic shower [5].

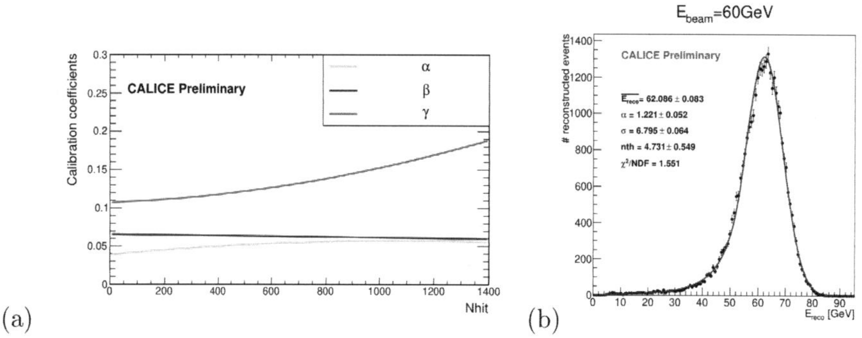

Figure 4. (a): α (green line), β (blue line) and γ (red line) evolution in terms of the total number of hits. (b): Reconstructed energy distribution for a 60 GeV pion runs.

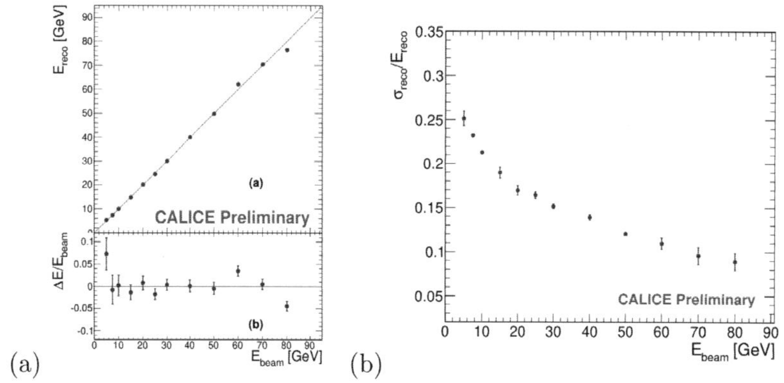

Figure 5. (a): Mean reconstructed energy for pion showers and relative deviation of the pion mean reconstructed energy with respect to the beam energy as a function of the beam energy. (b): $\frac{\sigma(E)}{E}$ of the reconstructed pion energy E as a function of the beam energy

4.1. Hough Transform method

Hough Transform is a simple method that allows to distinguish aligned points from others. The principle is to replace the point coordinates (x,z) by curves (ρ,θ) using the following transformation :

$$\rho(\theta) = z\cos\theta + x\sin\theta \tag{3}$$

The aligned points have their curves intersecting in the same point (ρ_0,θ_0) in the (ρ,θ) plan. The point coordinates (ρ_0,θ_0) determine the distance from the origin to the track and the angle between the ordinate axis and the track. The previous method can not be applied directly to hadronic showers recorded in the SDHCAL. The spatial resolution and the Multiple Coulomb Scattering require an adaptation of the method. The (ρ,θ) plan is discretized into a histogram. The bins of this histogram are incremented when they are crossed by a curve associated to one point in (x,z) plan.

4.2. Hough Transform within hadronic showers in the SDHCAL

To apply the Hough Transform method to the SDHCAL data samples the hits are grouped into clusters if they share an edge. Then the clusters in the dense part of the shower are rejected if two other clusters are found in an area of 10×10 cm^2 around (in the same layer) or if one other cluster composed by more than 5 hits is found in this area. Hough Transform is then applied to the remaining clusters where the cluster coordinates are the geometrical barycenter of its hits. At first, the Hough Transform is applied using one projection of the cluster coordinates (for example (z,x)). The distance ρ is calculated for 100 values of θ covering the range $[-\frac{\pi}{2}, \frac{\pi}{2}]$ and used to fill the (ρ,θ) histogram. The bins with more than 6 entries are selected. With the clusters belonging to the selected bins, the procedure is perfomed again using the other projection to avoid accidentaly aligned clusters in one 2D projection. The selected clusters are used to build tracks. Fig. 6.a shows one event display of a 80 GeV hadronic shower with the Hough Transform selected hits. The detector efficiency has been estimated using the same method described in section 2.2 but with the reconstructed tracks from the Hough Transform method. This efficiency was found to be in good agreement with the previous study (Fig. 6.b).

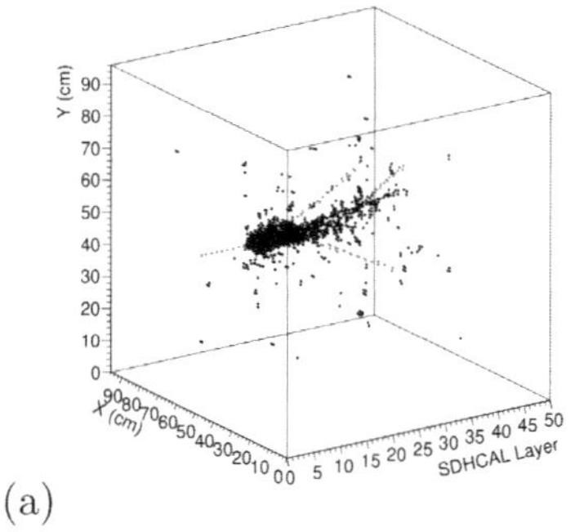

(a)

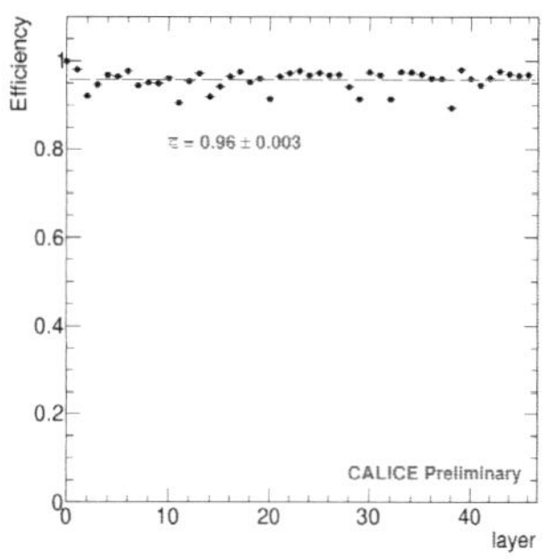

(b)

Figure 6. (a) 80 GeV hadronic shower with hits belonging to tracks in red. (b): Efficiency per layer estimated with tracks recontructed with Hough Transform technique.

5. Conclusion

The SDHCAL prototype achieved in 2012 was successfully tested at the SPS in CERN using trigger-less and power-pulsed modes. It has been exposed to beams of pions, electrons and muons on a large energy range (5-80 GeV). Encouraging results on the energy reconstruction and resolution have been reached without any gain correction. New algorithms to linearize the pion response have been developed. They allow to obtain a satisfactory linearity on the energy reconstruction and a good energy resolution which reaches 9% at 80 GeV. The Hough Transform technique has been applied succesfully and it was found to be a powerful tool for the track detection.

References

[1] I. Laktineh, Development of a semi-digital hadronic calorimeter using GRPC, Nucl.Instrum.Meth. A623 (2010) 231233

[2] F. Dulucq *et al.*, *HARDROC: Readout chip for CALICE/EUDET Digital Hadronic Calorimeter*, Nuclear Science Symposium Conference Record (NSS/MIC), 2010 IEEE, pp.1678-1683, Oct. 30 2010-Nov. 6 2010 doi: 10.1109/NSSMIC.2010.5874060

[3] The CALICE Collaboration, *First results of the CALICE SDHCAL technological prototype*, CALICE Analysis Note CAN-037, 30th November 2012

[4] E. Abat *et al.*, *Study of energy response and resolution of the ATLAS barrel calorimeter to hadrons of energies from 20 to 350 GeV*, NIM A, Volume 62, Issues 1-3, 1-21 September 2010.

[5] The CALICE Collaboration, *Tracking within Hadronic Showers in the SDHCAL prototype using Hough Transform Technique*, CALICE Analysis Note CAN-047, 04th December 2013

Hadron shower decomposition in a highly granular calorimeter

Marina Chadeeva, on behalf of the CALICE Collaboration

ITEP, Bolshaja Cheremushkinskaja 25, Moscow, 117218, Russia
MEPhI, Kashirskoje sh., 31, Moscow, 115409, Russia

E-mail: marina@itep.ru

Abstract. The spatial development of showers induced by positive hadrons with momenta 10-80 GeV in the highly granular CALICE scintillator-steel analogue hadronic calorimeter is analysed. The parametrisation of both longitudinal and radial shower profiles with the two-component functions are fit to the test beam data and simulations using the physics lists QGSP_BERT and FTFP_BERT from GEANT4 version 9.6 patch 01. The shower parameters, describing the longitudinal tail and radial halo, are in good agreement between data and simulations and are similar for pions and protons. For the longitudinal development, the most significant difference between data and simulations is in the relative containment of the separated components. For the radial development, the core slope parameter is underestimated by simulations. The physics list FTFP_BERT gives a very good description of proton showers in the studied energy range and gives better predictions of the pion shower development than QGSP_BERT.

1. Introduction

The CALICE collaboration has developed and constructed highly granular electromagnetic and hadronic calorimeter prototypes to evaluate detector technologies for future linear collider experiments. The linear collider physics program requires a very high jet energy resolution, which can be achieved using the particle flow approach based on the precise tracking and the possibility to disentangle showers induced by different particles in the calorimeter system [1, 2, 3]. The understanding of hadronic showers and their parametrisation is important for the estimation of leakage from the calorimeter, for the validation of hadronic shower models in simulations [4] and for the development of the particle flow algorithms [5].

The CALICE analogue hadronic calorimeter prototype (AHCAL) is the first experience of a large-scale application of silicon photomultipliers (SiPM) in the field of high energy physics [6]. The prototype was extensively studied during several test beam campaigns in two configurations: with tungsten (W-AHCAL) or steel (Fe-AHCAL) as an absorber material. The unprecedented granularity of the AHCAL allows to identify the position of the first inelastic interaction and deconvolve the distribution of the energy density inside the shower from the distribution of the shower starting point. The calorimeter response and resolution for pions are described in detail in [7] including the implementation of the developed software compensation techniques. The validation of several hadronic models from GEANT4 version 9.4 is discussed in [4], where the detailed comparison of the longitudinal centre of gravity, radius and profiles of pion-induced showers is shown.

Here we present the study of showers induced by positive pions and protons with initial momenta 10-80 GeV in the CALICE Fe-AHCAL, which is based on the parametrisation of shower profiles. The longitudinal and transverse shower profiles are decomposed into the core and halo contributions by fitting an empirical parametrisation to the energy density distributions extracted from both data and simulations. The fitted parameters and the extracted core shower fractions are compared to simulations using the physics lists QGSP_BERT and FTFP_BERT from GEANT4 version 9.6 patch 1 [8].

2. Data samples, event selection and simulations

The positive hadron data were collected during CALICE test beams campaigns at CERN in 2007 in the momentum range from 30 to 80 GeV/c and at FNAL in 2009 in the momentum range from 10 to 15 GeV/c. The CALICE calorimeter setup comprised the Fe-AHCAL and the scintillator-steel tail catcher and muon tracker (TCMT) [9]. The Si-W electromagnetic calorimeter (ECAL) [10] was placed in front of the AHCAL during CERN runs and is used as a tracker in this analysis. The information from a Čerenkov counter upstream of the calorimeter setup was used for off-line discrimination between pions and protons on an event-by-event basis.

The AHCAL is a sampling structure of 38 active layers interleaved with absorber plates. Each active layer is assembled from the small scintillator tiles (3×3, 6×6 and 12×12 cm^2) individually read out by SiPMs. The longitudinal depth of the Fe-AHCAL is $\sim 5.3\lambda_{\mathrm{I}}^{\mathrm{eff}}$ ($\lambda_{\mathrm{I}}^{\mathrm{eff}} = 231$ mm). The TCMT is also a sampling calorimeter with 16 active layers assembled from strips with SiPM readout. The first section (9 layers) of the TCMT has the same sampling as the Fe-AHCAL ($\sim 0.14\lambda_{\mathrm{I}}^{\mathrm{eff}}$ per layer) and is shown in the plots of the longitudinal profiles but is not used for fits. The effective radiation length X_0^{eff} of the Fe-AHCAL is estimated to be 25.5 mm.

The visible signal in each calorimeter cell is obtained in units of MIP (minimum-ionising particle) as described in [6]. Only cells with a signal above 0.5 MIP were used in the analysis and are called hits. The hadron sample cleaning includes the rejection of muons and positrons with the resulting purities of better than 99% and 98%, respectively. The additional selection of events was applied to minimise the leakage into the TCMT and to reduce the fraction of remaining positrons in the sample: the events with an identified shower start in the physical layers 2, 3, 4, 5, 6 (3, 4, 5, 6) of the Fe-AHCAL were selected for run taken with (without) ECAL. Hereinafter the profiles from the identified shower start are analysed.

The following sources of systematic uncertainties were found to affect the comparison of shower profiles: (a) layer-to-layer variations of the response in test beam data due to saturation correction issues (up to ~10% at 80 GeV); (b) accuracy of the identification of a shower axis for the data samples collected without ECAL (up to 10%); (c) pion contamination of the proton samples (estimated purity of the proton samples is from 74% to 95%). The uncertainties due to positron contamination of the samples taken without ECAL, accuracy of the shower start identification and leakage from the Fe-AHCAL were found to be negligible, provided the event selection described above. The statistical and systematic uncertainties are summed up in quadrature for the fit of the studied shower profiles.

The simulations were performed with GEANT4 version 9.6 patch 1, the test beam profile of each particular data run being closely reproduced. The simulated samples were digitised by taking into account the SiPM response, inter-tile crosstalk, map of calorimeter dead cells and noise from the corresponding data runs.

3. Parametrisation of hadron shower profiles

The parametrisation is an instrument for quantitative comparison of the observed shower development with the predictions of the Monte Carlo models. To obtain a stable fit procedure and reliable error estimates, no parameter limits were applied during the minimisation. Instead, a random variation of initial values for the minimisation procedure was used. From the sample of 100 attempts the best fit was chosen, the results with unphysical values were rejected.

3.1. Fit to longitudinal profiles

The parametrisation of the longitudinal development of hadronic showers with a sum of two gamma distributions was proposed in [11] as a natural extension of the parametrisation of electromagnetic shower profiles. The energy distribution $\Delta E(z)$ along the longitudinal coordinate integrated over the radial direction (longitudinal profile) is parametrised as a sum of the "short" and "long" components:

$$\Delta E(z) = A \cdot \left\{ \frac{f}{\Gamma(\alpha_{\mathrm{short}})} \cdot \left(\frac{z}{\beta_{\mathrm{short}}}\right)^{\alpha_{\mathrm{short}}-1} \cdot \frac{e^{-\frac{z}{\beta_{\mathrm{short}}}}}{\beta_{\mathrm{short}}} + \frac{1-f}{\Gamma(\alpha_{\mathrm{long}})} \cdot \left(\frac{z}{\beta_{\mathrm{long}}}\right)^{\alpha_{\mathrm{long}}-1} \cdot \frac{e^{-\frac{z}{\beta_{\mathrm{long}}}}}{\beta_{\mathrm{long}}} \right\}, \tag{1}$$

where z is the longitudinal distance from shower start, A is the scaling factor. The smaller slope parameter from the fit is called β_{short} with the corresponding shape parameter α_{short} and the fractional contribution f of the "short" component. The parameters α_{long} and β_{long} characterise the shape and slope of the "long" component. The fit range is from 0 to $\sim 4.5\lambda_{\mathrm{I}}^{\mathrm{eff}}$ with the step of $\sim 0.137\lambda_{\mathrm{I}}^{\mathrm{eff}}$. An example of the fit to the pion longitudinal profile is shown in Fig. 1.

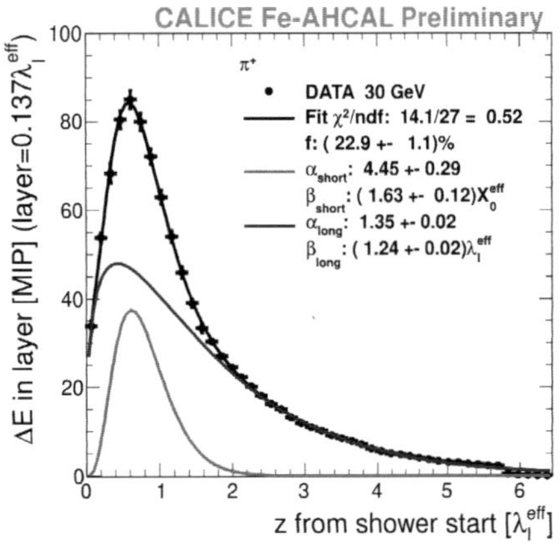

Figure 1. Fit (black curve) of the function (1) to the longitudinal profile (points) of showers initiated by the test beam pions with initial energy 30 GeV. The red and blue curves show the contributions of the "short" and "long" components, respectively. See text for details.

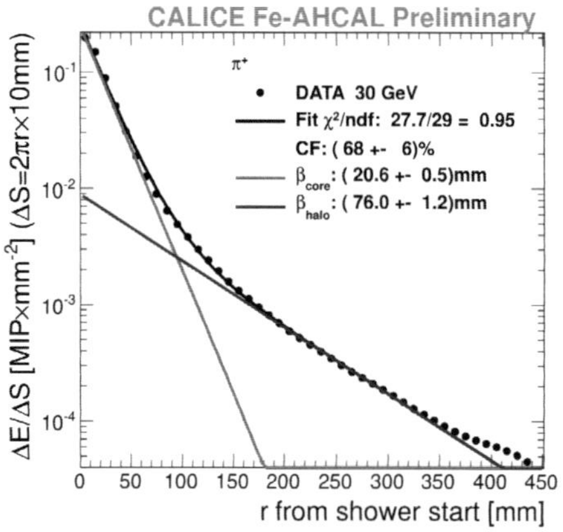

Figure 2. Fit (black curve) of the function (2) to the radial profile (points) of showers initiated by the test beam pions with initial energy 30 GeV. The red and blue curves show the contributions of the core and halo components, respectively. See text for details.

3.2. Fit to radial profiles

A transverse distribution of the energy density $\frac{\Delta E}{\Delta S}$ integrated over the longitudinal direction (radial profile) can be parametrised with a sum of the "core" component near the shower axis and the "halo" component far from the shower axis:

$$\frac{\Delta E}{\Delta S}(r) = A_{\text{core}} \cdot e^{-\frac{r}{\beta_{\text{core}}}} + A_{\text{halo}} \cdot e^{-\frac{r}{\beta_{\text{halo}}}} \tag{2}$$

where $\Delta S = 2\pi r \Delta r$ is the area of the ring of width Δr at the distance r from the shower axis, A_{core} and A_{halo} are the scaling factors, β_{core} and β_{halo} are the slope parameters. The fit range is from 0 to 340 mm with the bin width $\Delta r = 10$ mm (the perimeter 12×12 cm^2 cells of the AHCAL were excluded from the fit). The estimated accuracy of the determination of the position of the shower axis is $\sigma_r = 2$ mm. An example of the fit to the radial profile is shown in Fig. 2. The fractional contribution of the core component is called "core fraction" (CF) and is calculated as a ratio of the integral under the core component (the red curve in Fig. 2) to the total shower energy.

3.3. Comparison of shower parameters extracted from profiles parametrisation

The behaviour of the shape parameter α_{long} does not depend on the particle type, is well predicted by Monte Carlo and rises logarithmically with energy. The values and behaviour of the slope parameters β_{long} and β_{halo} are also predicted by simulations within uncertainties. They demonstrate no energy dependence and are very similar for pions and protons. Such a behaviour supports the general idea that both tail and halo of the hadronic shower consist of secondaries, which have already "forgotten" the energy and type of the initial particle.

The radial parameter β_{core} characterises the transverse shower development near the shower axis and is probably related to the angular distribution of secondary π^0s from the first inelastic interaction. The estimated value of this parameter is of the order of magnitude of one effective Molière radius of the Fe-AHCAL ($R_{\text{M}}^{\text{eff}} = 24.5$ mm), which is comparable with the spatial parameters of electromagnetic showers. It decreases with energy and is well predicted by both studied physics lists below 20 GeV and for protons by FTFP_BERT in the full studied energy range. The ratio of parameters β_{core} extracted from the test beam data and simulations is shown in Fig. 3. The underestimation of the parameter β_{core} above 20 GeV is $\sim$5% by FTFP_BERT for pions and $\sim$10% by QGSP_BERT for both types of particles.

The "short" component of the longitudinal profile describes the part of the shower, which develops immediately after the first inelastic interaction. The slope parameter of the "short" component is about $1.5 \cdot X_0^{\text{eff}}$. Both "short" parameters for pions are well predicted by Monte Carlo except for by FTFP_BERT at 10 GeV and are energy independent above 20 GeV. The fractional contribution of this component for protons is significantly smaller than for pions and cannot be reliably estimated for the energies below 20 GeV.

The position of the maximum of the "short" component $Z_{\text{max}}^{\text{short}}(\pi)$ can be calculated as follows: $Z_{\text{max}}^{\text{short}}(\pi) = (\alpha_{\text{short}} - 1) \times \beta_{\text{short}}$. Figure 4 shows the comparison of $Z_{\text{max}}^{\text{short}}(\pi)$, extracted from the "short" component of pion showers, with the estimation of the shower maximum position $Z_{\text{max}}(e)$, obtained from the pure electromagnetic showers induced by single electrons or positrons in the Fe-AHCAL [12, 13]. The position of the maximum of the "short" component for pions is calculated w.r.t. the shower start that corresponds to the estimates of $Z_{\text{max}}(e)$ from the calorimeter front for single electrons. In case of pions, the reconstructed energy, corresponding to the integral under the "short" component, $E_{\text{reco}}^{\text{short}}$ is calculated using the factor 42.3 MIP/GeV [13]. The maximums of the longitudinal profiles obtained for single electrons or positrons are shown versus the mean reconstructed energy E_{reco} of the corresponding particles, which coincides with the beam energy within 1-2%. The difference between $Z_{\text{max}}^{\text{short}}(\pi)$ and $Z_{\text{max}}(e)$ increases with decreasing energy for both data and simulations.

While the slope and shape longitudinal parameters extracted from the test beam data and simulations coincide within uncertainties, the fractional contribution f of the "short" component is overestimated by both studied physics lists above 20 GeV for pions and is slightly underestimated below 20 GeV. The behaviour of the parameter f is shown in Fig. 5a. The

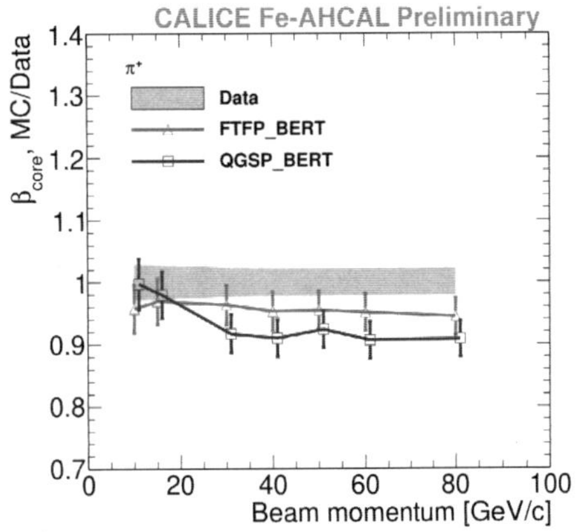

Figure 3. Energy dependence of the ratio of the radial slope parameter for pion showers β_{core} extracted from simulations with the FTFP_BERT (red) and QGSP_BERT (blue) physics lists to that extracted from data. The grey band corresponds to the uncertainties for data.

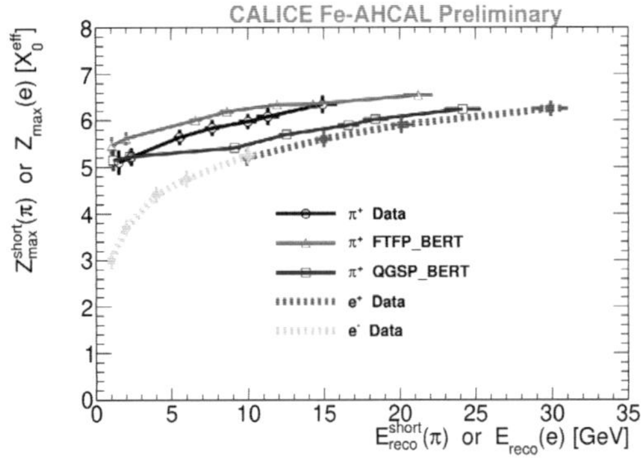

Figure 4. Solid lines: maximum of the "short" component versus energy of the "short" component estimated from the fit to pion shower profiles from data (black) and simulations with FTFP_BERT (red) and QGSP_BERT physics list (blue). Dashed lines: maximum of the longitudinal profile obtained for electromagnetic showers induced by single electrons (green) [12] or positrons (magenta) [13] in the Fe-AHCAL versus the mean reconstructed energy.

FTFP_BERT physics list gives a good prediction for protons while overestimates the parameter for pions at higher energies by 5-25%. The QGSP_BERT physics list significantly (up to 50%) overestimates the contribution of the "short" component above 20 GeV for both pions and protons. The fractional contribution of the "short" component in proton showers is roughly twice lower than in pion showers, which can be explained by a smaller electromagnetic fraction in proton showers due to lower amount of produced π^0s.

4. Conclusion

The spatial development of hadronic showers in the CALICE scintillator-steel analogue hadronic calorimeter was studied using the parametrisation technique. The positive hadron test beam data collected at beam energies from 10 to 80 GeV were analysed and compared with simulations performed using FTFP_BERT and QGSP_BERT physics lists from GEANT4 version 9.6. The energy density distributions within hadronic showers were deconvolved from the distributions of the shower starting point through the possibility to identify the position of the first inelastic interaction on an event-by-event basis.

The longitudinal profiles are parametrised with the sum of two gamma distributions. The parameters of the "short" component are comparable with those of electromagnetic showers induced by single electrons in the Fe-AHCAL. Hence the "short" component can be considered to represent the contribution of the electromagnetic showers from the decays of π^0s produced in the first inelastic interaction. The spatial parameters of the longitudinal tail are reproduced by simulations within uncertainties, do not depend on the type and energy of the initial particle. Proton profiles are characterised by a smaller fractional contribution of the "short" component. The parameter f for pions is overestimated by simulations above 20 GeV and exhibits a much steeper rise than observed in the test beam data.

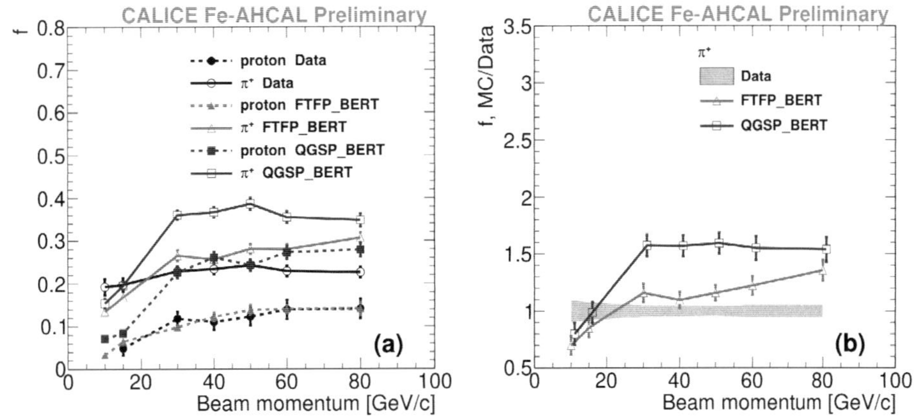

Figure 5. Energy dependence (a) of the "short" component fraction f for pions (solid curves) and protons (dotted curves) and (b) of the ratio of f extracted from simulations with the FTFP_BERT (red) and QGSP_BERT (blue) physics lists to those extracted from data for pion showers. The grey band corresponds to the uncertainties for data.

The radial profiles are parametrised with the sum of two exponential functions, which describe the behaviour near the shower axis ("core" region) and in the periphery ("halo" region). While the halo slope parameter is well reproduced by simulations, the core slope parameter is underestimated by ~5% for pions by FTFP_BERT and by ~10% by QGSP_BERT for both types of hadrons resulting in an underestimation of the shower width (shower radius) observed in the previous studies [4].

The tendencies observed in the current analysis for the models from GEANT4 version 9.6 are similar to those demonstrated in [4] for the version 9.4: the overestimation of the shower core and underestimation of the shower radius for pions increase with energy. The FTFP_BERT physics list from GEANT4 version 9.6 gives very good predictions of both longitudinal and radial shower development for protons over the full studied energy range and shows better agreement with data than the QGSP_BERT physics list for both pions and protons.

Acknowledgments
The work is supported by the Russian Ministry of Education and Science via grant P220 and RFBR grant 14-02-00873A.

References
[1] Brient J C and Videau H 2001 *eConf* **C010630** E3047
[2] Morgunov V L 2002 *Proc. 10th Int. Conf. CALOR 2002 (Pasadena)* (River Edge: World Scientific) p 70
[3] Thomson M 2009 *NIM* A **611** 25
[4] Adloff C *et al.* 2013 *JINST* **8** P07005
[5] Adloff C *et al.* 2011 *JINST* **6** P07005
[6] Adloff C *et al.* 2010 *JINST* **5** P05004
[7] Adloff C *et al.* 2012 *JINST* **7** P09017
[8] Agostinelli S *et al.* 2003 *NIM* A **506** 250; Dotti A *et al.* 2011 *J. Phys. Conf. Ser.* **293** 012022
[9] Adloff C *et al.* 2012 *JINST* **7** P04015
[10] Adloff C *et al.* 2008 *JINST* **3** P08001
[11] Bock R K, Hansl-Kozanecka T and Shah T P 1981 *Nucl. Instrum. Meth.* **186** 533
[12] Feege N 2011 Ph.D. thesis Hamburg University (*Preprint* DESY-THESIS-2011-048)
[13] Adloff C *et al.* 2011 *JINST* **6** P04003

The Time Structure of Hadronic Showers in Calorimeters with Scintillator and with Gas Readout

Marco Szalay on behalf of the CALICE collaboration

Max Planck Institut für Physik,
Föhringer Ring 6, 80805 München, Germany

E-mail: marco.szalay@mpp.mpg.de

Abstract. Hadronic showers are characterized by a rich particle structure in the spatial as well as in the time domain. The prompt component comes from relativistic fragments that deposit energy at the ns scale, while late components are associated predominantly with neutrons in the cascade. To measure the impact of these late components, two experiments, based on gaseous and plastic active layers with steel and tungsten absorbers, were set up. The different choice for the material of the active layers produces distinct responses to neutrons, and consequently to late energy depositions. After discussing the technical aspects of these systems, we present a comparison of the signals, read out with fast digitizers with deep buffers, and provide detailed information of the time structure of hadronic showers over a long sampling window.

1. Introduction

The timing resolution of calorimeters depends strongly on the time structure of hadronic showers. The study of this quantity is of particular interest for precision signal time stamping that would allow to reject pile-up background from hadrons produced in two-photon processes at future accelerators such as the Compact Linear Collider (CLIC[1, 2]).

In a calorimeter, the time distribution of energy deposition depends on several physical processes taking place within the hadronic shower. Relativistic hadrons and the electromagnetic fraction of the shower produce an intense prompt signal in the detector that extends for a few ns before fading out. After that, a rich set of nuclear interactions with the dense absorbing material releases neutrons that can be detected in the active layer via elastic scattering on hydrogen nuclei on a timescale that ranges over a few tens of ns. Even more delayed signals, up to ms, arise from energy released by neutron capture events and the decay of meta-stable nuclear states.

The sensitivity to neutron elastic scattering, as well as to the later parts of the hadronic cascade, also influences the spatial structure of the visible signal in the detector, especially when the absorbing material is tungsten, where the electromagnetic subshowers are denser, due to the shorter radiation length of the material. The extent of the cascade is of particular importance for the performance of particle flow algorithms[3, 4].

These algorithms, which are used for jet reconstruction with unprecedented precision in linear collider detectors, rely on two-particle separation in the calorimeters. The spatially resolved measurement of the time structure of showers and the comparison of detection media with different sensitivity to neutrons, such as plastic scintillators and gaseous detectors, is thus of high relevance for the development of calorimeter technologies for such a future collider.

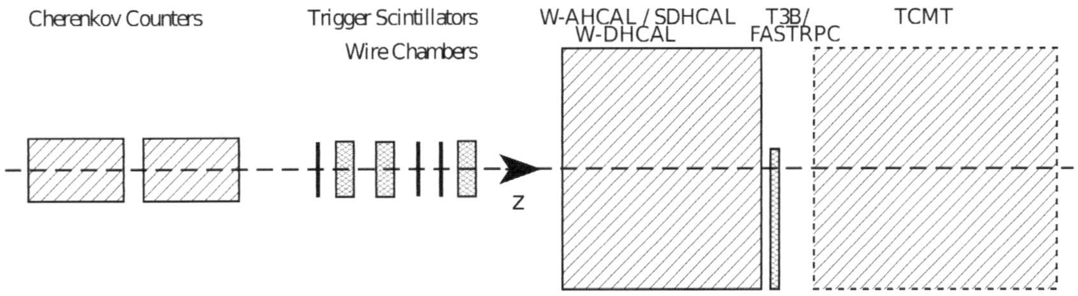

Figure 1. Setup schematics at the CERN SPS H8 beam line in the North Hall. The tail catcher TCMT was only present for runs with the CALICE W-AHCAL and W-DHCAL, and not installed for runs with the Fe-SDHCAL. Illustration not to scale.

2. Experimental Setup

Two complementary experiments, based on different signal detection technologies, have been developed to take data together with the CALICE HCAL prototypes and measure the average time structure of hadronic showers, namely *T3B* (Tungsten Timing Test Beam[5]) and *FastRPC*. The former consists of 15 3×3 cm^2 BC420 scintillating tiles, each coupled to a Hamamatsu SiPM, arranged in a stripe to sample the hadronic shower radially from its center. The latter uses a glass RPC active layer, operated in avalanche mode, with readout pads that have the same geometry as T3B.

The setups share the same readout based on 15 (one per channel) Infineon SiGe preamplifiers[1] and four 4-channel USB-oscilloscopes[2] which provide a sampling rate of 1.25 GS/s per channel with a buffer large enough to cover a 2.4 μs acquisition window for each event. This allows the study of time structure of the energy deposits in the active medium in detail over a long time window, while maintaining a sub-ns resolution.

The small number of readout channels is insufficient for event-by-event measurements, but is used to measure the average time structure of showers in large data samples. To produce the results presented here, sizable data sets, summarized in table 1, have been collected. T3B

Data Set	Number of Events
T3B 180 GeV μ	5.40 M
T3B 60 GeV hadrons (steel)	1.60 M
T3B 60 GeV hadrons (tungsten)	4.06 M
FastRPC 180 GeV μ	3.19 M
FastRPC 80 GeV hadrons (tungsten)	2.63 M

Table 1. Statistics recorded for each data sample at CERN SPS test beam facility.

has been installed behind the tungsten analog scintillator (W-AHCAL[6]) prototype as well as behind the steel semi-digital prototype (SDHCAL[7]). The FastRPC experiment was placed downstream to the tungsten digital RPC prototype (W-DHCAL[8]). All systems were installed at the CERN SPS facility and took data at beam energies up to 180 GeV. A schematic drawing of the test beam setup can be found in figure 1. The material budget in the beam line upstream of the experiment's layer differs for tungsten and steel runs. The total amount of material is $\sim$ 5.1 nuclear interaction lengths for tungsten data and $\sim 6.5\ \lambda_I$ for steel.

[1] Infineon BGA614 (http://www.infineon.com/)
[2] PicoTech PicoScope 6403 (http://www.picotech.com/)

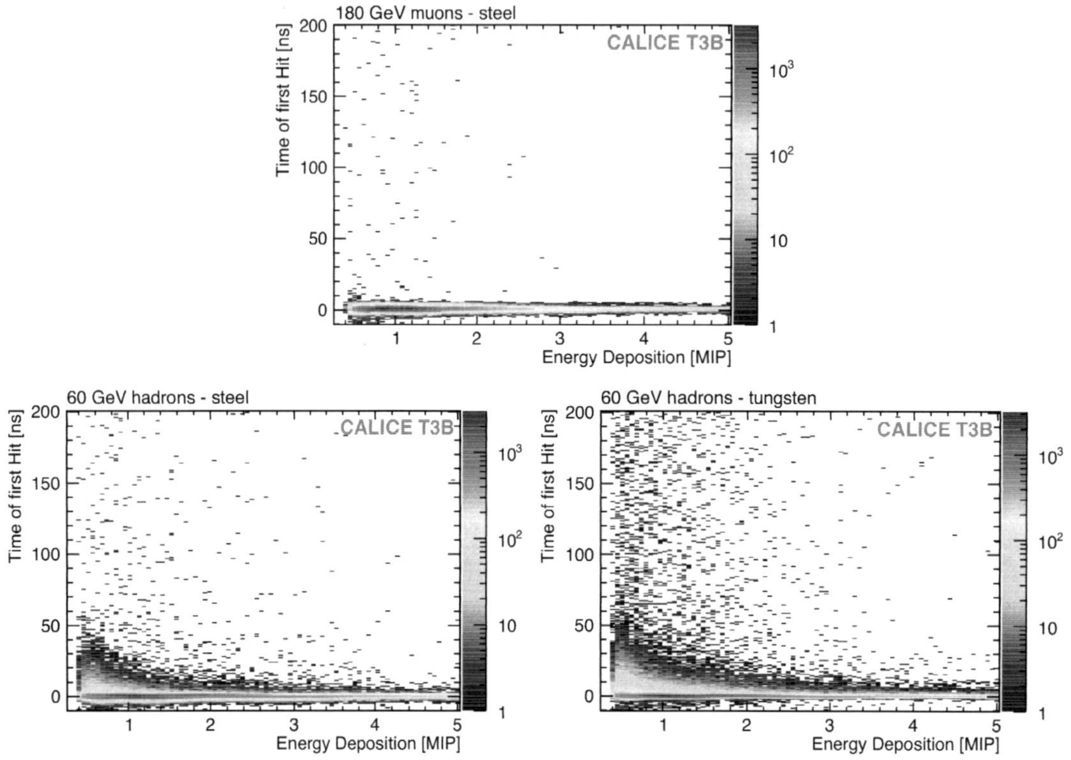

Figure 2. Distribution of reconstructed time of first hit in T3B as a function of deposited energy for muon data (top) and 60 GeV hadrons in steel (bottom, left) and in tungsten (bottom, right). With both absorbing materials, the time structure for hadrons extends over a longer time window, compared to muon data.

The trigger system is shared with the HCAL prototypes and consists of two 10×10 cm^2 scintillators in coincidence placed in front of the HCAL prototypes.

3. Results

For both experiments, a sophisticated calibration and reconstruction framework has been developed. In the case of scintillator readout, this system is capable of determining the arrival time of each photon on the photon sensor on the nanosecond level by iteratively subtracting single photon signals from the recorded waveform[5]. Further analysis is then performed on the photon time distribution. With RPC readout, the recorded pulse is used directly in the analysis.

To provide a robust base for comparison between the two systems and to eliminate effects from afterpulsing, the time of first hit is studied, which is defined by the time of the first energy deposit corresponding to at least the equivalent of 0.3 minimum-ionizing particles within 9.6 ns in a given cell in an event. Due to the high granularity of the readout, the probability for multiple hits in one cell in one event is on the percent level. The use of the time of first hit rather than using all observed hits thus does not result in an appreciable bias.

Figure 2 shows a comparison of the aforementioned time of first hit for T3B data[9]. Muon data show a prompt energy deposition compatible with relativistic particles depositing energy via ionization processes. In contrast, hadron data show a richer tail in the time domain due to slower components in the shower produced by particles interacting with the absorbing material.

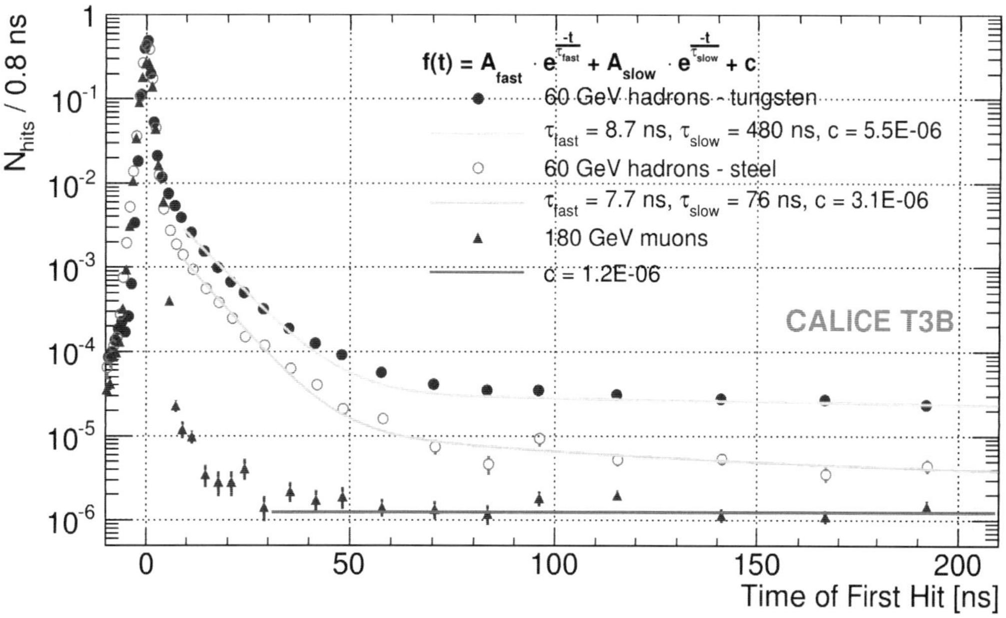

Figure 3. Time of first hit distribution of muon data with steel absorbers and hadron data with steel and tungsten absorbers in a time range of -10 ns to 200 ns. The histograms are normalized to the number of events in which at least one first hit could be identified and show the number of hits per T3B DAQ time bin of 0.8 ns.

Moreover tungsten shows a higher non-prompt energy deposition activity compared to steel, a hint that absorber materials with a high neutron content release an increased number of evaporation neutron that account for a more prominent tail. The discrepancy can be better appreciated in figure 3[9], where all T3B data sets are projected on the time axis. Here the sum of two exponential decays and a noise constant term are used to fit the signal tail. The two exponentials account for a faster tail component, in the time region from a few ns up to $\sim$ 50 ns, that transitions smoothly into a slower shower component that extends to hundreds of ns. The time range of the first one is compatible with fast neutrons that scatter elastically on hydrogen nuclei present in the active material. The second one can be associated primarily with neutron capture, which release detectable energy over a long period of time.

A similar comparison can be performed, for tungsten absorbers, between the two different experiments, as depicted in figure 4. It is apparent that, whereas for both active materials hadronic showers lead to substantial late signal components, in the range from 10 to 50 ns there is a large difference, up to a factor eight, between the two setups. In this time window MeV-scale neutrons scattering elastically on hydrogen account for the largest fraction of the signal. Due to the low density and the low hydrogen content in the RPCs, their sensitivity to this component is significantly reduced compared to plastic scintillator. This discrepancy is then again strongly reduced for more delayed signals, where the contribution of neutron elastic scattering in the active layer is negligible.

Finally, it is interesting to compare the prediction of simulations to experimental measurements. For the T3B data, detailed comparisons to GEANT4[10] simulations with different hadronic shower models have been performed for both absorbing materials. These show that while the distributions in steel are generally well modeled by all physics lists, the

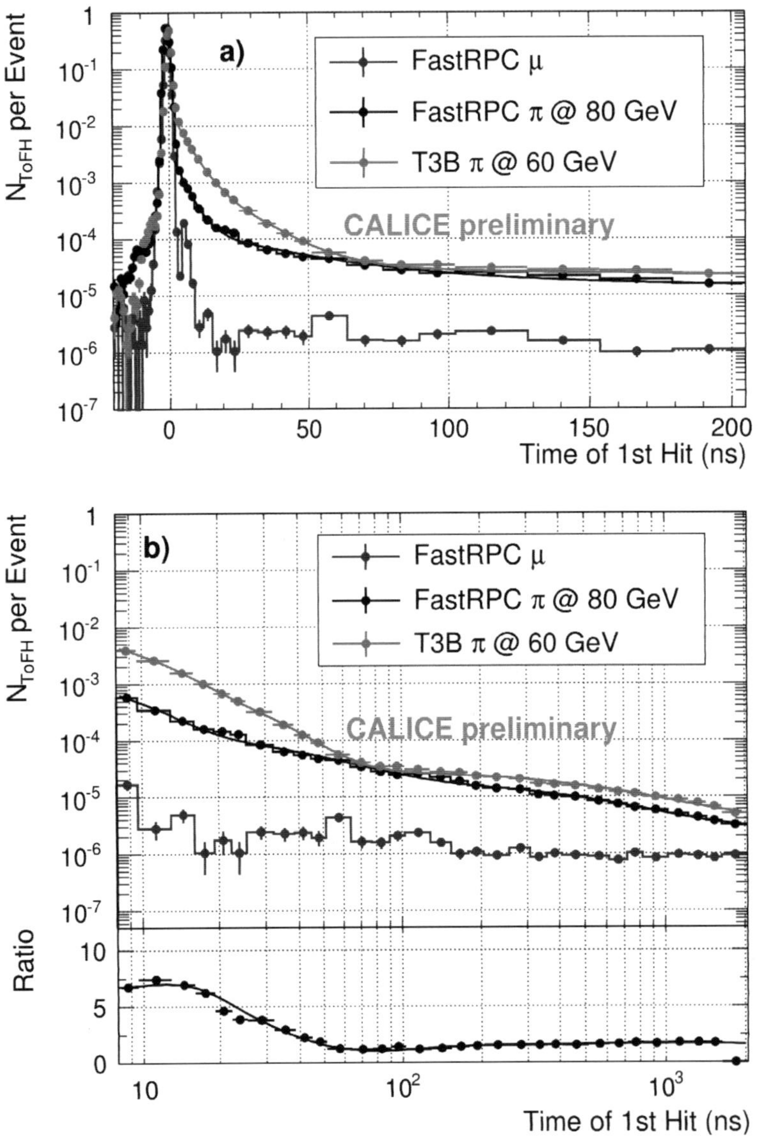

Figure 4. Comparison of the time of first hit with scintillator and RPC readout in tungsten in the [-20, 200] ns range (top) and on a logarithmic time scale for [8, 2000] ns (bottom). For reference, the time distribution for muons in FastRPC is also shown.

reproduction of the tungsten results requires a more precise neutron treatment. Indeed, as shown in figure 5, physic lists like QGSP-BERT fail to reproduce data over time windows longer than a few tens of ns, whereas QGSP-BERT-HP, an extension of the former list with tracking of neutrons down to thermal energies, describe experimental results over a broader range. For the FastRPC experiment the simulation codes are still being developed at present.

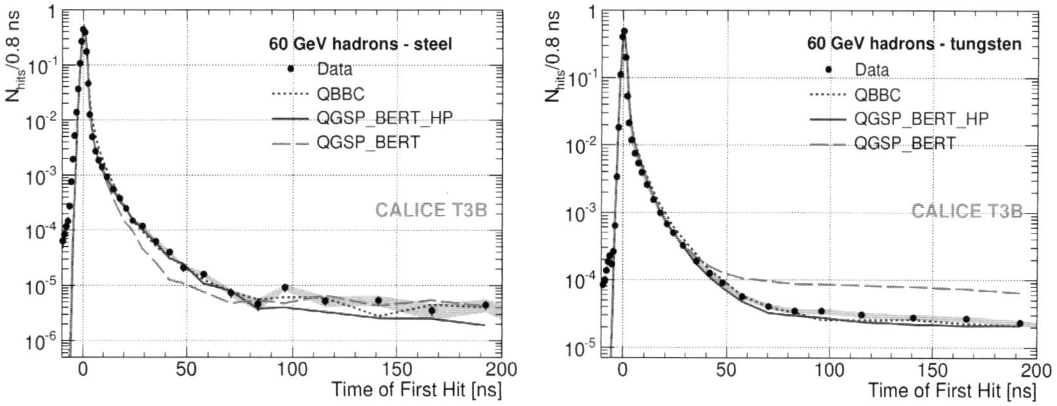

Figure 5. Distribution of the time of first hit in T3B for 60 GeV π^+ in steel (left) and tungsten (right), compared to Geant4 simulations with different physics lists.

References

[1] Lebrun P, Linssen L, Lucaci-Timoce A, Schulte D, Simon F, Stapnes S, Toge N, Weerts H and Wells J 2012 *arXiv preprint arXiv:1209.2543*

[2] Linssen L, Miyamoto A, Stanitzki M and Weerts H 2012 *arXiv preprint arXiv:1202.5940*

[3] Thomson M 2009 *Nuclear Instruments and Methods in Physics Research Section A: Accelerators, Spectrometers, Detectors and Associated Equipment* **611** 25–40

[4] Marshall J, Münnich A and Thomson M 2012 *Nuclear Instruments and Methods in Physics Research Section A: Accelerators, Spectrometers, Detectors and Associated Equipment*

[5] Simon F, Soldner C and Weuste L 2013 *Journal of Instrumentation* **8** P12001

[6] Adloff C, Karyotakis Y, Repond J, Brandt A, Brown H, De K, Medina C, Smith J, Li J, Sosebee M *et al.* 2010 *Journal of Instrumentation* **5** P05004

[7] Laktineh I 2011 *Journal of Physics: Conference Series* vol 293 (IOP Publishing) p 012077

[8] Bilki B, Butler J, Cundiff T, Drake G, Haberichter W, Hazen E, Hoff J, Holm S, Kreps A, May E *et al.* 2008 *Journal of Instrumentation* **3** P05001

[9] Adloff C *et al.* 2014 *arXiv preprint arXiv:1404.6454*

[10] Agostinelli S, Allison J, Amako K e, Apostolakis J, Araujo H, Arce P, Asai M, Axen D, Banerjee S, Barrand G *et al.* 2003 *Nuclear instruments and methods in physics research section A: Accelerators, Spectrometers, Detectors and Associated Equipment* **506** 250–303

Journal of Physics: Conference Series **587** (2015) 012038 doi:10.1088/1742-6596/587/1/012038

The CALICE digital hadron calorimeter: calibration and response to pions and positrons

B Bilki
On behalf of the CALICE Collaboration

Argonne National Laboratory, Argonne, IL 60439, USA
University of Iowa, Iowa City, IA 52242, USA

E-mail: burak-bilki@uiowa.edu

Abstract. In order to measure the jet products of the hadronic decays of electroweak bosons in a future lepton collider with 3-4% resolution, a novel approach named Particle Flow Algorithms is proposed. The Particle Flow Algorithms attempt to measure each particle in a hadronic jet individually, using the detector providing the best energy/momentum resolution. The role of the hadronic calorimeters is to measure the neutral component of the hadronic jets. In this context, the CALICE Collaboration developed the Digital Hadron Calorimeter, which uses Resistive Plate Chambers as active media. The 1-bit resolution (digital) readout of 1 x 1 cm^2 pads achieves a world record in the number of readout channels already at the prototyping stage. Here we report on the results from the analysis of pion events of momenta between 2 to 60 GeV/c collected in the Fermilab test beam with an emphasis on the intricate calibration procedures.

1. Introduction

The CALICE Collaboration develops calorimeters that are optimized for the application of Particle Flow Algorithms (PFAs) for future linear colliders [1]. The large CALICE Digital Hadron Calorimeter (DHCAL) prototype was built in 2009-2010 with a design based on the preliminary work done with a small-scale prototype, which underwent a rigorous test program in the Fermilab test beam and resulted in numerous publications [2-7]. The active media of the DHCAL are Resistive Plate Chambers (RPCs), which are read out by 1 x 1 cm^2 pads with a 1-bit resolution (digital readout). A single layer of the DHCAL measures roughly 1 x 1 m^2 and consists of 96 x 96 pads. During the Fermilab beam tests, the calorimeter consisted of a 38-layer structure (main stack) with 1.75 cm thick steel absorber plates and a 14-layer structure (tail catcher) with eight 2 cm thick steel plates followed by six 10 cm thick steel plates. In addition to the absorber plates, each layer of RPCs was contained in a cassette with a 2 mm thick Copper front plate and a 2 mm thick Steel back plate. The details of the DHCAL can be found in [8, 9]. Further information about the CALICE detectors can be found in [10-16].

The DHCAL is a calorimeter with the following unique features: RPCs for calorimetry (no other hadron calorimeter uses RPCs as active medium); Pad readout of RPCs (RPCs are usually read out with strips); Digital readout; Embedded front-end electronics; Large channel count (a world record of ~ 0.5M channels). Here, we describe the basics of the intricate calibration procedures and their implementation in the analysis of the Fermilab data.

2. Calibration procedures

The DHCAL data contain the hit position information, the time stamp of the individual hits and the time stamp from the trigger and timing unit. Additionally, discriminated signals from a beam Čerenkov counter and a muon tagger (a downstream scintillator 1 x 1 m^2 paddle) are integrated into the data stream by the data acquisition system.

The hits in each layer are combined into clusters using a nearest-neighbor algorithm. If two hits share a common edge, they are assigned to the same cluster. The cluster's x and y coordinates are calculated as the average of these coordinates over the constituent hits. Here, the x axis is the horizontal, the y axis is the vertical and the z axis is along the beam direction (a right-handed coordinate system with the origin at the center of the most upstream layer). The event selection requires at least five active layers (layers with at least one hit) in order to eliminate events with spurious triggers.

The calibration of the DHCAL involves several steps. To begin, the performance parameters of the individual RPCs, i.e. the efficiency and the average pad multiplicity, are measured. Here two methods are used: track fits and track segment fits. Dead or hot cells, if any, are identified on a run-by-run basis. In order to avoid a bias in the estimation of the performance parameters, regions within 1 cm of dead/hot cells or RPC edges are excluded from these measurements.

The track fits method uses dedicated muon calibration runs to assess the performance parameters of individual RPCs. This method starts with grouping the clusters that are laterally within a distance of 3 cm of each other in different layers for the events with tracks traversing the DHCAL. The group of clusters is then fit to the 3-dimensional parametric line $x=x_0+a_xt$; $y=y_0+a_yt$; $z=t$. For each layer, clusters within 2 cm of the point predicted by the fit are searched for. If a cluster is found, the layer is counted as efficient, and inefficient otherwise. If the layer is efficient, the pad multiplicity is given by the size of the found cluster. If multiple clusters are found in this search, the pad multiplicity is given by the size of the cluster that is closest to the fit point.

The track segment fits method is developed to measure the calibration parameters using the track segments within hadronic showers. With this method, the DHCAL provides another unique feature in calorimetry: For operation in a colliding beam environment, the DHCAL does not need a dedicated calibration system, as track segments can be used to monitor the performance of the RPCs. The method starts with searching for four clusters that are aligned within 3 cm in four different layers. The track segment is then fit to the parametric line defined above and this track segment is used to measure the performance parameters of a fifth layer (measurement layer). The measurement layer can either be within the layer span of the pick layers or outside, but only one measurement layer per track segment is allowed. In the measurement layer, clusters within 2 cm to the fit point are searched for. If a cluster is found, the layer is measured as efficient, and inefficient otherwise. If the layer is efficient, the pad multiplicity is measured as the size of the found cluster.

In the second step of the calibration, the number of hits measured in a given RPC is corrected for differences in its performance parameters using three different approaches: full calibration, density-weighted calibration, and hybrid calibration.

Using the track fits and track segment fits methods, the calibration factors per RPC per data taking run are obtained as $C_i= \varepsilon_i\mu_i/\varepsilon_0\mu_0$ where ε_i and μ_i are the efficiency and the average pad multiplicity of RPC i and ε_0 and μ_0 are the average RPC efficiency and pad multiplicity of the entire stack, 0.96 and 1.56 respectively [9].

Full Calibration: The hits in RPC i are weighted by $1/C_i$.

Density-Weighted Calibration: This approach takes into account that pads collecting charge from several nearby avalanches, for instance in the core of a shower, require a different calibration procedure than pads measuring single tracks. In other words, a pad in the core of a shower will register a hit with minimal dependence on the performance characteristics of this particular RPC and it should be calibrated in a different way than a pad e.g. along a MIP track in the same RPC. In this approach, the calibration factors of a pad may, in general, depend on the local hit density (as a measure of the number of avalanches contributing to the signal charge of that pad), the energy of the incident particle, and the type of incident particle, in addition to the performance parameters of the RPCs.

Hybrid Calibration: For the hits with 0 or 1 neighbor, the density effect is minimal. The hybrid calibration utilizes full calibration for density bins 0 and 1, and density-weighted calibration for the higher density bins.

Further details about the calibration approaches can be found in [17].

Figure 1 shows the application of the three calibration schemes to the 4 GeV π^+ (left) and 8 GeV e^+ (right) data. As expected, the uncalibrated data (black) show the largest amount of fluctuation in response between different runs. The full calibration (red), density-weighted calibration (green) and the hybrid calibration (blue) schemes all result in improved uniformity of the responses with significantly smaller fluctuations. All calibration schemes are successful in compensating for the slight differences in the RPC performance characteristics.

Figure 2 shows the χ^2/ndf for the constant line fits to the π^+ data collected in the Fermilab test beam. All calibration schemes improve the uniformity of the response across different runs and run periods. The three calibration schemes seem to perform at similar levels with no clear winner.

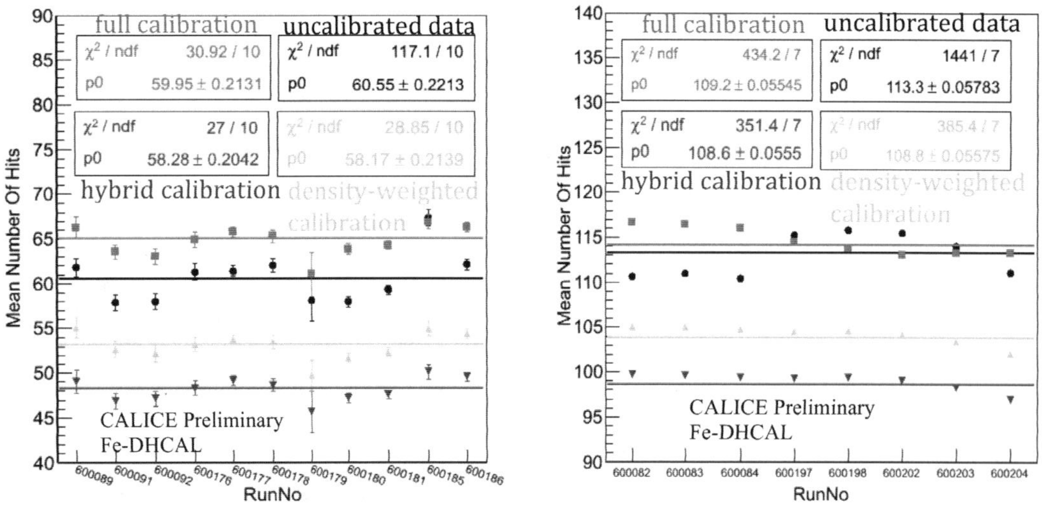

Figure 1. The results of the three calibration schemes applied to the 4 GeV π^+ (left) and 8 GeV e^+ (right) data. The uncalibrated data (black, 0), full calibration (red, 5), density-weighted calibration (green, -5) and the hybrid calibration (blue, -10) responses are all fit to a constant. The numbers in the parenthesis following the colors are the y-offsets applied to the data points in order to increase their visibilities. The error bars show the statistical errors on the means.

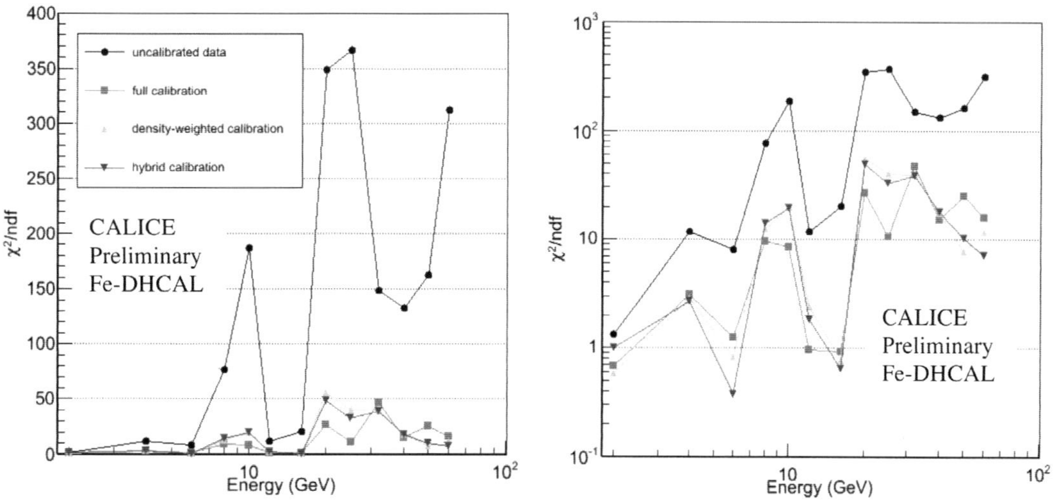

Figure 2. Normalized χ^2 of the fits of a constant to all the π^+ runs at a given energy (left), also in log-y scale (right).

3. Pion results

The event selection requires not more than 1 cluster in Layer 1 with at most four hits. This selection assures that upstream interactions are not included. The requirement of at least five active layers rejects events with spurious triggers. In order to separate the pions, positrons and muons in the beam, the Čerenkov counter in the beamline is utilized. In addition, a topological particle identification (PID) method is developed and implemented for 2, 4, 25 and 32 GeV data where the Čerenkov counter was not efficient. For details of the topological PID see [17]. The RPC performance parameters are calculated using the track segment fits method.

Figure 3 shows the mean response (a, c, e) and the energy resolution (b, d, f) for the uncorrected pion data (black in all plots), full calibration (red in a, b), density-weighted calibration (green in c, d) and hybrid calibration (blue in e, f). The mean response is fit to the power function $N=aE^m$ up to and including 60 GeV. The resolutions are fit to the generic $\frac{\sigma(N)}{N} = \frac{\alpha}{\sqrt{E}} \oplus C$ where α is the stochastic term and C is the constant term. The resolution fits are up to and including the 25 GeV point and they are extrapolated to 60 GeV. No additional corrections/selections are applied to the data (e.g. containment cuts, correction for response non-linearity). Therefore, the purpose of Figure 3 is to demonstrate the effect of the calibration schemes on the results.

All calibration schemes tend to normalize the mean response to the predefined DHCAL operating conditions. At lower energies, the methods agree with each other. However, at higher energies where the shower densities are large, the effect of employing the density weighting in the calibration procedure is clearly visible.

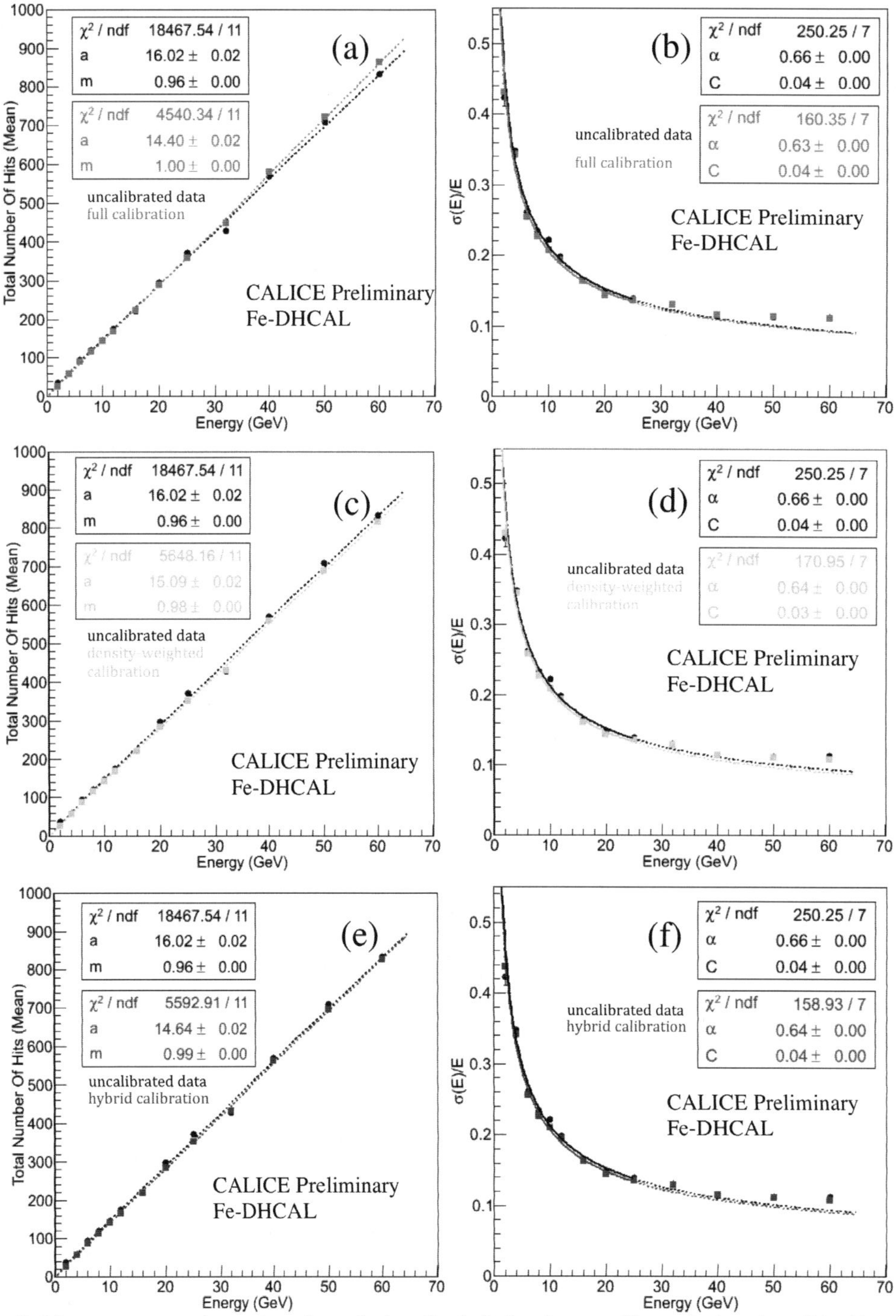

Figure 3. Mean response (a, c, e) and resolution (b, d, f) for the uncalibrated pion data (black) and the three calibration schemes (full calibration – red; density-weighted calibration – green; hybrid calibration – blue). For all calibration schemes, the fit quality is improved both for mean response (a, c, e) and resolution (b, d, f) compared to the fits to the uncalibrated data. The resolution fits (b, d, f) are up to 25 GeV (solid) and are extrapolated to 60 GeV (dashed).

4. Conclusions

The DHCAL recorded around 14 million secondary beam events over five test beam campaigns at Fermilab. The beam is a momentum-selected mixture of muons, pions and positrons. Data were collected at various energies between 2 and 60 GeV. The high granularity and the digital readout of the DHCAL enable the utilization of numerous topological event parameters for all purposes ranging from calibration to correcting the hadronic/electromagnetic response (software compensation) and improvements to the energy resolution measurements.

The calibration of the DHCAL is based on two performance parameters of the Resistive Plate Chambers: efficiency and average pad multiplicity. A simple multiplication of these parameters normalized to a reference value can serve as a calibration factor. However, the density of showering particles per pad impacts the calibration procedure in a complicated manner. As a result, the density-weighted calibration schemes provide better handles in understanding/manipulating response differences due to changes in individual RPC performances and operation conditions of the DHCAL.

All three calibration schemes, i.e. full calibration, density-weighted calibration and hybrid calibration, result in a more uniform response for all runs at each energy point when compared to the uncalibrated results.

5. References

[1] Thomson M A 2009 Nucl. Instrum. Meth. A 611 25; Brient J C et al. 2001 eConf C 010630 E3047; Morgunov V L 2002 Proc. of Calorimetry in Particle Physics (CALOR 2002) Pasadena USA
[2] Zhang Q et al. 2010 JINST 5 P02007
[3] Bilki B et al. 2009 JINST 4 P10008
[4] Bilki B et al. 2009 JINST 4 P06003
[5] Bilki B et al. 2009 JINST 4 P04006
[6] Bilki B et al. 2008 JINST 3 P05001
[7] Drake G et al. 2007 Nucl. Instrum. Meth. A 578 88
[8] http://www.hep.anl.gov/repond/DHCAL_FNAL_Nov_2010.pptx
[9] Repond J 2011 CALICE Analysis Note CAN-030; Repond J 2013 CALICE Analysis Note CAN-030A
[10] Poeschl R 2014 (CALICE Collaboration) *Hadronic interactions in the CALICE Si-W ECAL* talk given at this conference
[11] Poeschl R 2014 (CALICE Collaboration) *Construction and test of a large scale prototype of a silicon tungsten electromagnetic calorimeter for a future lepton collider* talk given at this conference
[12] Krueger K 2014 (CALICE Collaboration) *Prototype tests for a highly granular scintillator-based hadron calorimeter* talk given at this conference
[13] Sicking E 2014 (CALICE Collaboration) *Shower characteristics of particles with momenta up to 100 GeV in the CALICE Scintillator-Tungsten HCAL* talk given at this conference
[14] Steen A 2014 (CALICE Collaboration) *Results of the CALICE SDHCAL technological prototype* talk given at this conference
[15] Chadeeva M 2014 (CALICE Collaboration) *Hadron shower decomposition in a highly granular calorimeter* talk given at this conference
[16] Szalay M 2014 (CALICE Collaboration) *The time structure of hadronic showers in calorimeters with scintillator and with gas readout* talk given at this conference
[17] Bilki B 2013 *The DHCAL Results from Fermilab Beam Tests: Calibration* CALICE Analysis Note CAN-042

Journal of Physics: Conference Series **587** (2015) 012039 doi:10.1088/1742-6596/587/1/012039

Proof-of-principle of a new geometry for sampling calorimetry using inorganic scintillator plates

R. Becker, G. Dissertori, A. Gendotti, Q. Huang, D. Luckey[1], W. Lustermann, S. Lutterer, F. Nessi-Tedaldi, F. Pandolfi, F. Pauss, M. Peruzzi, M. Quittnat, R. Wallny

Institute for Particle Physics, ETH Zurich, Switzerland

E-mail: Francesca.Nessi-Tedaldi@cern.ch

Abstract. A novel geometry for a sampling calorimeter employing inorganic scintillators as an active medium is presented. To overcome the mechanical challenges of construction, an innovative light collection geometry has been pioneered, that minimises the complexity of construction. First test results are presented, demonstrating a successful signal extraction. The geometry consists of a sampling calorimeter with passive absorber layers interleaved with layers of an active medium made of inorganic scintillating crystals. Wavelength-shifting (WLS) fibres run along the four long, chamfered edges of the stack, transporting the light to photodetectors at the rear. To maximise the amount of scintillation light reaching the WLS fibres, the scintillator chamfers are depolished. It is shown herein that this concept is working for cerium fluoride (CeF_3) as a scintillator. Coupled to it, several different types of materials have been tested as WLS medium. In particular, materials that might be sufficiently resistant to the High-Luminosity Large Hadron Collider radiation environment, such as cerium-doped Lutetium-Yttrium Orthosilicate (LYSO) and cerium-doped quartz, are compared to conventional plastic WLS fibres. Finally, an outlook is presented on the possible optimisation of the different components, and the construction and commissioning of a full calorimeter cell prototype is presented.

1. Introduction

The detectors now in operation at the Large Hadron Collider (LHC) at CERN will have to face a challenging environment after the accelerator upgrade to High-Luminosity running (HL-LHC), that is currently being planned to start in 2022. The radiation levels expected there are typically a factor 10 larger, and the integrated luminosities — thus integrated radiation levels — a factor 6 more important than the design values at the LHC. The HL-LHC design parameters at 14 TeV foresee a luminosity of 5×10^{34} cm^{-2} s^{-1}, up to a total integrated luminosity of 30×10^5 pb^{-1} [1]. Typical exposure levels in the high-rapidity region of the calorimeters reach 30 Gy/h in ionising radiation and 2×10^{14} cm^{-2} in integrated fluences of high-energy hadrons. Calorimeters designed for an adequate performance through the years of running at the LHC may need to be partially upgraded, to withstand these increased exposure levels at the HL-LHC.

[1] also at Massachusetts Institute of Technology, Cambridge, Massachusetts, USA

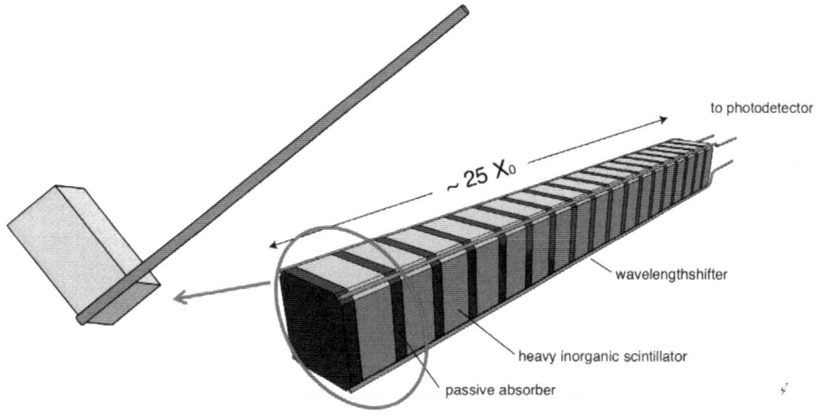

Figure 1. Concept drawing for an innovative sampling calorimeter cell geometry using inorganic scintillator plates as an active material, with wavelength-shifting fibres running along the four long cell edges, and enlarged view on the components used for proof-of-principle measurements.

2. A new sampling calorimeter geometry

Sampling calorimeters are being considered with a renewed interest as an option for HL-LHC detector upgrades. Some inorganic scintillators have been identified, that are likely to perform adequately in the strong radiation field and high particle fluences present during HL-LHC running ([2],[3]). A sampling calorimeter employing them might offer an interesting option in terms of costs and short light path. However, inorganic crystalline materials present mechanical challenges compared to plastic scintillators: this paper describes an innovative light collection geometry that minimises the mechanical processing complexity. It presents first test results demonstrating a successful signal extraction and evaluates possible WLS candidates.

The geometry (Fig. 1) consists of a sampling calorimeter cell made of passive absorber layers (e.g. W or Pb) interleaved with layers of an active medium made of inorganic scintillating crystals. Wavelength-shifting (WLS) fibres run along the four long edges of the stack, transporting the light to photodetectors at the rear. The chamfers along those edges have a width matching the WLS fibre dimensions. On the crystal samples they are depolished to maximise the amount of scintillation light escaping the scintillator and reaching the fibres. The simplicity of the design minimises the mechanical processing complexity, compared to other geometries with grooves or holes machined through the stack.

In this paper, a study has been performed, in order to to show that this concept is working for cerium fluoride (CeF_3) [2] used as the scintillating material. For the test, a CeF_3 crystal has been prepared, with a 3 mm wide and depolished chamfer, as visible in Fig. 2. Coupled to

Figure 2. The CeF_3 crystal sample ($20 \times 20 \times 10$ mm^3) used in this test, provided with a 3 mm wide, depolished chamfer along one of its 10 mm long edges.

it, several different types of materials have been tested as WLS medium, using cosmic muons and laboratory γ sources to excite the CeF_3 scintillation. In particular, materials that might be sufficiently resistant to the HL-LHC radiation environment, such as cerium-doped Lutetium-Yttrium Orthosilicate (LYSO) and cerium-doped quartz, are compared to conventional plastic WLS fibres. The most promising material being studied as a WLS candidate is in fact cerium-doped quartz, which is photo-luminescent in range of wavelengths of the CeF_3 scintillation emission [4].

The presented results provide a proof-of-principle that this kind of light extraction geometry via WLS fibres is viable. Finally, an outlook is presented on a further possible optimisation of the different components and the steps leading to a full calorimeter cell prototype construction are shown.

3. Cerium fluoride as a scintillator for HL-LHC calorimetry

Cerium fluoride is a scintillating crystal whose density (ρ = 6.16 g/cm^3), radiation length (X_0 = 1.68 cm), Molière radius (R_M = 2.6 cm), nuclear interaction length (λ_I = 25.9 cm) and refractive index (n = 1.68) make it a competitive medium for compact calorimeters. For Barium-doped crystals, the emission is centred at 340 nm, with decay time constants of 10 - 30 ns; it is insensitive to temperature changes (dLY/dT (20°C) = 0.08%/^{o}C) as well as bright (4 - 10% of NaI($T\ell$)) and thus suitable for high-rate, high-precision calorimetry [5, 6]. Minimum ionising particles deposit 7.9 MeV/cm in CeF_3, and cosmic μ are thus an interesting means of exciting its scintillation. It should be noted that cerium is a readily available rare earth, which would allow moderating raw material costs, were a mass production envisaged. Its melting point of 1460 ^{o}C allows applying well-known crystal growth technologies. A feature crucial for use in calorimetry at the HL-LHC, is its performance under exposure to considerable fluxes of high-energy hadrons. It has been demonstrated, that crystals can be grown, where the damage fully recovers [2], and thus never builds up as it is the case for lead tungstate [7] and for LYSO [8].

4. The test setup layout and its commissioning

For the presented proof-of-principle studies, a few variations of the same laboratory setup have been used, as visible in Fig. 3. The first setup (Fig. 3, left) foresees one of the big crystal faces to be coupled to a photomultiplier (PMT) of type Hamamatsu H7415 provided with a bialkali photocathode. On all other faces, the crystal was wrapped in Teflon, while the 3 mm chamfer was coupled, via an air gap, to a LYSO bar used here as a wavelength-shifter (WLS), due to its photoluminescence characteristics. The LYSO bar was parallelepipedic, with dimensions $3 \times 3 \times 100$ mm^3 yielding a volume of 0.9 cm^3, of the same order of magnitude as the crystal volume of 4 cm^3. At one end, the naked LYSO bar was coupled to a bialkali-photocathode PMT of type Hamamatsu H8443, while the opposite end was aluminised. Optical screens were introduced, to make sure that no direct scintillation light from the CeF_3 could reach the PMT coupled to the LYSO bar. The PMT output from CeF_3 scintillation light was split into two. One signal was used to trigger for cosmic muons traversing the crystal, in coincidence with the signal from a plastic scintillator placed on top of the setup, with 40×40 mm^2 active area. The other signal was integrated with an Analogue-to-Digital converter (ADC). The PMT signal from the LSYO light output was integrated by a different ADC, and data were acquired on an event-by-event basis. A correlation was observed from this setup, for a measurement with cosmic muons, which is visible in Fig. 4. It is apparent that a WLS signal due to the LYSO photoluminescence is detected.

However, the setup 1 suffers from the drawback that part of the scintillation light from the CeF_3 crystal is not available for producing a WLS signal, because it is directly detected. To overcome this limitation, setup 2 was adopted, with an identical, second CeF_3 crystal for direct scintillation light detection, to make all the light from the top crystal available for wavelength-

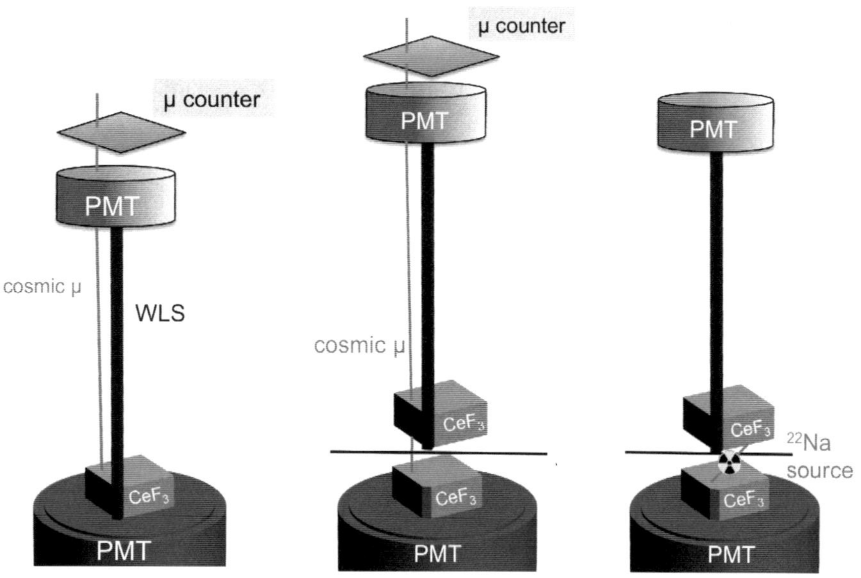

Figure 3. Test setup configurations: Left: setup 1, where a CeF_3 crystal is both, coupled to the bottom PMT for a direct scintillation light output measurement, and to a WLS bar seen by the top PMT. Center: setup 2, where the bottom CeF_3 crystal is coupled to a PMT, and the top crystal is coupled to a WLS bar. Right: setup 3 is used to trigger on annihilation photons from a ^{22}Na source placed between the crystals. Optical screens are used to optically insulate the crystals from one another and from the top PMT.

shifting. Both crystals were wrapped in teflon, except for the side coupled to the PMT or the WLS bar, and they were optically isolated from each other. The resulting correlation and spectra when triggering for cosmic muons are shown in Fig. 5. With a gain $G = 1.5 \times 10^7$, a PMT quantum efficiency of 25% in average over the CeF_3 emission spectrum, and an attenuation of 35 db on the signal, a light output of 300 p.e./MeV has been obtained, corresponding to a light yield $S_{CeF_3} = 1200\ \gamma$/MeV. For the PMT coupled to the LYSO bar, the gain used was 1.3×10^7 and the average quantum efficiency 20%, yielding a WLS light output for this setup of $S_{WLS} = 2.4$ p.e./MeV, thus a light yield of 12 γ/MeV and a WLS to direct light yield ratio $R_\gamma = 1\%$.

A study has also been performed placing a ^{22}Na source between the CeF_3 crystal, triggering

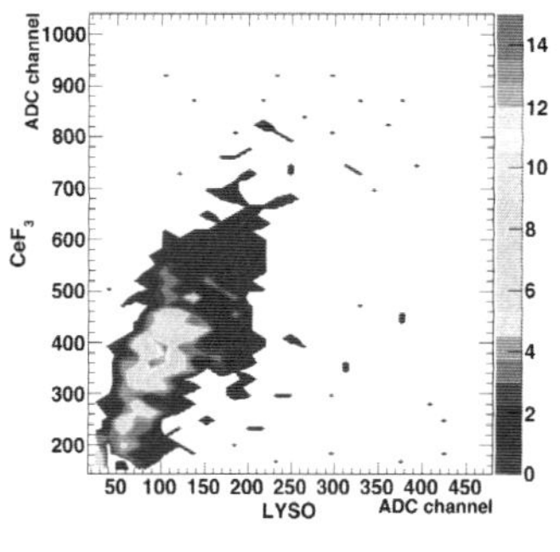

Figure 4. For cosmic muons in setup 1 (Fig. 3), correlation between direct scintillation signal from CeF_3 and WLS signal from the LYSO bar.

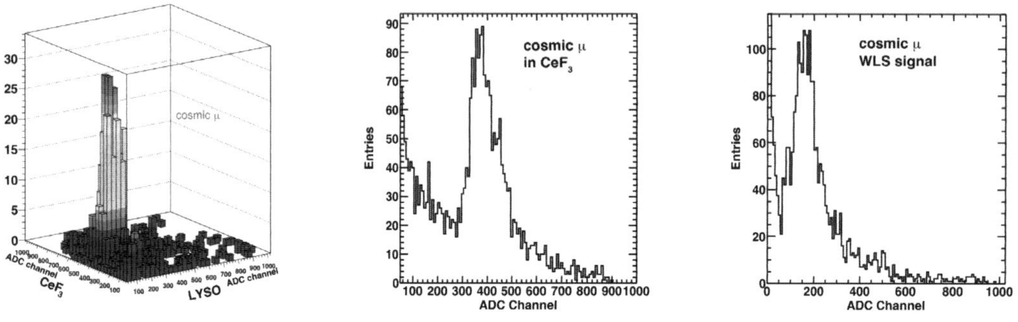

Figure 5. For cosmic muons in setup 2 (Fig. 3), correlation between direct scintillation signal from the bottom CeF_3 and WLS signal from the LYSO bar coupled to the top CeF_3 crystal, and one-dimensional projections of the distributions.

on back-to-back 511 keV annihilation γ quanta (setup 3 in Fig. 3). The correlation and spectra are found in Fig. 6. In the CeF_3 spectrum, one distinguishes well the photoelectric peak from 511 keV annihilation photons, and the Compton background and photoelectric peak for the 1.275 MeV emission which is occasionally detected in coincidence with a back-to-back annihilation photon. The distribution for WLS signals is less striking, due to the small amplitudes involved, but the photoelectric peak and the background at higher ADC channels are clearly distinguishable. One infers from this measurement a light output of 230 p.e./MeV and a light yield $S_{CeF_3} = 940\ \gamma$/MeV. For the PMT coupled to the LYSO bar, one obtains a WLS light output for this setup of $S_{WLS} = 1.4$ p.e./MeV, thus a light yield of 7 γ/MeV and a WLS-to-direct light yield ratio $R_\gamma = 0.8\%$. Taking into account possible systematic effects due to differences in light collection between the two cases, one can consider the results for cosmic muons and γ quanta reasonably compatible.

Direct scintillation signals from γ quanta depositing energy in the LYSO bar itself have also been studied, since the LYSO bar volume is $\sim 25\%$ of the CeF_3 crystal one. Indeed, they have been easily visualised. For this purpose, the signals from the PMT coupled to the LYSO bar

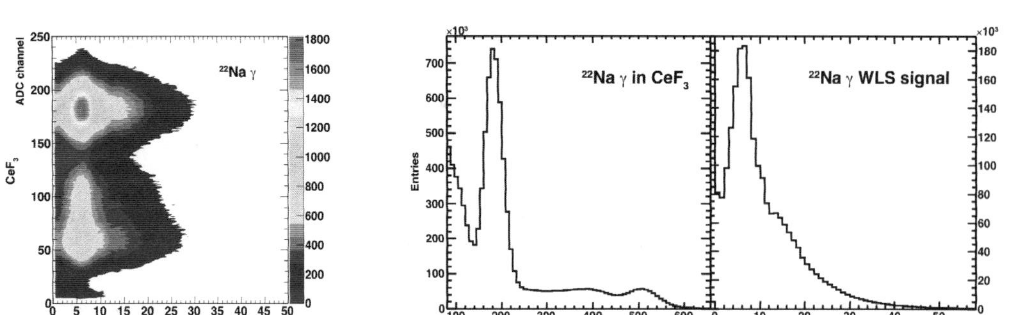

Figure 6. For a ^{22}Na source in setup 3 (Fig. 3), correlation between direct scintillation signal from the bottom CeF_3 and WLS signal from the LYSO bar coupled to the top CeF_3 crystal, and one-dimensional projections of the distributions.

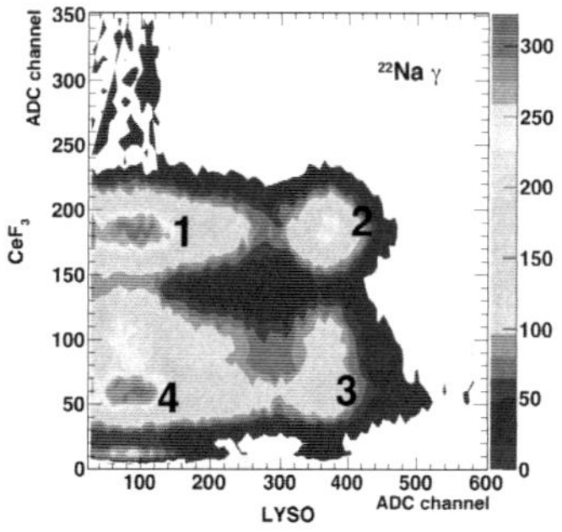

Figure 7. For a ^{22}Na source in setup 3 (Fig. 3), correlation between direct scintillation signal from the bottom CeF_3 and direct scintillation signal from the LYSO bar (see text for details).

have been attenuated by 12 db. The correlation is found in Fig. 7, where the numbers indicate the different topologies that are observed for 511 keV γ: (1) photoelectric effect in CeF_3 and Compton background in LYSO, (2) photoelectric effect in both, CeF_3 and LYSO, (3) Compton background in CeF_3 and photoelectric effect in LYSO and (4) Compton background in both, CeF_3 and LYSO.

Overall, the studies described in this section, making use of a relatively big LYSO bar coupled to a CeF_3 crystal, have allowed to commission the setup and to learn about the performance of the basic elements involved. Photoluminescence from LYSO, excited by the CeF_3 scintillation emission, has been observed, with cosmic μ and ^{22}Na γ quanta leading to a consistent WLS-to-direct signal amplitude ratio. Furthermore, as expected, a direct scintillation signal from ^{22}Na γ quanta interacting in the LYSO bar was observed. The lesson to be learned from these tests is, that for all candidate WLS materials and their volumina, the possible presence of direct scintillation needs to be always studied and its tolerable amplitude optimised in the design.

5. Cerium-doped quartz fibres as WLS candidates

As a material in calorimetry and dosimetry, cerium-doped quartz is known to be quite radiation-hard, a feature that makes it an interesting WLS candidate for HL-LHC applications. For this reason, a first attempt at extracting a WLS signal from such fibres has been performed. The photoluminescence of the fibres considered in this test matches the emission spectrum of CeF_3 [9], as visible in Fig. 8. A fast emission component, as needed for HL-LHC applications, had been previously observed [4]. A set of fibres was received, that were developed for dosimetry applications, and as such had been optimised for radioluminescence [10]. These fibres had not been optimised yet for photoluminescence, nor had its fast emission component been maximised.

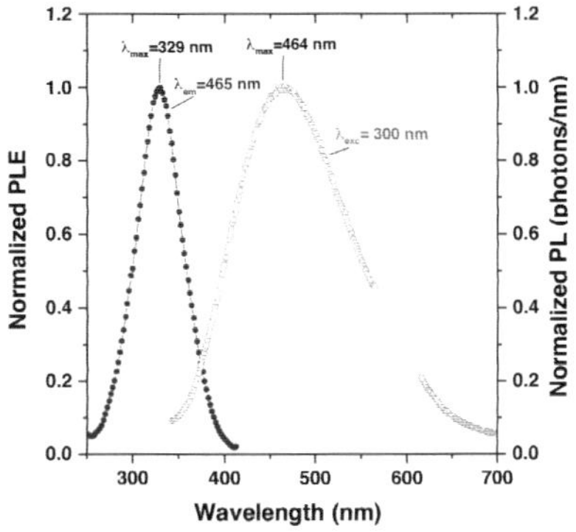

Figure 8. Photoluminescence emission and excitation curves for cerium-doped quartz fibres used in this test [4, 9].

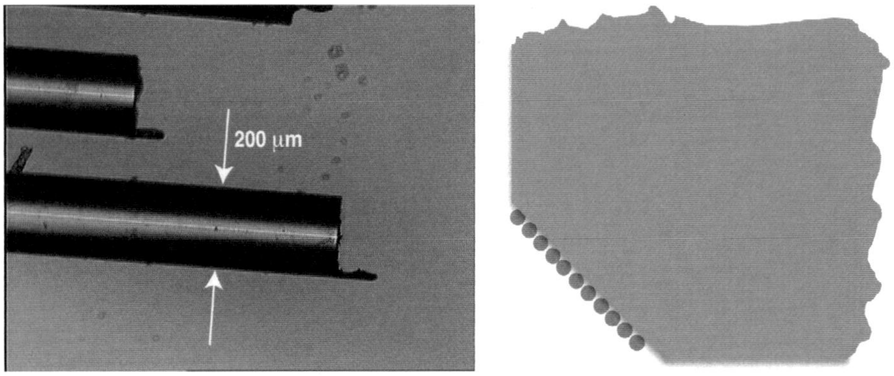

Figure 9. Image of the Ce-doped quartz fibre ends, as seen under the microscope (left) and concept view of the fibre arrangement along the 3 mm wide crystal chamfer (right).

Furthermore, they had been grown in an oxidising atmosphere, and were thus rich in Ce^{4+}, which is optically inactive but highly absorbing, while poor in photoluminescent Ce^{3+} content. The fibres received had a diameter of 200 μm, they had no cladding, its end were, out of necessity, simply cut with a razor blade and were not polished neither aluminised (Fig. 9). To maximise the light collection, 12 fibres, 10 cm long, were laid along the crystal chamfer in a single-layer ribbon (Fig. 9) replacing the LYSO bar in setup 2 (Fig. 3). An air gap of a few millimetres was left between the ends of the fibres and the PMT window. A modest, if at all, signal was expected.

The observed correlation for cosmic muons, between direct scintillation signal from CeF_3, and WLS signal from the ribbon of fibres, is visible in Fig. 10, together with one-dimensional projections. For the WLS signal, the direct scintillation amplitude was required to fall between ADC channels 300 and 550. The ratio of direct-to-WLS light detected was 0.3% for this setup. Although the ratio is very modest, taking into account all the limitations of the present test, it can be considered promising in view of an optimisation of such fibres characteristics. Due to the total fibre volume of only 0.037 cm^3, the direct scintillation signal in the fibres is very small,

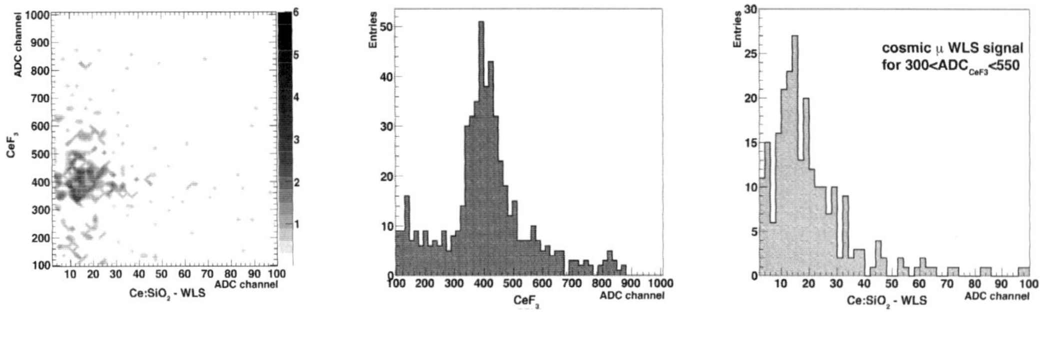

Figure 10. For cosmic muons in setup 2 (Fig. 3), correlation between direct scintillation signal from the bottom CeF_3 and WLS signal from a ribbon of Ce-doped quartz fibres coupled to the top CeF_3 crystal, and one-dimensional projections where, for the WLS signal spectrum, amplitudes corresponding to $300 < ADC_{CeF_3} < 550$ were required.

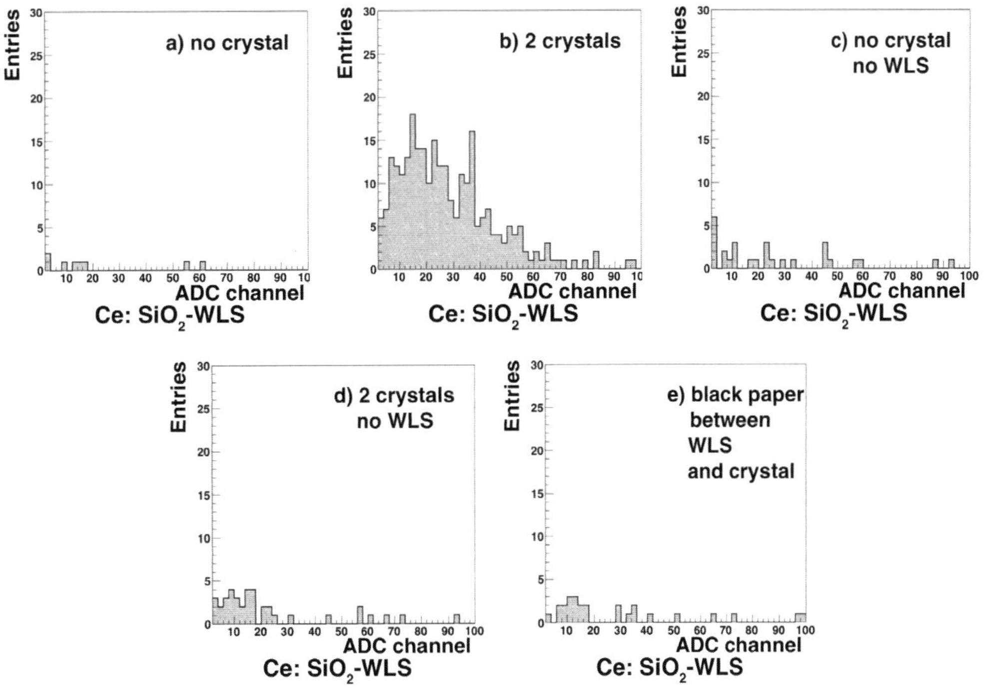

Figure 11. For the same duration of data taking for cosmic muons, WLS signal spectra for Ce-doped quartz fibres in setup 2 (Fig. 3), for several cross-check configurations to the results in Fig. 10: a) crystal removed; b) 2 crystals coupled to the fibres ribbon; c) no crystal, no fibres present; d) 2 crystals, but no fibres present and e) crystal optically isolated from the WLS fibres.

and remained undetectable in this setup.

A number of cross-check measurements were performed, always for the same duration of data taking, to ascertain that the observed correlation is indeed due to a WLS signal. The results match the expectations. In the test whose resulting WLS signal spectrum is visible in Fig. 11a), the crystal coupled to the WLS fibres ribbon was removed from the setup: the spectrum is compatible with no WSL signals detected. In Fig. 11b), the WLS signal spectrum was taken for two crystals, sitting on top of each other, both coupled to the WLS fibres ribbon. The mean of the distribution is higher and its width broader than for a single crystal, although slightly different from the exact convolution of two identical distributions as of Fig. 10, due to differences in chamfer dimensions and crystal light yield. When the crystals and also the WLS fibres ribbon were removed (Fig. 11c), the distribution was found to be compatible with no signal. Similarly, this is also the case when only the WLS fibres were removed (Fig. 11d) and when a black paper stripe was inserted between the crystal chamfer and the WLS fibres ribbon (Fig. 11e). The results of these cross-checks measurements confirm the observation of Fig. 10, that a WLS signal is observed, for a cerium fluoride crystal coupled to Ce-doped quartz fibres.

6. Tests of 3HF Kuraray fibres as WLS

The construction of a prototype calorimeter cell, to be tested in a particle beam, following the proposed readout concept, was being prepared. To allow for a WLS signal readout, while R&D is being performed on Ce-doped quartz fibres, the temporary use of conventional plastic WLS fibres was envisaged. A suitable candidate was found in 3HF fibres from Kuraray, that are known to be

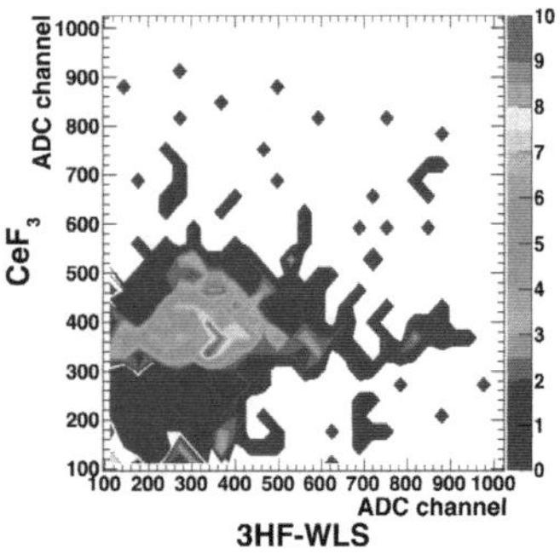

Figure 12. For cosmic muons in setup 2 (Fig. 3), correlation between direct scintillation signal from CeF_3 and WLS signal from a Kuraray 3HF fibre.

photoluminescent over the range of cerium fluoride scintillation emission wavelengths [12, 13]. Single-clad fibres, 1 mm in diameter, were chosen, since the cladding material, consisting of PMMA, is known to be at least partially transmitting over the wavelength range of the cerium fluoride scintillation emission [14], while no data were available for the proprietary outer cladding used on double-clad fibres, which consists of a fluorinated polymer. For this first test, a single 3HF fibre, 1 mm in diameter, single-clad, with no aluminised end, was used. An air gap was present between the fibre and the PMT. The correlation in Fig. 12 shows the detection of a WLS signal. The ratio of direct-to-WLS light output is at the level of 0.5%, an amplitude which corresponds to expectations, due to the measurement conditions, which have not been fully optimised yet.

7. Prototype cell construction and commissioning

Based on the successful proof-of-principle measurements described above, a prototype cell has been constructed for tests in high-energy particle beams, whose 3-dimensional drawing is presented in Fig. 13. The dimensions were optimised for a $\sim$ 500 MeV electron beam. Ten samplings were foreseen, with tungsten absorber plates of 3.1 mm thickness and cerium fluoride crystals of 10 mm thickness (Fig. 14), for a total of 17 X_0, while transverse dimensions of 24×24 mm^2 were chosen. It should be noticed that for this configuration the effective Molière radius is $R_M = 23$ mm, but a smaller value can be easily obtained through a different choice of relative absorber and scintillator thickness. For example, for 2.5 mm thick tungsten plates and 1.5 mm thick cerium fluoride crystals, $R_M = 14$ mm is achievable.

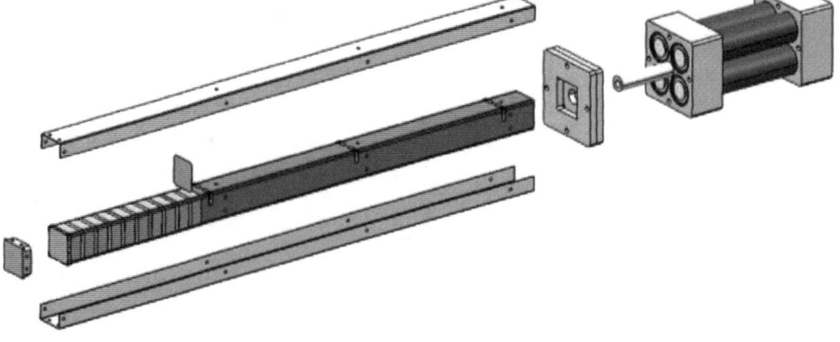

Figure 13. Concept drawing for a complete calorimeter sampling cell, in an exploded view.

Figure 14. Prototype calorimeter cell construction elements: single cerium fluoride crystal (left), set of tungsten plates (centre) and prototype during assembly, showing the ten samplings and the WLS fibres running along the edges (right).

The mechanical support structure was conceived to be easily expandable in depth for beam tests at higher energies. The samples were all provided with 3 mm chamfers along the four 10 mm long corners, and a Kuraray 3HF single-clad fibre was running along each one of the long cell edges. For the test, each fibre was coupled to a Hamamatsu PMT with a bialkali photocathode of type R1450, within a H6524 assembly, and it was aluminised at the end opposite to the photodetector.

Prior to an exposure to high-energy particle beams, the prototype was commissioned in the laboratory, by acquiring signals from cosmic muons required to traverse the cell along its length. This was achieved by placing the cell vertically upright, and triggering with scintillators placed on top and below the cell. Muons going only partially through the length of the cell were not excluded, due to a trigger counter size slightly larger than the transverse cell dimensions. The correlation between sums of signals coming from WLS fibres running along diagonally opposite edges is found in Fig. 15. This being raw data, the slight deviation from a unitary slope is due to the differences in quantum efficiency folded with the gain of the photomultiplier tubes. The measurement allowed to ascertain the functioning of all components and to make sure, that signals of detectable amplitude are generated.

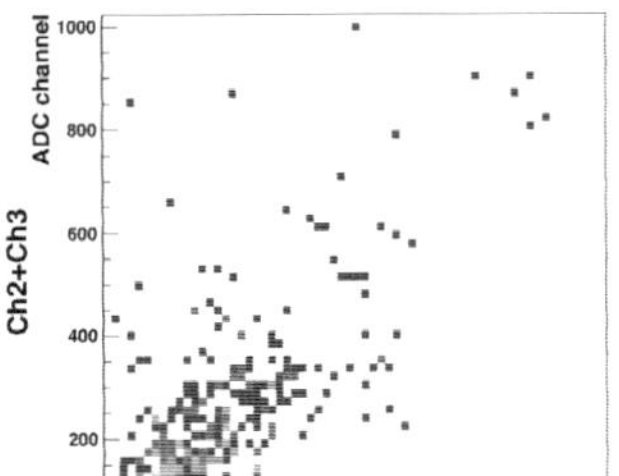

Figure 15. For cosmic muons going through the length of the prototype cell of Fig. 13, correlation between the two sums of WLS signals from fibres running along two diagonally opposite edges.

8. Conclusions

This paper summarises the work that has been performed so far, in pursuing cerium fluoride scintillating crystals used in a simplified sampling calorimeter geometry. The proposed concept has wavelength-shifting fibres running along the four edges of the stack, collecting light through slightly depolished crystal chamfers.

In proof-of-principle laboratory tests, wavelength-shifted signals have been detected for conventional fibres from Kuraray, and for Ce-doped quartz fibres. While the fibres need to be further optimised for HL-LHC calorimetry applications, a full prototype cell has been built and commissioned in the laboratory with cosmic muons, before tests in a high-energy beam.

The presented measurements demonstrate, that this principle with its components has the potential to be used for a high-rate, radiation resistant sampling calorimeter, as needed for HL-LHC upgrades, in a mechanical assembly that allows a simple construction.

Acknowledgments

The excellent support of the ETH Zurich technical staff is acknowledged, in particular M. Dröge, C. Haller and U. Horisberger, as is the one of CERN, by R. Dumps, T. Schneider and M. Van Stenis. We are grateful to A. Vedda, N. Chiodini and M. Fasoli (U. Milano Bicocca), and to Kuraray (Japan) for kindly providing us with samples for this test. We also acknowledge the effort made by Hamamatsu (Japan) and Tokuyama (Japan), to provide us with the needed parts in time for meeting a stringent construction schedule. This work was performed with the support of the Swiss National Science Foundation.

References

[1] M. Lucchini (CMS Collab.), "Evolution of the response of the CMS ECAL", Proc. IEEE Nuclear Science Symposium (Seoul, S. Korea, 2013) and CMS-CR-2013-401 (CERN, Geneva, Switzerland).
[2] G. Dissertori, P. Lecomte, D. Luckey, F. Nessi-Tedaldi, F. Pauss, Th. Otto, S. Roesler, Ch. Urscheler, Nucl. Instr. and Meth. Phys. Res. A 622 (2010) 41-48.
[3] J. Chen, R. Mao, L. Zhang, R.-Y. Zhu, IEEE Trans. Nucl. Sci. 54, (2007) 1319-1326.
[4] A. Vedda, N. Chiodini, D. Di Martino, M. Fasoli, S. Keffer, A. Lauria, M. Martini, F. Moretti, G. Spinolo, M. Nikl, N. Solovieva and G. Brambilla, Appl. Phys. Lett., Vol. 85 (2004) 6356.
[5] D. F. Anderson, Nucl. Instr. Meth A 287 (1990) 606-612.
[6] E. Auffray, S. Baccaro, T. Beckers, Y. Benhammou, A. N. Belsky, B. Borgia, D. Boutet, R. Chipaux, et al., Nucl. Instrum. and Meth. Phys. Res. A 383 (1996) 367-390.
[7] M. Huhtinen, P. Lecomte, D. Luckey, F. Nessi-Tedaldi, F. Pauss, Nucl. Instr. and Meth. A 545 (2005) 63-87.
[8] G. Dissertori, D. Luckey, F. Nessi-Tedaldi, F. Pauss, M. Quittnat, R. Wallny, Nucl. Instr. and Meth. Phys. Res. A 745 (2014), 1-6.
[9] A. Vedda, N. Chiodini, M. Fasoli, private communication.
[10] Courtesy of A. Vedda and N. Chiodini, University of Milano-Bicocca (Italy).
[11] M. Fasoli, A. Vedda, A. Lauria, F. Moretti, E. Rizzelli, N. Chiodini, F. Meinardi, M. Nikl, J. Non-Cryst. Sol. 355 (2009) 1140-1144.
[12] Kuraray, Japan, 3HF scintillation fibres technical specifications at http://kuraraypsf.jp/psf/sf.html, visited on May 15th, 2014.
[13] Kuraray, private communication.
[14] D. C. Miller et al., Proc. SPIE Solar Energy and Technology Conference, San Diego (USA, 2009).

Study of a 3x3 module array of the ECAL0 calorimeter with an electron beam at the ELSA

M Dziewiecki[1*], N Anfimov[2], V Anosov[2], J Barth[3], V Chalyshev[2], I Chirikov-Zorin[2], D Elsner[3], V Frolov[2], F Frommberger[3], A Guskov[2], F Klein[3], Z Krumshteyn[2], R Kurjata[1], J Marzec[1], A Nagaytsev[2], A Olchevski[2], I Orlov[2], A Rybnikov[2], A Rychter[1], A Selyunin[2], K Zaremba[1], M Ziembicki[1]

[1] Warsaw University of Technology, Institute of Radioelectronics, Nowowiejska 15/19, Warsaw 00-665, Poland
[2] Joint Institute for Nuclear Research, Joliot-Curie 6, Dubna 141980, Russia
[3] Physikalisches Institut der Universität Bonn, Nussalee 12, Bonn 53115, Germany

E-mail: M.Dziewiecki@ire.pw.edu.pl

Abstract. ECAL0 is a new electromagnetic calorimeter designed for studying generalized parton distributions at the COMPASS II experiment at CERN. It will be located next to the target and will cover larger photon angles (up to 30 degrees). It is a modular high-granularity Shashlyk device with total number of individual channels of approx. 1700 and readout based on wavelength shifting fibers and micropixel avalanche photodiodes. Characterization of the calorimeter includes tests of particular sub-components, tests of complete modules and module arrays, as well as a pilot run of a fully-functional, quarter-size prototype in the COMPASS experiment. The main goals of the tests on low-intensity electron beam at the ELSA accelerator in Bonn were: to provide energy calibration using electrons, to measure angular response of the calorimeter and to perform an energy scan to cross-check previously collected data. A dedicated measurement setup was prepared for the tests, including a 3x3 array of the ECAL0 modules, a scintillating-fibre hodoscope and a remotely-controlled motorized movable platform. The measurements were performed using three electron energies: 3.2 GeV, 1.6 GeV and 0.8 GeV. They include a calibration of the whole detector array with a straight beam and multiple angular scans.

1. The detector

1.1. ECAL0 at COMPASS II

ECAL0 is a new electromagnetic calorimeter to be built for COMPASS II experiment at CERN for Generalized Parton Distribution (*GPD*) study. It's going to be mounted as one of the most upstream detectors, right next to the fixed target and before SM1 magnet (see figure 1). It will cover particle angles from 5 to 30 degrees and energy range from 0.2 up to 30 GeV.

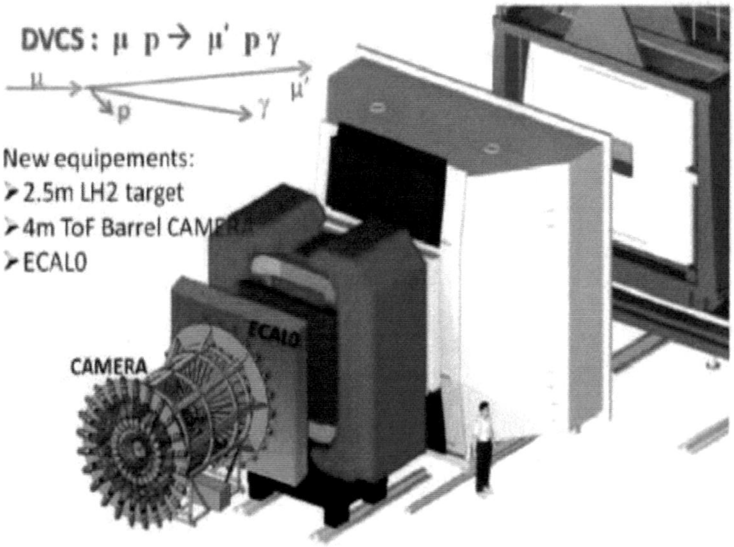

Figure 1. Upstream detectors of COMPASS II experiment. The target is installed inside the CAMERA detector.

1.2. Detector design

ECAL0 is a modular, multi-channel, sampling electromagnetic calorimeter with external dimensions of approximately 2x2m. It consists of 194 9-channel modules, giving a total of 1746 channels. The modules are aligned in an almost round pattern with rectangular beam window (see figure 2).

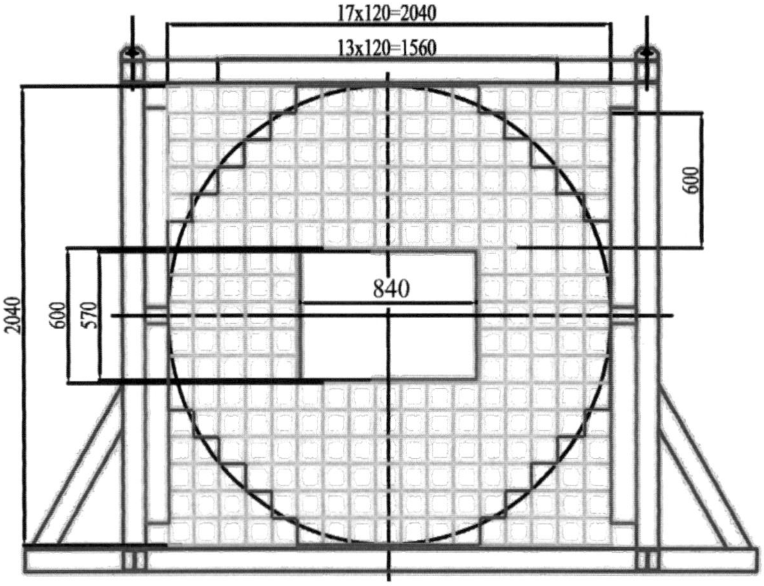

Figure 2. Front view of the detector.

The position of the detector in the experiment gave some constraints on its construction:

- due to proximity to SM1 magnet, the detector must be insensitive to magnetic field;
- the depth of the device is limited to 50 cm;
- the power consumption should be as low as possible due to limited cooling capabilities.

1.3. Colorimeter module

A single detector module consists of two main parts: calorimeter part and readout assembly (figures 3, 4, 5). They are detachable for service and for transportation.

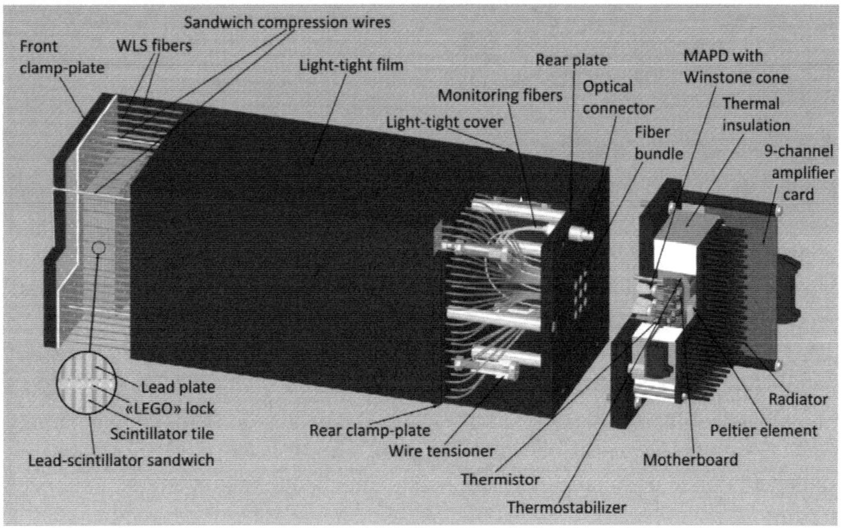

Figure 3. Overview of the detector module.

The calorimeter part (figure 4) has nine independent channels having a form of a shashlyk with 1.5 mm thick polystyrene scintillators and 0.8 mm thick lead plates, giving a total thickness of 25.2 cm with 109 layers. The Moliere radius of the detector is 3.5 cm while a single channel has a cross-section of 3 x 3 cm. The light from the scintillator plates of each channel is collected by 16 Bicron BCF91AMC WLS fibres (1.2 mm), forming a 6.5 mm bundle at the output, along with 1 clear fibre for LED monitoring.

The readout part (figure 5) incorporates photodetector and accompanying electronics as well as Peltier-based temperature stabilization system. All electronics is divided into three PCBs:

- the motherboard with photodetectors and some passive components for power supply
- slow control board with bias voltage generator (approx. 80 V), power monitoring and temperature control circuits;
- analog front-end for signal shaping and transmission over twisted pairs.

The temperature of the photodetector is kept at a constant level set around 16-18°C to keep the noise on a sufficiently low level and maintain a constant, well defined electron gain.

Figure 4. Calorimeter part of the module (unsealed).

Figure 5. Readout part of the module with Winston cones, photodetectors, electronics and Peltier cooler.

1.4. Photosensors

The scintillation light is detected by multipixel avalanche photodiodes. Generally, Zecotek MAPD-3A and MAPD-3N were used for various tests of the detector, with MAPD-3N intended to be used in the final design. However, Hamamatsu MPPC S12572-10P is also considered to be used in the detector due to its low noise and reasonable price. Table 1 presents the comparison of these two products.

Table 1. Basic parameters of MAPD-3N and S12572-10P MPPC.

	Zecotek MAPD-3N	Hamamatsu S12572-10P
Size, mm^2	3 x 3	3 x 3
Pitch, μm	8 (size: 5)	10
Number of pixels	135 000	90 000
Bias voltage, V	90	70
Gain · 10^5	3-5	14
PDE, % (λ=520 nm)	30	10

Generally, MPPC has slightly lower pixel density and much lower photon detection efficiency (PDE), but it's still competitive to MAPD-3N due its high gain, low noise and short recovery time.

1.5. Signal read-out

The readout path of the detector consists of two main parts: front-end amplifiers and ADC converters (figure 6). The amplifiers are located directly on the readout modules. They provide 4th order signal shaping and transmission over a 110Ω twisted pair. This is done by two op-amps (the second one is symmetric and acts as the line driver) powered by a single +5V supply. The peaking time of the output pulses is 40 ns and is selected to be optimal for 80MHz sampling.

The analog-to-digital conversion is performed by external Mezzanine Sampling ADC (MSADC) boards installed in a 9U VME crate. Each board has 4 16-channel modules with 12-bit resolution and 80 MHz sampling rate. One MSADC board covers 7 calorimeter modules which occupy 63 channels. The remaining channel is used for debug purposes – either as analog sum of 7 modules or as LED monitor.

A patch board is designed between amplifiers and MSADC for the final design. Its main aim is to interface round, 10-pair cables from modules to 17-pair ribbons to be connected to MSADCs. It will also allow easier detector disconnecting and provide analog sum of all detector channels for trigger generation. For the prototype, slightly different patch boards were used, which were installed directly on MSADC connectors, without ribbon cables.

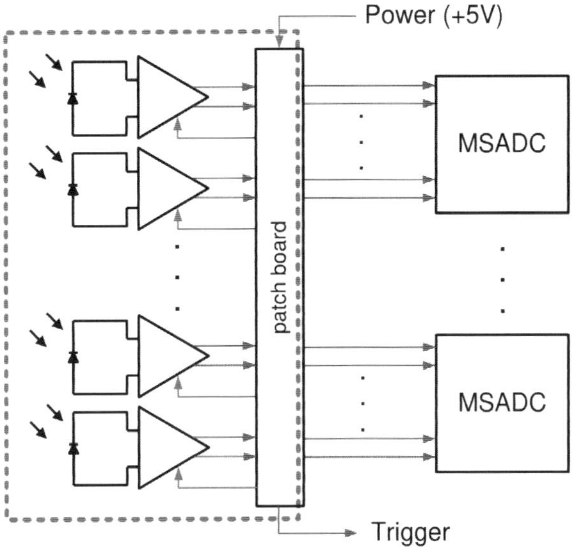

Figure 6. Block diagram of the readout electronics.

2. Previous detector tests

2.1. Prototypes and test setups

Two types of detector setup were used for various tests of ECAL0 modules. The first one is a 81-channel, 9-module (3 by 3 modules) matrix used for measuring various parameters of the modules, including the tests performed in Bonn (figure 7). The second one is a fully functional, quarter-size detector which was installed in 2012 on Compass beam to check overall detector capabilities.

Figure 7. 3 x 3 (81-channel) setup used for various tests, including the tests in Bonn (left) and quarter-size detector for tests on Compass beam in 2012 (right).

2.2. Calibration at CERN

Before starting the experiment in Bonn, some experiments in CERN were already completed to test the detector. They included:

- tests on cosmics and muons (9-module matrix) to get energy calibration, noise performance and energy resolution;
- a pilot run of a quarter-size detector prototype within COMPASS experiment in 2012 with π^0 calibration;

Also, extensive tests of photodetectors, readout electronics, slow-control as well as tests of single modules were performed.
Some data used for energy calibration are shown in figures 8 and 9.

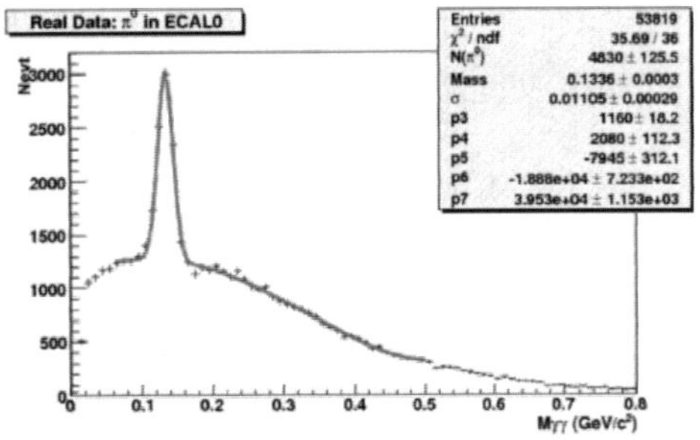

Figure 8. π^0 peak in ECAL0.

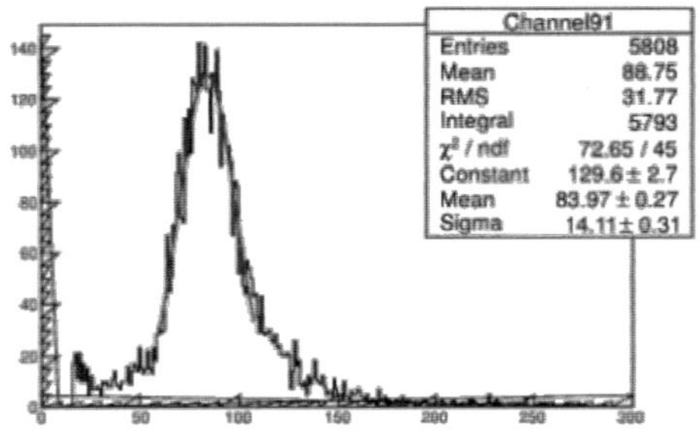

Figure 9. The muon peak.

3. Detector tests on electron beam in Bonn

3.1. Reasons and goals

The main goals for the tests to be done in Bonn at the ELSA were:

- to calibrate the energy response of the detector for electrons;
- to check the angular response of the detector over a range of 0 to 36 degrees with 6-degree step;
- to perform an energy scan over three points (3.2, 1.6 and 0.8 GeV) to cross-check previously collected data.

3.2. Measurement setup

The measurement setup for tests in Bonn (figure 10) consisted of three main parts:

- a low-intensity electron beam source, the ELSA accelerator;
- a Sci-Fi hodoscope for system triggering, high-resolution electron position readout and beam monitoring;
- a 9-module ECAL0 matrix installed on an automated moving platform for spatial and angular scanning.

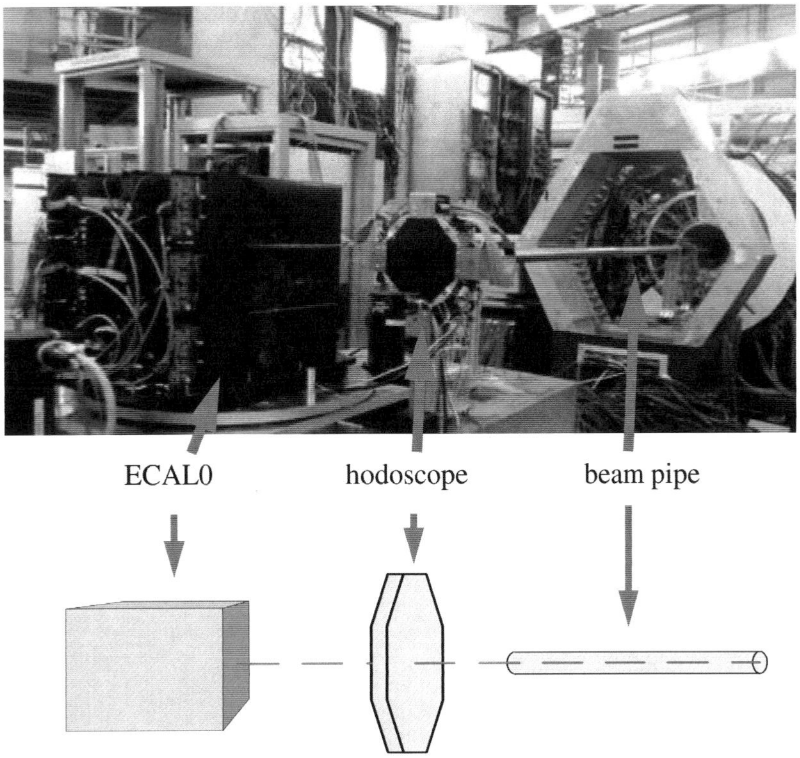

Figure 10. The measurement setup.

The first element, namely the ELSA accelerator, is capable of supplying an electron beam with energy of up to 3.5 GeV for various aims. For ECAL0 tests, the accelerator was prepared to provide a low intensity beam (approx. $2\cdot10^3$ particles per second, which gives as little as one electron per 1000 beam rotations). The experimantal setup was installed on a position normally occupied by the Crystal Barrel (CB) experiment. In order to make space and provide beam, some parts of the Crystal Barrel experiment were removed. Also, some parts of the beamline were removed or re-ordered. The main changes were:

- removing parts of CB experiment: TAPS, target system, photon flux detector;
- removing parts of beamline (cleaning magnet, collimator system);
- adding an evacuated beam pipe between the tagging system and the hodoscope.

The main problem was that normal beam monitoring was not possible during run due to low beam intensity and the fact that ECAL0 was installed between the beam output and the beam monitor. Thus, a special beam positioning scheme was applied:

- first, the accelerator was tuned on high intensity beam and with ECAL0 removed from the setup; after getting right beam shape, appropriate settings of the accelerator control were frozen;
- after tuning, the accelerator was switched off and re-programmed for lower intensity, wile ECAL0 was inserted into its working position;
- for the normal run, the accelerator was running on frozen settings, the beam intensity was cotrolled on-line by means of the hodoscope's trigger output, and the beam size and position was only manually checked by observing beam profiles collected by the hodoscope.

Journal of Physics: Conference Series **587** (2015) 012040 doi:10.1088/1742-6596/587/1/012040

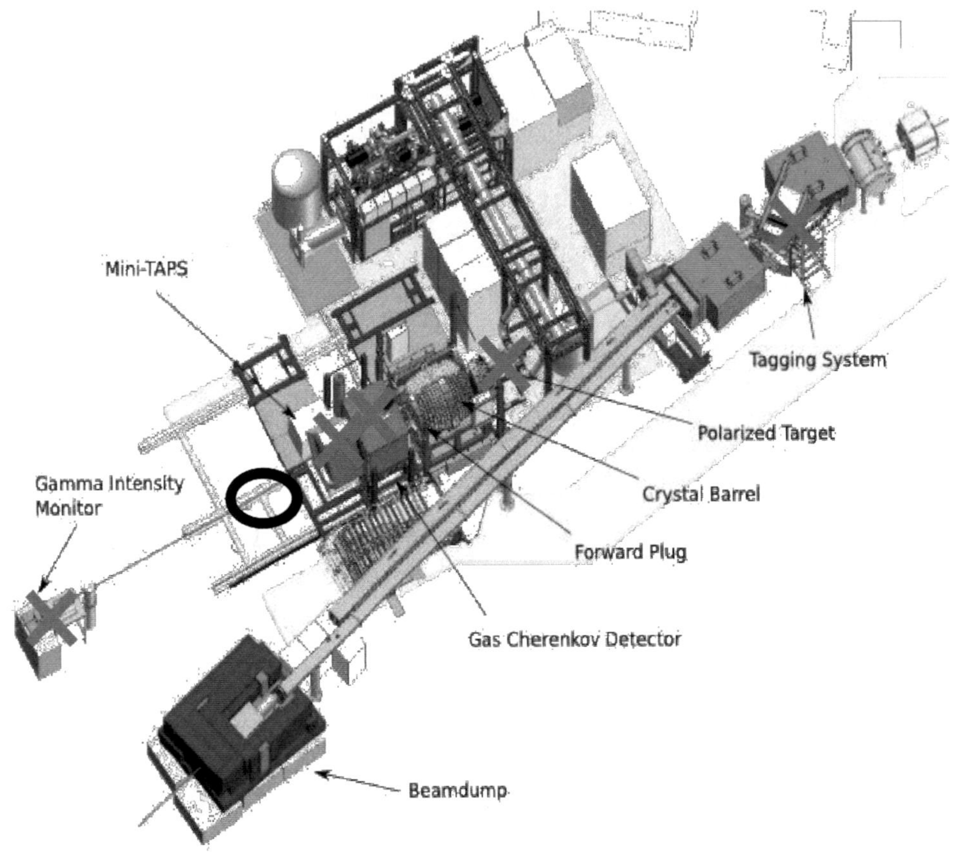

Figure 11. CB-part of the ELSA system; removed components are shown with crosses; the ECAL0 position is marked with black circle.

The Sci-Fi hodoscope was designed specially for the tests and will be described in more detail in the following paragraph.

The moving platform (kindly supplied by Giessen University) allowed high-precision positioning over two axes. It was remotely controlled via a Windows PC with Lab-View software and remote desktop. For detector rotation, a manual rotating stage was designed in Bonn and mounted on the top of the moving platform. The rotating stage allowed to fix the detector at an angle of -36 to +36 degrees with 6-degree step, as well as at +/-90° and 180°. Thanks to mechanical locking system, the positioning error was negligible.

3.3. Program of the tests

The program of the tests contained two parts. The first part was to scan the whole matrix (81 channels) with 3.2 GeV beam. Generally, one measurement point per one channel was collected (40 mm step), although few channels were scanned with 10 mm step to check the spatial uniformity of the response. The second part was the angular scan, which was performed for three different energies: 3.2, 1.6 and 0.8 GeV. Seven angle settings (0-36 degrees with 6° step) and 15 various beam positions were scanned for each energy. There were about 400 runs performed in total, which gave approx. 250 GB of raw data.

4. The Sci-Fi hodoscope

4.1. Design
The main aims to use the hodoscope were accurate read-out of the position of incident electrons and system triggering. It was used also for accelerator control, as beam intensity and shape monitor.
It was designed as two-plane Sci-Fi detector with approx. 23 x 23 mm active area and 32 channels in each plane, giving position resolution of 0.7 mm with fibre diameter of 1.2 mm. Each detector channel consists of one fibre only, so aluminium mirrors were made at the blind ends of the fibres to increase light output (the actual light output is 16 photons/MIP). Four 16-channel photomultipliers (Hamamatsu H-8711) were used as photodetectors. PMT power is supplied by an external HV power supply.
All heavier detector parts (PMTs, electronics) were mounted to a rigid aluminium frame (figure 13), while the fibre array has a support made of extruded polystyrene. Plastic covers with thin beam windows made of kapton were used to get light tightness of the whole frame. Additional covers were made to protect electronic circuits (see figure 12).

Figure 12. Complete Sci-Fi detector.

Figure 13. The detector without covers and cables. Fibres are visible inside.

4.2. Detector read-out
Two kinds of output signals are provided at the output of the Sci-Fi detector:
- channel signals for position recognition (analog);
- output trigger signal (digital).

The channel output is designed to be compatible with MSADC in terms of amplitude and signal shape, which corresponds to approx. 1 V amplitude, 40 ns pulse peaking time and differential output with 34-pin ribbon cables. It was decided to use MSADC for Sci-Fi readout to simplify the DAQ system, as the same boards were used for ECAL0.
The trigger signal is provided via LEMO connector in NIM standard to be directly connected to NIM-based trigger system of the experiment.
All the electronics for both outputs is located directly on the detector's frame to make it possibly compact. All circuits are powered from a single +5 V line consuming approx. 0.5 A.

5. Measurement Results

5.1. Beam Profiles

The first measurements concerned not the ECAL0 detector itself, but the beam spot. It turned out that the beam is very narrow for lower energies (figures 16 and 17) while a tail on X-axis is visible at 3.2 GeV, giving an increase of beam size to approx. 3 mm (σ). This is due to accelerator's intensity controlling method, which includes moving the beam relative to its nominal position in the synchrotron ring. Nevertheless, the beam size was still within desired range. The effect of the beam tail takes place also at lower energies, but is relatively smaller.

Table 2 presents beam sizes for all used energies.

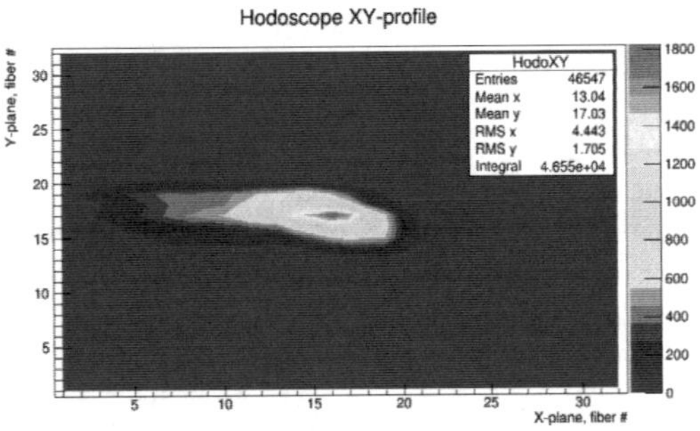

Figure 14. Beam shape at 3.2 GeV.

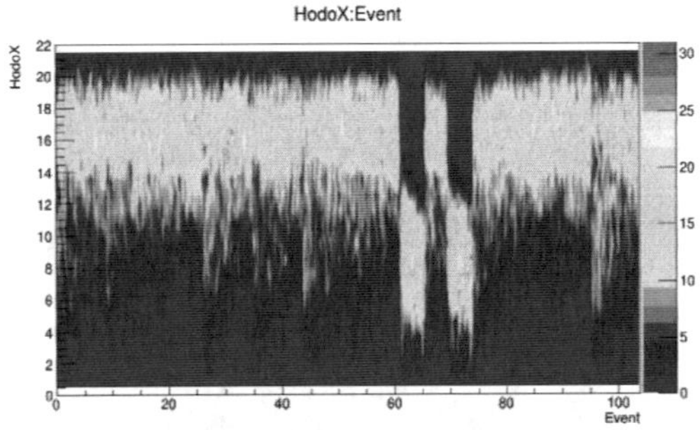

Figure 15. Beam position changes over time.

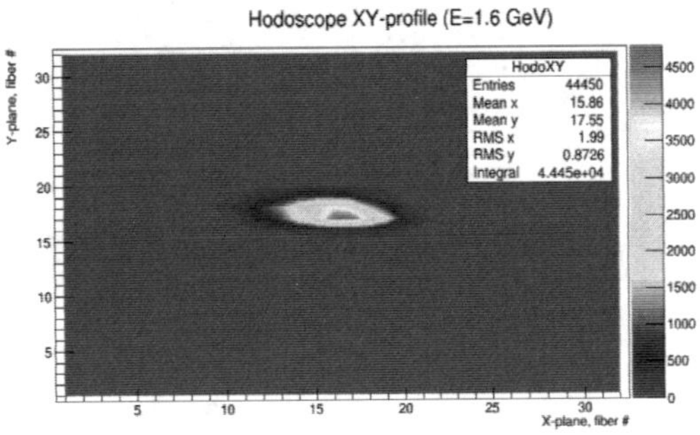

Figure 16. Beam shape at 1.6 GeV.

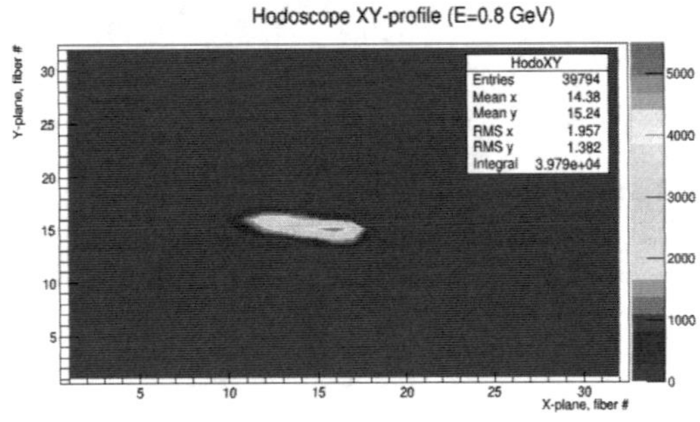

Figure 17. Beam shape at 0.8 GeV

Table 2. Beam size for different energies.

	3.2 GeV	1.6 GeV	0.8 GeV
X profile σ [mm]	3.1	1.4	1.3
Y profile σ [mm]	1.3	0.5	0.5

5.2. Energy calibration and linearity

Figure 18 presents the energy response of the detector on electrons. Three measurement points (0.8, 1.6 and 3.2 GeV) are consistent with Monte-Carlo simulations. The energy resolution (figure 19) is bigger than the expectations by approx. 20%, but it's still within desired limit of $10\% / \sqrt{E}$ [GeV].

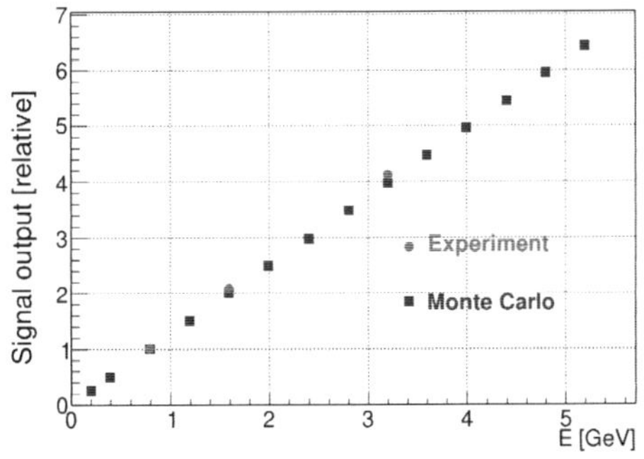

Figure 18. Energy response of the detector.

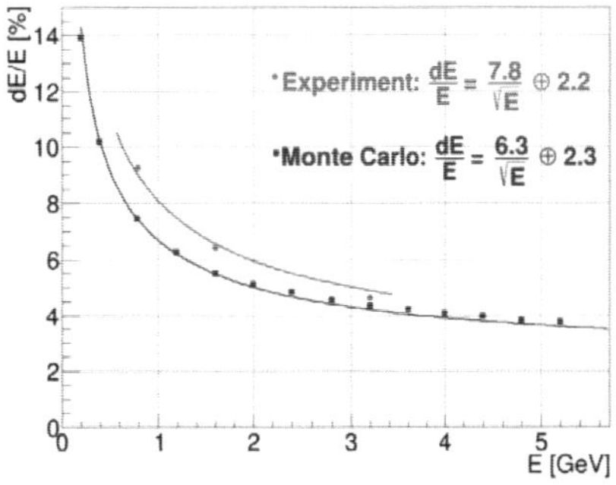

Figure 19. Measured energy resolution

5.3. Angular response

The results of the angular scan are sumarized in figure 20. The main conclusions are that:

- the energy deposit is almost independent of the incident angle;
- the energy resolution remains virtually constant over the angle.

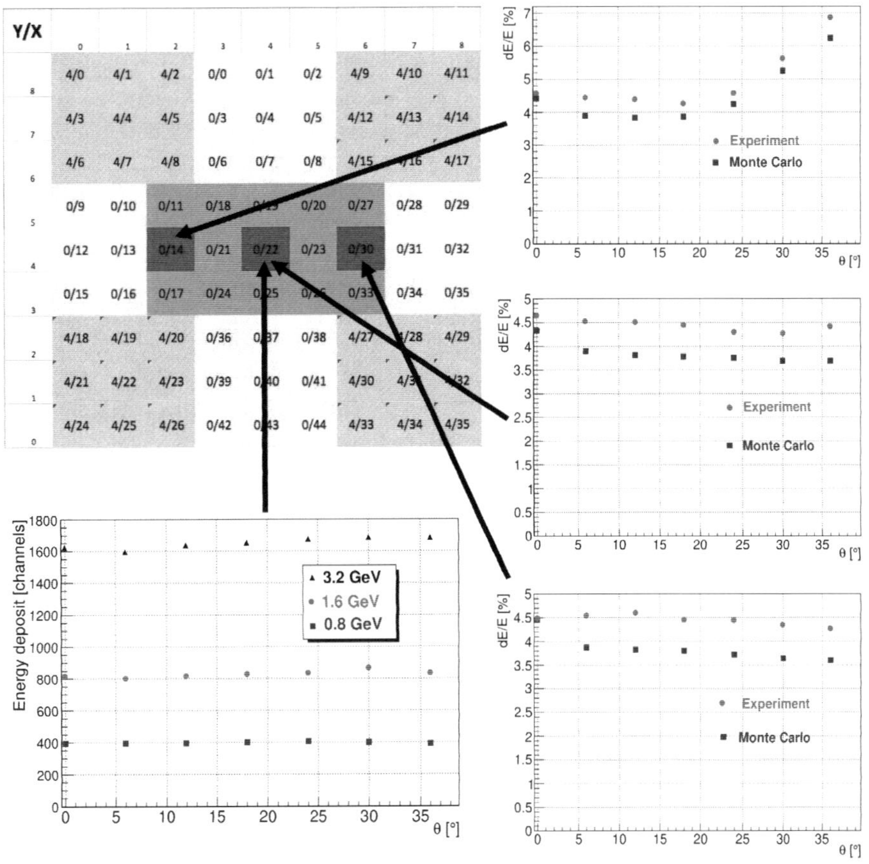

Figure 20. Energy response and resolution of the detector for different angles.

The increase of the noise in the upper right figure is an apparatus effect. The data presented comes from a cell which is close to the edge of the detector and there's a leakage of the particle shower for higher energies. The resolution decrease is due to fluctuations in the leakage from event to event.

6. Conclusions and further actions

The main outcome from the tests in Bonn is that:

- the calibration results from previous tests from 2011 were confirmed;
- measured energy resolution is within desired level and close to expectations based on Monte-Carlo simulations;
- the angle dependence of energy response and resolution is small and also consistent with Monte-Carlo.

Also, it was confirmed that a low-intensity electron beam for detector testing can be smoothly provided by ELSA. A special experimental area is considered to be developed for various detector tests not to interfere with ELSA's everyday research program.
A very good performance of built Sci-Fi tracker was confirmed. It will be used for further ECAL0 tests and probably also as a part of low-intensity beam monitoring system at the ELSA.

The next ECAL0 test to be done is planned for October 2014 at CERN T10 test beam. For these tests, installing MPPC photosensors instead of MAPD-3N is being considered, which needs minor hardware changes in the detector modules. The tests of new photodetectors are in final phase.

7. References

[1] Anfimov N et al. 2010 *Nucl. Instr. and Meth.* A **617** 78
[2] Anfimov N et al. 2013 *Nucl. Instr. and Meth.* A **718** 75
[3] Chirikov-Zorin I, Krumshteyn Z, Olchevski A, Utkin V and Zhmurin P *The Design of a Module of a new Electromagnetic Calorimeter for COMPASS II, SPIN-Praha-2013* (Prague)
[4] Arefev A V et al. 2008 *Instr. Exp. Tech.* V **51** 4
[5] *COMPASS-II Proposal* CERN-SPSC-2010-014 SPSC-P-340 (Geneve)

Acknowledgments

We would like to express our gratitude to the people of the ELSA team, who provided us with an electron beam for the test, as well as the 2nd Physics Institute of Giessen University for providing the moving table, especially Rainer Novotny and Peter Drexler.

Current Status and Performance of the Crystal Ball and TAPS Calorimeter

Andreas Neiser for the A2 collaboration

Institute of Nuclear Physics, JGU Mainz, J.-J.-Becher-Weg 45, 55128 Mainz, Germany

E-mail: neiser@kph.uni-mainz.de

Abstract. In the A2 experiment at the Mainz microtron (MAMI) electron accelerator, the production of nucleon resonances and light mesons off protons and nuclei using energy-tagged bremsstrahlung photons is studied. Decay products are measured with two electromagnetic calorimeters: Crystal Ball (CB), whose 672 NaI crystals cover almost the entire solid angle, and TAPS, which consists of 366 BaF2 and 72 $PbWO_4$ crystals in the forward direction. Here, we report on the current performance and status of both detectors after 10 years of operation in Mainz. In addition, we present the new CB high-voltage system, a new fast readout scheme for TAPS, and the planned upgrade of the CB data-acquisition system.

1. Introduction

This contribution gives a short overview of the goals of the A2 collaboration in Mainz and the experimental setup to achieve them. Subsequently, it focuses on the two deployed calorimeters Crystal Ball and TAPS in corresponding sections, which have been in use for over a decade now at A2.

2. The A2 experiment at MAMI

The A2 collaboration investigates strongly interacting particles in the low-energy regime, where quantum chromodynamics as the underlying theory is not perturbatively solvable. Using energy-tagged bremsstrahlung photons, we currently study Compton scattering $\gamma p \to p\gamma$ to obtain the scalar and vector polarizabilities of protons [1]. Furthermore, we photo-produce mesons and investigate their their mainly uncharged decay channels to test effective field theories and to obtain information about the nucleon excitation spectrum, eventually leading to an unambiguous partial wave analysis [2].

MAMI is an electron accelerator located at the University of Mainz. It consists of four staged racetrack microtrons and provides a 100 % duty cycle electron beam. Maximum beam currents of 100 μA and 20 μA can be reached with unpolarized and 85 % polarized electrons, respectively. With three stages, energies from 180 to 883 MeV are available. Since 2007, a fourth stage provides 1604 MeV. Besides A2, the A1 collaboration uses the electron beam directly with excellent magnetic spectrometers.

A2 creates photons via the bremsstrahlung process off various radiators to obtain circularly and longitudinally polarized photons with the characteristic $1/\omega$ spectrum, where ω is the photon energy. The momentum of the post-bremsstrahlung electrons are determined in the Glasgow-Mainz electron tagging spectrometer [3]. It has 384 logical channels covering ω from 6 to 93 %

(a) (b) (c)

Figure 1. (a) The detector setup at the A2 experiment. The target is surrounded a the particle identification detector (PID), two multi-wire proportional chambers (MWPC) and the calorimeter Crystal Ball (some crystals intentionally not drawn). In forward direction, TAPS with its hexagonal modules is depicted. (b) and (c) show Crystal Ball and TAPS, respectively, inside the A2 experimental hall. For TAPS, the installed veto detectors (plastic scintillators) are visible, see text.

of the initial electron energy. The electron rate can be up to 1 MHz per channel, corresponding to a total photon flux after collimation of roughly $10^6 - 10^7$ Hz. Additionally, the electron beam polarization can be determined online via Moeller scattering by measuring both electrons in time coincidence. For the energy range $\omega \approx 1450 \ldots 1600$ MeV, which is largely not covered by the Glasgow-Mainz tagger, an end-point tagger is available since 2012 to study in particular the η' photo-production, which falls in this energy regime.

The photons impinge on different targets, most notably the "frozen spin" target providing longitudinally or transversely polarized nucleons [4]. It consists of radical-doped butanol beads cooled down to 20 mK with a ^{3}He/^{4}He dilution cryostat, allowing typical polarization degrees of 85 % with relaxation times of 2000 h. Additionally, liquid hydrogen or helium targets in thin containers with different lengths are typically used for unpolarized measurements with very low background.

In general, A2 is a fixed target experiment with full angular coverage aiming at a good two-photon invariant mass resolution,

$$m_{\gamma\gamma}^2 = 2E_{\gamma 1}E_{\gamma 2}(1 - \cos\phi_{\gamma\gamma}), \tag{1}$$

in a large energy range from a few MeV to one GeV. Therefore, an excellent energy and angular resolution and a good particle identification is required. This is achieved with the two calorimeters Crystal Ball, covering 94 % of the solid angle, and TAPS in forward direction. It is complemented by two multi-wire proportional chambers for tracking and plastic scintillators for particle identification. See figure 1 for the general detector setup.

3. The NaI calorimeter Crystal Ball

The Crystal Ball NaI calorimeter was built in the mid-1970s for the e^+e^- collider SPEAR at SLAC for Charmonium spectroscopy. With this detector, the η_c was discovered. Later, it moved to DESY, and then in 1996 to BNL. Eventually, it was installed in Mainz in 2004 and equipped

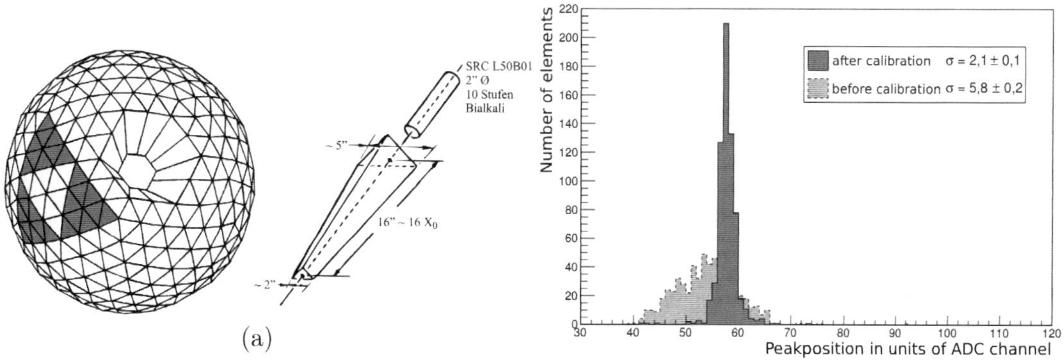

(a) (b)

Figure 2. (a) The icosahedron geometry of the Crystal Ball, showing one colored major triangle, subdivided in four minor triangles. Additionally, one single NaI(Tl) crystal with its PMT is depicted [7]. (b) The distribution of Am/Be photopeak positions before (lighter color) and after (darker color) adjusting the PMT voltages of the Crystal Ball individually [9].

with new electronics [5]. ADCs with 40 MHz sampling rate are used to achieve a high dynamic range of > 400 and a good linearity in combination with TDCs with 117 ps ticks for timing. Thus, the data-acquisition system is not limiting the energy resolution of the Crystal Ball and, for example, the 4.4 MeV photons from ^{12}C, excited by incoherent π^0 photo-production, can be directly measured with sufficient resolution [6]. Since the rate is only limited by the VME cycle time of 1.5 μs, the number of VME CPUs on the Crystal Ball and tagger side were doubled during a major DAQ upgrade in October 2013. This increased the data acquisition rate by a factor of 2 to roughly 2 kHz under typical experimental conditions.

As shown in figure 2a, the Crystal Ball has an icosahedron geometry as major triangles, further divided into four minor triangles, which consist of 720 truncated pyramids in total. Excluding the beam entrance and exit holes, there are in total 672 Tl-doped NaI crystals, $15.7X_0$ long, still readout with the original PMTs [7]. They cover 94 % of the 4π solid angle. At 1 GeV, the energy resolution is about 1.7 %. The angular resolution is about 2.5° in polar and $2°/\sin\theta$ in azimuthal direction [8]. After over thirty years of operation, 8 channels are dead and 9 channels show bad ADC spectra. This is mostly due to insufficient light collection, but also due to three broken PMTs.

The electronics provide an analog sum of all PMT signals, being proportional to the total energy deposited in the Crystal Ball, which makes the calorimeter a self-triggering detector. Since 2012, there is a flexible FPGA-based trigger system yielding decisions on coplanarity and multiplicity of Crystal Ball clusters within a few hundred nanoseconds.

3.1. Hardware gain calibration

Owing to the very stable environmental conditions of the A2 experiment hall, a Crystal Ball gain calibration is only necessary every few years to achieve stable operation and thus trigger conditions concerning the already mentioned analog sum. To this end, an Am/Be radioactive source emitting 4.438 MeV photons is put in the center of the Crystal Ball and the PMT gain is adjusted such that the photon peak above an exponential neutron background is always located at the same position in the raw ADC spectrum. Previously, since the high voltage is provided by only four very stable supplies for all PMTs, each PMT gain was calibrated by manually adjusting variable resistor, which is a tedious procedure. In 2013, voltage regulator boards were developed in Mainz and replaced the simple distributor boards, allowing to set the PMT voltage

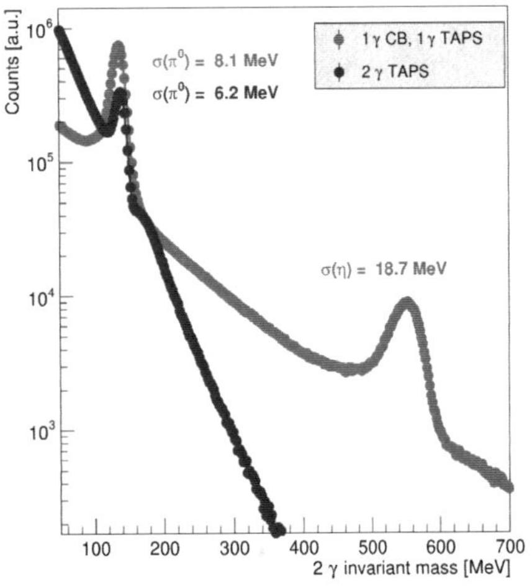

(a)

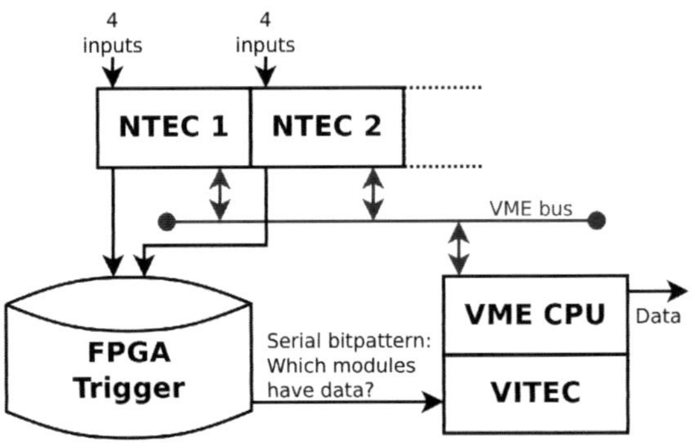

(b)

Figure 3. (a) The two photon invariant mass spectrum for one photon detected in Crystal Ball, one photon in TAPS (red), and for two photons detected in TAPS (blue). The observed peaks correspond to the decays of the uncharged pion and the eta, see text. (b) Schematic representation of the readout speed improvement by using trigger information, see text. The "New TAPS Electronics Card" (NTEC) is a CAEN V874 VME module with customized piggybacks.

individually but still using the very stable high-voltage supplies [9]. This cost-effective upgrade allows for a faster and more precise calibration method, see figure 2b, remotely controlled and monitored with EPICS [10]. The stability is roughly 0.3 V over 100 hours, corresponding to a negligible variation in measured energies of 0.02 MeV. After the calibration with the upgraded high-voltage system, π^0 peak widths of 8.6 MeV in the invariant two-gamma spectrum are measured, close to the resolution limit of the detector system, since offline analysis calibration does not reduce this width significantly.

3.2. Upgrade of the data-acquisition system

After one decade of operation, the data-acquisition system will be replaced by a readout system based on the TRB3 FPGA board, developed at the GSI, Darmstadt, for the HADES experiment [11]. This will increase performance further, but is also intended to sustain the maintainability of the system since spare parts for the current system are not available anymore. In the far future, the TRB3, as a flexible 4 + 1 FPGA board with customizable piggybacks, can also be part of a hardware-triggerless data acquisition, allowing multiple experiments at the same time. Up to now, first tests with different front-ends have been carried out. The main challenge is the large energy range of 1 . . . 400 MeV of the NaI signals. Additionally, the A2 collaboration profits from the large TRB3 user base, including the collaborations PANDA and MUSE, providing auxiliary software, documentation and debugging [12].

4. The BaF_2 and $PbWO_4$ calorimeter TAPS

The "Two-Arm Photon Spectrometer" TAPS is a BaF_2 calorimeter consisting of hexagonal modules, which can be arranged in different configurations tailored to various experiments.

Since 1988, the number of modules increased from 190 to 600 in 2006. Then, the detector was split and 384 modules were installed in Mainz and 216 in Bonn, both as forward wall detectors. In 2001, plastic scintillators in front of each module were installed as a charged particle veto system, and in 2004 a new readout system was commissioned [13]. In 2007, the modules of the two inner rings were replaced by 72 $PbWO_4$ crystals in Mainz to cope with higher hit rates in the very forward region.

In Mainz, the detector covers the important forward region $\theta = 0° \dots 20°$. The BaF_2 crystals are $12X_0$ long, and the inner $PbWO_4$ $17X_0$. In addition to the veto detectors, photons and protons can be separated by a pulse shape analysis of the BaF_2 scintillation light. This is reflected in the electronics by integration gates of different lengths. The energy resolution is 2 % at 1 GeV with all 366 BaF_2 crystals working and stable. The $PbWO_4$ crystals are operated at room temperature, and their calibration is more difficult due to large backgrounds. Currently, 6 modules have a broken PMT and another 8 show bad ADC spectra. However, they become increasingly important for the upcoming η' measurements with the end-point tagger in 2014 since most of the outgoing protons will be scattered into the very forward region. Hence, the 6 currently broken $PbWO_4$ modules will be replaced in June 2014 and the reason for the 9 bad spectra will be investigated.

In figure 3a, the overall performance of the calorimeter system is demonstrated with the inclusive reaction $\gamma p \to X\gamma\gamma$, where the analysis identified the photons as uncharged clusters with no further cuts applied. The initial electron energy was 1557 MeV, and liquid deuterium was used as a target. Although there is a significant amount of combinatorial background, signals from the decays $\pi^0 \to \gamma\gamma$ and $\eta \to \gamma\gamma$ can be nicely identified.

During the hardware upgrade in October 2013, each VME CPU was additionally equipped with a multi-purpose I/O card, the VITEC, which received the serialized event number and handles the readout synchronization between VME crates. It was measured that the TAPS VME readout system was the bottleneck after the upgrading the Crystal Ball and tagger systems. Since the VITEC was developed in Mainz, its FPGA was easily reprogrammed such that it receives a bit pattern from the TAPS FPGA-based trigger subsystem, enabling a fast pedestal suppression on the VME CPU by skipping empty modules, see figure 3b. Since the hits over the whole TAPS system are typically sparse, this improved the readout time by a factor of 3 from 176 μs to 56 μs.

5. Conclusion

In summary, both calorimeters Crystal Ball and TAPS showed stable operation and excellent performance during the various tagged photon experiments carried out at A2. In the next years, upgrading the Crystal Ball and tagger hardware is a major task to ensure maintainability and improve performance. Moreover, the performance of TAPS can be increased even further by splitting its VME crates similar to the Crystal Ball and tagger crates.

Acknowledgments

A.N. receives a Fellowship through GRK Symmetry Breaking (DFG/GRK 1581).

References

[1] E. J. Downie, H. Fonvieille, Eur. Phys. J. Special Topics **198**, 287, 2011
[2] B. Krusche, Eur. Phys. J. Special Topics **198**, 199, 2011
[3] J. C. McGeorge *et al*, Eur. Phys. J. **A 37**, 129, 2008
[4] A. Thomas, Eur. Phys. J. Special Topics **198**, 171, 2011
[5] D. Krambrich, "Aufbau des Crystal Ball-Detektorsystems und Untersuchung der Helizitätsasymmetrie in $\gamma p \to p\pi^0\pi^0$", PhD thesis, Mainz, 2007
[6] C. M. Tarbert *et al*, Phys. Rev. Lett. **100**, 132301, 2008
[7] M. Oreglia, "A Study of the Reactions $\Psi' \to \gamma\gamma\Psi$", PhD thesis, SLAC-R-236, 1980

[8] S. Prakhov *et al*, Phys. Rev. **C 79**, 035204, 2009
[9] J. Wettig, "Aufbau und Inbetriebnahme einer neuen HV-Versorgung für den Crystal Ball Detektor am MAMI", Diploma thesis, Mainz, 2013
[10] EPICS website. `http://www.aps.anl.gov/epics/`
[11] TRB3 website. `http://trb.gsi.de`
[12] A. Neiser *et al*, JINST **8** C12043, 2013
[13] P. Drexler *et al*, Nuclear Science Symposium Conference Record, 585, 2002

The BGO Calorimeter of BGO-OD Experiment

B. Bantes [1], D. Bayadilov [2], R. Beck [2], M. Becker [2], A. Bella [1], P. Bielefeldt [1], J. Bieling [1], M. Bleckwenn [1], S. Böse [2], A. Braghieri [3], K.-Th. Brinkmann [4], D. Burdeynyi [5], F. Curciarello [6,7], V. De Leo [6,7,*], R. Di Salvo [8], H. Dutz [1], D. Elsner [1], A. Fantini [8,9], O. Freyermuth [1], S. Friedrich [4], F. Frommberger [1], V. Ganenko [5], D. Geffers [1], G. Gervino [10,11], F. Ghio [12,13], G. Giardina [6,7], B. Girolami [12,13], D. Glazier [14], S. Goertz [1], A. Gridnev [15], E. Gutz [4], D. Hammann [1], J. Hannappel [1], P.-F. Hartmann [1], W. Hillert [1], A. Ignatov [16], R. Jahn [2], R. Joosten [2], T.C. Jude [1], F. Klein [1], K. Koop [2], B. Krusche [17], A. Lapik [16], P. Levi Sandri [18], I. Lopatin [15], G. Mandaglio [6,7], P. Mei [1], F. Messi [1], R. Messi [8,9], V. Metag [4], D. Moricciani [8], M. Nanova [4], V. Nedorezov [16], D. Novinskiy [15], P. Pedroni [3], M. Romaniuk [8], T. Rostomyan [17], N. Rudnev [16], C. Schaerf [8,9], G. Scheluchin [1], H. Schmieden [1], V. Sumachev [15], V. Tarakanov [15], V. Vegna [1], D. Walther [2], D. Watts [14], H.-G. Zaunick [2], T. Zimmermann [1]

[1] Universität Bonn, Physikalisches Institut, Nußallee 12, Bonn, D-53115, Germany
[2] Helmholtz-Institut für Strahlen- und Kernphysik, Nußallee 14-16, Bonn, D-53115, Germany
[3] INFN, sezione di Pavia, Via Agostino Bassi 6, Pavia, 27100, Italy
[4] Justus-Liebig-Universität Gießen, II. Physikalisches Institut, Heinrich-Buff-Ring 16, Gießen, D-35392, Germany
[5] National Science Center, Kharkov Institute of Physics and Technology, Akademicheskaya St. 1, Kharkov, 61108, Ukraine
[6] INFN, sezione di Catania, via Santa Sofia 64, Catania, 95123, Italy
[7] Università degli Studi di Messina, Dipartimento di Fisica e di Scienze della Terra, v.le F. Stagno d'Alcontres 31, Messina, 98166, Italy
[8] INFN, Roma Tor Vergata, Via della Ricerca Scientifica 1, Roma, 00133, Italy
[9] Università di Roma "Tor Vergata", Dipartimento di Fisica, Via della Ricerca Scientifica 1, Roma, 00133, Italy
[10] INFN, sezione di Torino, Via P.Giuria 1, Torino, 10125, Italy
[11] Dipartimento di Fisica, Università di Torino, via P. Giuria 1, 10125, Torino, Italy
[12] INFN, sezione di Roma, c/o Dipartimento di Fisica - Università degli Studi di Roma "La Sapienza" P.le Aldo Moro 2, Roma, 00185, Italy
[13] Istituto Superiore di Sanità, Viale Regina Elena 299, Roma, 00161, Italy
[14] The University of Edinburgh, James Clerk Maxwell Building, Mayfield Road, Edinburgh EH9 3JZ, United Kingdom
[15] Petersburg Nuclear Physics Institute, Gatchina, Leningrad District, 188300, Russia
[16] Russian Academy of Sciences Institute for Nuclear Research, prospekt 60-letiya Oktyabrya 7a, Moscow 117312, Russia
[17] Institut für Physik, Klingelbergstrasse 82, Basel, CH-4056, Switzerland
[18] INFN - LNF, Via E. Fermi 40, Frascati, 00044, Italy

E-mail: vdeleo@unime.it

Abstract. The BGO Rugby Ball is a large solid angle electromagnetic calorimeter now installed in the ELSA Facility in Bonn. The BGO is operating in the BGO-OD experiment aiming to study meson photoproduction off proton and neutron induced by a Bremsstrahlung polarized gamma beam of energies from 0.2 to 3.2 GeV and an intensity of 5×10^7 photons per second. The scintillating material characteristics and the photomultiplier read-out make this detector particularly suited for the detection of medium energy photons and electrons with very good energy resolution. The detector has been equipped with a new electronics read-out system, consisting of 30 sampling ADC Wie-Ne-R modules which perform the off-line reconstruction of the signal start-time allowing for a good timing resolution. Performances in linearity, resolution and time response have been carefully tested at the Beam Test Facility of the INFN National Laboratories in Frascati by using a matrix of 7 BGO crystals coupled to photomultipliers and equipped with the Wie-Ne-R sampling ADCs.

1. Introduction

A detailed and precise knowledge of the nucleon spectroscopy is undoubtedly one of the cornerstones for our understanding of the strong interaction in the non-perturbative regime. Todays privileged way to get information on the excited states of the nucleon is the meson photo- and electroproduction[1]. The attempt is to extract, from the phoproduction, the electromagnetic couplings and furthermore the hadronic properties of the excited nucleon states that cannot be accessed via pion scattering, either because the resonances largely overlap, or because of a weak coupling to the single pion-nucleon channel. Polarization degrees of freedom in photoproduction processes play a crucial role. Polarization observables arising from interference mechanisms, are highly sensitive to the details of the interaction revealing resonance properties that are difficult to extract from a cross-section measurement, where a single contribution often dominates[2]. The corresponding database has considerably expanded over the last years thanks to a combined effort of dedicated facilities worldwide. Not only did the recent experiments brought a quantitative improvement by measuring cross-sections with unprecedented precision for a large number of channels but they also allowed a qualitative leap by providing high-quality data on polarization observables.

The BGO-OD experiment at the ELSA facility in Bonn involves the use of a Bremsstrahlung tagged and polarized photon beam of energy between 0.2 and 3.2 GeV, a large solid angle high resolution BGO calorimeter combined with the plastic scintillator barrel and two Multi-Wires Proportional Chambers (MWPC) and the Open Dipole spectrometer equipped with tracking detectors. This apparatus will be used to measure polarization observables and cross sections in the photoproduction of pseudo-scalar and vector mesons off a Hydrogen or Deuterium target.

2. Experimental apparatus at BGO-OD

The BGO-OD apparatus is composed of two main parts: a central detector system and a forward spectrometer for charged particles, completed by a photon tagging system. The tagger, consisting of 120 channels, will cover the energy range $(0.1\div 0.9)E_{e^-}$ with variable resolution $(10 \div 40$ MeV)[3]. It is designed to handle a high intensity beam (up to $5 \cdot 10^7 s^{-1}$). One of the central detectors of the experimental setup is the high energy resolution and large solid angle (0.9 π) BGO electromagnetic calorimeter that will be described in detail later. The calorimeter is combined with two multi-wire proportional chambers (MWPC) for inner tracking and a plastic scintillator barrel for particle identification through the measurement of dE/dx. The region between the acceptance of the central detector and the forward spectrometer will be covered by an azimuthally symmetric Multi-gap Resistive Plate Chamber (MRPC) detector.

The forward spectrometer is based on a large open dipole magnet that uses a dipolar field of about 0.5 T for the separation, identification and measurement of the momentum (3% resolution) of charged particles emitted in the photoproduction processes. For this purpose, the

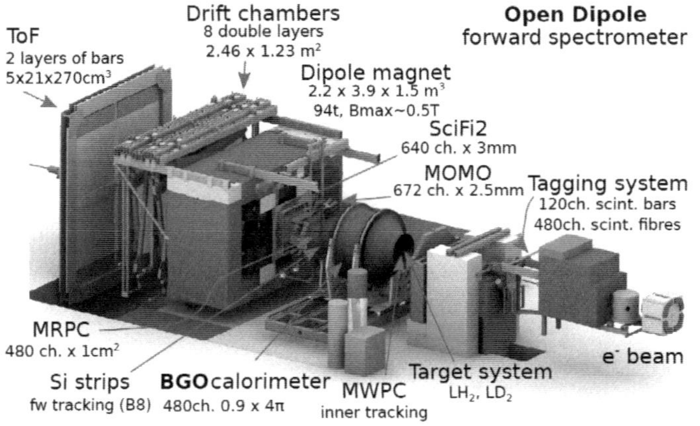

Figure 1. BGO-OD experimental apparatus.

spectrometer is equipped with MOMO and SciFi2 detectors, eight double layer drift chambers (DCs) and a time-of-flight (TOF) detector [3]. The two fiber hodoscopes MOMO and SciFi2, located in front of the magnet, allow the tracks reconstruction before the curvature caused by the magnetic field. Behind the magnet, tracking is done by a set of 8 double layer drift chambers in four different orientations, vertical wires to measure the x -coordinate, horizontal for the the y -coordinate, and tilted by 9 from vertical, for a u - and v -coordinate. The forward spectrometer is completed by a set of time-of-flight walls that are used for charged particle identification in combination with the momentum reconstruction provided by the tracking detectors[4].

3. The BGO calorimeter

The calorimeter is designed to study photonuclear reactions up to about 2 GeV, in particular exclusive channels where neutral products (γ and neutral mesons decaying in two or more photons, as π, η, η', K^0, ϕ) and charged particles (detectable to some extent) are emitted[5].

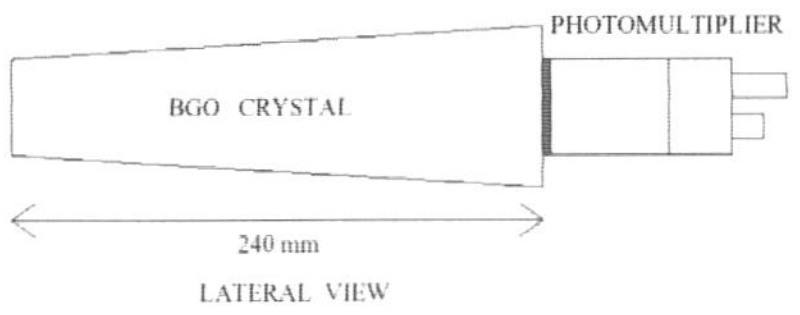

Figure 2. Schematic views of a BGO crystal.

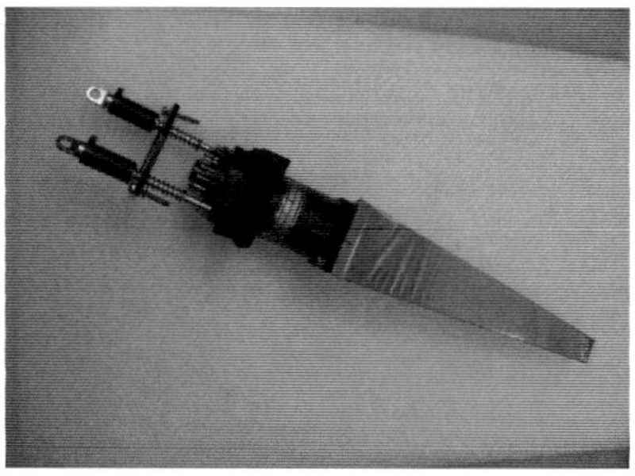

Figure 3. BGO crystal coupled to the photomultiplier.

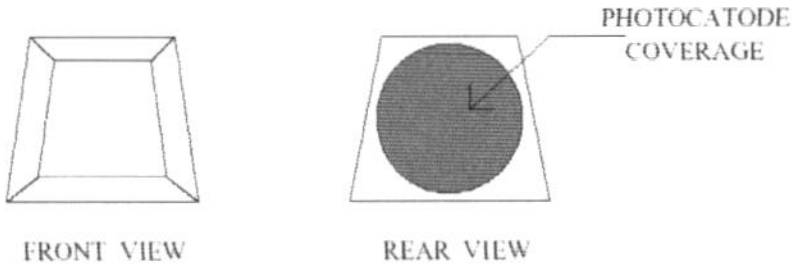

Figure 4. Schematic views of a BGO crystal.

The features of the apparatus are a large solid angle (0.9π), extended granularity, good detection efficiency for neutral products and also a good energy resolution to low momentum (<0.8 GeV/c) protons. Moreover, the high density, the short radiation length ($X_0 = 1.12$ cm), the high atomic number made this inorganic scintillating material ($Bi_4(GeO_4)_3$) particularly suited for the construction of an electromagnetic calorimeter with limited sizes. The crystal is not hygroscopic and the wavelength range of the emitted light (480 nm) can be easily matched with standard photomultipliers. The calorimeter is composed of 480 BGO crystals. The crystals are of 8 different dimensions and are shaped like pyramidal sectors with trapezoidal bases. They define 15 angular regions (θ) in the plane containing the symmetry axis of the calorimeter, coincident with the beam axis, and 32 (ϕ) in the plane orthogonal to the beam axis[6]. The 480 crystals have all the same length of 24 cm ($\simeq$ 21 radiation lengths), for a good confinement of photon showers in the GeV region, and are arranged in such a way that the reaction products encounter the same thickness of BGO. Each crystal is wrapped up in a thin (30 μm) aluminized mylar reflector, and its back side is optically coupled to a photomultiplier (PM). The crystals have been treated by the manufacturer in order to guarantee a uniformity in the light collection along the crystal length better than at least 95% (it is even better than 98% for some crystals) and this will represent a deterioration of the energy resolution by a constant term of the order of 0.3%[6].

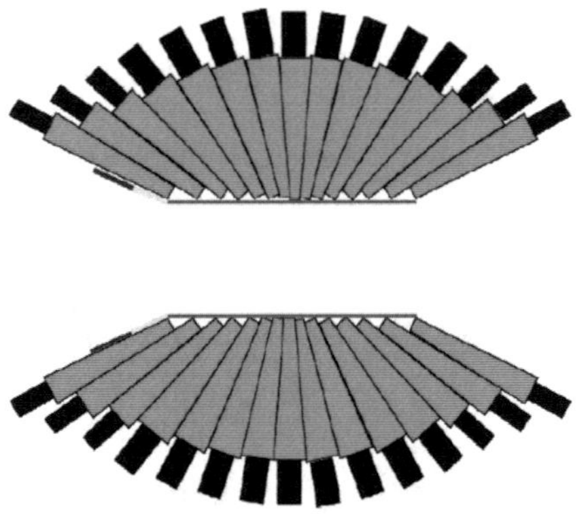

Figure 5. Scheme of the BGO sectors. The crystals cover the polar angle between 25° and 155° in the laboratory reference frame (15 sectors), while the azimuthal angle is totally covered (32 sectors).

The mechanical support structure consists of 24 baskets of Carbon fiber composite material supported by an external steel frame. Each basket is divided into 20 cells with very thin walls, 0.38 mm for the inner and 0.54 mm for the outer walls, to keep the crystals optically and mechanically separated. The Carbon fiber has been preferred to other materials like Aluminum for its higher rigidity and lower gamma ray attenuation due to its low Z number. The support frame is divided into two halves which can be taken apart by 1.5 m to allow access to the target and central detector region. When closed the structure leaves a 20 cm diameter hole along the beam-line for the insertion of the target, the cylindrical wire chambers and the plastic scintillator barrel[5].

4. BGO readout system and calibration procedure

The BGO acquisition is performed by using thirty Wiener Sampling ADC modules. The ADC modules AVM16-Mambo operate with a sampling frequency of 160 MHz (corresponding to 6.25 ns interval between two samples) with 12 bit resolution (i.e. 4096 channels). The signal is sampled inside a time interval defined by the user, that can start after or before the trigger signal. The AVM16 modules have 16 channels and one trigger input but they are provided also of an internal trigger (level trigger). We use this option for calibration purposes. The modules allow the extraction of the main parameters of the pulse as the total and partial integrals of the signal, amplitude and time of relative maxima and minima and the start time of the signal.

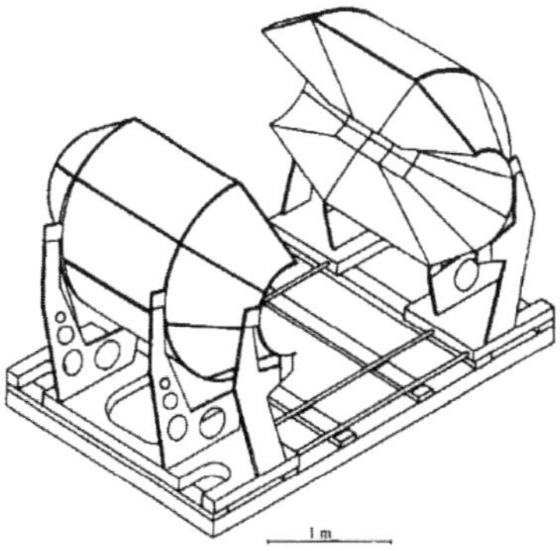

Figure 6. Sketch of the BGO calorimeter showing the carbon fiber baskets mounted on the external support frame separable into two halves.

Figure 7. Top view of the BGO Calorimeter.

An example of the main features for a signal with pile-up events is illustrated in the figure 8: the window start point (P0), which represents the user defined latency time; amplitude and time of minima (Pi, Ppi, ...) and maxima (Pa, Ppa, ...), extracted when a rising slope is detected; the pulse start time (Pz, Ppz, ...), extrapolated from the crossing of the slope and the pedestal value (or the current pedestal value for pile-up signals) and the total and partial charge integrals (Pq, Ppq).

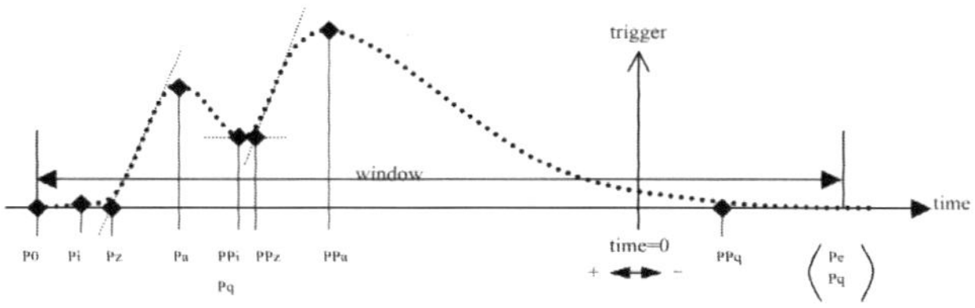

Figure 8. Feature extraction parameters.

The absolute calibration of the Bgo Rugby Ball is performed by using the more energetic photons (1.275 MeV) emitted by three ^{22}Na sources located inside the BGO cylindrical hole. The equalization of the response of the 480 crystals is obtained by setting all the photomultiplier (PM) gains varying their high voltages.

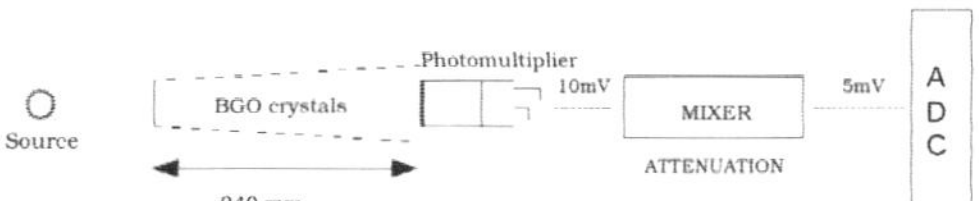

Figure 9. Scheme of the electronic chain used for the calibration procedure.

This procedure is strictly necessary, because a threshold on the hardware sum of the energy released in the BGO is used as a trigger for the experiment. After the equalization, for the same energy released, the response of the PM associated to each crystal will be located at the same ADC channel. For the calibration procedure an internal trigger which is provided by the ADC, when a signal overcomes a certain threshold, was used; it was chosen in order to guarantee the same statistics on all the crystals, which a global trigger on the hardware sum of the energies of all the crystals could not insure. In Figure 9 a schematic representation of the experimental chain of calibration is shown: the output signal from the phototube, coupled to the crystal, is sent to a mixer that delays, in case attenuates, splits the signals into two parts, one sent to the ADC for the readout and the other to the sum over the signals. To obtain the calibration constants for each channel, we fix the PM voltages so that the response to the energy of the second peak of the source is located at the channel 60 of the ADCs. The channel is fixed within a tolerance of $\pm 3\%$, which, in any case, does not affect the calibration of the data, since the calibration constant of each crystal is registered and used for the offline energy conversion[7]. It is necessary to repeat the calibration procedure two times per day because the calibration constants of each crystal may change during time. The reasons are essentially two: variation in the crystal light output due to temperature effects and variation in the PM gain. Gain variations of the photomultipliers may occur as a consequence of the following effects[5]:

-photocathode temperature variation;
-instability in the high voltage power supply;
-aging of the cathode and dynode materials;
-voltage dividers instability.

In figure 10, a typical BGO calibration spectra acquired with the AVM-16 ADC modules is represented, where both the peaks of the ^{22}Na source are visible.

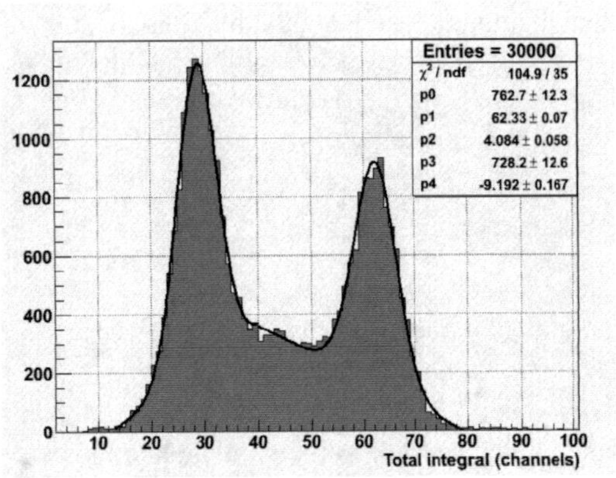

Figure 10. BGO calibration spectrum where the two peaks of the ^{22}Na source are visible.

5. Performances of the BGO calorimeter

The linearity and resolution in the energy reconstruction and the performances in the time response of the BGO crystals in combination with the new readout system (AVM16 ADC modules) have been recently studied at the BTF (Beam Test Facility) in Frascati.

The BTF is part of the DAΦNE ϕ-factory complex which includes a high current electron and positron LINAC, a 510 MeV e^- and e^+ accumulator and two 510 MeV storage rings[8]. Before the high-intensity electron or positron beam pulses produced by the 60 m long LINAC are injected into the double storage ring, they can be extracted to a transfer line that is dedicated to the calibration of particle detectors (the BTF transfer line). Here, the number of particles can be reduced to a single electron per pulse by means of a variable thickness copper target. Moreover, a dipole magnet and a system of slits, located before and after the dipole, allow the selection of a particular energy of the beam and its collimation. In this way, it is possible to keep under control not only the energy of the beam but also the number of electrons that enter in the experimental area (the multiplicity), allowing to reach energy values up to 2-3 GeV when several electrons simultaneously impinge on the detector.

During the test we used 7 BGO crystals arranged as a matrix inside a PVC basket with removable walls. The position of the walls could be adjusted with Aluminum screws in order to guarantee a better tightness among the crystals. No inert material was present between them because they were not mechanically separated.
The trigger to the acquisition has been provided by a plastic scintillator plate, located in front of the crystals, when the signal is in coincidence with an over threshold deposited energy in the BGO crystals.

5.1. Electronics readout

A scheme of the electronics readout used during the tests is shown in the figure 11. Here, the signal coming from the PM of each crystal was split and attenuated and sent to the ADC module for digitization, or sent to a linear Fan-In-Fan-out together with the signals coming from the other crystals for trigger purposes. The linear sum of the seven signals was amplified and sent to a threshold discriminator and then to a coincidence unit, together with the discriminated signal of the scintillator pad. A veto to the coincidence unit, to avoid further triggers during the dead time of the acquisition, is provided by a Dual Timer module activated by an output of the coincidence unit.

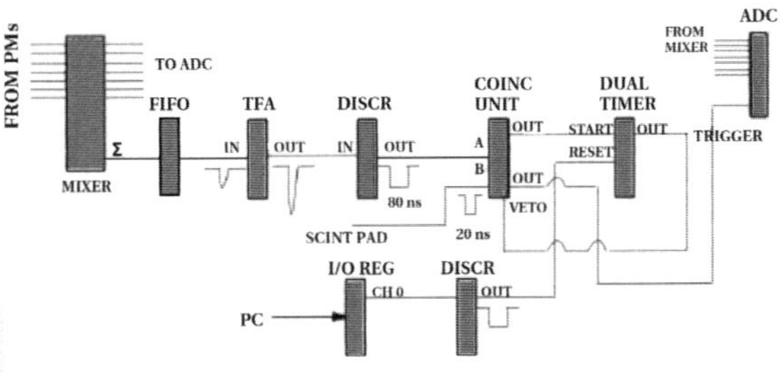

Figure 11. The electronics scheme of the crystals read-out: signals from the BGO are split and (i) attenuated and sent to the ADC module or (ii) summed with the other signals, amplified and discriminated for trigger purposes.

The reset of the Dual Timer is externally provided by the acquisition program via an I/O-Register.

5.2. Analysis and results

We collected data at six different settings of the beam energy with different multiplicity values: 100 MeV (4 peaks of multiplicity=1÷4), 150 MeV (6 peaks), 200 MeV (7 peaks), 300 MeV (7 peaks), 400 MeV (9 peaks) and 500 MeV (4 peaks). For each event, we considered a cluster of all the crystals, summing up the energies deposited in all of them, since the e.m. shower spreads over several crystals.

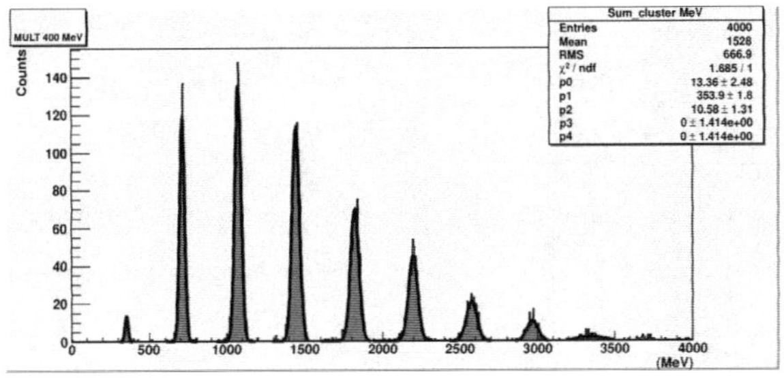

Figure 12. Energy spectrum of the beam when the single electron has an energy of $400\,MeV$. The visible peaks correspond to the different value of multiplicity of electrons, i.e. correspond the energies which are multiple of E_{NOM} (E_{NOM}, $2E_{NOM}$, ... nE_{NOM}).

The energy spectrum of the beam for the energy setting of 400 MeV is shown in figure 12. The peaks correspond to the different values of multiplicity of electrons and then correspond to the energies which are multiple of E_{NOM} (E_{NOM}, $2E_{NOM}$, ... nE_{NOM}).

As one can see, the distributions are well distinguished so they have been fitted with a gaussian function (plus a background approximated with a straight line) with mean value of the energy E_{meas} and resolution $\sigma_{E_{meas}}$.
Thus, it was possible to extract the linearity curve of the detector plotting the mean value of the energy E_{meas} as a function of the nominal energy E_{NOM}.

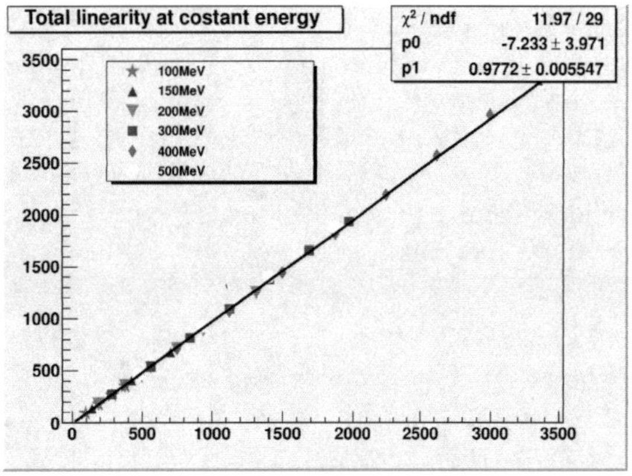

Figure 13. Linearity curve of the detector: the mean value of the energy (E_{meas}) obtained from the fit is plotted as a function of the nominal energy (E_{NOM}). The error bars correspond to the resolution σ of the gaussian fit. Points of the same colour refer to the data of the same energy setting with increasing value of multiplicity.

In figure 13 we reported the fitted values of the energy with the error coming from the fitted resolution as a function of the nominal energy for all energy settings and multiplicities. Points of the same colour refer to the data of the same energy setting with increasing value of multiplicity.
They follow a linear behaviour:

$$E_{meas} = p_0 + p_1 * E_{nom} \tag{1}$$

with
$p_0 = (-7.233 \pm 3.971)MeV$
$p_1 = -0.977 \pm 0.0055MeV$
and a satisfactory reduced χ^2_{red} value: $\chi^2_{red} = 0.4$. Only the three points at the highest energies, 2400MeV, 2800 MeV, and 3200 MeV (corresponding to 6,7 or 8 electrons of energy 400MeV each) are slightly higher than expected, probably denoting small non-linearities in the PM of the central crystal of the matrix or of the ADC module.
The linearity curve was extracted also plotting the fitted vs. the nominal energy for the different multiplicities (up to four) at each energy setting in order to see if differences arise with the previous case.
Since the calorimeter is a linear device, we expect that if a single electron of energy E_0 deposits its energy in the detector or n electrons each of energy $E_i = E_0/n$, the response of the detector should be essentially the same in the two cases.
We compared the two results and we found that they are in good agreement. In the last case the fit provides a lower value of the χ^2_{red} (0.25) and:
$p_0 = (-1.868 \pm 4.845)MeV$
$p_1 = 0.9628 \pm 0.01027$
because we have not considered the peaks of higher multiplicity that are characterized by a greater spread.

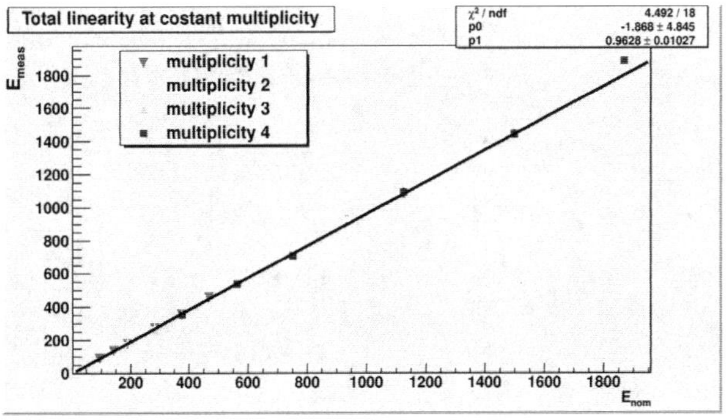

Figure 14. Linearity curve of the detector: the mean value of the energy (E_{meas}) obtained from the fit is plotted as a function of the nominal energy (E_{NOM}). The error bars correspond to the sigma of the gaussian fit. Only data for multiplicities less or equal four are reported and points with the same colour represent data obtained at the same multiplicity.

The BGO resolution in combination with the new ADC module, was extracted subtracting the BTF resolution, which is $\simeq 1\%$(RMS) at all the energies, from the measured resolution:

$$(\sigma_{BGO})^2 = (\sigma_{meas})^2 - (\sigma_{BTF})^2 \tag{2}$$

The energy resolution was parameterized in the usual way:

$$\frac{\sigma}{E}(\%) = a_{CONST} \oplus \frac{a_{STAT}}{\sqrt{E_{NOM}}} \oplus \frac{a_{NOISE}}{E_{NOM}} \tag{3}$$

where the terms are summed in quadrature.
The first term, a_{CONST} is the constant term. It gives the asymptotic behaviour and it is connected to the intercalibration precision. It is related also to the longitudinal non-uniformities ($\simeq 0.3\%$) in the collection of the emitted light and to the temperature fluctuations.
The second term, a_{STAT} is the statistical term. It containes sampling and fluctuations of all kinds.
The last term, a_{NOISE} is generally negligible at high energies and it is connected to the electronic noise.
The data were fitted according to the previous formula and the result of the fit is displayed in figure 15.

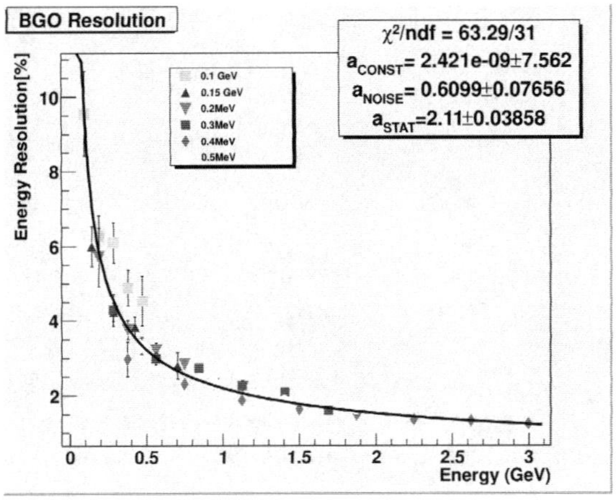

Figure 15. The BGO energy resolution, extracted from the fit of the peaks with a gaussian distribution and subtracted of the BTF energy resolution, as a function of the nominal energy. Data for all energy settings and multiplicities are reported and points with the same colour represent data obtained at the same energy settings.

We can see that:

- the constant term a_{CONST} is resulted about zero. Probably the reason is that the data taking was very near in time to the equalization procedure, so the fluctuations in calibrations are negligible as well as the fluctuations in temperature. Another possible reason is that we are overestimating the BTF resolution.
- a_{STAT} is reasonable and comparable with other published results[9].
- a_{NOISE} is small as it could be expected from read-out devices such as the photomultipliers.

As before, we have compared the energy resolutions concerning energies obtained in different ways (one single electron with energy E_0 or n electrons with energies E_0/n) and we found reasonable agreement between the different cases. Data of the same colour represent the results obtained at the same multiplicity for different energy settings.

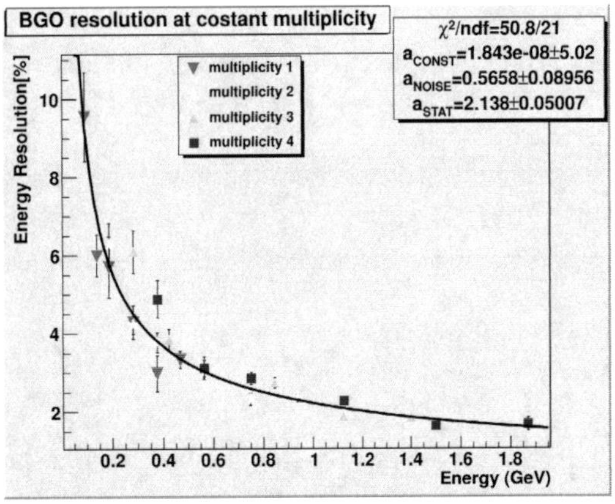

Figure 16. The BGO energy resolution, extracted from the fit of the peaks with a gaussian distribution and subtracted of the BTF energy resolution, as a function of the nominal energy. Only data for multiplicities less or equal four are reported and points with the same colour represent data obtained at the same multiplicity.

We have not observed clear difference with respect to the simplest case of one single electron. However, as it's possible to see from the plot 15, the better fit is given from the energy setting of $400MeV$. The fact has an explanation. In fact, studying the spot of the beam has been found that only in the case of this energy setting the beam was perfectly centered and above all, it has a spread lower than the other settings of energy. In the figure 17 the spatial distribution of the beam, overlapped to the matrix of the seven crystals, is shown for the energy settings of 400 MeV and 500 MeV.

5.3. Time response of the detector

For the first time the time response of the detector has been studied with a test beam and its time resolution has been estimated. In figure 19, the distribution of the start time for a central and a lateral crystal of the BGO matrix is illustrated.

As it is possible to see from the figure 19 for a central crystal (as the crystal number 3) we have a typical time resolution less than $3ns$; for a lateral crystal (as the crystal number 0) the measured resolution is about $6 - 8ns$. The reason for the different measured resolution is that since the crystal number three received much more energy than the other ones, most of the times the acquisition has been triggered by this crystal.
A difference in typical mean value is visible between the large and the small crystals: in the first case the t_{lrg} is about $-234ns$, instead in the second case the t_{sml} is about $-222ns$ with a $\Delta t \simeq 10 - 12ns$ and $|t_{lrg}| > |t_{sml}|$. The different electron transit time in the photomultipliers of different sizes is the explanation for the Δt.

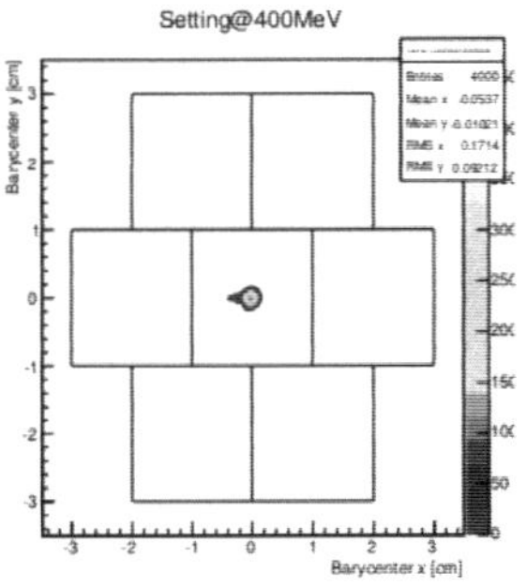

Figure 17. Bidimensional distribution of the x and y coordinates of the centroid for the energy setting of 400 MeV.

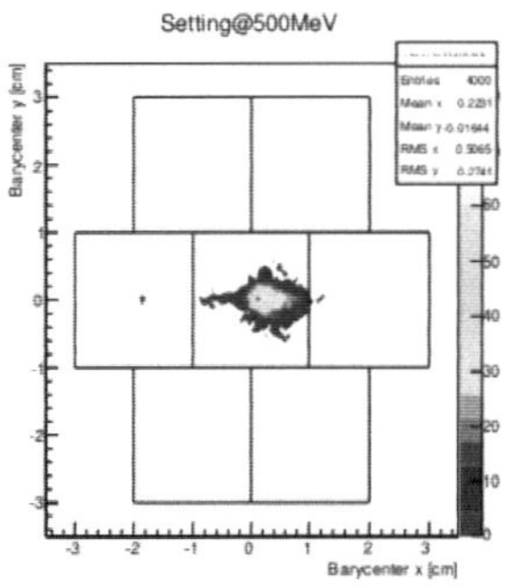

Figure 18. Bidimensional distribution of the x and y coordinates of the centroid for the energy setting of 500 MeV.

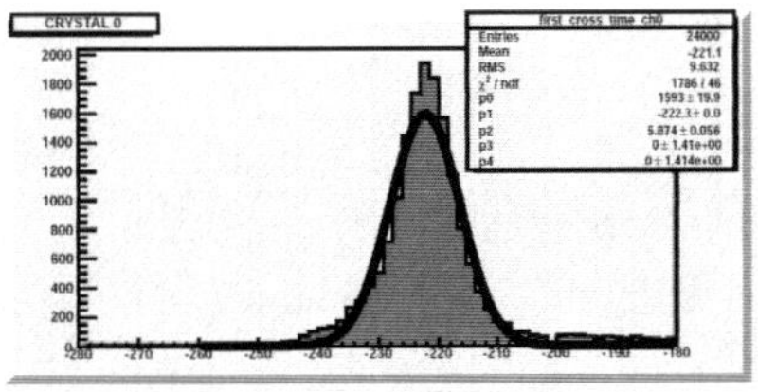

Figure 19. The start time distribution of a lateral crystal of the matrix. Distributions has been fitted with a gaussian; the fit parameters are reported in the plot.

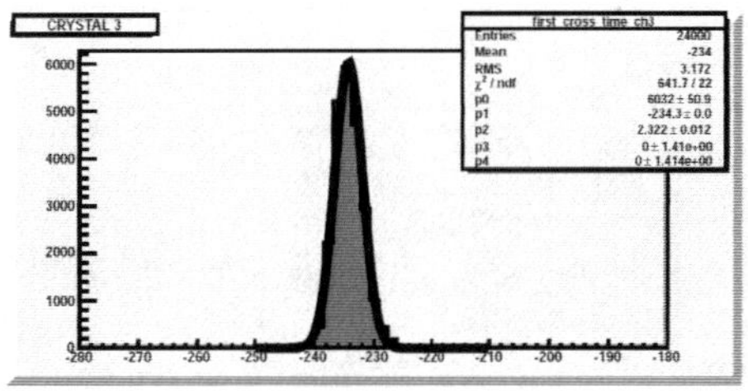

Figure 20. The start time distribution of a central crystal of the matrix. Distributions has been fitted with a gaussian; the fit parameters are reported in the plot.

6. Some preliminary results at BGO-OD

The good status of the BGO calorimeter allows the invariant mass reconstruction of the π^0 meson and the missing mass from two gammas that are visible in the figures 21 and 22.
They have been obtained by using the geometrical coincidence between the BGO calorimeter and the barrel scintillator.

7. Conclusions

The BGO calorimeter operating in the BGO-OD experiment has been described. We reported the calibration procedure of the BGO calorimeter that we used.
The performances of the detector in combination with the new readout system of AVM16 Sampling ADC modules have been studied at the BTF of Frascati. The global resolution of the BGO is confirmed to be about 3% at 1 GeV, which is compatible with the requests at the BGO-OD experiment. For the first time, we have studied the time response of the detector with a test beam and we have estimated its time resolution ($\sim$ 3ns (RMS)). The excellent status of the BGO Ball has allowed to obtain promising results during the beam time tests, as the reconstruction of the invariant mass of the π^0 meson and the missing mass from two gammas.

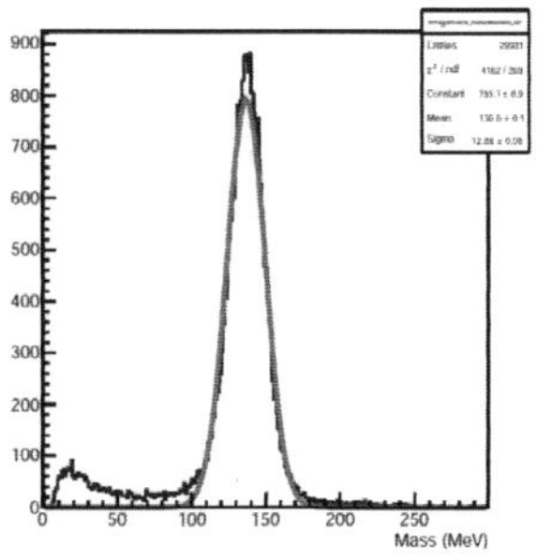

Figure 21. Invariant mass reconstruction of the π^0 meson from two gammas where the particle identification is obtained by using the geometrical coincidence between the BGO calorimeter and the barrel scintillator[10].

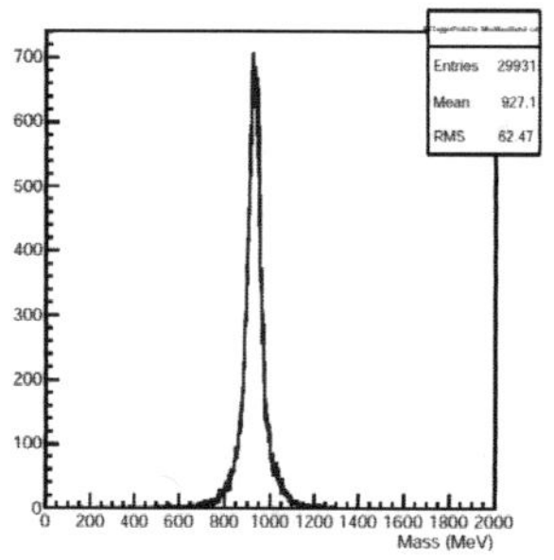

Figure 22. Missing mass reconstruction from two gammas where the particle identification is obtained by using the geometrical coincidence between the BGO calorimeter and the barrel scintillator[10].

References

[1] Lleres A et al. 2009 *Eur. Phys. J. A* **39** 149161
[2] Di Salvo R et al. 2009 *Eur. Phys. J. A* **42**, 151157
[3] http://bgo-od.physik.uni-bonn.de
[4] Levi Sandri P et al. 2013 *Int. J. Mod. Phys.* **26** 1460093
[5] Ghio F et al. 1998 *Nucl. Instr. and Meth. in Phys. Res. A* **404** 71-86
[6] Zucchiatti et al. 1992 *Nucl. Instr. and Meth. in Phys. Res. A* **317** 492-497
[7] De Leo V et al. 2013 *J. Phys.: Conf. Ser.* **424** 012008
[8] Mazzitelli G, Ghigo A, Sannibale F, Valente P and Vignola G 2003 *Nucl Instr. Meth. A* **515** 516.
[9] Kampert K -H et al. 1994 *Nucl. Instr. and Meth. A* **349** 81.
[10] Magnisi L 2014 Bachelor Thesis *Private Communication.*

The New APD Based Readout for the Crystal Barrel Calorimeter

M. Urban[1], Ch. Honisch[1] and M. Steinacher[2]
for the CBELSA/TAPS collaboration

[1] HISKP, University of Bonn, 53115 Bonn, Germany
[2] Physikalisches Institut, University of Basel, Switzerland

E-mail: urban@hiskp.uni-bonn.de, honisch@hiskp.uni-bonn.de, michael.steinacher@unibas.ch

Abstract. The CBELSA/TAPS experiment at ELSA measures double polarization observables in meson photoproduction off protons and neutrons. To be able to measure purely neutral reactions off polarized neutrons with high efficiency, the main calorimeter has to be integrated into the first level trigger. This requires to exchange the existing PIN photo diode by a new avalanche photo diode (APD) readout. The newly developed readout electronics will provide an energy resolution compatible to the previous set-up and a fast trigger signal down to 10 MeV energy deposit per crystal. After the successful final tests with a 3x3 CsI crystal matrix in Bonn at ELSA and in Mainz at MAMI all front-end electronics were produced in fall 2013. Automated test routines for the front-end electronics were developed and the characterization measurements of all APDs were successfully accomplished in Bonn.

The project is supported by the Deutsche Forschungsgemeinschaft (SFB/TR16) and Schweizerischer Nationalfonds.

1. Introduction

The systematic investigation of the baryon excitation spectra demands an extension of the experimental program from the photoproduction off protons towards the photoproduction off neutrons, and gives rise to the necessity of measuring entirely neutral final states. The world data on all-neutral final states is sparse [1], which makes an experiment able to obtain such data very desirable. The planned instrumental improvements of the Crystal Barrel calorimeter, the central component of the CBELSA/TAPS experiment, will allow a study of such processes (e.g. $\gamma n \to n\pi^0$, $\gamma n \to n\eta$) with a high efficiency and full solid angle coverage. A new readout based on avalanche photo diodes (APDs) has been developed and tested for the CsI(Tl) crystals of the Crystal Barrel calorimeter in order to replace the existing PIN diode readout, which was optimized for a good energy resolution but did not provide a timing information of the analog signal [2]. With the new APD readout a fast timing signal in addition to the energy information for every single CsI(Tl) crystal is available, allowing an integration of the main calorimeter into the fast trigger decision of the CBELSA/TAPS experiment.

2. Experimental Setup

The setup of the CBELSA/TAPS experiment located at the ELSA electron accelerator in Bonn [3] is shown in Fig. 1. Photons are produced by scattering electrons off different radiators (e.g. a

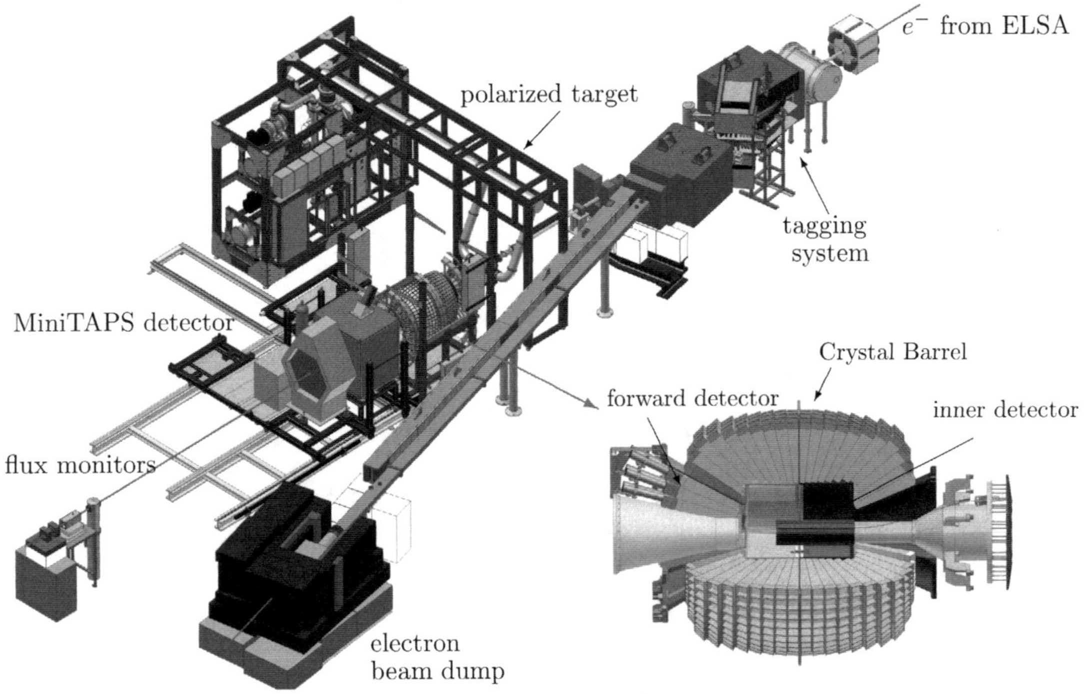

Figure 1. The experimental setup of the CBELSA/TAPS experiment at ELSA. Inset: the main crystal barrel detector with forward cone and inner detector.

diamond crystal) receiving unpolarized, linearly or circularly polarized photons. Several different reaction targets are available, e.g. transversely or longitudinally polarized butanol targets. The according target is surrounded by the Crystal Barrel main and forward detector consisting of 1320 CsI(Tl) crystals and an inner detector which is used to identify charged particles. In forward direction the angular coverage is closed down to 1° by the MiniTAPS detector. The complete setup offers a close to 4π angular coverage with a high detection efficiency for neutral particles. It is therefore ideally suited to measure crosssections and polarization observables for the photoproduction of final states with neutral mesons decaying into photons.

3. Avalanche Photo Diodes

Several devices can be used in order to convert the light output from a scintillator into a analog signal. PIN photo diodes, photomultipliers, silicon photomultipliers (SiPM) and avalanche photo diodes (APD) are the most common devices. A PIN photo diode shows a small temperature dependency of the generated charge per photon. This devices are used in strong magnetic fields. The drawback is the signal to noise ratio (SNR) which is smaller compared to the other devices, if all devices are operated at room temperature.

APDs show a better SNR than PIN photo diodes due to their intrinsic amplification and they can also be used in strong magnetic fields of a few Tesla, but their temperature dependence is quite large. We investigated the characteristics of the S11048(X3) APD from Hamamatsu [4] which will be used in the upgrade of the Crystal Barrel calorimeter to provide a fast trigger signal down to an energy of 10 MeV per crystal.

3.1. *Gain Temperature Dependence of APDs*

Fig. 2(a) depicts the gain temperature dependence of the S11048(X3) APD from Hamamatsu. It shows a coefficient of roughly $-2.3\,\frac{\%}{\mathrm{K}}$ around 27.5 °C. Its behavior between 20 °C and 40 °C

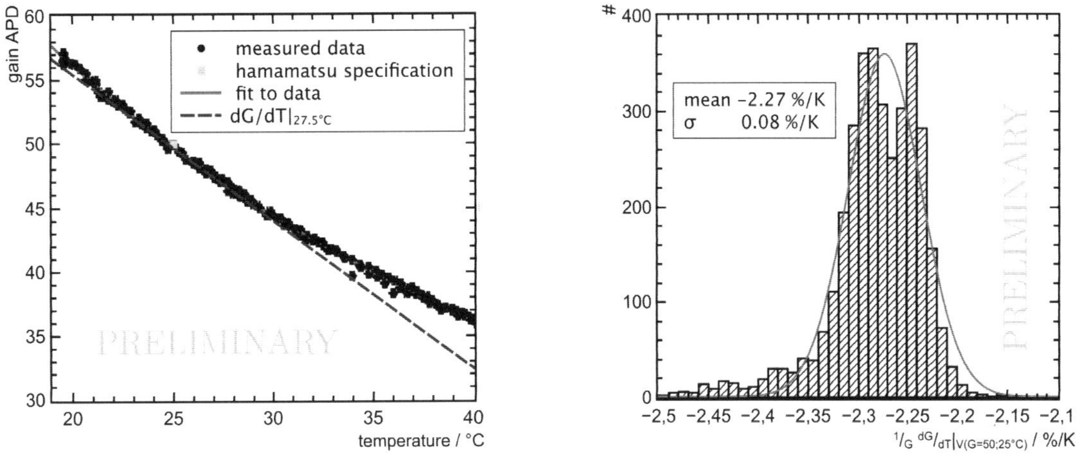

(a) Gain temperature dependence

(b) Distribution of gain temperature dependence

Figure 2. Gain temperature dependence of S11048(X3) APDs; 2(a): temperature dependence of one APD; red: fit using eqn. 1. 2(b): Gain temperature dependence at $G(V_{\mathrm{bias}}) = 50$ at 25 °C for a sample of 3500 APDs.

is described by the function

$$G(T) = a + b \cdot (1 + \frac{c}{100})^T \tag{1}$$

which is plotted in red in Fig. 2(a) alongside the data. The temperature dependence of the S11048(X3) APD was measured for 3500 pieces to investigate the variations within the sample used to upgrade the CB detector. The result is shown in Fig. 2(b).

A more detailed description of the measurement of the APD gain and its temperature dependence will be given in section 4. There are various approaches to stabilize the gain of an APD. This is mainly done by automatically adjusting the bias voltage of the APD.

3.2. *Gain Voltage Dependence of APDs*

The APD needs an external bias voltage for the internal amplification mechanism. The gain of the amplification depends on the applied bias voltage, shown in Fig. 3(a) for a S11048(X3) APD of Hamamatsu. The measured data is described very well by the following modified Miller function in a region where the gain is about 50:

$$G(V_{\mathrm{bias}}) = \frac{a}{(1 - \frac{V_{\mathrm{bias}}}{b})^c} \quad . \tag{2}$$

If the bias voltage is changed according to the temperature of the APD, a stable gain behavior is achievable. For such a compensation not only the temperature dependence but also the gain voltage dependence needs to be well known. Therefore this dependence of all 3500 S11048(X3) APDs were measured. To compare their behavior and to design a simple compensation algorithm the gain voltage dependence around $G = 50$ was approximated linearly. It was found to be on average $2.95\,\frac{\%}{\mathrm{V}}$. The values obtained for all 3500 APDs can be seen in Fig. 3(b).

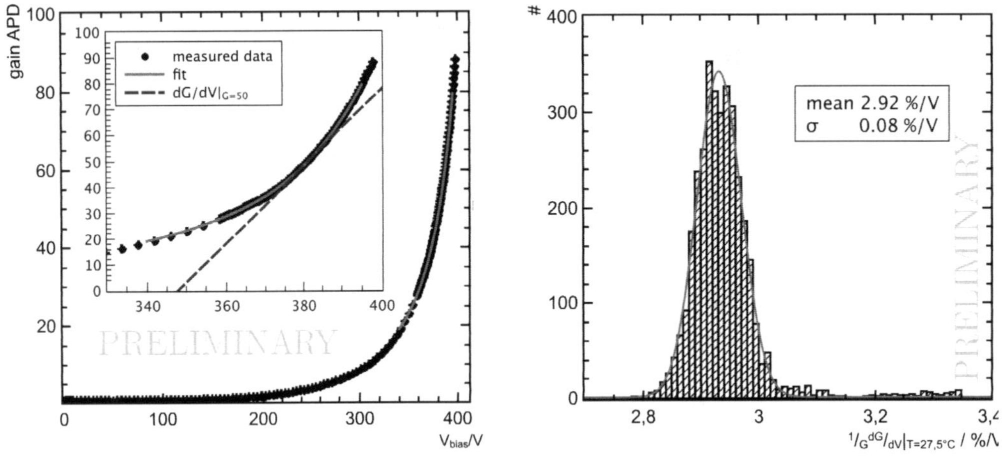

(a) Gain voltage dependence (b) Distribution gain voltage dependence

Figure 3. 3(a):Gain voltage dependence of a S11048(X3) APD; red: fit using eqn. 2. 3(b): Gain voltage dependence at 27.5 °C for the sample of 3500 S11048(X3) APDs.

4. Gain Monitoring and Light Pulser Measurements

One possibility to overcome the gain temperature dependence is to continuously measure the gain of the APD. A common method is to illuminate the APD with a continuous light source and to measure the current flowing through the APD with and without an applied bias voltage. A drawback of this method is its requirement that the current flowing through the APD needs to be measured quite precisely. Utilized in a calorimeter setup this demands a second read-out chain.

This disadvantage can be overcome if pulsed light is used for the gain measurement. Using the light pulse as fixed reference peak seems to be obvious, but a reference measurement is still necessary to get rid of long term variations in the light pulser setup. A reference measurement at $V_{\text{bias}} = 0$ can be used. At this bias voltage no internal amplification takes place inside the APD. Furthermore the generated charge per photon shows no strong dependence on the temperature. However, this needs extra efforts because the capacitance of the APD at about 7.1 nF without bias voltage is quite large compared to 280 pF at a bias voltage for an amplification of 50. On the one hand the noise increases due to the high capacitance, on the other hand the signal is lower due to the missing internal amplification. In addition, there is a charge division between preamplifier and APD which results in further reduction of the signal. This can be compensated by a modified preamplifier. Still the SNR is bad compared to the case where the bias voltage is applied. This can be avoided with high statistics in the measurement and a sufficiently high intensity of the light pulser. As the dark current only manifests as noise but not as a signal seen by a charge sensitive preamplifier, a background subtraction is not necessary when pulsed light is used. The gain is proportional to the quotient of the light of the pulses with and without bias voltage. Knowing about the charge devision between APD and preamplifier and therefore choosing a sufficiently high coupling capacitance, the gain of the APD can directly be calculated from these two measurements.

This method is used to determine the gain temperature and voltage dependence of the 3500 S11048(X3) APDs, illuminating the APDs with 2 μs pulses. A small setup (shown in Fig. 4) consisting of an insolated temperature controlled box with four APD slots was designed. The APDs are read out by charge sensitive preamplifiers SP 917 developed by the Physics Department Basel and supplied with a constant bias voltage from a precise ISEG HV.

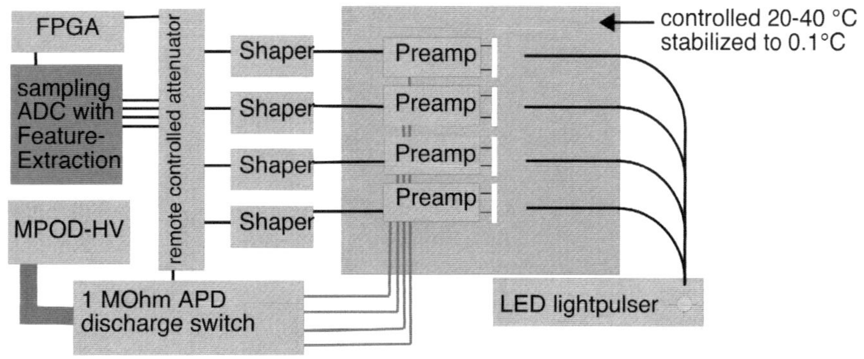

Figure 4. Setup used for characterizing APDs.

The 4 signals are shaped outside the box and read out by a sampling ADC, located on a VFB 2.2 FPGA board of ELB-Elektroniklaboratorien Bonn, with a provided firmware to extract baseline and area of the pulses. These values are also directly provided as multi channel analyzer spectra. Measurements were carried out supplying the APD with a defined bias voltage and also without bias voltage, short circuit the HV supply with 1 MΩ to ensure a complete discharge.

A automated routine taking approximately 4 hours to retrieve gain temperature and voltage dependence was carried out for 3500 APDs. The gain is continuously determined, while the box is heated from 20 to 27.5 °C and the APD is supplied with a bias voltage specified from Hamamatsu for a gain of 50 at 25 °C ($G = 50$). Stabilized at 27.5 °C, the most probable temperature for operating the calorimeter, the gain voltage dependence of the APD is determined. Therefore a voltage from 0 V up to 5 V above the specified voltage for $G = 50$ varying between 335 V and 415 V is applied, leading to a gain of 0 up to approximately 100. Afterwards the temperature dependence is further measured from 27.5 °C up to 40 °C and then down to 20 °C, securing the equality of the temperature of the APD to one of the measuring sensor. Otherwise hysteresis effects of the measurement might arise by changing the temperature to fast. Hence the temperature variation was limited to 17.5 °C per hour.

The described method for the determination of the gain will also be used to monitor the gain of the APDs online in the CB detector and is already implemented in the prototype setup. Light guides connected to each crystal offer the possibility to insert short light pulses from a LED which is placed in a stabilized surrounding. Reference measurements at $V_{\text{bias}} = 0$ V are only necessary to eliminate external changes like e.g. a varying light intensity. With this measurement the gain determination is a pure relative measurement. The gain is directly determined by inserting marked light pulser events into the data stream while data taking or using a spill break of the ELSA accelerator.

5. New Front End Electronics - the High Voltage Supply

One important part of the new front end electronics is the high voltage supply of the APDs. Since each APD needs a certain bias voltage varying between 335 V - 415 V to achieve a defined amplification, a board for the remote adjustment of the incoming bias voltage to the desired value [5] was designed. Additionally this board includes, aside from an identification chip and monitoring ADCs, a temperature compensation circuit. By using a negative temperature coefficient thermistor (NTC), the temperature of the APD is approximated and a generalized average correction is applied to the bias voltage supply. This correction is held as simple as possible and was designed with only passive elements. Its aim is only to reduce the gain temperature dependence, offering a temperature region where the dependence is minimized.

5.1. Performance of the Temperature Compensation

The temperature compensating high voltage board was tested using a light pulser designed as prototype for the upgraded Crystal Barrel experiment. In addition there were simulations [6] to adjust the circuit according to the measured gain temperature and voltage dependence, choosing a very flat gain dependence region to occur around 27.5 °C. Fig. 5 shows the result of a test

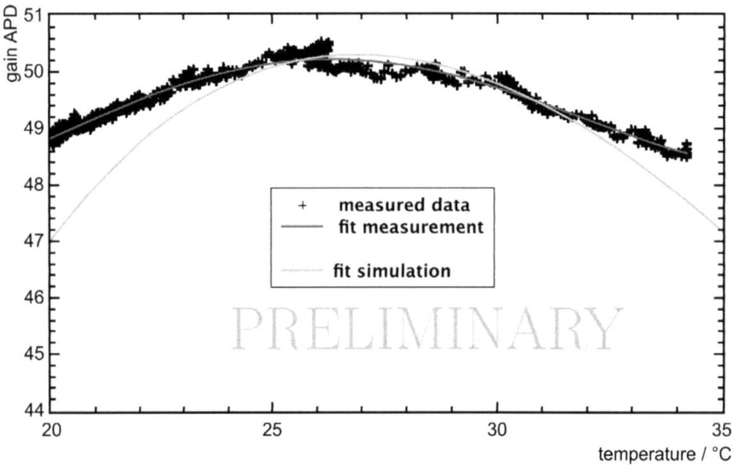

Figure 5. Simulation and measurement of the HV board temperature compensation. The discrepancy arises from a simplified simulation.

measurement between 20 °C and 35 °C. The measured APD gain using the compensation circuit shows even a flatter behavior than in the simulation. This is most probably caused by assuming a linear gain and temperature dependence around gain 50 and 27.5 °C and neglecting gain and temperature dependent changes of the dependencies in the simulation.

5.2. Testing of the Final Readout

The complete new readout was tested in several different experiments. The latest test was carried out at MAMI in Mainz in January 2014. An array of 3×3 crystals equipped with the final electronics was set up and a readout as it will be used in the experiment. The tagged photon beam of MAMI created by bremsstrahlung of an electron beam with an energy of 1.6 GeV and 180 MeV was directly focussed on the central crystal. The new electronics provide a fast trigger signal of each crystal. The OR of all crystals in coincidence with the OR of 32 used tagger bars of the tagging system formed our trigger. Measurements were carried out to determine the energy and time resolution at different photon rates to ensure a high rate capability of the electronics.

5.3. Performance of Final Upgrade Electronics

To analyze the energy resolution, the energy sum measured with all 9 crystals of the detector array is plotted against the energy of the incoming photon. The projection of each used tagger bin shows an energy resolution information for the specific tagger bar energy. The result is shown in Fig. 6 for different rates for both electron beam energies. The rate in the central

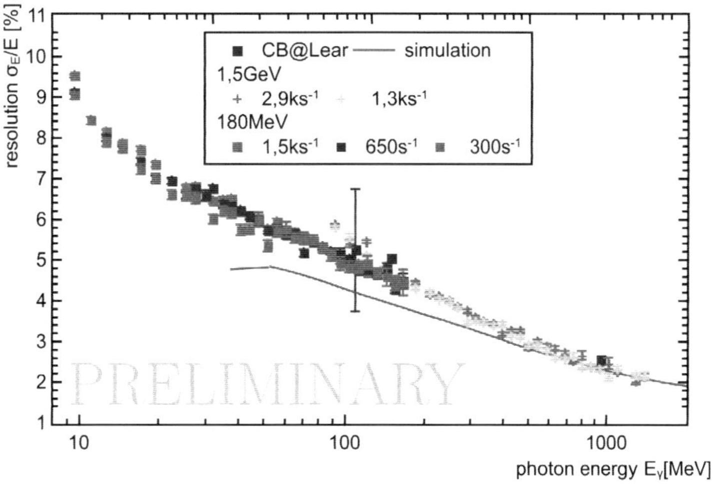

Figure 6. Energy resolution of a 3×3 subunit with new front end electronics compared to PIN diode readout [2] and simulations.

crystal above a threshold of 6.5 MeV is used as reference. In the current setup the crystals in the forward angle show a maximum rate of approximately $1.5\,\mathrm{ks}^{-1}$ above 6.5 MeV during production beamtimes. Three reference points of CB at LEAR are also shown.

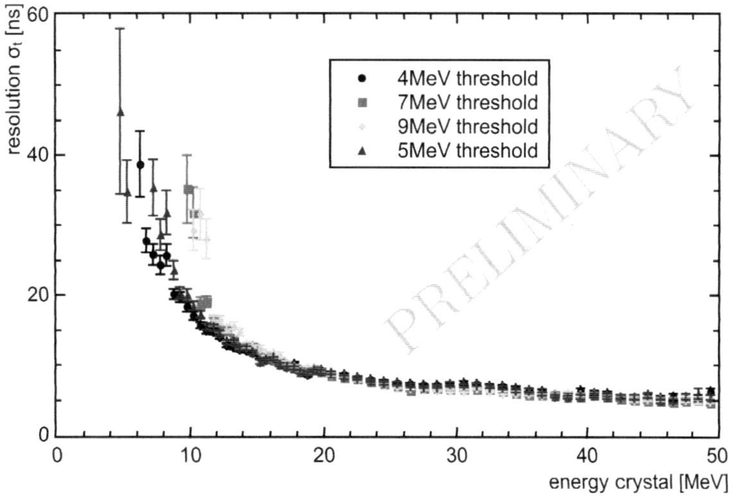

Figure 7. Time resolution of the central crystal in a 3×3 matrix with new front end electronics.

Due to the upgraded front end electronics it is now possible to retrieve a timing signal of each single crystal. These signals were read out for testing purposes using a TDC of CAEN. The resulting time resolution is shown in Fig. 7.

6. Conclusion

The upgrade of the Crystal Barrel calorimeter readout is currently in progress. The 3500 S11048(X3) APDs of Hamamatsu show a average temperature dependence of $(2.27\pm0.08)\,\frac{\%}{\mathrm{K}}$ and voltage dependence of $(2.9\pm0.08)\,\frac{\%}{\mathrm{V}}$. With a specially designed high voltage supply board, equal for all APDs, a circuit to compensate the gain variations with temperature was introduced, which results in a reduction of the gain dependence to $\pm1.5\%$ in a region from $20\,^{\circ}\mathrm{C}$ to $35\,^{\circ}\mathrm{C}$. The gain of the APDs is determined with a light pulser using a relative measurement without applied bias voltage, a method which will be implemented for monitoring purposes in the calorimeter. Tagged photon beam tests showed that the energy resolution of the new readout electronics also achieves 2.5 % at 1 GeV. Due to the upgrade it is possible to also achieve a time information of each crystal with a resolution of $\sigma_t \leq 10\,\mathrm{ns_{RMS}}$ for an energy deposit $E_{\mathrm{crystal}} \geq 20\,\mathrm{MeV}$. The timing signals, included in the first level trigger, will lead to a improved trigger efficiency on neutral reactions. This upgrade will allow detailed studies of double polarized photoproduction off the neutron at the CBELSA/TAPS experiment.

References

[1] Arndt R.A. et al.: Analysis of pion photoproduction data *Phys. Rev. C* 66 (2002) 055213
[2] Aker, E. et al.: The Crystal Barrel Spectrometer at LEAR. *Nuclear Instruments and Methods in Physics Research* A (1992) Nr. 321, S. 69–109
[3] Hillert, W.: The Bonn Electron Stretcher Accelerator ELSA: Past and future.*European Physical Journal* A(28):139–148, 2006
[4] Hamamatsu Photonics K.K.: Delivery Specification Sheet, Doc. No. K30 B70082
[5] Design by C. Honisch and M. Steinacher
[6] Simulations carried out by M. Steinacher

Journal of Physics: Conference Series **587** (2015) 012044 doi:10.1088/1742-6596/587/1/012044

Readout concepts for the suppression of the slow component of BaF_2 for the upgrade of the TAPS spectrometer at ELSA

Stefan Diehl[1], Rainer W. Novotny[1], Benjamin Wohlfahrt[1] and Reinhard Beck[2]

[1] 2nd Physics Institute University Giessen, Heinrich-Buff-Ring 16, Giessen, 35392, Germany
[2] Helmholtz-Institut für Strahlen und Kernsphysik University Bonn, Nussallee 14-16, Bonn, 53115, Germany

E-mail: stefan.diehl@exp2.physik.uni-giessen.de

Abstract. For the measurement at extremely high interaction rates with fast scintillators, pile-up of consecutive events is a limiting factor. With a decay time of 600 ps of the fast cross-luminescence component, Barium Fluoride (BaF_2) is one of the fastest inorganic scintillators known today. However, the dominating slow component with a 3 orders of magnitude longer decay time of 630 ns limits the rate capability. To circumvent this limit, different approaches have been made in the past. The slow component can be suppressed for example by doping the crystals with rare earth ions like La^{3+}. The paper will give an overview over the various concepts investigated in the past and present the suppression via optical band pass filters. This method has been chosen for the upgrade of the BaF_2 crystals in the most forward region of the TAPS-spectrometer at ELSA in Bonn. It allows to reuse the existing crystals and to achieve a high degree of suppression of the slow component. The focus of the paper will be on the selection of the filters, the achievable rate capability and the energy resolution of the fast component.

1. Introduction

The Mini-TAPS, an hexagonally shaped forward wall of BaF_2 detector modules, is used together with the Crystal Barrel based on CsI(Tl) to operate as a nearly 4π electromagnetic calorimeter in experiments with real photons at the Electron Stretcher Accelerator (ELSA) at Bonn. Due to the high hit rates in the most forward region of the detector setup, an upgrade of the inner parts of the spectrometer will be performed to increase the rate capability, which is limited due to pileup of the slow component of BaF_2. The implementation of optical bandpass filters was chosen to suppress the slow component of BaF_2 peaking at an emission wavelength of 310 nm.

2. The CB-ELSA/TAPS Experiment

The CB-ELSA/TAPS Experiment [1] is used to investigate photonuclear reactions with energy marked photons of an energy up to 3.5 GeV, produced form a primary electron beam by tagging photons due to Bremsstrahlung. The experimental setup is shown schematically in Fig. 1. The photons interact with a fixed target in the center of the Crystal Barrel detector comprising 1380 CsI(Tl) crystals arranged in a barrel shape covering the angles between 12 and 168, respectively. Due to the kinematics of a fixed target experiment one needs a fast and efficient detector in the

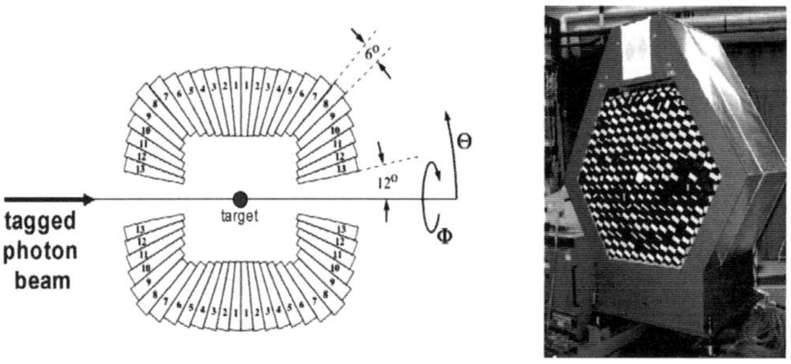

Figure 1. Setup of the Crystal Barrel / TAPS experiment.

forward direction. This role is accomplished by the Mini-TAPS forward spectrometer covering the angles below 12. It consists of 216 BaF_2 crystals with a length of 12 X_0, a hexagonal shape of 5.9 cm diameter and readout individually via photomultiplier tubes (2 inch diameter, Hamamatsu R2059-01 with fused silica window). The crystals are arranged annularly in 8 rings around the beam axis [2]. One removed single detector in the center allows the photon beam to pass.

2.1. The Scintillation Mechanism of BaF_2

Barium fluoride is among to the group of scintillator materials providing the fastest mechanism presently known. For illustration, Fig. 2 shows schematically the band model of BaF_2 explaining the origin of the two different luminescence processes. As an insulator, the conduction band is normally empty, while the valence band and the upper core band are nearly completely filled. In the excitation phase due to ionizing radiation, electrons are excited from the lower bands to the conduction band. Holes in the upper core band can relax very fast to the upper end of the core band and recombine via a core-to-valence transition with an electron from the valence band. This so called "cross luminescence", a special type of intrinsic luminescence within two bands with a decay time of 600 - 800 ps, is enhanced since the electron energy is not sufficient for excitation of Auger electrons from the valence band. The energy of the scintillation photons peaks at 195 nm and 220 nm, respectively, differing by the width of the valence band The excited and slowly relaxed electrons in the conduction band can recombine via self trapped exciton formation with hole in the valence band. These luminescence processes are significantly slower at the level of 620 ns. However, this process has a much higher probability and dominates the emission spectrum. The emission spectrum of BaF_2 resulting from this two processes is shown in Fig. 2 with the maxima of the fast component at 195 nm and 220 nm and the slow component at 310 nm, which has in contrast to the fast component is strongly temperature quenched Fig. 3 gives an overview over the most relevant parameters of BaF_2.

2.2. Hit Rate Distribution for TAPS

To illustrate the rate limitation of TAPS, Fig. 4 shows the hit rate distribution for the different regions of the TAPS spectrometer in a typical photonuclear reaction with a fixed target. The figure clearly shows, that the rates are significantly enhanced in the central region and peaking at the first and second ring leading to an increase by a factor of 3 to 5. To avoid pile-up of the slow component, which requires on average an integration time of $2\mu s$, the single crystal rate has to be limited to $\simeq 100\ kHz$, which has a strong impact on the detection efficiency of weak processes. As a solution, the suppression of the slow component at least in the inner 2 to 3 rings could exclude the rate limitation due to the scintillator itself.

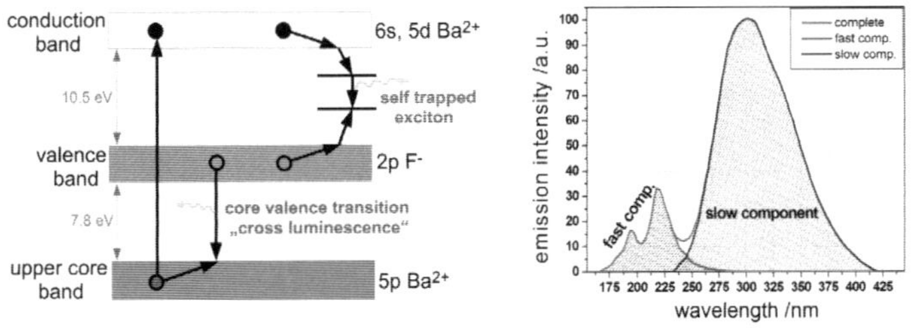

Figure 2. Scintillation mechanism (left) and emission spectrum (right) of BaF_2.

property	value
density	4.89 g/cm^3
rad. length X_0	2.05 cm
Moliére-Radius R_M	3.4 cm
light yield for γ-rays	11300 ph/MeV
	fast: 1300 ph/MeV
	slow: 10000 ph/MeV

Figure 3. Selected physical and scintillation parameters of BaF_2.

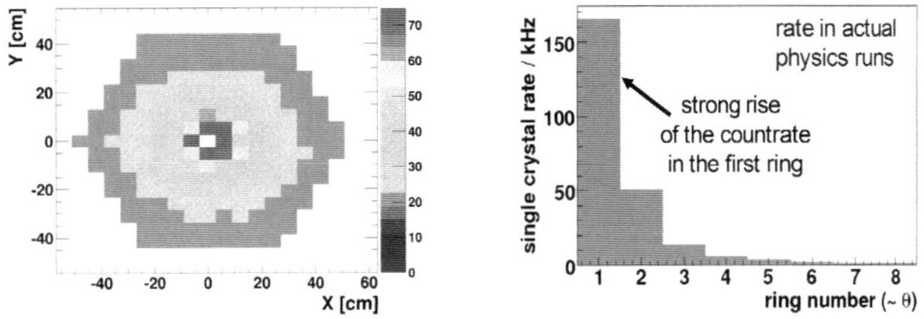

Figure 4. Hit rate distribution for photons for the complete TAPS detector (left) and single crystal rate in a typical physics experiment as a function of the number of detector rings (right)

3. Concepts for the Suppression of the Slow Scintillation Component

In the past several concepts have been investigated to suppress the slow component of BaF_2. Several attempts have been made to change the scintillation process by appropriate co-doping. Alternative approaches are based on a selective sensitivity of photo sensors or the use of optical filters, which has been chosen for the upgrade of the TAPS modules.

3.1. Use of Doped Crystals

The doping of BaF_2 crystals, especially with rear earth elements like La^{3+} ions was studied in detail by Schottanus et al. [3]. Fig. 5 shows the impact of different La^{3+} concentrations on the emission spectra of BaF_2. The created color centers lead to a strong but not sufficient suppression primarily of the slow component. However, even for the largest concentration (13.3 mol%, both components are nearly of similar intensity, since the fast component is reduced to a level of around 50 % in parallel.

3.2. Use of Special Photo Sensors

Another possibility to suppress the slow component would be the use of photo sensors which are insensitive in the spectral range of the slow component. Solar blind PMTs, such as the (Hamamatsu R4679), exploit a CsTe photocathode with a peaking quantum efficiency at 225 nm and a sensitivity above 300 nm reduced by almost 3 orders of magnitude. Fig. 6 (left) shows the quantum efficiency of such a PMT (Hamamatsu R3197) in comparison to the actually used Hamamatsu R2059-01 PMTs and compared to the emission spectrum of BaF_2. Test measurements showed, that the finally detected light consists of 50 % fast and 50 % slow component which can be explained by the quantum efficiency (QE) in Fig. 6 and the relative intensity of the two luminescence components. A second better adapted possibility is the use of diamond photo sensors, which were developed for astrophysical applications [5]. Due to the

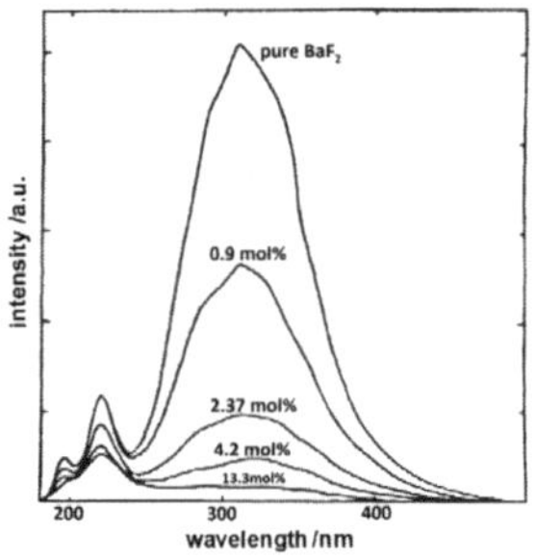

Figure 5. Emission spectrum of doped BaF_2 for different concentration of La^{3+} [3].

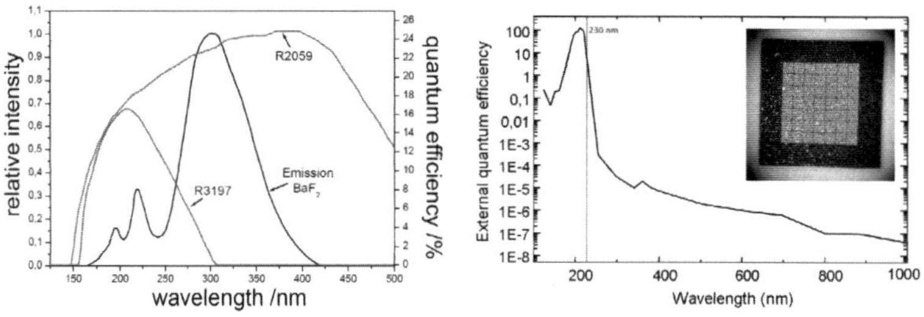

Figure 6. Quantum efficiency of a Hamamatsu R3197 solar blind PMT [4] in comparison to a Hamamatsu R2059-01 PMT (bialkali photocathode) [4] and to the relative emission intensity of BaF_2 (left). Quantum efficiency of a single cell of a diamond photo sensor [5] and a picture of a pixel array of diamond sensors [5] (right).

relatively wide band gap of diamond of 5.5 eV these sensors have a high sensitivity in the narrow range from 180 nm to 225 nm with a large drop for wavelengths $\lambda > 230$ nm (compare Fig 6 (right)). For spectroscopic applications an arrangement in pixel arrays similar to silicon photomultipliers is needed. So far, there are no practical experiences in spectroscopy and first prototypes provide only an option for future detector projects.

3.3. Use of Optical Filters

The use of optical filters became the most promising and finally chosen concept for the upgrade of the existing TAPS setup with the advantage that all existing components can be reused. Fig. 7 (middle) illustrates schematically the position of the filter within an existing TAPS module consisting of a BaF_2 crystal, a PMT and a voltage divider base. The main requirements for an optical filter are a high transmittance for the fast component and at the same time a high grade of suppression for the slow component. An ideal filter would be a short pass filter with a sharp edge between 230 nm and 235 nm. Fig 7 (left) shows a theoretically calculated transmission performed with a special simulation program provided by LOT-Oriel. However, there are no practical experiences in producing such filters and a costly development would be required.

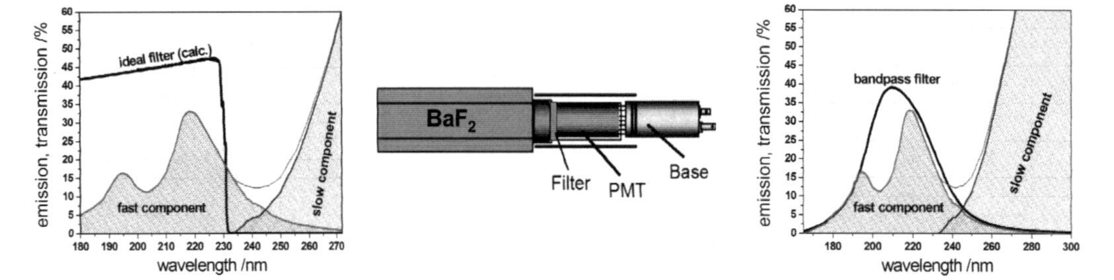

Figure 7. Calculated transmission of an ideal filter compared to the emission spectrum of BaF_2 (left) and the measured transmission spectrum of a band pass filter compared to the emission of BaF_2 (right). The schematic setup of an individual TAPS BaF_2 detector with the implemented filter, PMT and voltage divider is shown in the middle part of the figure.

Finally, standard optical band pass filters consisting of hard coated quartz glass, as shown in Fig. 7, will be used after an optimization of the most probable wavelength. As a remaining disadvantage, a small but tolerable part of the tail of the slow component towards short wavelengths causes a LY_{fast} / LY_{slow} ratio > 5. Since the remaining slow component is emitted with a significantly longer decay constant of 620 ns, only a small background contribution will be left over underneath setting a short integration gate of typically 20 ns.

4. Response Measurements Using Optical Band Pass Filters

To verify the applicability of the optical filters, different response tests have been performed.

4.1. Quality Control of the Filters

For the upgrade 40 filters have been delivered from Princeton Instruments. Fig. 8 (left) shows the overlay of the measured transmission curves and a picture of a typical filter. The Gaussian transmission distribution peak is on average at a wavelength of 214 ± 3 nm with an absolute transmission value for λ_{max} between 36% and 42%, respectively. Under the assumption of the spectral distribution of the emission spectrum, as shown in Fig. 7 (right), the contributions to both scintillation components have been calculated for each of the measured transmission curves of the available filters considering in addition the quantum efficiency of the photomultiplier R2059-01 and the transmission of the crystal. However, the kinetics of the luminescence processes has been ignored. The two correlations shown in Fig. 8 (right) relate the normalized intensity of the slow component to the fast component or the integral light yield collected after the filter, respectively. Unfortunately, both figures illustrate the consequence from the nearly identical transmission shape of the filters that a small shift of the transmission curve to higher wavelengths increases the total light yield after the filter but allows also a larger contribution from the slow scintillation component.

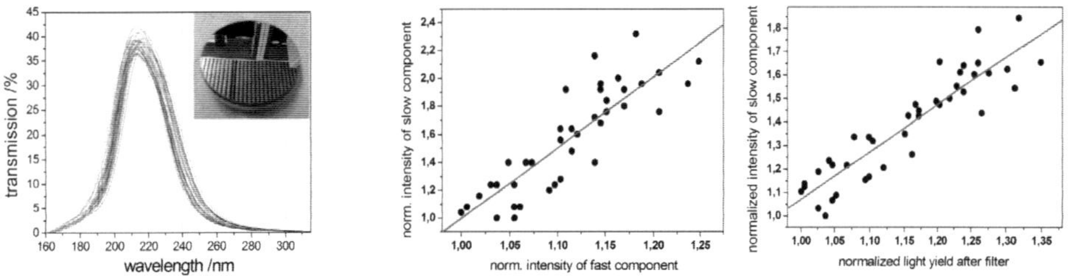

Figure 8. Measured transmission spectra of 40 band pass filters and a picture of a typical band pass filter (left). Correlations between the calculated intensity of the slow component and the fast component (middle) and the total light yield detected after the filter (right), respectively. The assumptions are explained in the text.

4.2. Signal Shapes with and without a Filter

Fig. 9 shows the measured signal shape for a low energy γ-source of a BaF_2 detector, read out with a R2059-01 PMT with and without an optical band pass filter in between, recorded with a digital oscilloscope at high sampling rate. Without a filter, the slow component is dominating after 10 ns with a significant contribution up to 2.5 μs. Using a filter, there remains nearly no visible contribution of the slow component. Only on a larger scale one can identify a small contribution of the remaining slow component slightly above the baseline in the first 50 ns. However, the contribution underneath the fast component is well below 10%. From this observations, a rate capability of at least 1-2 MHz can be estimated for the modified detector modules without major limitations due to any pile-up. In contrast to this, with an ideal short pass filter rates of 10 MHz should be possible.

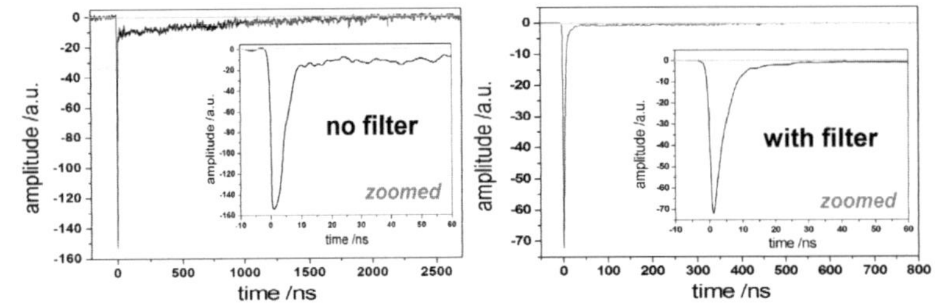

Figure 9. Signal shape of BaF_2 recorded without (left) and with a band pass filter (right), respectively.

4.3. *Response to Low Energetic Gamma Rays and Cosmic Muons*

To get a first estimate for the light yield and the energy resolution using optical band pass filters, measurements with radioactive gamma sources (^{137}Cs, ^{22}Na, ^{60}Co) and cosmic muons have been performed. Fig. 10 (left) shows that a broad photo peak can be clearly distinguished above the noise threshold and a linear response is confirmed. In these tests, a light yield of 20-25 pe/MeV and a time resolution of σ_t = 400-450 ps have been obtained. Based on the calibration with low energy photons the most probable energy deposition for passing cosmic muons is shown Fig. 10 (right). The detector was placed horizontally and the diameter between two opposite sides of the hexagonal cross section amounts to 5.9 cm. The measured most probable energy loss comes close to the energy deposition of minimum ionizing particles of 37.6 MeV for 5.9 cm BaF_2. The TAPS calorimeter is used for reconstructing electromagnetic showers up to 3.5 GeV energy. The obtained individual detector threshold well below 500 keV will not limit the achievable energy resolution of the inner detectors relying on the fast light component only.

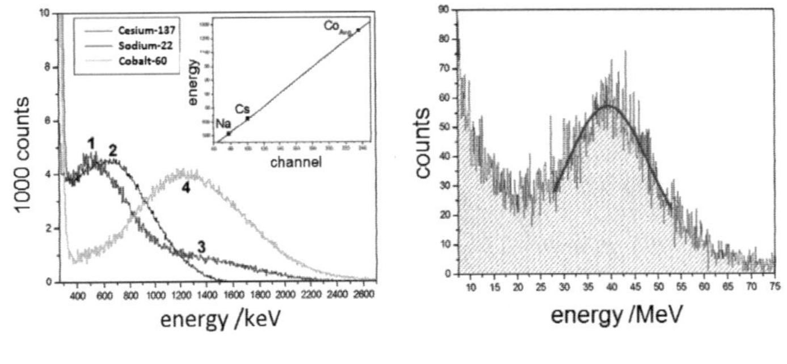

Figure 10. Response of a BaF_2 detector equipped with a band pass filter measured for low energetic gamma rays (left) and the energy loss of cosmic muons (right).

4.4. *Response Expected for High Energy Photons*

To get an estimate for the overall energy resolution for electromagnetic showers up to several GeV energy equipping the detectors with optical band pass filters, a calculation was made based on the known energy resolution of the TAPS calorimeter measured for the total light output as well as the fast component separately by A. Gabler et al. [6]. The contribution of the scintillation light of the fast component was obtained by charge integration of the photomultiplier signal within an integration gate of 50 ns. Assuming an optical transmittance of the UV-filters of 30 - 35 % and considering an additional light loss of 10-20 % due to the missing slow component within the first 50 ns, an energy resolution given by $\sigma/E = (1.41\% - 1.61\%)/\sqrt{E_\gamma} + 1.80\%$ can be estimated, which is comparable to the energy resolution of the target calorimeter of the

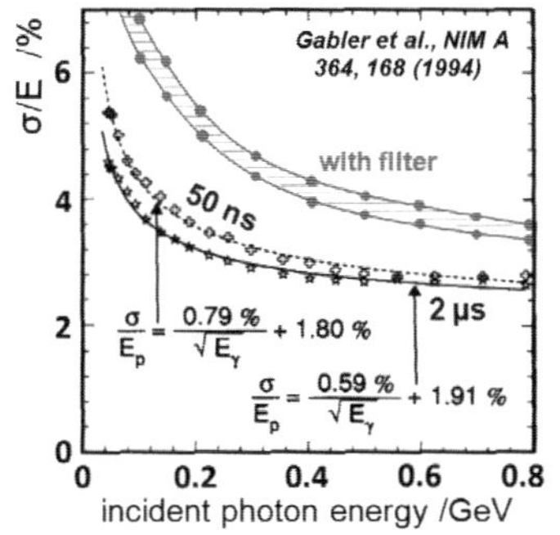

Figure 11. Energy resolution of a matrix of 19 BaF_2 scintillation detector in TAPS geometry read out with a PMT with bialkali photocathode. The emission light in total was collected within an integration time of 2 μs (solid line) and the fast component was extracted separately by integrating over the first 50 ns only (dotted line) [6]. For comparison, the expected energy resolution using a filter with reduced transmittance is indicated by the dashed area.

future PANDA detector at FAIR based on lead tungstate when operating at room temperature. The measured and expected energy resolutions, respectively, are compared in Fig. 11.

5. Discussion and Outlook

The fast component of BaF_2 represents the fastest scintillation mechanism in inorganic scintillators, presently known. Therefore, the selective readout of the fast component only (10 times faster kinetics than PWO) would enable very high rate capabilities without giving up the required energy resolution for electromagnetic showers in the energy range between a 50 MeV and 3.5 GeV, respectively. Even commercially available optical band pass filters provide a high grade of suppression of the slow component and the reduced photon statistics is still sufficient to guarantee a high resolution electromagnetic calorimeter. Presently, only the most forward part of the TAPS-wall is considered to be modified. However, there is one serious drawback in the applicability of BaF_2. Due to the very different origin of the two luminescence components, the line shape of a BaF_2-detector allows particle/photon separation via pulse shape analysis exploiting the ratio of the two components. For highly ionizing charged particles appears the fast component significantly reduced compared to electron/positrons generated in the electromagnetic shower. Therefore, the modified detector modules will provide only information on energy and timing. A direct proof of the energy resolution of a sub-matrix of modified detectors is scheduled in summer 2014 using energy marked photons up to 800 MeV at the tagging facility at MAMI, Mainz (Germany).

References

[1] Crystal-Barrel/TAPS at ELSA, *http : //www1.cb.uni − bonn.de/index.php?id* = 6 (2014)
[2] R. Novotny, The BaF2 Photon Spectrometer TAPS, IEEE Transaction on nuclear science, 38(2) (1991).
[3] P. Schottanus, P. Dorenbos, C.W.E. Van Eijk and H.J. Lamfers, Suppression of the slow scintillation light output of BaF2 crystals by La3+doping, NIM A281, 162 (1989).
[4] Hamamatsu Photonics, http://www.hamamatsu.com (2014).
[5] E. Pace and A. De Sio, Innovative diamond photo-detectors for UV astrophysics, Mem. S.A.It. Suppl. Vol. 14, 84 (2010).
[6] A.R. Gabler et al., Response of TAPS to monochromatic photons with energies between 45 and 790 MeV, Nuclear Instruments and Methods in Physics Research A 346, 168-176 (1994).
[7] M. Laval et al., Barium Fluoride Inorganic Scintillator for Subnanosecond Timing. Nuclear Instruments and Methods in Physics Research 206, 169-176 (1983).

Journal of Physics: Conference Series **587** (2015) 012045 doi:10.1088/1742-6596/587/1/012045

Electromagnetic calorimeter for Belle II

Belle-ECL[1],V Aulchenko[1,5], A Bobrov[1,5], A Bondar[1,5], B G Cheon[2], S Eidelman[1,5], D Epifanov[6], Yu Garmash[1,5], Y M Goh[2], S H Kim[2], P Krokovny[1,5], A Kuzmin[1,5], I S Lee[2], D Matvienko[1,5], K Miyabayashi[4], I Nakamura[3], V Shebalin[1,5], B Shwartz[1,5*], Y Unno[2], Yu Usov[1,5], A Vinokurova[1,5], V Vorobjev[1,5], V Zhilich[1,5], V Zhulanov[1,5]

[1] Budker Institute of Nuclear Physics SB RAS, Novosibirsk 630090, Russia
[2] Hanyang University, Seoul 133-791, Korea
[3] High Energy Accelerator Research Organization (KEK), Tsukuba 305-0801, Japan
[4] Nara Womens University, Nara 630-8506, Japan
[5] Novosibirsk State University, Novosibirsk 630090, Russia
[6] Department of Physics, University of Tokyo, Tokyo 113-0033, Japan

E-mail: * `shwartz@inp.nsk.su`

Abstract. The electromagnetic calorimeter of the BELLE II detector for experiments at Super B-factory SuperKEKB is briefly described. The project of the calorimeter upgrade to meet severe background conditions expected at the upgraded KEK B factory is presented.

1. Introduction

Experiments with the Belle detector [1] at KEKB B-factory [2], energy-asymmetric collider with high luminosity, were conducted from 1999 until 2010. In these experiments the world highest luminosity of the collider, $2 \times 10^{34}\mathrm{cm}^{-2}\mathrm{s}^{-1}$, was achieved that provided a basis to obtain ample results in several fields of particle physics. Although the most known Belle achievements concern the CP symmetry violation in the quark sector, very important results were also obtained in the heavy quarkonium spectroscopy, tau lepton decays and two-photon physics.

The Belle detector is a forward/backward asymmetric detector with high vertex resolution, precise magnetic spectrometry, very good calorimetry and sophisticated particle identification ability. It should be stressed that high energy resolution and detection efficiency for photons in a wide energy range is expremely important since one third of B-decay products are π^0's and other neutral particles providing photons. Thus, a calorimeter able to detect photons in a energy range from 20 MeV to 4 GeV with high characteristics, is a very important part of the detector. The CsI(Tl) scintillation crystals were chosen as a material for the Belle calorimeter due to their high light output, short radiation length, good mechanical properties and moderate price. During more than 10 years of operation electromagnetic crystal calorimeter (ECL) of the Belle detector demonstrated its high quality and performance [4].

Initiated by the success of the KEKB/Belle experiment, the new advanced project, KEKB II/Belle II, was accepted. The KEKB II luminosity will exceed the previous one by about 40 times. Belle II should have considerably higher performance than Belle. This upgrade will open exciting possibilities in a search and study of new physics phenomena in the heavy quarkonia, lepton flavour violation in tau decays as well as in other particle physics fields.

The first priority of the electromagnetic crystal calorimeter (ECL) upgrade is the calorimeter electronics modification following the general strategy of the Belle upgrade. The main idea is to shorten the shaping time from 1 μsec to 0.5 μsec and to use a pipe-line readout with waveform analysis.

The potential improvement at this way is limited by the long scintillation decay time of CsI(Tl) – 1 μsec. Then for further inprovement we consider the next stage of upgrade with a replacement of the slow CsI(Tl) crystals in endcaps to fast pure CsI scintillation crystals. This work is at the R&D stage now.

2. Belle/Belle II electromagnetic calorimeter

The Belle II detector project is described in the Technical Design Report [3]. Schematic view of the Belle II detector (top half) in comparison to the previous Belle detector (bottom half) is presented in Fig. 1.

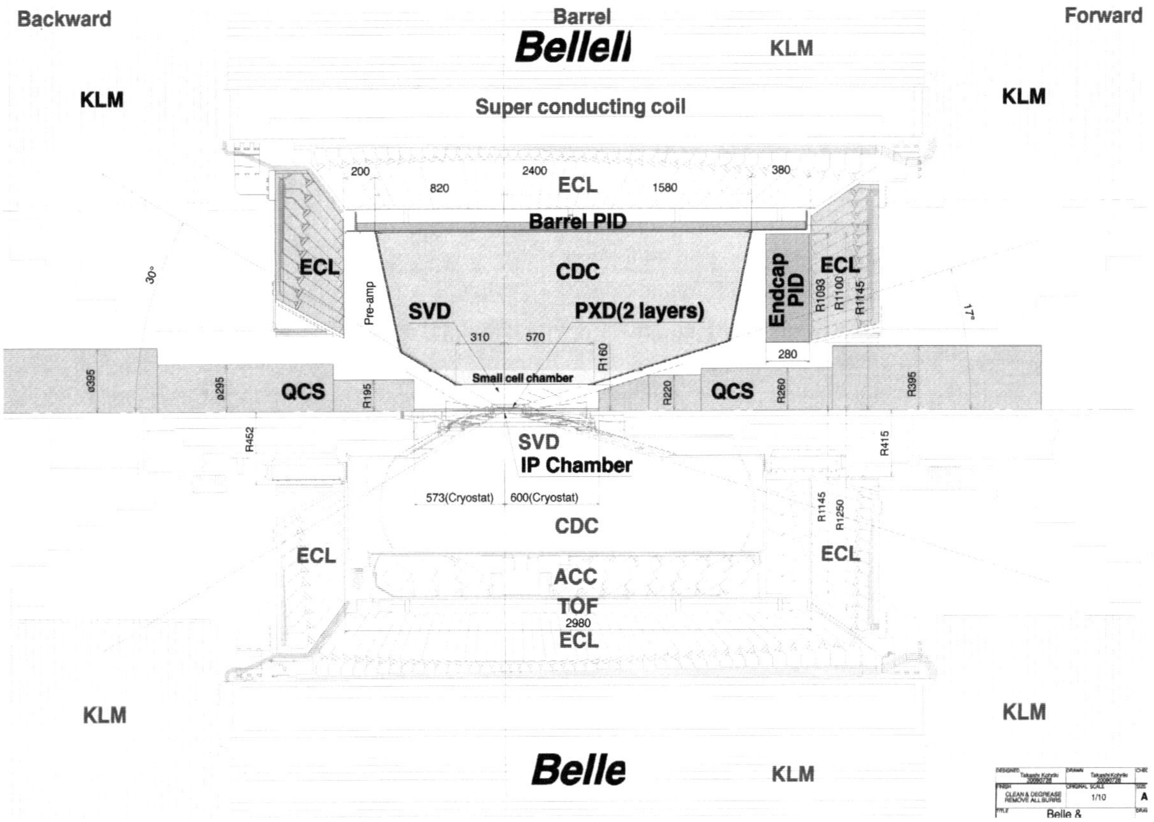

Figure 1. Schematic view of the Belle II detector (top half) in comparison to the previous Belle detector (bottom half).

The main tasks of the calorimeter are:

- detection of γ-quanta with high efficiency,
- precise determination of the photon energy and coordinates,
- electron/hadron separation,
- K_L detection together with KLM,
- generation of the proper signal for trigger,

- on-line and off-line luminosity measurements.

The electromagnetic calorimeter (ECL) consists of a 3 m long barrel section with an inner radius of 1.25 m and the annular endcaps at $z = 2.0$ m (Forward part) and $z = -1.0$ m (Backward part) from the interaction point. The calorimeter covers the polar angle region of $12.4^\circ < \theta < 155.1^\circ$ except two gaps $\sim 1^\circ$ wide between the barrel and endcaps.

The barrel part has a tower structure projected to the vicinity of the interaction point. It contains 6624 CsI(Tl) elements of 29 types. Each crystal is a truncated pyramid of the average size about 6×6 cm^2 in cross section and 30 cm ($16.2X_0$) in length. The endcaps contain altogether 2112 CsI crystals of 69 types. The total number of the crystals is 8736 with a total mass of about 43 tons.

Each crystal is wrapped with a layer of 200 μm thick Gore-Tex porous teflon and covered by the 50 μm thick aluminized polyethylene. For light readout two 10×20 mm^2 Hamamatsu S2744-08 photodiodes are glued to the rear surface of the crystal via an intervening 1 mm thick acrylite plate. The LED attached to the plate can inject the light pulses to the crystal volume to control the optical condition stability. A preamplifier is attached to each photodiode providing two independent output lines from each crystal. These two pulses are summed at the shaper board. For the electronics channel control the test pulses are fed to the inputs of the preamplifier. The average output signal of the crystals measured by calibration with cosmic rays is about 5000 photoelectrons per 1 MeV while a noise level is equal to about 200 keV. The energy resolution for photons, $\sigma E/E$, slightly changed from about 2.5% at 100 MeV to 1.7% at 5 GeV.

However, the experiment at SuperKEKB with a luminosity up to 8×10^{35} cm^{-2}s^{-1} puts new, severe requirements on the detector. Since KEKB collider is an e^+e^- machine with circulating beams up to 3–4 A, an obvious concern is a probable degradation of the crystal parameters due to a high radiation dose. The measured integrated dose absorbed by Belle ECL crystals by summer 2009 (i.e at integrated luminosity 900 fb^{-1}) is about 100 rad for the barrel part and about 400 rad in the most loaded part of the endcaps. The light output of the crystals is around 7% in the barrel and up to 13% in the endcap parts close to the beam pipe. These results are in a good agreement with previous measurements of the crystal radiation hardness [5]. Since these studies show the loss of the light output to be less or about 30% at 3.6 krad, an increase of the absorbed dose by one order of magnitude will not provide serious problems.

Another effect of the radiation background is the increase of the dark current of photodiodes induced by the neutron flux. At present the increase of the dark current in the barrel part is not substantial, less than 10 nA, while it is rather large in the endcap calorimeter – up to 200 nA. Thus, at the further high luminosity, we may encounter some problems caused by the high dark current.

The issue we have to treat more carefully is so called pile up noise caused by the soft background photons with average energy of about 1 $\lesssim$MeV. The fluctuation of the number of these photons coming during the integration time contributes to the total noise level. The noise level, σ, measured at the luminosity of $L = 10^{34}$ cm^{-2}s^{-1} already approaches to 1 MeV in the end caps that substantially exceeds the electronics noise.

High-energy photon background produces random clusters in the calorimeter. At present each event contains in average 6 background clusters (3 in barrel and 3 in the both endcaps) of the energy exceeding 20 MeV. We can expect 10 times high rate at Belle II which will produce a huge combinatorial background at the event reconstruction unless we upgrade calorimeter seriously.

3. ECL electronics upgrade

The main idea of the ECL electronics upgrade is to shorten the shaping time and to use a pipe-line readout with waveform analysis. The shaping time becomes 0.5 μs and a counter signal is digitized continuously with 0.5 μs clock interval. Initiated by the trigger pulse, 16 points within the signal are fit to the signal shape function $F(t)$: $F(t) = A_0 \times f(t - t_0)$ where A_0 is a pulse height and t_0 is an event time. The signal shape function, $f(t)$, is evaluated from separate measurements. Both amplitude (A_0) and time (t_0) are obtained by the on-line shape fit:

$$\chi^2 = \sum_{i,j}(A_i - F(t_i))S_{ij}^{-1}(A_j - F(t_j)), \qquad (1)$$

where A_i is an amplitude at the time t_i and S_{ij} - covariance matrix.

Time information allows to suppress the fake clusters by 7 times by rejecting wrong time clusters. At the described scheme the pileup noise will be reduced by a factor 1.5

A block diagram of the new ECL electronics is presented in Fig. 2. At least in the first years

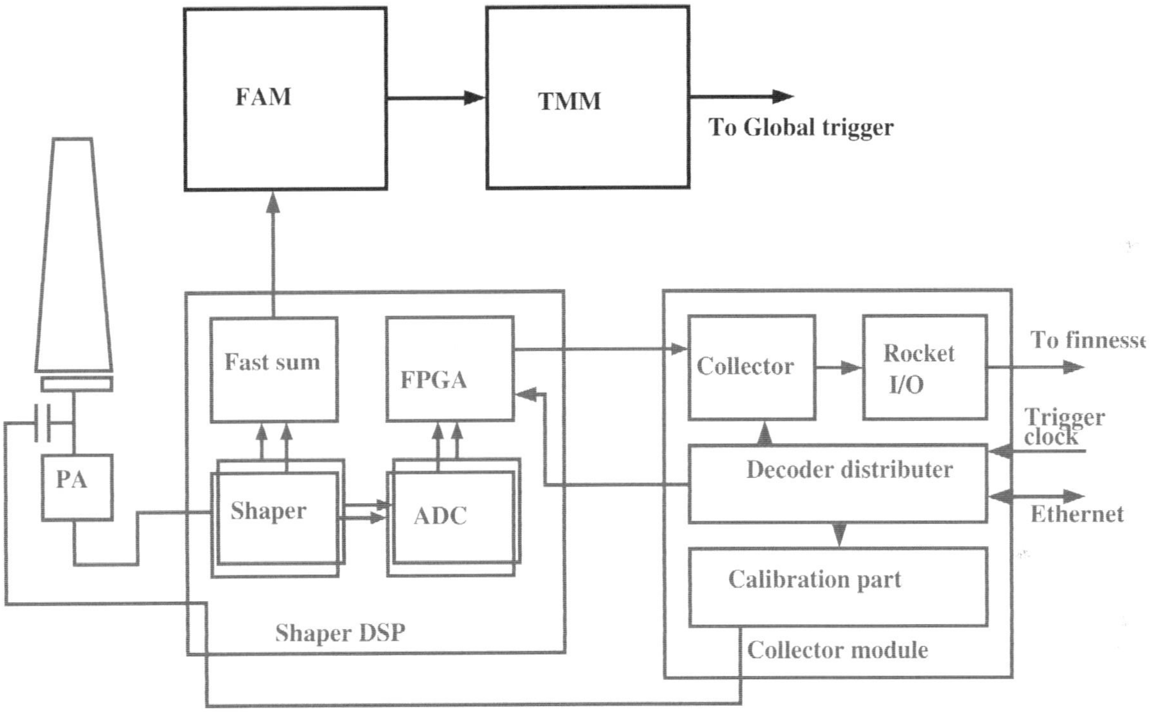

Figure 2. Block diagram of ECL electronics to read out CsI(Tl) crystals at Belle-II experiment.

of of Belle-II experiment, the existent CsI(Tl) crystals, both in the barrel and endcaps will be used together with the same PIN-Photo-diodes, preamplifiers and the cables to reach the shaper are planned to be used. The Shaper-QT modules used for the Belle ECL are replaced by the Shaper-DSP modules.

New shaper-digitizer boards will keep the density of 16 channels per module. Each channel contains a shaping amplifier CR-(RC)^4 ($\tau_s = 0.5\mu$s) and 18-bit flash ADC (Analog Devices AD7641) which digitizes the signal with 2 MHz clock frequency. The ADC data are read out by the digital processor realized in the XILINX FPGA. The data processing is initialized by the trigger signal and yields three parameters (amplitude, time and a quality of the data) which are calculated according to the algorithm described above.

The data from the Shaper-DSP are sent to the Collector module which packs the data of 8-12 Shaper-DSP modules and sends this by ROCKET-I/O line to COPPER FINNESSE module of Belle II DAQ. The collector module contains also a part for the test pulse generation to calibrate the response of the channel.

Each Shaper-DSP module also includes fast shaping ($\tau_d = 0.2\,\mu$s) and gain adjustment circuits per each channel for a trigger output. A fast analog sum signal which is combined with 16 gain-corrected fast shaping channels is delivered to the FAM (Flash-ADC trigger Module) where a trigger cell signal is generated after comparing with a threshold ($\sim$ 100 MeV). These trigger cell signals are collected and processed by the TMM (Trigger and Monitoring Module) where ECL subtrigger arguments are formed and sent to GDL.

The simplified block diagram of the pulse analizing part is shown in Fig. 3. The digital processor is built in XILINX FPGA XC3S1500-FG456 which has enough resources to realize waveform analysis in real time. The SDRAM chips are used to store precomputed coefficients for the digital signal processor (DSP). The idea of the waveform analysis algorithm is following.

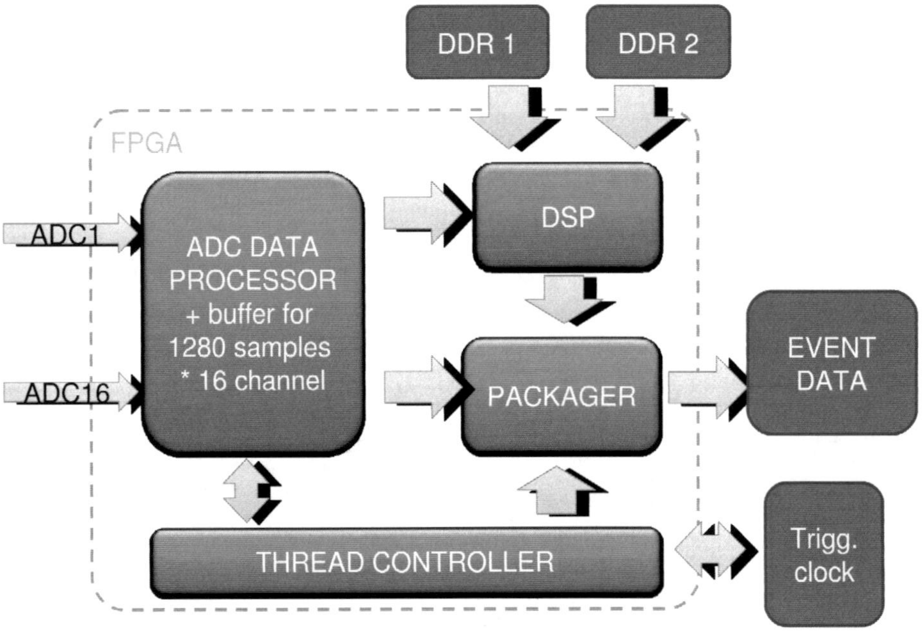

Figure 3. Schematic diagram of digital circuit realized by the FPGA for the energy and time measurements.

First, the signal function $f(t)$ as well as its deriviative $f'(t)$ are tabulated to obtain the tables of values $f_{ki} = f(t_{ki}) = f(k\delta t + i\Delta t)$, where Δt is the digitizing interval, and δt is the timing bin width of tabulation that must be much smaller than Δt, i and k are the corresponding bin indexes. Linearizing $A_0 f(t)$ in Eq. 1

$$Af(t_{ki} - t_0) = A_0 f(t_{ki}) - t_0 A_0 f'(t_{ki}) = A_0 f_{ki} + B f'_{ki}\,,$$

we can rewrite Eq. 1 taking a pedestal p into account:

$$\chi^2 = \sum_i (A_i - A_0 f_{ki} - B f'_{ki} - p)(S^{-1})_{ij}(A_j - A_0 f_{kj} - B f'_{kj} - p)\,. \tag{2}$$

The minimization of this functional is equivalent to the linear system solution:

$$\sum_{i,j} f_{ki} S^{-1}_{ij} (S^{-1})_{ij}(A_j - A_0 f_{kj} - B f'kj - p) = 0 \qquad A_0 = \sum_i \alpha_i^k A_i$$

$$\sum_{i,j} f'_{ki} S^{-1}_{ij} (S^{-1})_{ij} (A_j - A_0 f_{kj} - B f' kj - p) = 0 \quad \Rightarrow B = \sum_i \beta^k_i A_i,\ t_0 = -B/A_0$$

$$\sum_{i,j} S^{-1}_{ij} (S^{-1})_{ij} (A_j - A_0 f_{kj} - B f' kj - p) = 0 \qquad p = \sum_i \gamma^k_i A_i \tag{3}$$

where α^k_i, β^k_i, γ^k_i are expressed via tabulated value f_{ki}, f'_{ki} and covariation matrix S^{-1}. Three parameters are obtained by the solution: amplitude (A_0), peak time (t_0) and pedestal (p). The obtained t_0 allows to calculate new k-index for the next iteration. Three iterations are enough for the convergence of the solution. For the first iteration the trigger time is used as the initial pulse time. As one can see from Eq. 3 one division and multiply-and-accumulate (MAC) operations are needed at each iteration. Fixed-point arithmetic and precomputed array of coefficients (α^k_i, β^k_i, γ^k_i) are sufficient to realize the algorithm at FPGA in real time computing.

For some fraction of data both input information (at 16 or 32 points) are sent to DAQ together with output parameters for algorithm and electronics monitoring.

4. Current status

The current activity on the ECL upgrade is going in several directions.

- New electronics production and tests.
- ECL trigger modules (FAM and TMM) development.
- On-line luminosity moniror development.
- Development of the software for the calorimeter DAQ and optimisation of the signal processing in the DSP.
- R&D works towards further endcap calorimeter upgrade.

By now the Shaper-DSP and Collector modules has been developed and these are in a production stage. All 432 barrel Shaper-DSP modules were delivered to KEK and tested with the specially developed test bench. In the test procedure following characteristics of the modules are checked: signal shape; FAM signal shape; slope and linearity; electronics noise level; FPGA logic and the attenuation range. All tested modules showed the proper characteristics. The measured nonlinearity of the Shaper-DSP is less than 2×10^{-3} within the full 260 000 bins of the ADC range.

All 144 Sh-DSP modules for the endcap parts will be delivered to KEK and test during 2014.

After the Great Japan Earthquake of 2011 it was important to check the status all channels of the barrel calorimeter (endcap parts were dismounted by that time). A special stand-alone test setup containing 12 Shaper DSP (prototype) modules, collector and FAM modules in one VME crate controlled by PC were prepared. Then all 6624 channels were tested batch by batch using cosmic ray particles. Fortunately all channels appeared to be operable.

5. Pure CsI in the endcaps

The drastic way to improve the characteristics of ECL is a replacement of the slow CsI(Tl) crystals with faster ones, at least in the end cap parts of the calorimeter. In the frame of this option pure CsI crystals of the same shape and size as the presently used CsI(Tl) counters are proposed for the end cap. The advantages of this crystal are a short scintillation decay time and moderate cost in comparison to other fast scintillation crystals. Since physical properties of the pure CsI are the same as CsI(Tl) the present sizes of the calorimeter element as well as the mechanical structure can be kept. Properties of pure CsI in comparison with CsI(Tl) scintillation crystals are listed in the Table 1.

However, the light output of pure CsI crystals is approximately ten times lower than that of doped crystals. Thus, to keep the electronics noise value at the same level one needs to use photo detectors with internal gain. A suitable solution is to use vacuum photopenthodes (PP), i.e. PM tubes with three dynodes. Since pure CsI emission wavelength is about 300 nm the photosensor should be UV sensitive. Such a device of 2-inch diameter with a low output capacity C≈10 pF has been recently developed by the Hamamatsu Photonics. The low capacity is important since the noise level depends on this value. The new developed PP have a quantum efficiency of about 20-25% and internal gain factor of 120-200 at zero magnetic field which reduces by factor 3.5 times for the B=1.5 T (see Fig. 4).

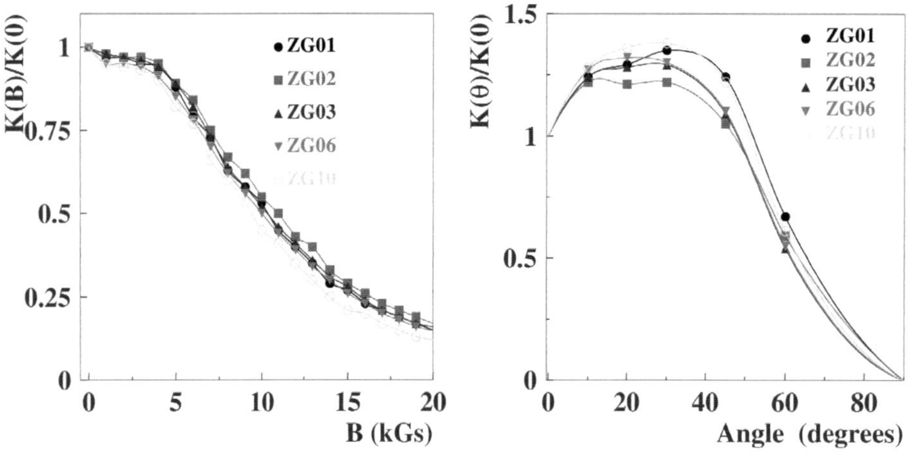

Figure 4. The photopenthode gain in dependence on the magnetic field and the angle device axis to the field direction for the five devices.

The electronics channel will be built in the same general line as for the barrel part (see previous Section). The shaping and digitizing part is proposed to be implemented on VME boards. Each VME board contains 16 channels. Each channel includes a differential receiver, CR-(RC)^4 shaper with shaping time τ = 30 ns and two 14-bits fast flash ADCs. Usage of two ADCs allows to have effective 18-bit digitization for a full dynamical range. Digitization of the signals from the shaper outputs will be performed under the control of the common clock of 43 MHz. The digitized data are processing in FPGA to obtain the amplitude(A_0), time(t_0) and quality bits(Q) for each hit.

In the tests with the counter and electronics prototypes we obtained the signal of about $2 \cdot 10^4$ electrons per 1 MeV (without magnetic field) and the noise of about 900 electrons. These results can be extrapolated to the 1.5 T magnetic field as 6000 e/MeV and 200 keV of the noise energy equivalent.

Shorter scintillation decay time with matched shorter shaping time of the electronics channel allows to suppress the pile-up noise by a factor of about 5.5 in comparison to the present end cap parts of the calorimeter. The fake photon rate is suppressed by a factor of more than 100 due

Table 1. Properties of pure CsI and CsI(Tl) scintillation crystals

	ρ, g/cm^3	X_0, cm	λ_{em}, nm	$n(\lambda_{em})$	N_{ph}/MeV	τ_d, ns	dL/dT, %/° 20°C
pCsI	4.51	1.85	305	2.0	2000	20/1000	- 1.3
CsI(Tl)	4.51	1.85	550	1.8	52000	1000	0.4

to a good timing providing by the mentioned scheme. Thus, even in the case of 20 times larger background the resulting fake rate is expected to be lower than in the current barrel calorimeter.

We performed a study of radiation hardness of the full size pure CsI crystals produced in Kharkov (Ukraina). In these study 14 full size crystals were irradiated by the wide gamma beam with the average energy of about 1 MeV generated by the 1.5 MeV electron accelerator ELV-6 at BINP. The absorbed dose increased step by step up to 14 krad. The most of the crystals (11) showed a light output reduction less than 20% while for 3 crystals the light output drops to 25-30% Thus, we consider the radiation hardness of the pure CsI crystals as acceptable for Belle II encaps.

The beam test with 20 pure CsI crystals coupled with PP and electronics provided waveform analysis was carried out at BINP photon beam [6].

The photon beam is produced by the Compton back scattering of the laser photons by the high energy electron beam circulating in the VEPP-4M storage ring. The resulting Compton spectrum is approximately uniform with a sharp edge corresponding to the maximum energy $\omega = 4\gamma^2\omega_0/(1+4\gamma\omega_0/m_e)$, where γ is a electron beam relativistic factor, ω_0 is the energy of the laser photon and m_e – electron mass. The energy resolution can be studied both by analyzing of the smearing of the Compton edge of the measured spectrum using the scattered electron tagging system.

Measured energy resolution (Fig. 5) is consistent with the CsI(Tl) ECL energy resolution [7] and MC predictions. The time resolution better than 1 ns for energy more than 20 MeV has been obtained.

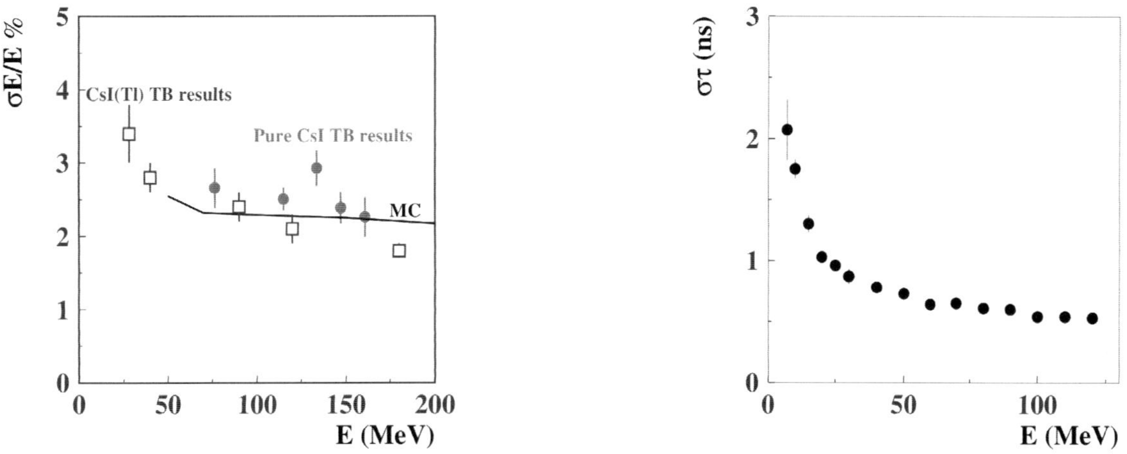

Figure 5. Energy (left plot) and time (right) resolution for photons obtained at the beam test (circles) and the results from beam test of the Belle ECL CsI(Tl) prototype [7] (squares).

6. Conclusion

- To keep good performance of the calorimeter at high background conditions of the Belle II experiment we upgrade the calorimeter electronics. This work is in progress. Most of the modules are produces and tested. The tests demonstrated proper performance of the new electronics.
- R&D works on the endcap calorimeter based on pure CsI counters with modified electronics is going on. This modification should provide drastic suppression both pile-up noise as and fake clusters rate in the endcaps.

7. Acknowledgments
We acknowledge support from the Ministry of Education, Culture, Sports, Science, and Technology (MEXT) of Japan; National Research Foundation of Korea and the Ministry of Education and Science of the Russian Federation

References
[1] Abashian A *et al* . 2002 *Nucl. Instr. and Meth.*, A **479** 117
[2] Kurokawa S *et al* . 2003 *Nucl. Instr. and Meth.* A **499** 1
[3] Abe T *et al* . 2010 Belle II technical design report, KEK Report 2010-1
[4] Cheon B G *et al* . 2002 *Nucl. Instr. and Meth.* A **494** 548
[5] Kazui K *et al.* 1997 *Nucl. Inst. and Meth.* A **394** 46
Beylin D M *et al.* 2005 *Nucl. Inst. and Meth.* A **541** 501
[6] Shwartz B 2009 *Nucl. Inst. and Meth.* A **598** 220
[7] Ikeda H *et al* . 2000 *Nucl. Instrum. and Meth.* A **441** 401

Journal of Physics: Conference Series **587** (2015) 012046 doi:10.1088/1742-6596/587/1/012046

Current Status and Performance of the BESIII Electromagnetic Calorimeter

Florian Feldbauer for the BESIII Collaboration
Johannes-Gutenberg Universität/Helmholtz-Institut Mainz, 55128 Mainz, Germany
E-mail: feldbaue@kph.uni-mainz.de

Abstract. The BESIII experiment is located at the Beijing Electron Positron Collider (BEPCII) in China. Its electromagnetic calorimeter (EMC) consists of 6240 CsI(TI) crystals, each read out by two Photodiodes (PD) at the end of the crystal. Changes in the response of the calorimeter due to radiation damage in the crystals or changes in the photo detector output are monitored with a light pulser system.

1. Introduction

The Beijing Electron Positron Collider II (BEPCII) is a symmetric collider with a double storage ring. The center of mass energy of the e^+e^- system can be varied from 2.0 to 4.6 GeV with an energy spread of 1.1 MeV at 3.686 GeV. Inside the Beijing Spectroscopy III (BESIII) experiment [1] the electrons and positrons collide under a crossing angle of 11 mrad.
The detector is composed of several subsystems. The Multilayer Drift Chamber (MDC) is the inner most detector. It provides excellent tracking with momentum resolution of $\sigma_p/p = 0.5\%@1\,\mathrm{GeV}$ and a spacial resolution of $\sigma_{xy} = 130\,\mu\mathrm{m}$. The next detector is the Time of Flight (TOF) system used for particle identification. The TOF is composed of two plastic scintillator bars providing a time resolution of $\sigma_T = 100\,\mathrm{ps}$. Behind the TOF the Electromagnetic Calorimeter (EMC) is located followed by the 1 T magnet and the muon system. The muon system uses 9 layers of resistive plate chambers for the detection of the muon tracks.
Overall BESIII has about 40000 read out channels with a data rate of 5 kHz and 50 MB/s, respectively.

2. BESIII Electromagnetic Calorimeter

The BESIII EMC is based on CsI(Tl) crystals [2] with a designed energy resolution of 2.5% and a position resolution of 6 mm for 1 GeV showers. The EMC consists of a barrel and two endcaps as shown in figure 1. The barrel is segmented in 44 rings in theta each consisting of 120 crystals. The two endcaps are arranged in 6 rings in radial direction. A total of 6240 crystals are used. The crystals are 28 cm ($15.1X_0$) long with an average front size of 5.2 by 5.2 cm^2 and an average rear size of 6.4 by 6.4 cm^2. In order to minimize the dead material and gaps between neighbring crystals, they are suspended from a support girder, i.e. there is no supporting walls between the crystals. There is however a 5 cm gap between the last rings of the barrel and the endcaps of the EMC.
Two Hamamatsu S2744-08 photodiodes are glued to the back face of every crystal. The

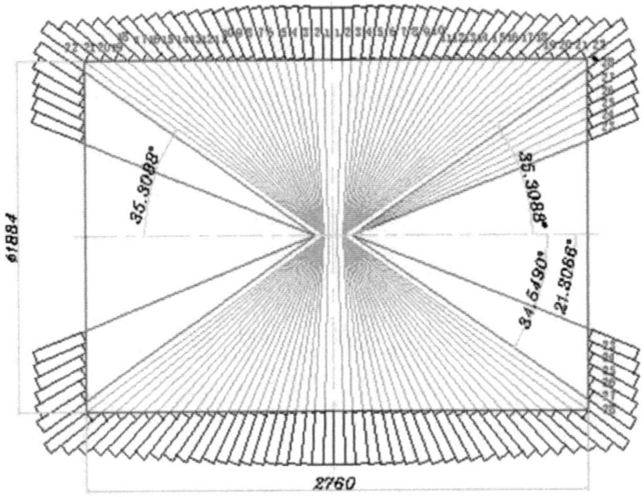

Figure 1. The structure of the BESIII EMC.

photodiodes have a size of 1 cm×2 cm and are readout by a preamplifier [3]. The signals from the preamplifier are send to the main amplifier with 1 μs shaping time and then digitized by a Q module. The noise of the read out chain is required to be less than 1100 electrons which corresponds to the signal of a 220 keV shower. For RF shielding the read out units are housed in an aluminium box. This read out box is attached to the crystal by four screws drilled into the crystal and then mounted to the support structure. In order to prevent the CsI(Tl) crystals from deliquescence and to maintain the humidity inside the calorimeter in a suitable range, the EMC is flushed with freeze-dried compressed air.

3. Operation Experience

According to the design, the allowed radiation dose per year should be less than 200 rad at the crystals. In order to monitor the applied radiation dose, Radiation-sensing Field-Effect Transistors (RadFET) based on a metal-oxide-silicon p-channel structure are used. RadFETs are integrated dosimeters which measure the dose by the virtue of the field effect caused by space charge trapped in an inorganic insulator. The RadFETs are distributed at a radius close to the crystal position. Inside the barrel four rows with 12 probes along the z direction at $\varphi = 0°, 90°, 180°$ and $270°$ are placed. Inside both endcaps 12 additional probes are placed. The integrated dose accumulated from the beginning of operation in 2009 to December 2013 is shown in figure 2. In six years of operation the maximum integral dose is about 1754 rad. It is striking to note that the integral dose at the west endcap is much larger than at the east endcap. This is due to the fact, that BEPCII is also used for synchrotron radiation experiments. In this operation mode only electrons are used, accelerated from east to west. Thus the west endcap is exposed to high synchrotron radiation while the east endcap is not.

To check the quality of a module and monitor radiation damages LED-fiber units consisting of a LED, light mixer and fiber bundle are used. In order to monitor the crystal's light output and responce of the photodiodes each crystal is connected to one fiber. The light intensity of the LED-fiber system can be adjusted to simulate the energy deposite in a crystal from 10 MeV to 1.5 GeV. Two independent references using an Am^{241} source are used to monitor the variances of each LED. The stability of the LED-fiber system is less than 1% in a run.

Overall 76 units are used in the EMC, 60 units in the barrel each serving 88 crystals and 16 units

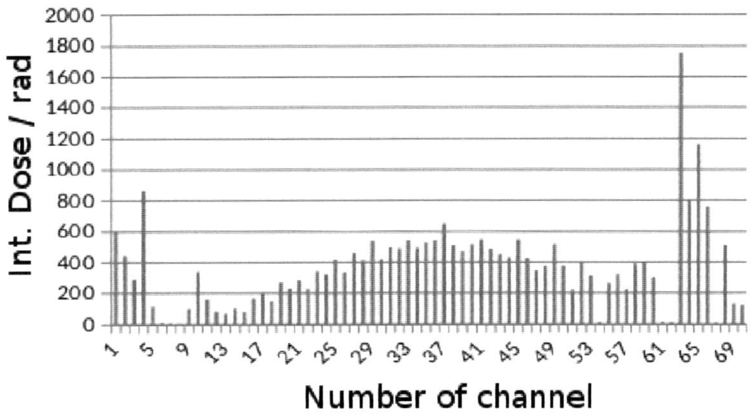

Figure 2. The integral dose of RadFET over six years.

in the endcaps each serving 60 crystals. During data tacking the system is used to determine bad channels and monitoring of the performace of the EMC, including the gain reduction due to radiation damages. Figure 3 shows the relative decrease of the light output measured by the LED-fiber system. In most cases the decrease is less than 10%. Only 30 crystals show a decrease larger than 10% and among them only 10 crystals exceed 20%. The results are in agreement

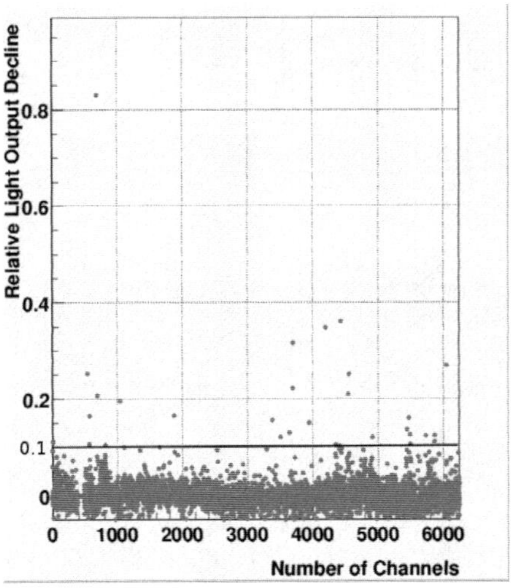

Figure 3. Relative decrease of light output measured by LED-fiber system

with the offline calibration using Bhabha scattering events. Also these numbers have not been increased in the last two run periods.

In the time from April 2009 to September 2010 17 crystals showed a sudden decrease of the light output by about 15%. This was caused by an ungluing of the photodiodes from the crystal. There is one crystal with a relative light output of less than 20%. This may be caused by a total drop of the photodiodes from this crystal. These loses in light output can be recovered through offline calibration.

Each crystal is read out by two PDs and two preamplifiers. The signals from the preamplifiers are fed to differential amplifiers and then go through a switching stage. By default the selection

"$A + B$" is used, which builds the sum of both signals. After 2011 however nine crystals were only read out by one PD. For those channels the selection "$2A \vee 2B$" is used to boost the signal of the remaining PD by a factor of two.

4. Performance of the EMC

For calibration of the gain of each crystal BESIII uses Bhabha events. From this data the energy peak and resolution of each channel can be determined at 1.5 GeV. Figure 4 shows the measured energy resolution. It can be seen that the results from data tacking in 2009 and 2012 are in very good agreement with each other and with the results from Monte Carlo simulations at a center of mass energy of 3.097 GeV. After six years of operation no distinctly decrease of the crystals light output can be observed. The energy resolution in the barrel part of the EMC is about 2.3% and 4.1% in the endcaps. In this plot also the gap between barrel and endcaps is clearly visible.

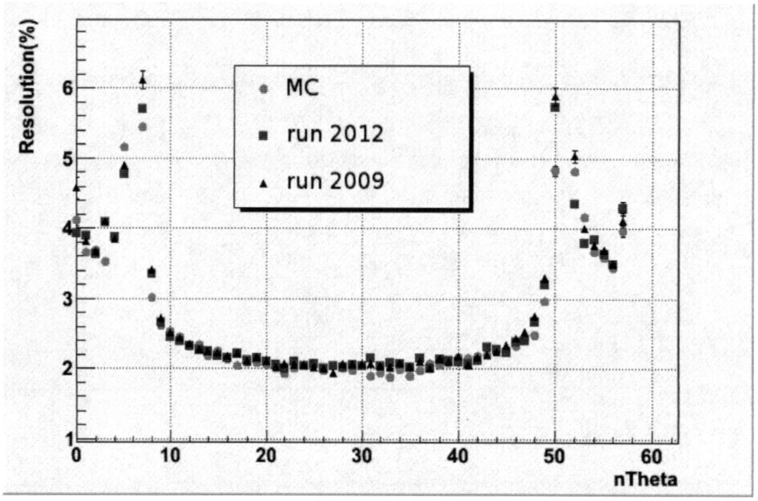

Figure 4. Energy resolution of crystals for Bhabha at 1.5 GeV, $E_{CMS} = m(J/\psi)$.

5. Conclusion

The status and performance of the BESIII calorimeter is very good. All 6240 channels are working well. the total integral radiation dose is still in an acceptable region. The decrease of the light output monitored with a LED-fiber system is for most of the crystals below 10% and the losses due to ungluing of the photodiodes and failures of the electronics could be compensated. The BESIII EMC is taking data smoothly and there is no degradation of the performance after six years of operation.

References

[1] M. Albikim, *et al* (BESIII Collaboration), "Design and construction of the BESIII detector", Phys. Rev. D81, 052005 (2001).

[2] Dong Ming-Yi, *et al*, "Design and assembly of the CsI(Tl) crystal module of the BESIII electro-magentic calorimeter", Chinese Physics C. Vol. 32, No.11, Nov. 2008.

[3] Chang Jin-fan, *et al*, "The design of BESIII EMC charge measurement with three ranges", Nuclear Electronics and Detection Technology, Vol.29, No.4, Jul. 2009.

Progress status for the Mu2e calorimeter system

Gianantonio Pezzullo[1,2], J. Budagov[3], R. Carosi[2], F. Cervelli[1,2], C. Cheng[4], M. Cordelli[5], G. Corradi[5], Yu. Davydov[3], B. Echenard[4], S. Giovannella[5], V. Glagolev[3], F. Happacher[5], D. Hitlin[4], A.Luca[5], M.Martini[4], S. Miscetti[5], P. Murat[7], P. Ongmonkolkul[4], F. Porter[4], A. Saputi[5], I. Sarra[5], F. Spinella[2],V.Stomaci[5], G. Tassielli[6]

[a]Departement of Physics, University of Pisa, Largo B. Pontecorvo 3, Pisa , Italy
[b] INFN sezione di Pisa, Italy, Largo B. Pontecorvo 3, Pisa , Italy
[c] Joint Institute for Nuclear Research, Joliot-Curie 6, Dubna, Russia
[d] Departement of Physics, California Institute of Technology, 1200 E California Blvd, Pasadena (CA), USA
[e] INFN Laboratori Nazionali di Frascati, via Enrico Femri 40, Frascati, Italy
[g] INFN sezione di Lecce, Via Arnesano 73100, Lecce, Italy
[h] Fermi National Accelerator Laboratory, Main Entrance Rd, Batavia (IL), USA

E-mail: pezzullo@pi.infn.it

Abstract. The Mu2e experiment at FNAL aims to measure the charged-lepton flavor violating neutrinoless conversion of a negative muon into an electron. The conversion results in a monochromatic electron with an energy slightly below the muon rest mass (104.97 MeV). The calorimeter should confirm that the candidates reconstructed by the extremely precise tracker system are indeed conversion electrons while performing a powerful μ /e particle identification. Moreover, it should also provide a high level trigger for the experiment independently from the tracker system. The calorimeter should also be able to keep functionality in an environment where the background delivers a dose of $\sim$ 10 krad/year in the hottest area and to work in the presence of 1 T axial magnetic field. These requirements translate in the design of a calorimeter with large acceptance, good energy resolution O(5%) and a reasonable position (time) resolution of $\sim$ <1 cm (<0.5ns). The baseline version of the calorimeter is composed by two disks of inner (outer) radius of 351 (660) mm filled by 1860 hexagonal BaF_2 crystals of 20 cm length. Each crystal is readout by two large area APD's. In this paper, we summarize the experimental tests done so far as well as the simulation studies in the Mu2e environment.

1. Introduction

Several experiments of the last decades proved that lepton-flavor-violation (LFV) for neutral leptons (neutrinos) is a fact. This implies that also charged LFV (cLFV) is possible, but with a rate not yet reached by the current experimental sensitivity. An observation of a cLFV signal would be an unambiguous evidence of new physics. A method to search for cLFV exploits muons; so far two processes involving muons have been studied: $\mu \to e\gamma$ and $\mu\mathrm{N} \to e\mathrm{N}$. The current upper limits for these two processes are: BR($\mu \to e\gamma$)< 5.7×10^{-13} [1] and $R_{\mu e}$(Au targets) < 7×10^{-13} [2]. Compared to $\mu \to e\gamma$, the rate of $\mu\mathrm{N} \to e\mathrm{N}$ is scaled by a factor of $\sim \alpha Z^2$, as the photon becomes virtual and couples with the nucleus. Due to the interaction with the nucleus, additional new physics processes can contribute to $\mu\mathrm{N} \to e\mathrm{N}$ compared to $\mu \to e\gamma$, the ones with a contact term or a four-fermion interaction, such as compositeness or

Journal of Physics: Conference Series **587** (2015) 012047 doi:10.1088/1742-6596/587/1/012047

lepto-quarks[3]. The search for muon conversion has a clear signature to look for. Indeed in the muon conversion process the initial state is a muonic atom that makes a transition to a 2-body state consisting of a mono-energetic electron recoiling to the atomic nucleus, with no neutrinos in the final state and no nucleus breakup. The nucleus is not observed while the conversion electron (CE) has an energy of the muon rest mass minus small corrections for the recoil and the binding energy of the muon.

2. The Mu2e experiment set-up

The Mu2e experiment [4] is composed by three main superconducting magnet systems (see figure 1). The first is the production solenoid (PS) where an intense 8 GeV proton beam (pulsed with a period of $\sim$ 1800 ns) strikes a tungsten target producing mostly π 's. A graded magnetic field (from 2.5 to 4.6 T) then collects most of the produced charged particles and moves them through the second magnet, the transport solenoid (TS). The TS is dedicated to move and select

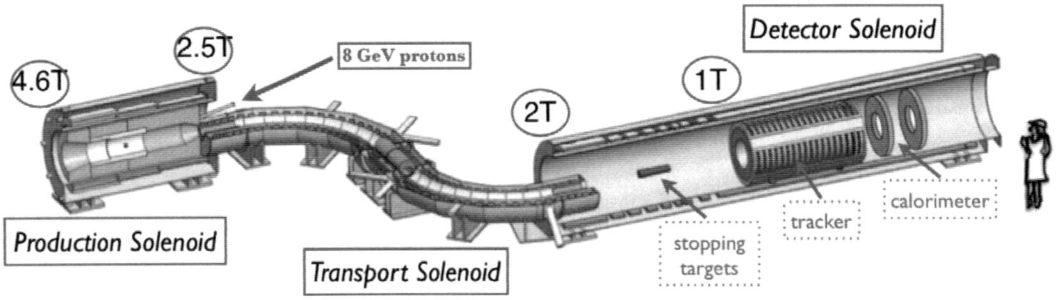

Figure 1. Mu2e experimental layout. The Cosmic ray veto is not shown.

only negative low-momentum muons through the last solenoid, the detector solenoid (DS). Such a selection is done by means of a system of collimators at the sides and in the middle of the TS. The S shape of the transport solenoid allows also to avoid that all neutral particles generated in the PS reach the DS. The DS is the solenoid where the muon beam arrives, stops on the target and the reaction products are analized by the detectors. In order to do that, it houses the muon stopping targets (Al) and the detection system devoted to identify and analyze the conversion electrons (CE). The Al targets reside in a graded field region that varies from 2 to a 1 T, so that electrons emitted upstream are reflected downstream through the detectors region. Downstream the targets a proton absorber, made of high density polyethylene is located to reduce the proton flux (produced in nuclear reactions) to the tracker. The tracking system consists of 20 straw tubes stations designed for maximizing the acceptance and the precise reconstruction of the CE. This is the main detector of the experiment and it is expected to get a resolution of $\sim$ 120 keV/c for 100 MeV/c momentum. The second detector adopted is a crystal electromagnetic calorimeter (EMC) that will be described in details in the next section 3. A Cosmic Ray Veto (CRV) system surrounds the DS on three sides (the ground is not covered) and extends up to the midpoint of the TS; it consists of four layers of extruded scintillator bars with embedded wavelength shifting fibers read out with Silicon photomultipliers (SiPM's). The Veto signal corresponds to coincident hits in three out of four layers. It is designed to reduce the number of expected cosmic induced background events to 0.05 events during the entire running period (three years). In the region of the muon stopping target the CRV is expected to be 99.99% efficient. Neutron shields, made of concrete blocks, are also present between the DS and the

CRV to reduce the neutron flux (coming from the muon stopping targets) which otherwise can compromise the efficiency and functionality of the CRV.

3. The Mu2e Calorimeter

3.1. Calorimeter requirements

The Mu2e calorimeter should confirm that the candidates reconstructed by the extremely precise tracker system are indeed conversion electrons while performing a powerful mu/e particle identification. Moreover, it should provide a high level trigger for the experiment, independently from the tracker system, to bring the rate on disk to $\sim$ few kHz. The calorimeter should also be able to keep functionality in an environment where the background delivers a dose of $\sim$ 12 krad/year in the hottest area and to work in the presence of 1 T axial magnetic field. The calorimeter will also help on selecting correct hits in the tracker. All these requirements translate in the following parameters to be satisfied by the Mu2e calorimeter:

- Provide a large acceptance for electron and positron tracks.
- Determine the energy with a resolution of O(5 MeV).
- Reconstruct the time with a resolution $\leq$ 0.5 ns.
- Reconstruct the impact position with a resolution better than $\sim$ 1 cm

The calorimeter should survive in the Mu2e environment where a radiation dose of 100 Gy/year/crystal is expected in the hottest areas. Moreover, the calorimeter is embedded in the DS where a 1 T magnetic axial field exists.

3.2. Calorimeter design

The calorimeter design consists of two disks whose dimension have been optimized by simulating signal electron events with the official Mu2e framwork [8]. Simulation results (see table 1) shows that the calorimeter acceptance gets its maximum with an inner (outer) radius of 35.1 (66) cm.

Table 1. Calorimeter configurations studied for optimizing the geometry.

Crystal apothem [cm]	Disk radii [cm]	# crystal	Crystal volume [cm^3]	Efficiency [%]
1.55	35.91/64.3	996	168830	90.5 ± 0.6
1.55	35.91/67.23	1110	193998	90.4 ± 0.6
1.6	34.01/66.35	1044	194424	92.2 ± 0.6
1.6	37.1/66.35	966	179898	90.2 ± 0.6
1.65	35.1/66.0	930	184188	92.2 ± 0.6
1.7	36.12/64.73	798	167769	90.4 ± 0.6

Figure 3 shows that the optimum disks separation is 75 cm, which is $\sim \frac{1}{2}$ wave-length of the signal electron helix.

Each disk is composed by $\sim$ 930 BaF_2 hexagonal crystals, see figure ??, with a length of 20 cm and a distance among flats of 1.65 cm. The readout system is constitued by two large area APD's, in order to grant a better light yield while providing redundancy of the readout system.

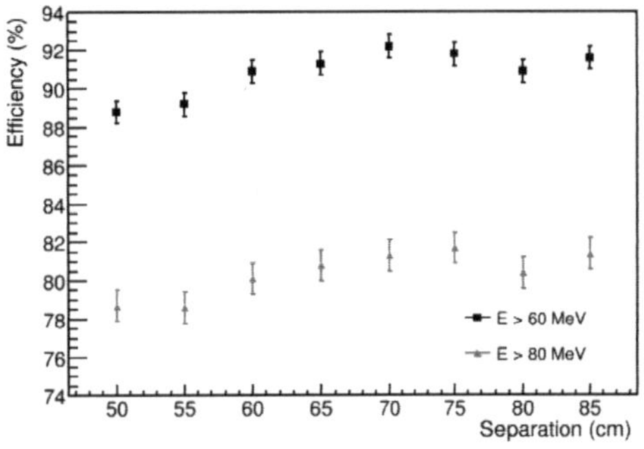

Figure 2. Calorimeter acceptance versus the disk separation.

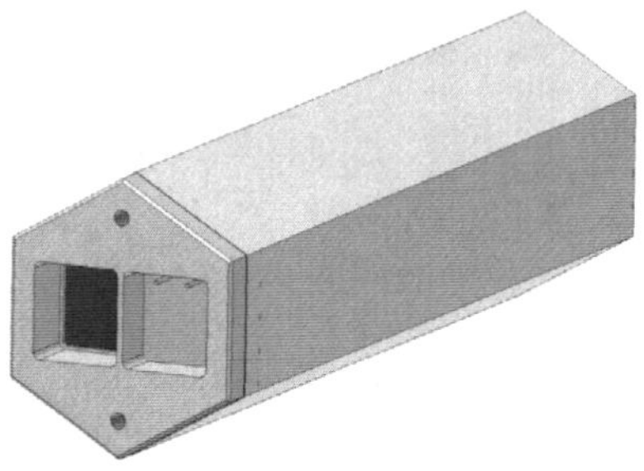

Figure 3. Crystal shape with APD's boxes colored on blue and green.

Table 2. Several configurations studied for optimizing the geoemtry.

Crystal	BaF_2	LYSO	CsI	$PbWO_4$
Density [g/cm^3]	4.89	7.28	4.51	8.28
Radiation length [cm] X_0	2.03	1.14	1.86	0.9
Molieère radius [cm] R_m	3.10	2.07	3.57	2.0
Interaction length [cm]	30.7	20.9	39.3	20.7
dE/dx [MeV/cm]	6.5	10.0	5.56	13.0
Refractive Index at λ_{max}	1.50	1.82	1.95	2.20
Peak luminescence [nm]	220, 300	402	310	420
Decay time τ [ns]	0.9, 650	40	26	30, 10
Light yield (compared to NaI(TI)) [%]	4.1, 3.6	85	3.6	0.3, 0.1
Light yield variation with temperature [%/°C]	0.1, -1.9	-0.2	-1.4	-2.5
Hygroscopicity	None	None	Slight	None

3.3. Crystal choice

In the Conceptual Design Report [4] the calorimeter baseline choice was based on LYSO crystals. Since then, estensive R&D program has been carried out on this option and test beams [5] and electronic design were tuned accordingly. However, during the last year, and despite an active R&D program at Caltech, in cooperation with SICCAS and SIPAT, aiming to reduce the commercial price of LYSO crystals, the large increase in Lu2O3 salt price over the past two years has made the cost of a LYSO calorimeter unaffordable.

Table 2 shows the alternatives which have been studied. At the end, we have selected BaF_2 as baseline choice due to the following advantages:

- it has a small decay time of the fast component;
- it is not hygroscopic;
- It is rad hard.

However BaF_2 presents also some drawbacks and difficulties. In particular, the fast component is emitted in the deep UV region at 220 nm (see figure 5) and, moreover, there is a large component (peaked at 300 nm) that has a long decay time of 650 ns. Name of the game for this crystal is to find an adequate photosensor.

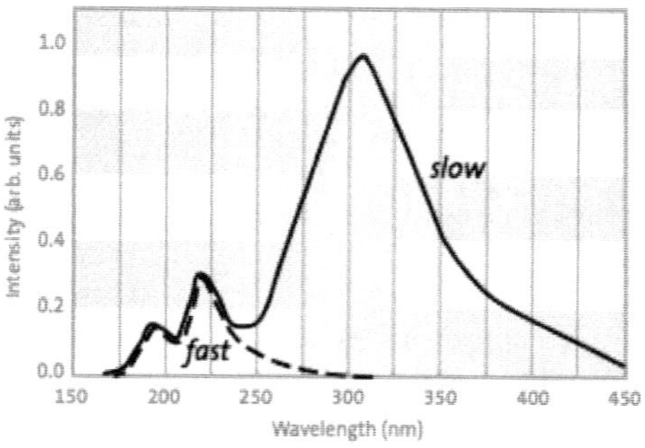

Figure 4. BaF_2 emission spectrum.

3.4. Photosensor choice

There are not existing photosensor candidates for BaF_2 readout that could have a high quantum efficiency at 220 nm while working in the presence of 1 T magnetic field. Photomultipliers are excluded. Channel plate PMTs are at present far too expensive, although spinoffs from the LAPPD project are still being pursued. Our main thrust, however, is to use solid state photosensors, either APD's, SiPMs or MPPCs, with extended UV response. APD's and MPPCs from Hamamatsu and RMD made without the normal protective epoxy coating, and therefore somewhat fragile, can have quantum efficiencies in the 200 nm region of $\sim$ 17%, but do not discriminate between the 220 nm fast component and 300 nm slow component of BaF_2 . The presence of the slow component limits the rate capability of the calorimeter, and can therefore be an issue in high background conditions. We are pursuing an experimental R&D approach to both improving the photosensor quantum efficiency and the slow component discrimination. A Caltech/JPL/RMD consortium has been formed to develop a modified RMD large-area APD into a delta-doped superlattice APD. This device will also incorporate an atomic layer deposition antireflection filter that will provide 60% quantum efficiency at 220 nm and $\sim$ 0.1% efficiency at 300 nm, thereby enabling us to not only have a larger number of photoelectrons/MeV (x3), but also to take full advantage of the fast decay time component of BaF_2 . The greatly reduced undepleted region of this device will also result in substantially improved rise time of the device.

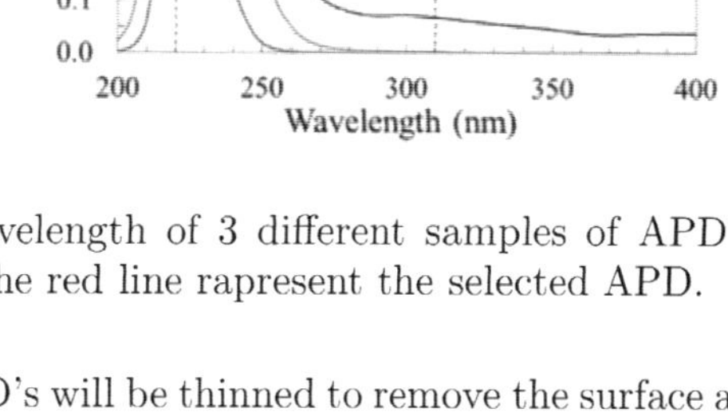

Figure 5. QE versus wavelength of 3 different samples of APD's with different number of atomic layer deposition. The red line rapresent the selected APD.

Conventional RMD APD's will be thinned to remove the surface and undepleted region before the avalanche layer, and the superlattice structure and optimized antireflection coating will then be deposited at the JPL Microdevices Lab. Figure 5 shows the calculated QE response of the resulting APD as a function of wavelength. For a five-layer AR coating, the QE at the fast component of BaF2 is nearly 70%, and the extinction at the slow component wavelength is nearly complete.

3.5. Energy resolution

The study of the expected energy resolution has been factorized in two steps:

(i) Estimate of the light yield via measurements done in laboratory;

(ii) Simulation of the calorimeter performance in the Mu2e framework including educated guess on light yield and electronic noise with the final setup.

The measurement of the light yield has been performed using a modified version of the LNF QA station developed for determining the response and the longitudinal uniformity of the LYSO crystals. The station was modified by adding a UV extended PMT from EMI while the mechanics for the movement of the Na^{22} source was kept unchanged. The EMI photo tube has a QE $\sim$ 30% @ 220 nm and an operation gain of 3.8×10^6 for an high voltage set at $V_{\rm bias}$ = 1800 V. With this setup we measured the response of the BaF_2 to the source in two different wrapping conditions (Teflon foil and ESR 3M reflector) and by coupling the PMT to the crystal in air, i.e. without the application of any optical grease. Figures 6 shows the integated charge in a time window of 50 ns (left) and of 650 ns (right) respectively. The shorter time window has been selected to give a first estimate of the expected prompt component due to the total signal width of the PMT and of the RMD APD's (that is expected to be $\sim$ 40-50 ns).

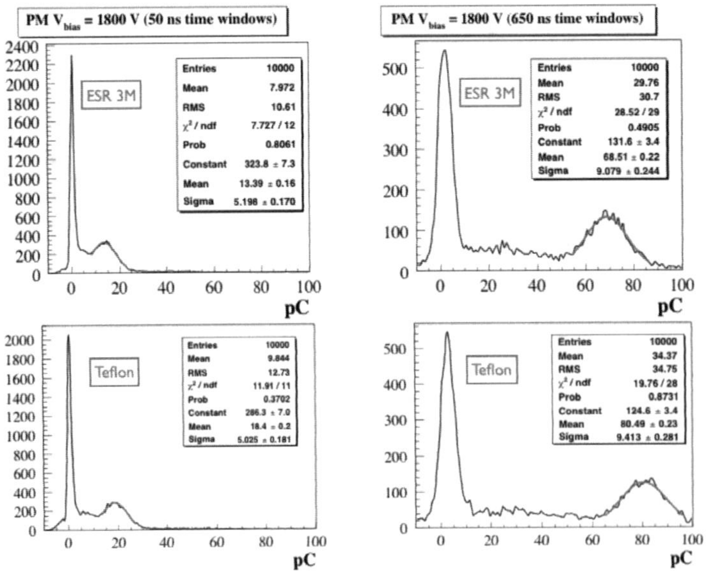

Figure 6. Left: integrated chrge distribution in 50 ns time window. Right: integrated charge distribtuion in 650 ns.

The best light output result was obtained with the Teflon wrapping. In this configuration the measured light yield, in 50 ns integration window, is 15.0$\pm$3.8 pe/MeV. Scaling this result to the expected area and quantum efficiency of the RMD APD, and applying a factor of two correction for the usage of optical grease or glue, we expect the light yield to be of $\sim$ 30 pe/MeV. A dedicated measurement of the light yield obtained with the first RMD prototypes will be carried out during next autumn.

We have also estimated the expected electronic noise (in MeV) for the BaF_2 crystal starting from the measurement done for the LYSO crystals and related FEE electronics. The electronics, see figures 7 and 8, consists of a discrete preamplifier chip and a local HV regulator that has been developed by the LNF electronic departments to grant an amplification of G$\sim$ 300, a precise stability on voltage supply and a ENC of 1000 e^- without input source connected. For the LYSO case readout by a S8664 APD from Hamamatsu, that has a capacitance of 270 pF, we have measured an ENC of 13000 e^- at an APD gain of 150. This noise corresponds to a ENE of 30 keV. For the BaF_2 case, the RMD APD has two improved characteristics with respect to the S8664: (i) a capacitance of 60 pF and (ii) an operation gain of 500. The ENC(e^-) will be of $\sim$ 5000 e^-. Assuming a light yield of 30 pe/MeV, the expected noise level is ENE=ENC(e^-)/(G $\times$ LY) = 5000/(30 $\times$ 500) = 0.33 MeV.

Figure 7. Amp-HV prototype.

Figure 8. Amplification HV side.

We have then simulated the calorimeter performance for reconstructing CE events by including the final optimized geometry, all the environmental background surces in the experiment (photon, protons and neutrons from muon capture and related secondaries, DIO events) and the estimated values of the light yield and of the electronic noise. To be realistic and in order to test the dependence of the resolution to these experimental inputs we have varied them up to a factor of two.

Figure 9 shows the distribution of the energy residuals. The 20 cm ($\sim$ 10 X_0) length of the BaF_2 crystal does not allow the total containment of the electron shower, which results in the long tail on the right of the distribution. The value of the FWHM/2.35 of the distribution is $\sim$ 4.3 MeV. For testing the stability of the energy resolution several simulations (with and without the background mixed) have been performed including the electronic noise contribution and assuming different photo detection efficiencies. Figure 10 shows that scaling to $\sim$ 20 p.e./MeV the expected FWHM/2.35 is less than 5 MeV.

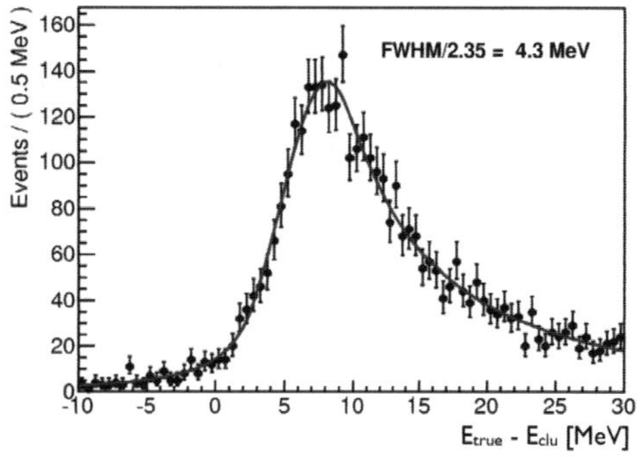

Figure 9. Distribution of the difference between the energy at the entrance of the EMC (E_{true}) and the energy deposited in the EMC (E_{clu}). The blue line is the result of the fit with a "Crystal-ball" function [9] able to describe the asymmetry due the non-perfect energy containment.

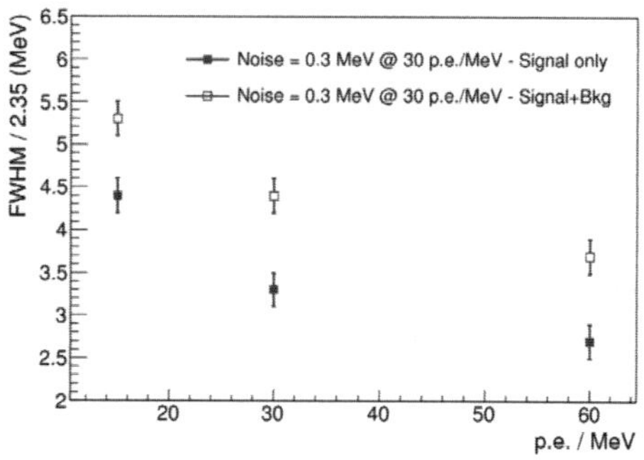

Figure 10. Dependence of the energy resolution for a BaF_2 based calorimeter as a function of the light yield for different values of electronic noise, and with and without nominal background..

3.6. Absolute calibration and system monitoring

The absolute energy scale, the equalization in response of the crystal, as well as a check of response linearity are provided by a liquid source system (C_8F_{18}) which is pumped through aluminum pipes (3 mm of diameter) which cover both disk surfaces. This liquid source can be activated by a neutron source, producing the following decay chain:

$$\begin{aligned} {}^{19}F + n &\rightarrow {}^{16}N + \alpha \\ {}^{16}N &\rightarrow {}^{16}O^* + \beta^- + \bar{\nu}_e \; (\tau_{1/2} = 7\mathrm{s}) \\ {}^{16}O^* &\rightarrow {}^{16}O + \gamma(6.13 \text{ MeV}) \end{aligned}$$

The fluorine activated by the neutron source produces the ^{16}N isotope, which decays with an half-life time of 7 s in to an exited state of the ^{16}O, which finally decays to its ground state emitting a 6.13 MeV photon. The source spectrum, as seen with a CsI(Tl) crystal [6] with PIN diode readout is shown in figure 12. There are three principal contributions to the overall peak, one at 6.13 MeV, another at 5.62 MeV, and the third at 5.11 MeV, the latter two representing escape peaks. It is important to note that all three peaks have well-defined energies and thus all are useful in the calibration, providing both an absolute calibration and a measurement of linearity. Figure 11 showes the liquid source circuit which is planned to be used. It is constitued of 12 Al pipes, 0.5 mm thick, with a diameter of 3/8 inches and a length which ranges from 1.5 to 1.7 m. The selected geometry allows to have an uniform illumination of the disk with variation on the intensity less then 5%.

The variation of the crystal optical transmittance and photosensors' gains will be monitored in a continuos way by means of a laser system, following a scheme similar to the one used for CMS [7]. To achieve this goal, each crystal is illuminated by UV light transported it, with fused silica optical fibers, to the back of each BaF_2 crystal. A reference monitor station based on Pin Diodes will also be used to control the variation of the input laser source and of the fibers at the end of the distribution system. In Figure 13 the basic scheme of the laser system in Mu2e is shown. Prototype of the distribution system has been developed and tested in preparation for the upcoming test beam periods.

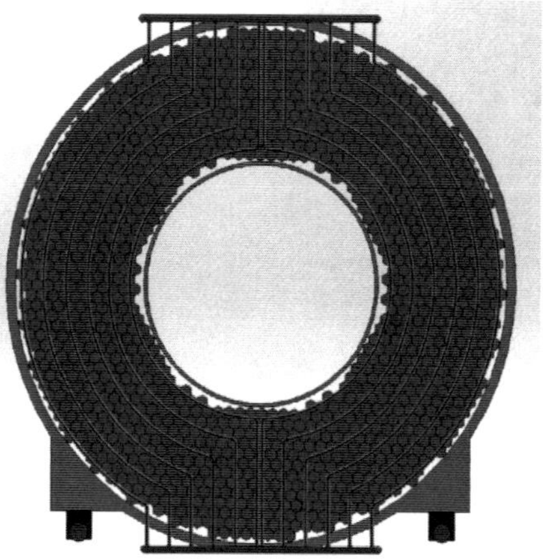

Figure 11. Liquid source circuit scheme.

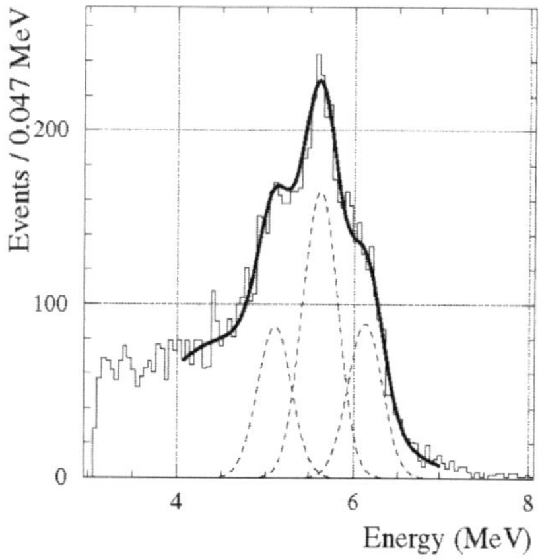

Figure 12. Energy spectrum in a *BABAR* CsI(Tl) crystal irradiated with 6.13 MeV photons from an $^{16}O^{*}$ source. The solid curve is a fit to the data, including Gaussian contributions at 6.13 MeV, 5.62 MeV, and 5.11 MeV, indicated by the dashed curves.

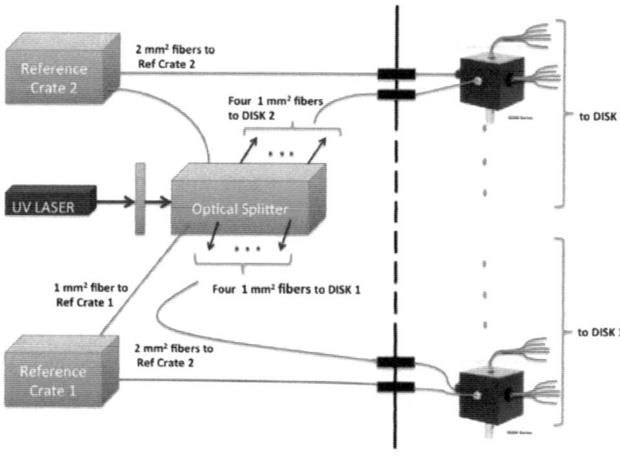

Figure 13. Laser system scheme.

In the final setup, one UV pulsed lasers will funnel light in a integrating sphere of 2" diameter through a dichroic prism outside of the DS.

3.7. Calorimeter driven pattern recognition

The first stage of the track reconstruction is the pattern recognition of the hits in the tracker. The straw tube has a diameter of 5 mm and the maximum drift time is $\sim$ 50 ns while a CE crosses all the detector region in $\sim$ 15 ns. Once a calorimeter cluster is identified as "candidate" (just looking at the deposited energy), its time can be used for collecting a pattern of hits in the tracker which are close in time with the calorimeter cluster in a window of $\sim$ 50 ns.

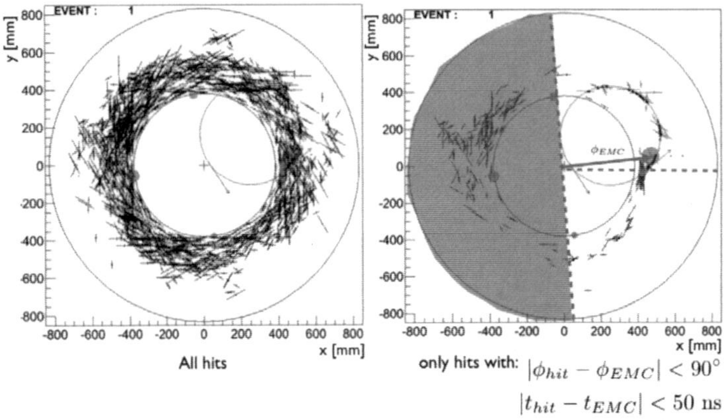

Figure 14. CE with background overlapped (left) and after time selection (right).

Moreover also the calorimeter cluster position allows to reduce significantly the tracker occupancy. Figure 14 shows that setting a cut on the angular distance between the straw hits (ϕ_{hit}) and the calorimeter cluster (ϕ_{EMC}) at 90° helps on reducing the tracker occupancy by a factor > 2. When simulating the CE with all background hits in overlap, the described procedure is able to increase the reconstruction efficiency by a factor $\sim 12\%$. Such a result improves when increasing the background rate.

3.8. Calorimeter particle identification

Rejection of background due to cosmic rays (CR) requires either the identification of muons mimicking the CE or the removal of electrons directly produced by CR. At low energies, a strong particle identification technique is based on the fact that at the same momentum, particles with different masses travel at different speed. In case of Mu2e signal, 100 MeV/c electrons could be considered ultra-relativistic ($E/m_e \sim 200$), while 100 MeV/c muons travel at a speed of about 0.7 c. Extrapolating the reconstructed track, it is possible to predict the particle arrival time and impact position on the calorimeter surface. Using the correct assumption about the particle mass, the predicted time will be close to the reconstructed time of the calorimeter cluster. All the way around, if a muon track has been extrapolated into the calorimeter assuming it was an electron, the predicted arrival time of the muon will be systematically lower than the reconstructed time of the calorimeter cluster associated to the track.

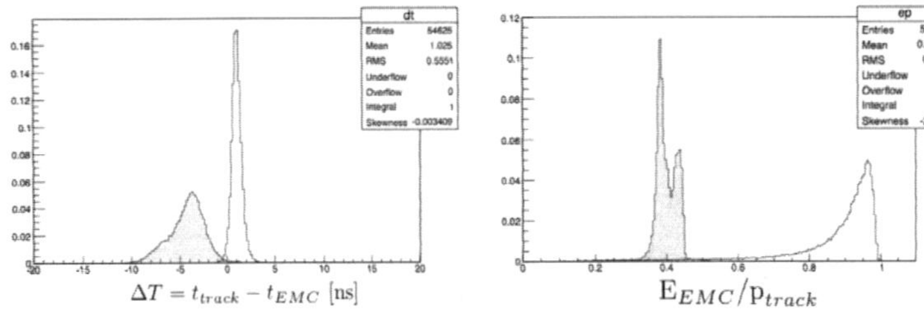

Figure 15. (Left) Distribution of $\Delta T = T_{track} - T_{cluster}$ [ns] for 100 MeV/c electrons (open histogram) and muons (blue filled histogram). (Right) Distribution of E / P for 100 MeV/c electrons (open histogram) and muons (blue filled histogram). E is the cluster energy from the calorimeter and P is the reconstructed momentum.

Figure 15 (left) shows the distributions in $\Delta T = T_{track} - T_{cluster}$ for 100 MeV/c electrons and muons; T_{track} is the predicted time of the particle at the calorimeter, calculated from the reconstructed track, and $T_{cluster}$ is the reconstructed cluster time. The separation between the maxima of the electron and muon distributions is $\sim$ 5 ns.

The distribution of the deposited energy in the calorimeter is also very different between electrons and muons. Figure 15 shows distributions in E/P, where E is the cluster energy and P is the associated track momentum. The kinetic energy of a 100 MeV/c muon is about 40 MeV. The structure and the tail above 50 MeV in the muon energy deposition are due to the capture and beta decay of the stopped muons, as modeled by Geant-4. A likelihood-based approach allows to reach an overall muon rejection factor of $\sim$ 500 (with an electron identification efficiency $\sim$ 99.9%), which is above the requirement of the experiment set at $\sim$ 200. The latter requirement is driven by the consideration that, when simulating a very large sample of Cosmic Ray, equivalent to the full elapsed time of the experiment, 11 muons were able to fake the CE signal in the tracker while were not vetoed by by the CRV system. Since the experiment wants to keep the CR related background contained to 0.05 event, a factor 200 of rejection is needed.

3.9. Calorimeter based trigger

So far, the developed calorimeter trigger algorithm is based on the application of an energy threshold on the reconstructed cluster energy. To study the efficiency and the performances of this method CE and DIO electrons were simulated with the Mu2e official framework [8]. DIO electrons were simulated since they represent the most intense source of background that can create calorimeter clusters. In order to study the behavior of trigger efficiency and DIO rejection on the calorimeter performances, we have convoluted the energy simulated by Geant-4 with additional Gaussian functions to deteriorate the resolution.

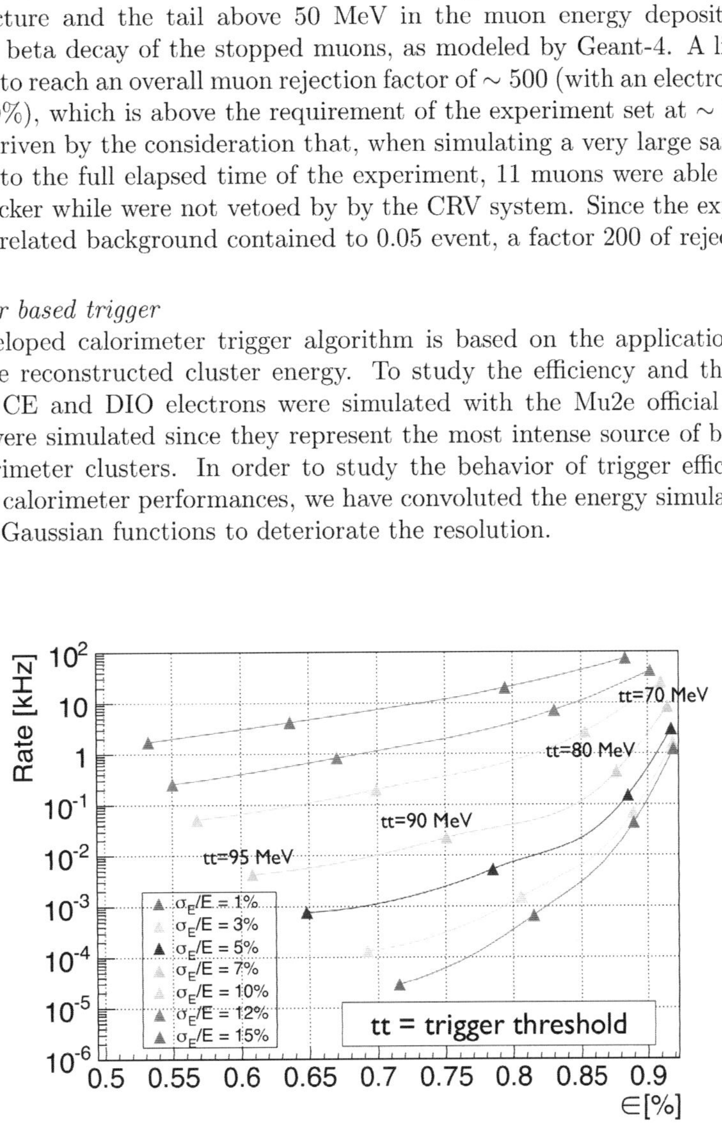

Figure 16. DIO's rate versus the signal efficiency. The different colors shows results from different assumptions on the energy resolution indicated relatively to 100 MeV.

Figure 16 shows the DIO's rate as a function of the signal efficiency for all the configurations tested. In the case of an energy resolution of $\sim$ 5%, it demonstrates that when setting a trigger threhold at 70 MeV we can get a signal efficiency $\sim$ 92% while reducing the DIO rate down to $\sim$ 2 kHz.

3.10. Calorimeter dose

To evaluate the total dose absorbed by the calorimeter all the processes related to the muon beam arrival in the detector solenoid have been studied within the Mu2e framework:

- The beam flash. This component is constitued by all particles wich reach the DS practically in time with the μ beam (π 's, e^- 's, ecc);
- Electrons from μ decay-in-orbit (DIO) processes;
- μ out-of-target (OOT);
- All secondary products related to the μ nuclear capture: photons, neutrons, protons;

The calorimeter dose have been evaluated separetly for the two disks. For the first disk (the one closer to the target), in figure 17, we show the crystal dose as a function of the radial distance, evaluated with respect to the beam axis. The beam flash is the most relevant contribution for the ionization dose. The inner ring is the hottest region with $\sim$ 15 krad/year. in the rest of the disk the dose/year is almost constant at a value of $\sim$ 4 krad/year. To validate this result

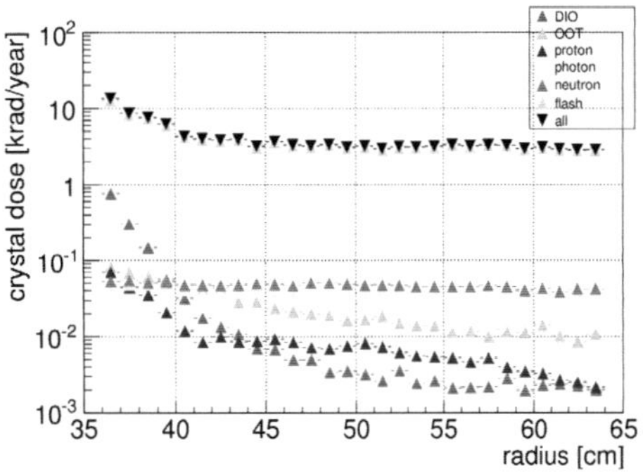

Figure 17. Disk 1: Crystal dose (krad/year) versus the radial distance (cm).

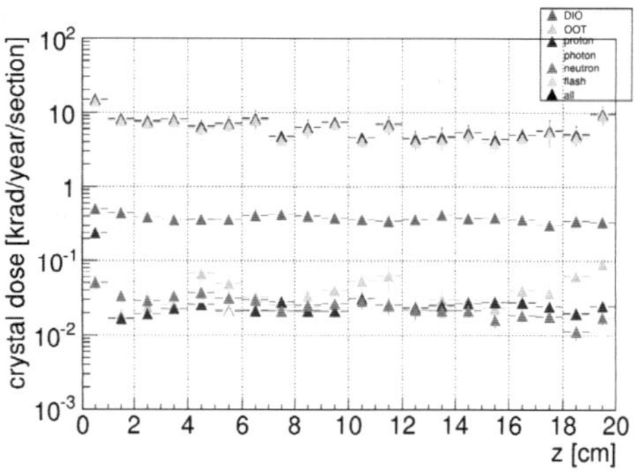

Figure 18. Disk 1: Crystal dose (krad/year/section) versus the crytal depth (cm).

also the dose along the crystal has been studied. Figure 18 shows the dose/year/section for the crystals in the inner ring of the first disk as a function of the crystal depth. It shows that

As one naively expects, the dose in the second disk is smaller, thanks to the shielding effect provided by the first one. Figure 19 shows the dose as a function of the radial distance. The hottest area, as in the previous case, is the inner ring, where the dose is $\sim$ 8 krad/year, than the dose scales down at 0.2 krad/year for radial distances higer then 50 cm.

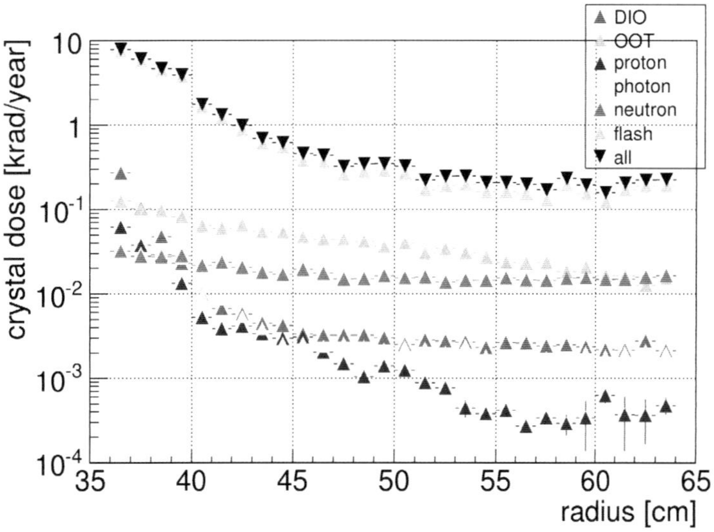

Figure 19. Disk 2: Crystal dose (krad/year) versus the radial distance (cm).

4. Conclusion

In this paper the calorimeter project for the Mu2e experiment has been presented. We have shown that the calorimeter design is able to satisfy the Mu2e requirements, providing excellent particle identification, an efficient trigger and is also a helpfull tool for improving the tracking pattern recognition. Simulation studies, integrated with eperimental measurements of light yield and electronic noise, indicated that we can achieve an energy resolution of O(5 MeV) at 100 MeV. Test beam with a small prototype of the calorimeter is under construction for validating the simulation results in this energy region.

References

[1] J. Adam et al, *New constraint on the existence of the $\mu^+ \to e^+\gamma$ decay. Phys. Rev. Lett.* 110 (2013), 201801 .
[2] W. Bertl et al, *A search for $\mu - e$ conversion in muonic gold. The European Physical Journal C - Particles and Fields 47(2) (2006), 337-346.*
[3] William J. Marciano et al, *Charged lepton flavor violation experiments. Annual Review of Nuclear and Particle Science* 58 (2008), 315-341.
[4] J. R. Abrams et al, *Mu2e conceptual design report*
[5] M. Cordelli et al, *Test and Simulation of a LYSO+APD matrix with a tagged Photon Beam from 40 to 300 MeV J. Phys.: Conf. Ser.* 404 (2012) 012027
[6] B. Aubert et al, *The BABAR detector. Nucl. Instr. Meth. A 479 (2002), 1.*
[7] M. Anfreville et al, *Laser monitoring system for the CMS lead tungstate crystal calorimeter. Nucl. Inst. Meth. A* 594 (2008), 292.
[8] Robert K. Kutschke, *Offline Software for the Mu2e Experiment J. Phys.: Conf. Ser.* 396 (2012), 022028
[9] T. Skwarnicki, *A study of the radiative cascade transitions between the υ′ and the υ resonances. Ph.D Thesis,* DESY F31-86-02(1986) Appendix E

16th International Conference on Calorimetry in High Energy Physics (CALOR 2014) IOP Publishing
Journal of Physics: Conference Series **587** (2015) 012048 doi:10.1088/1742-6596/587/1/012048

A $PbWO_4$-based Neutral Particle Spectrometer in Hall C at 12 GeV JLab

Tanja Horn

Catholic University of America, Washington, DC 20064, USA

E-mail: hornt@cua.edu

for the JLab Neutral Particle Spectrometer Collaboration

Abstract. The Neutral Particle Spectrometer is a standalone electromagnetic calorimeter capable of detecting high energy photons from, for instance, DVCS or π^0 decay with good energy and spatial resolution in a high rate environment. It can be used together with the Hall C high-momentum spectrometers for a suite of experiments with the underlying scientific objective of studying quark dynamics through exclusive and semi-inclusive reactions.

1. Introduction

Over the last decade, hard exclusive processes have emerged as a class of reactions providing novel information on the quark and gluon distributions in hadrons. This information is more complete than what can be obtained from studies of inclusive and elastic scattering alone [1, 2, 3]. QCD factorization theorems allow for expressing amplitudes of hard exclusive processes in terms of the Generalized Parton Distributions (GPDs) [4, 5], which are expected to provide a universal (process-independent) description of the nucleon, and have a known QCD (Q^2) evolution. GPDs are hybrid distributions that combine aspects of the usual collinear Parton Distribution Functions and elastic form factors. As such, GPDs simultaneously encode information on parton distributions and correlations in both momentum (in the longitudinal direction) and coordinate (in the transverse direction) spaces. An interesting aspect of GPDs is also their connection to the form factors of the energy-momentum tensor, which, among other things, establishes the decomposition of the proton spin in terms of the quark and gluon contributions to the total orbital momentum [6]. GPDs provide an unprecedented means to describe nucleon structure. The GPD program thus lies at the heart of the scientific motivation of 12 GeV JLab.

The Hall C 12 GeV base experimental equipment consists of two magnetic spectrometers: the High Momentum Spectrometer (HMS) and the Super High Momentum Spectrometer (SHMS). Important features of these magnetic spectrometers include their ability rotate around a pivot on a rigid support structure and a well-shielding detector enclosure. The acceptance of magnetic spectrometers is limited, but within this acceptance they provide excellent control over systematic uncertainties allowing for high precision cross section measurements [1]. Depending on the specific requirements of the experiments, the HMS and SHMS can detect either negatively

[1] This has been demonstrated by more than 200 Rosenbluth separations (close to 1000 kinematics).

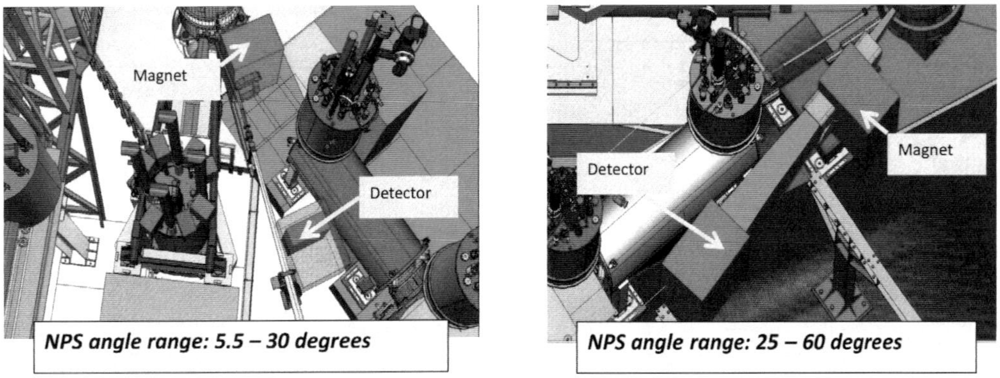

Figure 1. The Neutral Particle Spectrometer in Hall C at JLab.

or positively charged particles by choosing the magnetic field and the trigger configuration. However, additional instrumentation is needed to carry out the Hall C scientific program with neutral particle final states. The program consists of Exclusive Deeply Virtual Compton and π^0 precision cross section measurements, measurements of the SIDIS π^0 production as validation of factorization, and Wide-Angle Compton Scattering and exclusive photoproduction of π^0 mesons at large angles.

The Neutral Particle Spectrometer (NPS) is envisioned as a facility utilizing the well-understood HMS and the infrastructure of the SHMS to allow for precision (coincidence) cross section measurements of neutral particles. It can be cantelevered off the SHMS platform covering spectrometer angles between 5.5 and 30 degrees and be installed on the SHMS platform to cover angles between 25 and 60 degrees as illustrated in Fig. 1.

2. Physics with neutral final states

The best studied hard exclusive process is Deeply Virtual Compton Scattering (DVCS). Present analyses assume validity of twist-2 dominance, and a leading order formalism. The GPD program at JLab with 12 GeV has the ambition to go beyond these analyses. The PAC40 approved DVCS experiment in Hall C [7] is in the unique position, using the spectrometer capabilities in Hall C, to exploit the full kinematical dependence of the DVCS cross section in order to isolate all Fourier moments of the cross-section on an unpolarized target and separate the DVCS-BH interference and DVCS^2 contributions to each of them. Each individual term has a distinct Q^2 dependence that will be measured allowing one to quantify the size of higher-twist corrections. Notice that so far only the Q^2-dependence of the imaginary part of the DVCS amplitude has been studied experimentally [8]. The data showed indications of leading twist dominance. However, the test on several other observables including the real part of the DVCS amplitude, which is thus far not well understood, will be important for our global understanding of the GPD formalism. Precise study of Q^2 dependences and higher twists are of great interest for our global understanding. Twist-three GPDs can give access to a longitudinal proton spin sum rule [9] in a similar way as twist-2 GPDs probe the transverse spin through the well-know Ji sum rule [6]. The potential of DVCS to provide additional constraints on the real part of CFFs is important for developing more accurate GPD models and these data could also be used in global fits of CFFs and for dispersion relation analysis.

The energy separation requiring the measurements of conjugate settings (same Q^2 and x_B) at different beam energies is key to the overall DVCS program. With the CEBAF upgrade Hall C will receive electron beams of maximum energy of 11 GeV and luminosity of up to 10^{38} cm^{-2}

Table 1. Experiment requirements on $PbWO_4$.

Parameter	**Ideal**	**Acceptable**
Light Yield (pe/MeV)	10-15	5-6
Transmittance (20cm) at $\lambda \sim$420nm	>60%	~40%
Uniformity of optical properties (%)	~10	<20
Radiation hardness LY loss for 1.5 krad at dose rate 15 rad/h	5%	~10%
Uniformity rad. degradation at the same dose (20%)	~20	<30
Tolerance in dimensions (μm)	±50	±100
Timing property (ns, %)	30-50, 90%	

s^{-1} providing a great facility for the study of electron-nucleon scattering and for the separation of the DVCS cross section. Energy separation can only be performed with the high momentum spectrometers in Hall C due to the high momentum of scattered electrons increasing the Q^2 reach to even higher values at fixed x_B. Besides the requirement on momentum, efficient detection of photons is demanded over the entire kinematic range for the Hall C DVCS measurements to provide the highest precision data in the kinematic domain accessible with a 11 GeV beam extracting all possible independent observables on an unpolarized proton target.

The flexible design of the NPS allows other experiments to use its new capability of detecting neutral particles in Hall C as well. For instance, the PAC40 approved, A-rated SIDIS experiment [10], which probes the foundation of all SIDIS experiments at energies available at 12 GeV JLab, would be part of the same run group as DVCS and also use the NPS.

The neutral pion electroproduction SIDIS reaction plays a unique role in validating the SIDIS framework at JLab energies, and is thus an important tool in our study of hadron structure. The main advantages of neutral pion production include the lack of diffractive ρ contributions, the lack of pole contributions and thus radiative tail contributions at large z, the reduced nucleon resonance contribution (as for example compared to ep), and the proportionality to an average fragmentation function, are all points in favor to validate low-energy (x, z) factorization required to substantiate the SIDIS science output.

3. The Neutral Particle Spectrometer (NPS)

The basic concept for the NPS is a highly segmented electro-magnetic calorimeter preceded by a compact sweeping magnet. The experiments it enables, require detection of neutral particles with energies ranging between 0.5-7.6 GeV with good energy resolution (1-2%), and good coordinate (2-3 mm) and angular resolution of 0.5-0.75 mrad, which is comparable to the resolutions of the focusing spectrometers in Hall C. The neutral particle scattering angles cover 6.7-25 degrees at distances from the target ranging between 3 m and 6 m. For example, the minimum angle required at a distance of 4 m from the target is 7.2°.

The NPS is an efficient and economical way to meet all of the experimental requirements. It will consist of the following components:

- $PbWO_4$ crystals in a temperature controlled frame;
- a set of high voltage distribution bases with built-in amplifiers for operation in high-rate environments;
- essentially deadtime-less digitizing electronics to independently record the pulse amplitudes from each crystal;
- a sweeping magnet and a modified beam pipe with a large critical angle to minimize backgrounds that still reaches small angles and could be used as general purpose beam

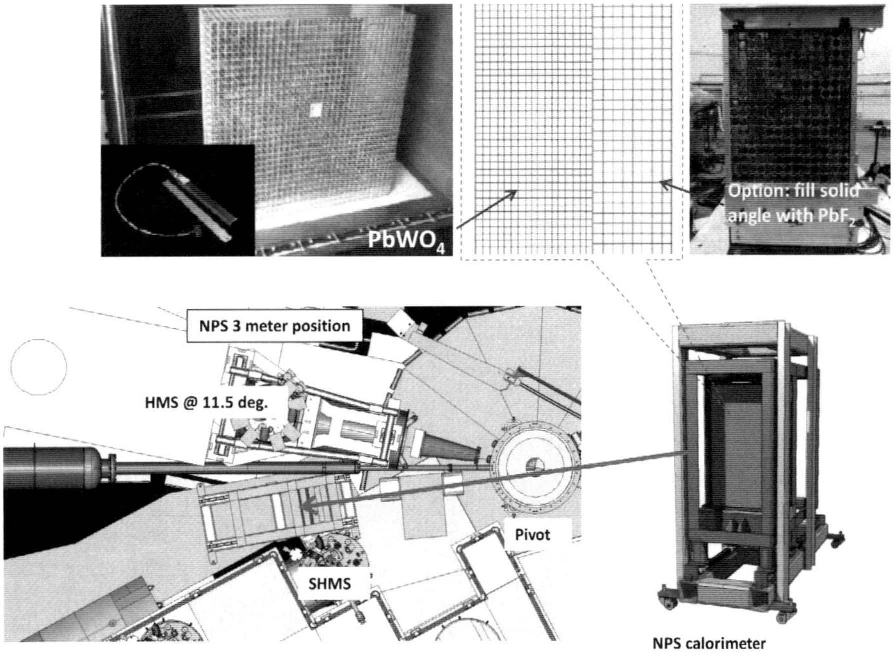

Figure 2. NPS crystal matrix and frame installation in Hall C.

pipe in Hall C ;

- a cantelevered platform of the SHMS carriage to allow precise, remote rotation around the Hall C pivot over an angular range between 6 and 25 degrees;
- a light monitoring and blue light curing system to monitor and restore crystal optical properties;

3.1. Crystals

The NPS will consist of an array of up to 1116 scintillating $PbWO_4$ and up to 208 PbF_2 crystals. Both types of crystals are fast ($PbWO_4$: 5-14 ns and PbF_2: < 30ns) and so suitable for the experiments, which require fast signals with short tails to minimize pile-up at high rates, e.g., timing resolution of better than 100 ns [2]. The requirements of the NPS on PWO are summarized in Table 1. In general, the NPS needs crystals with high transparency, high light yield, good timing where 90% of the light is emitted within 30-50 ns, and good radiation hardness. Also important are crystal geometry and integrity.

In the ideal case the NPS calorimeter will consist of a set of brandnew $PbWO_4$ crystals. Taking advantage of the existing $PbWO_4$ crystals [3] of the high-resolution inner part of the Hybrid Electromagnetic Calorimeter (HYCAL) [12] used for PRIMEX /PRIMEX-II experiments one arrangement is an arrangement of 1080 scintillating $PbWO_4$ crystals in a 36 by 30 matrix covering 25 msr at a distance of 4 meters from the target. We could use the PbF_2 crystals from the Hall A DVCS calorimeter to fill out the solid angle if, in the worst case scenario, only around

[2] Bismuth Germanate (BGO), another commonly used scintillation crystal has a timing property of 300 ns and is thus not suitable for the NPS. LSO/LYSO is a crystal with acceptable timing, but is not an economically favorable option.
[3] and accompanying photomultiplier tubes (PMTs)

600 $PbWO_4$ crystals should be available due to scheduling conflicts. The remainder could then be replaced by 208 PbF_2 crystals in a hybrid configuration, one option of which arranging the faster $PbWO_4$ along the beam side is shown in Fig. 2 [4]. Such a hybrid spectrometer configuration with a combined crystal matrix was successfully used in the PrimEx HYCAL calorimeter, which had an inner core of $PbWO_4$ and an outer ring of lead glass crystals. Experience with the HYCAL yielded energy and coordinate resolutions of $\sigma/E = 1.3\%$ and $\sigma_x \sim$1.28-2.10 mm at a neutral-pion energy of 5 GeV, giving an invariant mass distribution with a width of 2.3 MeV/c^2, which fully satisfies the needs of the DVCS and SIDIS π^0 experiments.

3.2. PMTs and amplifiers

The phototubes available with the $PbWO_4$ crystals from PrimEx, are the 10-stage Hamamatsu R4125 model with 19-mm diameter. For operation in the high-rate environment of the Hall C experiments, a new set of HV distribution bases with built-in amplifiers is needed. To increase the gain stability at the high rates anticipated with the NPS setup, the R4125 photomultiplier will be equipped with an active base [13]. The active base design improves gain stability with an efficient output signal range up to $\sim$160 μA. In this range, the PMT base system gain, or pulse amplitude and pulse width, remains stable to within 1%. The measured pulse rate at this current corresponds to a rate of about 1.2 MHz of 300 mV output pulses. The total count rate advantage of the newly designed active base is a factor of $\sim$25 as compared to the PrimEx base. It also indicates a noticeable improvement of the gain stability of the complete PMT plus base system versus count rate (or anode current). Additionally, radiation tests of the active PMT base were performed in Hall C during the Qweak operations in 2011/12 and no degradation of the base for a radiation dose of 100 kRad was found.

3.3. Trigger and electronics

The singles rate of electrons in the HMS for the experiment will be sufficiently low (<1kHz) to allow using a minimum-bias electron trigger and reading out the NPS in each event. In this way, exclusive, semi-inclusive, and inclusive cross sections can be compared directly at each kinematic point. To take full advantage of the high-resolution crystals while operating in a high-background environment, JLab-developed flash ADCs (fADCs) will be used to digitize the signal [5]. They continuously sample the signal every 4 ns, storing the information in an internal FPGA memory. When a trigger is received, the samples in a programmable window around the threshold crossing are read out for each crystal that fired. Since the readout of the FPGA does not interfere with the digitizations, the process is essentially deadtime free. If needed, the DAQ system will support windows up to 200-300 ns at 1 kHz and 100% occupancy in the 1200 channel calorimeter ($\sim$ 200 MB/s), but projected data rates will be smaller by orders of magnitude for all kinematics even if the thresholds are set very low. The sampled signals can then be fitted and integrated off-line, effectively eliminating issues with pile-ups, baseline shifts, etc. The fADC timing resolution was measured to be $<$ 1 ns for all signals [22]. For low-Q^2 measurements, where the electron singles rate could be high, the fADC-based system can also support a coincidence trigger. Such a trigger would take advantage of the ability of the fADC to perform the integration of the pulse and pass it along to the trigger for cluster finding. The appropriate conditions for the latter can then be used to select, for instance, DVCS or π^0 events. The integration and cluster finding will delay the trigger decision, but this can be easily accommodated in a pipelined system without any need for delay cables or analog delay modules. In summary, the system will provide a low dead time, precision signal processing off-line, and support both high-rate operations in singles mode as well as advanced, trigger-level

[4] This option has the minimum number of transitions between the different block types

[5] The fADC250 modules [23] collect and process data with 10 or 12 bit resolution at 250 MHz sample rate.

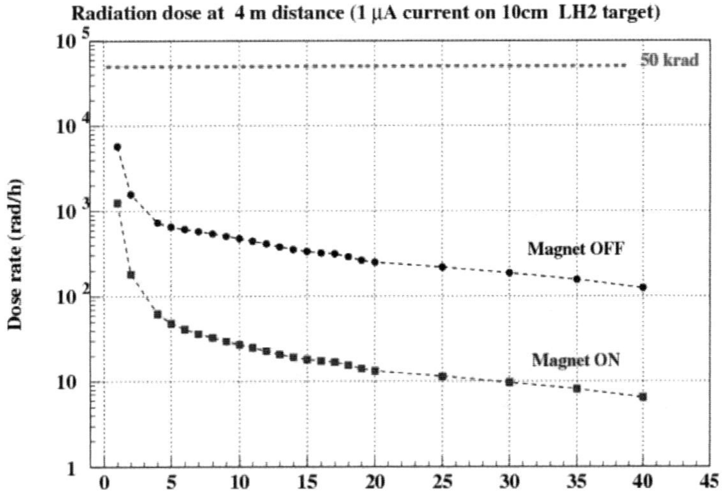

Figure 3. NPS crystal matrix and frame installation in Hall C.

cluster finding in coincidence mode.

3.4. Temperature controlled frame, cantilevered platform and positioning mechanism
The crystals will be installed in a frame based on that from the HyCal as shown in Figure 2. The detector frame will be mounted on a base frame with a rail system allowing the spectrometer to be moved in z to positions between 3 m and 12 m from the target.

The calorimeter box will be installed off a cantilevered platform on the SHMS (see Figure 2). It will rotate with the spectrometer allowing for precise positioning and eliminating the need for a separate motion control system, which is essential given the space constraints between the spectrometer and the beam line at small angles. In collaboration with the JLab engineering/design group we have analyzed the NPS positioning together with the other equipment in the experimental hall. The minimum central angles the NPS can reach fulfill all the requirements of DVCS experiment and SIDIS, and are also suitable for future experiments. The spectrometer will be designed so that it can be easily moved and removed from the SHMS platform.

The scintillation light output of the $PbWO_4$ crystals is temperature sensitive, and thus the entire calorimeter must be kept at a constant temperature (to within 0.1° to guarantee 0.5% energy stability for absolute calibration and resolution). Furthermore, the high-voltage dividers on the PMTs may dissipate a total of several hundred Watts, and this power must not create temperature gradients or instabilities in the calorimeter. The NPS will thus be thermally isolated and surrounded on all four sides by water cooled copper plates. This design is based on that of the HYCAL temperature controlled frame optimized with more recent experience from CMS [14], which has shown stability to 0.05°. The final choice of temperature will be a trade-off between light yield and radiation resistance. For $PbWO_4$ the light yield increases at lower temperatures, but luminescence decay time and radiation resistance decreases. The properties of PbF_2 are nearly independent of temperature.

4. Studies of backgrounds and radiation dose

To detect particles produced at small scattering angles, the Neutral Particle Spectrometer (NPS) must be installed close to the beamline. It is thus important to pay attention to the effect of

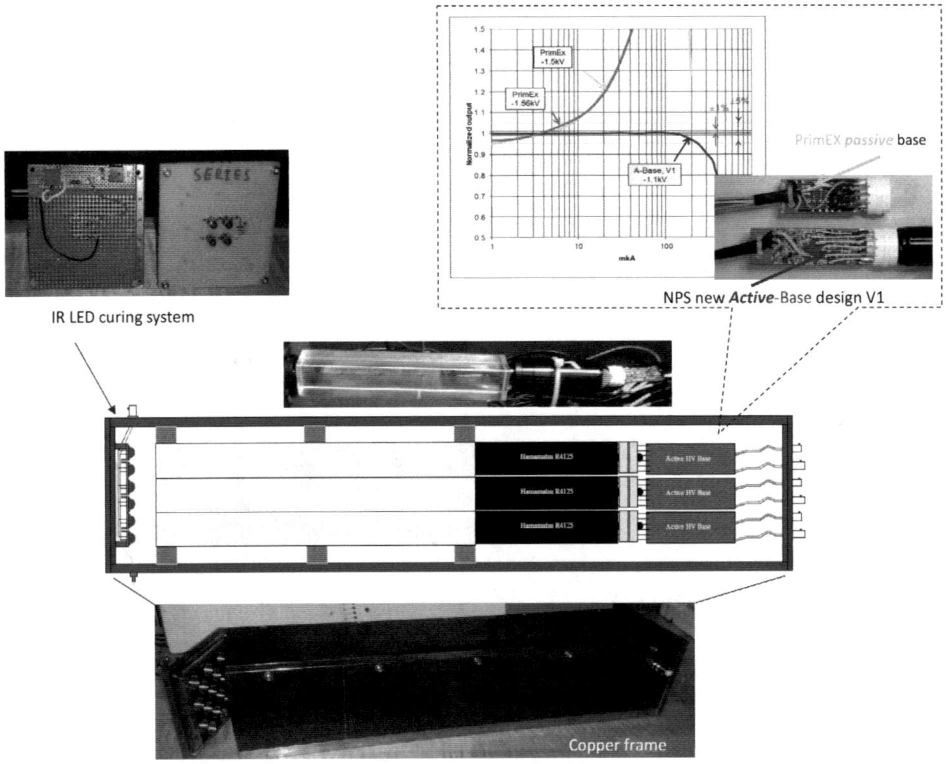

Figure 4. The NPS prototype. It consists of $PbWO_4$ and PbF_2 crystals in a copper frame. The readout is performed with R4125 with active bases. These active bases outperform the passive bases from PrimEx by about a factor of two. The setup also contains a prototype of the curing system.

radiation dose on the performance of the spectrometer. In particular, attention must be paid to degradation of the crystal optical properties and recovery thereof, and damage to the photosensors. The background rates could also directly affect the resolution of the spectrometer.

To select the optimal configuration for the NPS, simulations of the background rates and radiation doses were carried out. The baseline simulation is a well calibrated GEANT simulation used for all JLab offical dose rate estimates. It includes the 10 cm LH_2 target, a 0.3 T-m sweeping magnet, and the NPS, with its front face located at a distance of 4 meter from the target, covering an angular range between 5 and 25 degrees. The simulation showed a strong angular dependence of the background, and that the major fraction of the rate comes from Moeller electrons, which can be eliminated by the 0.3 Tm sweeping magnet. With the magnetic field added, the dominant parts of the beam-induced background rate become photons with an energy above 10 MeV and electrons with an energy above 300 MeV.

Radiation dose rates from the JLab official background simulation have been used to estimate the exact radiation condition as seen by the $PbWO_4$ detector as part of the NPS facility. Fig. 3 shows the total dose rates as a function of NPS angle, with magnet on and off, compared to a 50 krad dose limit before curing of the crystals is required. This number based on LHC studies [11] is conservative (i) even at much higher integrated doses only small (few %) effects are seen; (ii) additional shielding materials are not yet taken into account; and (iii) actual dose rates out-of-plane are much reduced. Additional and more detailed information on our background and radiation dose studies can be found in Appendix VII of reference [10].

4.1. Light monitoring and curing system

The expected doses for the NPS experiments is dominated by small-angle operations, where the integrated dose is 1.7 Mrad in the center and 3.4 Mrad at the edge. To control the stability of the NPS and degradation of the crystals due to accumulated radiation dose a light monitoring system is needed. It will periodically inject light into detector modules between the real events during data taking or during special calibration runs with a frequency 10-20 Hz. Such a system will measure variations of the crystal transmittance over the course of the experiment and provide calibration in situ.

To restore the crystal optical properties, a curing system will be constructed with minimum impact on the experiment running. Our baseline method is to use blue light of wavelengths between 400-600 nm (optical bleaching [18]). The required light intensity is of an order of 1-2 mW/cm^2, and thus for the NPS 2x2 cm^2 ($PbWO_4$) or 3x3 cm^2 (PbF_2) crystals a curing system with power 5-10 mW/crystal (at wavelengths of 400-600 nm) is needed. This is a well established technique and has been used by many other experiments, e.g., GEP-III in Hall C and DVCS in Hall A. However, this technique is invasive and based on experience with the DVCS experiment can affect PMT operation. We will thus also develop a curing system with permanent infrared illumination based on those from, e.g., Refs. [20, 19, 21]. This method has been observed by the PANDA collaboration to yield significant improvement of the crystal optical transparency after being exposed to modest radiation doses.

5. NPS prototype studies

To optimize the technical aspects of the calorimeter before finalizing the NPS design a prototype has been constructed. As illustrated in Fig. 4 it has an active area of about 6x6 cm^2 including a crystal matrix of $PbWO_4$ (and PbF_2 to test the hybrid configuration) in a copper frame. The readout is done by 19 mm Hamamatsu R4125 PMTs with our new active HV base. The prototype will test the light monitoring system as well as two approaches for the curing system, i.e., standard with a blue light source and IR curing of wavelengths $>$ 900 nm. The prototype IR curing system was constructed using superbright LEDs like the OSRAM LD274 with peak wavelength 950nm and Vishay TSAL7400 with peak wavelength 940nm.

The efficiency of IR curing is reduced by a factor 20-50 relative to blue light. This must be compensated by using high intensity ($\sim 10^{16}$ photons/s per crystal). We have measured the intensity of IR LEDs LD274 and TSAL7400 whose peak wavelenghts are around 950 nm. The measurements were done using a calibrated Hamamatsu silicone photodiode S2281 with quantum efficiency of 67% at 950nm to measure the intensity of the emitted light [6]. The distance betwen LED and photodiode was varied between 0.5 cm and 20 cm. The LED driving current was measured with a multimeter and the photodiode current was measured with a picoammeter. The results for the LED emission intensity as a function of driving current for distances from the photodiode of 3 cm and 7cm are shown in Fig. 5. The output is almost linear with driving current for both LEDs. Beyond 60 mA the output begins to curve slightly indicating the onset of saturation. Additional measurements with attenuation of the LEDs output light intensity were made with a neutral density filter with attenuation factor 9.25 installed in front of the LED. The data showed good agreement of the data without filter and with filter when scaled by the attenuation factor. The saturation does thus not seem to be due to the photodiode. The measured photon flux at LED driving current of 100 mA and 3 cm distance between the LED and the photodiode is equivalent to 2 x 10^{16} photons/s depositing a power of 4.2 mW/cm^2. The

[6] its current is nearly proportional to the LED intensity

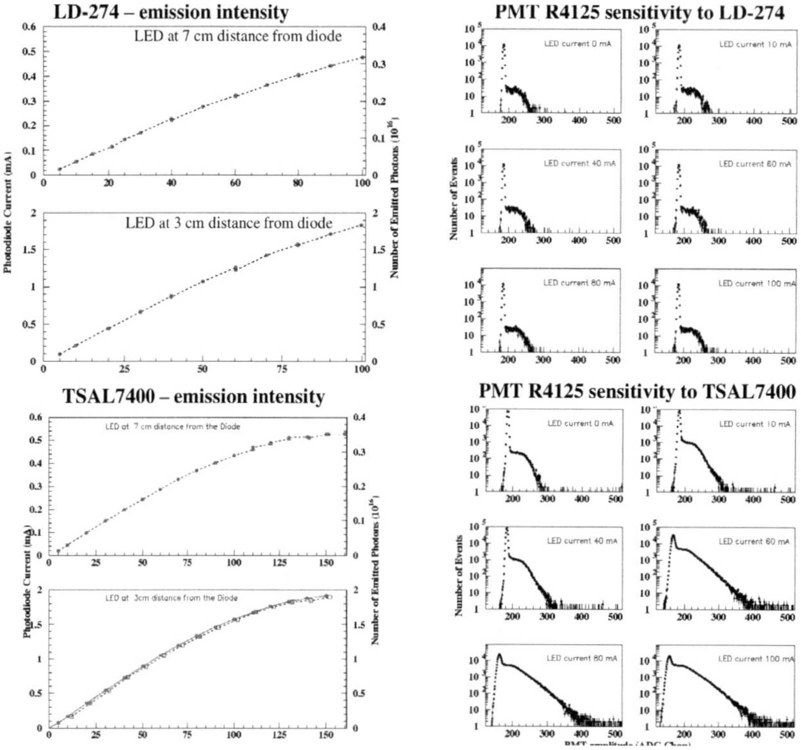

Figure 5. LEFT: the emission intensity of the IR LEDs LD274-3 (top figure) and TSAL7400 (bottom figure) vs. driving current at distances 7cm (top panels) and 3cm (bottom panels) and as indicated in the figure. The PMT HV for the measurements was 1600 V. The blue open symbols in the bottom figure denote data taken with a neutral density filter with attenuation factor 9.25 in front of the LED; RIGHT: the amplitude distributions of the signals detected by PMT R4125 at different values of the infrared LEDs driving currents (top: LD-274-3, bottom: TSAL7400).

photon flux at 7cm distance is 0.4 x 10^{16} photons/s. The results of the intensity tests suggest that both LEDs LD274 and TSAL7400 would be suitable for curing applications.

The possibility to perform continous curing of $PbWO_4$ using permanent illumination with infrared light when the PMT HV is on depends on the sensitivity of the PMTs to the IR light source. We have measured the PMT sensitivity to IR light. This is effectively a PMT quantum efficiency measurement at wavelength far beyond the sensitivity range shown in any of manufacturer data sheets. We have measured the PMT sensitivity at different driving LED currents between 0 and 100 mA for different configurations, e.g., distances between LED and PMT ranging between 0.5 cm and 18 cm, with and without the crystal in front of the PMT, and different PMT gains. The output signal was analyzed with an ADC of gate width 150 nsec, and 1 channel equivalent to 100 fC. The resulting signal amplitude distributions for different LED driving current (at different intensity of the IR light) are shown in Fig. 5. The data suggest that the PMT R4125 has a very low, but non-negligible efficiency to IR light. One thought about the PMT sensitivity is that the IR LED emission spectrum may be contamination by short wavelength light. Tests were thus repeated with a 900 nm long-pass filter cutting all wavelengths but IR. No difference between the measurements with and without filter were found.

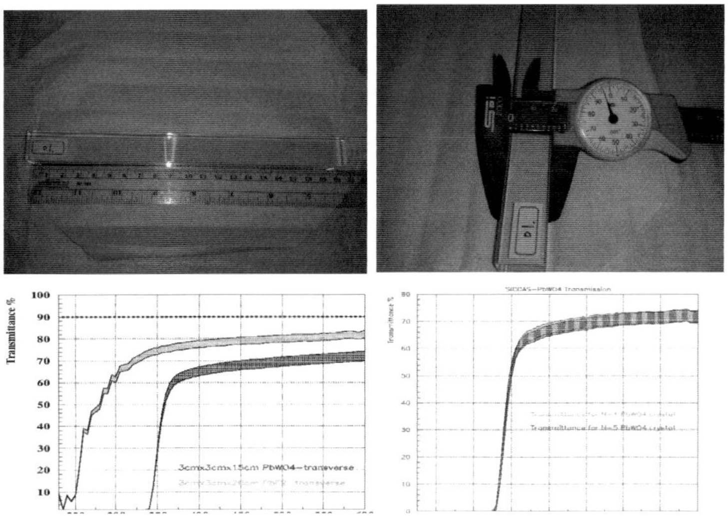

Figure 6. TOP LEFT/RIGHT: $PBWO_4$ crystal from SIC obtained in spring 2014. The measured dimensions are 20±0.15 mm x 20 ±0.15mm x 200mm. LEFT BOTTOM: The transmittance of PbF_2 from the Hall A DVCS calorimeter (green) and $PbWO_4$ from the HYCAL crystals vs. wavelength. The color band denotes the spread between the data measured at different points on the crystal. RIGHT BOTTOM: the transmittance for a $PbWO_4$ crystal from SIC obtained in spring 2014.

This suggests that the majority of the light has IR wavelengths, that the PMT has a real non-zero efficiency relative to IR light, and that the PMT could receive long term damage if left on during IR curing.

To estimate the effect of IR light on the PMT lifetime the relation between PMT anode current and LED driving current was analyzed. Measurements were made for 4 IR LEDs at 19cm distance from the PMT and an LED driving current of 50 mA. The PMT high voltages varied between 1400-1600 V, the typical operating voltages for these PMTs. For a 50mA LED driving current the 4-IR LEDs will have about $4x2x10^{16}$ photon/cm^2/s. The results are 309 nA for 1400 V, 492 nA for 1500 V and 758 nA for 1600 V. This is below the R4125 maximum anode current of $\sim$ 0.1 mA even for the highest voltages, and thus there will likely not be a large effect on the PMT lifetime. However, additional studies are ongoing to improve the PMT lifetime using LEDs with larger peak wavelengths, e.g., >1000 nm. The efficiency of curing with these IR LEDs will be tested with about 20krad irradiated crystals.

The $PbWO_4$ crystals used in the prototype were obtained from SIC in spring of 2014. The results of a transmittance measurement compared to the transmittance of crystals from the Hall C photon detector and PbF_2 crystals is shown in Fig. 6. The $PbWO_4$ and PbF_2 transmittance curves are very different in the short wavelength region (λ=250-350 nm), but for both types of crystals the transparency saturates for wavelengths above $\lambda \sim 400$ nm. A common blue light source should thus be acceptable for both types of crystals for the NPS light monitoring system. The efficiency of the blue light monitoring system will be tested with the prototype.

A prototype for electronics and triggering tests is being constructed to see how the fADC based electronics would work with PbF_2. These pulses rise/fall faster than $PbWO_4$, which could result in fluctuations in the integral depending on the timing of the pulse with respect to the fADC clock. Furthermore, studies of existing elastic data from JLab are ongoing to check for possible impact on the resolution in timing and amplitude when doing a waveform analysis for PbF_2.

6. Acknowledgments

This work is supported in part by NSF grant PHY-1019521 and PHY-1036227, and Jefferson Science Associates (JSA).

7. Bibliography

[1] K. Goeke, M.V. Polyakov and M. Vanderhaeghen, Prog. Part. Nucl. Phys. **47**, 401 (2001)
[2] M. Diehl, Phys. Rept. **388**, 41 (2003).
[3] A.V. Belitsky and A.V. Radyushkin, Phys. Rept. **418**, 1 (2005).
[4] J.C. Collins, L. Frankfurt, M. Strikman, Phys. Rev. D **56**, 2982 (1997).
[5] J.C. Collins, A. Freund, Phys. Rev. D **59**, 074009 (1999).
[6] X. Ji, A. Freund, Phys. Rev. Lett. **78**, 610 (1997).
[7] A. Ahmidouch, I. Albayrak, D. Androic, J.R.M. Annand, A. Asaturyan, M. Ben Ali, M. Boer, A. Camsonne, M. Canan, M. Carmignotto, C. Chen, S. Covrig, S. Danagoulian, D. Day, M. Defurne, J. Denes-Couto, C. Desnault, R. Dupre, M. Elaasar, R. Ent, A. Fradi, M. Garcon, B. Garillon, D. Gaskell, L. Ghedira, F.X. Girod, P. Gueye, M. Guidal, D.J. Hamilton, Y. Han, M. Hattawy, N. Hlavin, T. Horn, G. Huber, C. Hyde, H.S. Jo, M. Jones, C. Keppel, P. King, D. Keller, F. Klein, P. E. C. Markowitz, A. Marti Jimenez-Arguello, M. Mazouz, A. Mkrtchyan, H. Mkrtchyan, C. Munoz-Camacho, P. Nadel-Turonski, B. Nepal, S. Niccolai, R. Paremuzyan, A.J.R. Puckett, M.N.H. Rashad, J. Roche, O. Rondon, F. Sabatie, B. Sawatzky, M.H. Shabestari, S. Sirca, P. Solvignon, V. Sulkosky, V. Tadevosyan, L. Tang, F. Wesselmann, S.A. Wood, B. Wojtsekhowski, S. Zhamkochyan, and the Neutral Particle Spectrometer Collaboration, *Approved Jefferson Lab experiment E12-13-010.*
http://www.jlab.org/exp_prog/proposals/13/PR12-13-010.pdf
[8] C. Muñoz Camacho, A. Camsonne, M. Mazouz, C. Ferdi, G. Gavalian, E. Kuchina, M. Amarian, K. A. Aniol, M. Beaumel, H. Benaoum, P. Bertin, M. Brossard, J.-P. Chen, E. Chudakov, B. Craver, F. Cusanno, C. W. de Jager, A. Deur, R. Feuerbach, J.-M. Fieschi, S. Frullani, M. Garçon, F. Garibaldi, O. Gayou, R. Gilman, J. Gomez, P. Gueye, P. A. M. Guichon, B. Guillon, O. Hansen, D. Hayes, D. Higinbotham, T. Holmstrom, C. E. Hyde-Wright, H. Ibrahim, R. Igarashi, X. Jiang, H. S. Jo, L. J. Kaufman, A. Kelleher, A. Kolarkar, G. Kumbartzki, G. Laveissière, J. J. LeRose, R. Lindgren, N. Liyanage, H.-J. Lu, D. J. Margaziotis, Z.-E. Meziani, K. McCormick, R. Michaels, B. Michel, B. Moffit, P. Monaghan, S. Nanda, V. Nelyubin, M. Potokar, Y. Qiang, R. D. Ransome, J.-S. Réal, B. Reitz, Y. Roblin, J. Roche, F. Sabatié, A. Saha, S. Sirca, K. Slifer, P. Solvignon, R. Subedi, V. Sulkosky, P. E. Ulmer, E. Voutier, K. Wang, L. B. Weinstein, B. Wojtsekhowski, X. Zheng, and L. Zhu [Jefferson Lab Hall A Collaboration], Phys. Rev. Lett. **97**, 262002 (2006)
[9] X. Ji, X. Xiong, F. Yuan, Phys. Rev. Lett. **109**, 152005 (2012).
[10] A. Ahmidouch, I. Albayrak, D. Androic, A. Asaturyan, F. Benmokthar, A. Camsonne, M. Carmignotto, C. Chen, M. E. Christy, S. Covrig, S. Danagoulian, D. Day, P. Degtiarenko, J. Denes-Couto, D. Dutta, J. Dunne, M. Elaasar, R. Ent, D. Gaskell, D.J. Hamilton, Y. Han, D.W. Higinbotham, M. Guidal, N. Hlavin, T. Horn, G. Huber, C. Hyde, M. Jones, N. Kalantarians, D. Keller, C. Keppel, M. Khachatryan, P. King, F. Klein, M. Kohl, V. Kubarovsky, W. Li, P. E. C. Markowitz, M. Metz, A. Mkrtchyan, H. Mkrtchyan, C. Munoz-Camacho, P. Nadel-Turonski, B. Nepal, R. Paremuzyan, M.N.H. Rashad, J. Roche, O. Rondon, B. Sawatzky, M.H. Shabestari, S. Sirca, P. Solvignon, V. Tadevosyan, L. Tang, J. Taylor, F. Wesselmann, S.A. Wood, B. Wojtsekhowski, S. Zhamkochyan, and the Neutral Particle Spectrometer Collaboration, *Approved Jefferson Lab experiment E12-13-007.*
http://www.jlab.org/exp_prog/proposals/13/PR12-13-007.pdf
http://www.jlab.org/exp_prog/proposals/09/PR12-09-011.pdf
[11] R.Y. Zhu et al., *Nucl. Inst.Meth* A**413**, 297 (1998).
[12] M. Kubantsev et al., Performance of the Primex Electromagnetic Calorimeter, arXiv:physics/0609201, 22 Sep. 2006; A. Gasparyan, Performance of PWO crystal Detector for a High Resolution Hybrit Electromagnetic Calorimeter at Jefferson Lab., Proceed. X Int. Conf. Calorimetry in Particle Physics, Perugia, Italy, 29 March-2 April 2004, pp. 109-115.

[13] V. Popov and H. Mkrtchyan, New potomultiplier active base for Hall C Jefferson Lab Lead Tangstate Calorimeter, NSSS2012-1098.
[14] G. Franzoni et.al, *Nucl. Instrum. Meth.* A **628**, 90 (2011).
[15] V. A. Batarin et.al, (BTeV electromagnetic calorimeter group), Nucl. Instrum. Meth. A512, 488 (2003), hep-ex/0210011.
[16] P. Adzic et.al, The CMS Electromagnetic Calorimeter Group Report, http:iopscience.iop.org1748-0221503P0310, 2010.
[17] Qu Xiangdong, A Study of Lead Tungstate Crystals and Improvement of Radiation Hardness. Ph.D. Thesis, Shanghai Institute of Ceramics. The Chinese Academy of Sciences, Shanghai, China, January 16, 2001.
[18] R.Y. Zhu et al., *Nucl. Instrum. and Meth.* A **376**, 319 (1996); C. Woody et al., Proc. of SCINT95 Int'l Conf. Delft, August 1995 and IEEE-NUCL-S V 43 (1996) 1585.
[19] M. Anfreville et al., Nucl. Inst.Meth A**594**, 292 (2008).
[20] R.W. Novotny, PANDA collaboration, IEEE-2009-05402124.
[21] D.A. Mao et al., Nucl. Inst.Meth A**356**, 309 (1995).
[22] J.V. Bennett, M. Kornicer, M.R. Shepherd, M.M. Ito, "Precision timing measurements of phototube pulses using a flash analog-to-digital converter", JLab technical report (2013).
[23] H. Dong, C. Cuevas, D. Curry, E. Jastrzembski, F. Barbosa, J. Wilson, M. Taylor, and B. Raydo, IEEE Nuclear Science Symposium Conference Record, N15-375 (2007).

iPhos, a new technique for the CALIFA CsI(Tl) calorimeter

M. Bendel[1], R. Gernhäuser[1], M. Winkel[1], H. Alvarez-Pol[2], D. Cortina-Gil[2], B. Heiss[1], W.F. Henning[1], P. Klenze[1], T. Le Bleis[1], C. Pfeffer[1] and B. Pietras[2]

[1] Technische Universität München, Physik-Department E12
James-Franck-Str. 1, 85748 Garching bei München, Germany
[2] Universidade de Santiago de Compostela, E-15782, Spain

E-mail: Michael.Bendel@tum.de

Abstract. Within this paper we will present an entirely new energy reconstruction method iPhos for charged particles detected in CsI(Tl) crystals. Based on pulse shape analysis of CsI(Tl) scintillation signals [1] protons that cannot be stopped within the active detector material can be distinguished from those fully stopped. Due to a special signature of punch-through protons, they can be identified and their full energy is reconstructed from their specific energy loss. In a dedicated experiment at the Cyclotron Center Bronowice, Krakow the first proof of this method was achieved.

1. Introduction

The target region of the R^3B experiment at the new accelerator facility FAIR (Darmstadt, Germany) will be surrounded by the large volume calorimeter CALIFA. This calorimeter consists of about 3000 CsI(Tl) crystals that are read out by large area avalanche photodiodes (LAAPD). It is a multipurpose detector that plays a key role in the realization of kinematically complete measurements. The main requirements are a high efficiency, a good energy resolution of 6% at 662 keV γ-ray energy and a huge dynamic range that allows simultaneous measurement of γ-radiation of a few 100 keV up to scattered particles of several 100 MeV. In nuclear reactions with relativistic beam energies of up to 700 AMeV high energy secondary particles are emitted mainly in forward direction. For their detection the main feature of a calorimeter namely the stopping of these particles cannot be fulfilled due to geometrical constraints.

2. RPID

It is known for decades that CsI(Tl) scintillation light is dominated by two different components, a fast one with a decay time constant of $\tau_f \approx 600$ ns and a slow one with $\tau_s \approx 3.5$ μs. Evaluating these different scintillation amplitudes allows for a particle identification [2; 3]. This has already been performed in several experiments like CHIMERA [4] and INDRA [5]. For the development of a simplified algorithm for this kind of particle identification (PID) that could be implemented in high density fast sampling ADC modules a first experiment in 2010 at the Maier-Leibnitz-Laboratorium (MLL) in Garching, Germany to produce γ-rays up to 15.1 MeV and protons in the energy region up to 21 MeV was performed. The experimental method, the detector and electronics setup as well as the implementation of the algorithm for PID of fully stopped particles are discussed in detail in [1].

For the separation of scattered protons and γ-rays the so called RPID algorithm [1] was used to extract the amplitudes of the fast component $\mathrm{N_f}$ and the slow component $\mathrm{N_s}$ of the CsI(Tl) signal from the traces of the preamplifier output signals. In that context $\mathrm{N_f}$ respectively $\mathrm{N_s}$ are the total amounts of emitted light in the particular component.

γ-rays and protons down to energies of about 1 MeV were clearly separated. The linear distribution of the amplitude of the fast component N_f versus the slow component N_s shows a constant and energy independent ratio for γ-rays.

In contrast the proton distribution at low energies differs considerably from the linear characteristics. For stopped protons an analytical relation between the amount of light emitted in the slow component and the fast component $\mathrm{N_{s,p}(N_f)}$, now with index p for protons, fitted the data very well.

$$N_{s,p}(N_f) = a\left(e^{-bN_f} - 1\right) + cN_f \tag{1}$$

As a byproduct also the pattern of protons not fully stopped $\mathrm{N_{s,p,frac}(N_f)}$ in the active detector material has been identified [1, p. 5].

3. Intrinsic phoswich - iPhos

Equation (1) shows that $\mathrm{N_{s,p}(N_f)}$ is a nonlinear function especially for low energy protons. It was shown in [6; 7] that the total light output L for protons is in good approximation linear over a wide energy range

$$L(E) \propto E = c_L \cdot E. \tag{2}$$

Additionally the assumption $\mathrm{L = N_f + N_s}$ is a good approximation for the light production in CsI(Tl). So it can be concluded that $\mathrm{\frac{N_s}{N_f}(L)}$ is also nonlinear or the derivative $\mathrm{\frac{dN_s}{dN_f}(L)}$ is not a constant. From experimental data described in the following section $\mathrm{N_s(L)}$ and $\mathrm{N_f(L)}$ can be extracted by fitting a similar function as Eq. (1) to the accessible energy range

$$N_f(L) = a\left(e^{-bL} - 1\right) + cL \tag{3}$$

The derivative then describes the amount of fast component that is emitted within a infinitesimal amount of light production or path length of a proton through the active material

$$\frac{dN_f}{dL}(L) = -ab\,e^{-bL} + c \tag{4}$$

Knowing the energy loss in the detector material the amount of $\mathrm{N_f}$ can now be calculated by integration

$$\int_{L_1}^{L_0} \frac{dN_f}{dL}(L)\,dL \overset{Eq.(2)}{=} \int_{L(E_1)}^{L(E_0)} \frac{dN_f}{dL}(L)\,dL \tag{5}$$

Here L_0 corresponds to the total amount of light that would be produced if the incoming proton is stopped. Due to Eq. (2) this can be described in terms of the energy E_0 of the incoming proton. If the crystal is not long enough to stop the proton, E_1 corresponds to its energy when leaving the crystal and L_1 would be the total amount of light produced by stopped proton of this energy. For stopped protons ($E_1 = 0$) the integration starts at $L(0) = 0$ for punch-through protons at $L(E_1) \neq 0$. The amplitude of the slow component N_s is treated accordingly.

If we assume a detector with a fixed thickness of the CsI(Tl) crystals, the integration limits for punch-through protons are correlated due to the dependence of the energy loss within the crystal on the incoming energy.

$$E_1 = E_0 - \Delta E = E_0 - f(E_0) \tag{6}$$

where ΔE is the energy loss in the CsI(Tl) crystal. The function f is analytically not accessible. To overcome this problem the energy loss calculation and the integration is done numerically within the simulation package GEANT4 [8]. In that way the behaviour of the punch-through protons in the N_s vs. N_f plot can be nicely reproduced.

4. Experimental verification

To verify this finding and to define the potential use of this effect, an experiment to explore the "intrinsic phoswich properties" of CsI(Tl) was performed in spring 2013 at the Cyclotron Center Bronowice at the Henryk Niewodniczański Institute of Nuclear Physics [9] (IFJ PAN) in Krakow, Poland. The proton cyclotron "Proteus C-235" is used for different purposes like nuclear physics experiments and radiation therapy for cancer patients. The accelerator facility allows for protons with energies from 70 MeV up to 226 MeV using a fast and easy energy and intensity adjustment for efficient use. To cope for the minimum beam intensity of about 1 pnA and to allow for a synchronous measurement in several different experimental setups, scattered protons from a 50 μm thick titanium target which also acted as an exit window of the vacuum beam pipe were used.

4.1. Experimental setup

The detector setup for this experiment was placed in a distance of about 2 m and an angle of Θ =10 ° with respect to the beam axis. It consisted of the prototype detector *DemoZero* [10] of the CALIFA Barrel that is equipped with 32 CsI(Tl) trapezoidal-conical shaped crystals. The crystal shapes and sizes correspond to the forward part of the CALIFA Barrel between $55.5° < \Theta < 70.4°$ (see [11, p. 15]). Each crystal was wrapped in one layer of high reflective foil [1] (see Fig. 1, left bottom) and four of these were grouped within one carbon fibre alveoulus. The scintillation light was read out by Hamamatsu S8664-1020 LAAPDs [12]. The LAAPD signals were preamplified by a 32-channel Mesytec [2] MPRB preamplifier at a dynamic range of 30 pC.

The detector was placed vertically to the incoming particles (see Fig. 1, left top). To limit the scattering angles of the protons, but also the hit position along the crystals most of the active volume was covered by a massive lead shielding. Protons could reach the active volume only through a vertical slit of d = 1 cm. Here the average thickness of the crystal layers were between 13.0 mm and 15.3 mm (see Fig. 1, right).

For a systematic investigation of the proposed effect eight different beam energies (90 MeV, 100 MeV, 105 MeV, 110 MeV, 120 MeV, 130 MeV, 155 MeV and 180 MeV were used. In the following only the first crystal row with an effective thickness of 15.3 mm of active detector material is used for the energy reconstruction.

[1] Vikuiti^{TM} Enhanced Specular Reflector by 3M

[2] http://www.mesytec.com

Journal of Physics: Conference Series **587** (2015) 012049 doi:10.1088/1742-6596/587/1/012049

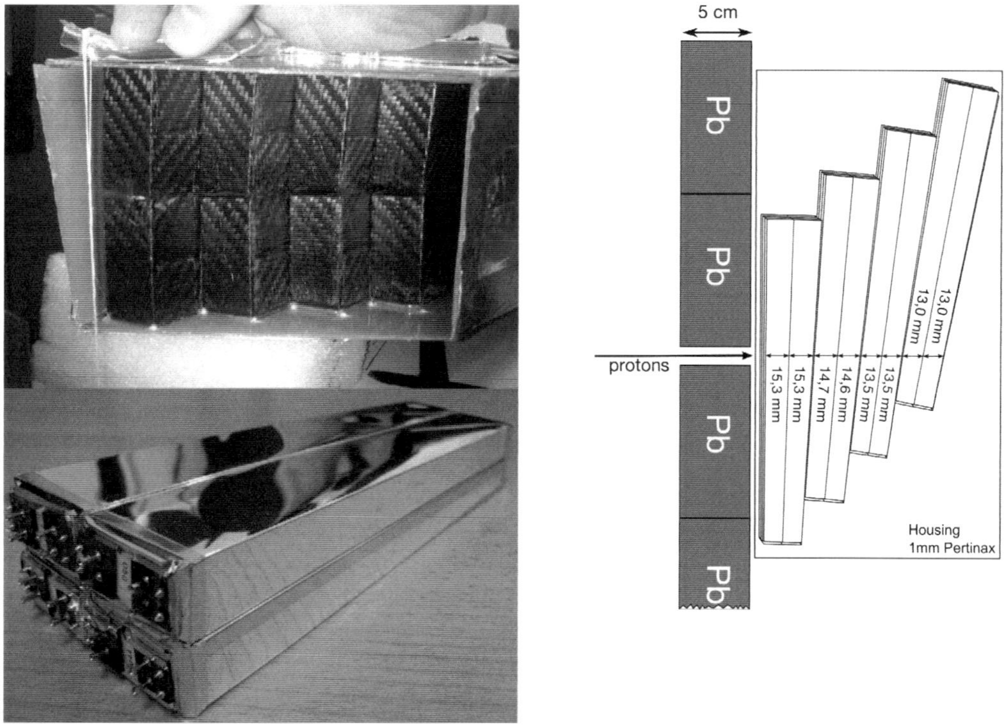

Figure 1. The detector setup consisted of 32 CsI(Tl) crystals that were wrapped with high reflective foil and read out by by two 10x10 mm^2 LAAPDs (bottom left). Four of these crystals were grouped within one carbon fibre alveoulus and surrounded by a 2 mm thick Pertinax housing (top left). More details can be found in [10]. A detailed schematic drawing of the setup is shown on the right. The effective thickness of the first two crystals is about 15.3 mm and is decreasing for those following.

4.2. Identification and reconstruction of punch-through protons

The results of the RPID pulse shape analysis, which separates the two components N_f and N_s are shown in Fig. 2.

In that two dimensional spectrum three different distributions (1-3 in Fig. 2) can be identified clearly. The structure marked with (2) corresponds to the low energy protons that are stopped within the crystals. For increasing energies and therefore protons that punch through the crystal, this line is strongly bent in the distribution (1). Due to angular uncertainties and therefore slightly varying effective detector thickness and also due to multiple scattering, the distribution of the stopped protons exceeds the punch-through energy in some cases. So in descending order the four peaks from 90 MeV to 180 MeV can be identified in the distribution (1), which corresponds to the reduced light output due to a reduced energy loss for punch-through protons. A simulation of the reaction in TALYS [3] showed that deuterons are also produced in the reaction $^{48}\mathrm{Ti(p, d)}^{47}\mathrm{Ti}$ with a cross section of about 10 % compared to $^{48}\mathrm{Ti(p, p')}^{48}\mathrm{Ti}$. They can be found in the RPID histogram marked as (3).

The function $N_{s,iPhos}(N_f)$ in Fig. 3 (drawn in black) shows the expected signal distribution of the punch-through protons. As first step of the reconstruction of the full energy all the events that are identified as punch-through protons are projected on this function to reduce

[3] http://www.talys.eu

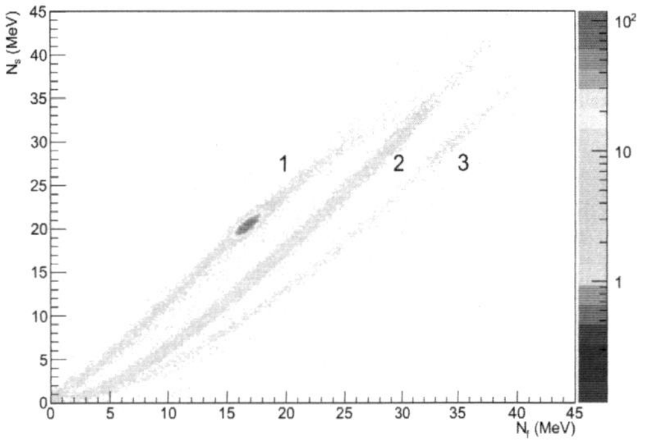

Figure 2. Result of the RPID pulse shape analysis for one crystal and 90 MeV protons. The different distributions correspond to punch-through protons (1), stopped protons (2) and deuterons (3).

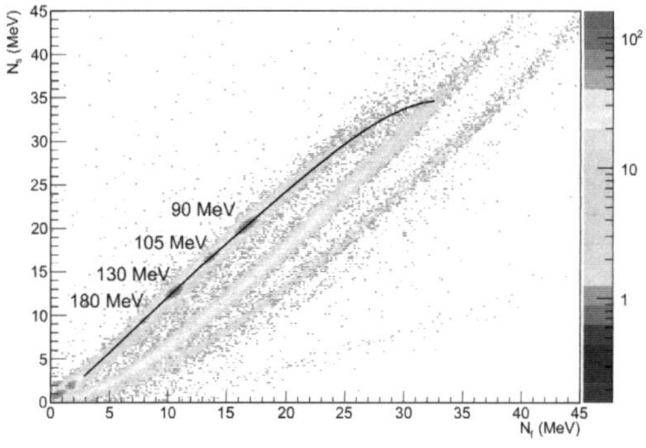

Figure 3. Result of the RPID pulse shape analysis for one crystal and the four energies 90 MeV, 105 MeV, 130 MeV and 180 MeV. The black drawn function is the projection function for punch-through protons. More explanations can be found in the text.

noise contributions. The second step is the determination of the energy loss ΔE in the active detector material. Therefore the two amplitudes of the *projected events* in the RPID histogram are summed up.

$$\Delta E = N_f + N_s \tag{7}$$

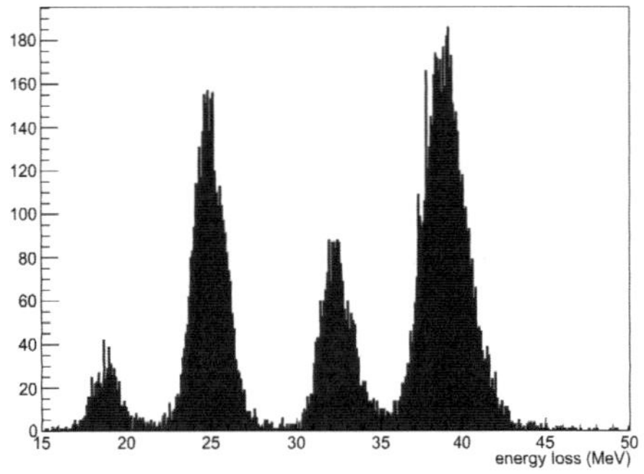

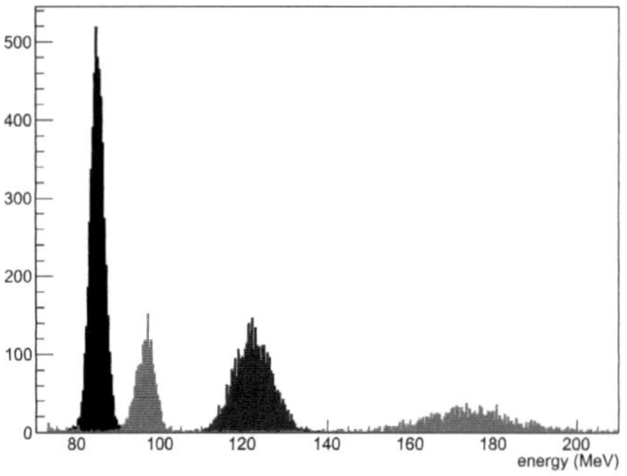

Figure 4. Energy loss of protons with 90 MeV, 105 MeV, 130 MeV and 180 MeV in 15.3 mm of CsI(Tl) (left). The energy loss is extracted from the projected events of the RPID. The least energy loss corresponds to the highest energy and vice versa. The reconstructed proton energies of these four energies are shown on the right.

The last step is a standard evaluation of the full energy from the measured energy loss. This could be done numerically with the help of energy loss calculators like ATIMA [4] or LISE++ [5].

[4] http://web-docs.gsi.de/~weick/atima/

[5] http://lise.nscl.msu.edu/lise.html

For the present data the full energy is known from the accelerator setting with a precision of $\frac{\Delta E}{E} < 1\%$ and is used for calibration.

After the reconstructed full energy is plotted in a 1D histogram (see Fig. 4). While protons with an incoming energy of 90 MeV can be reconstructed with a resolution of $\frac{\Delta E}{E} = 4.9$ %, it is decreasing for 180 MeV to $\frac{\Delta E}{E} = 13.0$ %. All evaluated energy resolutions can be found in Table (1). For comparison the energy straggling (FWHM), calculated by LISE++, divided by the energy loss is summarized in the right column. From Table (1) we conclude that the resolution of the reconstructed energy is dominated by the energy straggling inside the material, which is expanded by the extrapolation between ΔE and E_{iPhos}. For energies near the punch-through energy of about 60 MeV the iPhos energy reconstruction provides a resolution close to the straggling limit. For increasing energies the extrapolation leads to a degradation of the achieved resolution.

E	Resolution (FWHM)	Energy straggling (FWHM)
90 MeV	4.9 %	4.3 %
100 MeV	6.3 %	4.4 %
105 MeV	5.6 %	4.5 %
110 MeV	7.2 %	4.7 %
120 MeV	7.8 %	4.9 %
130 MeV	8.9 %	5.2 %
155 MeV	9.6 %	5.9 %
180 MeV	13.0 %	6.5 %

Table 1. Energy resolution of the iPhos reconstruction technique using 15.3 mm CsI(Tl) crystals in comparison to the relative energy straggling.

5. Acknowledgement

We would like to acknowledge the IFJ PAN, Krakow and especially Adam Maj for their kind support during the beamtime. We also want to thank Katharina Nowak for the fruitful discussions and the creative neologism. This work is support by BMBF (contract no. 12P05WOFNF and 12P05WONUE), Spanish MICCIN (FPA2012-39404-C02-01 and PRI-PIMNUP-2011-1357) and EraNet NupNet.

References

[1] Bendel M, Gernhäuser R, Henning W F, Krücken R, Le Bleis T and Winkel M 2013 *The European Physical Journal A* **49** 1–7 ISSN 1434-6001

[2] Storey R S, Jack W and Ward A 1958 *Proceedings of the Physical Society* **72**

[3] Valentine J D, Moses W W, Derenzo S E, Wehe D K and Knoll G F 1993 *Nuclear Instruments and Methods in Physics Research Section A: Accelerators, Spectrometers, Detectors and Associated Equipment* **325** 147–157 ISSN 0168-9002

[4] Alderighi M, Anzalone A, Basssini R, Berceanu I, Blicharska J, Boiano C, Borderie B, Bougault R, Bruno M, Calí C, Cardella G, Cavallaro S, D'Agostino M, D'Andrea M, Dayras R, Filippo E d, Fichera F, Geraci E, Giustolisi F, Grzeszczuk A, Guardone N, Guazzoni P, Guinet D, Iacono-Manno C, Kowalski S, La Guidara E, Lanchais A, Lanzalone G, Lanzanó

G, Le Neindre N, Li S, Maiolino C, Majka Z, Manfredi G, Nicotra D, Paduszynski T, Pagano A, Papa M, Petrovici C, Piasecki E, Pirrone S, Politi G, Pop A, Porto F, Rivet M, Rosato E, Saccá G, Sechi G, Simion V, Sperduto M, Steckmeyer J, Trifiró A, Trimarchi M, Urso S, Vannini G, Vigilante M, Wilczynski J, Wu H, Xiao Z, Zetta L and Zipper W 2002 *Nuclear Instruments and Methods in Physics Research Section A: Accelerators, Spectrometers, Detectors and Associated Equipment* **489** 257–265 ISSN 0168-9002

[5] Pouthas J, Borderie B, Dayras R, Plagnol E, Rivet M F, Saint-Laurent F, Steckmeyer J C, Auger G, Bacri C O, Barbey S, Barbier A, Benkirane A, Benlliure J, Berthier B, Bougamont E, Bourgault P, Box P, Bzyl R, Cahan B, Cassagnou Y, Charlet D, Charvet J L, Chbihi A, Clerc T, Copinet N, Cussol D, Engrand M, Gautier J M, Huguet Y, Jouniaux O, Laville J L, Le Botlan P, Leconte A, Legrain R, Lelong P, Le Guay M, Martina L, Mazur C, Mosrin P, Olivier L, Passerieux J P, Pierre S, Piquet B, Plaige E, Pollacco E C, Raine B, Richard A, Ropert J, Spitaels C, Stab L, Sznajderman D, Tassan-got L, Tillier J, Tripon M, Vallerand P, Volant C, Volkov P, Wieleczko J P and Wittwer G 1995 *Nuclear Instruments and Methods in Physics Research Section A: Accelerators, Spectrometers, Detectors and Associated Equipment* **357** 418–442 ISSN 0168-9002

[6] Quinton A R, Anderson C E and Knox W J 1959 *Physical Review* **115** 886–887

[7] Horn D, Ball G C, Galindo-Uribarri A, Hagberg E, Walker R B, Laforest R and Pouliot J 1992 *Nuclear Instruments and Methods in Physics Research Section A: Accelerators, Spectrometers, Detectors and Associated Equipment* **320** 273–276 ISSN 0168-9002

[8] CERN 05112013 Geant4 URL `http://geant4.web.cern.ch/geant4/`

[9] Institute of Nuclear Physics PAN URL `http://www.ifj.edu.pl`

[10] Pietras B, Gascón M, Álvarez-Pol H, Bendel M, Bloch T, Casarejos E, Cortina-Gil D, Durán I, Fiori E, Gernhäuser R, González D, Kröll T, Le Bleis T, Montes N, Nácher E, Robles M, Perea A, Vilán J A and Winkel M 2013 *Nuclear Instruments and Methods in Physics Research Section A: Accelerators, Spectrometers, Detectors and Associated Equipment* **729** 77–84 ISSN 0168-9002

[11] R3B 2012 *Technical Report for the Design, Construction and Commissioning of the CALIFA Barrel: The R3B CALorimeter for In Flight detection of gamma-rays and high energy charged pArticles*

[12] Hamamatsu Photonics KK Solid State Division 2005 Datenblatt Si APD S8664-series URL `http://www.hamamatsu.com/resources/pdf/ssd/s8664_series_kapd1012e04.pdf`

The Forward Endcap of the Electromagnetic Calorimeter for the PANDA Detector at FAIR

Malte Albrecht for the PANDA collaboration

Institut für Experimentalphysik I, Ruhr-Universität Bochum, Universitätsstr. 150, 44801 Bochum, Germany

E-mail: malte@ep1.rub.de

Abstract. The versatile 4π-detector PANDA will be built at the Facility for Antiproton and Ion Research (FAIR), an accelerator complex, currently under construction near Darmstadt, Germany. A cooled antiproton beam in a momentum range of $1.5 - 15\,\mathrm{GeV}/c$ will be provided by the High Energy Storage Ring (HESR). All measurements at PANDA rely on an excellent performance of the detector with respect to tracking, particle identification and energy measurement. The electromagnetic calorimeter (EMC) of the PANDA detector will be equipped with 15744 $PbWO_4$ crystals (PWO-II), which will be operated at a temperature of $-25°$C in order to increase the light output. The design of the forward endcap of the EMC has been finalized. The crystals will be read out with Large Area Avalanche Photo Diodes (LAAPDs) in the outer regions and with Vacuum Photo Tetrodes (VPTTs) in the innermost part. Production of photosensor units utilizing charge integrating preamplifiers has begun. A prototype comprised of 216 $PbWO_4$ crystals has been built and tested at various accelerators (CERN SPS, ELSA/Bonn, MAMI/Mainz), where the crystals have been exposed to electron and photon beams of 25 MeV up to 15 GeV. The results of these test measurements regarding the energy and position resolution are presented.

1. Introduction

The future accelerator complex FAIR will be built as an extension to the existing facilities of the *GSI Helmholtzzentrum für Schwerionenforschung* near Darmstadt, Germany. The PANDA experiment is one of the approved key experiments at this facility, which will make use of a secondary antiproton beam. The detector will be mounted at an internal target position of the HESR, which will provide a cooled antiproton beam with a precise definition of the beam energy in a momentum range of 1.5 - 15 GeV/c. Due to the interaction of the antiprotons with protons of an internal target a luminosity of up to $2 \cdot 10^{32}\,\mathrm{cm}^{-2}\mathrm{s}^{-1}$ will be achieved, while the beam momentum resolution will be as good as $\Delta p/p = 4 \cdot 10^{-5}$. The main focus of the experiment lies on hadron spectroscopy, especially the search for exotic states in the charmonium mass region. Other goals from the broad spectrum of accessible physics cover the interaction of hadrons with nuclei as well as electromagnetic processes, which will lead to a better understanding of the nucleon structure [1].

Since PANDA is a fixed target experiment, a large fraction of the scattered and produced particles will be emitted under small azimuthal angles with respect to the beam axis and direction. Therefore the PANDA spectrometer is split into a target spectrometer based on a 2 T superconducting solenoid magnet, which is surrounding the interaction point, and a forward

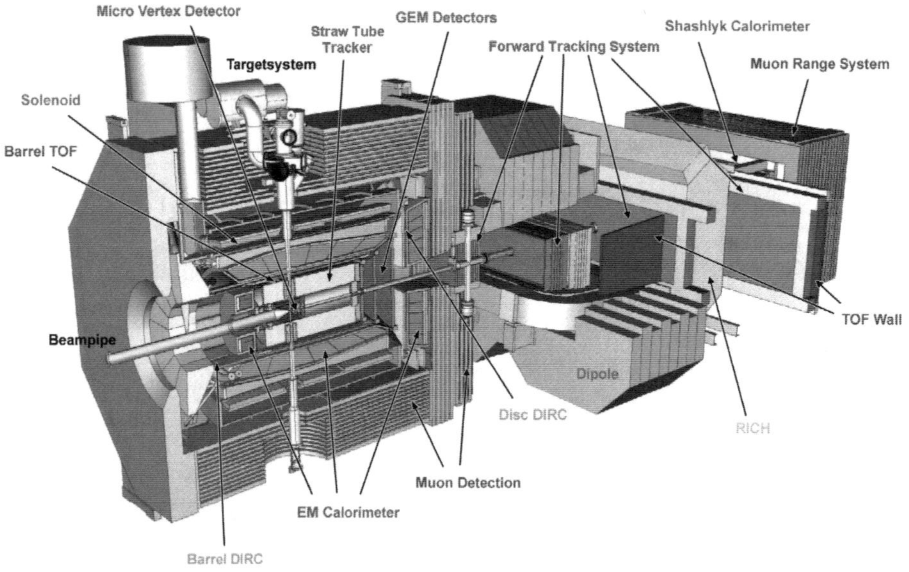

Figure 1. Technical drawing of the PANDA detector. The electromagnetic calorimeter systems in the target spectrometer are depicted in a violet color.

spectrometer, which contains a 2 Tm dipole magnet. A sketch of the PANDA detector is shown in Fig. 1. Both the target and the forward spectrometer contain subdetectors for charged particle identification, tracking, electromagnetic calorimetry and muon identification.

1.1. The PANDA Electromagnetic Calorimeter

The physics program of the PANDA experiment sets high demands on the performance of the electromagnetic calorimeter systems. A photon reconstruction threshold of 10 MeV is a central requirement, which leads to a single crystal threshold of about 3 MeV and a corresponding noise level equivalent to 1 MeV. These requirements are justified by the need to efficiently measure low-energy photons in order to make the reconstruction of final states involving many photons possible. This is especially important for the envisaged precision spectroscopy of charmonium states and exotic hadrons in the charmonium region. The most abundant sources of final state photons are π^0 and η mesons, which must be clearly identified in order to seperate them from single photons occuring e.g. in radiative charmonium decays.

In the Forward Spectrometer a sampling-type "shashlyk" calorimeter using lead absorbers and plastic scintillator tiles, that are read out via wavelength-shifting fibres with photomultipliers will be installed in order to cover the most forward part of the solid angle. This part of the EMC is described in more detail in [2].

The EMC in the Target Spectrometer is a homogeneous crystal calorimeter utilizing 15744 lead tungstate crystals ($PbWO_4$). This material has been chosen due to its compactness, fast response, short decay time and radiation hardness. The material used for the PANDA crystals is denoted as PWO-II, a second generation of lead tungstate material in comparison to the first generation material used e.g. in the calorimeter of the CMS experiment at the LHC. The main improvement is the significantly higher light yield that one can gain for PWO-II crystals [3]. A comparison of the two materials is given in Table 1.

The PANDA EMC will consist of a cylindrical barrel part comprised of 11360 crystals, a forward (3856 crystals) and a backward endcap (528 crystals). While all crystals will have a

Table 1. Characteristics of PWO-I and PWO-II crystals [4]

Characteristic	PWO-I (CMS)	PWO-II (PANDA)
La, Y concentration level (ppm)	100	40
Light yield of full size (20cm) crystal with PMT-readout at room temperature (phe/MeV)	8 − 12	17 − 22
Light yield temperature coefficient at $T = +20°C$ (%/K)	−2.0	−3.0
EMC operating temperature (°C)	+18	−25

length of 200 mm, which is equivalent to 22 times of the radiation length of PWO-II, the tapering of individual crystals differs with the position of the crystal in the calorimeter. The crystals in the endcaps are only slightly (forward endcap) or not tapered at all (backward endcap), while crystals foreseen for the barrel part are produced in eleven different geometrical shapes. In order to increase the light output of the crystals roughly by a factor of four compared to operation at $T = +25°C$, the EMC will be operated at a temperature of −25°C. All crystals are pointing off the asymmetrically located target position. The crystals in the barrel and backward endcap part of the EMC will be read out with each two Large Area Avalanche Photodiodes (LAAPDs), while a readout partly with LAAPDs and partly with Vacuum Photo Tetrodes (VPTTs) is foreseen for the forward endcap, which is discussed in more detail in Section 2. One reason for the use of these types of photosensors is their tolerance regarding an external magnetic field, since the EMC will be placed inside the 2 T superconducting solenoid in the target spectrometer. An energy resolution of $\sigma_E/E = 1\% \oplus 2\%/\sqrt{E[GeV]}$ and a time resolution better than 2 ns should be achieved with the PANDA EMC. A technical drawing of the barrel and forward endcap part of the EMC is shown in Fig. 2. [6]

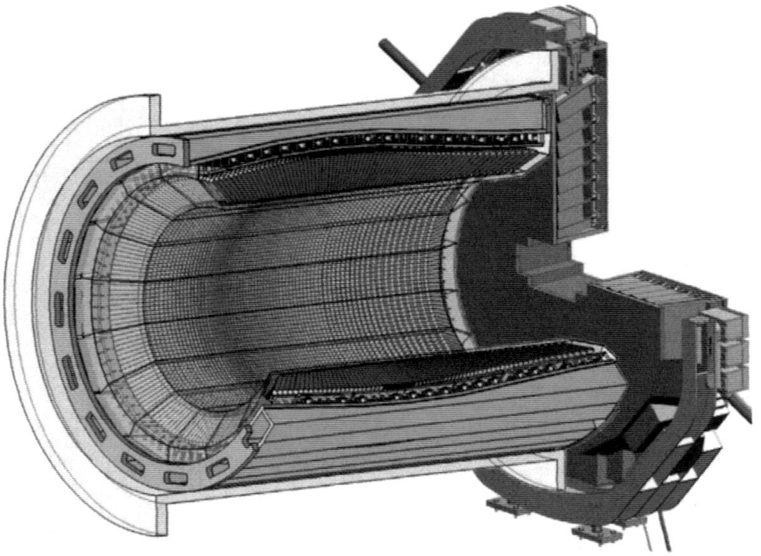

Figure 2. Schematic drawing of the barrel- and forward endcap part of the PANDA EMC. The crystals in the forward endcap are depicted in orange, covered by a thermal insulation shown in green.

2. The Forward Endcap of the PANDA-EMC

The geometrical shape of the Forward Endcap is that of a disc with an asymmetrical opening in the center, through which the beam pipe will be led. Particles passing the opening will be detected in the forward spectrometer. Thus, the forward endcap covers the angular range between $5°$ and $23.6°$ in vertical, and $10° - 23.6°$ in horizontal direction. The components used in this part of the detector have to withstand some extremely challenging environmental conditions. Due to the Lorentz-Boost in a fixed target experiment, the single crystal hit rate in the forward endcap is expected to be much higher (up to $1 \cdot 10^6\,\mathrm{s}^{-1}$ for the innermost crystals) than in the barrel and backward parts of the detector. A radiation dose of up to 125 Gy per year will be accumulated when running at the full design luminosity of $2 \cdot 10^{32}\,\mathrm{cm}^{-2}\mathrm{s}^{-1}$. Furthermore, the magnetic field in this region of the target spectrometer will be as high as 1.2 T. These conditions, together with the operating temperature of $-25°\mathrm{C}$ require a robust design of the electronics and photosensors placed in the cold volume. The proper detection of signals from the energy threshold of 3 MeV up to a 12 GeV maximum photon energy deposition per single crystal must be ensured. Due to the large dynamic range, radiation dose and rate a low-power, low-noise charge integrating preamplifier has been designed for the forward endcap, which will be attached directly to the photosensors at the back of each crystal [7].

Since not only the light yield of the PWO-II crystals, but also the characteristics of the photosensors, especially those of the LAAPDs, are strongly temperature dependant, a precise temperature monitoring system has been developed for the PANDA-EMC, called the THMP (Temperature and Humidity Monitoring board for Panda)[8]. Ultrathin platinum-wire temperature sensors (thickness $< 200\mu\mathrm{m}$) are placed inbetween the crystals and read out using the microcontroller driven THMP board in order to monitor the temperature inside the forward endcap at the location of the crystals [9]. Three independantly controllable cooling circuits are used for the temperature stabilization of the volume that contains the PWO-II crystals. The front insulation contains plates with meander-shaped grooves, through which dried and

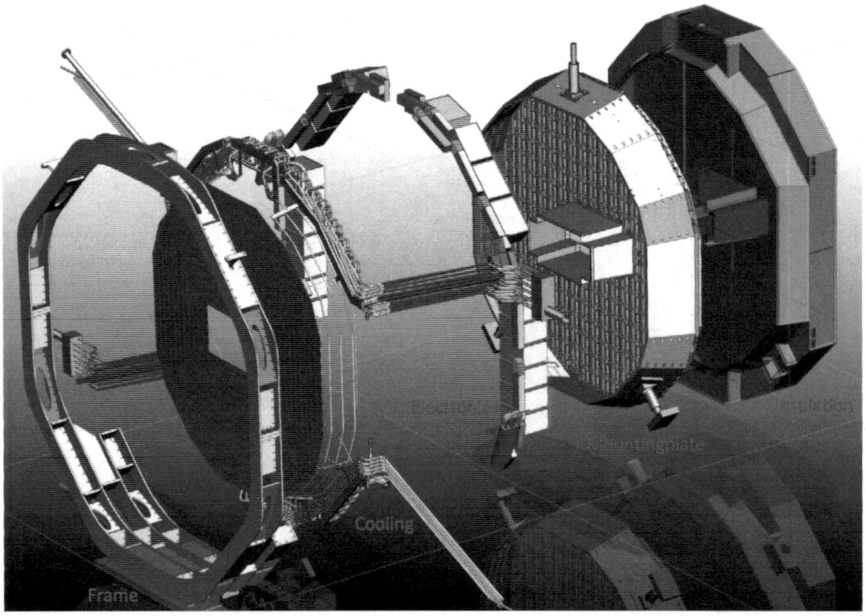

Figure 3. This exploded assembly drawing shows the main constructional parts of the forward endcap. The PWO-II crystals are not depicted in this drawing.

cooled air will be pressed. This provides cooling of the crystals from the front side, whereas the outermost ring of crystals will be cooled by pipes that follow the outer shape of the endcap and carry a cooled liquid (methanol-water mixture). The largest fraction of cooling power is being transported by the so called backplate, an aluminium plate that has a thickness of 30 mm and a diameter of 2 m. All crystal units, containing in total 3856 crystals, will be mechanically attached to this backplate using aluminium mounting structures, so that heat transport through conduction can be used here. A liquid methanol-water mixture is pumped through long vertical holes inside the backplate. This provides the most efficient cooling of the crystals from their back side. All cooling systems and the backplate can be seen in Fig. 3.

2.1. Crystal subunits

Each PWO-II crystal is wrapped in DF2000MA [5], a highly reflective foil, in order to reflect as much of the scintillation light as possible back into the crystal. At the back face of the crystal a cutout in the reflective foil with the size of the active area of the photodetector is left for the scintillation light to leave the crystal. LAAPDs as well as VPTTs are glued to the crystal using the room temperature vulcanizing silicone based adhesive Dow Corning 3145. The preamplifiers are directly soldered to the photodetectors. In total 16 of these crystal-photodetector-preamplifier units are mechanically held together by a carbon fibre alveole. The ultrathin temperature sensors are affixed to the side faces of selected crystals before they are sled into the alveole. The electrical connections of the temperature sensors are fed out at the back of the subunit, together with the signal and supply cables for the photodetector units (see Fig. 5). Aluminium mounting structures are pushed into the carbon fibre alveoles at their rear end in order to mechanically stabilize the positions of the photodetectors and enable the subunit to be mounted to the backplate. This 4 × 4 matrix is called a subunit. A technical drawing as well as a photograph of a fully equipped prototype subunit are depicted in the Figures 4 and 5.

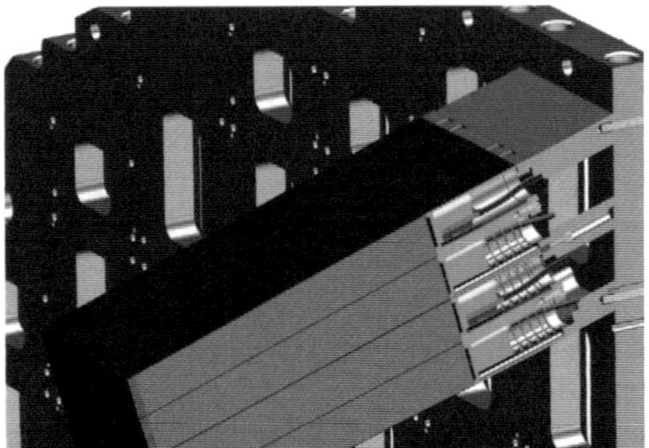

Figure 4. Technical drawing of a 16-crystal subunit mounted to the aluminium backplate. For illustration of the mounting mechanics the side face of the subunit has been removed.

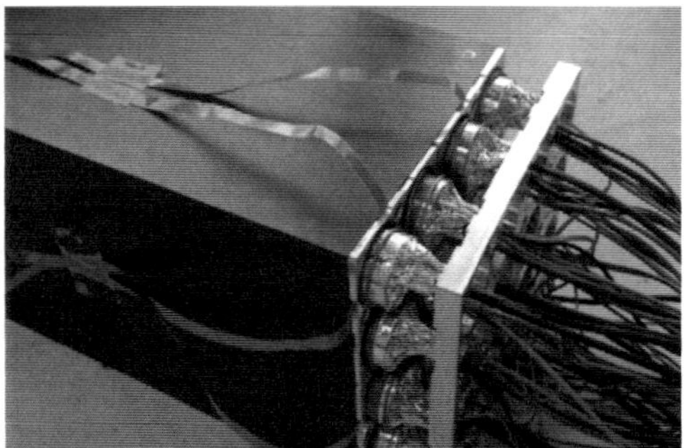

Figure 5. Photograph of a prototype subunit equipped with 16 VPTT readout units. The mounting structure has not yet been fully attached.

2.2. Photodetectors

The innermost 768 crystals of the forward endcap will be equipped with Vacuum Photo Tetrodes (R11375-MOD) of the Japanese manufacturer Hamamatsu. VPTTs are small, two-stage photomultipliers with an active area of $\approx 200\,\text{mm}^2$, a diameter of 23.9 mm and a length of 40 mm. All electrodes and dynodes are oriented in parallel to the front face of the glass tube, so that one dynode as well as the anode had to be designed as a mesh in contrast to a solid electrode (see Fig. 6). After delivery different parameters for each VPTT are being measured in order to ensure proper operation and reproduction of the manufacturer's values. The absolute amplification of the VPTTs is determined by an explicit measurement of the anode and cathode current while the VPTT is being illuminated with constant light. Furthermore the dark current for each tube is measured. The magnetic field strength will be about 1 T at the position of the VPTTs, which will result in a drop of the gain by about 50% (see Fig. 7). The extremely good radiation hardness in combination with the small detector capacitance of $\approx 22\,\text{pF}$ led to the choice of this photodetector type for the innermost crystals, for which the highest radiation doses and peak rates are expected.

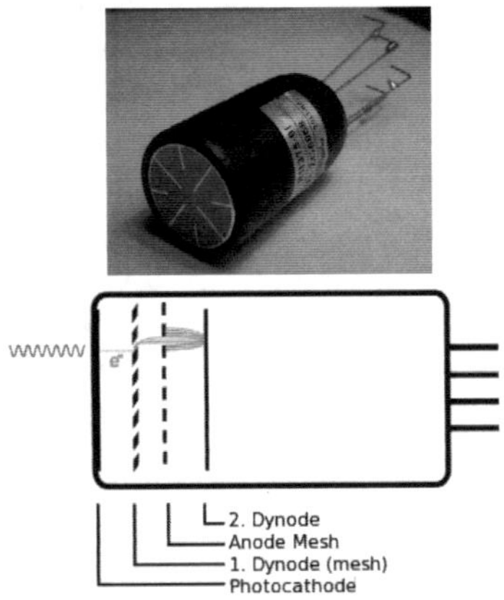

Figure 6. Photograph and functional drawing of a VPTT

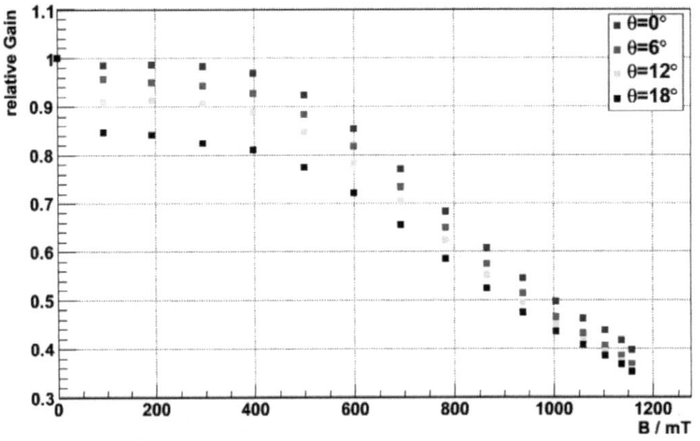

Figure 7. Dependance of the relative VPTT gain from the applied external magnetic field for different mounting positions given in terms of the azimuthal angle Θ.

The outer 3088 crystals will be read out with two rectangular Large Area Avalanche Photodiodes (S11048, Hamamatsu) each. The active area of one PANDA-LAAPD is $(6.8 \times 14)\,\text{mm}^2 = 95.2\,\text{mm}^2$, so that the sensitive area of two APDs is comparable to that of a VPTT. The functionality of APDs is largely independant of the strength of an external magnetic field, but the capacity of these photodetectors is about ten times higher ($\approx 270\,\text{pF}$) than for a VPTT. Due to the avalanche effect, an APD has an internal signal amplification, dependant on the applied bias voltage. Thus, each APD has to be operated at the proper bias voltage to realize operation at the same gain (M). However, the gain of these semiconductor devices is strongly temperature dependant, so that the response curve for each APD has to be measured for different temperatures prior to glueing the photosensor to a crystal. The response curve of one APD measured at $T = +25°\text{C}$ and $T = -25°\text{C}$ is shown in Fig. 9. In the forward endcap the APDs will be operated at $T = -25°\text{C}$ and a gain of $M = 200$. At this point the slope of the response curve reaches a value of $\frac{dM}{M}/dU \approx 6.5\,\%/\text{V}$, while the dependance of the operating temperature is $\frac{dM}{M}/dT \approx 1.5\,\%/\text{K}$. Due to this dependance it is of utmost importance also for the operation of the photodetectors to have a precise temperature measurement inside the forward endcap. Fig. 8 shows a photograph of the front face of two LAAPDs and a functional diagram.

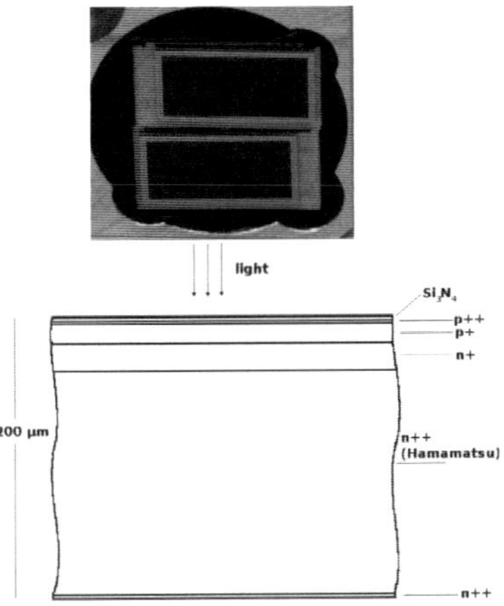

Figure 8. Photograph and functional drawing of an APD

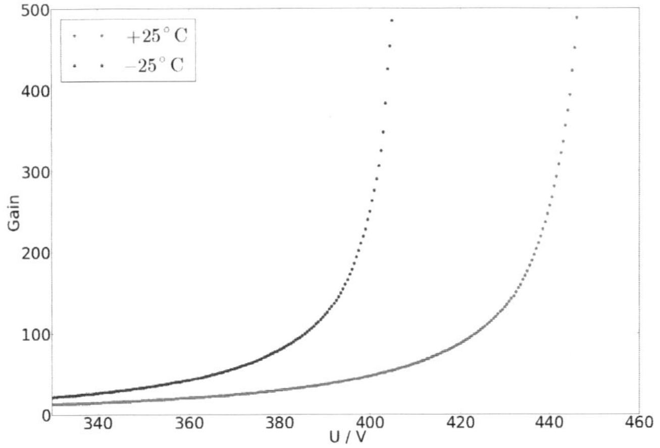

Figure 9. Gain vs. applied bias voltage for one PANDA-LAAPD at $T = +25°\text{C}$ and $T = -25°\text{C}$

3. Forward endcap prototype setup

A prototype of the forward endcap comprised of 216 full-sized PWO-II crystals has been built. Geometrically the prototype resembles a cutout of the forward endcap near to the asymmetrical hole in the center of the detector (see Fig. 10). Both detector types foreseen for the forward endcap have been used to read out crystals in the prototype, which have been cooled down to the operating temperature of $T = -25°\text{C}$. The prototype is being used to estimate the expected performance of the PANDA-EMC in a most realistic scenario. Tests of the mechanical components and assembly procedure, the cooling and temperature readout systems, the photosensor readout electronics and the slow control could be performed in the lab as well as at various accelerator facilities. At testbeams the crystals have been exposed to electron and photon beams in order to determine the achievable performance in terms of noise, energy resolution, minimal energy threshold and spatial resolution. An overview over the different testbeams is given in Table 2. All photodetectors are attached to the PANDA low noise preamplifiers discussed in Section 2. The signals are then fed to custom VME shaper modules with a shaping time of 100 ns and finally digitized using commercial VME sampling ADCs

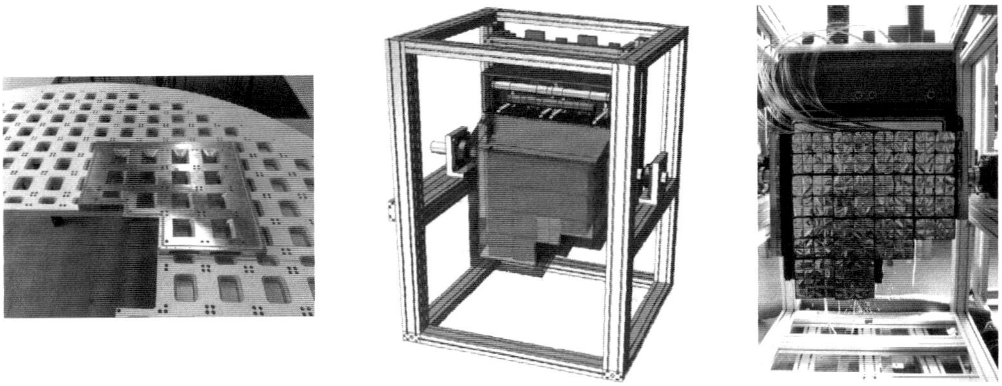

Figure 10. Left: Prototype backplate on wooden 1:1 model for the forward endcap backplate; Middle: CAD drawing of the forward endcap prototype; Right: Photography of the prototype

Table 2. Testbeams with the forward endcap prototype

	Beam particles	E_{Beam} or p_{Beam}	Specialties
CERN/SPS	e^+	10, 15 GeV/c	max. PANDA energy;
	μ^+	150 GeV/c	dep. energy $\approx$ 230 MeV
ELSA/Bonn	Tagged γ	1, 2.1, 3.1 GeV	Rates up to $2 \cdot 10^6\,\text{s}^{-1}$
MAMI/Mainz	Tagged γ	20 – 415 MeV	excellent beam energy resolution
CERN/SPS	e^-	5 – 15 GeV/c	Fibre / Si-strip
	$\pi^+, K^+, \overline{p}$	15, 50 GeV/c	TrackingStation

(WIENER AVM16, Struck SIS3302). A detailed model of the prototype was implemented in Geant 4 [11], including dead material (aluminium mechanics, alveoles, insulation, cooling tubes + liquid, ...) and the correct alignment of the prototype axis to the beam axis, according to measurements performed at the testbeams. Whenever possible, the beam profile was also measured using fibre hodoscopes ($e^{\pm}$ beam) or a beam camera (photon beam), and properly included into the simulation.

Single crystal energy deposits for centrally impinging particles at different beam energies have been used to calibrate each single readout unit. The most probable value for the energy deposition has been determined from the Monte Carlo simulation. After applying the same cuts to testbeam data the corresponding ADC conversion for the most probable value has been determined. This procedure is repeated for each available beam energy. Afterwards a polynomial fit is performed in order to obtain a calibration function (E_{dep}(ADC Conversion)) for each readout unit.

3.1. Results from testbeams

3.1.1. Noise and Threshold

The best performance regarding energy equivalent noise has been measured for the Vacuum Photo Tetrodes. A mean noise of $\langle\sigma_{\text{Noise}}\rangle_{\text{VPTTs}} = 1.6\,\text{MeV}$ has been determined, which is near the envisaged value of 1 MeV in the Technical Design Report of the EMC [6]. For the APDs a slightly worse value of $\langle\sigma_{\text{Noise}}\rangle_{\text{APDs}} = 2.2\,\text{MeV}$ has been measured.

In order to determine the optimal value for the single crystal energy threshold (E_{thr}), the dependance of the energy resolution for a 3×3 crystal matrix has been studied in dependance of E_{thr}. For this scan the threshold has been varied in the range of $1 - 15\,\text{MeV}$. Fig. 11 shows threshold scans exemplatory for two different beam energies. The best resolution can be obtained by applying a single crystal threshold of 5 MeV, which corresponds roughly to $3 \cdot \langle\sigma_{\text{Noise}}\rangle_{\text{VPTTs}}$.

3.1.2. Energy resolution

For the PANDA PWO-II crystals about 80% of the incident energy is deposited in the crystal that is centrally hit by a particle. In order to reconstruct the incident energy, the deposition in many crystals is summed up. For the prototype studies the deposition in symmetrical 3×3 or 5×5 crystal matrices is used. Due to the fact that electromagnetic showers can have asymmetrical shapes and large fluctuations, later an algorithm will be used to select crystals that contribute to the shower, starting from the crystal with the largest deposition and subsequently moving outwards until crystals are reached, for which the deposition lies below the threshold E_{thr}.

For each beam energy the resolution is determined from a fit to the distribution of the energy sum. This distribution has an asymmetrical shape, which can be described by a gaussian function with an additional tail towards lower energy deposits. The empicirally determined function for

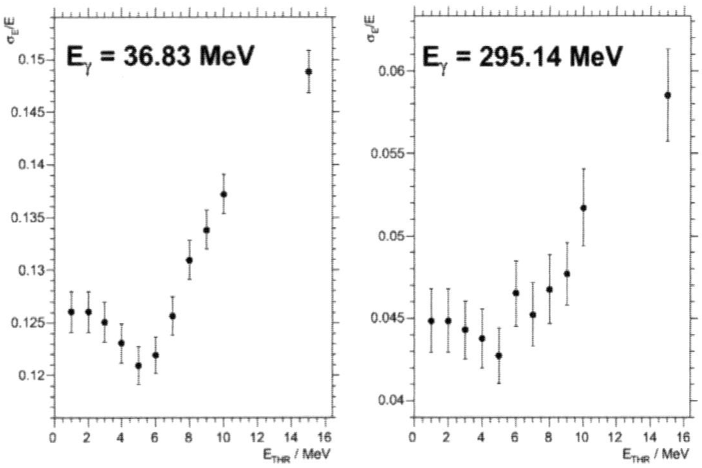

Figure 11. Energy resolution vs. applied single crystal threshold shown for two different beam energies (tagged photons, MAMI testbeam). The optimal energy thresold determined by this is $E_{\text{thr}} = 5\,\text{MeV}$ (measured with VPTTs).

this distribution is described in more detail in [10]. The resolution is determined from the width σ_E of the fitted function and the energy of the incident particle E as σ_E/E. The width is defined as $\sigma_E = \text{FWHM}/(2\cdot\sqrt{2\ln 2})$. Examples for the energy deposition in the central and surrounding crystals as well as the sum are shown for a low ($E = 295.14\,\text{MeV}$) and a high ($E = 10.0\,\text{GeV}$) beam energy in Fig. 12 and 13, respectively.

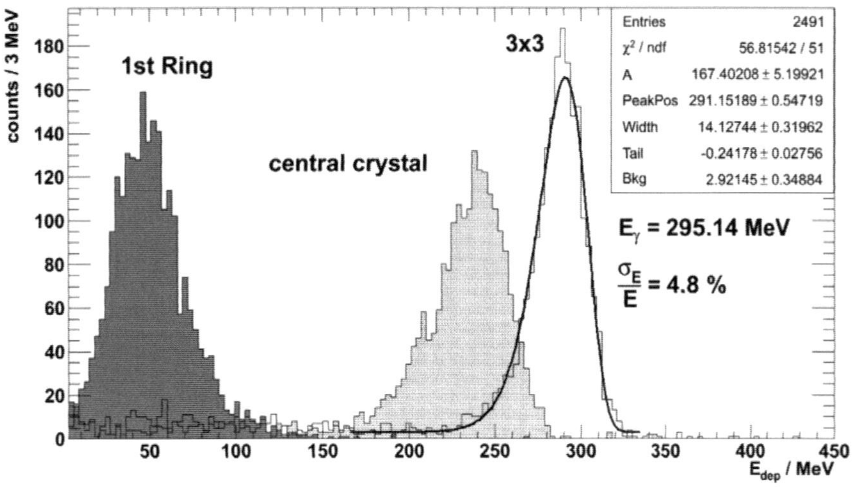

Figure 12. Energy deposition in the central crystal, the surrounding ring of crystals and the sum of all nine crystals at a beam energy of $E = 295.14\,\text{MeV}$. The energy sum has been fitted to obtain the energy resolution (measured with VPTTs).

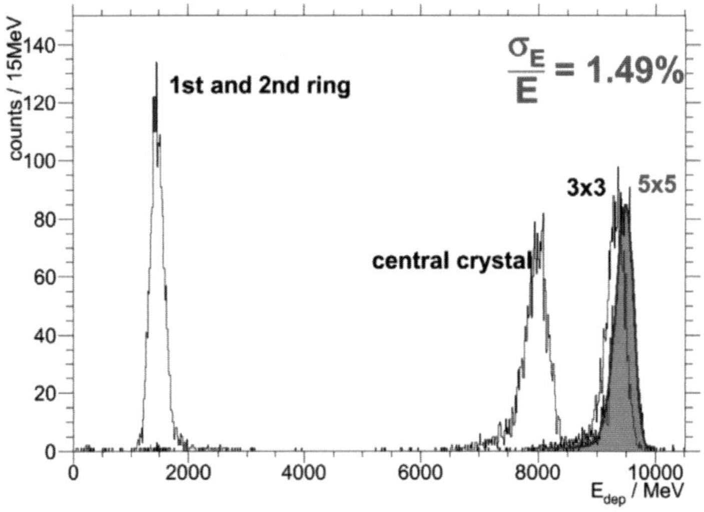

Figure 13. Energy deposition in the central crystal, the surrounding two rings of crystals and the sum of all 25 crystals at a beam energy of $E = 10.0\,\mathrm{GeV}$. The energy sum has been fitted to obtain the energy resolution.

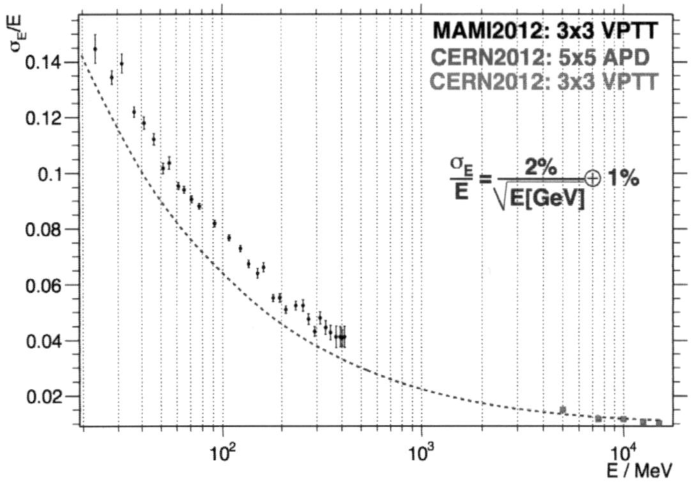

Figure 14. Energy resolution as a function of beam energy. The blue curve shows the envisaged energy resolution.

The resolution has been determined for incident energies in the range of 25 MeV up to 15 GeV for the different photodetector types. All results are summarized in Fig. 14 together with the envisaged energy resolution function given in the Technical Design Report [6] (blue curve). For small energies the resolution is slightly worse than the targetted value, while for the highest energies expected in the PANDA experiment the expectations can be met very well.

3.1.3. Spatial resolution

The position resolution of the forward endcap prototype could be studied at the CERN testbeam, due to the use of Si-microstrip detectors with a pitch of 50 μm and a fibre hodoscope, that were set up in front of the prototype. The point on the surface of the central crystal, at which a beam particle hit the crystal, can be reconstructed from the information of the tracking detectors as well as from the measured energy depositions in the prototype. A logarithmic weighting method was used to calculate the point of impact from the energy deposition in the crystals. The difference between the point of impact calculated from the energy deposition and tracking detectors, denoted as Δx, shows the spatial resolution (Fig. 15). The width of this gaussian shaped distribution, σ_x, was determined for all four beam energies. The resolution was determined to be $\sigma_x = 1.6$ mm at $E = 5\,GeV$ (black curve in Fig. 15) and subsequently decreases to 0.9 mm at an Energy of 15 GeV (blue curve). The resolution for the intermediate energies were measured as $\sigma_x = 1.3$ mm at $E = 7.5\,GeV$ (red curve) and 1.1 mm at an energy of 10 GeV. These results are in good agreement with the design value for the forward endcap of $\sigma_{x,\text{TDR}} < 3.5$ mm. Additionally the dependance of the position resolution from the point of impact along the front face of a crystal has been studied. It was found, that the resolution improves towards the borders of the crystal. In these cases the largest fraction of the energy is deposited in two crystals instead of only the central crystal. This behaviour could also be qualitatively reproduced with Monte Carlo simulations (see Fig. 16: red curve: simulation, black: testbeam data). Along the front face of the crystal the position resolution is about a factor of three better than the targetted design value for all points that were studied.

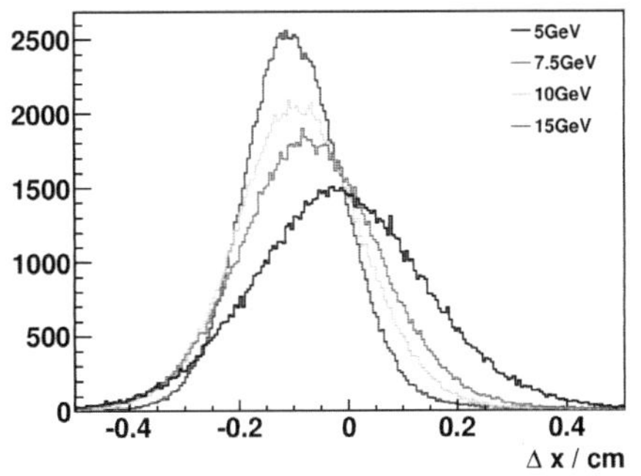

Figure 15. Position resolution for different beam energies. The distribution is shifted from the center for higher beam energies due to the non-zero angle between the crystal- and beam-axis.

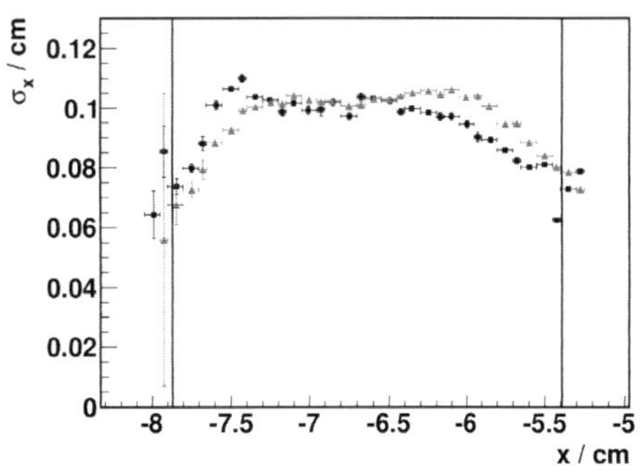

Figure 16. Position resolution as a function of the point of impact on the crystal surface for testbeam data (black) and Monte Carlo simulation (red). The vertical lines indicate the borders of the central crystal.

Acknowledgments

We wish to thank the staff of the SPS (CERN), ELSA (U Bonn) and MAMI (U Mainz) accelerators for their support and help during the forward endcap prototype testbeams.

References

[1] Erni W et. al. (PANDA Collaboration), Physics Performance Report for PANDA: Strong Interaction Studies with Antiprotons, *arXiv:0903.3905v1*

[2] Semenov P, Status on the PANDA Forward Shashlyk Calorimeter, *contribution to this conference*

[3] Novotny R W, Doring W M, Hjelm F, Melnychuk D, Makonyi K, Reiter A, Salz C, Steinacher M, Thiel M and Zwieglinski B 2005 *Proc. IEEE Nuclear Science Symposium*, Puerto Rico, Conference Record, ISBN: 0-7803-9222-1, N12-4

[4] Borisevich A et al., 2008, PWO-II scintillation crystals for the PANDA electromagnetic calorimeter, *Nuclear Science Symposium Conference Record, 2008. NSS '08. IEEE*, http://dx.doi.org/10.1109/NSSMIC.2008.4774932

[5] 3M Light Management Solutions 2007, Daylighting Film DF2000MA, Product Bulletin DF2000MA

[6] Erni W et. al. (PANDA Collaboration), Technical Design Report for: PANDA Electromagnetic Calorimeter (EMC), *arXiv:0810.1216v1*

[7] Keshelashvili I, Development of Low Noise/Low Power Preamplifier for the Readout of Inorganic Scintillators and its Mass Production Test System, *contribution to this conference*

[8] Feldbauer F, 2012, Dissertation Ruhr-Universität Bochum, Analyse des Zerfalls $\chi_{cJ} \to K^+K_s^0\pi^-\pi^0$ bei BES-III und Entwicklung der slow control für das PANDA-Experiment, *urn:nbn:de:hbz:294-35182*

[9] Schulze J, 2012, Dissertation Ruhr-Universität Bochum, Analyse des Zerfalls $\chi_{c0} \to K^+K^-\pi^0\pi^0$ bei BESIII und Entwicklung von mechanischen Komponenten für einen Prototypen des PANDA-EMC, *urn:nbn:de:hbz:294-35247*

[10] Aubert B, 2005 *Phys. Rev.* Lett. **94** 221803

[11] Agostinelli S et.al., Geant4a simulation toolkit, Nuclear Instruments and Methods in Physics Research Section A: Accelerators, Spectrometers, Detectors and Associated Equipment, Volume 506, Issue 3, 1 July 2003, Pages 250-303, ISSN 0168-9002, http://dx.doi.org/10.1016/S0168-9002(03)01368-8.

The backward end-cap for the PANDA electromagnetic calorimeter

L Capozza[1,3,*], **F E Maas**[1,2,3], **O Noll**[2], **D Rodriguez Pineiro**[1,3], **R Valente**[1,3]

[1]Helmholtz-Institut Mainz, J.-J.-Becherweg 36, 55128 Mainz, Germany
[2]Institut für Kernphysik, Johannes Gutenberg-Universität, 55099 Mainz, Germany
[3]GSI Helmholtzzentrum für Schwerionenforschung GmbH, Planckstraße 1, 64291 Darmstadt, Germany
[*] corresponding author

E-mail: capozza@kph.uni-mainz.de

Abstract. The PANDA experiment at the new FAIR facility will cover a broad experimental programme in hadron structure and spectroscopy. As a multipurpose detector, the PANDA spectrometer needs to ensure almost 4π coverage of the scattering solid angle, full and accurate multiple-particle event reconstruction and very good particle identification capabilities. The electromagnetic calorimeter (EMC) will be a key item for many of these aspects. Particle energies ranging from some MeVs to several GeVs have to be measured with a relative resolution of $1\% \oplus 2\%/\sqrt{E/\mathrm{GeV}}$. It will be a homogeneous calorimeter made of $PbWO_4$ crystals and will be operated at -25°C, in order to improve the scintillation light yield. With the exception of the very forward section, the light will be detected by large area avalanche photodiodes (APDs). The current pulses from the APDs will be integrated, amplified and shaped by ASIC chips which were developed for this purpose. The whole calorimeter has been designed in three sections: a forward end-cap, a central barrel and a backward end-cap (BWEC). In this contribution, a status report on the development of the BWEC is presented.

1. Introduction

The physics programme covered by the PANDA experiment will be focused on strong interaction studies using an antiproton beam and an internal target. The physics topics range from hadron structure and spectroscopy to nuclear physics and are summarised in ref. [1], together with an account of the expected performances, achievable by the designed apparatus. For many physics channels, electromagnetic final states including electrons, positrons and gammas have to be reconstructed and these particles need to be detected within a solid angle of approximately 4π. In addition, particle identification is needed by many of the planned measurements. For these reasons, the electromagnetic calorimeter (EMC) [2] will play an important role for the successful operation of the spectrometer.

The Panda Collaboration decided in favour of a $PbWO_4$ homogeneous calorimeter which will be operated at -25°C and has a goal relative energy resolution of $1\% \oplus 2\%/\sqrt{E/\mathrm{GeV}}$. The EMC will be divided into three parts, a forward end-cap, a central barrel and a backward end-cap (BWEC) which is the subject of this contribution. The total solid angle acceptance will be about 99% of 4π and the large polar scattering angles from 140° to 170° will be covered by the

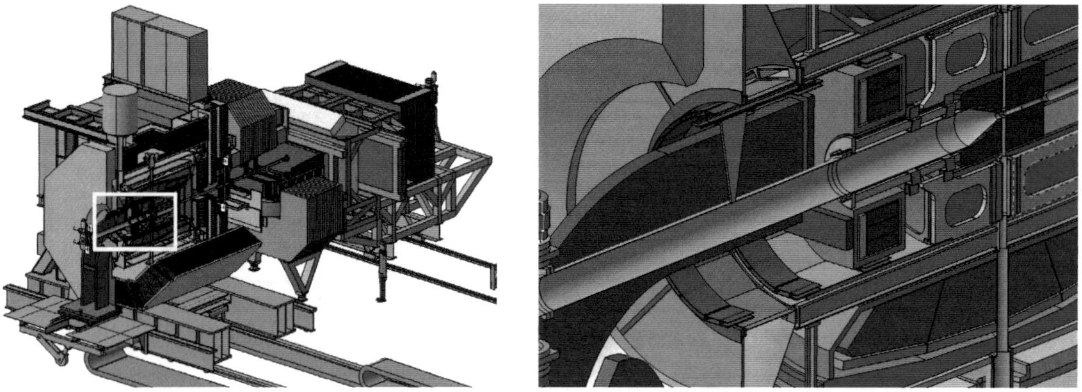

Figure 1. Left: technical drawing of the PANDA spectrometer. The white box shows the position of the backward end-cap of the electromagnetic calorimeter. A close-up of this region is shown on the right. The beam direction is from left to right and the target position is where the beam pipe becomes narrower. The target region is surrounded by the micro-vertex detector (in dark grey) and the straw tubes tracker (STT, in orange). The ring-shaped volume upstream of the STT (in light blue) is the backward end-cap of the calorimeter.

BWEC. This range is important for those physics channels which are particularly sensitive to the forward and backward kinematics, e.g. the time-like form factor measurement from the exclusive proton-antiproton annihilation into leptons. For other proposed experiments, like the extraction of transition distribution amplitudes from the $\bar{p}p \rightarrow e^+e^-\pi^0$, where the pion should be detected only at the very forward and very backward polar angles, the BWEC will be essential.

The most severe constraints to the development of the BWEC are imposed by the geometrical boundary conditions. The BWEC will be situated inside the barrel and all services to the inner detectors will be routed through its central hole or through the narrow gap between BWEC and barrel (Fig. 1).

On the other hand, the requirements in terms of signal readout are favourable. In this scattering angle region, counting rates up to only 1 kHz are expected and particle energies ranging from about 10 MeV to 1 GeV have to be measured, which is one order of magnitude less than the dynamic range required to the forward end-cap (10 MeV to 10 GeV).

2. Mechanical construction

An overview on the mechanical design of the BWEC is shown in Fig. 2. The detector will hang from an arm-shaped support structure which will lie on a platform mounted on rails, in order to be able to move out the BWEC from the spectrometer and grant access to the inner detectors. In order to attach the crystals to the support structure, two aluminium mounting plates are foreseen. One of them will be at the border between the cold volume and the other one completely inside of it. The mounting plates will be held together by basalt-fibre-reinforced plastic feet which have enough mechanical strength for supporting the weight of the full structure (about 1.5 ton) but have a low thermal conductivity.

The crystals will be kept inside carbon-fibre-reinforced alveoli which will be glued to aluminium parts, called inserts, which in turn can be screwed to the cold mounting plate (see Fig. 3).

Because of the tight geometrical constraints imposed to the BWEC very high accuracy is needed for the realisation of all mechanical parts.

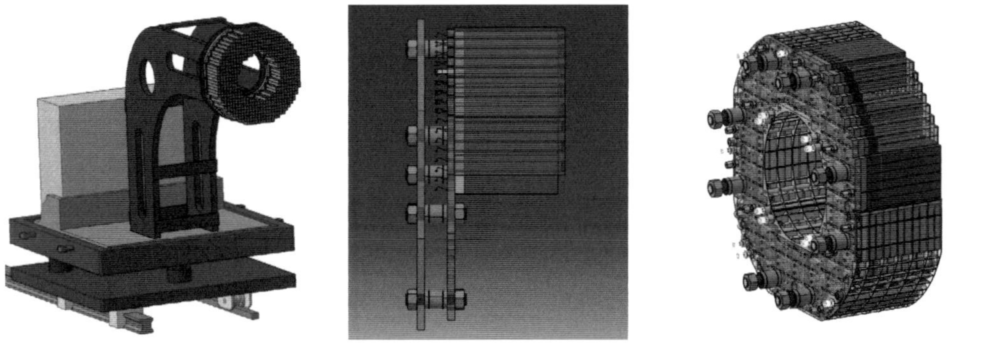

Figure 2. Left: technical drawing of the BWEC, mounted on its support structure (cooling shell and insulation cover are not shown). Centre: side view of the BWEC. The two mounting plates are visible and held together by the basalt-fibre-reinforced plastic feet (in yellow). Right: axonometric view of the BWEC without the outer mounting plate. On the upper-right quarter the cooling shells are removed and the carbon fibre alveoli are shown.

Figure 3. Left: carbon fibre alveoli in two different shapes with their aluminium insert (left one) and crystal stopper (right one). Centre: four $PbWO_4$ crystals wrapped in reflective foil and equipped with thin Pt-100 thermal sensors. Right: the crystals inside the carbon fibre alveole.

3. Cooling system

In order to improve the scintillation light yield of $PbWO_4$ the calorimeter will be operated at $-25°C$. Because of the dependence of the light output on the temperature, it is necessary to avoid strong temperature gradients and fluctuations inside the detector volume, in order to improve the linearity of the detector response and the energy resolution.

One important item for having a good thermal stability will be the insulation, because relevant heat sources (i.e. the read-out electronics of the inner detectors) will be placed in the vicinity of the BWEC. Vacuum insulating panels (VIP) will be used for this purpose.

Another important task will be to remove uniformly the heat produced inside the cold volume (e.g. by the front-end electronics) as well as the heat coming from the outside (through cables and the not perfect insulation). To this end, a system of pipes for a cooling medium are being developed. The constraints to this system, apart from having a compact design, are: it needs uniformly distributed pipes, the heat exchange through the pipe walls should be maximised, and the pressure drop along the system should be lower than 0.5 bar, in order to be integrated into the leak-less cooling plant foreseen for PANDA. In order to match these requirements, cooling shells encasing the detector will be manufactured by laser sintering. A prototype for such a shell was already designed and produced (Fig. 4). Finite element simulations are being performed for finding the most appropriate design in terms of pressure drop and cooling efficiency.

Figure 4. Laser sintered alumide/polyamide cooling shell.

The temperature will be monitored at some significant points in the cold volume by self-manufactured flat Pt-100 sensors (Fig. 3).

4. Readout electronics

The scintillation light pulses are detected by large area avalanche photodiodes (APD), like in most of the PANDA calorimeter (see [2] for details). Each crystal will be equipped with two APDs. The APD output pulses are read out by means of a customised ASIC, called APFEL (ASIC for Panda Front-end Electronics), which is developed at the GSI [3]. The integrated circuit implements a charge sensitive preamplifier as input stage, followed by a shaper and a main amplifier with two different gains and two output signals (see the schematics in Fig. 5) which have to be sent to a sampling ADC for digitisation. With the two amplification gains it is possible to enlarge the dynamic range of the measurement matching the input range of the ADC.

Since it is not yet decided, whether the sampling ADCs will be placed on the support structure, close to the detector or rather some metres away, outside of the spectrometer, booster boards have been developed for driving the APFEL-ASIC output signals over several metres of cable.

In Fig. 5, the output pulse of an APFEL-ASIC is shown. One can see the smooth shape of the pulse which have a FWHM of about 1 μs. The signal was obtained by reading out an APD which was illuminated with a short LED pulse of about 70 ns of duration. The test was performed at room temperature and no bias voltage was applied to the APD. The signal was transmitted by the booster boards over 10 m of cable.

5. Tests with prototypes

Before constructing the final setup of the BWEC, prototype tests are being performed. A first arrangement with four $PbWO_4$ crystals has been set up and tested with cosmic rays. Since the APFEL chips were not yet available at the time of these tests, the APD pulses were read out with charge sensitive preamplifiers from the University of Basel which were also developed for PANDA [4]. A sample result from these tests is shown in Fig. 6. The improvement obtained by the cooling is quite clear. At room temperature the bias voltage of the APDs has to be raised, in order to obtain visible pulses, up to a value at which the dark pulses are amplified making the signal baseline very unstable. Cooling down the crystals, one has to lower the bias voltage in order to keep the pulse within the input range of the ADCs (sampling ADC of type SIS3301 by Struck GmbH were used to digitise the signals). This way the baseline is much more stable and the pulses are cleaner.

Currently, a new prototype with 16 crystals is being built. In the new setup, the APD pulse readout is performed by APFEL ASICs. Almost all components will be the same as in the final BWEC setup. Besides the tests with cosmic rays, a test beam at the MAMI facility at Mainz is foreseen for July 2014. The test measurements will be performed in the so-called tagger hall, where experiments with real photons are usually run. In this hall the MAMI electron bin is scattered on a radiator producing a secondary gamma beam. The momentum of the electrons after the radiation is analysed in a dipole magnet, providing an energy measurement of each single gamma. These "tagged" photons are a very well suited tool for performing a calorimeter characterisation. The main objective of the beam tests is studying the dependence of

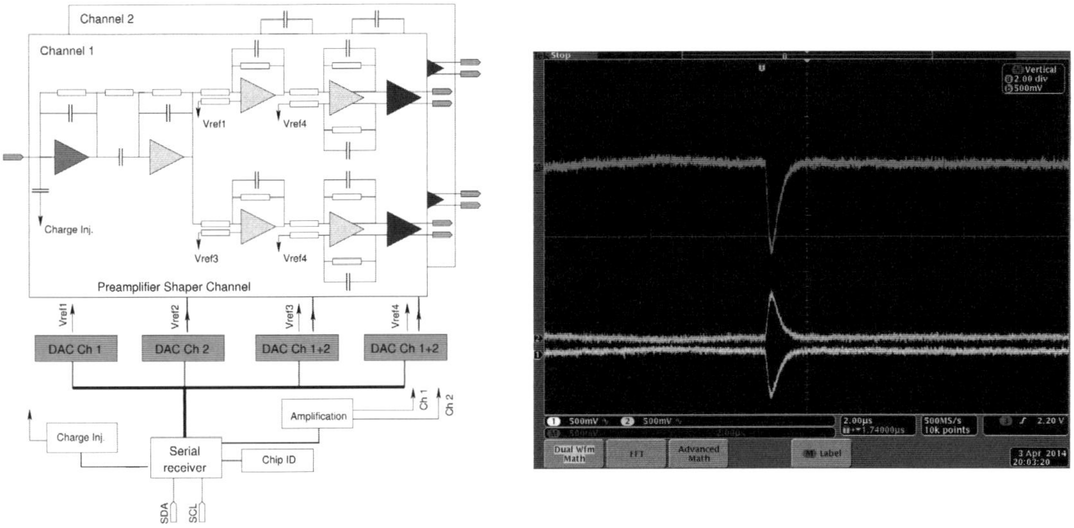

Figure 5. Left: schematics of the printed circuit implemented in the APFEL ASIC (see [3] for details). Right: output signal of an APFEL ASIC, where the input pulse was produced by an APD which was irradiated with a short (~70 ns) LED pulse. The yellow and cyan waveforms are the positive and negative differential outputs, respectively, whereas their difference is shown in red.

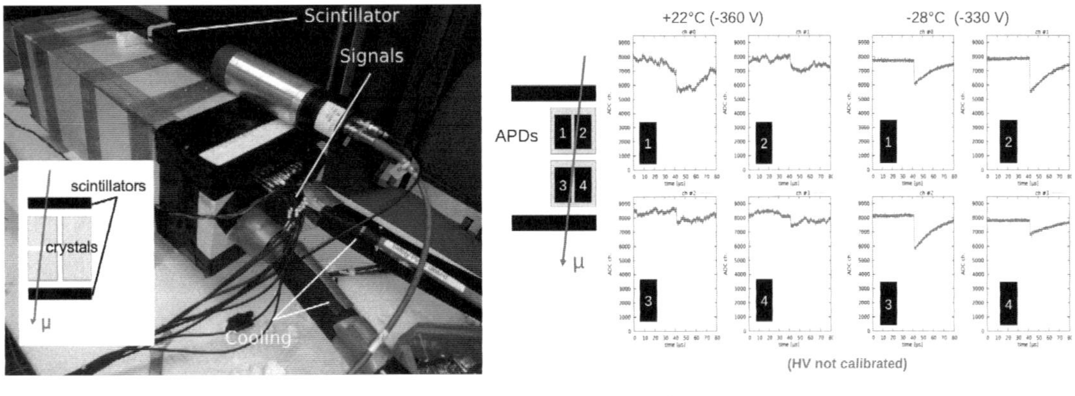

Figure 6. Left: test prototype for cosmic ray measurement. Inside the light-blue insulating box four $PbWO_4$ crystals are kept at $-28°C$ (this temperature differs from the design operation temperature of the PANDA EMC just for convenience operating the available chiller). The APD pulses are amplified by charge sensitive preamplifiers SP883a02 [4] and a 10-fold linear amplifier. Plastic scintillators are place above and below the prototype and are used for triggering. Right: examples of cosmic ray events at different temperatures. The improvement in the pulse height is clearly visible.

the calorimeter response on several parameters like incoming particle energy, angle and position, temperature of the crystals, APD bias voltage, and count rate. It will be the first time that the APFEL ASICs are tested in a full arrangement with a gamma beam.

6. Conclusions

The EMC will be an essential ingredient for the physics programme of the future PANDA experiment. Since a full solid scattering angle coverage is required for many measurements, the calorimeter end-caps play an important role. In this contribution the design, development and construction of the backward end-cap have been addressed. Many of the activities related to these topics are being put forward at the Mainz University within the Helmholtz-Intitut and the Institut für Kernphysik and at the GSI. Currently, the design of the equipment is mostly settled and tests with prototypes are being performed for finalising all needed subsystems. Within the next year, the construction of the final setup of the BWEC is planned to begin and the advancement of the project is, as of now, within the schedule required by the PANDA Collaboration.

Acknowledgments

This project is supported by the German Bundesministerium für Bildung und Forschung through the grant 05P12UMFP9.

References

[1] PANDA Collaboration, *Physics Performance Report for PANDA: Strong Interaction Studies with Antiprotons*, `arXiv:0903.3905 [hep-ex]`, 2009
[2] PANDA Collaboration, *Technical Design Report for PANDA Electromagnetic Calorimeter (EMC)*, `arXiv:0810.1216 [physics.ins-det]`, 2008
[3] Wieczorek P. and Flemming H., *IEEE Nucl.Sci.Symp.Conf.Rec. 2010*, 1319-1322
[4] Erni W., Datasheet of the SP883a02 preamplifier , 2012
`http://jazz.physik.unibas.ch/panda/uploads/2012-07-24--11-45-10_SP883d_Datasheet_Manual_PreampV0.6.4.pdf`

Performance of prototypes for the PANDA barrel EMC

D A Bremer[1], T Eissner[1], P Drexler[1], M Moritz[1], R W Novotny[1] for the PANDA collaboration

[1] II. Physikalisches Institut, JLU Gießen , Heinrich-Buff-Ring 16, D-35392 Gießen, Germany

E-mail: daniel.a.bremer@exp2.physik.uni-giessen.de

Abstract. The PANDA experiment will be part of the future Facility for Antiproton and Ion Research (FAIR) and aims for the study of strong interaction within the charm sector via antiproton proton collisions up to antiproton momenta of 15 GeV/c. Reflecting the variety of the physics program the PANDA detector is designed as a multi-purpose detector able to perform tracking, calorimetry and particle identification with nearly complete coverage of the solid angle. The Electromagnetic Calorimeter (EMC) contained inside its Target Spectrometer is based on cooled $PbWO_4$ scintillator crystals. In order to ensure an excellent performance throughout the large dynamic range of photon/electron energies ranging from a few MeV up to 15 GeV an extensive prototyping phase is mandatory. This contribution describes the measured response of the EMC barrel part prototype PROTO60 at the largest design energy to secondary beams provided by the SPS at CERN. In addition to PROTO60 a tracking station was deployed, providing precise position information of the 15 GeV/c positrons. For calibration purposes a 150 GeV/c muon beam and cosmic radiation, in combination with estimations from GEANT4 simulations were used. The obtained performance concerning energy, position and time information is presented.

1. Introduction

The PANDA experiment will be located at the future Facility for Antiproton and Ion Research (FAIR), which is an extension of the existing GSI Helmholtzzentrum für Schwerionenforschung GmbH (GSI) at Darmstadt, Germany. The High Energy Storage Ring (HESR) will provide antiproton beams with momenta ranging from 1.5 GeV/c - 15 GeV/c at a precision of $\frac{\Delta p}{p} \approx 10^{-4} - 10^{-5}$. This regime enables access to QCD bound states containing open and hidden charm as well as exotic states like gluonic excitations. Furthermore the fields of hypernuclei physics or nuclear structure experiments are accessible.

1.1. The PANDA detector

The PANDA detector is designed as a multi-purpose detector and is, adhering to fixed target kinematics, divided into two parts: Target- and Forward Spectrometer, as indicated in the schematic overview shown in Fig. 1. Each spectrometer contains several tracking and Cherenkov detectors for particle identification. The detection of electromagnetic probes in the Forward Spectrometer is performed by a Shashlyk type lead / plastic scintillator sampling calorimeter, while the Target Spectrometer contains an electromagnetic calorimeter based on second generation lead tungstate scintillator crystals (PWO-II) of improved quality, compared

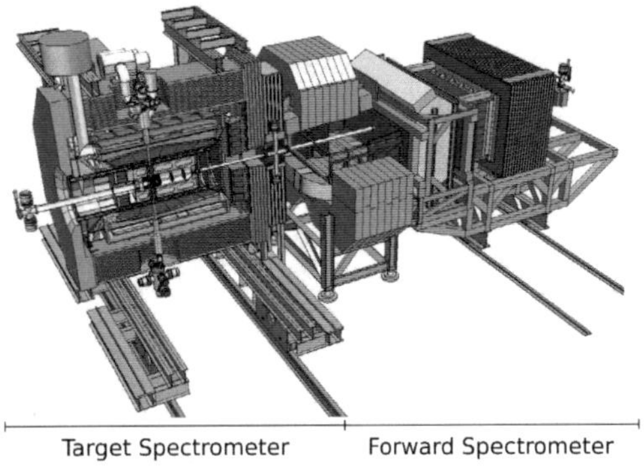

Figure 1. Schematic overview of the PANDA detector.

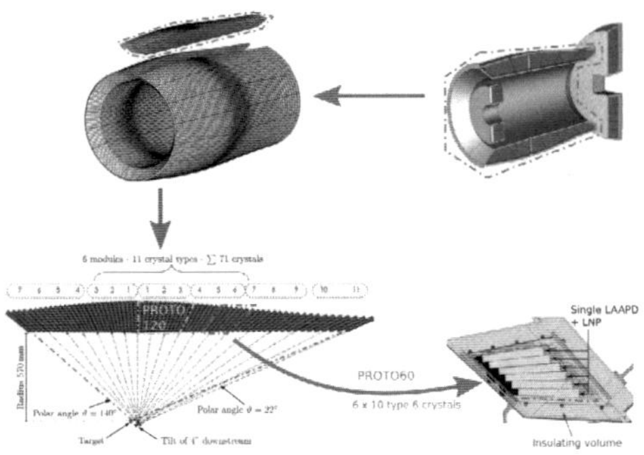

Figure 2. Schematic view of the Barrel EMC subsection represented by PROTO60.

to the crystals employed at the CMS experiment at LHC. Cooled operation at $-25\,^{\circ}$C enhances the light yield by approximately a factor of 4 compared to room temperature, which entails an improved energy resolution and enables measurements over the large dynamic range from 10 MeV to 15 GeV [1], respectively. The Target Spectrometer electromagnetic calorimeter (EMC) in turn is divided into three parts, named according to their position with respect to the beam direction. The Forward Endcap covers polar angles from $5\,^{\circ}$ to $22\,^{\circ}$, the Barrel EMC angles from $22\,^{\circ}$ to $140\,^{\circ}$ and the Backward Endcap angles larger than $140\,^{\circ}$. With this design a virtually complete coverage of the solid angle can be achieved, which is mandatory for the intended physics cases. Due to the fact that the EMC is contained inside the magnetic field of the solenoid and necessarily has to be compact, readout is based on two rectangular Large Area Avalanche Photo Diodes (LAAPDs) per crystal. Merely in the low polar angle regions of the Forward Endcap, the use of one Vacuum Photo Tetrode per crystal is foreseen, which ensures a higher rate capability.

1.2. The Barrel EMC Prototype - PROTO60

The first prototype for the PANDA EMC, the so called PROTO60 consists of a matrix of 6×10 PWO-II crystals. As indicated in Fig. 2, geometrically it resembles a subsection of one of the 16 geometrically identical Barrel EMC slices. All 60 crystals possess the same shape and represent frusta of pyramids with a trapezoidal base with an intermediate degree of tapering (geometry type 6). Each crystal is read out by a single quadratic LAAPD with $1\,\mathrm{cm}^2$ sensitive area (type Hamamatsu S8664-79 1010SPL). The signals of groups of four crystals are amplified by the Low-Noise-low-Power charge preamplifier (LNP, Version Quad)[2] and routed outside the insulated containment by backplane-printed-circuit-boards (PCBs), which are also used for the high voltage distribution. During operation, the complete insulated volume is cooled to $-25\,^{\circ}$C and flooded with slightly over-pressured dry nitrogen to reduce humidity and avoid ice formation. An detailed description of PROTO60's mechanical layout and front end electronics is given in [3].

2. Test Beam at CERN-SPS

2.1. Setup and Readout

Previously several successful beam tests with energy marked photons in the range from 50 MeV to 1.5 GeV were performed and are described in detail in [3]. The test experiment described in this work was performed with a positron beam of 50 $\frac{\mathrm{GeV}}{c}$ momentum, representing the highest design energy of the PANDA EMC. This secondary beam was provided by the H4 beamline at CERN, using a primary protons delivered by the Super-Proton-Synchrotron (SPS). In addition

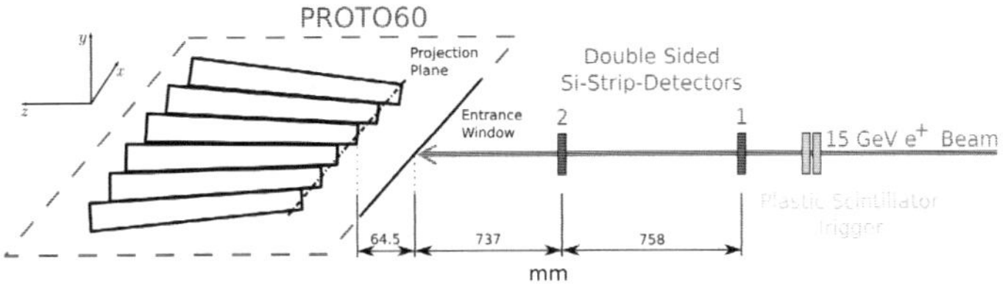

Figure 3. Schematical drawing of the test beam setup at CERN SPS. The depiction is not true to scale.

to PROTO60, the so called Bonn Tracking Station [4] was used for charged particle tracking. In the used configuration it consists of two double sided silicon microstrip detectors (50 μm strip pitch) triggered by a plastic scintillator cross with photomultiplier tube (PMT) readout. A schematic drawing of the setup is displayed in Fig. 3. Each Si-strip detector delivers information on the x- and y- position of the traversing positron, which is linearly extrapolated to the front face of the PROTO60 crystal matrix to obtain the point of impact of the individual positrons. The raw signals provided by the PROTO60 crystals are digitized by Sampling ADCs (Type SIS 3302) with 50 MHz sampling rate. These raw traces were stored to hard disc without further treatment. However, energy and time information are extracted offline via feature extraction algorithms, which are elaborately described in [5]. The energy is extracted from the pulse height after a moving window deconvolution and moving average filter, while for time extraction, the moving average is replaced by a constant fraction timing algorithm.

2.2. Analysis

Due to the expected high energy depositions for crystals located at the centre of the electromagnetic shower, the expected signal output exceeded the limited input charge of the LNP. Therefore, for the 5 innermost crystals of the matrix, the LAAPD gain was reduced to avoid saturation effects, while the outer crystals used gain settings consistent with earlier beam times at lower incident energies. The scheme of the crystal matrix depicted in Fig. 4 shows an overview of the settings applied to the individual crystal as well as the method employed for calibration. The energy calibration via cosmic minimum ionizing particles (MIPs) established in previous PROTO60 beam times is due to the significantly higher energy deposition only applicable for the outermost crystals of the matrix. Crystals exposed to the core of the electromagnetic shower, calibration was performed according to the energy deposited by a muon beam of 150 GeV/c momentum. This method provides a reliable relative calibration even at high energies. However, for an exact absolute energy calibration one has to rely on GEANT4 simulations for determining the absolute amount of energy deposited. Nevertheless, because of the high kinetic energy of the muon beam, which is close to the critical energy of muons inside lead tungstate as measured by [6], these absolute values are prone to uncertainties.

3. Results

3.1. Energy Resolution

The line shape obtained for the energy sum with a single crystal threshold of 1 MeV is depicted in Fig. 5. To extract the energy resolution in an appropriate manner, taking into account the tailing to lower energies, a fit of the Novosibirsk-function [7] was used. In case of the 15 $\frac{\text{GeV}}{c}$ positron beam of approximately 1 cm diameter impinging at the crystal center, a relative energy resolution of $\frac{\sigma}{E} = 1.7\,\%$ was measured. However, to investigate the position dependence of the energy response, the tracker information was used to restrict the beam size to 4 mm diameter, which represents a reasonable compromise between position precision and the number of collected

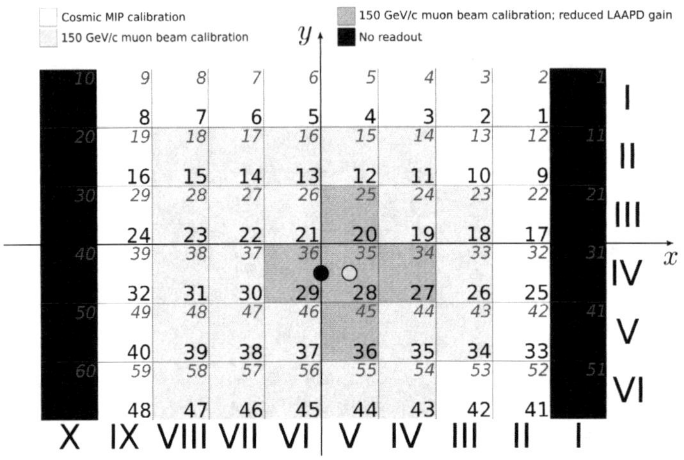

Figure 4. Schematic view of the PROTO60 crystal matrix in beam direction. Grey italic numbers indicate the physical number of the crystal matrix, while black numbers represent the crystal indices used in readout and analysis. The positron beam was impinging either at the center of crystal 28 (run 1) or in between crystal 28 and 29 (run 2), as marked by an open and filled circle, respectively. The schematics are not true to scale.

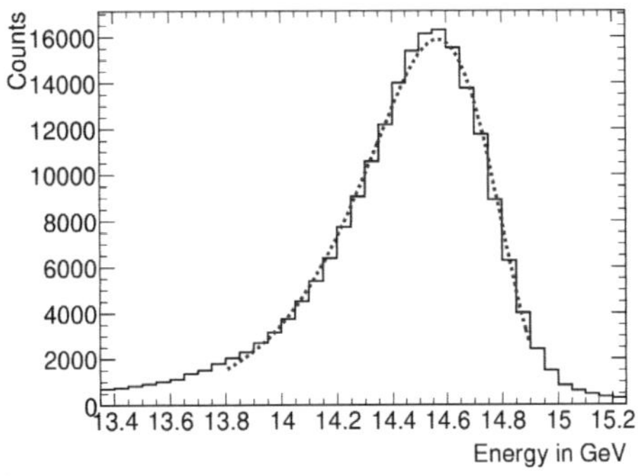

Figure 5. Energy sum line shape obtained at central position and fitted with a Novosibirsk function.

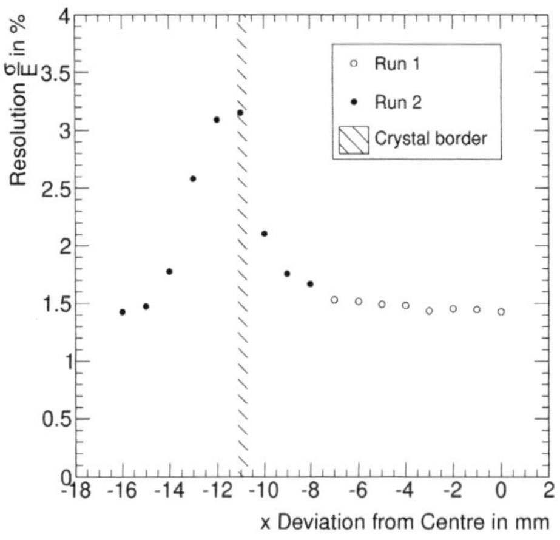

Figure 6. Energy resolution as a function of the deviation of the beam position form the crystal center. The beam size was restricted to 4 mm diameter by software conditions.

events. Fig. 6 shows the obtained relative resolution in dependence on the deviation of the beam position from the crystal center. Apparently, the resolution is significantly deteriorated in the crystal junction region, while shower containment and therefore resolution is achieved at the crystal center. Nevertheless, correction algorithms to reduce these effects were studied and are elaborately described in [8].

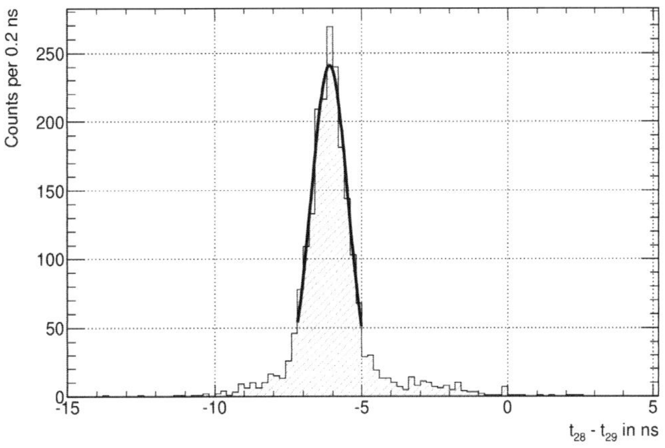

Figure 7. Time difference of the response of crystal 28 and 29 in case of an energy deposition equal within 10 %.

3.2. Time Resolution

Time resolution is extracted from the fluctuations in the time response of two detectors with comparable signal rise times. Therefore, events were selected, where the energy deposition inside the two neighboring crystals 28 and 29 was amounting to the same value within a margin of 10 %. The obtained distribution of their time difference Δt is presented in Fig. 7 and shows a Gaussian shape with a width of $\sigma_{\Delta t} = (631 \pm 14)$ ps. Under the assumption that both crystals contribute equally, the single crystal time resolution at on average 5.5 GeV deposited energy is obtained to be

$$\sigma_t = \frac{\sigma_{\Delta t}}{\sqrt{2}} = (441 \pm 15)\,\text{ps}. \tag{1}$$

3.3. Position Resolution

For position reconstruction a center of gravity algorithm with logarithmic weighting, as given in Eq. 2, was employed, which takes into account the exponential transversal profile of the electromagnetic shower.

$$x_{log} = \frac{\sum\limits_i w_i x_i}{\sum\limits_i w_i}, \qquad w_i = \max\left\{0, W_0 + \ln\left(\frac{E_i}{\sum\limits_i E_i}\right)\right\}. \tag{2}$$

If selecting the appropriate weighting parameter W_0 a linear position response is achieved, as displayed in Fig. 8. The position resolution is obtained by measuring the width σ_x of the distribution of the difference between reference position x_{Si} provided by the tracker and reconstructed position x_{log}. For slices of 0.5 mm in x_{Si} the extracted position resolution is plotted within Fig. 9. Best resolution of 0.5 mm is observed at the crystal junctions, while in the crystal center the position resolution is limited by the unequal sharing of the electromagnetic showers energy and amounts to 1.3 mm. In y- direction, a comparable behavior of the energy response is obtained. Consequently, the overall position resolution including all events regardless of x_{Si} amounts to $\sigma_x \approx \sigma_y \approx 1.1$ mm, for both x- and y-direction.

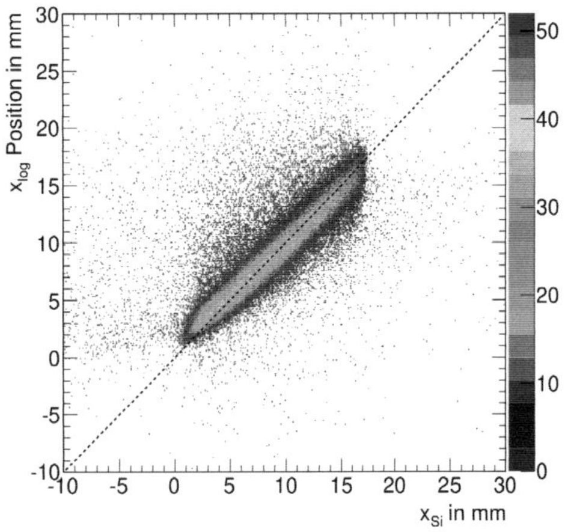

Figure 8. Correlation of reconstructed position x_{log} and reference position x_{Si}. The dashed line corresponds to a linear correlation.

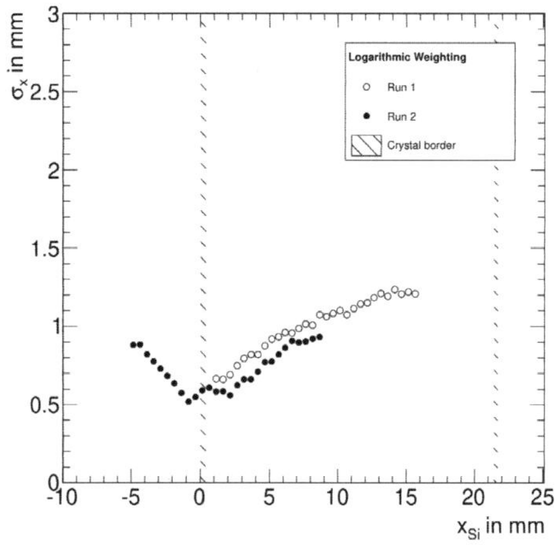

Figure 9. Position resolution in dependence of the reference position x_{Si}.

4. Conclusion

The performance of the first prototype for the PANDA Barrel EMC was measured at its highest design energy with a positron beam of 15 GeV/c momentum. Readout was based on sampling ADCs and offline feature extraction of the sampled detector signal, as described in detail in [3]. A 150 GeV/c muon beam and MIPs from cosmic rays were used for calibration of the detector matrix. Obtained energy, position and time resolution are compliant to the specification given in the EMC's technical design report. An energy resolution at 15 GeV of 1.5 % at the crystal center could be achieved. For all positron impact positions a position resolution better than 1.3 mm was measured. In addition, the obtained time resolution of $\sigma_t = (441 \pm 15)$ ps is in agreement with previous measurements [3] and satisfies the demands of the triggerless readout system for precise timing information. Towards the final design, the LNP will be replaced by an ASIC dual gain output. Therefore, the preamplifier will set no limitations on the LAAPD gain and results in an increased dynamic range. The mandatory tests of the modified readout chain and close to final mechanical integration will be performed with a new prototype consisting of a matrix of 120 strongly tapered PWO-II, which is currently under commissioning.

5. References

[1] PANDA Collaboration 2008 Technical Design Report for: PANDA Electromagnetic Calorimeter (EMC) (*Preprint* `0810.1216v1 [physics.ins-det]`)

[2] Erni W 2012 Basel LNP, a discrete Preamplifier for VPT/T (and APD) readout, model SP883d, Preliminary Version

[3] Kavatsyuk M *et al.* 2011 *Nucl. Instr. and Meth. A* **648** 77 – 91 ISSN 0168-9002

[4] Bianco S, Becker M, Brinkmann K T, Kliemt R, Koop K, Schnell R, Würschig T and Zaunick H G 2011 *PoS(RD11)025*

[5] Kavatsyuk M, Guliyev E, Lemmens P J J, Lohner H, Poelman T P and Tambave G 2009 *Nucl. Science Symp. Conf. Record (NSS/MIC), 2009 IEEE* pp 210–213 ISSN 1095-7863

[6] CMS Collaboration and others 2010 *JINST* **5** P03007 (*Preprint* `0911.5397 [physics.ins-det]`)

[7] Strother P D 1998 *Design and Appplication of the Reconstruction Software for the BaBar Calorimeter* Ph.D. thesis University of London

[8] Eißner T 2013 *The new PWO Crystal Generation and Concepts for the Performance Optimisation of the PANDA EMC* Ph.D. thesis Justus-Liebig-Universität Gießen

Development of a forward calorimeter system for the STAR experiment.

O D Tsai[1], E Aschenauer[2], W Christie[2], L E Dunkelberger[1], S Fazio[2], C A Gagliardi[3], S Heppelmann[4], H Z Huang[1], W W Jacobs[5], G Igo[1], A Kisilev[2], K Landry[1], X Liu[1], M M Mondal[3], Y X Pan[1], M Sergeeva[1], N Shah[1], E Sichtermann[6], S Trentalange[1], G Visser[5] and S Wissink[5]

[1] University of California, Los Angeles, California 90095
[2] Brookhaven National Laboratory, Upton, New York 11973
[3] Texas A&M University, College Station, Texas 77843
[4] Pennsylvania State University, University Park, Pennsylvania 16802
[5] Indiana University, CEEM, Bloomington, Indiana 47408
[6] Lawrence Berkeley National Laboratory, Berkeley, California 94720

E-mail: tsai@physics.ucla.edu

Abstract. We present results of an R&D program to develop a forward calorimeter system (FCS) for the STAR experiment at the Relativistic Heavy Ion Collider at BNL. The FCS is a very compact, compensated, finely granulated, high resolution calorimeter system being developed for p+p and p+A program at RHIC. The FCS prototype consists of both electromagnetic and hadron calorimeters. The electromagnetic portion of the detector is constructed with W powder and scintillation fibers. The hadronic calorimeter is a traditional Pb/Sc-plate sandwich design. Both calorimeters were readout with Hamamatsu MPPCs. A full-scale prototype of the FCS was tested with a beam at FNAL in March 2014. We present details of the design, construction technique and performance of the FCS prototype during the test run at FNAL.

1. Introduction

STAR is a large acceptance detector at RHIC. In the coming years the STAR detector system will continue to evolve through a series of upgrades towards superior detection capabilities and broader kinematic coverage, which will position STAR well for the p + p, p + A and A + A measurements and at the initial phase of an Electron Ion Collider (EIC) [1]. The STAR forward upgrade is mainly driven to explore QCD physics in the very high or low region of Bjorken x. At the core of the forward detector upgrade is the Forward Calorimeter System. This system, as shown in figure 1, will provide superior detection capability for neutral pions, photons, electrons, jets and leading hadrons covering a pseudo-rapidity region of 2.5-4.5. The design of the FCS is driven mainly by consideration of detector performance, integration into the STAR system and cost optimization. The FCS consists of a Spaghetti Electromagnetic (EM) Calorimeter (SPACal), followed by a lead and scintillator plate sampling Hadronic Calorimeter (HCal). The EM and HCal sections of the detector both are compensated calorimeters. Compensation (which requires small sampling fraction) provides the best possible energy measurement for single hadrons and at the same time keeps the system compact. To achieve good EM energy resolution in the calorimeter the sampling frequency in the EM section is kept

Journal of Physics: Conference Series **587** (2015) 012053 doi:10.1088/1742-6596/587/1/012053

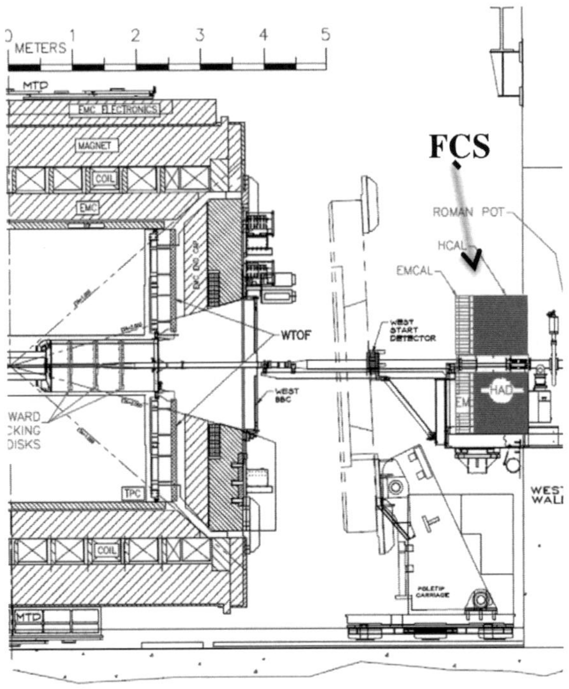

Figure 1. Planned upgrade of the STAR detector with a Forward Calorimeter System. Only the West side of the STAR detector is shown.

sufficiently high with the choice of Scintillation Fiber (ScFi) technology. The SPACal section will be made of tungsten powder and scintillating fibers in a spaghetti geometry. In order to do this, we have developed a novel construction technique through an EIC generic detector R&D project [2]. A prototype module was tested at a FNAL test beam in early 2012 and satisfactory performance of the detector was achieved [2]. Over the last two years we continued development of this technique through the STAR forward upgrade R&D and at present this technology is mature enough for mass production of W/ScFi modules. The proposed SPACal has 120x80 towers. Active volume of a tower is approximately 2.5 cm x 2.5 cm x 17 cm corresponding to $23X_0$ radiation lengths. The HCal section will be made of lead and scintillating tiles with each tower of 10 cm x10 cm x 81 cm corresponding to 4 interaction lengths. The sampling fraction and frequency were copied from the ZEUS Pb/Sc prototype [3]. We developed a novel construction technique for HCal to build the detector by stacking the tiles in-situ. This allows smooth integration in the STAR configuration. There will be 30x20 towers of HCal covering the same area as the SPACal. Whenever possible, we also minimize the number of mechanical components and the construction and operation resources needed from the collaboration in order to carry out the forward upgrade project under the expected constraints for budget and other resources.

2. Description of FCS prototypes

A detailed description of the construction technique for EM sections can be found in [2]. A brief description of our technique is the following: first we form a matrix of scintillation fibers, then a powdered absorber is poured into this matrix. Residual air in the absorber is then replaced with epoxy, which turns the assembly into a rigid object. Different combinations of fiber arrangements and very fine sampling structures can be easily produced. Mechanical properties of the resulting W/ScFi assembly are close to that of structural steel, which allows hermetic, self-supporting structures to be built from this material. For the FCS EM section, we developed a method of building superblocks, each consisting of four individual towers, at once. This technique is mature enough to start the process of industrialization.

Figure 2. HCal Mechanical Structure.

The HCal section is a stack of layers of absorber and scintillation plates. The easiest way to describe the assembly process is to imagine building an entire HCal block from LEGO style parts layer-by-layer. Figure 2 shows the basic mechanical structure of the HCal mechanical prototype. Holes in the bottom base plate of

Figure 3. A Sixteen channels HCal prototype.

the detector provide locations of the absorber plates. Each absorber plate has four holes for dowel pins, two at the bottom and two at the top. Steel dowel pins (5 mm in diameter) position absorber plates with respect to the bottom base and top steel master plates. A single master plate covers one and a half rows of HAD towers, providing interlinks: between all absorber plates within one tower, between front and back steel plates of the HAD section and between adjacent rows of the HAD towers. The thickness of the absorber plates is 10 mm. They are made of lead-antimony alloy (4% Sb) and painted with white diffusive reflective paint (Sherwin Williams F63WC134). The gap between two adjacent absorber plates is 3.1 mm. Scintillation plates of thickness 2.5 mm (EJ-212) are placed inside these gaps. There are 63 absorber plates and 64 scintillation plates in a single HCal tower. Scintillation light from a single tower is collected with a 3 mm thick wavelength shifting (WLS) plate (EJ-280), which is placed in the gap between the two adjacent HCal towers as shown in figure 2. All scintillation and WLS plates are "floating" within each layer (there are no mechanical loads on these elements). Figure 3 shows an assembled HCal prototype. We assembled this prototype in place at the FNAL test beam to validate the construction technique. It took about eight hours for four people to build the sixteen-channel HCal prototype from the component parts at the test beam site.

The light collection scheme of the HCal was optimized to provide uniform and efficient light collection from all scintillation tiles along the depth of the HCal tower. All optical connections (except for coupling of the silicon photo-multipliers to the WLS) in the HCal were made through a thin air gap. We found that the best combination of reflective materials for the WLS plate is white diffusive reflector (Bicron BC-620) at the far end (from the photo-detector) and aluminized mylar at the back side (opposite to the edge of the scintillation tiles) of the WLS. The mylar film also serves as an optical isolator between HCal towers within one layer of the HCal. To achieve uniform light collection (within 10%) along the depth of the HCal tower, we placed a variable density filter printed on a clear mylar sheet inserted between scintillation tiles and the WLS plate. MC calculations show that without such a filter, the energy resolution of the HCal degrades by about factor of two for the energy range above 20 GeV compare to an ideal detector. Variation of light from tile-to-tile in the tower within 10% has negligible effect on energy resolution. Bench test measurements show that variation of the thin air gap between the scintillation tiles and WLS plate (due to mechanical tolerances required for HCal assembly) has a negligible effect on energy resolution of the detector as well. We found no degradation in light collection efficiency for unwrapped scintillation tiles placed between painted absorber plates compared to scintillation tiles wrapped in Tyvek.

We developed a new compact readout for the FCS. For both EM and HCal sections we decided to use silicon photo-multipliers (Hamamatsu Multi-Pixel Photon Counters (MPPC) S10931-025p). They are very compact, fast and insensitive to the magnetic field and sufficiently radiation hard for FCS readout in STAR [4]. The measured light yield (with a very efficient light collection scheme and PMTs) from the EM prototype in the test run in 2012 was 2000 p.e./GeV. We estimated that with 4 MPPC per tower for the EM section and with 8 MPPC per tower for the HCal section we will collect enough light to keep the contribution from photo-statistics to the energy resolution of the detector at a negligibly small level. The geometrical efficiency (ratio of active area of eight MPPCs to the output surface area of the WLS plate) of the light collection scheme for HCal is 8.2% and ~21% for EMcal towers.

A light collection scheme with WLS plates for HCal towers provides perfect light mixing. The situation is different for the EM section where light from the scintillation fibers is collected by the MPPC through a short (25 mm long) light guide/mixer. A bench test measurement prior the test run showed that with this scheme, non-uniformity of the light collection might be as high as 20% (difference between the hottest spots just under the MPPCs and at the corners of the towers). We decided to proceed with this scheme anyway to measure the absolute light yield with readout based on MPPCs and later redesign the light collection scheme for the EM section depending on results of the test run. The optical, mechanical and electrical integration of the readouts for the EM and HCal sections are given below.

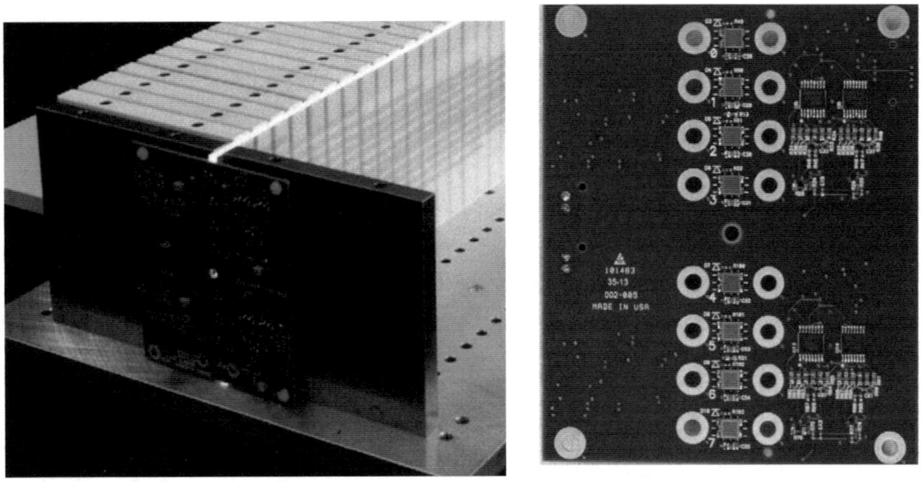

Figure 4 A compact readout for the HCal section. The HCal FEE board is coupled with a WLS plate through Dow Corning 3145 RTV (left). Eight MPPC sensors mounted on the HCal FEE board (right).

We used two different sets of front end electronics to readout the FCS with the MPPC during the test run. The front end electronics for the HCal section is shown in figure 4. The HCAL front end board is designed for low power and reliable, low cost integration into a large detector system. The unregulated 90VDC input is regulated to the required SiPM bias voltage set by two DAC channels, one incorporating a thermistor. In this way, both the voltage and temperature compensation slope (to maintain a constant gain of the SiPM) are programmable. Signals from four SiPM's are directly summed at the input of a single preamplifier, which is a regulated common-base stage using BFR92A transistors. The amplifier input impedance is very low, a few Ohms; as a result the high capacitance (1.3nF) of the four parallel SiPM's does not limit the charge collection time as it would with a 50 Ohm input. The preamplifier is followed by a differential output driver providing a 4 V peak-to-peak signal to be used with low cost, low mass twisted pair cables to an external ADC system. The large signal swing and differential interface enables robust high dynamic range performance even if there is external EMI or ground noise in the system.

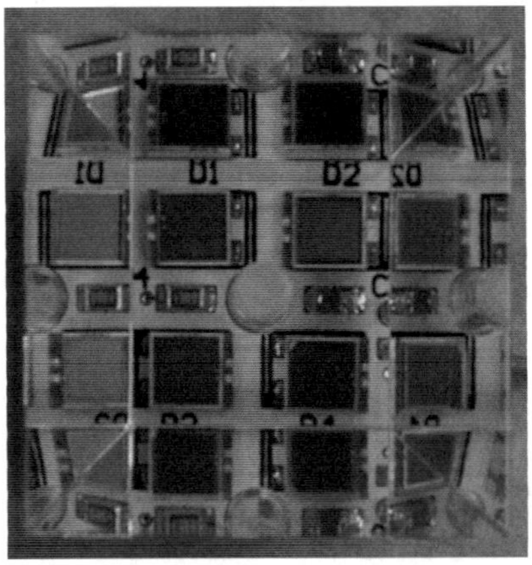

Figure 5. Compact readout for EM section. EM FEE board with four MPPCs attached to a light guide.

For readout of the EM towers, four MPPCs per tower were grouped so that all of them had the same gain at the same bias voltage. A single bias voltage was used for all sixteen channels. A simple resistive divider was used to set the required bias voltage for individual towers. There was no temperature compensations of bias voltage for the EM prototype. Analog signals from these MPPCs were summed with the OPA691 operational amplifier. MPPC sensors for the EM prototype were first mounted on the FEE board, calibrated, and then potted with Dow Corning Sylgard 184 silicone. This silicone layer (~ 2 mm thick) served as an optical cookie and as a spring to keep the FEE board firmly attached to the light guide with a single plastic screw (visible at the center of the light guide in figure 5). The design of the light guide was not completely optimized prior to the test run. The light guide is an acrylic truncated pyramid with a height of 25 mm, base 25 mm x 25 mm and 13 mm x 13 mm at top. The distance between the centers of the MPPCs is 7 mm. Geometrical efficiency of the light collection scheme for the EMcal is approximately 21%.

3. Preliminary results of the test run.

The FCS system and a prototype of the EM central barrel calorimeter for a dedicated EIC detector were tested at the FNAL (experiment T1018) test beam facility in March of 2014. The data analysis and comparison with MC predictions are still in progress. The EM prototype of the central barrel calorimeter was built using the same technique as the forward EM prototype for the FCS. Towers for the EIC prototype had wedge shapes as required for a cylindrical geometry of the barrel central calorimeter. The readout of the FCS prototype was placed upstream of the incoming beam, while for the EIC EM prototype it was downstream during the test run. We used the same MPPC FEE assemblies to read out both EM prototypes during the test run.

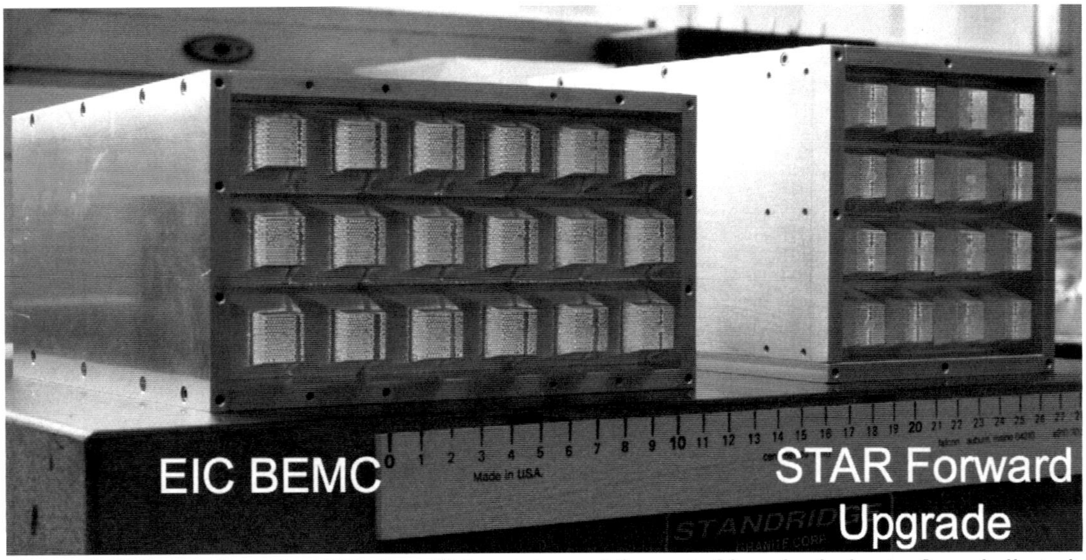

Figure 6 Eighteen channels, 18 X_0 deep EM prototype of the central barrel calorimeter for a dedicated EIC detector (left). A sixteen channels, 23 X_0 deep prototype of the forward EM calorimeter for FCS (right). Both prototypes used similar light collection schemes and frontend electronics.

We tested the response of the FCS system to hadrons, electrons and muons in the energy range 3 -32 GeV. Electrons were identified with a differential Cherenkov counter (standard equipment at the MTBF). Impact position was defined by a scintillator XY hodoscope (4.9 mm wide scintillator square rods readout by SENSL SiPMTs). We minimized the amount of material upstream of the calorimeters in the beam line to about 4 cm of scintillation counters. Additionally, FTBF personnel installed He-filled beam pipes between our apparatus and the upstream Cherenkov counter. The initial setup of our apparatus in the beam line is shown in figure 7. Two MTBF MWPCs (one is seen in figure 7) were used as additional monitoring devices during the beam energy scans to track reproducibility of the beam settings at different energies. The HCal was oriented with a fixed angle (2.5 degrees) between the beam and primary axis of the HCal towers. The EM prototype was attached to the front steel plate of the HCal. The angle between the axis of EM tower and beam was kept at 4 degrees. All channels of the FCS were equipped with an LED monitoring system. LED monitoring signals and pedestals events were continuously recorded with a rate about 1 Hz most of the time during the test run. Preliminary analysis of these data shows that stability of the gain for HCal and EMcal front end electronics was better than 1% during a typical twelve hour shift of data-taking.

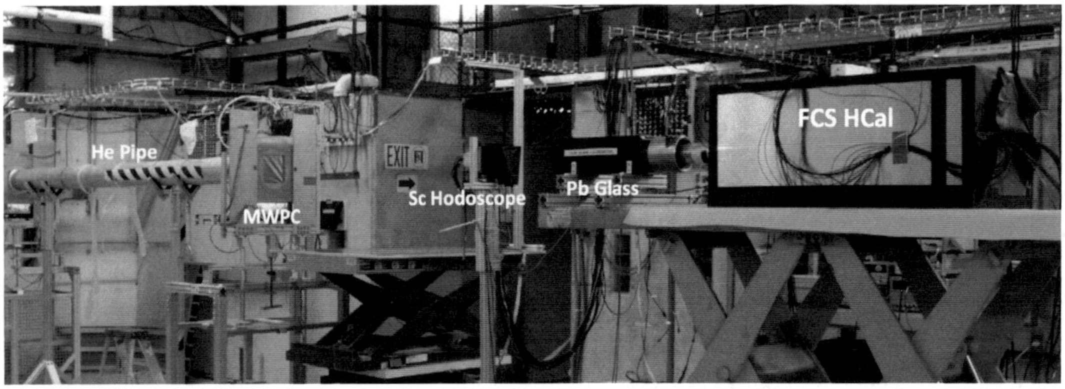

Figure 7 FCS HCal prototype at the beam line. Pb glass calorimeter in front of HCal was used for initial beam studies.

All MPPC's were tested with a laser system prior the test run. With this system we measured that the gain of the MPPC assemblies for both HCal and EM prototypes were set equal to within 1%. We found that no additional tower-by-tower calibration of the EM prototype with the beam was required. This was expected after our previous beam test in 2012 when we measured excellent internal homogeneity of the EM modules built with our technique. The HCal required additional tower-by-tower calibrations with MIPs. For that an absorber was inserted into the beam line (8 GeV muon mode for the MT6 test line). A MIP peak was selected in each HCal tower using an isolation requirement (a single muon hit in a tower with no other energy deposition in the entire HCal). For calibrations with MIPs the EM prototype was removed from the beam line. We found that quite large corrections at the level of approximately 20% were required on top of calibrations made prior to the test run. About 10% of this shift can be explained by the alignment of the WLS plate and the MPPCs (both have a 3 mm active area, about 250 microns misalignment is possible due to positioning of the MPPCs on the FEE board). The rest should be attributed to the quality of optical components; one possible source is a difference in the response of the WLS tiles used in different HCal towers (concentration of dopants and attenuation length has not been measured for every WLS tile used in the HCal, we assumed that they are all identical). The response of the FCS to hadrons is illustrated in figure 8. In an ideal, completely compensated calorimeter system, the reconstructed energy of the incoming hadron is a simple sum of the energy deposited in the EM and HCal sections (assuming that the response in both sections is the same and energy independent). To obtain the best energy resolution for hadrons in the FCS we found that a weighting factor for the EM section should be energy dependent. This changes from about 2 to 1.2 for a beam energy range from 3 to 20 GeV and stays flat after that. With this energy dependent weighting of the energy deposited in the EM section, we measured e/h ratio for the FCS to be close to 0.95 and almost constant above 10 GeV as seen in the right panel of figure 8.

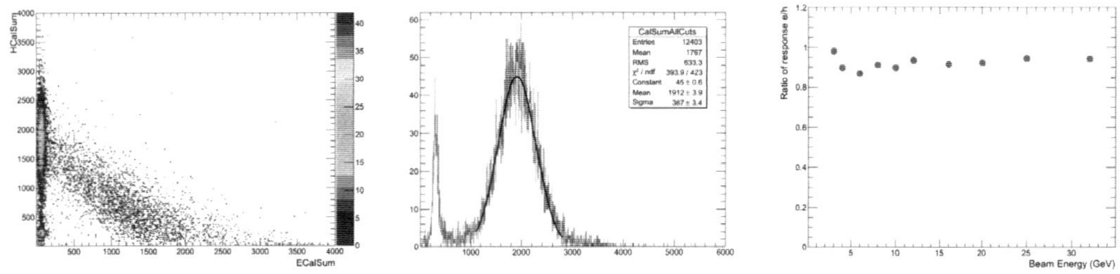

Figure 8 Response of FCS prototype to hadrons. Energy deposition in Hcal section (Y-axis) vs energy deposition in EMcal section (X-axis) for 12 GeV hadrons (left panel). A weighted sum of energy deposited in EM and HCal section for 12 GeV hadrons (center panel). Ratio of response of FCS to electrons vs hadrons (**e/h** ratio) vs beam energy (right panel).

We did not perform any corrections due to leakages in the transverse and longitudinal directions in the FCS. Qualitatively, this result is close to MC predictions, however in our MC model we did not include some of the structural elements between the EM and HCal sections as well as the limited size of the prototype tested at FNAL. The questions of optimal weighting and e/h ratio in the FCS still need to be clarified with a MC for the exact geometry of the detectors used in the test run.

The response of the FCS to electrons is illustrated in figure 9. Due to non-uniform light collection with MPPCs the response of the EM section is dependent on impact position. For the preliminary results we present in this article we corrected the energy deposition in the EM section according to impact position and restricted the impact area only to the circle with diameter 1.4 cm at the center of the EM tower.

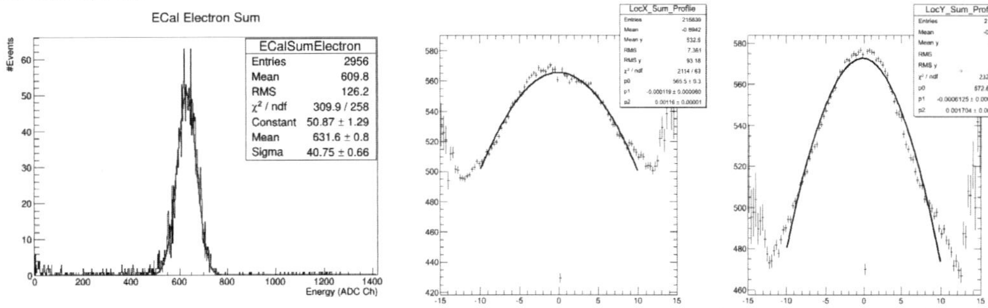

Figure 9 Response of FCS EM prototype to electrons. Energy deposition in the EM section for 4 GeV electrons, impact point restricted by the scintillation hodoscope to an area 5 mm x 5 mm (left panel). A dependence of energy deposited in EM section vs impact position in local X coordinate (center panel) and local Y coordinate (right panel).

Local coordinates of the impact position were determined using calorimeter information only. We used a logarithmic weighting method with the cut-off parameter set at 3.8. The difference in the shapes of the response of the EM section in the X and Y directions is due to a tilt of the EM prototype of 4 degrees around the Y axis.

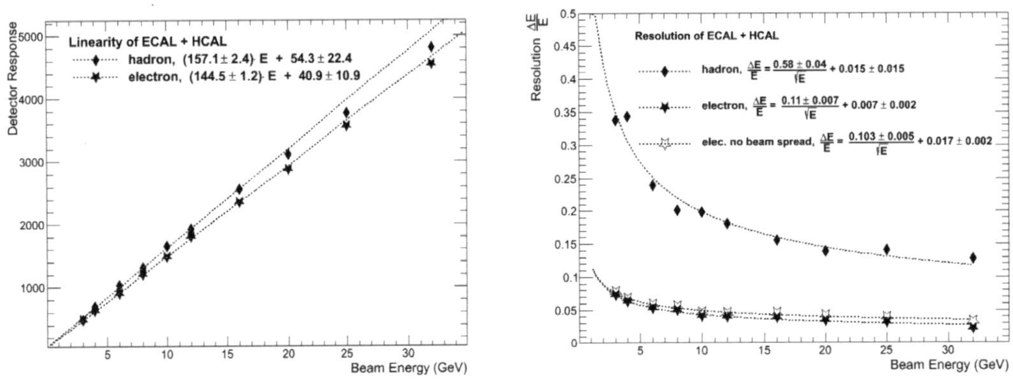

Figure 10 Response of the FCS prototype to electrons and hadrons vs energy (left panel). Energy resolution of the FCS for hadrons and electrons vs energy (right panel).

Summary plots illustrating the performance of the FCS prototype during the beam test are shown in figure 10. The response to electrons is almost linear, while the response to hadrons shows clear deviation from linearity above 15 GeV. The most likely reasons for this deviation are the weighting procedure of the fraction of energy deposited in the EM section and leakage from the FCS. We tested the HCal section alone (with EM section removed from the beam line) and did not observe a similar deviation from linearity in this energy range. The energy resolution of the FCS to hadrons, shown in figure 10, is about 15% worse compared to MC predictions for the FCS at 10 GeV. One of the reasons is a transverse leakage from the FCS prototype, which was not taken into account for the test beam results. We also should mention that the energy resolution of the FCS in MC depends on the physics list used in GEANT4. We used a LHEP physics list, which we believe provides the most accurate description of the FCS. The electromagnetic energy resolution of the FCS prototype is close to MC predictions. There are two fits of our experimental results shown in figure 10. One assumed that the momentum spread of the beam is zero. In this case the stochastic term is close to 10% and the constant term is 1.7%. If we use our earlier (2012) estimates for the momentum spread of the beam of 2.7 % below 4 GeV and 2.3% above this value, then the stochastic term becomes 11% and constant term is close to zero.

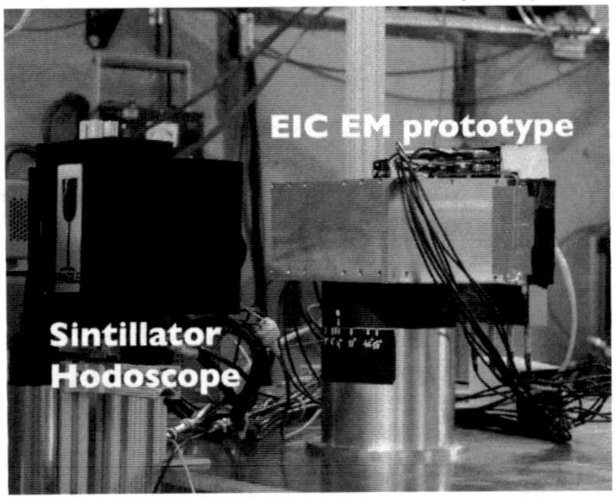

Figure 11 An eighteen channel EM prototype for EIC on the beam line at FNAL.

The absolute light yield measured in the EM section is about 400 pixels/GeV, with the front face of the EM prototype painted with the white diffusive paint BC-620. The measured absolute light yield for the HCal section is about 130 pixels/GeV. MPPCs for both EM and HAD sections should behave almost linearly with these types of light yields for the energy range used in the test run. The light yield measured for the EM prototype is sufficient so we can introduce a mask between the scintillation fibers and light guide to make the light collection more uniform in future. Given the fact that the most recent generation of MPPCs already have much better PDE compared to MPPCs used in the test run (and anticipated future improvements in SiPMs), we believe that there is no need for any type of reflector at the end of the scintillation fibers. This significantly simplifies the construction of the EM section. According to our measurements in 2012, with EM prototypes equipped with a good mirror and with black tape at the end of the fibers, all degradation in the energy resolution can be explained by photo-statistics, *i.e.*, degradation of light attenuation length in the scintillation fibers is not critical in this case.

Along with the FCS we tested a prototype of the central barrel EM calorimeter for EIC. A picture of this prototype on the beam line is shown in figure 11. The goal of the test run was to measure the response of this detector at different impact angles. The design of the prototype shares the same technology with the EM prototype for the FCS, except the geometry of the towers. A wedge-type geometry has always been technologically challenging for ScFi type calorimeters. We developed a simple construction technique for this type of geometry. The wedge shape is obtained by successively inclining meshes which form the matrix of scintillation fibers along the depth of the tower. This method is as simple as the method used for the traditional straight towers of the FCS. The sampling fraction and sampling frequency with inclined meshes slightly increases toward the front face of the towers. A Monte-Carlo calculation shows that this variation of sampling structure has a minor effect on the energy resolution of the detector. The goal of the test run was to measure response, energy resolution and absolute light yield in this detector as a function of impact angles. We performed beam energy scans for three different angles and tested three different light collection schemes. Since the light collection scheme for this prototype is similar to FCS EM detector similar problems with uniformity of light collection were observed. At shallow impact angles (detector was rotated along Y vertical axis) the non-uniformities in response along the X axis is almost vanished. For preliminary test beam results presented in this article we limited the impact area in the Y direction to 5 mm in order to minimize the effect of non-uniformity of light collection and leakage from the prototype due to its limited size.

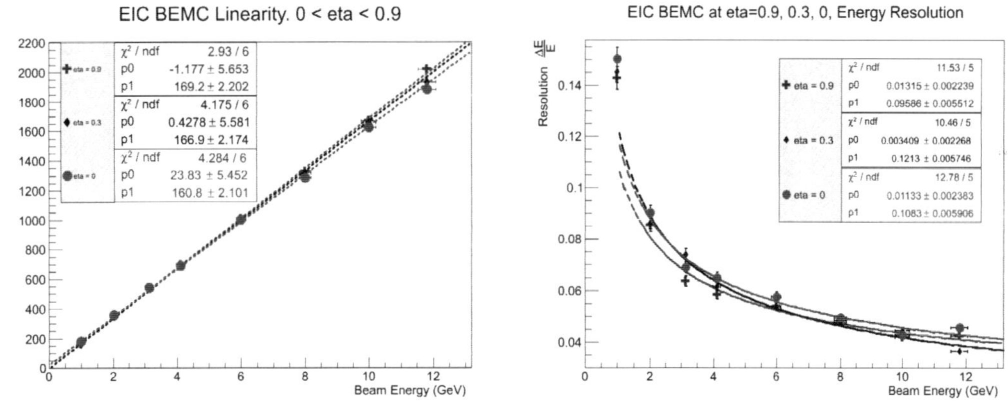

Figure 12 Response of the EM EIC prototype to electrons vs energy for different impact angles (left panel). Energy resolution (p0/√E + p1) of the EM EIC for electrons vs energy for different impact angles (right panel).

Figure 12 summarizes the performance of the EM prototype for a central barrel EIC calorimeter during the test run. As predicted by the MC, the response of the detector only slightly depends on impact angle. A change in the slope for the fits at the left side of figure 12 can be explained by a slightly increased sampling fraction of the detector at more shallow impact angles. The energy resolution will be practically the same for the entire pseudorapidity range covered by the central barrel calorimeter for an EIC detector as shown at the right side of figure 12, which is again in reasonable agreement with MC predictions. The light yield measured for this prototype was 430, 530 and 600 pixels/GeV depending on the type of reflector at the front face of the detector and materials surrounding the light guides. The highest light yield was achieved with white diffusive paint at the front face of the detector (BC-620) and diffusive reflectors surrounding the light guides. We found that multi-clad scintillation fibers (Kuraray SCSF 78M) used for the EIC EM prototype only slightly improves the light yield compare to single-clad fibers (Kuraray SCSF 78) used for the FCS EM prototype. There is no difference in performance of the EM prototypes when the readout was placed upstream (for FCS) or downstream (for EIC) of the detectors.

4. Summary

In 2012 we demonstrated a 'proof-of-principle' for a new simple and cost effective method of building compact scintillation fiber calorimeters utilizing tungsten powder. We continued development of this technique and tested two new EM prototypes designed for the STAR forward upgrade and future central barrel EM calorimeter for a dedicated EIC detector. Both prototypes were successfully tested in the test run at FNAL in March of 2014. We also developed a new construction technique for a high-resolution lead scintillation tile hadronic calorimeter. The FCS system designed specifically for the STAR forward upgrade is the perfect candidate for the forward calorimeter envisioned for the EIC dedicated detector for the outgoing hadron region. The performance of this system during test runs met our expectations. The compact readout scheme based on SiPM readout works well for the HCal prototype. For the EM sections, improvements in the uniformity of light collection have to be made in the near future. With the light yield measured at the test run, introduction of properly designed masks between scintillation fibers and light guides should solve the non-uniformity issue. A bench test measurements with existing prototypes, as well as final data analysis and MC simulation of the exact test beam configuration, presently is in progress.

Acknowledgments

We thank the STAR technical support group at BNL for help with preparation of the front end electronics for the test run at FNAL. We are grateful for the support of FNAL for giving us an opportunity to test our detectors with beam and the FNAL FTBF group for their help during the setup and running of the experiment. This study was supported by the United States Department of Energy.

References

[1] Boer D *et al.* 2011, Gluons and the quark sea at high energies: distributions, polarization, tomography, arXiv:1108.1713v2
[2] Tsai O *et al.* 2012, J. Phys. Conf. Ser.Vol **404**, 012023
[3] Bernardi E *et al.* 1987, Nucl. Instr. & Meth. **A262**, 229
[4] Fisyak Y *et al.* 2014, Nucl. Instr. & Meth. **A756**, 68

Design Studies of the Calorimeter Systems for the sPHENIX Experiment at RHIC and Future Upgrade Plans

C Woody and E Kistenev, for the PHENIX Collaboration

Physics Department, Building 510C
Brookhaven National Lab
Upton, NY 11973

E-mail: woody@bnl.gov, kistenev@bnl.gov

Abstract. The PHENIX Experiment at RHIC is planning a series of major upgrades that will enable a comprehensive measurement of jets in relativistic heavy ion collisions, provide enhanced physics capabilities for studying nucleon-nucleus and polarized proton collisions, and allow a detailed study of electron-nucleus collisions at the Electron Ion Collider at Brookhaven (eRHIC). The first of these upgrades, sPHENIX, will be based on the former BaBar magnet and will include a hadronic calorimeter and new electromagnetic calorimeter that will cover ±1.1 units in pseudorapidity and 2π in azimuth in the central region, resulting in a factor of 6 increase in acceptance over the present PHENIX detector. The electromagnetic calorimeter will be a tungsten scintillating fiber design with a radiation length ~ 7 mm and a Moliere radius ~ 2 cm. It will have a total depth of ~ 18 radiation lengths and an energy resolution ~ $15\%/\sqrt{E}$. The hadronic calorimeter will consist of steel plates with scintillating tiles in between that are read out with wavelength shifting fibers, It will have a total depth of ~ 5 interaction lengths and an energy resolution ~ $100\%/\sqrt{E}$. Both calorimeters will use silicon photomultipliers as the readout sensor. Detailed design studies and Monte Carlo simulations for both calorimeters have been carried out and prototype detectors have been constructed and tested in a test beam at Fermilab in February 2014. This contribution describes these design studies for the sPHENIX experiment and its future upgrade plans at RHIC.

1. Introduction

The PHENIX Experiment at RHIC is planning a series of major upgrades that will enable an extensive set of new physics programs over the next decade. This will involve replacing the current PHENIX Central Arm spectrometer with a new central spectrometer, sPHENIX, shown in Fig 1. It will utilize the former BaBar solenoid magnet and require the construction of two new large calorimeter systems, one electromagnetic and the other hadronic, covering ± 1.1 units in pseudorapidity and 2π in azimuth. These calorimeters, along with a silicon tracking detector, will form the basis of the new central spectrometer for sPHENIX. The hadronic calorimeter will be the first hadronic calorimeter ever used in a RHIC experiment. It will provide a measurement of the total jet energy and enable a comprehensive study of jets, di-jets and gamma-jet events in heavy ion collisions. With enhanced tracking capabilities, a detailed study of upsilon production in heavy ion collisions will also be possible. A detailed description of the sPHENIX experimental proposal can be found in [1].

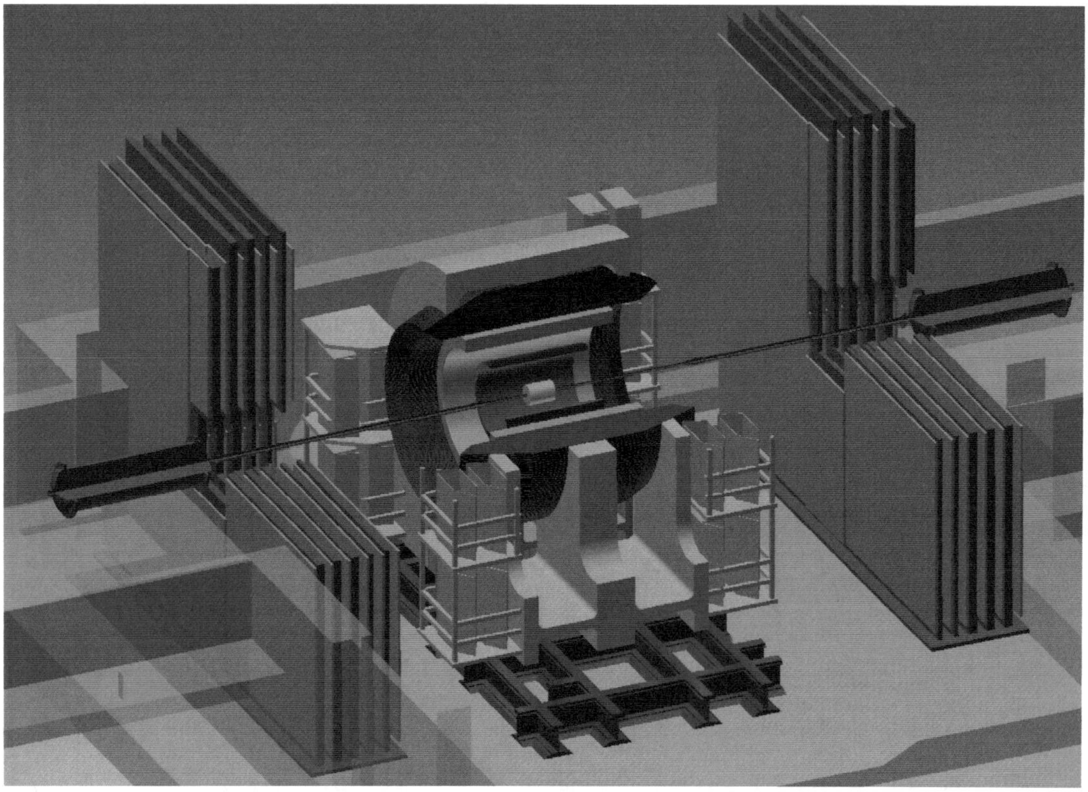

Figure 1. The sPHENIX detector utilizing the former BaBar solenoid magnet, new electromagnetic and hadronic calorimeters, and the current PHENIX silicon vertex tracker.

The new measurements from sPHENIX will allow an investigation of the properties of the Quark Gluon Plasma near the region of its critical temperature and provide detailed information about the phase transition. The flexibility of the RHIC collider also enables studies involving different ion species, comparisons with pp and pA collisions, including measurements of polarization and spin, and studying the energy dependence of these interactions over a wide energy range. The data obtained from RHIC will be complimentary to the data from the heavy ion program at the LHC, which operates at nearly fifteen times the center of mass energy and produces a much higher initial temperature, thus providing a complete and thorough investigation of the QGP and its properties.

The long range plan for RHIC is to transform the current heavy ion collider into an Electron Ion Collider (EIC) at BNL (eRHIC). It will have the capability for colliding polarized electrons with energies initially up to 21 GeV (and eventually up to 30 GeV) with hadrons up to 250 GeV and heavy ions up to 100 GeV/A. The greatly expanded kinematic coverage of eRHIC will allow a detailed study of the distribution of quarks and gluons, and their spins, inside the nucleus, and enable a broad new physics program to study nucleon structure and QCD in nuclei over a large range of x and Q^2.

A new suite of detectors would be added to sPHENIX to provide the additional measurement capabilities required to study ep and eA collisions, evolving it into a new spectrometer, ePHENIX, as shown in Figure 2. These would include additional tracking in the central region in the form of a fast, compact TPC and small angle GEM trackers, a central DIRC for particle id, a high resolution electromagnetic crystal calorimeter and additional GEM trackers in the electron going direction to precisely measure the scattered electron, and a new spectrometer in the hadron going direction. The hadron spectrometer would cover a pseudorapidity range of $1< \eta <4$ and would consist of more GEM trackers, a Gas RICH that utilizes a photosensitive GEM detector with a CsI photocathode for high momentum particle ID, an aerogel Cherenkov detector, and additional electromagnetic and hadronic calorimeters. A Zero Degree Calorimeter (ZDC) and Roman Pots would also be added in the forward direction. A more detailed description of these detectors, along with a discussion of the physics capabilities of ePHENIX, can be found in the ePHENIX Letter of Intent from the PHENIX Collaboration [2].

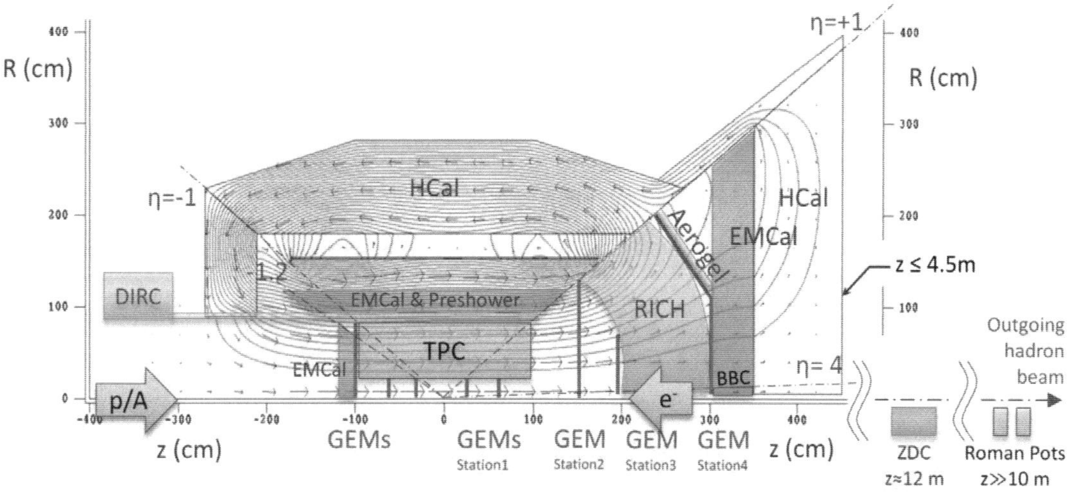

Figure 2. The ePHENIX detector for the Electron Ion Collider (eRHIC) at BNL. Additional detectors have been added to the sPHENIX spectrometer for studying ep and eA collisions.

Detector requirements

The main components of the sPHENIX spectrometer are the solenoid magnet, the two calorimeters and the silicon tracking detector. The BaBar magnet defines the magnetic volume of the spectrometer with a field of 1.5 T and also places constraints on where the electromagnetic and hadronic calorimeters can go. We are currently considering two designs, one with a portion of the hadron calorimeter inside the magnet and the rest outside, and another where the entire hadron calorimeter is outside. In the first configuration, the electromagnetic calorimeter would start at a radius of ~ 95 cm and extend to ~ 120 cm, including the space required for the readout. The absorber would consist of tungsten and scintillating fibers and would have a radial extent of ~ 13 cm (~18 X_0). The PHENIX group has been pursuing a design that would utilize tungsten plates and scintillating fibers oriented approximately parallel to the incoming particle direction, where the plates would either be tilted at a small angle, or have an accordion shape in order to avoid channeling of particles along the length of the absorber. The STAR group has been studying a similar design based on an absorber consisting of scintillating fibers embedded in a matrix of tungsten powder and epoxy. This design is reported on in a separate contribution to this conference [3]. Both designs are very similar. The absorber would have a radiation length of ~ 7 mm and a Moliere radius ~ 2 cm, but the sampling fraction for the PHENIX design is ~ 6%, whereas the STAR design has a sampling fraction ~ 2%. The energy resolution requirement for the EMCAL in sPHENIX is ~ $15\%/\sqrt{E}$, but is slightly more stringent for ePHENIX (~ $12\%/\sqrt{E}$). The segmentation requirement for sPHENIX is much higher than for ePHENIX in order to be able to measure jets in heavy ion collisions. The EMCAL would have a segmentation of approximately $\Delta\eta\times\Delta\phi$ ~ 0.025×0.025, leading to ~ 25K channels covering the central region of 2π in ϕ and ±1.1 in η.

Journal of Physics: Conference Series **587** (2015) 012054 doi:10.1088/1742-6596/587/1/012054

The primary purpose of both the calorimeters is to measure jets in heavy ion collisions. However, for events containing jets in heavy ion collisions, there is a significant amount of energy deposited in the calorimeter from the large number of particles produced in the underlying event of accompanying soft collisions, and the fluctuations on this energy ultimately limits the jet energy resolution. As a result, the calorimeter energy resolution need not be better than the limitations imposed on the jet energy resolution by these fluctuations. The requirement on the energy resolution for the HCAL is ~ 100%/√E for measuring jets over a momentum range from p_T > 20 GeV/c up to p_T ~ 50 GeV/c, and the segmentation is approximately $\Delta\eta\times\Delta\phi \sim 0.1\times0.1$.

The HCAL design has been described at the previous CALOR 2012 conference [4]. It utilizes steel plates and scintillating tiles that are read out using wavelength shifting fibers, where both the plates and the scintillating tiles are oriented parallel to the direction of the colliding beams. This allows the steel plates to serve as a flux return for the solenoid magnet. It also has the feature that the thickness of the plates (but not the scintillating tiles) increases with radius, and therefore the sampling fraction changes with radius. However, the calorimeter is divided into two longitudinal segments, which allows the ability to measure the longitudinal distribution of the hadronic shower, and therefore apply a correction for the longitudinal sampling fluctuations on an event by event basis. The plates will be tilted at an angle of approximately ± 15° in the ϕ direction for the front and back sections respectively in order to avoid channeling of particles passing through the scintillator alone. The total depth will be ~ 5 λ_{abs} (1 m) and the electromagnetic calorimeter will add another approximately 1 λ_{abs} in front. For the configuration where part of the HCAL is inside the magnet, there will be approximately 1 λ_{abs} inside and 4 λ_{abs} outside, and in the configuration with the entire HCAL outside, it would be divided with approximately 1.5 λ_{abs} in the front section and 3.5 λ_{abs} in the back section. Both sections will be divided into 22 towers in η and 64 towers in ϕ for a total of 2816 towers. The fibers from eight tiles will be bundled together to form a tower and will be read out with a single silicon photomultiplier.

The tracking system would initially consist of the existing PHENIX silicon vertex detector (VTX), but additional tracking layers would be added in order to measure heavy quarkonia, tagged heavy flavor jets and high-z fragmentation functions. These would be similar to the existing silicon tracking layers and would be placed at approximately 40 cm and 60 cm radius in order to provide the necessary momentum resolution for resolving the various upsilon states. In addition, a future preshower detector may be added that would enable the measurement of π^0's up to a p_T ~ 40 GeV/c.

2. Prototype Calorimeter Tests

Prototypes of both the HCAL tilted plate design and the EMCAL tilted plate design were constructed and tested in the MT6 Test Beam Facility at Fermilab in February of 2014 with beams of hadrons and electrons with momenta from 1 GeV/c up to 120 GeV/c. The 120 GeV/C beam was derived from the primary beam from the Fermilab Main Injector and was essentially purely protons, while a lower momenta, the beam consisted of a mixture of pions, electrons and muons. A Cherenkov detector upstream allowed for the identification of electrons and pions from 1 GeV/c up to 32 GeV/c. The detectors were tested individually and in tandem in order to study their performance in approximately the same configuration as they will be in sPHENIX. The layout of the detectors along the beam line is shown in Fig. 3. Each calorimeter was mounted on its own rotation platform that allowed it to be oriented at various angles with respect to the beam corresponding to both ϕ and θ. The EMCAL could be moved both vertically and horizontally, and the HCAL could also be moved vertically but not horizontally.

A common readout was used for the SiPMs for both calorimeters (Hamamatsu S10931-025P) that utilized a custom designed preamp and readout circuit that incorporated a programmable bias adjustment for each SiPM along with temperature sensors and feedback circuit that could automatically correct for the gain variation with temperature. Unfortunately, time did not permit fully testing this system during the beam test and the adjustments for the individual SiPM bias voltages to equalize the gains and the automatic temperature compensation was not applied during data taking.

The analog signals from the preamps were sent down ~ 15 m cables to a system of fast digitizing ADCs. The ADCs were previously used for the Hadron Blind Detector for the PHENIX experiment and digitized at a sampling rate of 65 MHz. An LED pulse system was also used for testing and calibration of all readout towers for both calorimeters.

2.1 Hadron Calorimeter

The hadron calorimeter prototype consisted of alternating layers of steel plates with scintillator tiles in between. The array of steel plates is shown in Fig. 3. The plate thickness varied from 2 cm in the front to 5 cm in the back with 8.5 mm gaps between the plates. The scintillating tiles were 7 mm thick and were coated with a UV reflective coating, and had a U-shaped groove in which a wavelength shifting fiber was embedded. The fibers from four tiles were bundled together into a small light mixing cavity and read out with a single SiPM, corresponding to one tower. They formed a 4x4 array of 16 towers for the front section and a similar array of 16 towers for the back section. The entire prototype measured 1.2 x. 1.7 x 1.3 m^3 and weighed approximately 4 tons.

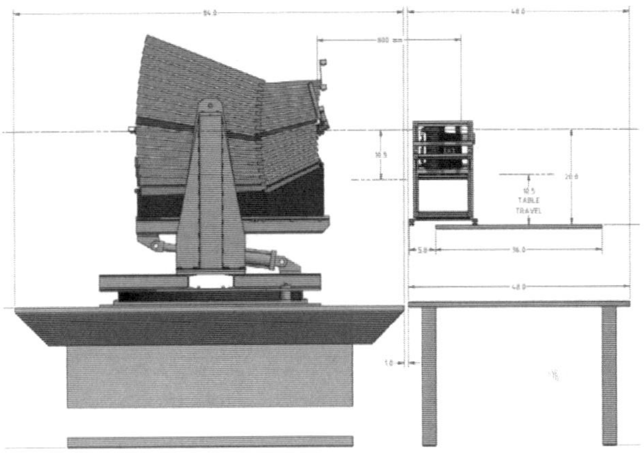

Figure 3. Left: Steel plate stack for the HCAL prototype calorimeter, Right: Layout of the prototype HCAL and EMCAL along the beam line in the test beam at Fermilab.

The analysis of the test beam data is still under way, but preliminary results indicate good performance from the calorimeter. Figure 4 shows the energy resolution for pions in the HCAL over a momentum range from 4 GeV/c to 60 GeV/c and gives a resolution of 88%/√E with a constant term ~ 13%. While this resolution is for single particles, it is within the acceptable range for measuring jets, and we expect the resolution to improve with further analysis. The rather large constant term we believe is due to uncertainties in the longitudinal center of gravity as measured by the ratio of the energy in the front and back sections which is used to correct for the longitudinal variation in the sampling fraction. This will be improved when the energy from the EMCAL is used in the analysis, and would also benefit from additional longitudinal segmentation of the HCAL itself. The energy resolution for electrons measured in the HCAL alone is also shown in Fig. 4 for momenta up to 32 GeV/c and gives a resolution of 53%/√E. The electron resolution would of course be significantly better using the EMCAL information, but measuring the electron energy in the HCAL allows an estimation of the e/π ratio, which gave approximately 1.2. The e/μ ratio was also found to be ~ 0.8. Finally, the right side of Fig. 4 shows the energy linearity for pions, electrons and muons and indicates good linearity over the measured energy range.

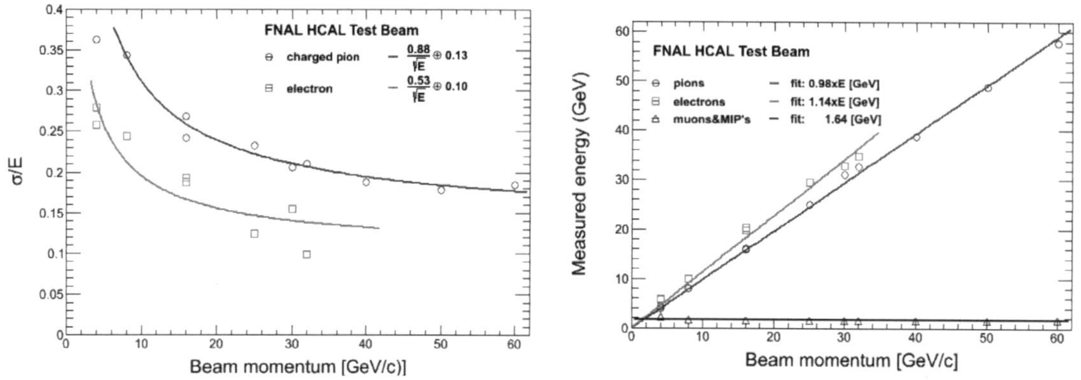

Figure 4. Preliminary data for the prototype hadron calorimeter. Left: Energy resolution for pions and electrons (note: electrons are measured only in the HCAL), Right: Energy linearity for pions, electrons and muons.

2.2 Electromagnetic calorimeter

The prototype electromagnetic calorimeter was constructed using a tilted plate design similar to the HCAL. It consisted of a stack of "sandwiches", where each sandwich was comprised of a pair of 0.5 mm thick flat tungsten metal plates glued together with a layer of 1 mm diameter scintillating fibers in between. When stacked together, this resulted in a tungsten absorber thickness of 1 mm and a nominal sampling fraction of 6.4%. A thin (~ 0.5 mm) spacer was imbedded at one end of each sandwich to produce a slight taper, which made the resulting absorber stack projective in one dimension (equivalent to the r-ϕ direction in sPHENIX). The fibers were also painted black over a length of ~ 2 cm at the readout end in order to produce a more uniform light output along the fiber, and the opposite end was covered with a reflecting mirror film (3M Enhanced Specular Reflector). Twelve sandwiches were stacked together to form a "tower module" and seven tower modules were stacked together to form the entire calorimeter.

Figure 5. EMCAL prototype calorimeter. Left: Absorber stack, Right: Completed calorimeter mounted in rotation stand for testing.

Figure 5 shows the absorber stack for the prototype EMCAL along with the completed calorimeter after it was mounted in its rotation stand. Segmentation into 7x7 readout towers (each 2.5 x 2.7 cm^2) was accomplished by placing an array of light collection cavities over the back of the absorber stack. Each cavity consisted of a small plastic dome that was painted with a white diffusing paint (Labsphere 6080 coating [5]) and was read out with either a single or pair of SiPMs. Figure 6 shows the light collection module with the array of 7x7 towers. Due to the small area of the SiPM (3x3 mm^2) relative to the size of the tower, the overall light collection efficiency was only ~ 4.7%, and the non-uniformity (given by the ratio of the maximum to minimum response across the tower) was ~ 1.7 with a single SiPM readout. The total light output was measured during the test beam run and found to be ~ 180 p.e./Gev with a single SiPM.

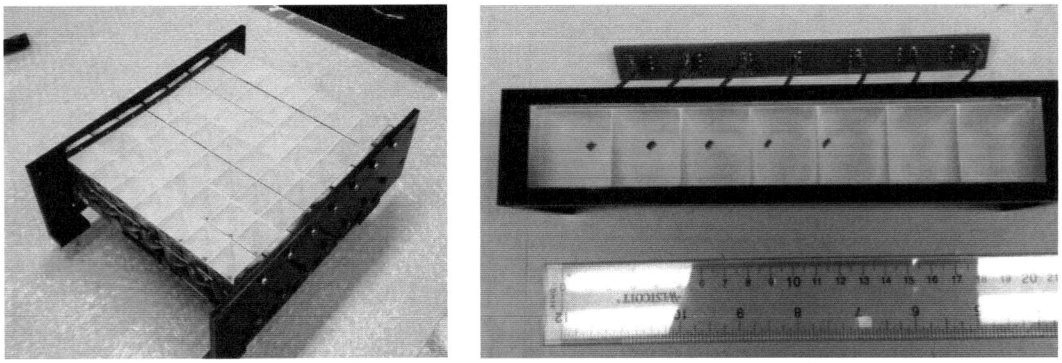

Figure 6. Left: Light collection module forming the 7x7 array of towers. Right: A detail of one row of 7 towers where the single SiPM is visible.

Analysis of the EMCAL data from the beam test is also still under way, but some preliminary results are shown in Figure 7. The plot on the left shows the sum of 3 rows of towers for a horizontal scan across the central row of towers with 8 GeV/c electrons, indicating a uniformity of ~ 5%. The plot in the middle shows the uniformity of the calorimeter measured along the vertical direction using 120 GeV/c protons and indicates a variation ~ 2%. The plot on the right shows the energy linearity of the calorimeter measured in two different orientations. The 90 degree position corresponds to the plates perpendicular to the incoming beam and the 20 degree position corresponds to the nominal orientation that the calorimeter would have in sPHENIX.

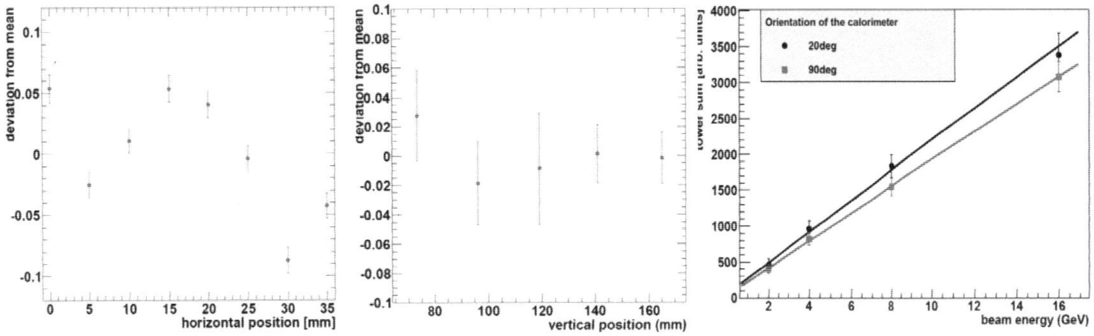

Figure 7. Preliminary data for the prototype EM calorimeter. Left: Uniformity for the sum of three rows of towers for a horizontal scan across the central row of towers with 8 GeV/c electrons, Middle: Longitudinal uniformity for a vertical scan along the calorimeter module with 120 GeV/c protons, Right: Energy linearity for the calorimeter module in two different angular orientations corresponding to the plates normal to the beam (90°) and in the nominal position the would be in for sPHENIX (20°).

3. Summary and Conclusions

The PHENIX Experiment at RHIC is planning a major upgrade, sPHENIX that will include two new calorimeter systems. One will be a hadronic calorimeter that will provide a measurement of the total energy of jets produced in heavy ion collisions for the first time at RHIC, and a new electromagnetic calorimeter that will provide an independent measurement of the electromagnetic energy. Both calorimeters will cover a region of $\pm$ 1.1 units in pseudorapidity and 2π in phi. The requirements on the energy resolution for the two calorimeters is not particularly stringent due to the large background in measuring the jet energy from the fluctuations in the energy deposited in the calorimeter from the underlying event. The energy resolution requirement for measuring jets in the hadron calorimeter is ~ $100\%/\sqrt{E}$ and ~ $15\%/\sqrt{E}$ for the electromagnetic calorimeter. Prototypes of both calorimeters were built and tested in a test beam at Fermilab in February of 2014. Preliminary results from the beam test have been obtained and the analysis of the data is continuing. The sPHENIX experiment will also form the basis for a new detector for eRHIC, a future Electron Ion Collider at Brookhaven, which would include a suite of additional detectors for measuring polarized ep and eA collisions.

4. References

[1] *sPHENIX: An Upgrade Concept from the PHENIX Collaboration*, arXiv:1207.6378v1 [nucl-ex].

[2] *Concept for an Electron Ion Collider (EIC) Detector Built Around the BaBar Solenoid*, arXiv:1402.1209v1 [nucl-ex].

[3] O. Tsai, *Development of a Forward Calorimeter System for the STAR Experiment*, contribution to these proceedings (CALOR2014)

[4] C.Woody and E.Kistenev, *The Calorimeter Systems for the sPHENIX Experiment at RHIC*, Proceedings of the XV International Conference on Calorimetry in High Energy Physics (CALOR2012), Journal of Physics Conference Series, December 2012.

[5] Labspehere, Inc., North Sutton, NH.

Journal of Physics: Conference Series **587** (2015) 012055 doi:10.1088/1742-6596/587/1/012055

The Next Generation of Crystal Detectors

Ren-Yuan Zhu
256-48, HEP, California Institute of Technology, Pasadena, CA 91125, USA

E-mail: zhu@hep.caltech.edu

Abstract. Crystal detectors have been used widely for decades in high energy and nuclear physics experiments, medical instruments and homeland security applications. Novel crystal detectors are continuously being found. Future HEP experiments require bright and fast crystal detectors with excellent radiation hardness. Cost-effectiveness is also a crucial issue for crystal detectors to be used in a large volume. To face these new challenges a thorough R&D program is required to investigate and develop crystal detectors for future HEP experiments in all frontiers.

1. Introduction

Crystal detectors have been used widely in high energy and nuclear physics experiments, medical instruments and homeland security applications. Novel crystal detectors are continuously being discovered and developed in academia and industry.

Table 1 Existing Crystal Calorimeters in High Energy Physics

Date	75-85	80-00	80-00	80-00	90-10	94-10	94-10	95-20
Experiment	C. Ball	L3	CLEO II	C. Barrel	KTeV	BaBar	BELLE	CMS
Accelerator	SPEAR	LEP	CESR	LEAR	FNAL	SLAC	KEK	CERN
Crystal Type	NaI:Tl	BGO	CsI:Tl	CsI:Tl	CsI	CsI:Tl	CsI:Tl	PWO
B-Field (T)	-	0.5	1.5	1.5	-	1.5	1.0	4.0
r_{inner} (m)	0.254	0.55	1.0	0.27	-	1.0	1.25	1.29
Crystal number	672	11,400	7,800	1,400	3,300	6,580	8,800	76,000
Crystal Depth (X_0)	16	22	16	16	27	16 to 17.5	16.2	25
Crystal Volume (m^3)	1	1.5	7	1	2	5.9	9.5	11
Light Output (p.e./MeV)	350	1,400	5,000	2,000	40	5,000	5,000	2
Photo-detector	PMT	Si PD	Si PD	WS + Si PD	PMT	Si PD	Si PD	Si APD
Gain of Photo-detector	Large	1	1	1	4,000	1	1	50
σ_N/Channel (MeV)	0.05	0.8	0.5	0.2	Small	0.15	0.2	40
Dynamic Range	10^4	10^5	10^4	10^4	10^4	10^4	10^4	10^5

Journal of Physics: Conference Series **587** (2015) 012055 doi:10.1088/1742-6596/587/1/012055

In high energy physics (HEP) and nuclear physics (NP) experiments, total absorption electromagnetic calorimeters (ECAL) made of inorganic crystals are known for their superb energy resolution and detection efficiency for photon and electron measurements [1]. A crystal ECAL is thus the choice for those experiments where precision measurements of photons and electrons are crucial for their physics missions. Examples are the Crystal Ball NaI:Tl ECAL, the L3 BGO ECAL and the BaBar CsI:Tl ECAL in lepton colliders, the kTeV CsI ECAL and the CMS PWO ECAL in hadron colliders and the Fermi CsI:Tl ECAL in space. Table 1 lists existing crystal calorimeters in high energy physics.

For future HEP experiments at the energy and intensity frontiers, however, the crystal detectors used in the above mentioned ECALs are either not bright and fast enough, or not radiation hard enough. Crystals have also been proposed to build a Homogeneous Hadron Calorimeter (HHCAL) to achieve unprecedented jet mass resolution by duel readout of both Cherenkov and scintillation light [2], where development of cost-effective crystal detectors is a crucial issue because of the huge crystal volume required [3]. This paper discusses several R&D directions for the next generation crystal detectors for future HEP experiments.

2. Performance of PWO Crystals

Table 1 shows that the CMS lead tungstate ($PbWO_4$ or PWO) crystal calorimeter, consisting of 76,000 crystals of 11 m^3, is the largest crystal calorimeter ever built. Because of its superb energy resolution and detection efficiency, the CMS PWO ECAL has played an important role for the discovery of the Higgs boson by the CMS experiment [4]. One crucial issue is crystal's radiation damage in the severe radiation environment at LHC, which requires precision monitoring to correct variations of crystal transparency [5]. After two years of operation up to 70% loss of light output was observed in the CMS PWO crystals at large rapidity *in situ* at LHC as shown in Figure 1 when the LHC was running at a luminosity of up to 5×10^{33} $cm^{-2}s^{-1}$ and a half of its designed energy [6]. The damage in PWO crystals increases when the luminosity increases, and recovers during LHC stops.

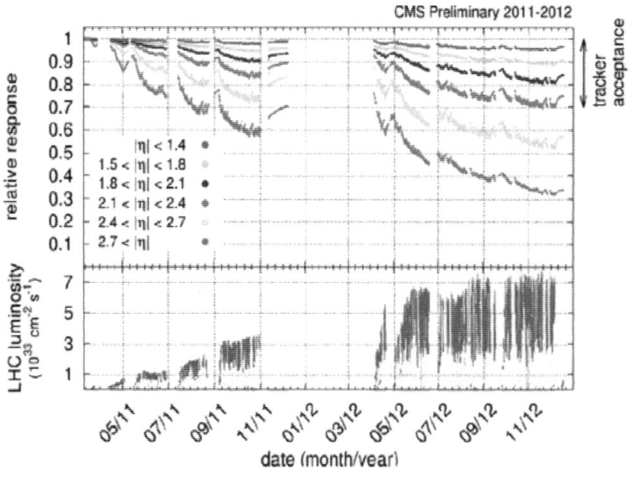

Figure 1. Monitoring response of CMS PWO crystals observed during LHC run I.

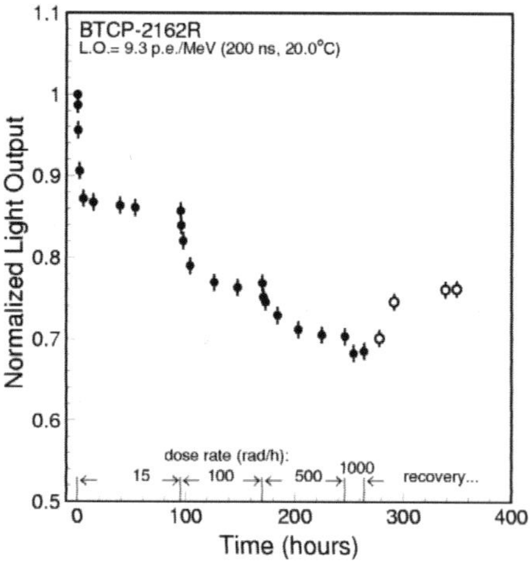

Figure 2. Dose rate dependent light output loss observed in a PWO crystal

The radiation damage of PWO crystals shown in Figure 1 is well understood. It is caused by radiation induces absorption, or color center formation, and is dose rate dependent [7]. Figure 2 shows that light output of a PWO crystal reached an equilibrium during irradiation under a defined dose rate, indicating a dose rate dependent radiation damage caused by colour center dynamics. At equilibrium the speed of the color center formation (damage) equals to the speed of the color center annihilation (recovery), so that the colour center density, or the radiation induced absorption, does not change unless the applied dose rate changes [7].

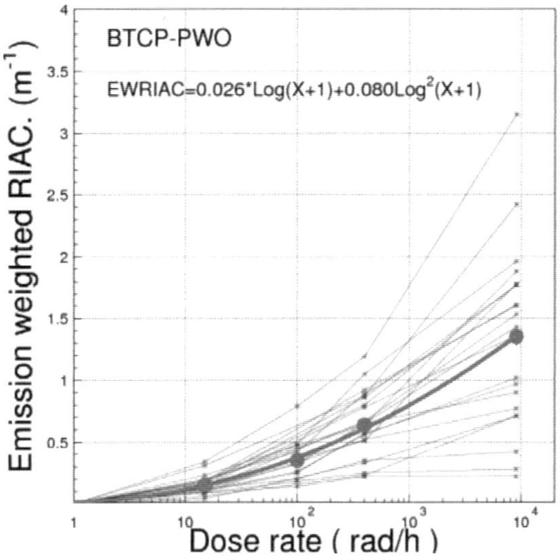

Figure 3. EWRIAC values are shown as a function of dose rate for mass produced PWO

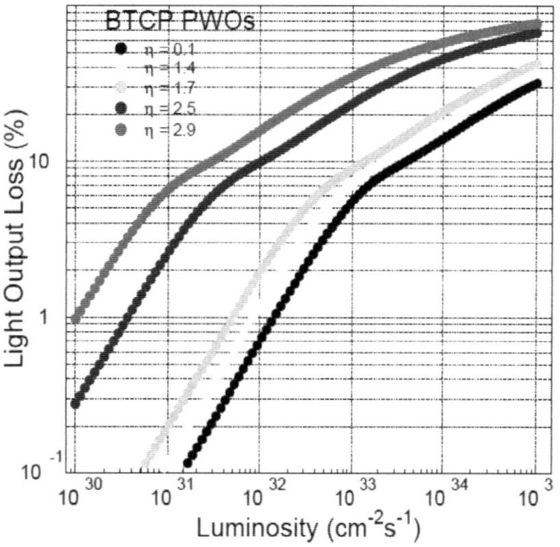

Figure 4. Predicated light output loss as a function of luminosity and rapidity

Figure 3 shows emission weighted radiation induced absorption coefficient (EWRIAC) as a function of the γ-ray dose rate measured for a batch of mass-produced BTCP PWO crystals [8]. The EWRIAC values of all crystals are less than 1 m^{-1} up to 400 rad/h, indicating no damage to the light response uniformity for PWO crystals used in the CMS ECAL barrel, where the maximum dose rate is expected to be a few hundreds rad/h even if the LHC luminosity is increased by a factor of ten, e.g. at HL-LHC. The EWRIAC values measured at 9,000 rad/h, however, are diverse. Some samples show up to 3 m^{-1}, indicating possible damages in the light response uniformity and thus the energy resolution. Figure 4 shows the expected light output loss as a function of luminosity for PWO crystals at different rapidity predicted in 2010 before LHC running [9]. This prediction was made according to the measured relation between light output and radiation induced absorption assuming the average EWRIAC values shown in red colour in Figure 3. A detailed comparison shows that this prediction agrees well with the data shown in Figure 1, indicating that radiation damage in PWO crystals observed so far is caused by ionization dose.

Additional damage caused by charged hadrons was also studied [10]. Because of these damages the CMS endcap PWO ECAL is proposed to be replaced by using more radiation hard technologies [11], one of which is a LYSO crystal based Shashlik sampling calorimeter [12].

3. Radiation Hard LYSO/LSO Crystals

Because of their high density (7.4 g/cm^3), short radiation length (1.14 cm), fast (40 ns) and bright (4 times BGO) scintillation, cerium doped lutetium oxyorthosilicate (Lu_2SiO_5:Ce, LSO) [13] and lutetium yttrium oxyorthosilicate ($Lu_{2(1-x)}Y_{2x}SiO_5$:Ce, LYSO) [14, 15] crystals have attracted a broad interest in the high energy physics community pursuing precision electromagnetic calorimeter for future high energy physics experiments [12, 16-19]. Their excellent radiation hardness against gamma-rays [15, 20], neutrons [17] and charged hadrons [10] also makes them a preferred material for calorimeters to be operated in a severe radiation environment, such as the HL-LHC.

The top plot of Figure 5 shows the photo-luminescence spectra of a LYSO crystal measured before (blue) and after (red) 1 Mrad γ-ray irradiation [20]. To facilitate a comparison, these spectra were normalized to the area between 380 and 460 nm under the spectra. The corresponding relative difference is shown in the bottom plot of Figure 5. The bin by bin average of the absolute difference between the spectra measured before and after the irradiation is found to be 0.6% in the normalization region, less than the systematic uncertainty for this measurement. This indicates that the γ-ray irradiation does not affect the scintillation mechanism in LYSO crystals.

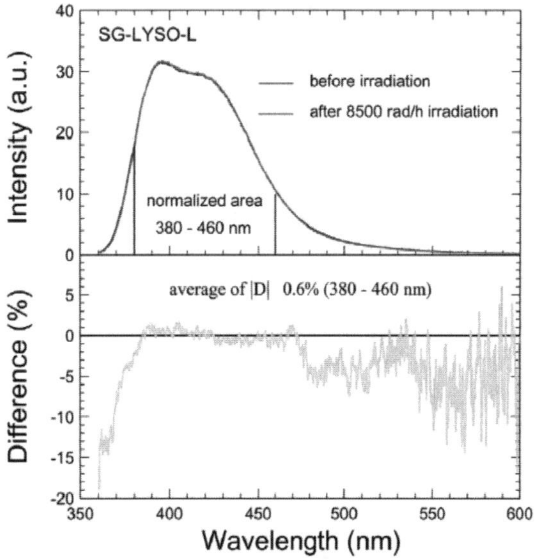

Figure 5. Photoluminescence of a LYSO crystal before and after 1 Mrad γ-ray irradiation

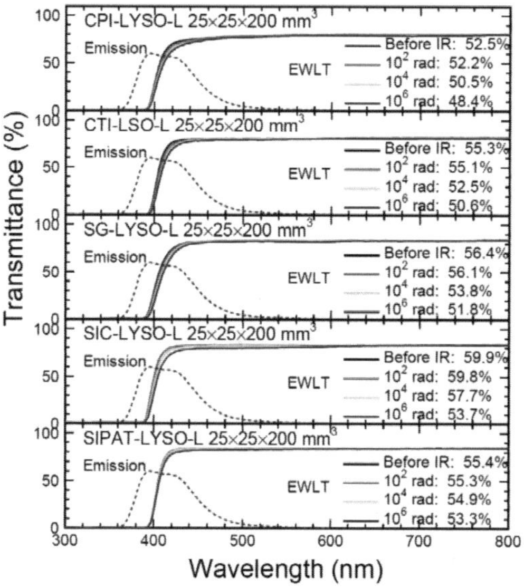

Figure 6. LT of 20 cm long LYSO crystals before and after irradiations

Figure 6 shows the longitudinal transmittance spectra for five 200 mm long LYSO crystals from different vendors before and after several steps of the γ-ray irradiation with integrated dose of 10^2, 10^4 and 10^6 rad. Also shown in the figure is the corresponding numerical values of the photo-luminescence weighted longitudinal transmittance (EWLT), which is defined as:

$$\mathrm{EWLT} = \frac{\int \mathrm{LT}(\lambda)\mathrm{Em}(\lambda)d\lambda}{\int \mathrm{Em}(\lambda)d\lambda}. \quad (1)$$

EWLT represents crystal's transparency better than the transmittance at the emission peak since it runs through the entire emission spectrum. This is particularly important for crystals with self-absorption nature, i.e. a part of scintillation is self-absorbed in the crystal [17]. Consistent damages are observed on the longitudinal transmittance of all LYSO crystal samples. It is noted that the degradation of longitudinal transmittance is small even after 10^6 rad.

Figure 7 shows light output (LO) and light response uniformity (LRU) for a 280 mm long LYSO crystal before and after several steps of the γ-ray irradiations with integrated dose of 10^2, 10^4 and 10^6 rad [21]. About 14% LO loss is observed after 1 Mrad irradiation with LRU maintained. Figure 8 shows LO and decay time for a LYSO plate of 25x25x5 mm grown by SIC before and after several steps of the γ-ray irradiations with integrated dose of 10^5, 10^6 and 10^7 rad. A LO degradation of 3.4% was observed after 10 Mrad, indicating significant reduction of radiation damage because of short light path. These results as well as the irradiation by 24 GeV protons of up to a fluence of 10^{14} cm^{-2} [10] demonstrate the excellent radiation hardness of LYSO crystals.

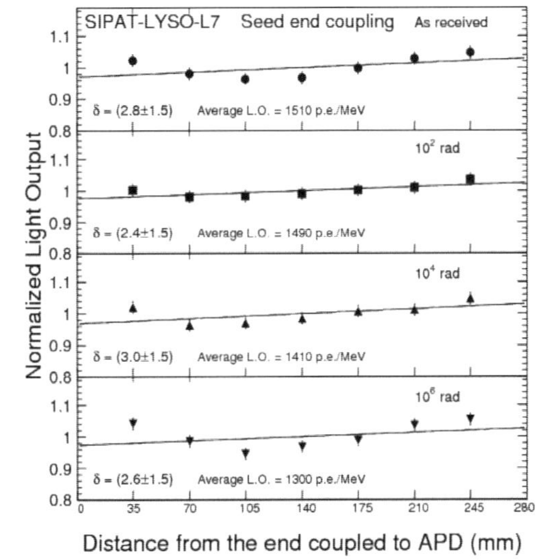

Figure 7. LO and LRU measured by APD for a 28 cm long SIPAT LYSO crystal

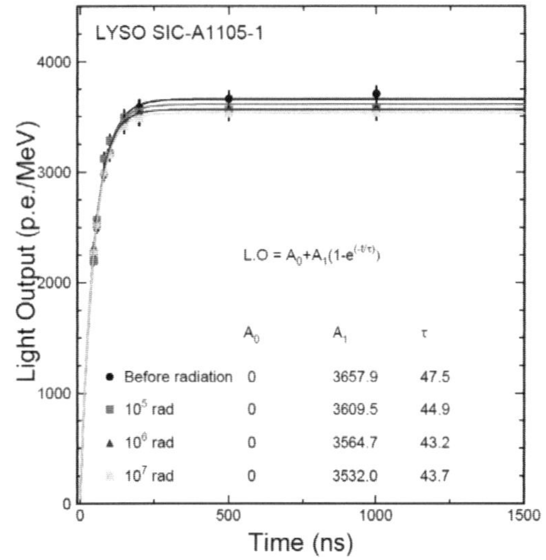

Figure 8. LO and decay kinetics measured by PMT for SIC LYSO plate

Figure 9 shows a LYSO crystal based Shashlik calorimeter detector concept with four evenly distributed wavelength shifting (WLS) fibers for readout and a monitoring fiber at the center [12]. This detector concept reduces the crystal volume and cost, and improves the radiation hardness of the calorimeter because of the much reduced light path. As shown in Figure 9, the initial design consists of 30 LYSO plates of 1.5 mm thick and 29 W plates of 2.5 mm thick. Each tower has a depth of 25 X_0 to accommodate electrons and photons with energies up to the TeV range. The sampling fraction was chosen to be around 20% to provide an adequate stochastic term of the energy resolution at a level of 10%. Because of the high density of both the LSO/LYSO crystals and the absorber materials, the average radiation length (0.51 cm) and Moliere radius (1.3 cm) are much smaller as compared to commonly used crystal scintillators. This detector concept thus also provides a very compact calorimeter to mitigate the pile-up effect expected at the HL-LHC. Figure 10 shows a photograph of LYSO/W Shashlik tower with four Y-11 WLS fibers coupled to a readout PMT. A beam test for a LYSO/W Shashlik matrix consisting of sixteen towers is being carried out at Fermilab.

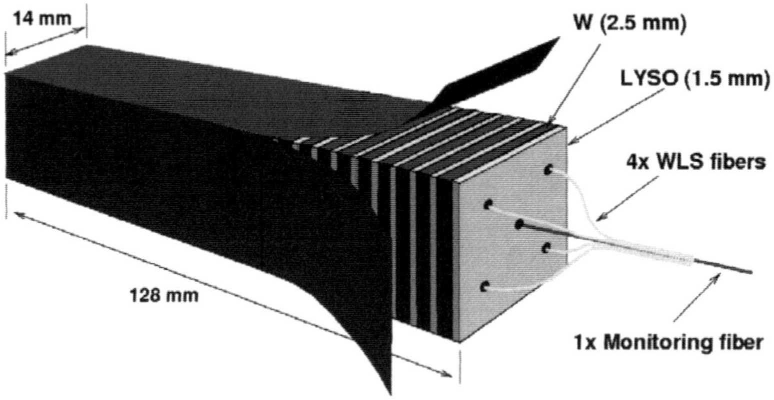

Figure 9. A schematic showing a LYSO/W Shashlik calorimeter concept consisting of LYSO plates of 1.5 mm thick and tungsten plates of 2.5 mm thick, 4 WLS fibers and one monitoring fiber.

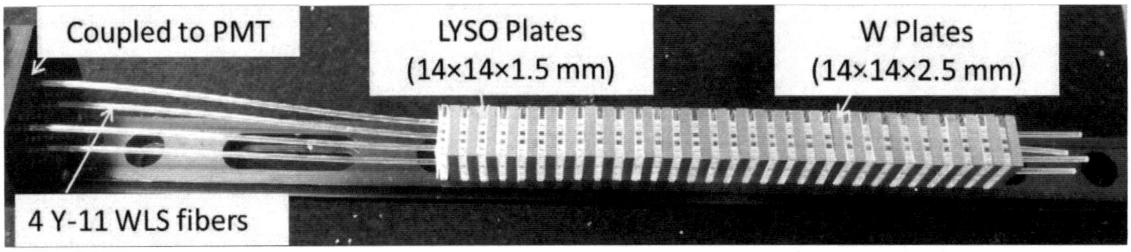

Figure 10. A photo showing a LYSO/W Shashlik tower with four WLS fibers inserted

Figure 11 shows a set-up used to measure longitudinal response uniformity for the LYSO/W Shashlik detector concept. A 14×14×1.5 mm^3 LYSO plate was moved along four Y-11 WLS fibers at five points of 2 cm apart. The LO was measured by using a γ-ray source shooting to the LYSO plate. Figure 12 shows the measured result and the corresponding fit for two straight lines using first and last three points. The corresponding slope of the fit is 0.2%/X_0 at the front, which is less than 0.3%/X_0 required to maintain good energy resolution [22]. The slope at the back is 0.7%/X_0, which is close to the optimum of 8% rise required in the last 10 X_0.

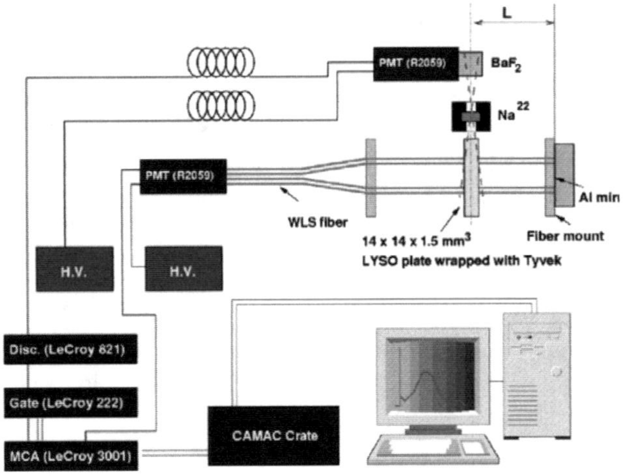

Figure 11. A setup used to measure the longitudinal uniformity for a LYSO/W Shashlik tower

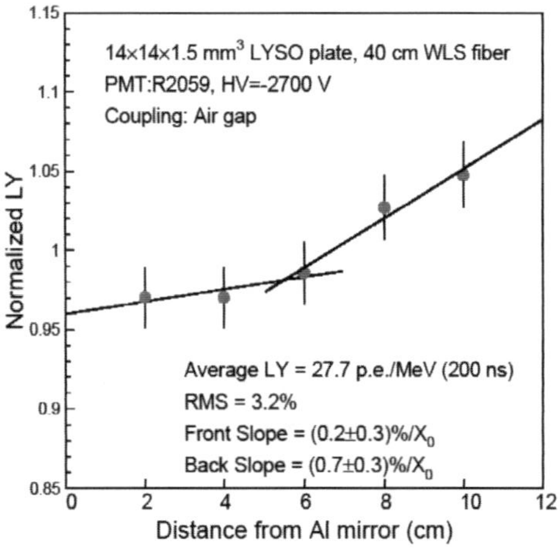

Figure 12. The longitudinal response uniformity of a LYSO/W Shashlik tower

4. Alternative Fast Crystals

The high cost of LYSO crystals caused by high Lu_2O_3 price, however, may limit their use in future HEP experiments. Table 2 lists basic optical scintillation properties for alternative fast crystal detectors with scintillation decay time ranged from sub-nanosecond to a few tens nanosecond, and compared to plastic scintillator. Among the fast crystals listed in Table 2 mass-production cost of barium fluoride (BaF_2) and pure CsI crystals is significantly lower than others because of their low raw material cost and low melting point. At this point BaF_2 is baselined for the Mu2e experiment with pure CsI as an alternative option [23]. For applications in severe radiation environment, such as HL-LHC, one of the crucial issues for their application is radiation hardness. Investigation of radiation hardness for alternative fast crystals would provide important input for future HEP experiments at the energy and intensity frontiers. On the other hand, cost-effective fast scintillating glass and ceramics may also be considered and developed.

Table 2 Basic Properties of Fast Crystal Scintillators

	LSO/LYSO	GSO	YSO	CsI	BaF_2	CeF_3	$CeBr_3$ [1]	$LaCl_3$	$LaBr_3$	Plastic scintillator (BC 404) [2]
Density (g/cm³)	7.40	6.71	4.44	4.51	4.89	6.16	5.23	3.86	5.29	1.03
Melting point (°C)	2050	1950	1980	621	1280	1460	722	858	783	70#
Radiation Length (cm)	1.14	1.38	3.11	1.86	2.03	1.70	1.96	2.81	1.88	42.54
Molière Radius (cm)	2.07	2.23	2.93	3.57	3.10	2.41	2.97	3.71	2.85	9.59
Interaction Length (cm)	20.9	22.2	27.9	39.3	30.7	23.2	31.5	37.6	30.4	78.8
Z value	64.8	57.9	33.3	54.0	51.6	50.8	45.6	47.3	45.6	-
dE/dX (MeV/cm)	9.55	8.88	6.56	5.56	6.52	8.42	6.65	5.27	6.90	2.02
Emission Peak[a] (nm)	420	430	420	420 310	300 220	340 300	371	335	356	408
Refractive Index[b]	1.82	1.85	1.80	1.95	1.50	1.62	1.9	1.9	1.9	1.58
Relative Light Yield[a,c]	100	45	76	4.2 1.3	42 4.8	8.6	141	15 49	153	35
Decay Time[a] (ns)	40	73	60	30 6	650 0.9	30	17	570 24	20	1.8
d(LY)/dT [a,d] (%/°C)	-0.2	-0.4	-0.1	-1.4	-1.9 0.1	~0	-0.1	0.1	0.2	~0

a. Top line: slow component, bottom line: fast component.
b. At the wavelength of the emission maximum.
c. Relative light yield normalized to the light yield of LSO
d. At room temperature (20°C)
#. Softening point

1. W. Drozdowski et al. ***IEEE TRANS. NUCL. SCI***, VOL.55, NO.3 (2008) 1391-1396
Chenliang Li et al, ***Solid State Commun***, Volume 144, Issues 5–6 (2007),220–224
http://scintillator.lbl.gov/
2. http://www.detectors.saint-gobain.com/Plastic-Scintillator.aspx
http://pdg.lbl.gov/2008/AtomicNuclearProperties/HTML_PAGES/216.html

4.1 Radiation hardness of large size BaF_2 crystals

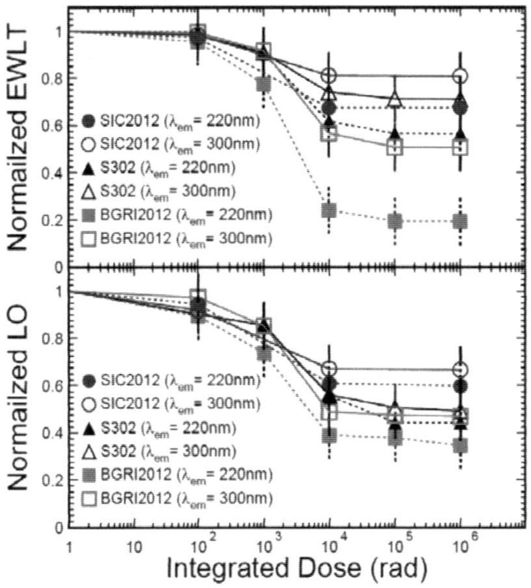

Figure 13. Normalized EWLT and LO as a function of integrated dose for 3 BaF_2 crystals

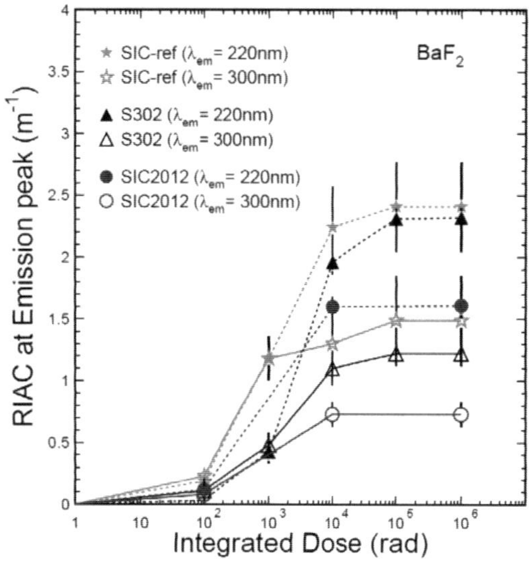

Figure 14. RIAC as a function of integrated dose for three long BaF_2 crystals

Figure 13 shows normalized emission weighted longitudinal transmittance (EWLT, top) and light output (LO, bottom) as a function of integrated γ-ray dose for three 250 mm long BaF_2 crystals. While SIC2012 and BGRI2012 were grown by SIC and BGRI respectively in 2012, the crystal S302 was grown by SIC twenty years ago during nineties for the SSC. The result shows that SIC2012 is more radiation hard than other samples, indicating an improvement of crystal quality. It was also found that the slow component (300 nm) of all crystals is more radiation hard than the fast component (220 nm). BaF_2 crystals also show stable damage after 10 krad, indicating limited defect density in these crystals which was fully exhausted after 10 krad. This promises a stable BaF_2 crystal calorimeter in a severe radiation environment. Figure 14 shows radiation induced absorption coefficient (RIAC) as a function of integrated dose for three BaF_2 crystals, indicating that RIAC of mass produced BaF_2 may be controlled to less than 1.6 m^{-1} for the fast component. Note, RIAC is independent of sample size, so can be compared to other crystals.

4.2 Radiation hardness of large size pure CsI crystals

Figure 15 shows normalized EWLT and LO as a function of integrated γ-ray dose for large size CsI crystals. SIC2013 (30 cm long) and SIC2011 were grown by SIC in 2013 and 2011 respectively. The crystals CsI-3 (20 cm long) and CsI-4 (19 cm long) were grown by Kharkov with data published in *Nucl. Ins. Meth. A* 326 (1993) 508-512. Consistent radiation damage between these crystals is observed. Figure 15 also shows that radiation damage in CsI is small below 10 krad, but degrades continuously with no sign of saturation at high dose, indicating high defect density in the crystals. Figure 16 shows RIAC as a function of integrated dose, which is longer than 3 m^{-1} after 1 Mrad.

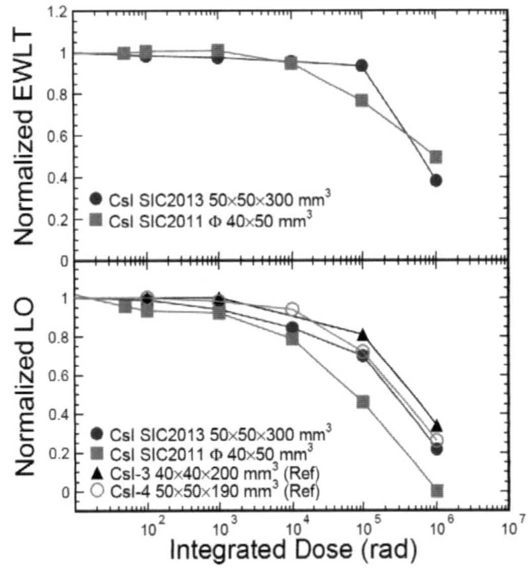

Figure 15. Normalized EWLT and LO as a function of integrated dose for 3 CsI crystals

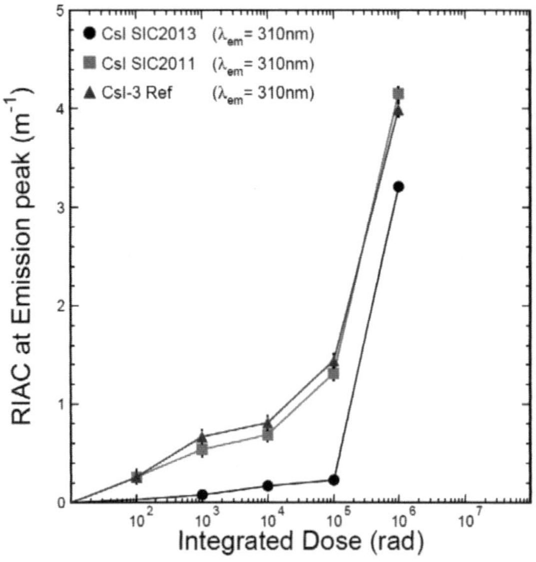

Figure 16. RIAC as a function of integrated dose for three CsI crystals

4.3 Comparison of radiation hardness between BaF_2, pure CsI and LYSO

Figure 17 and 18 show normalized EWLT, LO and RIAC as a function of integrated dose for crystals with dose rate independent radiation damage, i.e. BaF_2, pure CsI and LYSO. Because of no recovery these crystals promise a more stable calorimeter than that with dose rate dependent radiation damage, such as PWO. LYSO crystals show clearly the best radiation hardness. Among two crystals of low cost, the radiation hardness of BaF_2 is good at high dose and that of CsI is good at low dose.

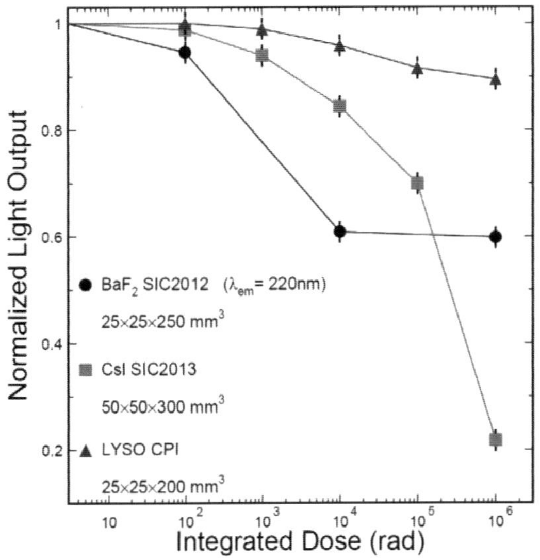

Figure 17. Normalized EWLT and LO as a function of dose for BaF_2, CsI and LYSO

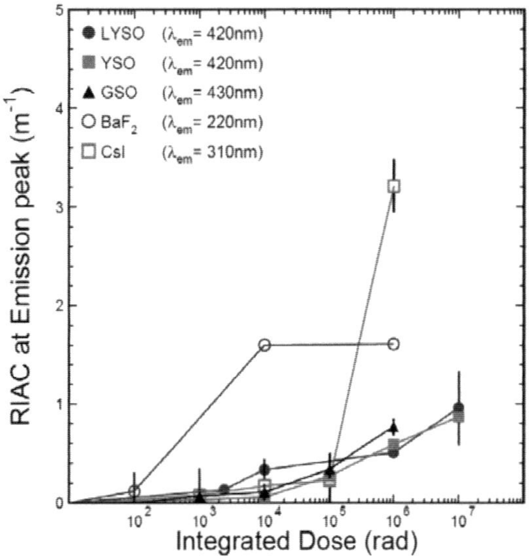

Figure 18. RIAC as a function of integrated dose for crystals with dose rate independent damage

4.4 Time resolution of crystal scintillation

Crystal time resolution is important for many applications. It depends on the signal to noise ratio for the rise time measurement. While scintillation light is known to have very different decay time for various crystals, the intrinsic rising time of most crystals is as fast as tens ps [24]. Figure 19 shows the rising time measured by using a Hamamatsu R2059 PMT for ten crystal samples of $1.5X_0$ size. The fast rise time of about 1.5 ns observed for BaF_2, LYSO, CeF_3 and BGO is dominated by the PMT rise time 1.3 ns (2500 V) and the rise time of 0.14 ns of the Agilent MSO9254A (2.5 GHz) DSO. The measured rise time values are also reduced for the same crystal with black wrapping, indicating effect of the light propagation in the crystal [25].

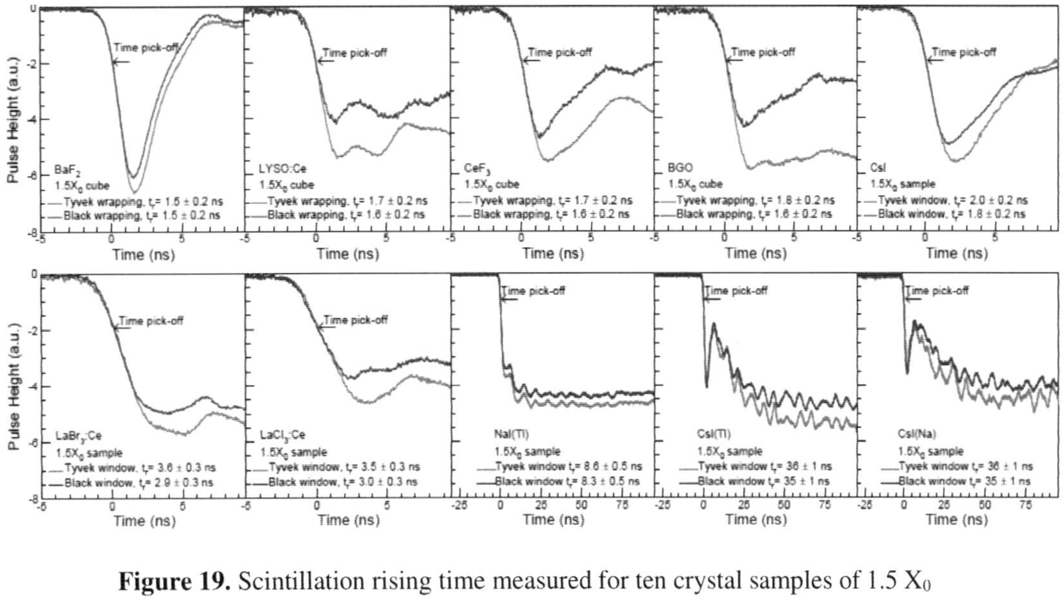

Figure 19. Scintillation rising time measured for ten crystal samples of 1.5 X_0

Table 3 lists the values of the figure of merit for time resolution for various crystal detectors, which is defined as the light output in the 1st, or the 1st 0.1, ns [25]. It is clear that the best crystal scintillators for ultra-fast timing are BaF_2, LSO:Ca,Ce and LSO/LYSO:Ce. $LaBr_3$ is a material with high potential theoretically, but suffers from scattering centers in the crystal as well as its intrinsic hygroscopicity.

Table 3 Figure of Merit for Time Resolution for Various Crystal Scintillators

Crystal Scintillators	Relative LY (%)	A_1 (%)	τ_1 (ns)	A_2 (%)	τ_2 (ns)	Total LO (p.e./MeV, XP2254B)	LO in 1ns (p.e./MeV, XP2254B)	LO in 0.1ns (p.e./MeV, XP2254B)	LY in 0.1ns (photons/MeV)
BaF_2	40.1	91	650	9	0.9	1149	71.0	11.0	136.6
LSO:Ca,Ce	94	100	30			2400	78.7	8.0	110.9
LSO/LYSO:Ce	85	100	40			2180	53.8	5.4	75.3
CeF_3	7.3	100	30			208	6.8	0.7	8.6
BGO	21	100	300			350	1.2	0.1	2.5
PWO	0.377	80	30	20	10	9.2	0.42	0.04	0.4
$LaBr_3$:Ce	130	100	20			3810	185.8	19.0	229.9
$LaCl_3$:Ce	55	24	570	76	24	1570	49.36	5.03	62.5
NaI:Tl	100	100	245			2604	10.6	1.1	14.5
CsI	4.7	77	30	23	6	131	7.9	0.8	10.6
CsI:Tl	165	100	1220			2093	1.7	0.2	4.8
CsI:Na	88	100	690			2274	3.3	0.3	4.5

5. Crystals for the HHCAL Detector Concept

Aiming at the best jet-mass resolution cost-effective inorganic crystal scintillators are being developed for a homogeneous hadron calorimeter (HHCAL) detector concept with dual readout of both Cherenkov and scintillation light for future high energy lepton colliders [2]. Because of the unprecedented volume (70 to 100 m^3) foreseen for the HHCAL detector concept cost-effectiveness is the most important requirement [3]. In addition, the material must be dense to reduce the calorimeter volume, UV transparent for effective collection of the Cherenkov light, and allow for a clear discrimination between the Cherenkov and scintillation light. The preferred scintillation light is thus at a longer wavelength, and not necessarily bright nor fast. Inorganic crystals being investigated are doped lead fluoride (PbF_2) [3, 26], lead chloride fluoride (PbFCl) [27-29] and bismuth silicate ($Bi_4Si_3O_{12}$ or BSO) [30-32].

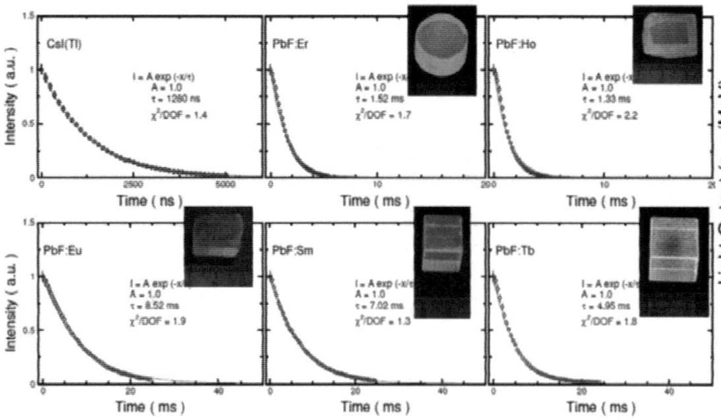

Figure 20. Decay kinetics of doped PbF_2 exited by UV light

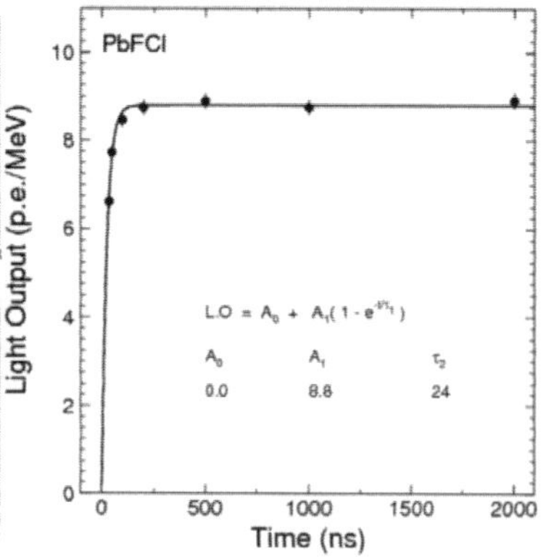

Figure 21. Decay kinetics of a PbFCl sample excited by γ-rays

Figure 20 shows UV light excited decay kinetics of doped PbF_2 samples [26]. The decay time of rare earth doped PbF_2 crystals is several milliseconds which is too long to be used for HHCAL. Figure 21 shows decay kinetics of PbFCl crystal excited by gamma rays. Its 24 ns decay time is appropriate for the HHCAL application. Because the layer structure, however, PbFCl crystal of large size is hard to grow.

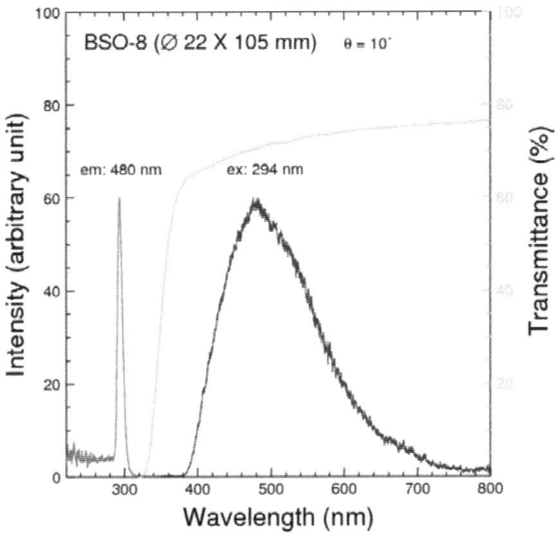

Figure 22. LT and PL of BSO crystal

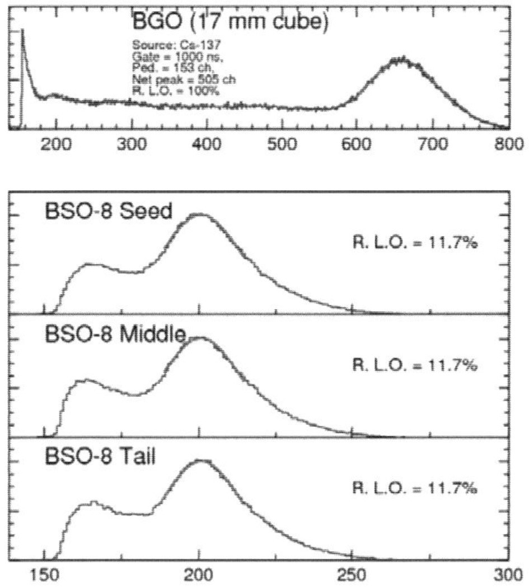

Figure 23. Relative LO of BSO crystal

Because of its low UV cut-off wavelength (300 nm) and low raw material cost (<50% of BGO), BSO crystals are under development at Shanghai Institute of Ceramics (SIC). Figure 22 shows longitudinal transmittance (LT) and photoluminescence (PL) spectra for a 105 mm long BSO crystal. Figure 23 shows relative light output of the BSO crystal comparing to BGO crystal. The LO of BSO with 100 ns decay time is approximately 12% of BGO crystal. See reference [33] for recent development for this crystal.

6. Summary

Bright, fast and radiation hard LYSO/LSO crystals may be used for a total absorption ECAL. LYSO/W Shashlik calorimeter is one of two options for the CMS FCAL upgrade technical report for the proposed HL-LHC. Crystal calorimeters with more than ten times faster rate/timing capability require very fast crystals, e.g. sub-ns decay time of the BaF_2 fast scintillation component. Crystals (PbF_2, PbFCl and BSO) may provide a foundation for a homogeneous hadron calorimeter with dual readout for both Cherenkov and scintillation light to achieve unprecedented jet mass resolution for future lepton colliders.

Acknowledgments
This work was supported in part by the US Department of Energy Grant DE-FG03-92-ER40701.

References

[1] Zhu R Y, 2012, *Physics Procedia* **37**, 372.
[2] Driutti A, Para A, Pauletta G, Briones N R, and Wenzel H, 2010, *J. Phys.: Conf. Ser.* (Beijing, China), **293** 012034.
[3] Mao R H, Zhang L Y, and Zhu R Y, 2012, *IEEE Trans. Nucl. Sci.*, **59**, 2229-2236.
[4] Chatrchyan S, Khachatryan V, Sirunyan A M, Tumasyan A, Adam W, Aguilo E, *et al.*, 2012, *Physics Letters B*, **716**, 30

[5] Zhu R Y, 2004, *IEEE Trans. Nucl. Sci.,* **51**, 1560
[6] T. T. de. Fatis on behalf of the CMS Collaboration, 2012, *J. Phys.: Conf. Ser.*(Sanat Fe, USA), **404**, 012002
[7] Zhu R Y, 1998, *Nucl Instrum Meth A,* **413**, 297
[8] Mao R H, Zhang L Y, and Zhu R Y, 2004, *IEEE Trans. Nucl. Sci.,* **51**, 1777
[9] Zhu R Y, *http://www.hep.caltech.edu/~zhu/talks/ryz_101209_pwo.pdf.*
[10] Dissertori G, Luckey D, Nessi-Tedaldi F, Pauss F, Quittnat M, Wallny R, *et al.*, 2014, *Nucl Instrum Meth A,* **745**, 1
[11] Biino C, *in these Proceedings.*
[12] Zhang L Y, Mao R H, Yang F, and Zhu R Y, 2014, *IEEE Trans. Nucl. Sci.,* **61**, 483
[13] Melcher C L and Schweitzer J S, 1992, *IEEE Trans. Nucl. Sci.,* **39**, 502
[14] Kimble T, Chou M, and Chai B H T, 2002, *Proc. IEEE Nuclear Science Symp. Conf.*, vol 3 (Norfolk, Virginia, USA), p 1434.
[15] Cooke D W, McClellan K J, Bennett B L, Roper J M, Whittaker M T, Muenchausen R E, *et al.*, 2000, *J Appl Phys,* **88**, 7360.
[16] Chen J M, Mao R H, Zhang L Y, and Zhu R Y, 2007, *IEEE Trans. Nucl. Sci.,* **54**, 718.
[17] Chen J M, Zhang L Y, and Zhu R Y, 2005, *IEEE Trans. Nucl. Sci.,* **52**, 3133.
[18] Eigen G, Zhou Z, Chao D, Cheng C H, Echenard B, Flood K T, *et al.*, 2013, *Nucl Instrum Meth A,* **718**, 107.
[19] Pezzullo G, Budagov J, Carosi R, Cervelli F, Cheng C, Cordelli M, *et al.*, 2014, *J Instrum,* **9**, C03018.
[20] Chen J M, Mao R H, Zhang L Y, and Zhu R Y, 2007, *IEEE Trans. Nucl. Sci.,* **54**, 1319.
[21] Mao R H, Zhang L Y, and Zhu R Y, 2012, *IEEE Trans. Nucl. Sci.,* **59**, 2224.
[22] Graham G and Seez C, *CMS Note 1996-002.*
[23] Pezzullo G, *in these Proceedings.*
[24] Derenzo S E, Weber M J, Moses W W, and Dujardin C, 2000, *IEEE Trans. Nucl. Sci.,* **47**, 860.
[25] Zhu R Y, *http://www.hep.caltech.edu/~zhu/talks/ryz_110428_time_resolution.pdf.*
[26] Mao R H, Zhang L Y, and Zhu R Y, 2010, *IEEE Trans. Nucl. Sci.,* **57**, 3841.
[27] Yang F, Zhang G Q, Ren G H, Mao R H, Zhang L Y, and Zhu R Y, 2014, *IEEE Trans. Nucl. Sci.,* **61**, 489.
[28] Zhang G Q, Yang F, Wu Y T, Sun D D, Shang S S, and Ren G H, 2014, *J Inorg Mater,* **29**, 162.
[29] Chen J M, Shen D Z, Ren G H, Mao R H, and Yin Z W, 2004, *J Phys D Appl Phys,* **37**, 938.
[30] Fei Y T, Fan S J, Sun R Y, Xu J Y, and Ishii M, 2000, *Prog Cryst Growth Ch,* **40**, 189.
[31] Barysevich A, Dormenev V, Fedorov A, Glaser M, Kobayashi M, Korjik M, *et al.*, 2013, *Nucl Instrum Meth A,* **701**, 231.
[32] Ishii M, Harada K, Senguttuvan N, Kobayashi M, and Yamaga I, 1999, *J Cryst Growth,* **205**, 191.
[33] Yang F, Yuan H, Zhang L and Zhu R Y, *in these Proceedings.*

New detecting techniques for a future calorimetry

E. Auffray[a], O. Buganov[b], A.Fedorov[c], M. Korjik[c], P. Lecoq[a], G.Tamulaitis[d], S.Tikhomirov[b], A.Vasil'ev[e]

[a]CERN, Switzerland
[b]Stepanov Institute of Physics, Belarus
[c]Institute for Nuclear Problems, Belarus
[d]Vilnius University, Lithuania
[e]Scobeltcyn Institute of Nuclear Physics, Belarus

E-mail: mikhai.korjik@cern.ch

Abstract. In the last forty years, application of crystalline materials in homogeneous Electromagnetic Calorimeters has played a crucial role in the discovery of matter properties and promoted a continuous progress in the detecting technique. The detection systems progressed from small detectors based on NaI(Tl), CsI(Na), BaF_2, PbF_2, and $Bi_4Ge_3O_{12}$ to giant Electromagnetic Calorimeters of CMS, ALICE Collaborations at LHC and PANDA Collaboration at FAIR, where the systems consisted of thousands lead tungstate $PbWO_4$ scintillation crystals. Lead tungstate (PWO) became the most extensively used scintillation material in high energy physics experiments. PWO possesses a unique combination of scintillation properties including high energy and time resolutions in the detection of high energy particles. Here, we report on the results of the two photon absorption in PWO crystals obtained by pump-probe technique using ultra short laser pulses. The results demonstrate that the relaxation processes in PWO offer capability of this material to be used in detection systems to make a time stamp with precision close to 10^{-12} s or even better.

1. Introduction

There is an upcoming demand for a new generation of detectors enabling a high time resolution in high energy physics experiments. To develop the new generation of ultrafast detectors, especially in experimental particle physics at high luminosity colliders, the processes permitting the measurement precision better than 10-20 ps for the interaction time between ionizing radiation and detecting medium need to be identified. In order to reduce the interaction region and have a compact experimental setup in high energy physics experiments, detectors with short radiation X_0 and nuclear interaction R_H lengths and small Moliere radius R_M are on demand. Among the variety of inorganic scintillation materials meeting these requirements, lead tungstate ($PbWO_4$, PWO) self-activated scintillation material possesses a unique combination of scintillation and other physical properties [1].

It ensures high energy and time resolutions, especially in electromagnetic calorimetry [2-4]. However, even the self-activated scintillation materials like PWO with fast scintillation decay (10 ns) and rise (0.1ns) time constants has limited time resolution due to stochastic origin of luminescence processes. The available data indicate that scintillation materials can not provide a better time resolution than 50 ps. The authors of [5] considered another options to improve the time resolution by exploiting of the transient phenomena, which appear at the early stages of the interaction between ionizing radiation and crystalline material. These processes occur when the density of states has peculiarities, such as a gap in the bottom of the conduction band. In such conditions, hot intra-band radiative and absorption transitions occur in the conduction band. These processes take place before free carriers are captured by the luminescent centers and/or shallow electron traps. The time scale of these processes is considerably shorter than the scintillation rise time of a self-activated scintillation crystal and does not exceed 10-30 ps. However, these transient phenomena can be reliably observed at high free carrier density, thus they can be utilized at the conditions of a large and high speed energy deposit in the material. Such conditions are achievable in collider experiments at LHC or ILC [6,7]. Thus, a search for fast processes with a weaker or indirect dependence on the free carrier density is a necessary. Two photon absorption in crystalline media is one of such processes. The effect is proportional to the laser pulse power squared and is relatively easy observed by picosecond laser spectroscopy in many media [8]. It is expected that electromagnetic shower might induce the polarization of the crystalline media and change the conditions for two photon absorption.

In the present work, we studied two photon absorption in PWO single crystals. The band gap E_g of PWO at room temperature is 4.33 eV, so a defect-free PWO crystal does not absorb in the visible range. Two photon absorption in the range of photon energy between 3 and 4 eV results in excitation to the electronic states in the bottom part of the conduction band. The bottom of the conduction band is formed by d orbitals of tungsten ions. Moreover, the density of electronic states has two peaks near 4 and 6 eV [9]. The first peak corresponds to the states of triplet terms $^3T_1$, $^3T_2$, whereas the peak near 6eV is formed with strong admixture of $^1T_1$, $^1T_2$ states. Thus, effective two photon absorption due to the transitions from the ground state which is the upper $^1A_1$ state of the valence band to the $^1T_2$ state of the conduction band is expected in the range of 6eV.

2. Experimental technique and results

A 10 mm thick sample of $PbWO_4$ crystal with perfect optical quality and oriented along the crystallographic axis *a* was selected for our measurements. The experiment was carried out using a pump-probe spectrometer based on a custom-made original femtosecond Ti:Al_2O_3 pulsed oscillator and a regenerative amplifier, both operating at 10 Hz repetition rate [10]. The laser pulse duration and energy after amplification were 140 fs and up to 0.5 mJ, respectively, while the wavelength was tunable over the spectral range from 770 to 820 nm. The pulses of the fundamental frequency (ω) at the output of the amplifier (790 nm wavelength was set for the present study) were divided into two parts in a 1:4 ratio (Figure 1). The beam of higher intensity was converted to the third harmonic ($\lambda \approx$ 263 nm, E up to 12 μJ) and used as the pump pulse. The pulse energy was chosen to prevent the sample surface from optical damage by the laser pulse. The second beam of smaller intensity was used as a probe. After passing the delay line, it was converted to a white supercontinuum generated by focusing the beam into a 1cm long water-containing cell. By using a semi-transparent mirror, the supercontinuum radiation (360-1500 nm) was subdivided into two pulses (reference and signal) of similar intensity. Both pulses were focused on the sample by mirror optics. The reference pulse is required to eliminate the impact of shot-to-shot instability of supercontinuum. It passes the sample always before the pump pulse. The pump-induced change of the optical density was calculated as:

$$\Delta D(\lambda, \Delta t) = \lg\left[\left(E_{sg}(\lambda)/E_{ref}(\lambda)\right)/\left(E^*_{sg}(\lambda, \Delta t)/E_{ref}(\lambda)\right)\right],$$

where E_{sg}, E_{sg}^*, and E_{ref} are the energies of signal pulses passed the sample before and after the pump pulse, and the reference pulse energy, respectively.

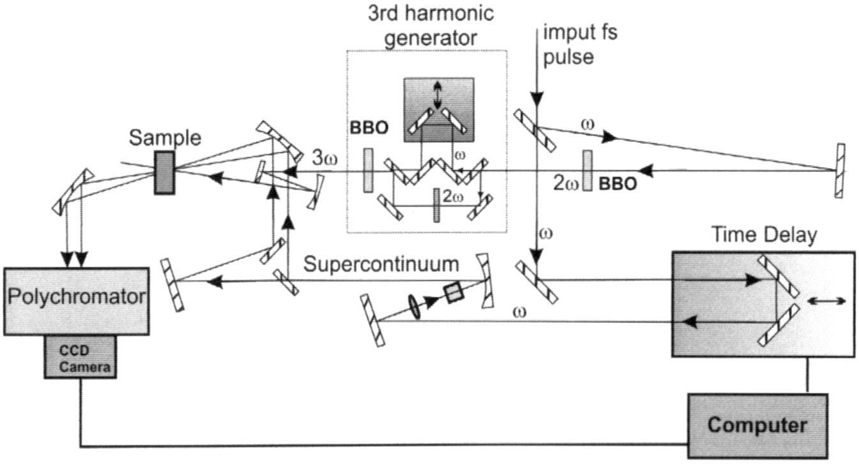

Figure 1: Set up of the femtosecond spectrometer used in this work.

The sample surfaces were carefully polished to prevent the surface damage during the illumination with femtosecond pulses. The beam spot diameter was 2 mm. The spectra of both pulses were recorded for each laser shot by a system including a polychromator equipped with a CCD camera and digitally processed. The absorption spectra of the white supercontinuum were measured in the spectral range from 400 to 730nm.

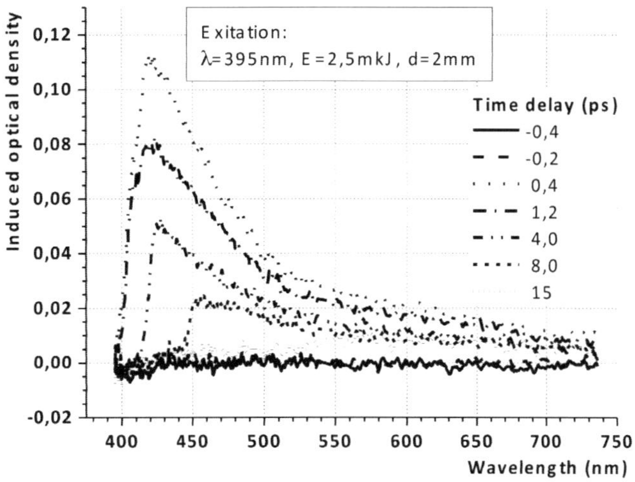

Figure 2: Transient induced absorption (ΔD) spectra of a PWO single crystal at different time delays between the pump and probe pulses. The pump pulse energy is 5 μJ.

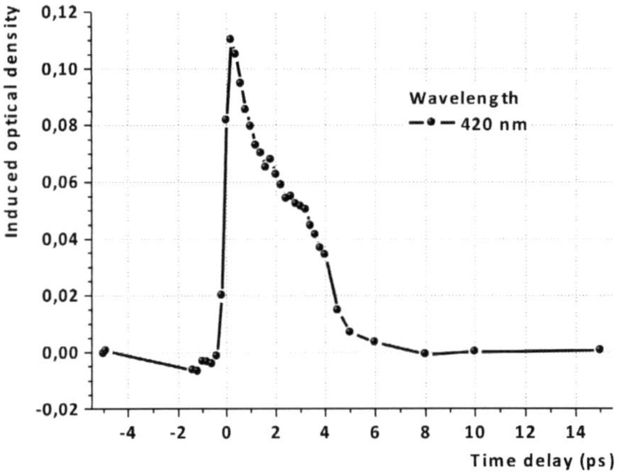

Figure 3: The two photon absorption kinetics of a PWO single crystal at probe wavelength of 420nm.

The simultaneous exposition of the sample to the pump pulse (the second harmonic at 395 nm, 3.13 eV) and the probe pulse with a supercontinnum spectrum results in strong two-photon absorption.. The effect is practically instantaneous. Thus, the time profile of the induced absorption determined in our experiments by the apparatus function of the detection system. Figure 2 shows the spectra of the induced absorption at different time delays. The time evolution of the probe at 420 nm is presented in Figure 3. This time evolution is connected with the dispersion of the group velocity of the probe pulse in the sample under study.

The data obtained show that the maximal efficiency of the two photon absorption in our experiments corresponds to the combination of the 395 nm (3.15 eV) pump pulse and the pulse probe with the closest to the pump pulse wavelength of 400 nm (3.13eV). This perfectly coincides with the second maximum in the spectrum of the density of electronic states described above. As the probe pulse wavelength is increased, the efficiency of the two photon absorption decreases and approaches zero at 500 nm. The energy corresponding to two photon absorption of photons with energies of 3.15 eV (395nm) and 2.5 eV (500 nm) corresponds to the minimum in the spectrum of the density of electronic states in the conduction band of PWO.

To exploit the observed ultrafast two photon absorption in development of precise time stamps related to the interaction of ionizing radiation with the PWO crystal for fast radiation detectors, further study of the fast response of two-photon absorption, especially under ionizing radiation, is required. Combination of the pump and probe experiments with the impact of picosecond X-ray tube emission or electron beam would be highly informative. The PWO scintillation could probably also be used as the probe. In this case, only the pump laser pulses would be required to ensure the time stamp. Moreover, the simultaneous absorption of two photons of the same energy optimally selected in the vicinity of 3 eV would strongly simplify the future application of this technique in the detector systems. In real applications, a set of several hundreds of short laser pulses, 1-5 ps in duration, within the total duration of 500-1000 ps is required, and has to be synchronized with the particle collisions to scale the time intervals between the probe pulses at the level of several picoseconds. The frequency of the laser pulse sets should be similar to the frequency of the particle collisions, i.e., approximately 40 MHz. Such detection systems can be assembled using commercially available lasers and optical components.

3. Conclusions

We observed two photon absorption of picosecond laser pulses in the extensively used PWO scintillation crystal. These processes are considered to be sensitive to polarization of the PWO single crystal under ionising radiation. An ultrafast change in the two photon absorption conditions due to the crystal polarisation are prospective to be used for development of precise time stamps of the interaction of ionizing particles with the crystal in high energy physics experiments at a picosecond time scale.

References

[1] Lecoq, P., Annenkov, A., Gektin, A., Korzhik, M., Pedrini, Inorganic Scintillators for Detector Systems, Springer, 2006.

[2] CMS Collaboration, Search for the standard model Higgs boson decaying into two photons in pp collisions at $\sqrt{s}$ = 7 TeV, Phys. Lett. B, 710 (2012) 403-425.

[3] R.Novotny et.al., Nucl. Instr. Methods A 648(2011)77

[4] Daniele del Re , Proceedings of CALOR2014, Report O1.04, 7-11 April 2014, Giessen, Germany

[5] P.Lecoq, M.Korzhik, A.Vasiliev, Can Transient Phenomena help improving Time Resolution in Scintillators?, IEEE Tras. Nucl. Sci., 61 (2014) 229-234.

[6] The CERN Large Hadron Collider: Accelerator and Experiments, Vol. 1-2, CERN, Geneva, 2009.

[7] International Linear Collider Technical Design Report, Tokyo, Geneva, Chicago – 12 June 2013.

[8] A. Dragonmir, Jh. G. McInerney, and D. N. Nikogosyan, Femtosecond Measurements of Two-Photon Absorption Coefficients at λ = 264 nm in Glasses, Crystals, and Liquids, Applied Optics, 41(2002) 4365-4376,

[9] R.Williams et.al., Inprganic scintillators and their applications / Ed.V.Mikhailin, Moscow State University, 2000, p.118

[10] .P. Blokhin, M.F. Gelin, O.V. Buganov, V.A. Dubovskii, S.A. Tikhomirov, G.B. Tolstorozhev, Femtosecond Dynamics of Optically Induced Anisotropy of Complex Molecules in the Gas Phase, J. Appl. Spectroscopy, 70 (2003)70–78.

Calorimeters for Precision Timing Measurements in High Energy Physics

Adolf Bornheim[1,2], Artur Apresyan[2], Javier Duarte[2], Cristian Pena[2], Anatoly Ronzhin[3], Maria Spiropulu[2], Si Xie[2]

[2] California Institute of Technology, 1200 E California Blvd., Pasadena, 91125,CA, United States of America
[3] Fermilab, P.O. Box 500, Batavia, IL 60510, USA

Abstract. Current and future high energy physics particle colliders are capable to provide instantaneous luminosities of 1034 cm-2s-1 and above. The high center of mass energy, the large number of simultaneous collision of beam particles in the experiments and the very high repetition rates of the collision events pose huge challenges. They result in extremely high particle fluxes, causing very high occupancies in the particle physics detectors operating at these machines. To reconstruct the physics events, the detectors have to make as much information as possible available on the final state particles. We discuss how timing information with a precision of around 10 ps and below can aid the reconstruction of the physics events under such challenging conditions. High energy photons play a crucial role in this context. About one third of the particle flux originating from high energy hadron collisions is detected as photons, stemming from the decays of neutral mesons. In addition, many key physics signatures under study are identified by high energy photons in the final state. They pose a particular challenge in that they can only be detected once they convert in the detector material. The particular challenge in measuring the time of arrival of a high energy photon lies in the stochastic component of the distance to the initial conversion and the size of the electromagnetic shower. They extend spatially over distances which propagation times of the initial photon and the subsequent electromagnetic shower which are large compared to the desired precision. We present studies and measurements from test beams and a cosmic muon test stand for calorimeter based timing measurements to explore the ultimate timing precision achievable for high energy photons of 10 GeV and above. We put particular focus on techniques to measure the timing with a precision of about 10 ps in association with the energy of the photon. For calorimeters utilizing scintillating materials and light guiding components, the propagation speed of the scintillation light in the calorimeter is important. We present studies and measurements of the propagation speed on a range of detector geometries. Finally, possible applications of precision timing in future high energy physics experiments are discussed.

[1] To whom any correspondence should be addressed. E-mail: bornheim@hep.caltech.edu

1. Introduction

Current and future high energy physics particle colliders are capable to provide instantaneous luminosities of 10^{34} $cm^{-2}s^{-1}$ and above. The high centre of mass energy, the large number of simultaneous collision of beam particles in the experiments and the very high repetition rates of the collision events pose huge challenges. They result in extremely high particle fluxes, causing very high occupancies in the particle physics detectors operating at these machines. To reconstruct the physics events, the detectors have to make as much information as possible available on the final state particles. Aside from the detailed spatial information of the final state particles as well as their momenta and energies one can make use of the relative time of arrival at a given location in the detector. The work presented in this document is inspired by the experimental conditions expected for the high luminosity upgrade of the Large Hadron Collider (HL-LHC) at CERN. At luminosities of 10^{34} $cm^{-2}s^{-1}$ and above at HL-LHC one expects 140 to 200 simultaneous proton-proton collisions at every bunch crossing. The proton bunches will cross every 25 ns inside the experiments. The luminous region inside the detector, the location where the protons collide, has a length of a few 10 cm. Due to the size of the proton bunches and the resulting time it takes them to pass through each other in the luminous region, the collisions are in addition spread out in time. For the LHC machine parameters as in the first LHC run in 2011 and 2012 this spread was about 200 ps. For HL-LHC this spread may increase as a consequence of an advanced beam optics of the interaction region (Fartouk). The multi-purpose detectors ALTAS and CMS are expected to undergo significant upgrades and modifications, the basic geometric layout and size will however remain unchanged. In this work we assume that a detector capable of measuring the time of arrival of a particle would be located at the outer perimeter of the tracking device or in the front part of the calorimeter, as for example in the electromagnetic part of the calorimeter. Under this assumption, the typical time of flight of a particle at the speed of light from the primary vertex of a proton-proton collision to the location of the precision timing device would be in the range of 4 ns to 11 ns in the case of the CMS detector. Finally, in this document we specifically study the possibility to measure the time of arrival of particles with scintillator based calorimeter. The size of the shower of electromagnetically interacting particles at high energies results in a typical time scale of order 1 ns over which the showering process evolves. In Figure 1 we summaries the relevant time scales discussed above.

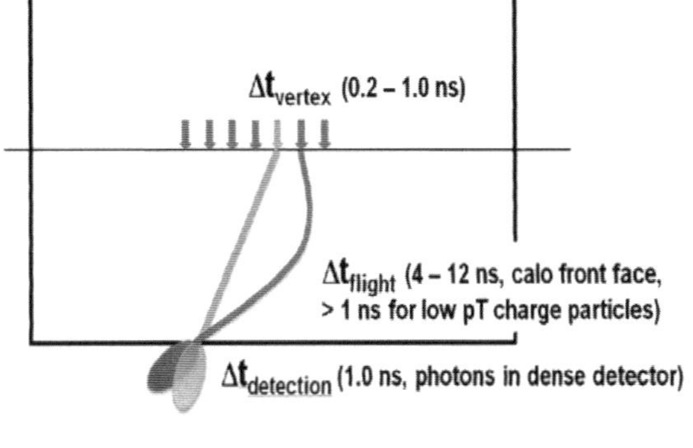

Figure 1 : The time of arrival of a high momentum particle created in a high energy proton collider such as the LHC can be decomposed in three major contributions. The creation time and location varies with respect to the nominal time and location (Δt_{vertex}). The time of flight from the vertex location to the location of the timing measurement may be referred to as the actual time of flight. The characteristic time of the detection process depends on the specific technology choice for the timing measurement

The reason that we focus on electromagnetically interacting particles is that the event reconstruction in the LHC detectors combines the information from tracking devices and calorimeter devices in a very sophisticated fashion. One can assume that the association of any charged particle to its origin in the collision region can be accomplished by the tracking system. Any calorimeter cluster originating from a charged hadron can be associated to an extrapolated track as well and hence to its origin in the collision region. Photons however, as the will not leave tracks, cannot be associated to their production vertex easily. This is where a very precise timing would benefit the event reconstruction tremendously if a precision of a few 10 ps, corresponding to a few mm resolution at the vertex location, can be achieved. To be utilized effectively it seems mandatory that the timing measurement is very closely linked to the energy measurement, ideally based on the same active detector element. It is in this context that we study the possibility to measure the time of arrival of electromagnetic particles with a calorimetric device.

2. Precision Timing Measurements with Scintillating Crystals

The study we present here is using scintillation crystals as an active medium to detect the high energy particles. This choice is driven by the potential use case of a precision timing detector for the Phase II upgrade of the LHC experiments. For this, the measurement of the time of arrival of high energy photons with energies of 1 GeV and above is of particular importance.

In Figure 2 we illustrate the major contributions to the precision of the timing measurement with a monolithic, large scintillating crystal. Upon entering the crystal the photon will need to convert and start showering to induce scintillation light in the crystal. This time span between entering the crystal and the first occurrence of scintillation light we refer to as t_C. The scintillation process may be described by two time constants, the rise time of the scintillation light and its typical decay constant. The rise time of the scintillation light signal from LYSO crystals has been measured to on the order of 75 ps [2]. The result was limited by the experimental setup of the respective measurement. The decay time constant for LYSO is on the order of 40 ns. For high energy particles in the multi GeV range the shower even for electromagnetic particles will extend over several centimetres. This process will take on the order of a nanosecond. With our current experimental setup we cannot de-convolute the time structure of the scintillation process and the propagation of the shower through the crystal. We denote the related characteristic time constant with t_S.

The scintillation light produced needs to be detected. Typically this is done with one or several photo detectors attached to the scintillator. The time t_T it takes the scintillation photons to travel through the crystal may vary substantially due to the isotropic angular distribution of the scintillation light and the possible internal reflection of the light inside the crystal before eventually reaching the photo detector. Finally, the photo detector and its readout system have characteristic time constants t_P and t_D which will affect the precision of the measurement as well.

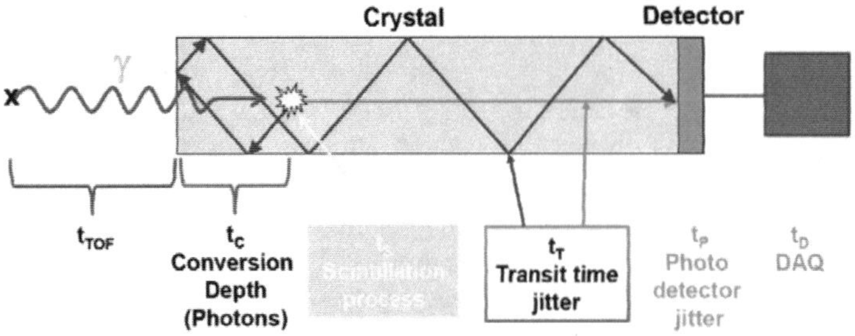

Figure 2 : Time difference between the two MCP photo sensors at the end of a 10 cm long LYSO crystal as measured in the transverse setup. The beam was hitting the crystal about 3cm from one end of the crystal. The time is extracted with a fit to the entire pulse shape. An electron beam was used for this measurement. The electrons were

In this study we focus on studying the effects of the shower development, scintillation light emission and light propagation inside the crystal with their characteristic time constants t_S and t_T. The contribution of the photo detectors and the readout system, t_P and t_D, was studied independently and reported at this conference [3]. In this reference the precision with which the time of flight can be measured with two reference detectors is better than 20 ps. In all results presented here which invoke the MCP based reference detector measurement we did not unfold its contribution to the final result. In the initial studies presented here we focus on measurements which allow to quantify the contribution of the showering process in the crystals, the scintillation light emission and the light propagation inside the crystal. These contributions will dependent on the type of scintillator used, its size and its surface properties. Future experiments will be carried out to study all these factors in detail.

3. Experimental Setup

In this document we report on measurements using cosmic muons, performed at Caltech Lauritsen Laboratory, and test beam measurements performed at the Fermi National Laboratory (Fermilab) M-test beam facility using high energy electrons of 4, 8 and 16 GeV as well as protons of 120 GeV.
For our test beam measurements we used commercially available multichannel plate photo multiplier tubes (MCP PMT or just MCP) from Photek 240 and Hamamatsu R3809U. In the test beam setup the MCP were read out with a wave form digitizer with an analogue bandwidth of 700 MHz and a sampling rate of 5 GS/s (DRS4). For the cosmic measurements at Caltech we used an Agilent MSO9404A oscilloscope with an analogue bandwidth of 4 GHz and up to20 GS/s. As a photo detector for the cosmic setup we used Philips XP2020 photo multipliers. A picture of the experimental setup as used in the Fermilab test beam is shown in Figure 3.
To extract the time of arrival information of a high energy particle from the scintillation light pulses we explored various methods, such as a constant fraction fit on the rising edge of the pulses as well as a fit of a parameterized fit function to the entire pulse. For this again we explored a range of functional forms. Results from a fit to the full pulse shape shown in this document are done with an exponentially modified Gaussian. At the precision we have tested so far we did not find a significant difference among the methods we tested. The observed rise time of the scintillation light signal as measured with the MCP and DRS4 is on the order of 1 ns. We did not find any significant dependency on the particle type inducing the signal, the size of the crystal used or the distance between the shower and the photo sensor. The signal line shape of the reference detector pulse is well described with a Gaussian. The observed sigma of the Gaussian line shape is about 0.5 ns. The details of the scintillation light pulse shape will be studied further in forthcoming measurement campaigns.

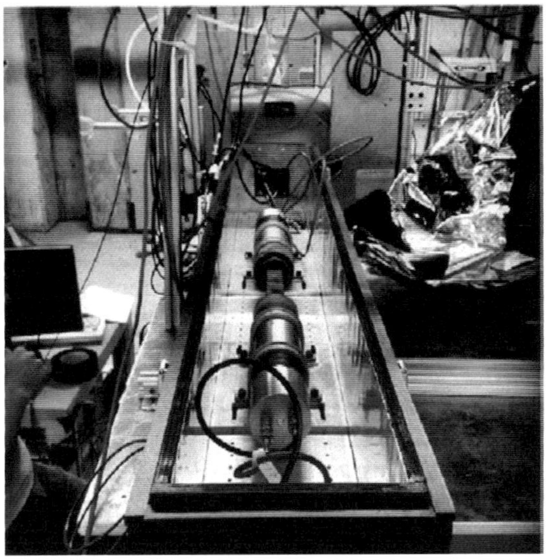

Figure 3 : Experimental setup for the test beam measurements. Inside a light tight box, shielded against environmental noise, a LYSO crystal with photo detectors mounted on both ends is installed. The setup shown is the one used for the Fermilab test beam. The setup for the cosmic test is largely identical. Depending on the configuration reference and trigger counter were installed. The different configurations are described schematically in the following figures.

In Figure 4 we show the conceptual setup of the cosmic ray test bench. The goal for measurements with this setup was to explore the effective propagation speed of the scintillation light inside in a solid LYSO crystal and to explore if there is any dependency on the pulse shape if the location and shape of the shower inside the crystal is varied. To achieve this, we spatially constrained the location at which scintillation light is created by cosmic muons passing through the crystal with two scintillation counters. The scintillation light will then propagate from where it was created through the crystal and be detected at both ends with a photo detector. By scanning along the axis of the crystal we extract the average propagation speed. We also used this setup to do initial studies of the pulse shape and test different time extraction methods.

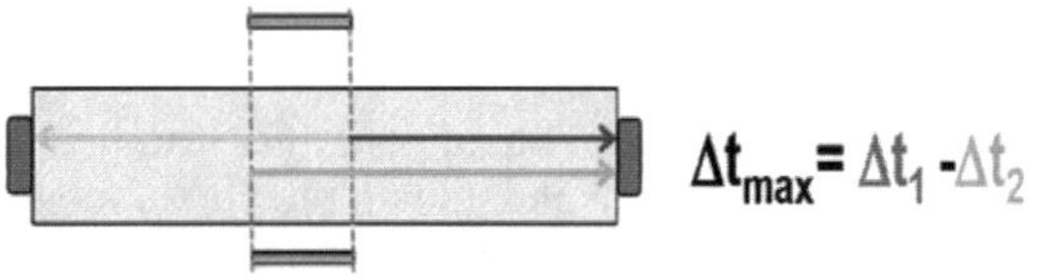

Figure 4 : Time difference between the two MCP photo sensors at the end of a 10 cm long LYSO crystal as measured in the transverse setup. The beam was hitting the crystal about 3cm from one end of the crystal. The time is extracted with a fit to the entire pulse shape. An electron beam was used for this measurement.

For measurements with high energy particles in the Fermilab test beam we used two different configurations. In Figure 5 we show what we refer to as perpendicular setup. The beam is triggered with two wire chambers which allow to spatially constrain the location where the beam particles hit the crystal. A MCP is used as a reference counter and the crystal is read out by two MPC as in the cosmic setup. The goal using this setup was to understand if the spatially much larger showers induced by electromagnetic particles compared to the energy deposition pattern of a muon would influence the pulse shape or the differential time distribution.

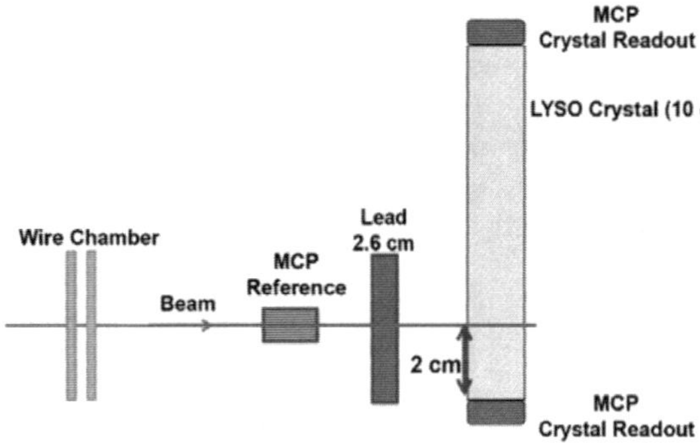

Figure 5 : Transverse setup for the test beam measurements. The LYSO crystal was equipped with two MCP photo detectors of both ends. The beam particles were triggered with two wire chambers, defining an area of about 1.6 cm x 1.6 cm. A third MCP detector was used as a reference counter in front of the crystal setup. A lead sheet with a thickness of 2.6 cm was used for some runs to have the particles create a shower before the crystal.

In Figure 6 we show what we refer to as the longitudinal setup. Here, the crystal is aligned along the axis of the incoming particle beam. The particles have to trespass a reference MCP and the MCP on one side of the crystal readout. We refer to the photo sensor closer to the reference counter as the near side counter, the one on the other side of the crystal as the far side counter. A key difference with this setup is that the two photo sensors are not symmetric with the respect to the location of the shower in the crystal any more. The shower of electromagnetic particles will develop closer to the near side sensor and then propagate through the crystal away from the near side and towards the far side detector. As the scintillation light created in any given location inside the shower has to propagate to the photo sensor, the effective time of arrival at the photo sensor – a convolution of the shower evolution time pattern and the light propagation time pattern – will be different for the far side and the near side.

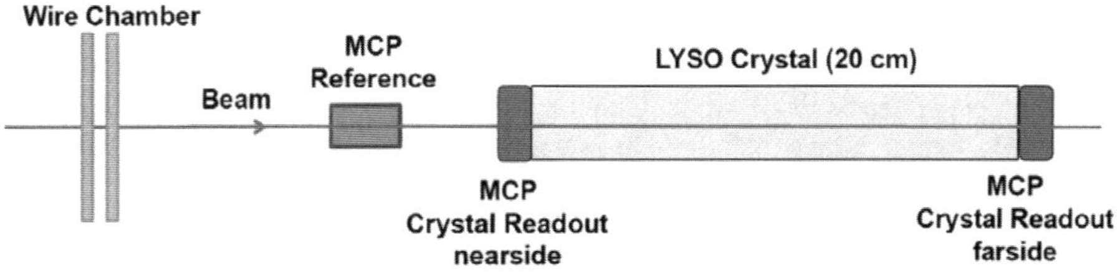

Figure 6 : The longitudinal setup for the test beam measurements. The LYSO crystal was equipped with two MCP photo detectors of both ends. The beam particles were triggered with two wire chambers, defining an area of about 1.6 cm x 1.6 cm. A third MCP detector was used as a reference counter in front of the crystal setup.

In this section we present initial results which reveal the importance of the contributions from the shower evolution and the scintillation light propagation to the measurement of the time of arrival. The results presented in this report are the first ones of a series of such measurements which we plan to carry out.

4.1 Results from the cosmic test setup.

A key result from the cosmic setup is the measurement of the propagation speed of the average scintillation light inside a large, monolithic LYSO crystal. Our cosmic setup was not equipped with a precise reference detector which defines the arrival of the particle inside the crystal. Instead we determine the propagation speed of the light by defining the distance the scintillation light had to propagate through the crystal. Using a setup in which we read out the crystal on two sides we eliminate the trigger jitter. As shown in Figure 7, we measure the propagation speed to be approximately 11 cm/ns or about one third the speed of light. Given the refractive index of LYSO of about 1.82 we conclude that there is a significant effect of the fact that the isotropic emitted scintillation photons do not reach the photo detector on the shortest possible path.
As can be seen in Figure 3 the observed relation of the mean time difference and the location of the scintillation light creation is linear. We did not observe any change in rise time or the pulse shapes if the distance of the source of the scintillation light to the photo detector is varied.

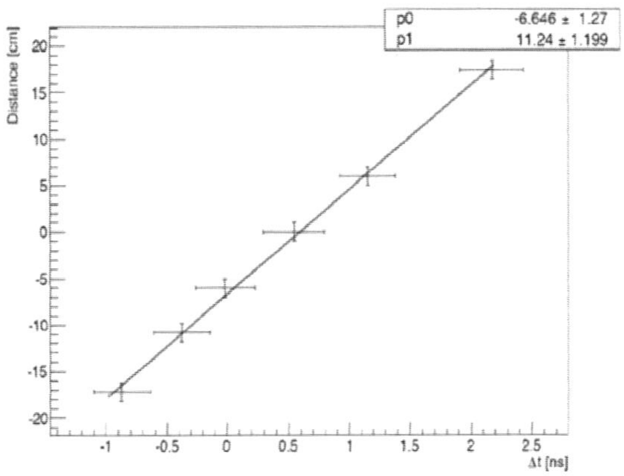

Figure 7 : Relative time difference measured between the two ends of long LYSO crystal. Two XP2020 photo multiplier was used as a photo detector, a high bandwidth oscilloscope as a data acquisition system. The scintillation signals were induced by cosmic muons traversing the crystal transversely in a small area defined with a plastic scintillator trigger counter. The time difference shown is the mean value of a large number of events recorded. From six measurements performed at different locations along the long axis of the crystal we deduce the average propagation speed of the scintillation light to be about 11 cm/ns inside the LYSO crystal.

The uncertainty of the Δt measurement is dominated by the size of the trigger counters which define the spatial precision with which the location of the scintillation light creation is defined. For cosmic muons we assume the energy deposit in the crystal to be uniform along the path of muon through the crystal and predominantly close to it. We have tested two different trigger counter sizes, with a width of approximately 3 cm and 1 cm. The spread of the measured time difference is 300 ps and 160 ps respectively. This dependency clearly reflects that the difference of the time measurement on the two ends of the crystal is very sensitive to the location where the scintillation light is created.

4.2. Results from the perpendicular test beam setup.

The perpendicular test beam setup allows testing the impact of the spatially much wider shower from an electromagnetically interacting particle. In Figure 8 we show the result of the time difference as measured with the perpendicular setup. We measure 120 ps differential time resolution. The size of the trigger counter is approximately 1.6 cm, slightly larger than the 1 cm wide counter used for the cosmic measurement in which we measured 160 ps differential time resolution. As can be seen from Figure 5 a lead sheet was placed in the beam before the crystal. This ensures that the shower is fully developed inside the LYSO crystal. Again we observe a further reduction of the spread in the measured time differential. With a Molière radius of about 1.2 cm (CHECK ME) the width of the shower is of similar width as the width of the trigger counter. We conclude that the dense nature of electromagnetic showers and the large number of scintillation photons created out weights the statistical nature of the showering process and we do not observe an increase in the measured differential time difference.

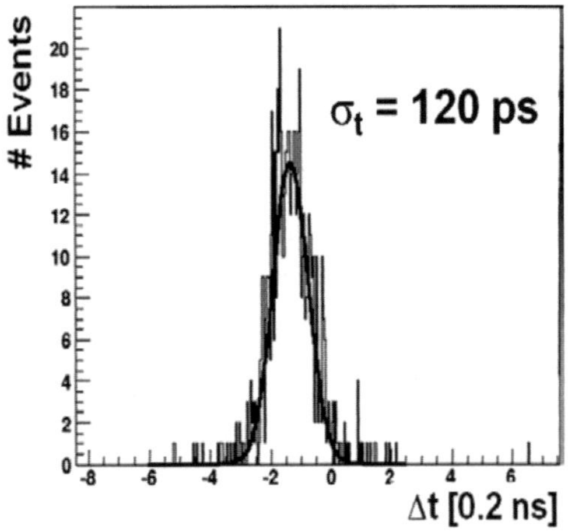

Figure 8 : Time difference between the two MCP photo sensors at the end of a 10 cm long LYSO crystal as measured in the transverse setup. The beam was hitting the crystal about 3cm from one end of the crystal. The time is extracted with a fit to the entire pulse shape. An electron beam was used for this measurement. The electrons were passing through a lead plate with 2.6 cm before hitting the crystal. We measure a time differential between the two ends of the crystal of 120 ps.

We note that the photo detectors and the photo detector readout were different in the perpendicular setup in the test beam compared to the cosmic setup. We used a full pulse shape fit to extract the time from the test beam data while we used a constant ratio fit for the cosmic data. Further we note that with the mean propagation speed measured for the scintillation light of 11 cm/ns the width of the 1.6 cm trigger counter in the test beam setup corresponds to a time difference of approximately 145 ps. We conclude that a significant portion of the observed differential dime difference can be attributed to the geometric spread of the mean shower position, as defined by the size of the trigger counter. Within the precision of or pulse shape sampling we did not observe any significant difference in the mean pulse shapes from the MCP detectors on the two sides of the crystals, even though the beam was not impinging in the middle between the two sensors. From this we conclude that the pulse shape does not depend on the average propagation distance of the scintillation photon inside the crystal. We conclude further that the observed differential time difference for the dual side readout in the cosmic setup as well as in the test beam setup is dominated by the geometric spread in the mean shower position, as defined by the size of the trigger counter. The photo sensor and data acquisition characteristics do not seem to impact the result.

We extracted the time of flight from the perpendicular setup as shown in Figure 9. Shown is the time difference between the reference MCP and the signal on one side of the crystal as detected by the MCP. For this result we use the constant fraction fit to the leading edge of the signal. We measure a resolution of about 60 ps. We do not observe any significant difference between measurements performed with and without an additional lead sheet in front of the crystal. This suggests that enhancing the shower activity in the crystal does not change the measured time resolution. We conclude from this that we are not sensitive to the photo statistics. The resolution we measure is better than the differential time resolution between the two sides of the crystal as shown in Figure 8. One difference is that for the time of flight measurements we use an MCP as a reference while the measurement in Figure 8 compares the two signals extracted from the crystal. Future measurements will be optimized to reduce the impact of the spatial uncertainties of the shower location contributing to the 120 ps measured as in Figure 8. The perpendicular setup will be optimized to better understand the contribution of charged shower particles hitting the MCP detectors. We know from our reference MCP detectors that direct hits of the MCP with particles allows to measure the time of arrival with a precision of around 15 ps. While this would allow a precise measurement of charged particles it is not applicable to photons which is the main target of our study as detailed in the introduction.

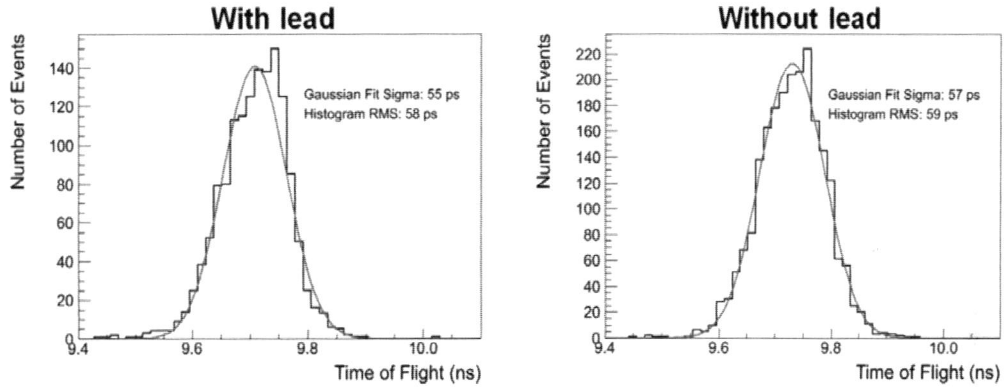

Figure 9 : Time of flight results for the perpendicular setup. The results show the time difference between the reference MCP and the signal extracted from the MCP attached to the crystal. The absolute time difference scale is arbitrary. We observe a resolution of about 60 ps. We perform the measurement with and without an additional lead sheet in the beam in front of the crystal. We do not see any significant change in the time resolution.

4.3. Results from the longitudinal setup

An important difference for the longitudinal setup is that there is no geometric effect due to the size of the trigger counter. However the shower development inside the crystal may fluctuate on an event by even basis. For the time measurement the shower depth is of importance. If a given shower develops deeper inside the crystal it will take longer for the shower reaching the point of maximal intensity and the mean path length of the scintillation light propagating to the near side photo detector will be longer. The inverse holds for the far side photo sensor. We recall that the mean speed of the scintillation light propagation inside the crystal is significantly less than the speed of light. As the shower development inside the crystal propagates with the speed of light the mean depth of the shower also impacts the time of arrival as measured at the far side detector. This effect is visualised in Figure 9 from a GEANT simulation of a single, 16 cm long LYSO crystal. The near and far side crystal face are at -8 cm and +8 cm respectively. A 100 GeV photon enters the crystal from the minus side. We show the mean arrival time of the scintillation and the Cherenkov photons on the near and far side of the crystal as a function of the location of the first conversion of the impinging photon inside the crystal. We see that the photon arrive later on the near side surface if the conversion happens deeper inside the crystal and earlier at the far side surface of the crystal. The effect is more pronounced on the near side crystal surface since the distance between the shower maximum and the crystal surface is smaller and the dispersion in the photon path is smaller. For the scintillation light we observe a delay of about 100 ps per 1 cm additional depth for the first conversion. This matches approximately the 11cm/ns we observed for the propagation speed of the scintillation light inside a long LYSO crystal.

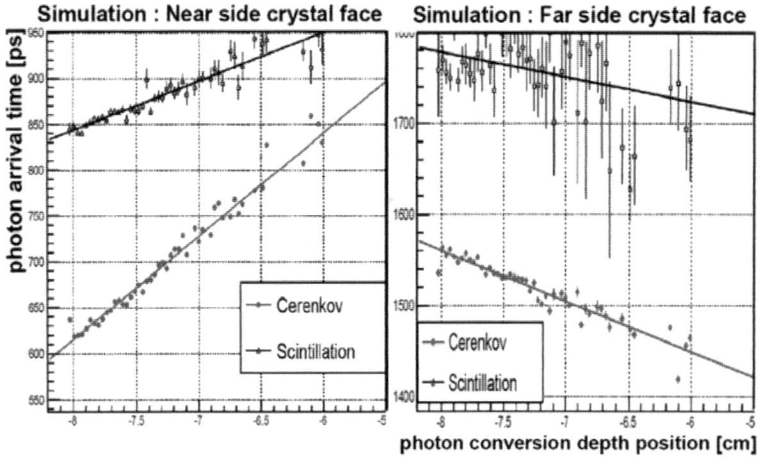

Figure 10 : Simulation of the mean arrival time of the scintillation light from a 100 GeV photon shower in a LYSO crystal. The way axis shows the depth of the first conversion of the photon. The crystal starts at -8 cm on the scale shown. One sees that the conversion depth is correlated with the mean arrival of the scintillation photons at the crystal face.

With the longitudinal test beam setup the goal was to explore if the expected differences with respect to the perpendicular setup can be reproduced. In Figure 11 we show the time difference between the near side and the far side photo detector in the longitudinal setup mounted on a 20 cm long LYSO crystal. We observe a differential time resolution of about 30 ps. As expected we see a further reduction of the differential time resolution compared to the corresponding measurement with cosmic muons and from the perpendicular setup. We attribute this to the fact that the geometric variation of the location of the shower relative to the photo sensor is reduced to the fluctuations of the shower depth. As these fluctuations are typically on the scale of one radiation length the resulting uncertainty is small.

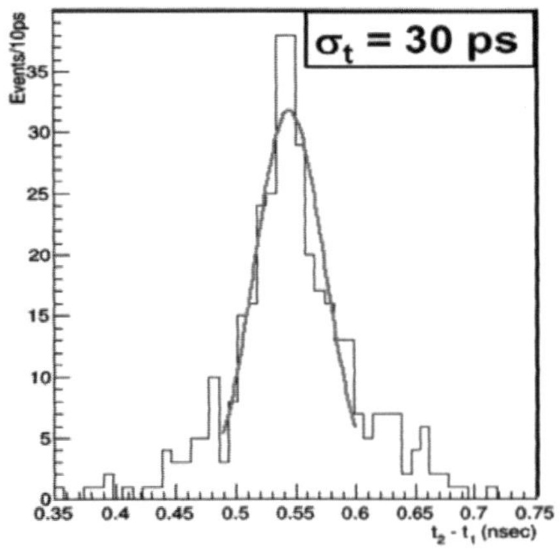

Figure 11 : Time difference between the two MCP photo sensors at the end of a 20 cm long LYSO crystal as measured in the longitudinal setup. The lateral size of the crystal is 2.5 cm x 2.5 cm. The beam was hitting the crystal approximately in the centre. The time is extracted with a constant fraction fit to the rising edge of the pulse shape. An electron beam was used for this measurement. In this setup the time difference between the two sides is not affected by geometric effects due the variations of the impact point. We measure 30 ps differential resolution. We observe some non-gaussian tails in this distribution which we attribute to particles showering before the near side MCP.

If indeed the shower depth is contributing to the measured resolution one would expect to find a correlation as shown in the simulation results in Figure 10. In our initial studies we found indeed indications for such a correlation. We are carrying out further measurements to study in more detail if this observation can indeed be solely explained by the effective path length of scintillation light propagating through the crystal.

5. Test with a small LYSO plate

A key feature of LYSO is its high light yield. This makes a sampling calorimeter with LYSO as active medium a feasible design. Even with a sampling fraction of around 0.1 and a readout of the LYSO plates via fibers the effective light yield for high energy showers is still be large enough to achieve an acceptable stochastic term in the energy resolution. The CMS collaboration at LHC is currently studying the feasibility of a sampling calorimeter with LYSO as active medium and tungsten as absorber. The results we presented above indicate that the achievable time resolution for high energy particles in a large LYSO crystal is significantly impacted by the optical transit time jitter. Shower fluctuations and photo statistics on the other hand are less relevant for multi-GeV electromagnetic showers. An important question is if this will remain true is the shower is sampled with a fraction of 0.1 and the light is guided from the active medium via fiber to the photo sensor. We carried out an initial test with a LYSO plate of 14 mm x 14 mm size and 1.5 mm thickness. Stacked with tungsten absorber plates of approximately 2.5 mm thickness the effective Molière radius of the Shashlik configuration is 14 mm. This is the current design baseline for the CMS ECAL Shashlik prototype. As shown in Figure 12 we read out a single such LYSO plate by bringing one of the sides of the plate into contact with the MCP window. We then exposed the LYSO plate to electron and proton beams.

Figure 12 : Setup for the test measurement of a single LYSO plate. The plate dimensions are 14 mm x 14 mm x 1.5 mm. It is mounted in a plastic holder which fixes the scintillator plate against the window of the MCP.

Results from the initial tests suggest that the rise time observed on the pulses from the LYSO plate is not significantly different from the one observed on larger LYSO crystals. We were able to extract time resolutions on the order of 50 ps with respect to the reference MCP detector. With our initial setup it was not possible to control the shower position on the LYSO plate very well. Also, the signal from the MCP may have contributions from shower particles hitting the sensor directly. Future tests with a refined setup will allow to disentangle the contribution of the signal from direct hits and from scintillation light. We also plan to test the time response of individual LYSO plates inside a full Shashlik configuration as well as the time response of a full Shashlik cell where the light is extracted from the LYSO/tungsten stack via wavelength shifting fibers.

6. Summary

We perform measurements on the timing properties of scintillation light from high energy electromagnetic showers in LYSO scintillating crystals of various sizes. For our measurements we use commercially available Photek and Hamamatsu multichannel plate photo multipliers, read out with DRS4 sampling digitizer. We found that the rise time observed on LYSO samples of different size, for particles of different type and energy does not vary significantly within the precision of our measurements. We measure the average propagation speed of scintillation light to be around 11 cm/ns inside a large monolithic LYSO crystal of 20 cm length. Differential time resolution measurements on a large LYSO crystal, read out by two identical photo sensors, suggest that the scintillation light propagation inside the crystal is a very significant contribution to the measured time of arrival extracted from the scintillation light. Time of arrival measurements with high energy electrons 4 GeV, 8 GeV and 16 GeV suggest that time resolutions of a few 10 ps can be achieved.

[1] A. Bornheim,"Evolution of the CMS ECAL response, R&D studies on new scintillators and possible design options for electromagnetic calorimetry at the HL-LHC", Proceedings of the CHEF 2013 (Calorimetry for the High Energy Frontier) International Conference, Eds. J.-C. Brient, R. Salerno, and Y. Sirois

[2] S Seifert, J H L Steenbergen, H T van Dam and D R Schaart, 2012 JINST 7 P09004. doi:10.1088/1748-0221/7/09/P09004

[3] A. Apresyan, "Development of a New Ultra-fast Shower Maximum Detector Based on Micro Channel Plates as an Active Element", CALOR2014, these Proceedings

[4] A. Ronzhin, S. Los, E. Ramberg, M. Spiropulu, A. Apresyan,, S. Xie, H. Kim, A. Zatserklyaniy; NIM A (doi: 10.1016/j.nima.2014.05.039)

New Fast Shower Max Detector Based on MCP as an Active Element

A. Ronzhin[1], S. Los[1], E. Ramberg[1], M. Spiropulu[2], A. Apresyan[2], S. Xie[2], H. Kim[3], A. Zatserklyaniy[4]

[1] Fermilab, Batavia, IL, USA
[2] California Institute of Technology, Pasadena, CA, USA
[3] University of Chicago, Chicago, IL, USA
[4] University of California, Santa Cruz, CA, USA

E-mail: Artur.Apresyan@cern.ch

Abstract. One possibility to make a fast and radiation resistant shower maximum (SM) detector is to use a secondary emitter as an active element. We present below test beam results, obtained with different types of photo detectors based on micro channel plates (MCP) as secondary emitter. The SM time resolution - we obtained for this new type of detector is at the level of 20-30 ps. We estimate that a significant contribution to the detector response originates from secondary emission of the MCP.

1. Introduction

The first proposal to use secondary emitter materials as an active element in sandwich type calorimeters can be found in the reference [1]. This approach allows the development of a new type of detector that is highly radiation resistant and fast. The measurements with such a type of element were performed with a micro channel plate (MCP) as an electron multiplier. With the recent progress of the Large Area Picosecond Photo Detector collaboration [2], the cost may decrease sufficiently to allow practical construction of this type of detectors. An alternative approach to achieve a calorimeter with presicion timing capabilities is presented in [3].

The principle of operation of the shower maximum detector (or calorimeter) is the following. High-energy particles passing through absorber material in sandwich type calorimeters produce secondary particles (e.g. positron electrons pairs, gammas, etc.) that are detected by MCPs. The generation of secondary particles in the calorimeter will be referred to as a shower. The response of the MCPs to the secondary particles is proportional to their number, which is in turn proportional to the energy of the initial particle.

Most of the secondary particles in the showers have low energy, where the MCPs detection efficiency is high. But the low energy component of the showers is not well investigated, and even reliable Monte Carlo simulation is limited by the minimal energy of the secondary particles to 100 keV. Therefore, to explore the feasibility of a shower maximum detector based on MCPs as the secondary emitter we started with experimental measurements. The goal of the measurements was to obtain time resolution (TR) of the MCP, placed inside of the absorber material. We also performed measurement at 120 GeV proton beam with start and stop counters. The goal was

Journal of Physics: Conference Series **587** (2015) 012058 doi:10.1088/1742-6596/587/1/012058

to estimate what time resolution we can get for the minimum ionizing particles. We did not use any absorber material between the start and the stop counters in the measurement.

2. Description of the MCPs used in the measurements

We used two types of photo detectors based on MCPs. One of them was a Photek 240 with superb time resolution (TR). Main parameters of the Photek 240 were measured and presented in [4]. The Photek 240 has a 41 mm diameter circular sensitive area. The input window of the phototube is made of 9 mm thick quartz which enables extended UV quantum efficiency. The single photoelectron time resolution (SPTR) was measured to be 40 ps. The output signal time shift is limited to 3.9 ps across the sensitive area of the Photek 240.

The second type of the MCP-PMT used was a Photonis 85011-501 which is composed of 64 pads, arranged as an 8x8 matrix. The size of each pad is 6x6 mm^2. 16 central pads (4x4 matrix) were connected together making a sensitive area of 24x24 mm^2. The input window is made of UV Grade Fused Silica with a thickness of 2 mm.

3. Test beam setup

We used the Fermilab Test Beam Facility (FTBF), which provided proton beams from Fermilab's Main Injector accelerator at 120 GeV/c, as well as secondary positron beams with 12 GeV/c and 32 GeV/c. Detectors were located inside of a dark box lined with copper foil for RF shielding (Fig. 1). Trigger was based on a 10x10 mm^2 scintillation counter. The three detectors (two Photek 240 and one Photonis between them) were placed in line. A stack of lead generating a shower, when high energy particle pass through it, was placed before MCP-PMT (Photonis 85011-501 in this case, Fig. 1). The Photonis 85011-501 was swapped with the second Photek 240 in some of the measurements, without any modification to the cabling.

Photek 240 (start counter), upstream Photonis 85011-501 and downstream Photek 240 were placed on the optical table (for normal particle incidence). A stack of lead plates can be seen to the left from Photonis 85011-501 detector. A Cherenkov counter was located before the dark box (in the beam path, not shown in the picture).

Two DRS4 waveform digitizer units [5] performed the main readout. The DRS4 were triggered by TTL level signals originating from the trigger counter. Signals from the four detectors, two Photek 240, the Photonis 85011-501 and the Cherenkov counter were split by high frequency Mini-Circuits ZFRSC-42-S+ splitters (4.2GHz BW). The outputs were connected in the same order to two DRS4 units. We attenuated the input signals to one of the two DRS4 units to cover the full dynamic range of the measurements. The other DRS4 did not have any additional attenuation. The schematic of the readout with the DRS4 units is shown in Fig. 1.

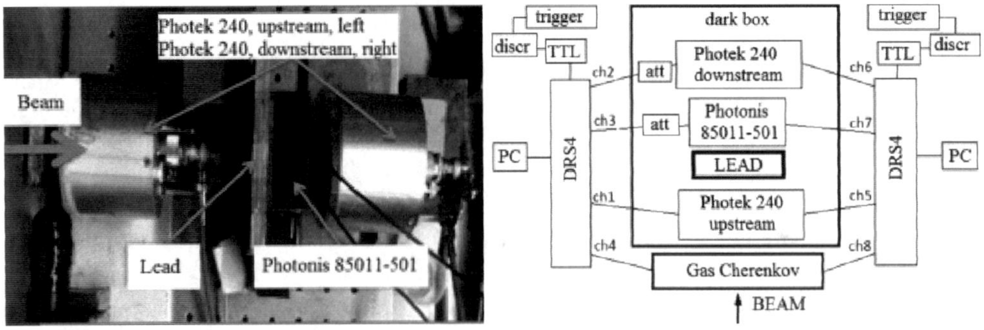

Figure 1. (Left) Test beam setup, located in the dark box. A Cherenkov counter was located before the dark box (not visible in the picture). (Right) A schematic of the readout. Attenuators were placed on channel 2 (CH2) and channel 3 (CH3).

4. Event selection and analysis

To assign a time stamp for each signal pulse, we first determine the time position of the pulse peak. A Gaussian function is fitted to the pulse maximum using three points before the maximum of the pulse peak and four points after the maximum (sample pulse is shown in Fig. 2). The mean value of the Gaussian was used as the time stamp for each pulse. Large signals above 500 mV were also rejected because they saturated the DRS4 inputs. Pulses with an irregular peak profile were rejected, as well as pulses which experienced a sudden reversal of polarity that is occasionally observed with our readout. We selected the pulses with amplitude larger than 20 mV for future analysis.

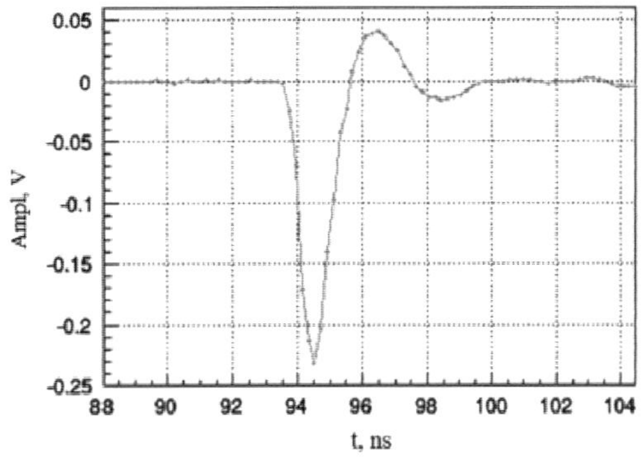

Figure 2. Signal of the Photek 240 recorded with a DRS4 during a run with 120 GeV/c protons passing through the PMT input window.

5. Test beam measurements

First we measured the electronic time resolution of the DRS4 units. The signal from one of the Photek 240 detectors was split into two equal parts, and connected to adjacent channels on the same DRS4 unit. The distribution of the time difference between these two channels was measured, and fitted with a Gaussian. The standard deviation of this Gaussian is defined as the time resolution. The measured electronic time resolutions of the two DRS4 units were different, but it was small with respect to the time resolution obtained later on. The result was dependent on the signal amplitude and was 4.8 ps (sigma, Gaussian fit) for one DRS4 and 6.7 ps for the other one at the same signal amplitude. We attribute the difference to the higher input noise in one of the two DRS4 units, which was an earlier version of the device. Distribution illustrating electronic time resolution of the DRS4 is shown in Fig. 3.

Next we measured the resolution of the two Photek 240 placed in a line close to each other. Signal of the one of them was used as "start" and another one as "stop". No lead plates were used between the start and stop counters (both of them were Photek 240) in the measurements. The time difference between the start and stop signals was measured. The Cherenkov light from 120 GeV protons in the Photek 240 input window created the signals in the MCP-PMT. The standard deviation of the Gaussian fit to the time difference between the two Photek 240, obtained with 120 GeV/c protons, was found to be 15.0 ps, as shown in Fig. 3. Assuming that the resolution of both Photek 240 is the same, one can derive the time resolution of a Photek 240 to be $15.0/\sqrt{2} \approx 10.6$ ps.

Next measurements were performed with positively charged particles of the 12 GeV/c secondary beam. We refer to these measurements as "shower-max" (SM) below. The goal of the measurements was to measure the dependence of the detector signal amplitude and time resolution on the lead thickness. We made some measurements with different quartz plates installed in optical contact with the Photonis 85011-501 detector input window to estimate the contribution of Cherenkov light into the MCP-PMT response. The plates were made of UV

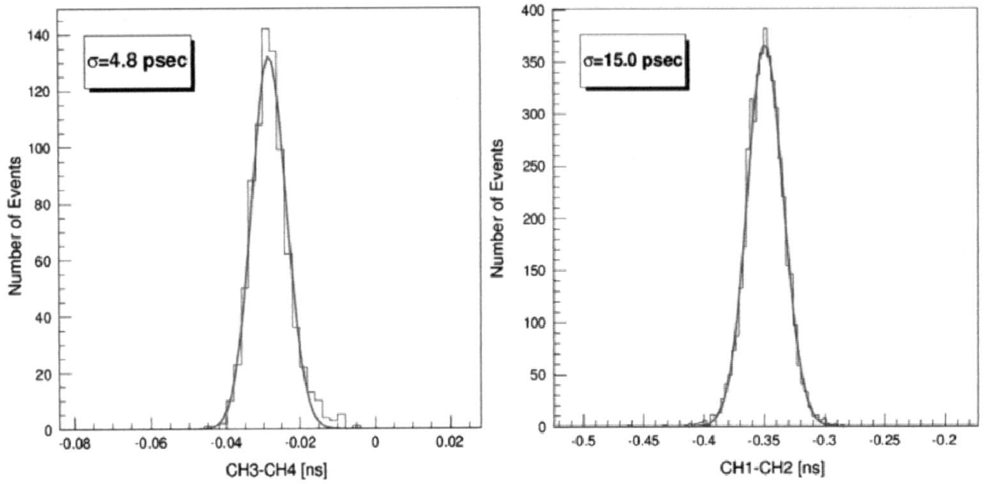

Figure 3. (Left) Distribution presenting electronic time resolution of the DRS4, and a Gaussian fit for it. In this setup the signal in channels 3 and 4 is identical, originating from one of the Photek 240 detectors, and is split into two equal parts. (Right) Spectrum presenting time resolution for the two Photek 240 in line, obtained with the 120 GeV/c proton beam. No lead plates between start and stop counters were used in the measurement.

Grade Fused Silica with the thickness of 3.5 mm and 7 mm. The stack of the lead plates with total thickness of either 10 mm or 20 mm was placed close to the Photonis 85011-501 input window. The transverse size of each set of the lead well covered the transverse size of the MCP-PMT.

We found that it is also possible to identify positrons using the signal of the downstream Photek 240. For positron events large signals near the end of the dynamic range of the unattenuated channel were produced. Using positron events selected by the gas Cherenkov counter we found the efficiency of the amplitude selection with the downstream Photek 240 to be 97%. We present the pulse height (PH, attenuated channel) and time distribution for zero and 3.5 mm quartz plates thickness placed upstream of the Photonis 85011-501 in Fig. 4.

Fig. 5 shows dependences of the PH and TR distributions on the quartz plate's thickness.

The measurements described above were repeated with Photek 240 as the SM detector in the 12 GeV/c secondary beam. No additional quartz was attached to the input window. We measured TR of the SM as a function of the lead thickness. The obtained TR values equal to 27.3 ps, 22.2 ps and 23.5 ps for the lead thickness of zero mm, 10 mm and 20 mm, Fig. 6.

The next series of measurements with the SM was performed at 32 GeV/c. We used the downstream shower maximum detector (Photek 240) to select positrons with the 32 GeV/c beam. According to our estimation the positrons component in the beam was less than 10%. We carried out the measurements with 10 mm, 20 mm of lead thickness and without lead. The TR dependence on absorber length is shown in Fig. 6.

6. Discusssion

The obtained data indicated that the shower particles were registered through the Cherenkov radiation in the input window as well as through direct interaction with the MCP. The secondary emission component of the response looks significant. Under the assumption that the Cherenkov response increases linearly with the radiator thickness (Fig. 5), one can extrapolate the dependence of the PH on the thickness of the Photonis 85011-501 input window down to

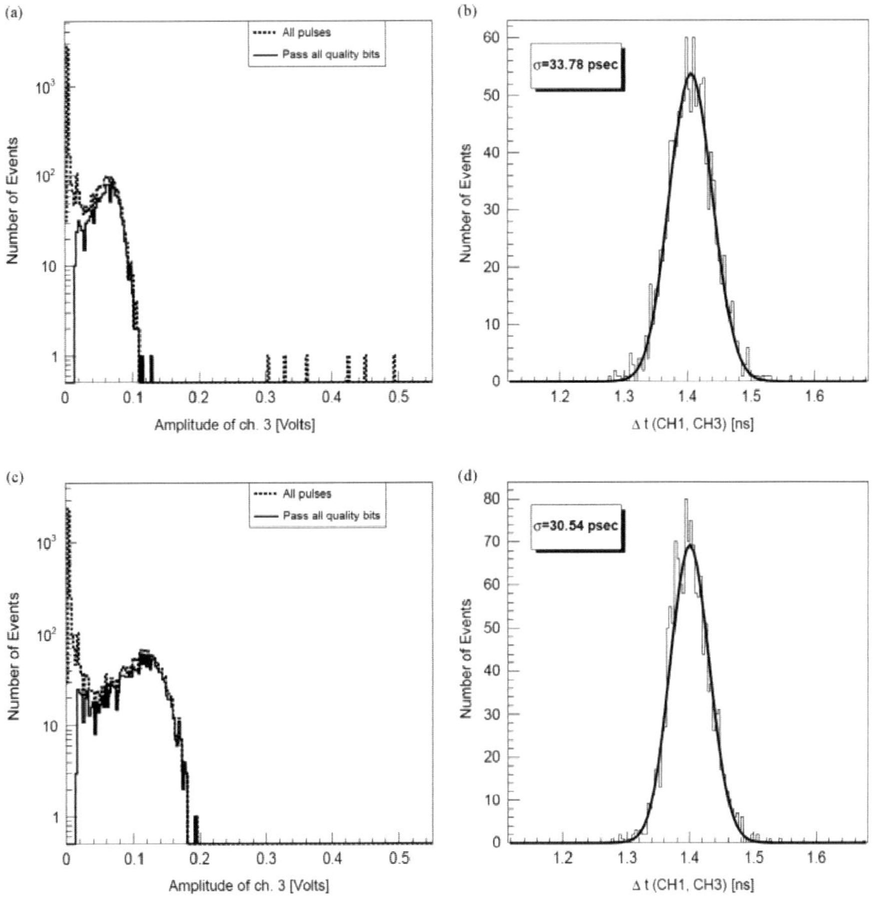

Figure 4. Distributions of the Photonis 85011-501 pulse height and the time difference between the start Photek 240 and Photonis 85011-501. Two pictures at left (a and b): no quartz plate is attached to the Photonis 85011-501 window. Two right pictures (c and d): 3.5 mm quartz plate is attached to the Photonis window. The other conditions are: 12 GeV/c, 10 mm of lead before the Photonis 85011-501.

zero to get an estimate for the secondary emission. We can estimate that 70% of the Photonis 85011-501 response is due to the MCP secondary emission and 30% is due to Cherenkov light produced in the 2 mm thick input window. The TR of the secondary emission component can be estimated to be around 35-40 ps for the Photonis 85011-501.

The Photek 240 shows an even better SM TR, at the level of 20-30 ps. The Photek 240 has an input window thickness of 9 mm. We did not make measurement with different input window thickness, but we have seen the response increase with additional radiator optically coupled with the Photek 240 input window. Nevertheless we can assume that the low energy component of the shower detected by an MCP will be significant, no matter which device is used.

The measured main parameters of the electromagnetic shower are well described. The SM detector, (1-100 GeV electron energy) is usually located in the beginning of the shower, after less than $4X_0$ - absorber thickness. We used lead thickness in correspondence with the SM locations in described detectors. The transverse size of the shower is very small at such depth and allows good separation from other showers. Our SM detectors cover this transverse size for lead absorber. The presented results show that the TR is at the level of 20-30 ps. The SM based

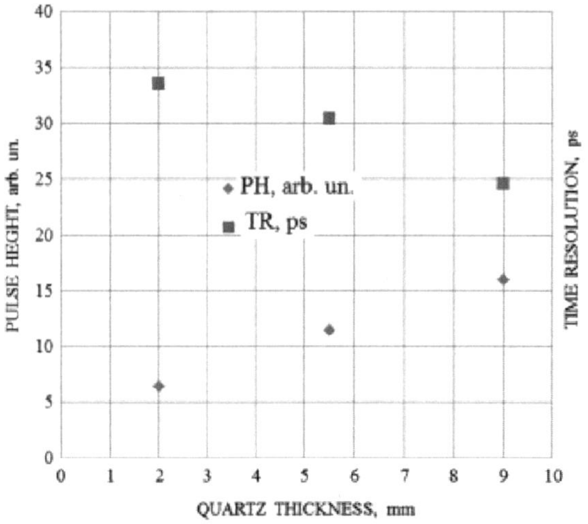

Figure 5. PH and TR dependences on absorber thickness in radiation length units. Photonis 85011-501 as SM, beam momentum is 12 GeV/c.

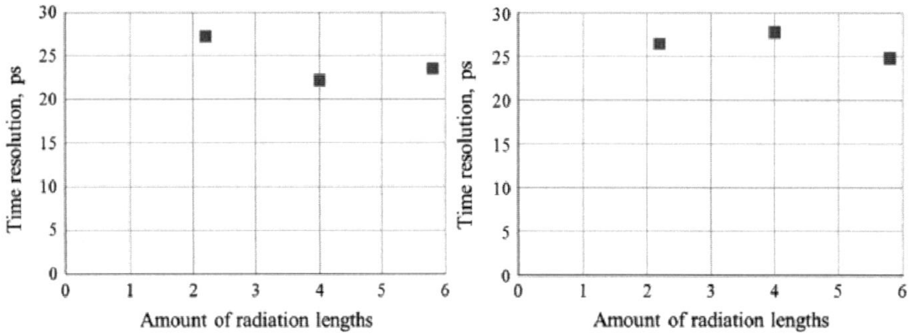

Figure 6. Dependence of the TR on the absorber thickness (in units of radiation length). Photek 240 as SM, 12 GeV/c (left) and 32 GeV/c (right) positrons beam.

on the approach could be very economical. A sandwich type calorimeter based on the same principle (MCP for all active layers) is also possible to build, but the cost is still a question in such a design.

References

[1] A. Derevshchikov, V. Khodyrev, V. Kryshkin, V. Rakhmatov, A. Ronzhin, "On possibility to make a new type of calorimeter: radiation resistant and fast". Preprint IFVE 90-99, Protvino, Russia, 1990.

[2] B. Adams, A. Elagin, H. Frisch, R. Obaid, E. Oberla, A. Vostrikov, R. Wagner, M. Wetstein. "Measurements of the gain, time resolution, and special resolution of a 20x20 cm^2 MCP based picosecond detector". Nuclear Instruments and Methods. A 732 (2013) 392-396.

[3] A. Bornheim, C. Pena, A. Apresyan, M. Spiropulu, J. Duarte, S. Xie, A. Ronzhin "Calorimeters for precision timing measurements in high energy physics", proceedings in CALOR 2014.

[4] A. Ronzhin, M. Albrow, M. Demarteau, S. Los, S. Malik, A. Pronko, E. Ramberg, A. Zatserklyaniy, "Development of a 10 ps level time of flight system in the Fermilab Test Beam Facility". Nuclear Instruments and Methods. A 623 (2010) 931.

[5] S. Ritt, R. Dinapoli, U. Hartmann, "Application of the DRS chip for fast waveform digitizing", Nuclear Instruments and Methods A 623 (2010) 486488.

FLUKA studies of hadron-irradiated scintillating crystals for calorimetry at the High-Luminosity LHC

Milena Quittnat for the CMS collaboration
Institute for Particle Physics, ETH Zurich, 8093 Zurich, Switzerland
E-mail: milena.quittnat@cern.ch

Abstract. Calorimetry at the High-Luminosity LHC (HL-LHC) will be performed in a harsh radiation environment with high hadron fluences. The upgraded CMS electromagnetic calorimeter design and suitable scintillating materials are a focus of current research. In this paper, first results using the Monte Carlo simulation program FLUKA are compared to measurements performed with proton-irradiated LYSO, YSO and cerium fluoride crystals. Based on these results, an extrapolation to the behavior of an electromagnetic sampling calorimeter, using one of the inorganic scintillators above as an active medium, is performed for the upgraded CMS experiment at the HL-LHC. Characteristic parameters such as the induced ambient dose, fluence spectra for different particle types and the residual nuclei are studied, and the suitability of these materials for a future calorimeter is surveyed. Particular attention is given to the creation of isotopes in an LYSO-tungsten calorimeter that might contribute a prohibitive background to the measured signal.

1. Motivation

The Compact Muon Solenoid (CMS [1]) detector will be upgraded for the High-Luminosity LHC (HL-LHC). The HL-LHC is scheduled for 2022 and will provide a five times higher instantaneous luminosity ($\mathcal{L} = 5 \cdot 10^{34}\,\mathrm{cm}^{-2}\,\mathrm{s}^{-1}$) than the LHC. Calorimetry at the HL-LHC will be performed in a harsh radiation environment with high hadron fluences, where typical doses in the forward region will be around 30 Gy/h at $|\eta| = 2.6$. The current forward electromagnetic calorimeter (ECAL endcaps, EE) is designed for LHC running with an integrated luminosity of $500\,\mathrm{fb}^{-1}$. In the forward region, this calorimeter consists of 2×61200 Lead Tungstate ($PbWO_4$) crystals and covers a region of $1.48 < |\eta| < 3.0$. The $PbWO_4$ crystals show a cumulative hadron-induced damage, which degrades the energy resolution [3] and makes a replacement of the EE for the HL-LHC desirable. One option for an upgraded forward ECAL would be a sampling calorimeter with alternating layers of absorber (tungsten), and active plates of a heavy inorganic scintillator. Suitable candidates are cerium fluoride (CeF_3), cerium-doped lutetium orthosilicate ($Ce_{2x}(Lu_{1-y}Y_y)_{2(1-x)}SiO_5$ - LYSO) and cerium-doped yttrium orthosilicate ($Ce_{2x}Y_{2(1-x)}SiO_5$ - YSO). These sampling calorimeter designs have to be studied in terms or radiation properties under HL-LHC conditions, using the Monte Carlo simulation program FLUKA ([6], [7]) which is optimized for such studies.

2. Comparison of FLUKA simulation to proton irradiated crystals

One goal of the following FLUKA radio-activation simulation is to validate the FLUKA description of the respective materials with measurements. The results of this study are

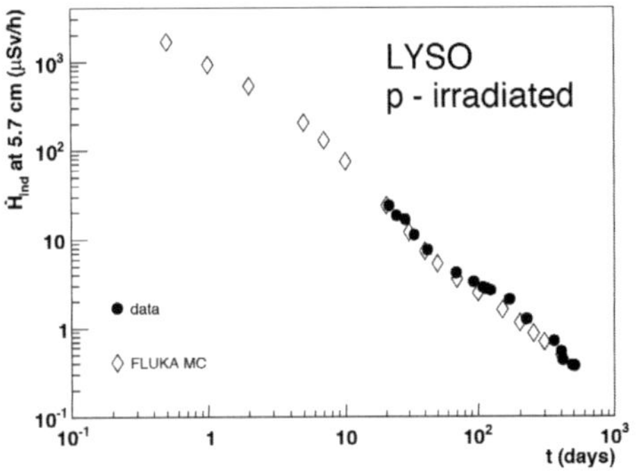

Figure 1. Measurements of induced ambient-dose equivalent rate $\dot{H}^*(10)_{\rm ind}$ at 5.7 cm from the side of the crystals as a function of time after irradiation, compared to FLUKA simulations for a fluence $\Phi_p = 10^{13}\,{\rm p/cm^2}$. The measurements are affected by a relative scale uncertainty of 7 % due to the precision of fluence determination [4].

described in detail in [4]. A LYSO sample with the dimension of $25 \times 25 \times 100\,{\rm mm}^3$, corresponding to $8.8\,{\rm X_0}$ in length, was irradiated with 24 GeV/c protons up to a fluence of $\Phi_p = (8.85 \pm 0.62) \times 10^{12}\,{\rm cm}^{-2}$ at the CERN PS T7 beam line. Since the fluence in this test has been delivered to the samples over few hours, the rate values measured here are not representative for what would be observed in situ [4]. However, the comparison between different crystal types is crucial to anticipate the expected exposure. The induced ambient-dose equivalent rate ("dose") $\dot{H}^*(10)_{\rm ind}$ has been measured at 5.7 cm distance from the middle of the LYSO sample at various times after irradiation. The measurements are compared with results from a FLUKA simulation (version 2011.2b.3), after rescaling to $\Phi_p = 10^{13}$ p/cm^2. A very good agreement is observed in Fig. 1, between FLUKA simulations and measurements, over one year and over two orders of magnitude in dose, with the simulation results falling slightly on the low side. Uncertainties in the comparison can arise from the limited knowledge of Lutetium cross-sections in FLUKA, and from non-uniformities of the beam intensity during irradiation combined with the self-shielding of the crystal. The comparison benchmarks the reliability and predictive power of FLUKA for such LYSO crystals. In a second step, FLUKA simulations have been performed to compare the expected remnant radioactivity among crystals of similar dimensions in terms of radiation lengths (26 X_0), while keeping the granularity the same ($24 \times 24\,{\rm mm}^2$). The description of CeF_3, based on [5], was added to the comparison. The simulation results for $PbWO_4$ are compared to existing measurements [3]. Accordingly, the crystals were assumed to be proton-irradiated with a 20 GeV/c squared beam profile for an integrated fluence of $\Phi_p = 10^{13}$ p/cm^2. The $\dot{H}^*(10)_{\rm ind}$ for each crystal was recorded as described in [4] and is depicted in Fig. 2. The FLUKA simulations for LYSO show a remnant dose similar to the one of $PbWO_4$, with the LYSO results being roughly a factor two above the $PbWO_4$ results. This is an indication that LYSO might become slightly more radioactive than $PbWO_4$ in a 26 X_0 deep calorimeter. As the transverse dimensions are kept the same, CeF_3 has effectively a smaller density. This smaller amount of material leads to a $\dot{H}^*(10)_{\rm ind}$ which is a factor two lower. In conclusion, the remnant dose rate for LYSO and CeF_3 is expected to be similar to $PbWO_4$ in a $26\,{\rm X_0}$ transversely extended calorimeter.

3. FLUKA studies for a forward ECAL for CMS at the HL-LHC

This section studies the behavior of an electromagnetic calorimeter in the forward region of CMS at the HL-LHC. The simplified FLUKA geometry is optimized to study the overall behavior of the sampling calorimeter with respect to radiation properties.

3.1. Implementation of CMS ECAL in FLUKA simulation

The implementation of a sampling calorimeter in the EE is based on the available baseline FLUKA geometry version 1.0.0.0, corresponding to the situation prior to the long shutdown 1

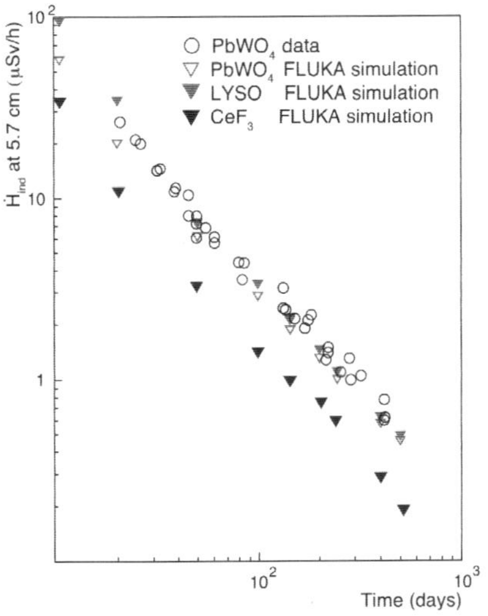

Figure 2. Induced ambient dose equivalent rate $\dot{H}^*(10)_{\rm ind}$ at distance 5.7 cm from the side of the crystal as a function of time calculated with FLUKA for a LYSO and CeF_3 crystal exposed to a fluence $\Phi_p = 10^{13}\,{\rm p/cm^2}$ of 20 GeV/c protons.These simulations are compared to data and simulations for $PbWO_4$ from [3] for the same irradiation conditions (all crystals 26 X_0 long). The uncertainties are discussed in the text. Based on Fig.8 in [4] with updated results from [5].

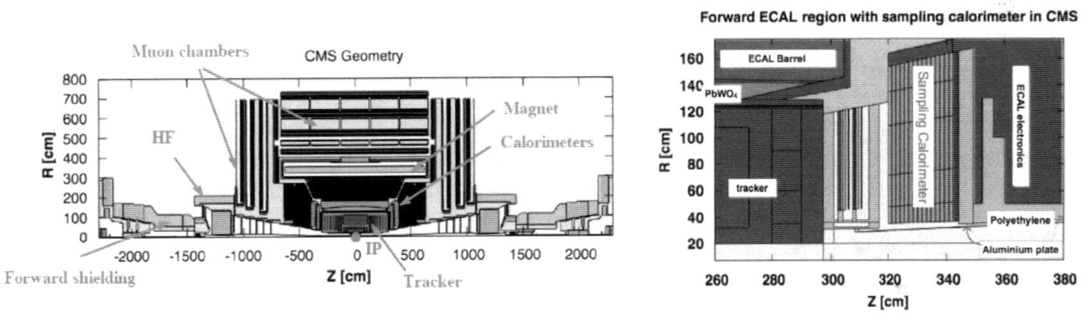

Figure 3. CMS geometry and sampling calorimeter in EE in FLUKA (CMS Beam Radiation, Instrumentation and Luminosity Project, BRIL, version 1.0.1-3).

Table 1. Dimensions of the sampling calorimeter options in terms of activation.

Calorimeter	L_{calo} [cm]	L_{abs} layer [cm]	L_{scint} layer [cm]	X_{0eff} [cm]	ρ [g/cm^3]
YSO/W	22.0	0.5	1.00	0.880	4.5
CeF3/W	15.4	0.5	0.56	0.616	6.2
LYSO/W	12.7	0.5	0.38	0.508	7.1

(LS1), that lasted between February 2013 until January 2015. Only the region of the current $PbWO_4$ crystals in the EE is modified (320 cm – 342 cm in z). The chosen sampling calorimeter starts with a scintillator layer to be sensitive to effects of the CMS environment. It consists of 15 scintillator and 14 tungsten layers. The 25 X_0 depth of the current $PbWO_4$ crystals in the EE is kept. For comparison, the dimensions in terms of radiation lengths are chosen to be the same for all options considered: 25 X_0 are split into 5 X_0 for the scintillator and 20 X_0 for the absorber. If the total length is smaller than 22 cm, the rest of the volume is filled with a region of low mass density, reflecting electronics and cooling devices. The dimensions of the calorimeter options are listed in table 1.

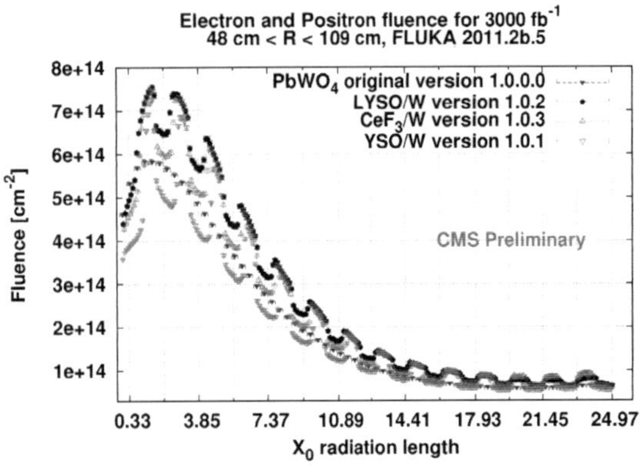

Figure 4. Electron and positron fluence in the EE for an integrated luminosity of $3000\,fb^{-1}$ as a function of depth expressed in X_0 using FLUKA. The fluence is averaged over a region of $48\,cm < R < 109\,cm$, corresponding to $1.8 < \eta < 2.6$ for $z = 320\,cm$. The average fluence is similar for all options.

3.2. Particle fluence in the sampling calorimeter at the end of HL-LHC

The primary proton-proton collisions with an energy of 7 TeV per beam were simulated with the DPMJetIII event generator and normalized to an inelastic collision cross section of 80 mb, using FLUKA version 2011.2b.5. Precise thresholds for the particle transport were applied. The transport and production thresholds were set to 100 keV for electrons and positrons, and to 50 keV for photons in the EE. Every other particle was transported down to 100 keV and the low energy neutron transport down to thermal energies (10^{-5} eV) was included. The electron and positron fluence in the EE for $3000\,fb^{-1}$ as a function of depth, expressed in X_0, is depicted in Fig. 4. The fluence is averaged over a region of $48\,cm < R < 109\,cm$, corresponding to $1.8 < \eta < 2.6$ for $z = 320\,cm$. The z-coordinate was converted to X_0 by scaling with the effective radiation length X_{0eff}, using a bin size of 10 % of X_{0eff}. For all options the average fluence is similar and comparable to the values of the $PbWO_4$ crystals. Most of the electromagnetically interacting particles in a minimum bias environment have an energy of a few MeV up to 1 GeV. The fluence reflects the development of the electromagnetic shower of such particles.The maximum electromagnetic particle flux develops in the first $7\,X_0$, particularly in the second scintillator layer. The effect due to the difference in density of the scintillator materials is clearly visible up to 11 X_0. In conclusion, the predicted electron-positron fluence will be comparable to the one of the current $PbWO_4$ crystals after $3000\,fb^{-1}$. Further studies show that the magnitude and overall shape of other particle fluences are also similar among the different sampling calorimeter options.

3.3. Absorbed dose in the sampling calorimeter at the end of HL-LHC running

The absorbed dose is the deposited energy per unit mass of the medium. Fig. 5 shows the dose absorbed in the EE for an integrated luminosity of $3000\,fb^{-1}$ as a function of the radiation length X_0 in the same region as above. Comparing the CeF_3 and YSO sampling calorimeters with the $PbWO_4$ calorimeter, the absorbed dose is overall comparable. The LYSO layers however show a significantly higher absorbed dose. This can be explained by the high thermal neutron capture cross section of ^{175}Lutetium ($\sigma_{th.n.} = 23.1$ barn [2]). In Fig. 6 an estimation of the maximum neutron fluence energy spectrum, which occurs in the second scintillator layer at $2.2 < \eta < 2.6$ (see Fig. 4) after $3000\,fb^{-1}$ is depicted. In the region of low energy neutrons, the number and width of the bins is given by the thermal neutron cross section library in FLUKA. Thermal neutrons carry an energy up to a few eV, followed by a resolved resonance region. This region shows different resonances for each option due to the different elements in the materials. Because of the high thermal neutron capture cross section of Lutetium, the LYSO/W option shows a

Journal of Physics: Conference Series **587** (2015) 012059 doi:10.1088/1742-6596/587/1/012059

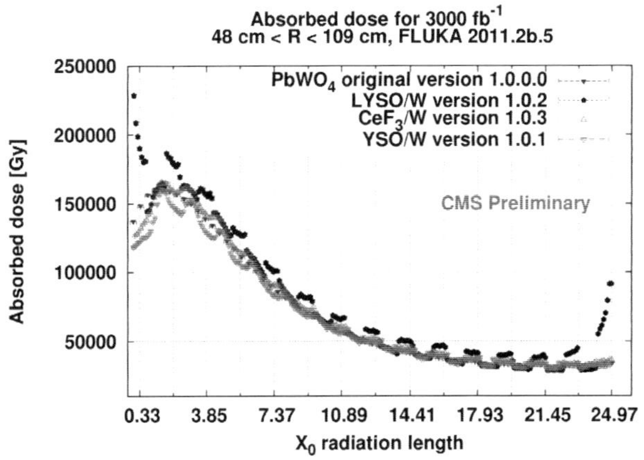

Figure 5. Dose absorbed in the EE for an integrated luminosity of $3000\,\text{fb}^{-1}$ as a function of depth expressed in X_0. Specifications as in Fig. 4. The average of the absorbed dose is comparable for all options.

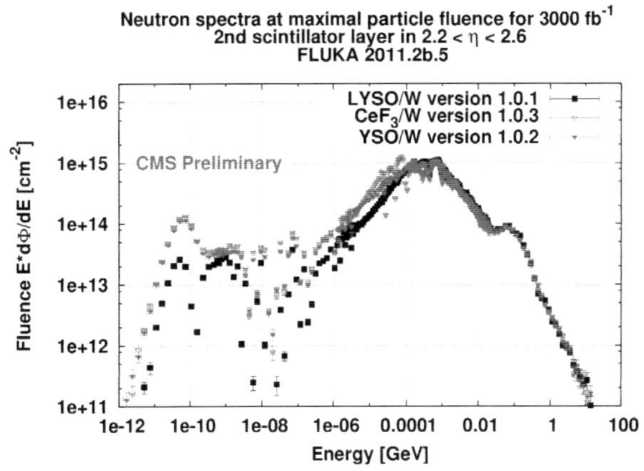

Figure 6. Fluence energy spectrum of neutrons for the maximal particle flux, in $2.2 < \eta < 2.6$ for $3000\,\text{fb}^{-1}$, simulated with FLUKA.

significantly lower fluence for thermal neutrons. This explains the higher absorbed dose in the LYSO layers in Fig. 5. Further FLUKA simulations show that the thermal neutron fluence in the low density regions adjacent to both sides of the sampling calorimeter is higher than the one in the scintillator materials. The enhancement is likely also due to the polyethylene layers at both sides, which moderate neutrons down to thermal energies, but do not absorb them. Due to the high $\sigma_{th.n.}$ of Lutetium, the absorbed dose rate in LYSO increases significantly at both ends of the calorimeter and in the scintillator layers. A thin absorber layer made out of boron polyethylene at both ends is advisable to moderate neutrons and absorb the thermal ones.

3.4. Study of a potentially prohibitive background due to decaying isotopes

The energy emitted by decaying isotopes might lead to a prohibitive background to the main electromagnetic signal. The LYSO/W calorimeter shows the highest dose values and is thus surveyed in a conservative scenario. Fig. 7 compares the absorbed dose rate ($\dot{D}$) during HL-LHC running (1.44×10^{13} pp-int./h) with the $\dot{D}$ by decaying particles at the irradiation stop after $2475\,\text{fb}^{-1}$. The irradiation schedule follows the official recommendations for a radio-activation simulation from LS3 until LS5. For the activation study, coalescence and evaporation of heavy fragments were enabled. $\dot{D}$ is visualized for two regions in R: 48 cm - 71 cm and 109 cm - 135 cm, corresponding to $\eta = 1.6$ - 1.8 and 2.2 - 2.6 at $z = 320$ cm. The inelastic cross section of the prompt electromagnetic shower is normalized to running conditions whereas the decay products are normalized to 100 mb, a conservative limit adopted for activation studies.

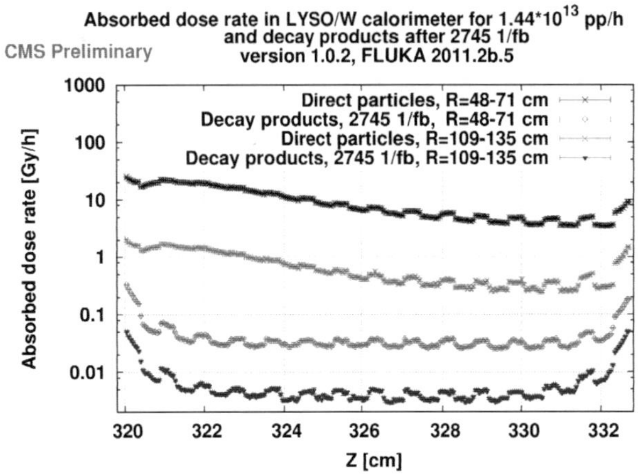

Figure 7. Absorbed dose rate due to the prompt electromagnetic shower and the decay products for an LYSO/W sampling calorimeter using FLUKA. The absorbed dose rate is projected to the calorimeter length in z, from 320 cm - 332.7 cm.

In Fig. 7, the dose rate $\dot{D}$ in the higher $\eta-$region is on average one order of magnitude higher than the one of the lower $\eta-$region. In comparison, the dose rate of the decay products is at a constant 1 % level of the prompt particles. This offset of deposited energy by decaying isotopes is still at a manageable level for HL-LHC running. The overall activity of the created residual nuclei after 2475 fb^{-1} in the first scintillator layer at $2.2 < \eta < 2.6$ is $A = (4.53 \pm 0.07) \times 10^{10}$ Bq. A percentage of 96 % of this activity comes from residual nuclei created by low energy neutrons. Furthermore, the dose rate of the decay products is dominated by short lived radio nuclides. The deposited energy by the decaying isotopes after 10 h is one third of the one after 2475 fb^{-1}. Responsible for this slight build-up is mainly the long lived isotope ^{177}Lutetium with an activity of $A = (1.91 \pm 0.03) \times 10^{10}$ Bq after 2475 fb^{-1} and a half life of $\tau = 6.64$ d.

3.5. Conclusions

A possible forward electromagnetic sampling calorimeter for the CMS experiment at HL-LHC was studied in terms of radiation exposure. The dedicated FLUKA study shows that the average particle fluence and the absorbed dose are similar for all options considered. As the ionizing dose in a LYSO-tungsten sampling calorimeter increases significantly towards both ends, a thin boron-polyethylen layer at both ends would be advisable. The deposited energy by decaying isotopes is at a constant 1 % level of the prompt electromagnetic shower and is dominated by short-lived radioisotopes. All three designs, using tungsten as an absorber and CeF_3, LYSO or YSO as the sensitive material are suitable candidates for a future sampling calorimeter at the HL-LHC in terms of activation levels.

References

[1] CMS Collaboration, **CERN/LHCC 2006-001**; **CERN/LHCC 2006-021**, 2006.

[2] ENDF/B-VII.1, M.B.Chadwick, M.Herman, P.Obloinsk, M.E.Dunn et al., *Nucl. Data Sheets* **112, 20112887**.

[3] M.Huhtinen, P.Lecomte, D.Luckey, F.Nessi-Tedaldi, F.Pauss, *Nucl. Instr. and Meth. Phys. Res.* **A545, 2005** 63-87.

[4] G.Dissertori, D.Luckey, F.Nessi-Tedaldi, F.Pauss, M.Quittnat, R.Wallny, M.Glaser, *Nucl. Instr. and Meth. Phys. Res.* **A 745, 2014** 1-6.

[5] G.Dissertori, P.Lecomte, D.Luckey, F.Nessi-Tedaldi, F.Pauss, T.Otto, S.Roesler, C.Urscheler, *Nucl. Instr. and Meth. Phys. Res.* **A 622, 2010** 41-48.

[6] G.Battistoni, S. Muraro, P.R.Sala, F.Cerutti, A.Ferrari, S.Roesler, A.Fasso, J.Ranft,*AIP Conf. Proc.* **896, 2007** 31-49.

[7] A.Ferrari, P.R.Sala, A.Fasso, J.Ranft, **CERN-2005-10, 2005, INFN/TC_05/11, SLAC-R-773.**

Journal of Physics: Conference Series **587** (2015) 012060 doi:10.1088/1742-6596/587/1/012060

Status of ADRIANO R&D in T1015 Collaboration

C Gatto[1], V. Di Benedetto[2] and A. Mazzacane[2]
On behalf of T1015 Collaboration[1]
[1] INFN, Sezione di Napoli, via Cinthia, 80126 Napoli, Italy
[2] Fermilab, Batavia (IL), USA
E-mail: corrado.gatto@na.infn.it

Abstract. The physics program for future High Energy and High Intensity experiments requires an energy resolution of the calorimetric component of detectors at limits of traditional techniques and an excellent particle identification. The novel ADRIANO technology (*A Dual-readout Integrally Active Non-segmented Option*), currently under development at Fermilab, is showing excellent performance on those respects. Results from detailed Monte Carlo studies on the performance with respect to energy resolution, linear response and transverse containment and a preliminary optimization of the layout are presented. A baseline configuration is chosen with an estimated energy resolution of $\sigma(E)/E \approx 30\%/\sqrt{E}$, to support an extensive *R&D* program recently started by *T1015* Collaboration at Fermilab. Preliminary results from several test beams at the *Fermilab Test Beam Facility* (FTBF) of a $\sim 1\lambda_I$ prototype are presented. Future prospects with ultra-heavy glass are, also, summarized.

1. Introduction

The physics program at future lepton and hadron colliders will be dominated by studies of processes involving multi-jets event. In such an environment, calorimeters will play a fundamental role as particle detector. A broad based *R&D* and Monte Carlo simulation activity is already in progress within the lepton colliders communities[2]. Similarly, the next round of precision experiments with High Intensity beams require an excellent particle identification capability in order to reduce background from beam contamination or by-products from interactions with shield materials (i.e., neutrons, photons, etc.).

A consensus has been established on the fact that the minimum hadronic energy resolution of the calorimetric systems needed to successfully distinguish the W from the Z signal in a high energy ($E_{cm} > 500GeV$) is $\sigma(E)/E \approx 30\%/\sqrt{E}$. Such a resolution is unprecedented for a detector at a collider and it is being reached only by massive compensating calorimeters with very small volume ratio between passive and active materials[3]. However, the large size needed to contain the showers, because of their relatively low density, make them impractical to build in experiments with high energy colliding beams. It would be challenging to achieve at the same time effective compensation and shower containment. Furthermore, the resolution of traditional calorimeters is limited, among other sources, by the fluctuation in the electromagnetic (EM) content of the hadronic shower and by the unequal response of such devices to the EM and hadronic components of the shower itself[4]. In recent years, an alternative technique has been developed in order to cope with such effects: the dual-readout calorimetry[7], based on an event-by-event measurement of the EM fraction of the shower. Such a technique is based on the simultaneous measurement of signals generated by different shower production mechanisms, thus providing complementary information on the composition of the hadronic shower.

The dual read-out calorimetry falls under two broad categories: sampling and integrally active. Sampling dual-readout techniques are currently investigated by the DREAM[7] Collaboration. However they introduce two important sources of fluctuations: Poisson fluctuations in the Čerenkov signal, induced by the low photo-electron statistics and sampling fluctuations. These fluctuations do not only impair the energy resolution for hadronic showers, but have also detrimental consequences on the detection quality of photons and electrons. Consequently, the detection of EM particles in high energy jets is similarly affected. An obvious solution to the latter problem would be to design a detector with two distinguished regions: a front EM section and a rear hadronic section. However, it is well known that such a configuration, most often consisting of media with very different properties, is sub-optimal in terms of energy measurement of hadronic particles due to the extra fluctuations introduced by the development of the shower into two different sections [4].

Integrally active dual-readout techiques are mostly free of the above limitations. Test beams an extensive simulations indicate that these techniques provide excellent energy resolution.

In this article we propose a novel Dual-Readout, Integrally Active and Non-homogeneous Option (*ADRIANO*), based on light signals produced in high transmittance optical glasses and scintillating fibers.

2. Description of ADRIANO techniques

The new detection technique named *ADRIANO* (*A Dual-readout Integrally Active Non-segmented Option*) is a part of an extensive *R&D* program in *T1015*[1] Collaboration. Two different layouts are currently under consideration, one being optimized for High Energy experiments, while a second adreesses the requirements of High Intensity experiemnts.

ADRIANO for High Energy. *ADRIANO* prototypes we have built for High Energy experiments are optimized for all have a modular structure, with the base unit consisting of an individual cell of parallelepiped shape with $40 \times 40\ mm^2$ cross-section and either 15 cm or 25 cm length. The cell consists of a sandwich of scintillating fibers and high density, optical grade heavy glass. The glass behaves as an absorber and as an active medium at the same time, generating almost exclusively Čerenkov light. The scintillation and Čerenkov sections of *ADRIANO* are optically separated. Therefore, the two generated lights are well separated, with minimal chance of cross-talk. For the results presented in this report, we have considered various techniques to optically separate the two regions: white and silver coating of the glass, white coating, silver coating and aluminum sputtering of the scintillating fibers and finally, a thin layer of Teflon between the glass plates and the array of scintillating fibers.

The scintillating fibers are either accommodated in grooves formed in the glass itself or in white plastic trays sandwiched among plates of glass. They run parallel to the longitudinal axis of the cell and are responsible for the generation of the scintillation component of the dual readout calorimeter. The pitch between nearby fiber is sufficiently small compared to the nuclear hadronic interaction length of the detector so that the shower sampling fluctuations are small. The Čerenkov light generated inside the glass is collected by WLS fibers running inside grooves parallel to the scintillating fibers and optically coupled to the glass by especially formulated optical compounds. The two light components are read out at the back of each cell with two distinct photodetectors. In some sense, *ADRIANO* is a spaghetti calorimeter with the passive absorber replaced by an active, transparent absorber made of heavy glass. Another advantage of *ADRIANO* relies on the fact that the heavy glass absorber can be used to detect electromagnetic showers in exactly the same ways as it has been done in the past with lead glass based electromagnetic calorimeters. Therefore, *ADRIANO* does not require a front electromagnetic section.

Figure 1. *Molding fabrication technology.*

Several heavy glasses (mostly lead and bismuth based) have been tested, with the intent of comparing the Čerenkov light yield and propagation. Their refractive index ranges from 1.85 through 2.24 while the densities range from $5.5g/cm^3$ through $7.5g/cm^3$. Several constructions techniques have been considered: diamond machining, precision molding, glass melting, laser drilling and photo-etching. However, only the former two have been considered for the production of the eleven *ADRIANO* prototypes presented in this report. A picture of a $8mm$ thick glass slice obtained with the precision molding technique is shown in Fig.1 .

A detailed description of the layouts of the eleven prototype and their corresponding construction techniques will follow in an upcoming article.

ADRIANO for High Intensity. *ADRIANO* prototypes intended for High Intensity experiments are optimized for larger light yield, rather than for high density, since they will operate mostly at lower energies and in the EM regime. They were, initially designed for the ORKA project at Fermilab, where they would operate in the 5 -100 MeV energy regime, but the layout can be properly optimized for any experiment with energy sensitivity above few MeV. In this case, we replaced the scintillating fibers with plates of thin (2 mm) and grooved extruded scintillator sandwiched between thicker (4.2 mm) heavy glass plates (Schott SF57), also grooved. Each plate is 10 cm wide and 37 cm long and it is formed with the molding technique described above. A picture of a the two plates is shown in Fig.2

.

The light readout of each plates uses WLS fibers: 6 for the plastic plates and 13 for the glass plates. The larger number of WLS fibers per unit surface of glass (0.031/cm^2 vs 0.012/cm^2 used in the High Energy version) captures more Cerenkov photons, at the expenses of a lower average density of the detector and an increased number of light sensors.

3. ADRIANO readout system

The scintillating and WLS fibers from each *ADRIANO* cell were bundled and routed each to a photodetector. In order to compare the performance of the light collection system in various situations we used three different photomultipliers (R647 and H3165 from Hamamatsu and P30CW5 from Sensetech) and two different SiPM ($4 \times 4mm^2$ square and $\O 2.7mm$ round from FBK) for WLS fibers and only one type of SiPM ($4\ times4mm^2$ square from FBK) for the scintillating fibers. When PMT were used, the fibers were routed through a plastice fixture and coupled to the photosensor window with a custom made optical grease. In the case of SiPM, we used either acrylic light concentrators (designed and produced by INFN Trieste) in

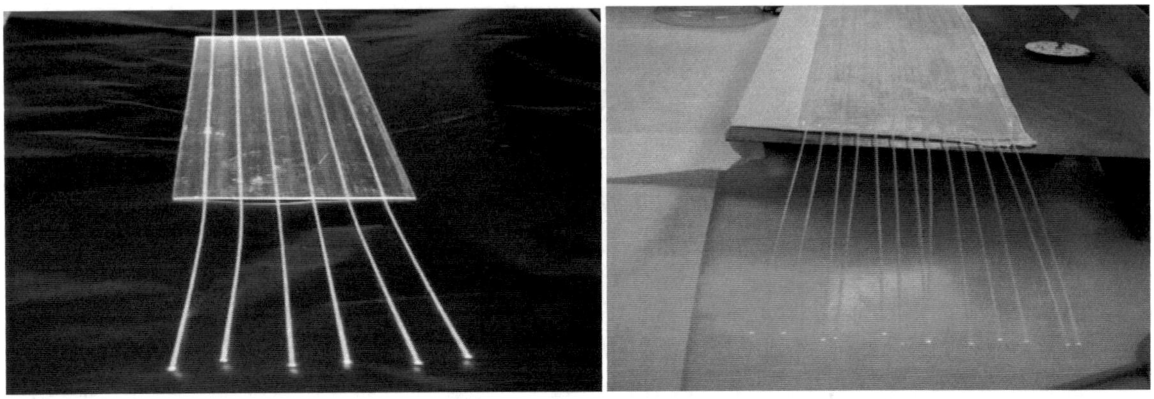

Figure 2. *Plastic scintillator (left) and glass (right) plates in the ADRIANO for High Intesity experiments.*

direct contact with the fibers on one side and spaced $0.1m\mu$ from the active SiPM surface or we routed the fibers directly to the SiPM up to $0.1m\mu$ from the active SiPM surface. A picture of a INFN-Trieste light concentrators is shown in top left of Fig.3. The output of the SiPM and PMT used was digitized by the TB4 DAQ system developed at Fermilab. Among the features of the DAQ, the most relevant for our application are:

- 50Ω inputs
- 14 bit ADC;
- $\sim$ 30 MHz bandwidth;
- $\sim$ 212 MSPS digitizer;
- Up to 16 channels per Motherboard;
- Bipolar, so both positive (from SiPM) and negative (from PMT's) signals can be acquired simultaneously;
- Slow-control over USB, readout over 100 Mbit Ethernet.

4. Preliminary Results From Test Beams

From March 2011 through December 2013 six test beam have been completed at FTBF Facility of Fermilab (Batavia, US). A total of thirteen *ADRIANO* cell prototypes with different dimensions and in different configurations and with different construction techniques have been illuminated by the secondary beam available at FTBF. A picture of a typical *ADRIANO* prototype for High Energy experiments, with scintillating fibers bundled together and WLS unbundled is shown in the bottom left of Fig.3

Similarly, Fig.4 shows a picture of two *ADRIANO* prototypes for High Intensity experiments, with WLS fibers from glass and scintillator plates bundled and routed to light sensors. One of the modules is intended for endcap applications, with fibers readout from one side and aluminized on the opposite side. The second prototype is intended for barrel applications, where the fibers are readout from both sides.

In summary we have tested:

- 5 glass types: lead and bismuth based.
- 3 glass coatings: TiO_2, Silver paint, Barium sulphate.
- 3 WLS fibers: Y11 (1.2 mm) & BCF92 (1.0 mm, 1.2 mm).

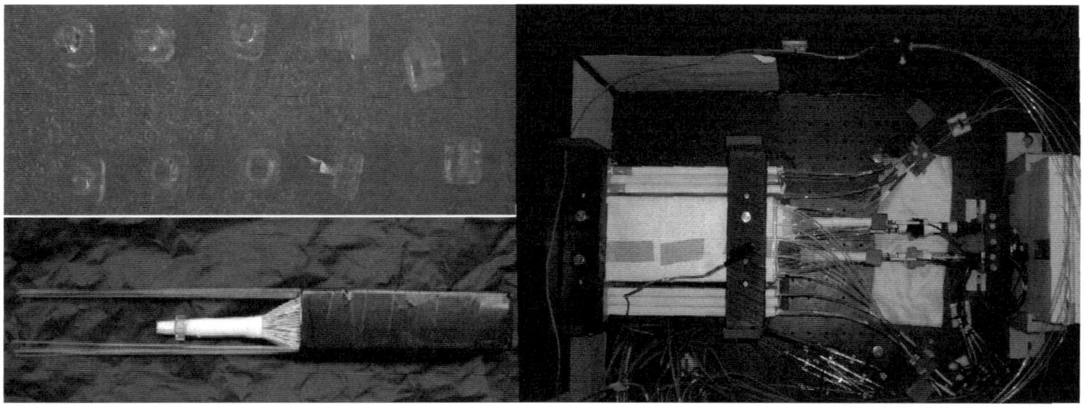

Figure 3. *INFN-Trieste light concentrators, a typical ADRIANO prototype, with scintillating fibers bundled together and WLS unbundled on the left and 2012 test beam setup at FTBF on the right.*

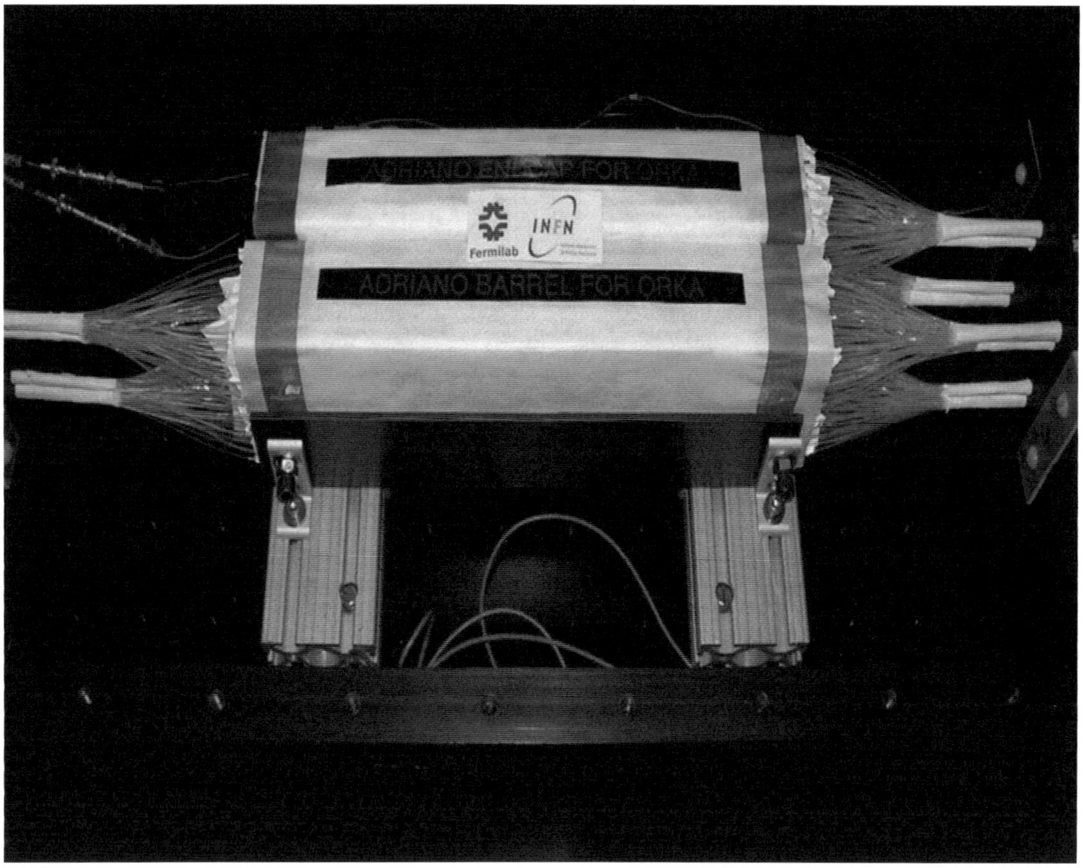

Figure 4. *ADRIANO for ORKA prototypes. Top is for endcap applications (one-side readout), while bottom is for barrel applications (two-side readout).*

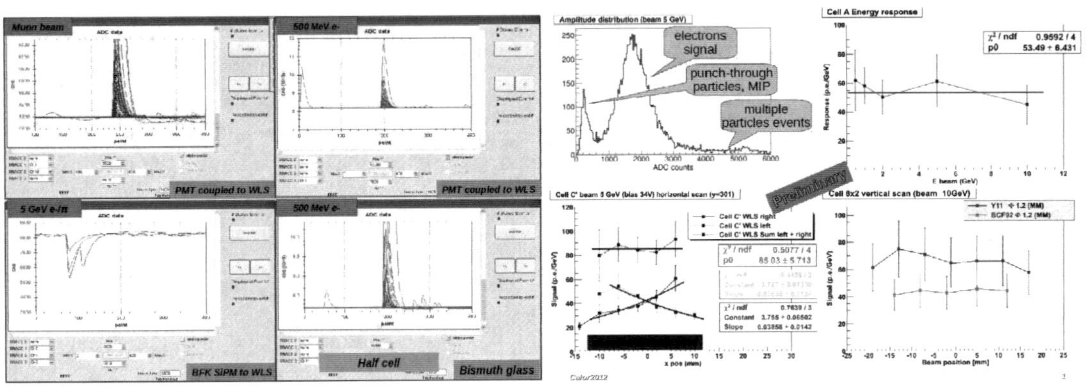

Figure 5. *Typical waveforms,the Čerenkov spectrum from various prototypes and the uniformity of Čerenkov response.*

- 1 Scintillating fiber: SCSF81.
- 4 scifi coating: TiO2, BasO4, Silver paint, Al sputter.
- Several optical glues (mostly custom made).
- 7 photodetectors: 4 SiPMs ($\o 2.7mm$ round and $4 \times 4mm^2$ square) manufactured by BKF and ST Microelectronics and 3 PMTs (P30CW5 , R647, H3165)
- 4 light coupling systems: direct glass + direct WLS + 4 light concentrators

The goal of these experiments were to optimize the construction technique and the materials used in order to maximize the Čerenkov light yield. We also measured some of the parameters needed for Monte Carlo simulations with ILCroot framework [8]. Due to the limited size of the cells, we do not expect to be able to test the dual-readout concept nor to evaluate the energy resolution of $ADRIANO$.

Typical waveforms for several beam configurations, as obtained by TB4 DAQ, the Čerenkov spectrum for various prototypes when exposed to a 5 GeV beam at FTBF and the uniformity of Čerenkov response across the cell are shown in Fig.4 (right side).

Similarly, Čerenkov (right) and scintillation (left) waveforms obtained from the High Intensity prototypes of $ADRIANO$ exposed to a 32 GeV muons beam are shown in Fig.4. The height of the pulse digitized by the TB4 DAQ systems (with 50Ω input impedence) is about 100 mV for the Čerenkov component and 290 mV for the scintillation component for the average energy deposited by a minimum ionizing muon in $ADRIANO$ is about 27 MeV. The peak value of electronic noise is about $2.5\,mV$ for the front-end electronics configuration in use. Therefore, assuming a threshold of three times that value, (namely 7.5 mV), we conclude that the sensitivity of the Čerenkov componenent of $ADRIANO$ for ORKA is about 2 MeV while the sensitivity of the scintillation component is about 0.7 MeV. A full analysys of test beam data to estimate the corresponding light yield is in progress.

Finally, the Čerenkov light yield from the eleven prototypes tested at FTBF is summarized in Table 1.

5. Summary

Since the start of ADRIANO $R\&D$ program, we were able to constantly improve the Čerenkov light yield by refining the construction techniques and the materials employed. The current limit for a $40 \times 40 \times 250mm^3$ prototype for High Energy is roughly 160 pe/GeV, obtained at a test

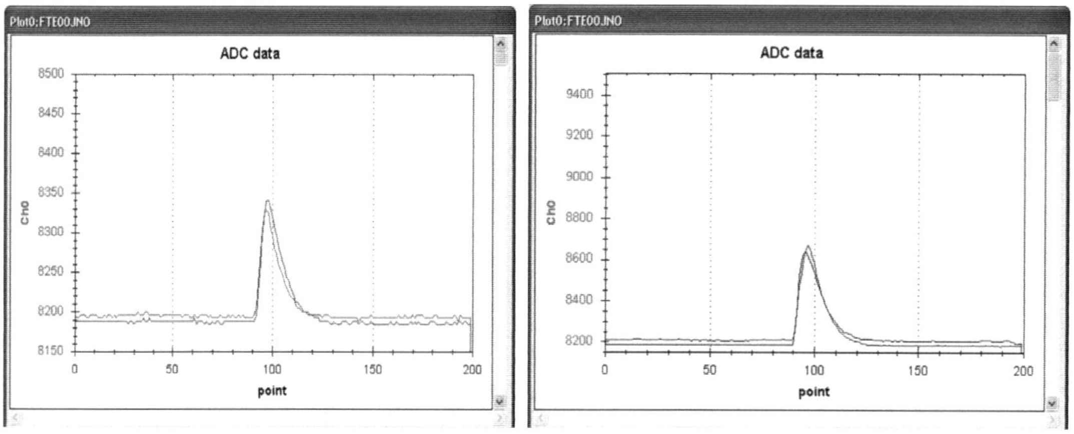

Figure 6. *Čerenkov (left) and scintillatio (right) waveforms ADRIANO for ORKA prototypes at a 32 geV muon beam (FTBF).*

Prototype #	Layout	Glass	g/cm^3	L.Y	Notes
1	5 slices, machine grooved, unpolished, white	Schott SF57HHT	5.6	82	SiPM readout
2	5 slices, machine grooved, unpolished, white, v2	Schott SF57HHT	5.6	84	SiPM readout
3	5 slices, precision molded, unpolished, coated	Schott SF57HHT	5.6	55	15 cm long
4	2 slices, ungrooved, unpolished	Ohara BBH1	6.6	65	Bismuth glass
5	5 slices, scifi silver coated, grooved, clear, unpolished	Schott SF57HHT	5.6	64	15 cm long
6	5 slices, scifi silver coated, grooved, clear, unpolished	Schott SF57HHT	5.6	120	improved version
7	10 slices, white, ungrooved, polished	Ohara PBH56	5.4	> 30	DAQ problems
8	10 slices, white, ungrooved, polished	Schott SF57HHT	5.6	76	
9	5 slices, scifi Al sputter, grooved, clear, polished	Schott SF57HHT	5.6	30	2 wls/groove
10	5 slices, ungrooved, polished	Schott SF57HHT	5.6	158	2012 version
11	2 slices, plain	Ohara experimental	7.5	–	DAQ problems
12	ADRIANO for ORKA - barrel (10 layers)	Schott SF57	5.6	-	Analysis in progress
13	ADRIANO for ORKA - endcap (10 layers)	Schott SF57	5.6	-	Analysis in progress

Table 1. Summary of light yields for ADRIANO modules for six test beams occurred from 2011 and 2013.

beam on 2012. Few modifications to the fibers-sensor coupling were made to the same prototype in 2013 and the test beam was repeated in October 2013. The data analysis is still in progress. We found that the Čerenkov light attenuation length inside the glass absorber critically depends on coating type and on the surface finishing of the glass slice. Coupling of fibers to SiPM's is critical as well; any thin air gap between the light concentrator and the active surface of an SiPM more than halves the light yield. Kuraray Y11 fibers produced almost 50% more light than Bicron BCF92 with the same diameter. Different optical glues, used to couple the glass to WLS fibers produced up to a factor of 2 in variation of light yield. This result should not surprise since we are attempting to collect light from an optical medium with considerably large refractive index (1.8 – 2.2) by transferring the Čerenkov light to plastic fiber with far lower refractive index (typical refractive index of the external cladding of WLS fibers is 1.45). Cold vs hot glass construction methods have shown no appreciable difference in Čerenkov light yield. Direct reading of Čerenkov light from glass with an acrylic light guide coupled to the back of a cell yielded consistently less light than when reading using the WLS fibers coupled longitudinally to the cell. This is a clear indication of the fact that, with thin layers of heavy glass having an absorption length of few centimeters, it is preferable to use light collection mechanisms based on WLS fibers distributed across the glass where the short path traveled by Čerenkov photons more than compensates the lower light collection efficiency. SiPM and PMT produce comparable signals, when their respective gains are taken into account. However, large noise from present generation of SiPM make them hard to use in low energy applications. This situation should quickly improve with the latest, low noise, generations of silicon devices.

Two protypes of *ADRIANO* for High Intensity experimens have been designed and built in 2013. Analysis of the data taken during a two weeks test beam at the Fermilab Test Beam Facility in December 2013 is still in progress. However, preliminary studies indicate that the light yield for this layout is at leas one order of magnitude better that for *ADRIANO* for High Energy experiments.

6. Future Prospects

The first four year of *ADRIANO R&D* have already produced clear directions. Precision molding technique is, at present, the preferred fabrication technique since it has the potential of making quick (less than 30 minutes) glass slices with optical surface finish and with appropriate grooves. Nonetheless, we will keep exploiting other fabrication techniques as they might have other potential advantages when compared to precision molding. Photo-etching techniques, for example, are expected to be equally fast and cheap, although they present quite severe chemical hazards.

A great boost in *ADRIANO* development is currently obtained with Ohara sponsorship/partnership as they provided bismuth glass strips of commercial optical glass ($6.6g/cm3$, $nd = 2.0$) and strips of an even denser experimental glass with density of $7.54g/cm^3$ and refractive index of 2.24.

Currently, two new and larger *ADRIANO* prototypes are under construction at Fermilab both addrssing High Energy experiments needs (high material density and moderate light yied). Each prototype measure about 10x8x105 cm^3 and is about 40 Kg in weight. The aim of these new modules is to study the dual-readout compensation mechanism by better containing the shower. We are planning, for the medium future, to continue adding new modules to the existing setup in order to increase the containement cabability of the detector by increasing its volume.

7. Conclusions

In this report we have presented the novel *ADRIANO* dual-readout technique along with preliminary results from several test beam at FTBF of eleven *ADRIANO* prototypes. The Čerenkov light yield we have obtained across the board is more than adequate for a hadronic calorimeter with an energy resolution of 30%/sqrt(E) or better. Our goal is to increase the light yield to a level that *ADRIANO* can be used as an electromagnetic and hadronic calorimeter at the same time. Hardware *R&D* and Monte Carlo simulations are in advanced state. More generally, current results from several test beams proved that Čerenkov light readout from heavy glasses with WLS collection system is feasible and provides equal or better results than traditional methods employing a large area PMT directly coupled to the glass. Correctly matching calorimetric techniques with SiPM and Front End Electronics is also crucial for a good performance of the detector. T1015 Collaboration will address these issues in the future and it will exploit new calorimetric techniques based on heavy glass, including scintillating heavy glasses.

8. Acknowledgments

We would like to thank Fermilab for its on-going support and for providing all the infrastructure for the construction and testing of *ADRIANO* prototype. In particular, we would like to thank Eileen Hahn and her team at Fermilab for her dedication in the assembly of *ADRIANO* prototypes and Aria Soha and her team for continuig assistence during the setup and data taking of T1015 at FTBF. We also would like to thanks Ohara for providing us with free samples of their BBH-1 bismuth based glass and of a new super-dense (about $7.5g/cm^3$) experimental glass which have been used for the construction of ADRIANO prototype #6 and #11.

This work has been financially supported by the Fermilab's Detector R&D program and the "TWICE project" of Istituto Nazionale di Fisica Nucleare (Italy).

References

[1] *wwwppd.fnal.gov/FTBF/MOUpDF/T1015_mou.pdf*
[2] *http : //www.linearcollider.org/ http : //clic − study.web.cern.ch/clic − study/*
[3] S. Buontempo et al., "Construction and test of calorimeter modules for the CHORUS experiment", Nucl. Instr. and Meth. A 349 (1994), p. 250.
[4] R. Wigmans, Calorimetry energy measurement in particle physics, in: International Series of Monographs on Physics, vol. 107, Oxford University Press, Oxford, 2000.
[5] 4th Concept Collaboration - Letter of Intent from the Fourth Detector (4th) Collaboration at the International Linear Collider, 2009 also at: http://www.4thconcept.org/4LoI.pdf
[6] N. Akchurin, et. al, "New Crystals for Dual-Readout Calorimetry", Nucl. Instr. Meth. A604(2007)512-526
[7] All DREAM papers are accessible at http://www.phys.ttu.edu/dream, and also http://highenergy.phys.-ttu.edu.particle physics, in: International Series of Monographs on Physics, vol. 107, Oxford University Press, Oxford, 2000.
[8] *http : //www.dmf.unisalento.it// danieleb/IlcRoot/*

Journal of Physics: Conference Series **587** (2015) 012061 doi:10.1088/1742-6596/587/1/012061

A particle counting EM calorimeter using MAPS

G. Nooren[1] **and E. Rocco**[1]

[1] SubAtomicPhysics Institute, Utrecht University and Nikhef, Princetonplein 5, Utrecht, 3584 CC, the Netherlands

E-mail: nooren@nikhef.nl, erocco@nikhef.nl

Abstract. The availability of full size MAPS sensors makes it possible to construct a calorimeter with pixelsize of a few tens of micrometers. This would be small enough to count individual shower particles and would allow a shower shape analysis on an unprecedented, small scale. Interesting features would be tracking capability for particle flow algorithms and a superior discrimination of single photons from neutral and charged pions at high momenta. A small Molière radius together with high transverse resolution would allow to separate close showers, induced by photons from neutral pion decay. A full scale (4 R_M, 28 X_0) prototype was constructed to demonstrate this. It features 30 micron pixelsize and a longitudinal sampling at 1 radiation length. We will show results from beam tests of this prototype at electron energies of 2 to 200 GeV.

1. Introduction

To cope with the availability of accelerators of increased energy, better detector performance, like spatial resolution, electronics readout speed and energy resolution are required. This might enable new measurement approaches. In calorimetry for example, instead of measuring the deposited energy of the showering particles it would be more convenient to count all the particles traversing the detectors. This would be an advantage in particle identification and in the separation of close showers. This entails fine detector granularity, namely a large number of channels. This is only feasible with digital signals.

We built a fully digital electromagnetic calorimeter equipped with Monolithic Active Pixels Sensors (MAPS) [1] with a longitudinal segmentation of 1 X_0 and an expected lateral resolution of few hundreds micron.

The main goal of such a device is the discrimination of direct photons from the photons produced by the π^0 decay at forward angles in high-energy physics experiments as at the LHC. The shower core density would be 10^3 mm^{-2} and to study the shower shape longitudinally and transversely, the pixel size is bound to be of the order of 30 - 50 μm for an average occupancy of 1 particle/pixel. Moreover to fully exploit such a small pixel size, a compact design of alternated tungsten planes with active planes has been realized with a calculated Molière radius of 11 mm.

The prototype was tested with particle beams proving the feasibility of a digital calorimeter [2] and demonstrating the integration of cooling without worsening the Molière radius.

Performance results from the test beams are reported.

2. The Detector

The prototype (Figure 1) has 24 layers with a thickness of 4 mm of which the sensor part, including the read-out board, is 1 mm. This leaves an air gap of only 0.1 mm between the overlapping sensors. The radiation thickness of one layer is 0.97 X_0. The active area of a layer (Figure 2) is 4 * 4 cm^2, composed of four sensors, while the absorber measures 5 * 5 cm^2. To compensate for some of the dead zones sensors overlap in one spatial direction. For ease of construction there remains a dead zone of 0.1 mm between each pair of chips in the other direction. The Molière radius calculated for this setup is 11 mm, which means that the tower is wide enough to study the lateral shower development. The first active layer has only 0.02 X_0 in front, to act as a charged particle detector. Between layers 21 and 22 6.7 X_0 of tungsten are placed to obtain a total depth of 28 X_0. At a dissipation of 70 W in 150 cm^3 a heat resistance of 0.3 K/W from chip to cooling water was measured.

The sensor chosen for the prototype is PHASE2/MIMOSA23 from IPHC [3]. This is the only full size MAPS (640 * 640 pixels), which allows the continuous read-out of all pixels, thanks to four outputs at 160 MHz. The 1 MHz rolling shutter implies an integration time of 640 μs. The resulting low event rate is no problem in the case of test experiments. The small pixel pitch of 30 μm allows very fine sampling of the shower core. In the detector there are sensors with different epilayer thicknesses and resistivity: 15 and 20 μm with resistivity of (400 Ωcm) and 14 μm with 10 Ωcm resistivity.

As each full detector read-out - a so-called frame - contains 39 M pixels, the raw data rate of the prototype is 61 Gb/s. Several FPGAs are used to manage this, as described in [4]. Incoming trigger bits are time-stamped with the read-out clock and stored separately.

Figure 1. The electromagnetic MAPS prototype measures 4 × 4 × 12 cm^3, hence it is a compact detector. In the picture the prototype is located in the centre and on the sides there are the printed circuit boards with the flat cables for the chip operation. At the time of this picture the cooling was not implemented yet.

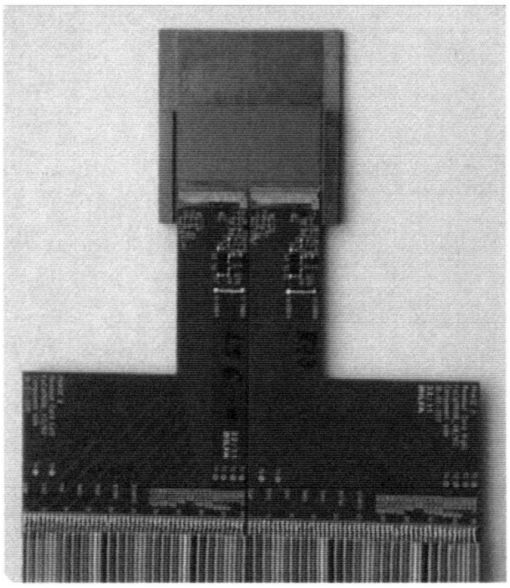

Figure 2. Picture of half module. On the tungsten plate two chips positioned side by side are connected to their printed circuit board.

3. Test beam results

Several test beam campaigns have been performed in the last couple of years: at DESY (in 2012 and 2014) with pure electron beam from 2 to 5.4 GeV and at CERN PS and SPS (in 2012) with mixed beams from few GeV till 200 GeV. In the meanwhile cosmic ray data have been taken and used to tune the sensors.

3.1. Energy Calibration

Due to diffusion, the charge created by a particle will lead to a cluster of pixel hits. The size of the cluster will depend on the charge created by the particle (Landau distribution) and the threshold of the discriminators. For application in trackers the discriminators are usually adjusted such that an acceptable fake rate, measured as clusters not belonging to a track, is achieved. In the case of a calorimeter it is a priori not known whether the clusters will be well-separated, especially in the core of the shower, Figure 3. Therefore it was decided to use the number of *hit pixels* as a measure of the energy, instead of trying to derive the number of *particles* from the hit distributions. This means however, that the discriminators should be set such that, in the absence of particles, an acceptable number of individual pixel hits is achieved, as opposed to clusters of hits. Figure 4 shows the average uncorrected response of the detector as a function of energy for the measurements at DESY. The figure also illustrates the total noise: this value is around 300 hits, which corresponds to a noise level of 10^{-5} per frame. Although the detector was not designed for such low energies a linear behaviour in number of hits per event is reported.

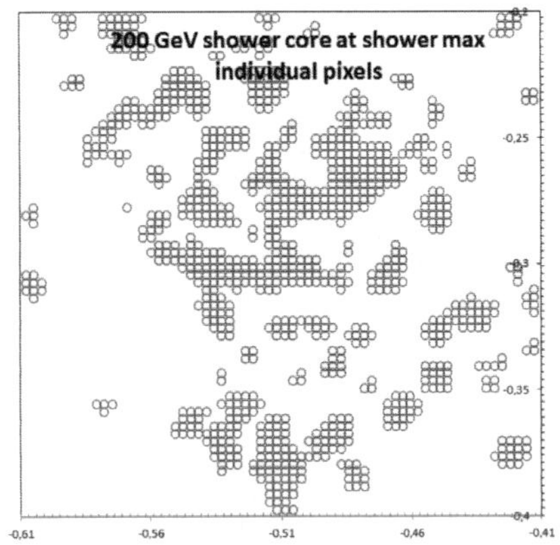

Figure 3. Close look at the shower core in 2 × 2 mm range. The individual pixel corresponds to a single circle.

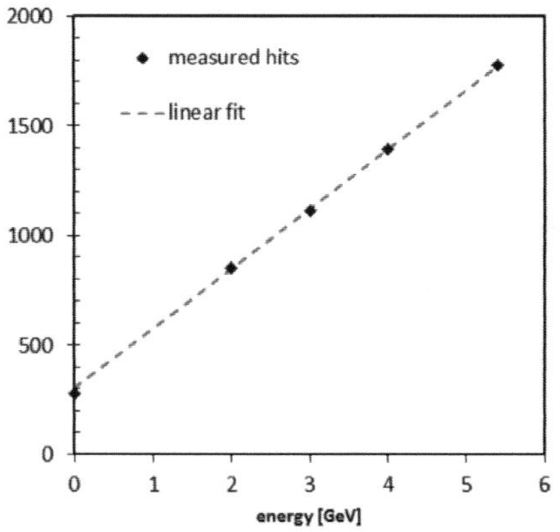

Figure 4. The average number of hits per event as function of the energy for raw data is shown. The point at zero corresponds to the integrated noise of the whole detector.

3.2. Longitudinal profile

The longitudinal profile averaged over more than 2000 shower events is depicted in Figure 5 for two energies: 2 and 5.4 GeV. The shower maxima are displaced according to the shower energy as expected. There are visible instrumental effects like, for example, in layer 5 where there is a decrease in number of hits for both energies. The decrease is caused by non working channels as illustrated in Figure 6. Unfortunately more than third of the detector is affected by such a problem that, as consequence, spoils the longitudinal profile and broadens the energy resolution.

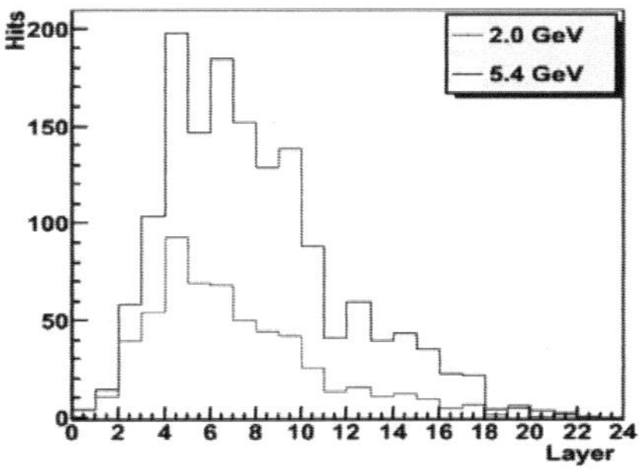

Figure 5. Longitudinal profile averaged over 2000 showers for 2 and 5.4 GeV electron beam. The not perfect longitudinal profile is due instrumental effects.

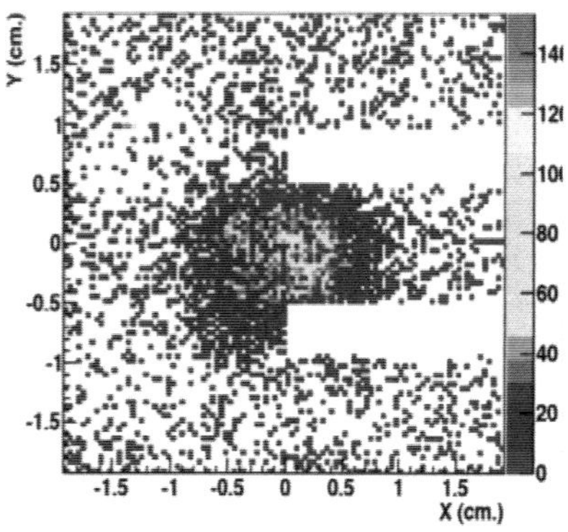

Figure 6. As example of not working channels, the hits map of layer 5 integrated over several shower events is reported. The white bands correspond to the missing channels that cause the decrease in the number of the counted hits in such a layer.

3.3. Lateral Profile

The lateral profile $\frac{dN}{2\pi r dr}$ has been estimated as the average over a few hundred showers. No calibration has been performed and only a preliminary alignment of the chips within 0.1 cm has been performed. In Figure 7 the average shower development for 5.4 GeV is plotted. The incoming particle leaves a track signal in layer 0 (there is no absorber in front of the sensor) and the shower develops leading to a maximum in layer 4. The distributions of layer 9 and layer 14 decrease at large r to an asymptotic value of 1 hit cm^{-2}, which is the noise level.

From the lateral distribution integrated over all the layers the Molière radius is calculated (see Figure 8). We obtain: $R_M = 11 \pm 1$ mm, which confirms the detector design expectations.

3.4. Energy Resolution

The energy resolution is a key quantity for a digital calorimeter. GEANT based simulation provides an estimate of the expected energy resolution, see Figure 9. Although the detector description is implemented to represent the real detector as closely as possible, e.g. also excluding the not working channels, there is still the approximation of identical sensors. This definitely contributes to a deteriorated energy resolution as measured from the data collected at the test beam at 30-50-100 GeV, which is about a factor of two worse than expected. In Figure 10 the detector response for data calibrated with MIP response are plotted. Further studies of the resolution including improvements of both the description in the simulation and a possible response correction of the experimental data are in progress.

3.5. Particle Identification

Despite the digital output it is possible to identify particles with unprecedented detail. In Figure 11 the number of hits as a function of the depth in units of radiation length (X_0) for two

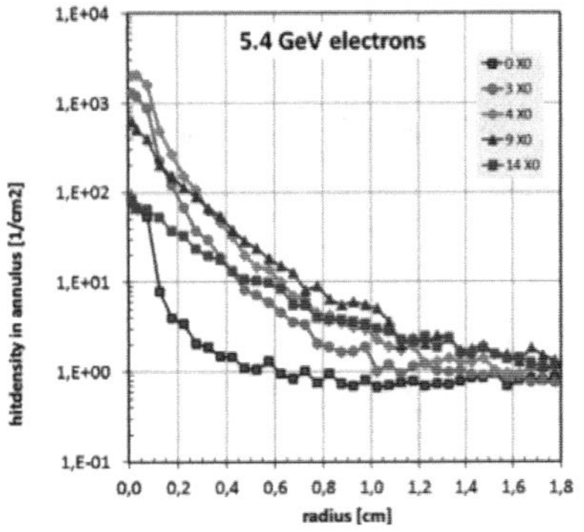

Figure 7. Lateral profile averaged over hundred showers at 5.4 GeV.

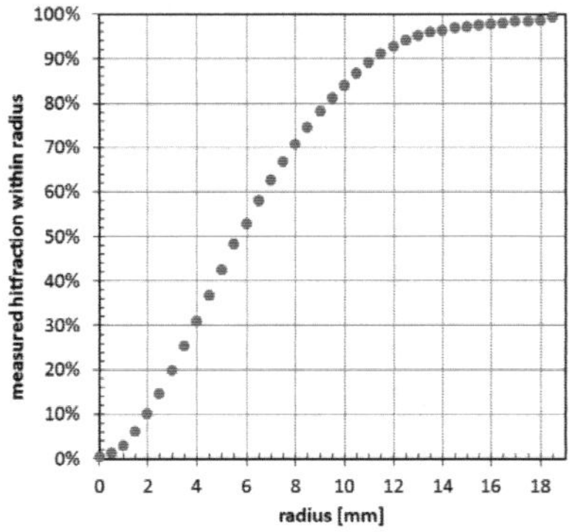

Figure 8. Hits fraction versus the distance from the shower core. From this measurement the estimate of the Molière radius for such a detector is 11 ± 1 mm as estimated from in the detector design.

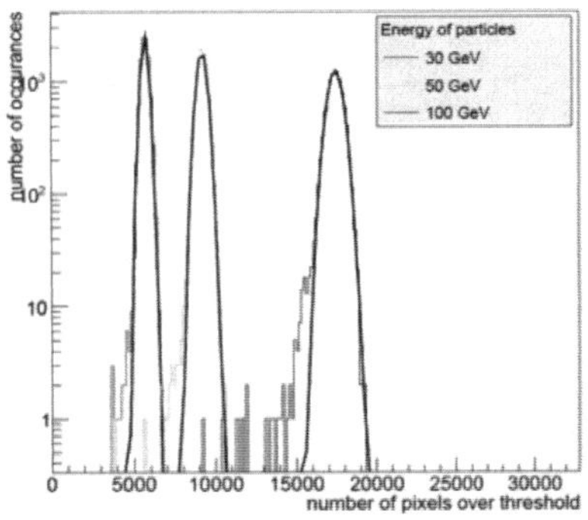

Figure 9. Simulations of the detector response for 30, 50 and 100 GeV with an energy resolution of 4.1%, 3.4% and 2.7% respectively.

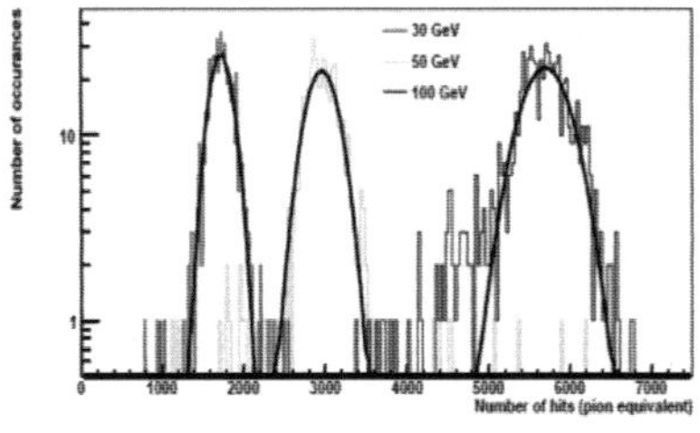

Figure 10. Calibrated detector response with pion for 30, 50 and 100 GeV. The calculated energy resolution for the same energies is: 8.1%, 6.8% and 5.3%.

showers at 50 GeV and 200 GeV, respectively is shown. These showers appear similar according to the total number of hits. The extremely high sensor granularity allows us to sample the lateral and longitudinal shower evolution in detail and to use this for the discrimination between electromagnetic and hadronic showers, as illustrated in Figures 12 and 13.

4. Conclusions

In this article the construction feasibility of digital electromagnetic calorimeter with sampling of $1X_0$ and spatial resolution down to few hundreds micron has been demonstrated. The sensors equipping the prototype are MAPS that have the front-end embedded in the sensor with very small pixel size, 30 μm. The 39 M pixels are read out at the GB/s rate. The sensor type is not the final one, but this large MAPS application shows the technology and integration feasibility,

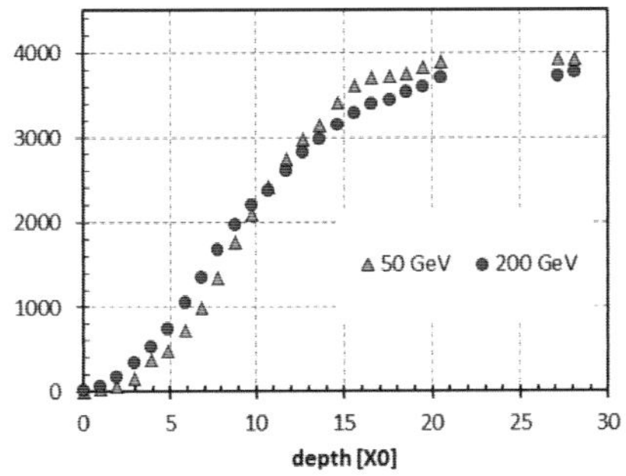

Figure 11. Cumulative number of hits as a function of the depth for an electromagnetic and a hadronic shower at 50 GeV and 200 GeV. The hits distribution for the same showers is displayed in figures 12 and 13.

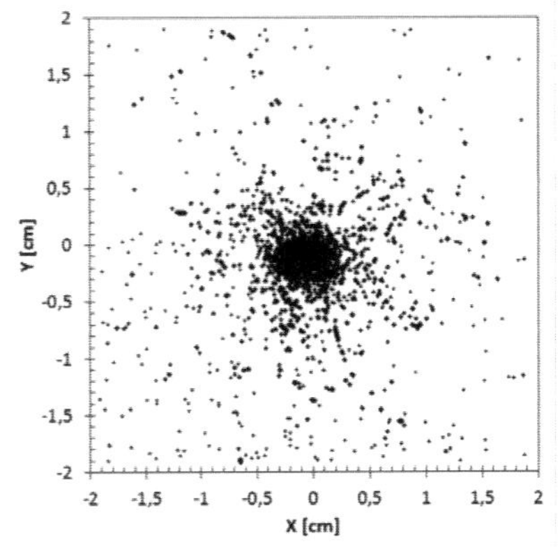

Figure 12. Event display of an electromagnetic shower at 50 GeV integrated over the full depth of the detector.

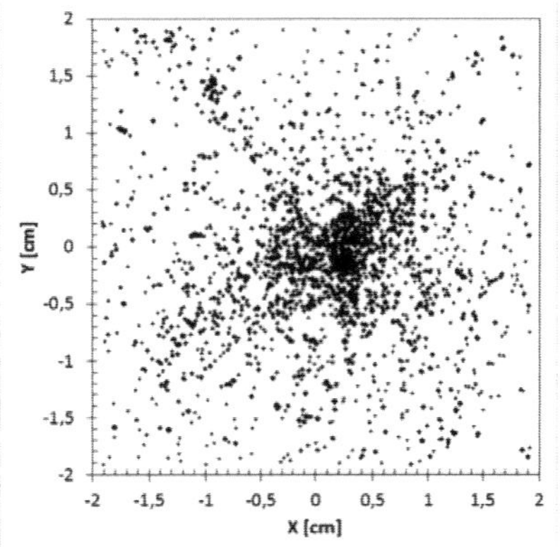

Figure 13. Event display of a hadronic shower at 200 GeV integrated over the full depth of the detector.

allows the MAPS characterisation study in terms of the signal saturation and study of the shower development.

The particle counting approach has been demonstrated to work for a fully digital electromagnetic calorimenter. Results on particle identification and the transverse and longitudinal shower development is reported. A very small Molière radius of 11 mm has been demonstrated, which is in perfect agreement with the design expectations.

The simulated energy resolution for 30, 50 and 100 GeV is a factor of two off from the measurements. This is partially explained by the missing implementation of the single chip response in the simulation. However it is work in progress.

Acknowledgments

The test beam performed at DESY in 2014 has received funding from the European Commission under the FP7 Research Infrastructures project AIDA, grant agreement no.262025.

References

[1] R. Turchetta *et al.*, *Nucl. Inst. Meth. A* **458** (2001), 677
[2] Th. Peitzmann, Prototype studies for a forward EM calorimeter in ALICE, Proceedings of CHEF 2013
[3] A. Brogna *et al.*, Manual PHASE1, IPHC Strasbourg
[4] D. Fehlker *et al.*, 2013 *JINST* **8** P03015

Journal of Physics: Conference Series **587** (2015) 012062 doi:10.1088/1742-6596/587/1/012062

DSB:Ce^{3+} scintillation glass for future

E. Auffray[a, 1], N. Akchurin[d], A. Benaglia[a], A. Borisevich[b], C. Cowden[d], J. Damgov[d], V. Dormenev[b], C. Dragoiu[d], P. Dudero[d], M. Korjik[b], D. Kozlov[b], S. Kunori[d], P. Lecoq[a], S.W. Lee[d], M. Lucchini[a,c], V. Mechinsky[b], K. Pauwels[a]
[a]CERN, Switzerland
[b]Institute for Nuclear Problems Minsk, Belarus
[c]University of Milano-Bicocca, Italy
[d]Texas Tech University, USA

E-mail: etiennette.auffray@cern.ch

Abstract. One of the main challenges for detectors at future high-energy collider experiments is the high precision measurement of hadron and jet energy and momentum. One possibility to achieve this is the dual-readout technique, which allows recording simultaneously scintillation and Cherenkov light in an active medium in order to extract the electromagnetic fraction of the total shower energy on an event-by-event basis. Making use of this approach in the high luminosity LHC, however, puts stringent requirements on the active materials in terms of radiation hardness. Consequently, the R&D carried out on suitable scintillating materials focuses on the detector performance as well as on radiation tolerance. Among the different scintillating materials under study, scintillating glasses can be a suitable solution due to their relatively simple and cost effective production. Recently a new type of inorganic scintillating glass: Cerium doped DSB has been developed by Radiation Instruments and New Components LLC in Minsk for oil logging industry. This material can be produced either in form of bulk or fiber shape with diameter 0.3-2mm and length up to 2000 mm. It is obtained by standard glass production technology at temperature 1400°C with successive thermal annealing treatment at relatively low temperature. The production of large quantities is relatively easy and the production costs are significantly lower compared to crystal fibers. Therefore, this material is considered as an alternative and complementary solution to crystal fibers in view of a production at industrial scale, as required for a large dual readout calorimeter. In this paper, the first results on optical, scintillation properties as well as the radiation damage behaviour obtained on different samples made with different raw materials and various cerium concentrations will be presented.

1. Introduction

In future high energy experiments at new particle colliders, the precise measurement of the jets and hadrons will be one of the main challenges. The dual readout technique is one of the approaches,

---

[1] To whom any correspondence should be addressed.

which allows the improvement of the hadronic energy resolution by measuring the electromagnetic fraction of the hadronic shower on an event-by-event basis [1]. This can be obtained by detecting simultaneously scintillation and Cerenkov signal produced in an active medium. In High Luminosity LHC (HL-LHC), the operational conditions in a high dose rate irradiation environment add a stringent requirement in terms of radiation hardness for the active material and therefore require search for new radiation tolerant materials. Among the different scintillating materials, scintillating glasses, despite their relatively simple and cost effective production, have, up to now, not been widely used in high energy experiments due to their poor radiation hardness and low light yield. The main cause of the low radiation tolerance of usual scintillating glasses is their amorphous structure. The "non-organized" structure of the atoms allows presences of structural defects, which favor the creation of colors center under irradiation.

In the case of binary systems based on SiO_2, the lattice structure is formed by layers of tetrahedral Si with metallic ions between them. Therefore monovalent or bivalent metallic ions can easily be replaced by bivalent or trivalent rare earth ions. In the last case, charge compensation occurs by reordering of cation systems. For instance, three divalent ions are replaced by two trivalent species. The disilicate of barium (BaO_2-SiO_2) doped with Ce (DSB:Ce) is one of the new scintillation materials made from binary composition and obtained by standard glass production technology with a successive thermal annealing. In this paper, we discuss the optical quality of DSB based on measurements performed on a first set of samples.

2. DSB production and properties

DSB is produced from a mixture of BaO_2 and SiO_2 by standard glass production technology at a temperature of 1400°C. The obtained glass is then thermally annealed at relatively low temperature (<1000°C) to achieve its nanostructure. It can be produced either in form of bulk or fiber shape with diameters of 0.3-2mm and lengths up to 2m as can be seen in Fig. 1.

As opposed to common heavy inorganic scintillators, DSB glass has a slightly lower stopping power. Its density is 3.8 g/cm^3, its effective Z is 51 and its radiation length (X_0) is 3.3 cm. Therefore applications of DSB in calorimetry will most probably require an additional absorber. On the other hand, on the contrary to crystalline materials, the production of large quantities is relatively easy and the production costs are rather low.

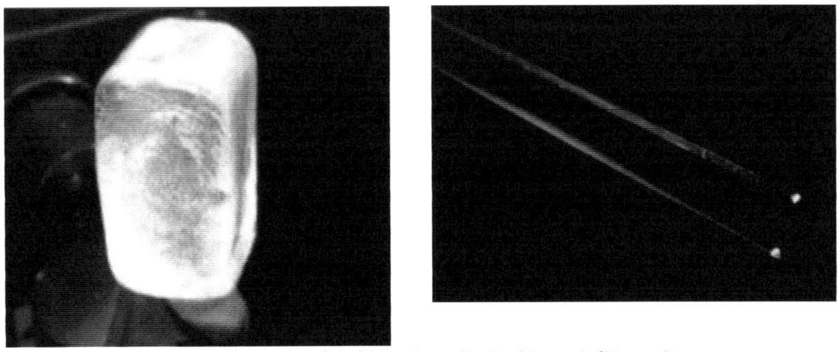

Figure 1: Picture of DSB glass in bulk and fiber shapes.

3. DSB:Ce samples

For the present study, five samples of bulk DSB glass material with different Cerium concentration were investigated (reference number #2743-2747). Samples #2743-2745 and #2746,2747 were produced from two different types of raw material purity. The Cerium concentration and raw material is provided together with a picture in Fig. 2.

Journal of Physics: Conference Series **587** (2015) 012062 doi:10.1088/1742-6596/587/1/012062

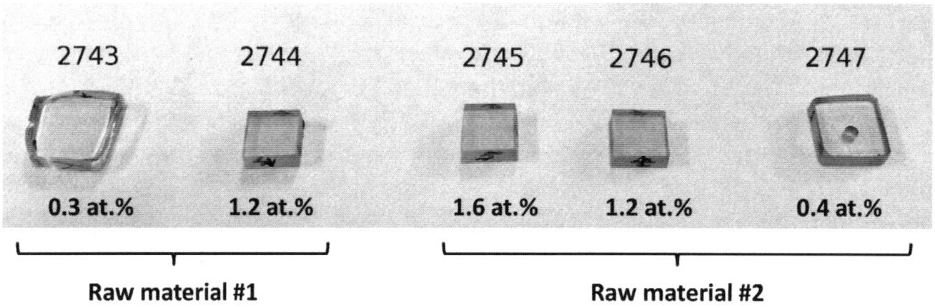

Figure 2: Picture of the five DSB samples studied here, with their respective Cerium concentration and raw material origin.

4. Measurements of the optical properties

A typical transmission spectra of DSB:Ce is shown in Fig. 3. It is characterized by an absorption edge around 365nm corresponding to the Cerium absorption band and a transmission above 80% at higher wavelengths. In Fig. 4, the excitation and emission spectra measured by photoluminescence at room temperature are shown. For these measurements, we selected the sample with the lowest Cerium concentration (#2743). The excitation spectrum (measured at an emission wavelength of 425 nm) is characterized by two peaks at 280 nm and 350 nm, due to the 5f-4d Ce^{3+} transition. The emission spectrum (measured at an excitation wavelength of 300 nm) is peaking at 425nm. It is noteworthy that the position of this peak depends on the concentration of Ce, as shown in Fig. 5.

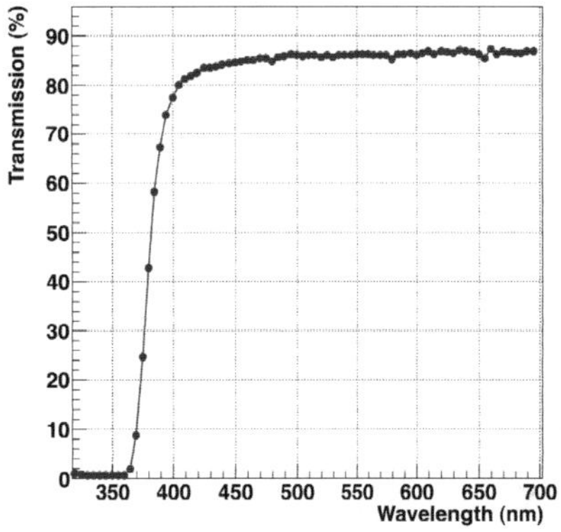

Figure 3: Optical transmission of DSB sample.

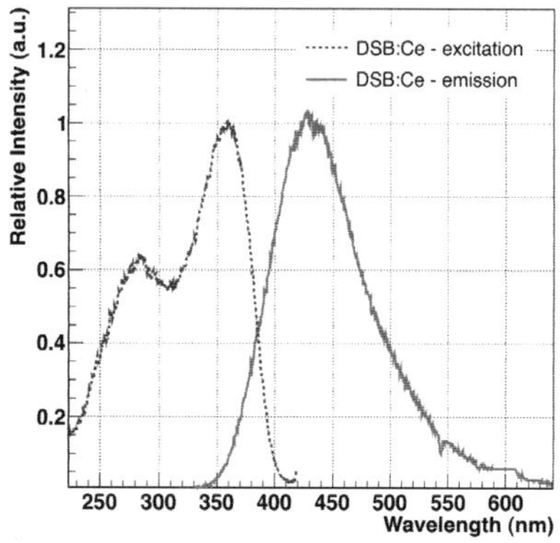

Figure 4: Excitation and emission spectra of DSB.

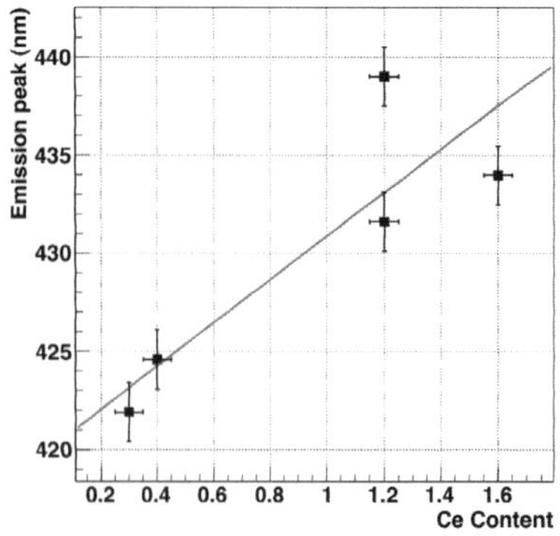

Figure 5: Correlation between position of peak emission and cerium concentration.

5. Measurements of the scintillation properties

All the samples show a very similar scintillation decay constants. Fig. 6 shows the scintillation decay time spectra of one of the DSB sample, measured at room temperature. The decay time is well approximated using a sum of three exponentials. The pulse shape is dominated by a fast 30-ns decay (~40% of total weight) and a slightly slower component (~50% of total weight) around 180 ns. The presence of this slower component in the scintillation depends both on Ce concentration and Ba/Si ratio in a final composition. It is noteworthy that ~74% of the light is contained in the first 100ns. Such a time response is compatible with high repetition rates, as those considered for HL-LHC. For what concerns the very fast component, the role of the instrumental response cannot here be completely excluded.

The light output of one of the DSB sample (volume ~ 1cm^3) was measured with γ-rays of 59.9 keV and 662 keV generated by respectively an ^{241}Am and a ^{137}Cs radioactive source. The obtained light output spectrum is presented on Fig 7. We estimate the light output to be around 100pe/MeV, thus five times larger than the one of a PWO crystal at room temperature [3]. We also investigated the temperature dependence of the light output from -25 to +25 °C. The temperature coefficient was found to be less that 0.04%/°C [3].

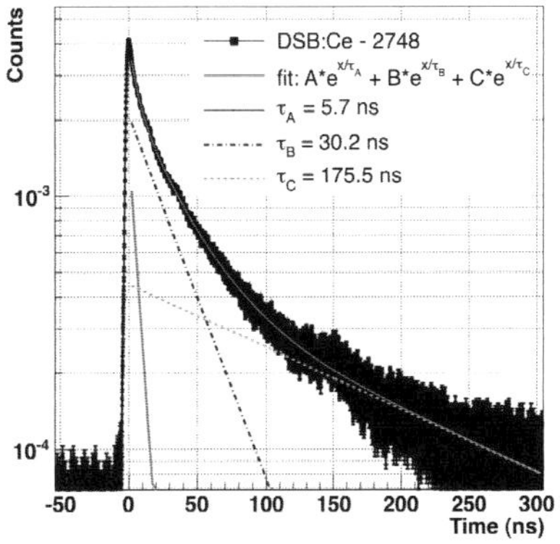

Figure 6: Typical decay time spectrum of DSB and its decomposition in three decay components.

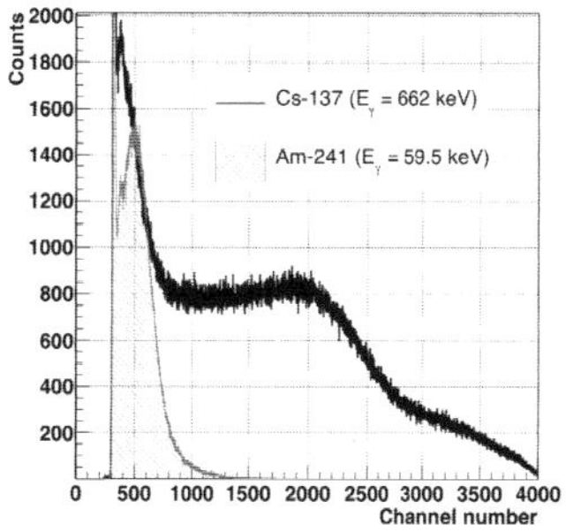

Figure 7: Light output spectrum obtained with γ-rays of 59.6 keV and 662 keV.

6. Preliminary measurements of the radiation hardness

First irradiation tests have been performed on DSB samples with gamma irradiation at a dose rate of 500Gy/h and a total dose of 1000Gy. Figure 10 shows the comparison of the induced absorption for two samples produced with two different types of raw material. A factor two in the damage intensity is observed between the two samples. These preliminary results indicate that by optimizing the raw material purity and the production parameter improvement of radiation damage could be obtained. Further optimization of raw material and production conditions are currently investigated.

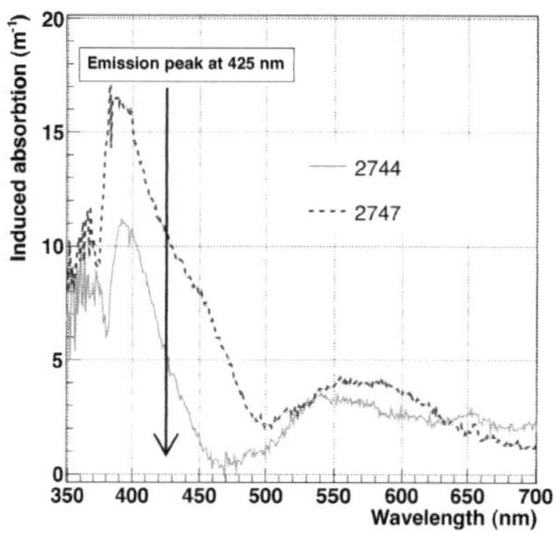

Figure 9: Comparison of induced absorption after 1000Gy irradiation (Co^{60}), on two DSB samples produced with two different raw materials.

7. Conclusions

DSB:Ce scintillating glass has been investigated for the first time. DSB is characterized by an emission at 440nm, which is well matching the sensitivity of the commonly used photodetectors like PMT, VPT, SiPM and APD. Moreover, DSB:Ce has a very small temperature dependence of the light output: 0.05 %/°C, thus about 40 times less than the one of PWO [3]. Therefore in the range of -20-+50°C no complex cooling system with be necessary to operate a detector built with DSB:Ce. Material. Undoped DSB will be a Cherenkov radiator.

The production of large quantities is relatively easy and the production costs are significantly lower compared to crystal fibers. Therefore, this material is considered as an alternative and complementary solution to crystal fibers in view of a production at industrial scale, as required for a large dual readout calorimeter. The possibility to produce also DSB in fiber shape at a rate of 2cm/s allows the production of large quantities of fibres relatively easily and a production cost significantly lower compared to crystal fibers. Therefore, this material is considered as an alternative and complementary solution to crystal fibers in view of a production at industrial scale, as required for a large dual readout calorimeter.

Acknowledgement

The authors thank for providing the samples Radiation Instruments and New Components LLC in Minsk.

References

[1] R. Wigmans et al. NIMA 617(2010)129–133

[2] V. Dormenev et al, Presentation at the calor conference 2014

[3] I. Dafinei, E.Auffray, P.lecoq, M. Schneegans, Mat. Res. Soc. Symp. PROC. Vol348 (1994)99 - 104

Journal of Physics: Conference Series **587** (2015) 012063 doi:10.1088/1742-6596/587/1/012063

Optical transmission radiation damage and recovery stimulation of DSB: Ce^{3+} inorganic scintillation material

A. Borisevich[a], V. Dormenev[b], M. Korjik[a], D. Kozlov[a], V. Mechinsky[a], R. W. Novotny[b]

[a] Institute for Nuclear Problems, Minsk, Belarus
[b] 2nd Physics Institute, Justus-Liebig-University, Giessen, Germany

E-mail: Rainer.Novotny@exp2.physik.uni-giessen.de

Abstract. Recently, a new scintillation material DSB: Ce^{3+} was announced. It can be produced in a form of glass or nano-structured glass ceramics with application of standard glass production technology with successive thermal annealing. When doped with Ce $^{3+}$, material can be applied as scintillator. Light yield of scintillation is near 100 phe/MeV. Un-doped material has a wide optical window from 4.5eV and can be applied to detect Cherenkov light. Temperature dependence of the light yield LY(T) is 0.05% which is 40 times less than in case of PWO. It can be used for detectors tolerant to a temperature variation between -20° to +20°C.
Several samples with dimensions of 15x15x7 mm^3 have been tested for damage effects on the optical transmission under irradiation with γ-quanta. It was found that the induced absorption in the scintillation range depends on the doping concentration and varies in range of 0.5-7 m^{-1}. Spontaneous recovery of induced absorption has fast initial component. Up to 25% of the damaged transmission is recuperated in 6 hours. Afterwards it remains practically constant if the samples are kept in the dark. However, induced absorption is reduced by a factor of 2 by annealing at 50°C and completely removed in a short time when annealing at 100°C. A significant acceleration of the induced absorption recovery is observed by illumination with visible and IR light. This effect is observed for the first time in a Ce-doped scintillation material. It indicates, that radiation induced absorption in DSB: Ce scintillation material can be retained at the acceptable level by stimulation with light in a strong irradiation environment of collider experiments.

1. Introduction

Up to date self-activated scintillation materials like BGO and PWO have found a wide application in detectors at high energy physics experiments. They possess a unique combination of the scintillation and physical properties, including high density, fast response and radiation hardness at least to the electromagnetic part of ionizing radiation. However, heavy materials demonstrate sufficient damage under the hadronic part of the ionizing radiation [1, 2] excluding consideration for high energy experiments at future particle colliders. Ce^{3+} doped scintillation materials are considered to be applied at calorimetric detectors. Among them, LYSO: Ce [3] shows a relatively high resistance to the damage of optical transmission. However, Lu-based crystals indicate after proton irradiation a high level of the radio-luminescence due to induced radio-nuclides and phosphorescence [4]. Therefore, we have paid our attention to medium heavy scintillation materials, particularly to a new scintillation material

Journal of Physics: Conference Series **587** (2015) 012063 doi:10.1088/1742-6596/587/1/012063

DSB: Ce^{3+} [5]. Scintillating glasses, despite their relatively easy and cost effective production, are so far not widely used in high energy experiments due to their poor radiation hardness and low light yield. The Di-silicate of barium (BaO-2SiO_2) doped with cerium (DSB: Ce) is one of the new scintillation materials made from binary stoichiometric composition and produced according to glass manufacturing technology with a successive thermal annealing to obtain nano-structuring of the material. In this paper, the first results of measurements of the radiation damage effects and its recovery are presented. For the first time, the effect of stimulated recovery has been observed in Ce-doped inorganic scintillation materials.

2. DSB samples and measurements.

DSB glass samples were provided by Radiation Instruments and New Components (Belarus) to evaluate their radiation hardness to γ-irradiation. The material can be produced as glass or nano-structured glass ceramics using the common production technology accompanied with successive thermal annealing. When doped with Ce^{3+}, it emits scintillation light at a yield of ~100 phe/MeV. Un-doped material has a wide optical window from 4.5eV and can be used as a Cherenkov radiator. The temperature dependence of the light yield LY(T) is on the level of 0.05 % which is 40 times less than in case of $PbWO_4$. Therefore, it becomes tolerant to temperature variation in the wide range of -20°C to +20°C. Therefore, DSB: Ce can be considered to be a prospective material for application in calorimetry. Several samples were irradiated at the Radiation Centre (Justus-Liebig-University Giessen, Germany) using a strong ^{60}Co source at a dose rate of 2Gy/min. Figures 1 and 2 present the transmission spectra of low- (less than 0.1 weight %) and heavy-doped (more than 1 weight %) DSB: Ce, before and after irradiation with γ-rays.

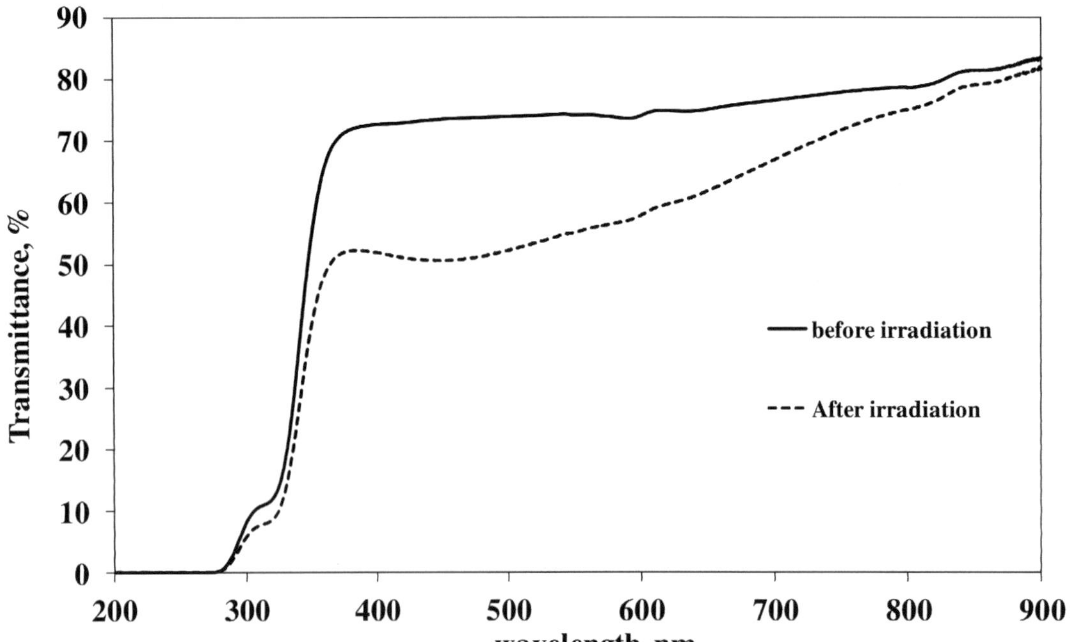

Figure 1. Comparison of the optical transmission spectra measured at room temperature of low-doped DSB: Ce samples before and after irradiation with a ^{60}Co source with an integral dose of 100Gy. The thickness of the sample is 8 mm.

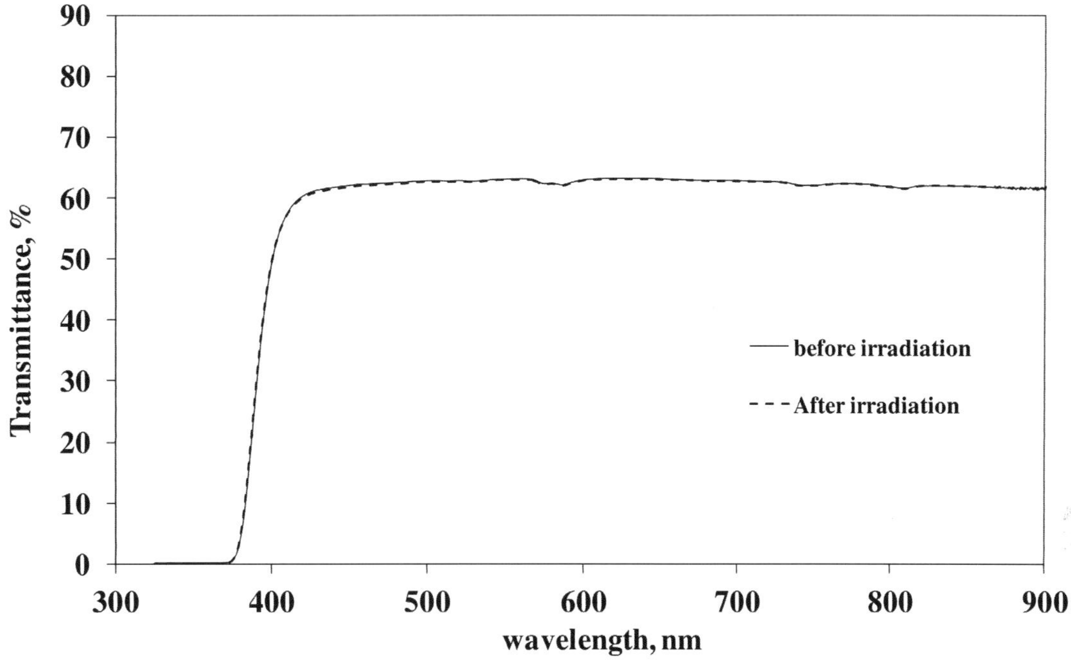

Figure 2. Comparison of optical transmission spectra measured at room temperature of a heavy-doped DSB: Ce sample before and after irradiation with a ^{60}Co source with an integral dose of 100Gy. The thickness of the sample is 4.45mm.

The low-doped sample shows several color centers peaking near 450nm and 650nm, respectively, whereas heavy-doped DSB samples show an extremely high radiation hardness against γ-irradiation.

3. Spontaneous and stimulated recovery of the gamma irradiation induced optical absorption

The effect of stimulated recovery of the radiation damage has been observed so far in self-activated scintillation material, namely $PbWO_4$ [6]. It was suggested that Ce-doped scintillation materials, due to the presence of cerium ions with a high capture cross-section of free carriers, should show fast recombination of color centers, especially created due to shallow traps. However, our study shows that even at high concentration of Ce^{3+} in the inorganic material, spontaneous recovery of the induced absorption is a relatively slow process. Figure 3 shows spontaneous recovery at room temperature of the normalized radiation induced coefficient of a heavy-doped DSB: Ce sample. Fast recovery is observed only in the initial. Only up to 25% of the reduced optical transmission is recuperated after 6 hours. Afterwards, the value of induced absorption remains practically constant if the samples are kept in the dark. However, induced absorption is decreased by a factor of two due to annealing at 50°C and completely removed by a short annealing phase at 100°C. It appears that the fast component of the recovery is due to recombination of color centers with cerium ions. However, annealing of the majority of the color centers above room temperature and their slow spontaneous recombination at room temperature indicate deep traps which only weakly interact with cerium ions. Therefore, recombination can be significantly enhanced by illumination with external optical photons.

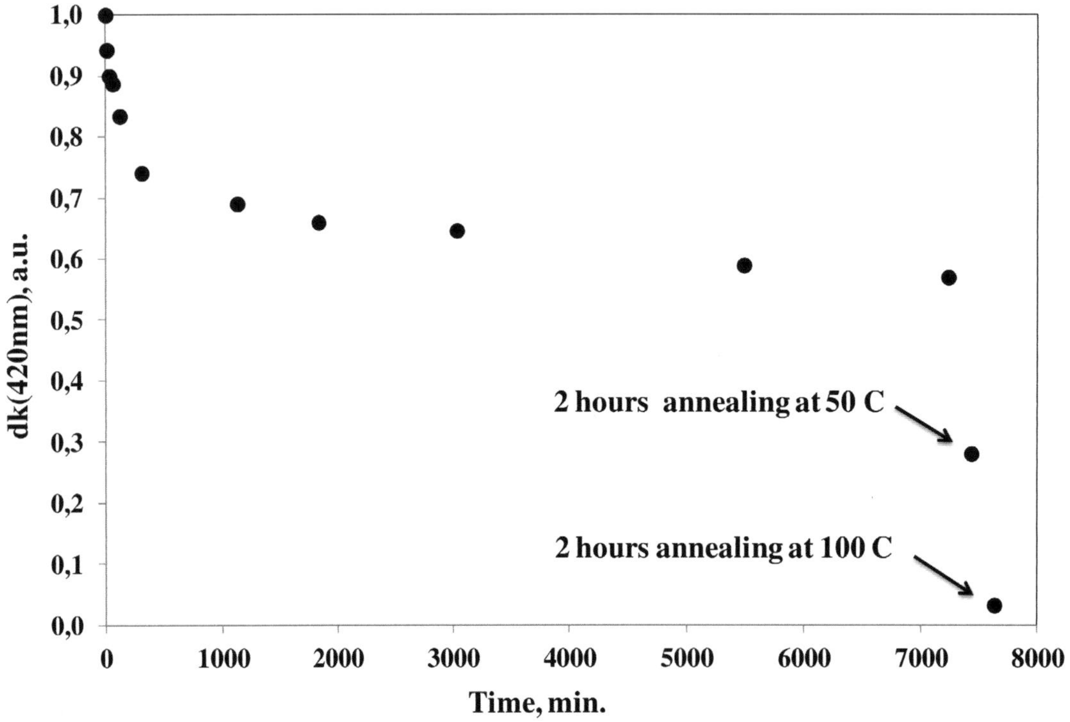

Figure 3: Spontaneous recovery at room temperature recovery of the normalized radiation induced coefficient at 420nm wavelength of a heavy-doped DSB: Ce sample after 500Gy of absorbed dose.

There are two mechanisms initiated by the photons: 1) ionization of color centers and 2) transport of the captured electron from the ground state of the color center to the radiating level or center. The first process depends on the energy width of the conduction band and the position of the ground state of the color center in the forbidden zone and may be initiated in a wide spectral region from UV to visible light. In Ce-doped material, ionization may be effective since electrons released from color centers can be captured by cerium ions preventing their re-capturing by other traps. The second type of process is an intra-center resonant transition among color centers or between a color center and a cerium ion. Due to the lower energy difference $E_f \sim E_{TA}$ even infrared (IR) photons can initiate the mechanism . Figure 4 illustrates the impact of light illumination on the recovery of the induced absorption coefficient in a heavy-doped DSB: Ce sample at room temperature.

The effect is achieved by the injection of photons of selected wavelengths with an intensity of 2 to $9 \cdot 10^{16}$ photons/s into the sample wrapped in a reflector foil. However, the time constants of the recovery progressively increase shifting towards the IR range. Stimulated relaxation provides relaxation within a wide spectral region. As expected, illumination with blue light, which can cause even ionization of the color centers, leads to the fastest recuperation. The observed results confirm that radiation induced absorption in cerium doped scintillation materials also can be promptly recuperated using stimulated recovery with optical photons.

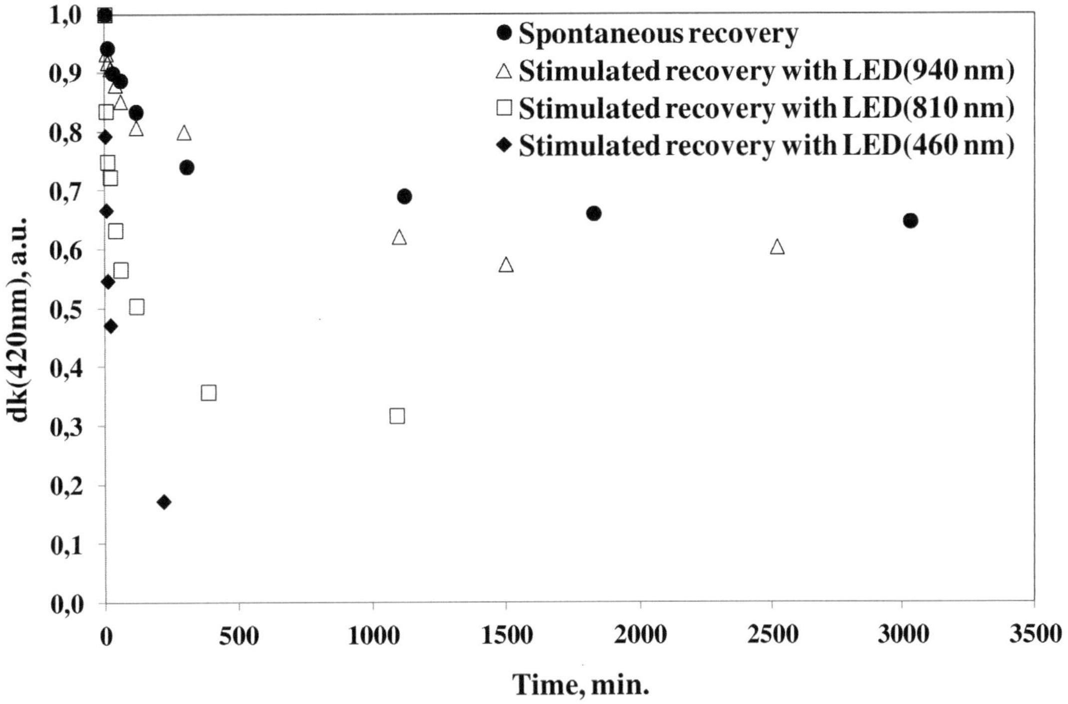

Figure 4. Stimulated recovery of the normalized radiation induced coefficient at 420nm for a heavy-doped DSB: Ce sample after an integral dose of 500Gy using optical photons of selected wavelengths measured at room temperature.

4. Conclusions

The radiation damage of new DSB: Ce scintillating glass quantified by the change of the optical transmission was investigated. It was found that heavy-doped glasses show a extremely high radiation resistance. It was observed for the first time that the radiation damage in Ce-doped scintillators can be recuperated by illuminating the samples with optical photons.

Acknowledgement

The authors thank Radiation Instruments and New Components LLC (Belarus) for providing the samples.

References

[1] E. Auffray, A. Barysevich, A. Fedorov, M. Korjik, M. Koschan, M. Lucchini, V. Mechinski, C.L. Melcher, A. Voitovich, Radiation damage of LSO crystals under γ- and 24 GeV protons irradiation. Nuclear Instr. and Meth. in Physics Res. **A721** (2013) pp.76-82.

[2] E. Auffray , M. Korjik, A. , Experimental Study of Lead Tungstate Scintillator Proton-Induced Damage and Recovery. IEEE Trans. on Nucl. Sci, **59** (2012) pp. 2219-2223.

[3] J. Chen, L. Zhang, R.-Y. Zhu, Large size LSO and LYSO crystal scintillators for future high-energy physics and nuclear physics experiments. Nuclear Instr. and Meth. in Phys. Res.

A572 (2007) pp.218-224.
[4] E. Auffray, A. Fedorov, M. Korjik, M. Lucchini, V. Mechinski, N. Naumenko, A. Voitovich, Radiation Damage of Oxy-Orthosilicate Scintillation Crystals Under Gamma and High Energy Proton Irradiation. IEEE Trans. on Nucl. Sci. **61** (2014) pp. 495-500.
[5] Baker Hughes International Information, 2013.
[6] A. Borisevich, V. Dormenev, T. Kuske, M. Korjik, S. Lugert, V. Mechinski, O. Missevitch, R. W. Novotny, A. Fedorov, Stimulated recovery of the optical transmission of $PbWO_4$ scintillation crystals for electromagnetic calorimeters after radiation damage. Nucl. Instr. and Meth. in Physics Res. **A623** (2010) pp. 1082-1085.

Journal of Physics: Conference Series **587** (2015) 012064 doi:10.1088/1742-6596/587/1/012064

BSO Crystals for the HHCAL Detector Concept

Fan Yang[1], Hui Yuan[2], Liyuan Zhang[1] and Ren-Yuan Zhu[1]
[1] 256-48, HEP, California Institute of Technology, Pasadena, CA 91125, USA
[2] Shanghai Institute of Ceramics, Shanghai, 264000, PRC

E-mail: zhu@hep.caltech.edu

Abstract. We report an investigation on optical and scintillation properties and radiation hardness of four 20 cm long BSO crystals grown at SIC for the HHCAL detector concept. Their optical and scintillation properties, such as longitudinal transmittance, light output and light response uniformity, were measured before and after γ-ray irradiation. Progresses are observed in optical quality, light output and radiation hardness. Their use for HHCAL concept is discussed.

1. Introduction

Aiming at the best jet-mass resolution cost-effective inorganic crystal scintillators are being developed for a homogeneous hadronic calorimeter (HHCAL) detector concept with dual readout of both Cherenkov and scintillation light for future high energy lepton colliders [1]. Figure 1 shows a schematic of a typical HHCAL cell with a pointing geometry with dual readout [2]. Because of the unprecedented volume (70 to 100 m^3) foreseen for the HHCAL detector concept cost-effectiveness is the most important requirement. In addition, the material must be dense to reduce the calorimeter volume, UV transparent for effective collection of the Cherenkov light, and allow for a clear discrimination between the Cherenkov and scintillation light. The preferred scintillation light is thus at a longer wavelength, and not necessarily bright nor fast. Inorganic crystals being investigated are lead fluoride (PbF_2) [2,3], lead chloride fluoride (PbFCl) [4-6] and bismuth silicate ($Bi_4Si_3O_{12}$ or BSO) [7-9]. Because of its low UV cut-off wavelength (300 nm) and low raw material cost (<50% of BGO), BSO crystals are under development at Shanghai Institute of Ceramics (SIC).

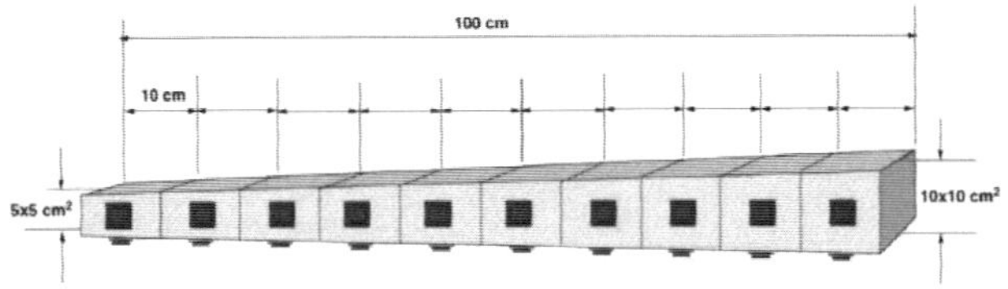

Figure 1. A schematic showing a typical cell for the HHCAL detector concept

Figure 2. Four 20 cm long BSO crystal samples grown at SIC

Figure 2 shows four BSO crystal samples grown at SIC by modified Bridgman method. Samples 12-2211 and 12-2223 were grown in 2012. Samples 13-05 and 13-09 were grown in 2013. While samples 12-2223, 13-05 and 13-09 are of a dimension of 20×20×200 mm^3, sample SIC 12-2211 is 15×15×200 mm^3. Their longitudinal transmittance (LT) was measured by using a PerkinElmer LAMBDA 950 UV/Vis spectrophotometer. The systematic uncertainty in repeated measurements is about 0.15%. Their scintillation light output (LO) was measured before and after irradiation by using a Hamamatsu R2059 PMT with a bi-alkali photo-cathode and a quartz window. For the light output measurement, one end of the samples was coupled to the PMT with Dow Corning 200 fluid while all other faces of the sample were wrapped with Tyvek paper. A collimated ^{137}Cs source was used to excite the crystals. The γ-ray peak positions were obtained by a simple Gaussian fit, and were used to determine photoelectron numbers by using calibrations of the single photoelectron peak.

2. Initial optical and scintillation properties

Figure 3 shows initial LT spectra before irradiation together with the emission spectrum for all samples. Also shown in the figure is the numerical values of the emission weighted longitudinal transmittance (EWLT), which is defined as [10]:

$$EWLT = \frac{\int LT(\lambda)Em(\lambda)d\lambda}{\int Em(\lambda)d\lambda} \qquad (1)$$

EWLT represents crystal's transparency better than the transmittance at the emission peak since it runs through the entire emission spectrum. Two observations are in order: (1) the longitudinal transmittance between 500 to 800 nm of the sample SIC 12-2211 approaches the theoretical limit, which was calculated assuming multiple bouncing without internal absorption [11]; and (2) the amplitudes of the permanent absorption bands observed between 300 and 500 nm are significantly reduced in the sample SIC 13-09 as compared to all other samples. These observations indicate that the optical quality of BSO crystals may be further improved.

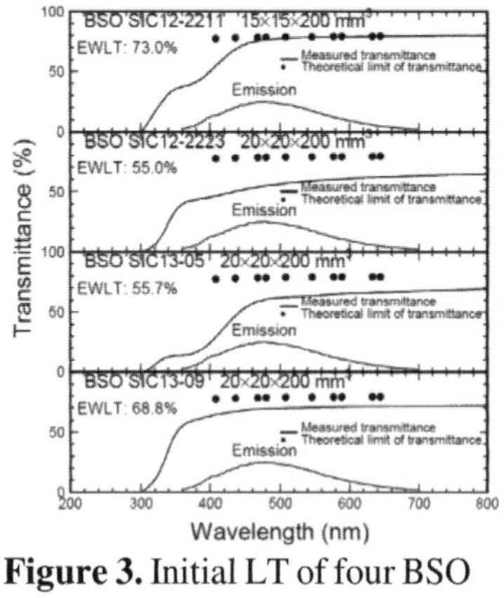

Figure 3. Initial LT of four BSO samples

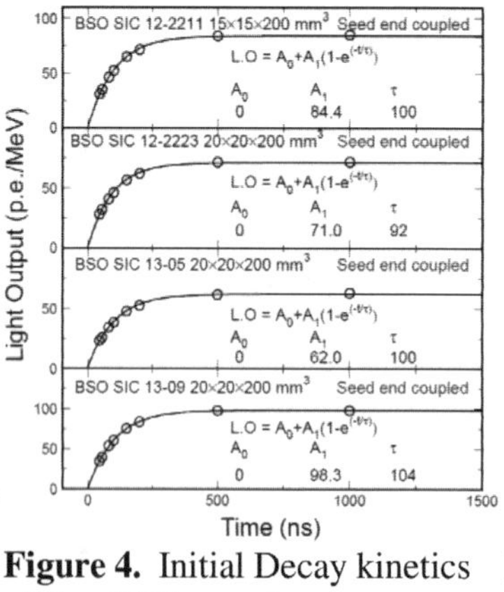

Figure 4. Initial Decay kinetics of four BSO samples

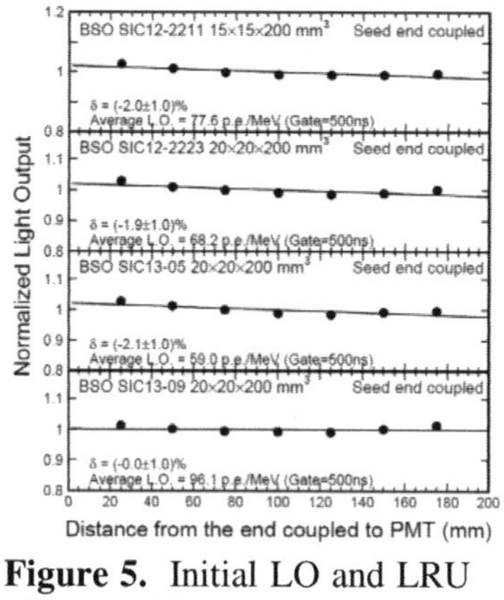

Figure 5. Initial LO and LRU of four BSO samples

Figure 4 shows the initial decay kinetics measured before irradiation for all samples. The scintillation decay time is about 100 ns. Figure 5 shows the initial light output (LO) and light response uniformity (LRU) measured before irradiation for all samples. The light response uniformity of long crystals was measured by moving a collimated γ-ray source along the longitudinal axis of the crystal at seven points evenly distributed along the crystal. The corresponding response (y) was fitted to a linear function [12]:

$$\frac{y}{y_{mid}} = 1 + \delta\left(\frac{x}{x_{mid}} - 1\right), \qquad (2)$$

where y_{mid} represents the light output at the middle of the crystal, δ represents the light response non-uniformity and x is the distance from one end of the crystal. Figure 5 shows that the non-uniformity observed in all samples are almost flat, and the sample SIC 13-09 has the best LO and LRU because of its best optical quality among all four samples.

3. Gamma ray induced radiation damage

Radiation damage in samples SIC 12-2211, 13-05 and 13-09 was found recover under room temperature, so they were irradiated by γ-rays at 2, 8, and 30 rad/h step by step until reaching equilibrium. Radiation damage in sample SIC 12-2223 does not recover under room temperature, so it was irradiated with integrated doses from 100 to 1 Mrad stepwise increased in an order of magnitude. The main consequence of γ-ray induced radiation damage in BSO crystals is radiation induced absorption which causes degradation of transmittance and light output. It may also affect crystal's light response uniformity and cause an unrecoverable degradation in the energy resolution if the light attenuation length is reduced significantly [13, 14].

3.1 γ-ray induced dose rate dependent damage in three BSO samples

Because of the recovery, the γ-ray induced radiation damage in samples 12-2211, 13-05 and 13-09 is dose rate dependent [14]. The optical transmittance and light output decreases when samples exposed to a certain dose rate until reaching equilibrium. At the equilibrium the speed of the color center formation (damage) equals to the speed of the color center annihilation (recovery), so that the color center density (radiation induced absorption) does not change unless the applied dose rate changes.

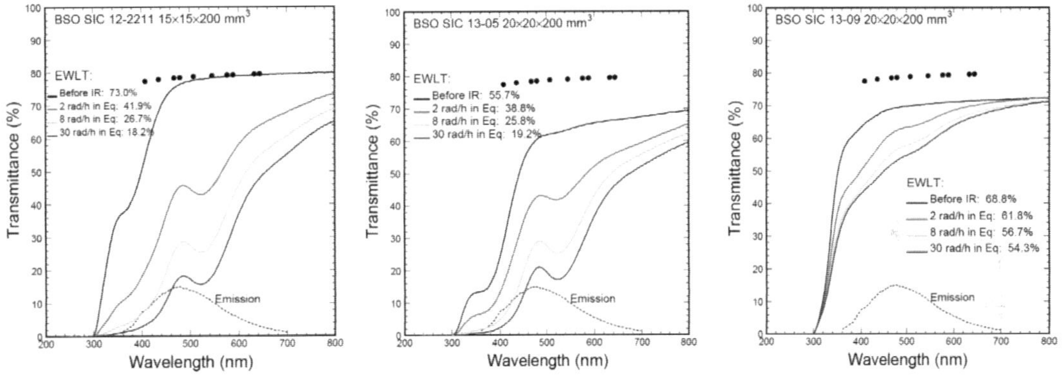

a b c

Figure 6. LT of three BSO samples in equilibrium under dose rates of 2, 8 and 30 rad/h:
a. SIC 12-2211; b. SIC 13-05; and c. SIC 13-09.

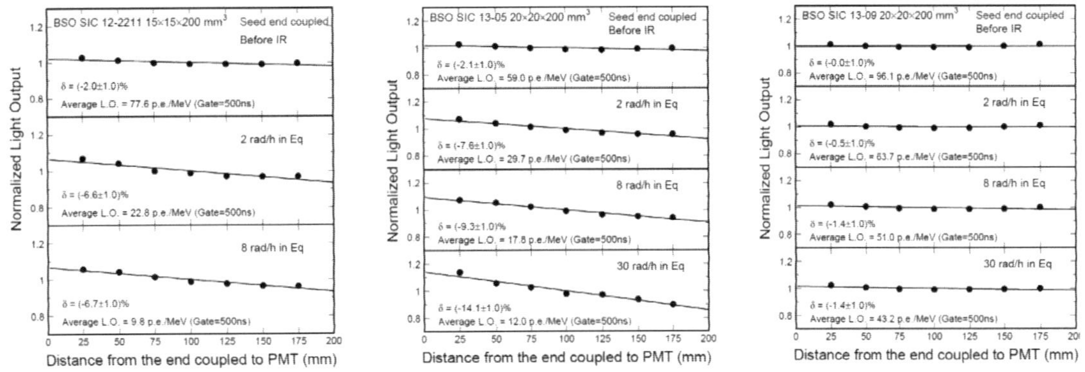

a b c

Figure 7. LO and LRU of three BSO samples in equilibrium under dose rate of 2, 8 and 30 rad/h:
a. SIC 12-2211; b. SIC 13-05; and c. SIC 13-09.

Figure 6 and 7 show LT and LO/LRU respectively of three BSO samples SIC 12-2211, 13-05 and 13-09 in equilibrium under dose rates of 2, 8, and 30 rad/h. The numerical values of EWLT are also listed in Figure 6, showing that the Sample SIC 13-09 has the lowest damage among three samples. The LO of sample SIC 12-2211 at 30 rad/h in equilibrium is not shown in Figure 7a since it is too low to be measured by using γ-rays. Figure 7c shows that the LRU of SIC 13-09 was not changed up to 30 rad/h, indicating sufficient long light attenuation length under this dose rate [14].

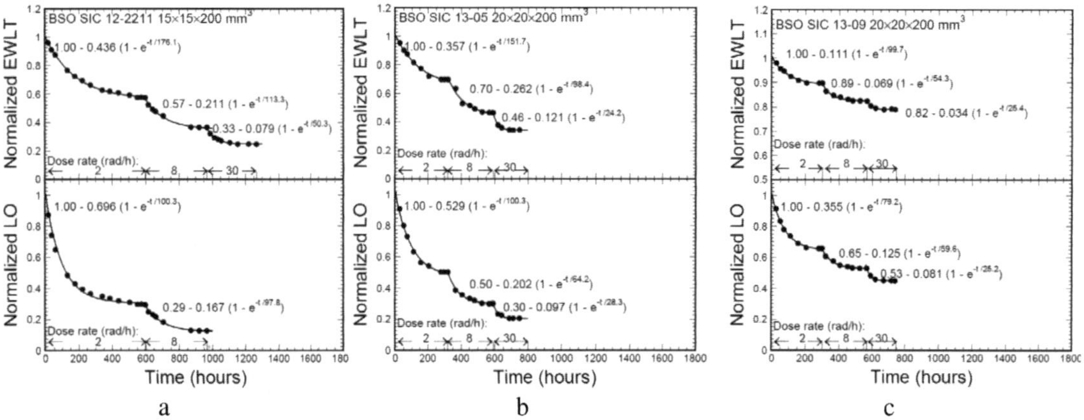

Figure 8. Histories of the normalized EWLT (top) and LO (bottom) under different dose rates: a. SIC 12-2211; b. SIC 13-05; and c. SIC 13-09.

Figure 8 shows histories of the normalized EWLT and LO for samples SIC 12-2211, 13-05 and 13-09 during γ-ray irradiations in three steps at dose rates of 2, 8 and 30 rad/h. Both EWLT and LO reached equilibrium when irradiated at a definite dose rate. The same phenomenon has also be observed in PWO [14] and BGO [15]. The damage time constants decrease when the dose rate increases, following well the color center kinetics described in [14].

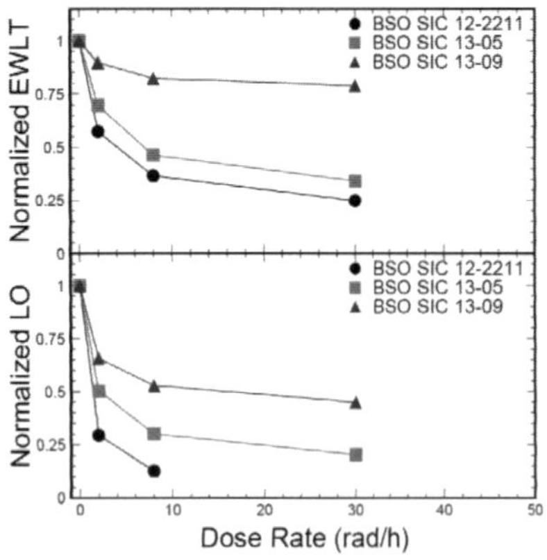

Figure 9. Normalized EWLT and LO are shown as a function of the dose rate

Table 1. EWLT and LO losses of three BSO samples in equilibrium under different dose rates

Samples		SIC 12-2211	SIC 13-05	SIC 13-09
EWLT Loss (%)	2 rad/h	43	30	10
	8 rad/h	63	54	18
	30rad/h	75	66	21
LO Loss (%)	2rad/h	71	50	34
	8rad/h	87	70	47
	30rad/h	-	80	55

Figure 9 shows the normalized EWLT (top) and LO (bottom) measured in equilibrium as a function of the dose rate for three BSO samples. The corresponding values of EWLT and LO losses are listed in table 1. It is clear that BSO samples grown in 2013 have better radiation hardness than that in 2012, and the sample SIC 13-09 has the best radiation hardness. This indicates that progress has been made in the radiation hardness of BSO crystals.

3.2 γ-ray induced dose rate independent damage in a BSO sample

Figure 10 shows that the LT of sample SIC 12-2223 does not recover after a γ-ray irradiation with a dose of 10^3 Rad. The γ-ray induced radiation damage in this sample thus is dose rate independent [14]. This sample thus was irradiated with integrated doses in steps of one order of magnitude from 10^2 to 10^6 rad. Figure 11 and 12 show LT and LO/LRU respectively for this sample at integrated doses from 10^2 to 10^6 rad. It is interesting to note that the LRU of this sample does not change although the loss of LO is significant. This indicates that energy resolution of a BSO calorimeter may be maintained up to 10^6 rad.

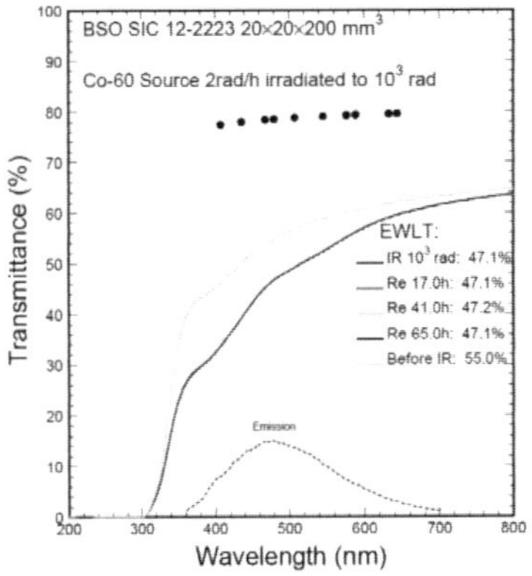

Figure 10. LT before and after irradiation of 10^3 rad for the BSO sample of SIC 12-2223.

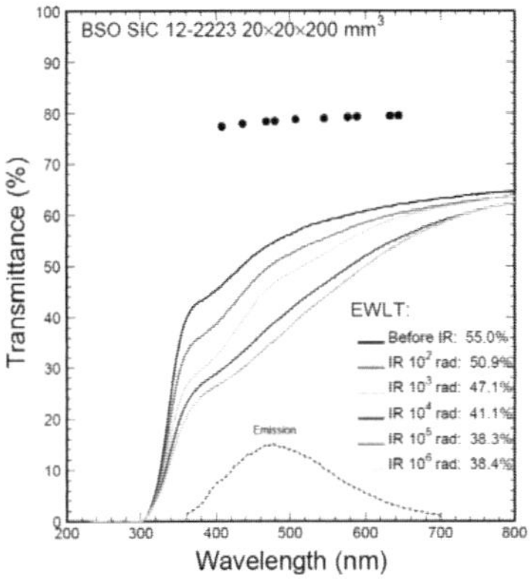

Figure 11. LT before and after irradiations of 100 to 10^6 rad is shown for sample SIC 12-2223.

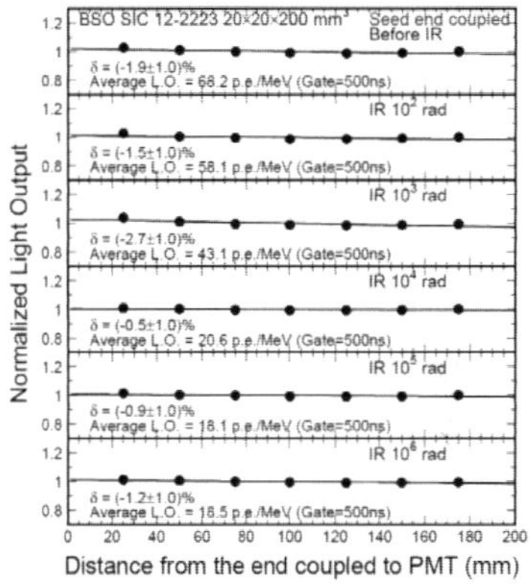

Figure 12. LO and LRU before and after irradiations are shown for sample SIC 12-2223.

Figure 13 shows the normalized EWLT and LO as a function of the integrated dose for the sample SIC 12-2223. The corresponding LO and EWLT losses after 1 Mrad are 30% and 73% respectively. The EWLT and LO losses are stable after 10^5 rad, indicating that the defect density in sample SIC 12-2223 was fully exhausted after 10^5 rad. Crystals with limited defect density have a potential to be improved for a stable detector in severe radiation environment after further reducing defect density.

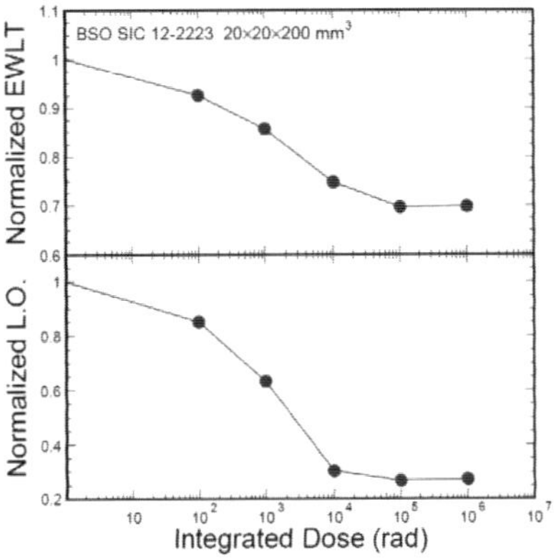

Figure 13. Normalized EWLT and LO are shown as a function of the integrated dose

Table 2. EWLT and LO losses of the BSO sample SIC 12-2223 after irradiations of different integrated dose

Integrated Dose (rad)	10^2	10^3	10^4	10^5	10^6
EWLT Loss (%)	7.5	14	25	30	30
LO Loss (%)	15	37	70	73	73

4. Summary

BSO crystals are a good candidate material for the HHCAL detector concept because of (1) high density of 6.8 g/cm3, (2) good UV cut-off edge at 300 nm, (3) emission peak at 470 nm, (4) 100 ns scintillation decay time, (5) good light output (20% of BGO) and (6) potential low cost. Among the four BSO samples grown at SIC in the last two years, SIC 12-2211 shows good transmittance approaching the theoretical limit at wavelength of longer than 500 nm, and SIC13-09 shows the improved LT, LO and LRU, indicating that progresses have been made in the crystal growth at SIC. BSO crystals satisfying HHCAL specifications, however, need to be developed after further improvement.

Gamma-ray induced radiation damage in three BSO crystals recovers at room temperature, so is dose rate dependent. The sample SIC 13-09 has better radiation hardness under gamma-ray irradiations, indicating further improvement of BSO quality is possible. Gamma-ray induced radiation damage in one BSO sample SIC 12-2223 does not recover at room temperature, so is dose rate independent. The radiation related defect density in this crystal, however, is limited, so there was no further damage after 100 kRad. Study on color centers in these crystals may help for further improvement of their radiation hardness.

Acknowledgments

This work was supported in part by the US Department of Energy Grant DE-FG03-92-ER40701 and the Natural Science Foundation of China (No. 11375250).

References

[1] Driutti A, Para A, Pauletta G, Briones N R, and Wenzel H, 2010, *J. Phys.: Conf. Ser.* (Beijing, China), **293** 012034
[2] Mao R H, Zhang L Y, and Zhu R-Y, 2012, *IEEE Trans. Nucl. Sci.*, **59**, 2229.
[3] Mao R H, Zhang L Y, and Zhu R-Y, 2010, *IEEE Trans. Nucl. Sci.*, **57**, 3841.
[4] Yang F, Zhang G Q, Ren G H, Mao R H, Zhang L Y, and Zhu R-Y, 2014, *IEEE Trans. Nucl. Sci.*, **61**, 489.
[5] Zhang G Q, Yang F, Wu Y T, Sun D D, Shang S S, and Ren G H, 2014, *J Inorg Mater,* **29**, 162.
[6] Chen J M, Shen D Z, Ren G H, Mao R H, and Yin Z W, 2004, *J Phys D Appl Phys,* **37**, 938.
[7] Fei Y T, Fan S J, Sun R Y, Xu J Y, and Ishii M, 2000, *Prog Cryst Growth Ch,* **40**, 189.
[8] Barysevich A, Dormenev V, Fedorov A, Glaser M, Kobayashi M, Korjik M, *et al.*, 2013, *Nucl Instrum Meth A,* **701**, 231.
[9] Ishii M, Harada K, Senguttuvan N, Kobayashi M, and Yamaga I, 1999, *J Cryst Growth,* **205**, 191.
[10] Chen J M, Mao R H, Zhang L Y, and Zhu R-Y, 2007, *IEEE Trans. Nucl. Sci.*, **54**, 718.
[11] Ma D A and Zhu R-Y, 1993, *Nucl Instrum Meth A,* **332**, 113.
[12] Zhu R-Y, Ma D A, Newman H B, Woody C L, Kierstead J A, Stoll S P, *et al.*, 1996, *Nucl Instrum Meth A,* **376**, 319.
[13] Zhu R-Y, 1997, *IEEE Trans. Nucl. Sci.*, **44**, 468.
[14] Zhu R-Y, 1998, *Nucl Instrum Meth A,* **413**, 297-311.
[15] Yang F, Mao R H, Zhang L Y, and Zhu R-Y, 2012, *J. Phys.: Conf. Ser.* (Santa Fe, New Mexican, USA), **404**, 012025.

Characterization of Three LYSO Crystal Batches

Fan Yang, Rihua Mao, Liyuan Zhang and Ren-Yuan Zhu
256-48, HEP, California Institute of Technology, Pasadena, CA 91125, USA

E-mail: zhu@hep.caltech.edu

Abstract. We report on three LYSO crystal batches characterized at the Caltech crystal laboratory for future HEP experiments: Twenty-five 20 cm long crystals for the SuperB experiment; twelve 13 cm long crystals for the Mu2e experiment and 623 14×14×1.5 mm plates with five holes for a LYSO/W Shashlik matrix for a beam test at Fermilab. Optical and scintillation properties measured are longitudinal Transmittance, light output and FWHM energy resolution. Correlations between these properties are also investigated.

1. Introduction

Because of their high density (7.4 g/cm^3), short radiation length (1.14 cm), fast (40 ns) and bright (4 times BGO) scintillation, cerium doped lutetium oxyorthosilicate (Lu_2SiO_5:Ce, LSO) [1] and lutetium yttrium oxyorthosilicate ($Lu_{2(1-x)}Y_{2x}SiO_5$:Ce, LYSO) [2,3] crystals have attracted a broad interest in the high energy physics community pursuing precision electromagnetic calorimeter for future high energy physics experiments [4-8]. Their excellent radiation hardness against gamma-rays [9, 10], neutrons [11] and charged hadrons [12] also makes them a preferred material for calorimeters to be operated in a severe radiation environment, such as the HL-LHC.

Figure 1. Twenty five 20 cm long LYSO crystals of tapered shape for SuperB

Figure 2. Twelve 13 cm long LYSO crystals for Mu2e

Figure 3. 623 14 x 14 x 1.5 mm LYSO plates for a LYSO/W Shashlik matrix

Figure 1 shows twenty five 20 cm long LYSO crystals of tapered geometry with about 20×20 mm^2 at the small end and 23×23 mm^2 at the large end for the SuperB experiment. They were grown by Czochralski method at Saint-Gobain (SG) Corporation, Shanghai Institute of Ceramics (SIC) and Sichuan Institute of Piezoelectric and Acousto-optic Technology (SIPAT). Figure 2 shows ten rectangular Mu2e crystals with dimension of 30×30×130 mm^3 and other two hexagonal crystals of 18.6 mm wide and 130 mm long. All the Mu2e crystals were grown at SIC. Figure 3 shows 623 14×14×1.5 mm^3 LYSO plates grown at SIC for a LYSO/W Shashlik matrix [6]. There are five holes in these plates to allow four wavelength shifters and one monitoring fiber to go through. All six faces of these crystals are polished.

The longitudinal transmittance (LT) was measured using a PerkinElmer LAMBDA 950 UV/Vis spectrophotometer with double beam, double monochromator and a large crystal compartment equipped with an integrating sphere. The systematic uncertainty in repeated measurements is 0.15%. The scintillation light output (LO) of SuperB crystals was measured by using a pair of Hamamatsu S8664-55 APDs. Their FWHM energy resolution (ER) was measured by a Hamamatsu R1306 PMT with a bi-alkali photo-cathode and a borosilicate glass window. The scintillation LO and ER were also measured by the Hamamatsu R1306 PMT for Mu2e crystals and Shashlik plates. In the LO measurement for long crystals the large (tapered crystals) or one (rectangular and hexagonal crystals) end of the crystals was coupled to the PMT with Dow Corning 200 fluid, while all other faces of the crystal were wrapped with Tyvek paper. A ^{22}Na source was used to excite the crystals, and provided the coincidence trigger. In the LO measurement for LYSO plates, one large face was coupled to the PMT with an air gap, while all other faces were wrapped with Tyvek paper. A ^{137}Cs source was used to excite the crystals. The γ-ray peak positions were obtained by a simple Gaussian fit. The systematic uncertainty in repeated measurements is less than 1%.

2. Properties of Long LYSO Crystals for the SuperB and Mu2e Experiments

2.1 Longitudinal Transmittance and Emission Weighted Longitudinal Transmittance

Figure 4 shows the LT at 420 nm for the SuperB LYSO crystals with divergences of less than 2% for crystals produced by each vendor. Among three venders SIPAT crystals have better average LT at 420 nm. Figure 5 shows the LT at 420 nm for the Mu2e LYSO crystals. Rectangular and hexagonal crystals have consistent LT with an overall divergence of less than 1%, where two crystals of different shape cut from the same ingot are shaded with red color. All crystals have LT at 420 nm better than the 75% specification, indicating good optical quality.

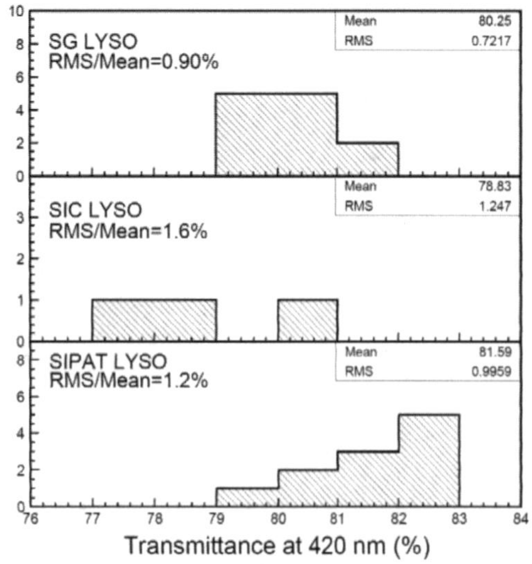

Figure 4. LT at 420 nm measured for twety five SuperB LYSO crystals

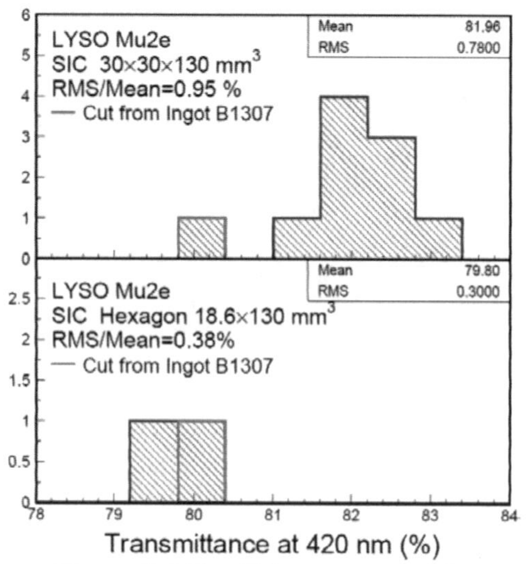

Figure 5. LT at 420 nm measured for twelve Mu2e LYSO crystals

Figures 6 and 7 show emission weighted longitudinal transmittance (EWLT) for SuperB and Mu2e crystals respectively, where EWLT is defined as [4]:

$$EWLT = \frac{\int LT(\lambda)Em(\lambda)d\lambda}{\int Em(\lambda)d\lambda} \quad (1)$$

EWLT presents crystal transmittance in the entire emission spectrum so is a better representation for crystal's transparency than the transmittance at 420 nm. Figure 6 shows that crystals grown at Saint-Gobain and SIPAT have better EWLT than that from SIC. Figure 7 shows that the rectangular and hexagon crystals have consistent EWLT, where two crystals cut from the same ingot are shaded with red color. A larger divergence for EWLT than the LT at 420 nm was observed since EWLT is affected by both transmittance and the UV cutoff wavelength, which is affected by the cerium concentration in the crystal [6]. A small divergence at a level of 1.7% was observed for the Saint-Gobain crystals as well as the Mu2e SIC crystals, indicating consistent cerium doping in these batches.

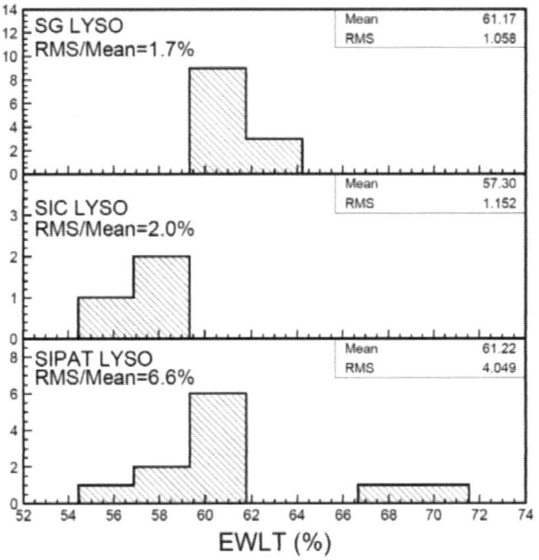

Figure 6. EWLT measured for twenty five SuperB LYSO crystals

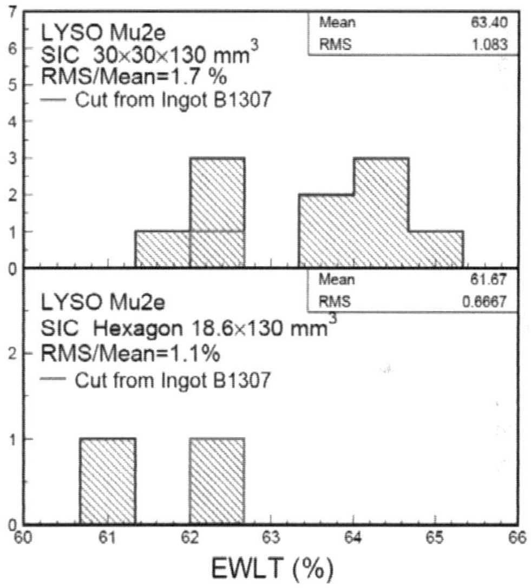

Figure 7. EWLT measured for twelve Mu2e LYSO crystals

2.2 Light Output and FWHM Energy Resolution

Figure 8 shows the LO measured by a pair of Hamamatsu S8664-55 APDs for twenty five SuperB LYSO crystals. While the SIPAT crystals have the best LO among three vendors, the Saint-Gobain crystals have the best consistency at about 3%. Figure 9 shows the LO measured by a Hamamatsu R1306 PMT with a divergence of less than 6% for twelve Mu2e crystals. The rectangular crystals appear to have higher LO than hexagonal ones cut from the same ingot. Further investigation, including a detailed ray-tracing simulation, however, is needed to understand this difference.

Journal of Physics: Conference Series **587** (2015) 012065 doi:10.1088/1742-6596/587/1/012065

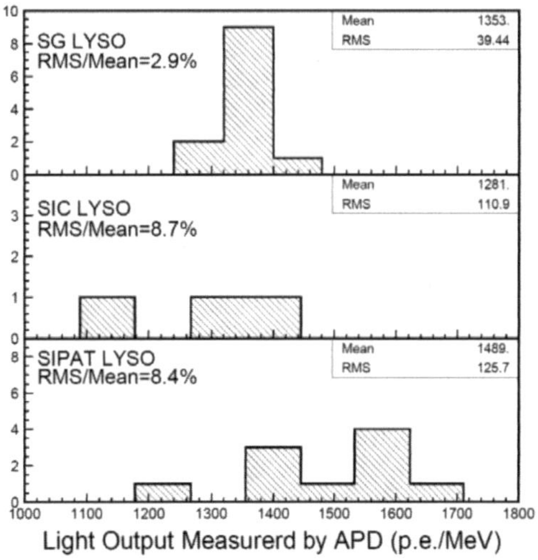

Figure 8. LO measured by APD for twenty five SuperB LYSO crystals

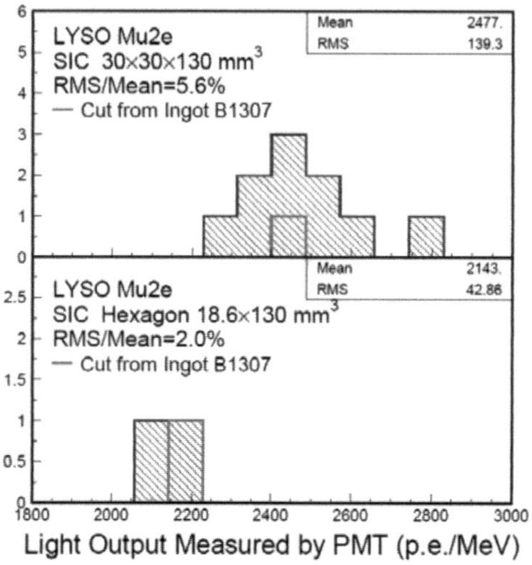

Figure 9. LO measured by PMT for twelve Mu2e LYSO crystals

Figure 10 shows the FWHM ER measured by a Hamamatsu R1306 PMT for twenty five SuperB crystals. The Saint-Gobain crystals have the best ER with a divergence of less than 2%. Figure 11 shows the FWHM ER measured by the Hamamatsu R1306 PMT for twelve Mu2e crystals. Consistent ER between the rectangular and hexagonal crystals is observed.

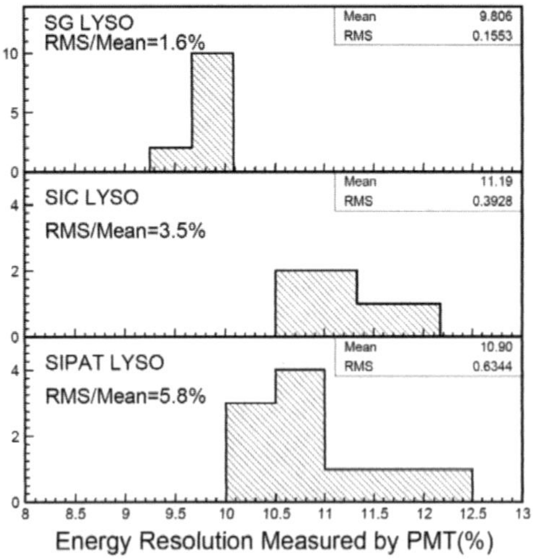

Figure 10. ER measured by PMT for twenty five SuperB LYSO crystals

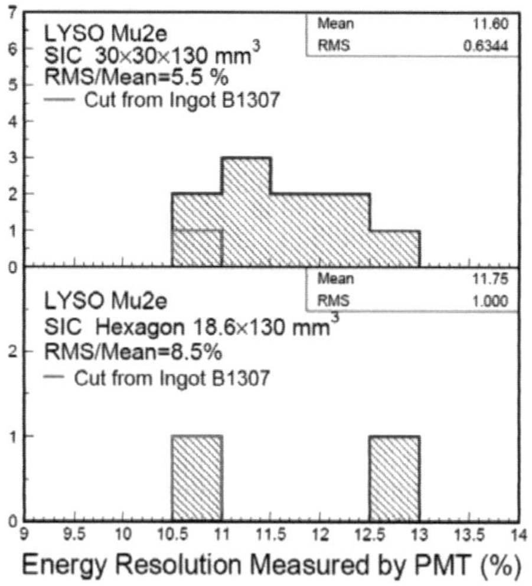

Figure 11. ER measured by PMT for twelve Mu2e LYSO crystals

2.3 Correlations between the Optical and Scintillation Properties

Figure 12 shows a slight correlation observed between the LO and the LT at 420 nm for twenty five SuperB crystals. Also shown in the figure is the corresponding linear correlation coefficients (CC), defined as [13]:

$$\mathrm{CC} = \frac{\sum(x-\bar{x})(y-\bar{y})}{\sqrt{\sum(x-\bar{x})^2 \sum(y-\bar{y})^2}} \qquad (2)$$

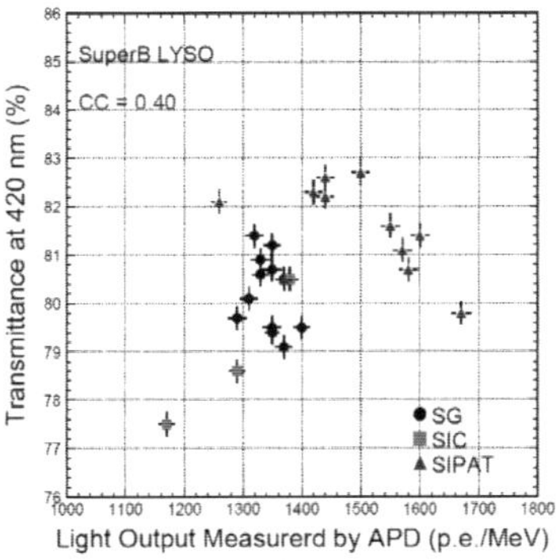

Figure 12. Correlation between the LO and the LT at 420 nm of 25 SuperB crystals

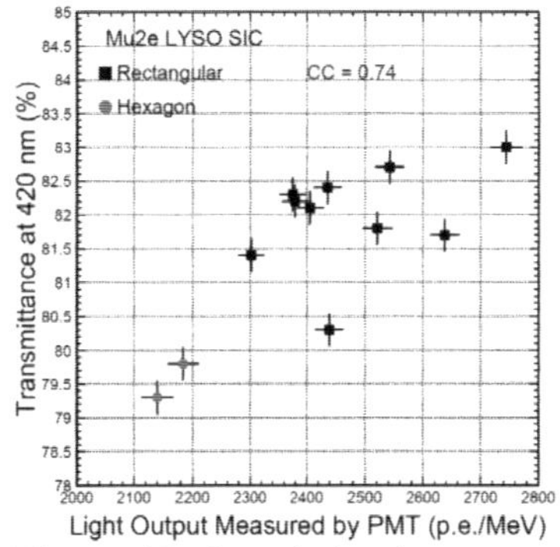

Figure 13. Correlation between the LO and the LT at 420 nm of 12 Mu2e crystals

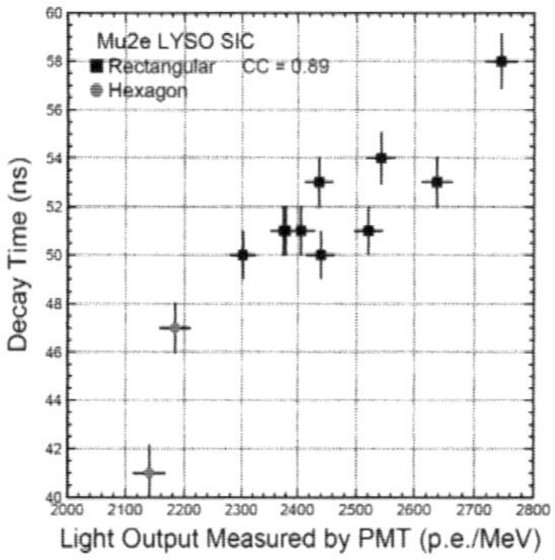

Figure 14. Correlation between the LO and the Decay time of 12 Mu2e crystals

Figure 13 shows a higher positive correlation coefficient at 74% observed between the LO and the LT at 420 nm for twelve Mu2e crystals than 40% observed for the SuperB crystals because of single crystal vendor. This positive correlation indicates that LO of long LYSO crystals is affected by their optical transparency through a light propagation process in the crystal. Figure 14 shows a strong positive correlation at 90% observed between the LO and the decay time for twelve Mu2e crystals. This positive correlation indicates that the decay time obtained by fitting the light output as a function of the integration time is affected by the light propagation in the crystal.

3. Properties of LYSO Plates for a LYSO/W Shashlik Test Beam Matrix

Figure 15 shows the LO measured by a Hamamatsu R1306 PMT for 623 LYSO plates for a LYSO/W Shashlik test beam matrix [6]. The plates can be divided into two groups according to their LO. While the high peak shows 3,284 p.e./MeV with a divergence of 3.6%, the 15% lower peak shows 2,793 p.e./MeV with a divergence of 3.2%. Plates in these two groups are actually from different ingots according to the manufacture. Figure 16 shows the FWHM ER measured for all plates, which can also be divided in two groups. The average ER value of the better group is 9.1% with divergence of 6.6% and that of the other group is 11.6% with divergence of 7.0%. Figure 17 shows a negative correlation between the light output and the energy resolution as expected.

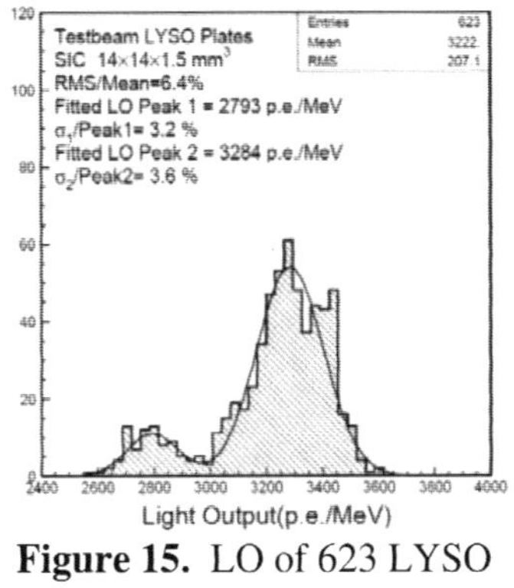

Figure 15. LO of 623 LYSO plates for a Shashlik matrix

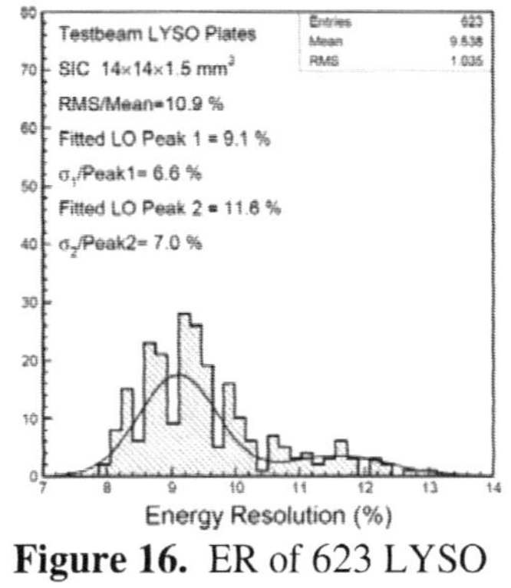

Figure 16. ER of 623 LYSO plates for a Shashlik matrix

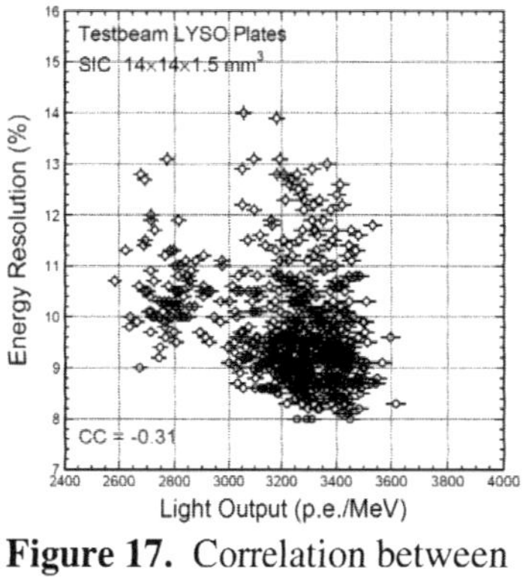

Figure 17. Correlation between the LO and the ER of 623 plates

Figure 18 shows very consistent transmittance at 420 nm for all LYSO plates except two with significantly lower transmittance. Figure 19 shows no correlation between the LO and the transmittance at 420 nm for these plates, indicating that the measured transmittance has negligible effect on the LO.

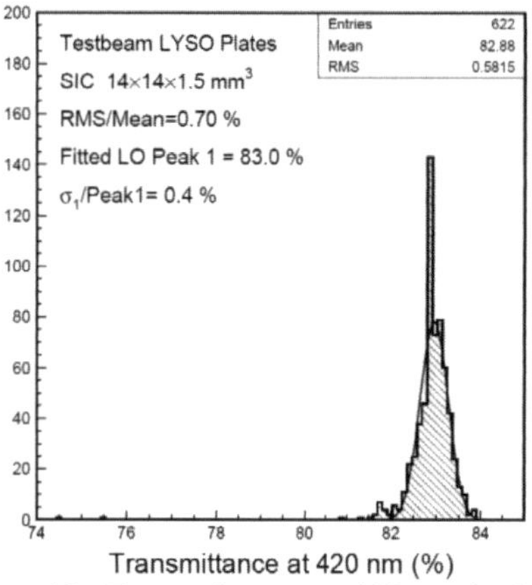

Figure 18. Transmittance at 420 nm is shown for LYSO plates foe a Shashlik matrix

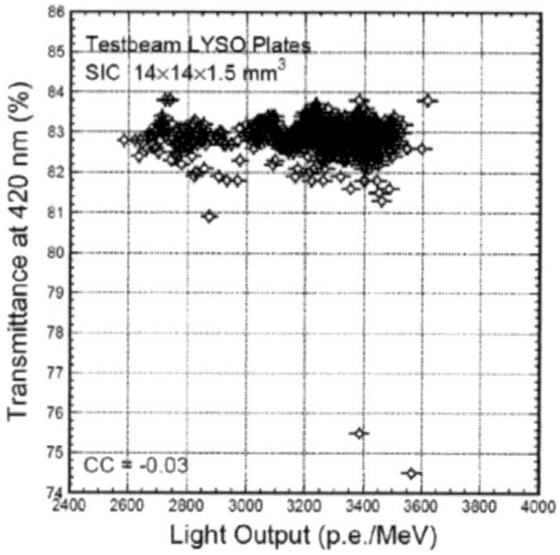

Figure 19. Correlation between the LO and the transmittance at 420 nm of 623 plates

4. Summary

Long LYSO crystals produced in industry show good transmittance exceeding 75% specification at 420 nm and good FWHM energy resolution better than 12.5% specification at 511 keV. Typical LO spread is at a level of 6% for long crystals, which may be reduced to about 3% in mass production. Correlations are observed between the LO and the LT at 420 nm as well as between LO and decay time for long LYSO crystals. A slight anti-correlation was observed between the LO and the energy resolution for 14×14×1.5 mm LYSO plates. Results of these investigations indicate that the quality of LYSO crystals grown in industry is adequate for future HEP calorimeters at both the energy and intensity frontiers.

Acknowledgments

This work was supported in part by the US Department of Energy Grant DE-FG03-92-ER40701.

References

[1] Melcher C L and Schweitzer J S, 1992, *IEEE Trans. Nucl. Sci.,* **39**, 502.
[2] Kimble T, Chou M, and Chai B H T, 2002, *Proc. IEEE Nuclear Science Symp. Conf.*, vol 3 (Norfolk, Virginia, USA), p 1434.
[3] Cooke D W, McClellan K J, Bennett B L, Roper J M, Whittaker M T, Muenchausen R E, *et al.*, 2000, *J Appl Phys,* **88**, 7360.
[4] Chen J M, Mao R H, Zhang L Y, and Zhu R-Y, 2007, *IEEE Trans. Nucl. Sci.,* **54**, 718.
[5] Chen J M, Zhang L Y, and Zhu R-Y, 2005, *IEEE Trans. Nucl. Sci.,* **52**, 3133.
[6] Zhang L Y, Mao R H, Yang F, and Zhu R-Y, 2014, *IEEE Trans. Nucl. Sci.,* **61**, 483.
[7] Eigen G, Zhou Z, Chao D, Cheng C H, Echenard B, Flood K T, *et al.*, 2013, *Nucl Instrum Meth A,* **718**, 107.
[8] SuperB Conceptual Design Report, INFN/AE-07/2, SLAC-R-856, LAL 07-15, Mar. 2007; and Cecchi C, 2010, *J. Phys.: Conf. Ser.*(Beijing, China), **293** 012066.
[9] Mao R H, Zhang L Y, and Zhu R-Y, 2012, *IEEE Trans. Nucl. Sci.,* **59**, 2224.
[10] Chen J M, Mao R H, Zhang L Y, and Zhu R-Y, 2007, *IEEE Trans. Nucl. Sci.,* **54**, 1319.
[11] Zhang L Y, Mao R H, and Zhu R-Y, 2009, *IEEE Nuclear Science Symp. Conf.* (Orlando, Florida, USA), p 2041.
[12] Dissertori G, Luckey D, Nessi-Tedaldi F, Pauss F, and Quittnat M, 2014, *Nucl Instrum Meth A*, **745**, 1-8.
[13] Chen J M, Mao R H, Zhang L Y, and Zhu R-Y, 2007, *IEEE Trans. Nucl. Sci.,* **54**, 375.

Monitoring LSO/LYSO Crystal Based Calorimeters

Fan Yang, Liyuan Zhang and Ren-Yuan Zhu
256-48, HEP, California Institute of Technology, Pasadena, CA91125, USA

E-mail: zhu@hep.caltech.edu

Abstract. Precision light monitoring is important for keeping excellent energy resolution promised by LSO/LYSO crystals in severe radiation environment. In this paper, we report an investigation on the wavelength choice for monitoring LYSO crystal based calorimeters. Gamma-ray induced absorption and light output loss were measured for 20 cm long crystals from five different vendors. Monitoring sensitivity and divergence between crystals from different vendors were investigated. The pros and cons of two monitoring approaches using emission and excitation light and their practical implementation for a LYSO/W Shashlik test beam matrix are discussed.

1. Introduction

Because of their high density (7.4 g/cm^3), short radiation length (1.14 cm), fast (40 ns) and bright (4 times BGO) scintillation, cerium doped lutetium oxyorthosilicate (Lu_2SiO_5:Ce, LSO) [1] and lutetium yttrium oxyorthosilicate ($Lu_{2(1-x)}Y_{2x}SiO_5$:Ce, LYSO) [2,3] crystals have attracted a broad interest in the high energy physics community pursuing precision electromagnetic calorimeter for future high energy physics experiments [4-8]. Their excellent radiation hardness against γ-rays [9, 10], neutrons [11] and charged hadrons [12] also makes them a preferred material for calorimeters to be operated in a severe radiation environment. As a result, LSO/LYSO crystals were chosen by the SuperB and Mu2e experiments to construct total absorption electromagnetic calorimeters. LSO/LYSO crystals are also proposed as active material for a Shashlik calorimeter which is currently one of the two options for the CMS forward calorimeter upgrade at the HL-LHC. Precision light monitoring is crucial for keeping excellent energy resolution promised by LYSO crystals in severe radiation environment. It is usually carried out by two approaches, either injecting light pulses around crystal's emission peak, e.g. CMS at LHC [13], or around the excitation peak, e.g. PHENIX at RIHC [14].

In this paper, we report an investigation on the wavelength choice for monitoring LSO/LYSO crystal based calorimeters. Five 20 cm long LSO/LYSO crystals from different vendors were irradiated step by step from 100 rad to 1 Mrad with γ-rays. Longitudinal transmittance, light output and longitudinal response uniformity were measured at each step. Monitoring sensitivities, defined as the ratio between the variations of longitudinal transmittance and light output, was determined as a function of wavelength for the emission approach, and was compared to that of the excitation approach. An OPOTEK Opolette 355II laser based light monitoring system with precision at a level of 0.1% was built for a LYSO/W Shashlik matrix beam test at Fermilab. It is expected that such a monitoring system will provide a solid foundation for the final monitoring system design for the proposed CMS forward calorimeter upgrade for the HL-LHC.

2. Determination of the monitoring wavelengths

2.1. Crystal transmittance and light output

Figure 1 is a photo showing 5 LSO/LYSO crystals of 20×20×200 mm^3 from Crystal Photonics Inc. (CPI), CTI Molecular Imaging (CTI), Saint-Gobain Corporation (SG), Shanghai Institute of Ceramics (SIC) and Sichuan Institute of Piezoelectric and Acousto-optic Technology (SIPAT). The UV excited emission spectrum was measured by using a Hitachi F4500 fluorescence spectrophotometer. The longitudinal transmittance (LT) was measured by using a Perkin Elmer Lambda-950 spectrometer equipped with double-beam, double-monochromator and a general purpose optical bench with light path up to 40 cm. The systematic uncertainty is about 0.15% [15]. Figure 2 shows the emission spectra, LT and emission multiplied longitudinal transmittance (EMLT) spectra (red) which represents the overall intensity of the monitoring signal as a function of wavelength in the crystal.

Figure 1. Five 20×20×200 mm^3 LSO/LYSO crystals from CPI, CTI, SG, SIC and SIPAT.

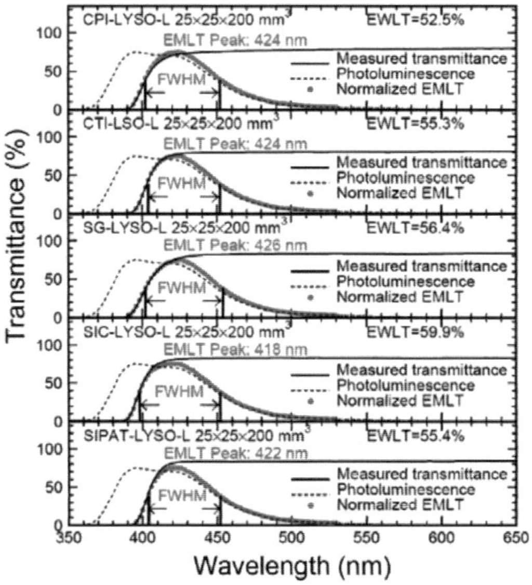

Figure 2. UV excited emission spectrum, longitudinal transmittance and EMLT.

Light output (LO) and longitudinal response uniformity (LRU) along the crystal were measured by injecting collimated γ-rays from a Na-22 source along the crystal with a coincidence trigger [4]. The light output (LO) is defined as the average of seven measurements uniformly distributed along the crystal [15]. The systematic uncertainty is less than 1% [15]. The LRU is parameterized as δ by a linear fit as follows,

$$\frac{LO}{LO_{mid}} = 1 + \delta\left(\frac{X}{X_{mid}} - 1\right), \qquad (1)$$

where LO_{mid} is the fitted value of the light output at the middle of the crystal (X_{mid}). Figure 3 shows that all crystals have good LO with LRU of better than 3% since the self-absorption effect caused by overlapping emission and transmittance is partially compensated by the cerium distribution along the crystal length [6].

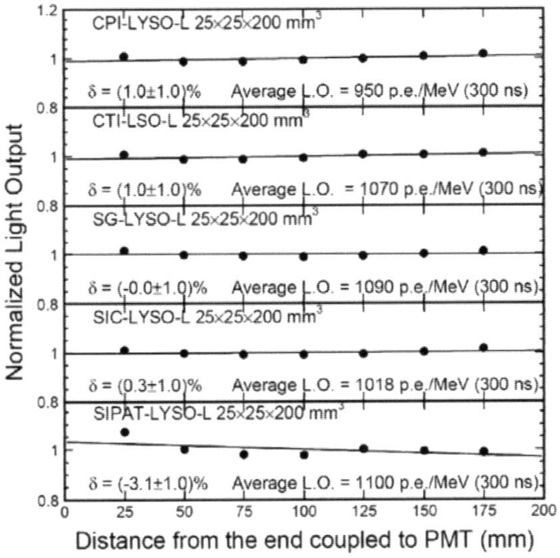

Figure 3. Initial light output and longitudinal response uniformity measured for 5 LYSO crystals.

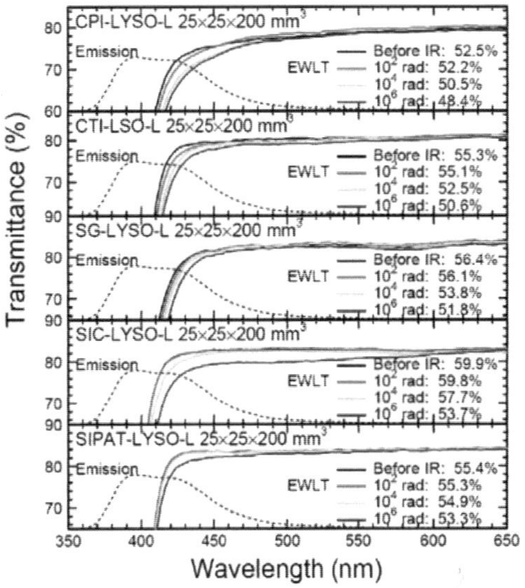

Figure 4. LT and EWLT before and after irradiation at 10^2, 10^4 and 10^6 rad.

2.2. γ-ray induced absorption, light output loss and monitoring wavelength

All crystals were irradiated by γ-rays in steps of one order of magnitude from 100 rad to 1 Mrad. Figure 4 shows LTs before and after irradiation at 10^2, 10^4 and 10^6 rad for each crystal. Also shown in the figure are the emission spectrum and the values of the emission weighted longitudinal transmittance (EWLT). One can see that all the crystals show consistent loss of about 10% at 1 Mrad in LT and EWLT. The longitudinal response uniformity (δ) is maintained in all five crystals up to 10^6 rad, as shown in figure 5 for crystal SG-LYSO.

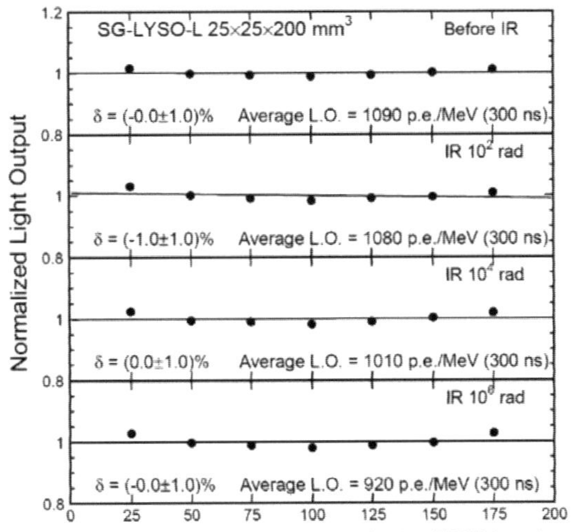

Figure 5. LO and LRU of SG-LYSO before and after irradiation at 10^2, 10^4 and 10^6 rad.

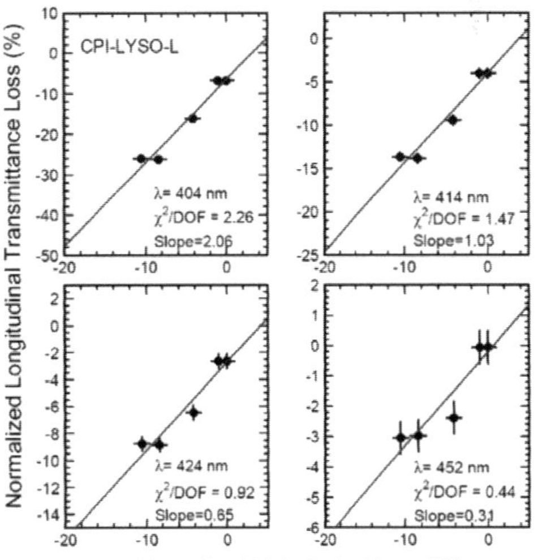

Figure 6. LO loss versus LT loss at 404, 414, 424 and 452 nm of crystal CPI-LYSO before and after irradiation at 10^2, 10^3, 10^4, 10^5 and 10^6 rad.

Figure 6 shows the relations between the losses in LO and LT at different wavelengths for CPI-LYSO. A linear fit determines the slope, which is defined as the monitoring sensitivity. Figure 7 shows a similar relation between the losses of LO and EWLT for five crystals. Figure 8 shows the average monitoring sensitivity (left scale) for five crystals and its divergence (right scale) defined as the rms over the average sensitivity as a function of wavelength. Also shown in the figure is the sensitivity and divergence obtained with EWLT (dotted lines). Taking into account the emission light propagation (EMLT in figure 2) in the crystal, the monitoring wavelength of choice is around 420 nm for the emission approach. On the other hand, 355 nm would be the choice for the excitation approach since the main excitation band of LYSO is between 340 and 380 nm. This particular wavelength is commercially available from Nd:YAG lasers operated in third harmonic generation. As shown in Figure 8, the excitation approach (EWLT) will have a similar monitoring sensitivity and divergence as that using emission wavelength at 420 nm. The only difference would be the requirement to the intensity of the monitoring light pulse since the former involves an addition process of converting the excitation light into the emission light.

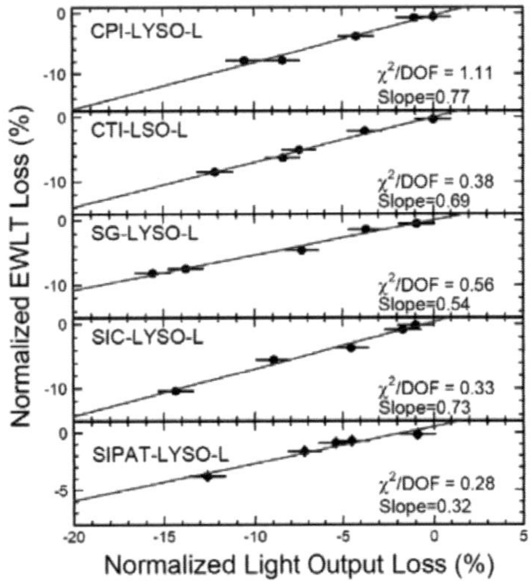

Figure 7. LO loss versus EWLT loss for 5 LYSO crystals before and after irradiation at 10^2, 10^3, 10^4, 10^5 and 10^6 rad.

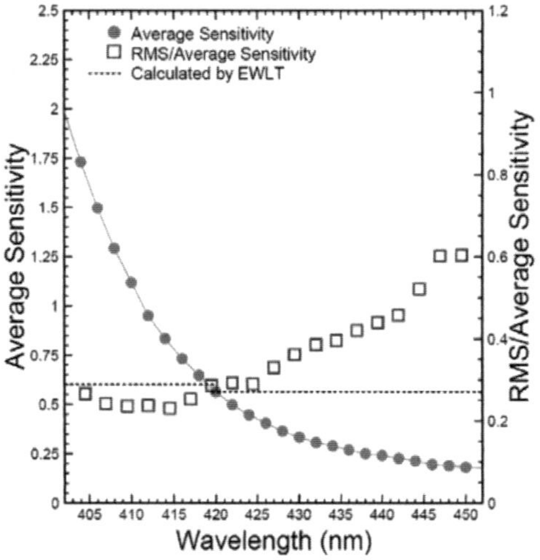

Figure 8. Average sensitivity and its divergence as a function of wavelength as well as the result using EWLT.

3. Monitoring LYSO-W Shashlik tower

Figure 9 shows a tunable OPO laser (OPOTEK Opolette 355+UV) based light monitoring system constructed for a LYSO/W Shashlik matrix beam test at Fermilab. After passing through a beam sampler, fast shutter and two neutral density (ND) attenuators, the laser pulses at 420 nm were coupled into a quartz fiber of 30 m long. Two ND attenuators, one in logarithm and other variable from 1 to 100%, allow an amplitude scan in 5 orders of magnitude. The laser pulses at the other end of the transportation fiber were distributed via a 2" integrating sphere to 16 LYSO/W Shashlik towers through monitoring fibers as shown in Fig. 10.

The monitoring light is delivered to the LYSO plates through natural scattering in the quartz monitoring fibers. The total efficiency for the laser source to the photon electrons of photo-detector was measured to be about 110 dB. More efficient leaky fiber [14] is being developed. Two reference channels from the integrating sphere are read out directly by the same photo-detectors and electronics used to read out Shashlik towers.

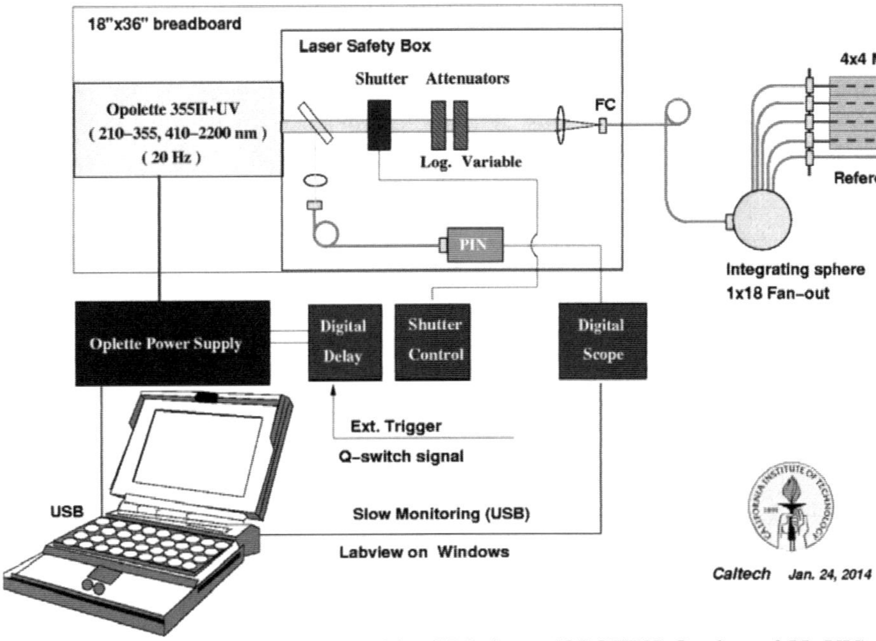

Figure 9. A schematic showing a tunable OPO laser (OPOTEK Opolette 355+UV) based light monitoring system for LYSO/W Shashlik beam test at Fermilab.

Figure 11 shows a distribution of the average ratio between two channels obtained with 500 pulses. They were readout by two identical Si Photo-detectors (Thorlabs DET10A) through a 500 MHz digital scope (Agilent 6052A). A precision at a level of 0.1% was achieved although the intrinsic rms fluctuation of the OPO laser pulse intensity was about 20%. The Opolette laser is being modified by the manufacturer to extract 355 nm directly. Monitoring performance using both 420 and 355 nm will be tested in the next run of the LYSO/W Shashlik matrix beam test at Fermilab.

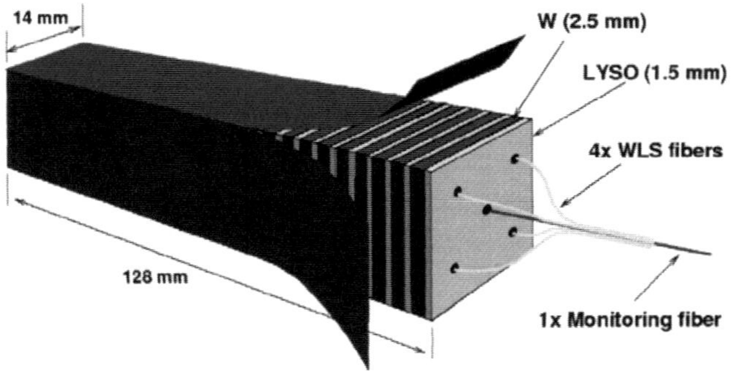

Figure 10. A schematic showing a LYSO/W Shashlik tower consisting of 30 LYSO plates of 1.5 mm thick and 29 tungsten plates of 2.5 mm.

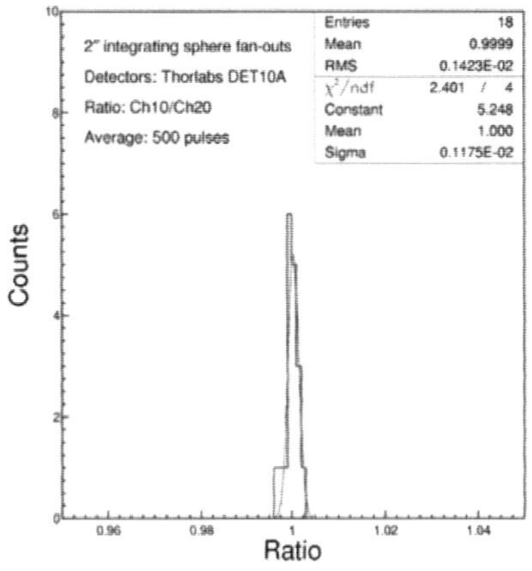

Figure 11. Average ratio of 500 pulses between two channels from the integrating sphere read out with the same Si photo-detectors.

4. Summary

LSO/LYSO crystals suffer from transparency loss, leading to light output loss. Variations of light output can be corrected by using variations of crystal response to monitoring light pulses. Precision light monitoring is important for achieving and maintaining excellent energy resolution promised by LYSO based calorimeters in severe radiation environment.

Two approaches may be used for monitoring LYSO crystal based calorimeters. One uses a wavelength around the emission peak. 420 nm is the choice for this approach. The other uses a wavelength at the excitation peak. 355 nm is the choice. The 2nd approach has three advantages: (1) crystal transparency is monitored with the entire emission spectrum; (2) crystal photo-luminescence production is also monitored and (3) cost-effective frequency tripled YAG laser at 355 nm is commercially available. The disadvantage of this approach is the requirement of high intensity monitoring light pulse.

A tunable OPO laser based light monitoring system is constructed and used for a LYSO/W Shashlik matrix beam test at Fermilab. The monitoring precision obtained with 500 pulse averages is about 0.1% although the instability of the laser pulse intensity is about 20%. It is expected that such a monitoring system will provide a solid foundation for the final monitoring system design for the proposed CMS forward calorimeter upgrade for the HL-LHC.

Acknowledgments

This work was supported in part by the US Department of Energy Grant DE-FG03-92-ER40701.

References

[1] Melcher C L and Schweitzer J S, 1992, *IEEE Trans. Nucl. Sci.,* **39**, 502.

[2] Kimble T, Chou M, and Chai B H T, 2002, *Proc. IEEE Nuclear Science Symp. Conf.*, vol 3 (Norfolk, Virginia, USA), p 1434.

[3] Cooke D W, McClellan K J, Bennett B L, Roper J M, Whittaker M T, Muenchausen R E, *et al.*, 2000, *J Appl Phys,* **88**, 7360.

[4] Chen J M, Mao R H, Zhang L Y, and Zhu R-Y, 2007, *IEEE Trans. Nucl. Sci.,* **54**, 718.

[5] Chen J M, Zhang L Y, and Zhu R-Y, 2005, *IEEE Trans. Nucl. Sci.,* **52**, 3133.

[6] Zhang L Y, Mao R H, Yang F, and Zhu R-Y, 2014, *IEEE Trans. Nucl. Sci.,* **61**, 483.

[7] Eigen G, Zhou Z, Chao D, Cheng C H, Echenard B, Flood K T, *et al.*, 2013, *Nucl Instrum Meth A,* **718**, 107.

[8] SuperB Conceptual Design Report, INFN/AE-07/2, SLAC-R-856, LAL 07-15, Mar. 2007; and Cecchi C, 2010, *J. Phys.: Conf. Ser.*(Beijing, China), **293** 012066.

[9] Mao R H, Zhang L Y, and Zhu R-Y, 2012, *IEEE Trans. Nucl. Sci.,* **59**, 2224.

[10] Chen J M, Mao R H, Zhang L Y, and Zhu R-Y, 2007, *IEEE Trans. Nucl. Sci.,* **54**, 1319.

[11] Zhang L Y, Mao R H, and Zhu R-Y, 2009, *IEEE Nuclear Science Symp. Conf.* (Orlando, Florida, USA), p 2041.

[12] Dissertori G, Luckey D, Nessi-Tedaldi F, Pauss F, and Quittnat M, 2014, *Nucl Instrum Meth A,* **745**, 1-8.

[13] Anfreville M, Bailleux D, Bard J P, Bornheim A, Bouchand C, Bougamont E, *et al.*, 2008, *Nucl Instrum Meth A,* **594**, 292-320.

[14] David G, Kistenev E, Stoll S, White S, Woody C, Bazilevsky A, *et al.*, 1998. *IEEE Trans. Nucl. Sci.,* **45**, 705-709.

[15] Chen J M, Mao R H, Zhang L Y, and Zhu R-Y, 2007, *IEEE Trans. Nucl. Sci.,* **54**, 1319-1326.

Journal of Physics: Conference Series **587** (2015) 012067 doi:10.1088/1742-6596/587/1/012067

Characterization and Applications of New High Quality LuAG:Ce and LYSO:Ce fibers

Stefan Diehl[1], Rainer W. Novotny[1], Nicolas Aubry[2], Sara Faraj[2], Gabriel Ferro[3] and for the HP3-WP22 and the Crystal Clear Collaboration

[1] 2nd Physics Institute University Giessen, Heinrich-Buff-Ring 16, Giessen, 35392, Germany
[2] Societe Fibercryst, Parc dctivite Wilson, 31 Rue Wilson, Descines Charpieu, France
[3] Universite Claude Bernard Lyon 1, CNRS, UMR 5615, Laboratoire des Multimateriaux (LMI), 43 Bd du 11 Nov. 1918, Villeurbanne, France

E-mail: stefan.diehl@exp2.physik.uni-giessen.de

Abstract. Inorganic scintillating fibers have a large potential as highly granular detector components in hadron and particle physics as well as in medical applications. With the micro-pulling-down-method a fast and cost efficient technique was developed over the last decades to grow such fibers from the melt. This paper will present the recent development of the quality of LuAG:Ce and LYSO:Ce fibers in terms of light output and light attenuation inside the fiber. For this purpose different steps in optimizing the fiber growth will be compared with the achieved fiber performance. In addition the response of the fibers to low energetic gamma rays will be studied and different readout concepts for the fibers will be discussed and compared.

1. Introduction

Inorganic scintillating fibers show significant advantages compared to organic fibers, like a higher light yield and a more effective interaction with electromagnetic probes due to the high Z elements contained in the fiber material. In addition one expects a higher radiation resistance, which is especially important in highly radioactive environments like the inner or forward parts of high energy physics detectors. These advantages in combination with the high granularity of the fibers open up a wide field of applications. In high energy physics the combination of doped and un-doped fibers will allow the design of compact hadron sampling calorimeters exploiting the dual-readout concept [1], [2]. In addition, the sub millimeter position resolution given directly by the fiber size can be exploited for beam monitoring and in combination with the good efficiency for γ-rays for medical applications like PET-scanners as well [3], [4].

2. Production Technology and Scintillator Materials

2.1. The micro-pulling-down method

Conventional methods for crystal growth like the Czochralski or Bridgeman method usually provide large crystal ingots. To produce highly granular detectors mechanical treatments like cutting and polishing impose a large workload. This can be avoided by producing crystals as fibers directly from the melt, like with the edge defined film-fed growth, the laser heated pedestal growth (LHPG) or the micro-pulling-down (μ-PD) method is needed. The μ-PD-method has been established over the last decades to be one of the best methods for the growth of single crystal fibers with a diameter below 1 or 2 mm, respectively, and with a length up to several tens of centimeters [5], [6]. Fig. 1 shows a schematic view of the fiber growth with the micro pulling down method. The crystal melt is heated up in a crucible by an external radiofrequency

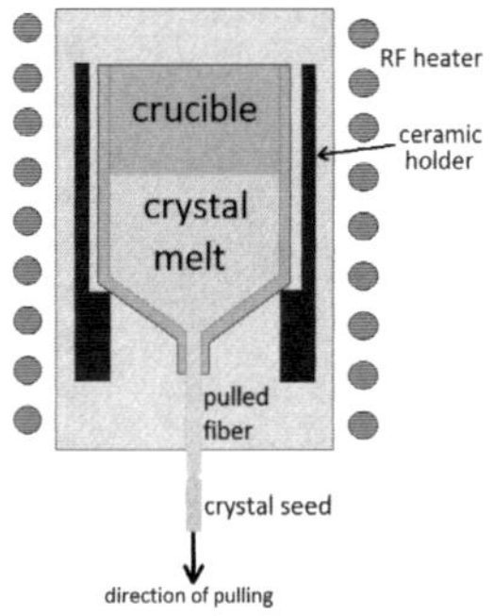

Figure 1. Schematic view of the fiber growth with the micro pulling down method.

heater. The fiber is pulled down on a small seed through an specially shaped opening on the bottom of the crucible. During this process several critical parameters, like the temperature gradient, and the speed of pulling down have a strong influence on the fiber quality.

2.2. Properties of LuAG:Ce and LYSO:Ce
For the fiber production the materials LuAG:Ce and LYSO:Ce have been chosen. Table 1 summarizes the most relevant physical and scintillation properties of the two materials. It shows that both fulfill the requirements for high energy physics detectors. However, LYSO:Ce appears most favorable due to the short radiation length and due to the high light yield. On the other hand, LuAG:Ce is much easier to grow due to the lower melting point.

Table 1. Properties of LuAG:Ce and LYSO:Ce single crystals. *The light yield strongly depends on the Ce-concentration.

	LuAG:Ce	LYSO:Ce
density	6.7 g/cm^3	7.1 g/cm^3
Z_{eff}	63	66
radiation length	1.41 cm	1.2 cm
decay time	(50-60) ns	45 ns
$\lambda_{max.emission}$	530 nm	420 nm
refractive index	1.85 @ 530 nm	1.82 @ 420 nm
light yield	12000 - 25000 ph/MeV*	26000 - 32000 ph/MeV*
melting point	1980C	2100C

3. Characterization - setup and procedure
The readout of the fibers is performed with silicon photomultipliers (SiPMs). Sensors with a pixel size of 25 μm (S12572-025C) and 50 μm (S12572-050C), respectively, from Hamamatsu [7] have been used. The measurements confirmed, that for fibers with 2 mm diameter the 50 μm version has a sufficient numbers of pixels to detect the total light yield with a significantly increased photon detection efficiency compared to the 25 μm version. The signals of the SiPMs are amplified by a special preamplifier (photonique AMP-0611, AMP-0604) and digitalized with commercial NIM and CAMAC electronics using a charge integrating ADC. The fiber is coupled at one or both ends to the SiPMs using optical grease. Fig. 2 shows the experimental setup with the movable collimated source on the top of the fiber.

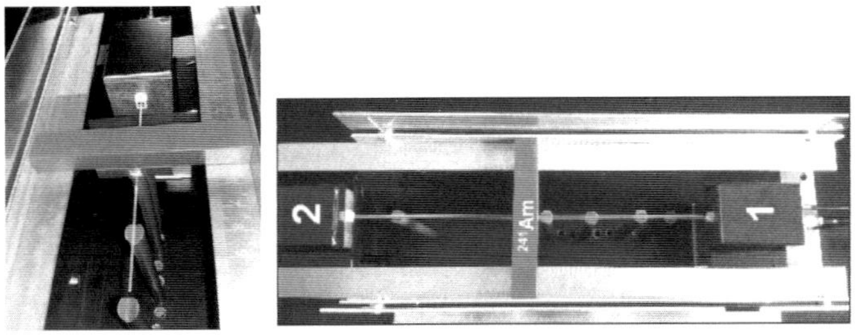

Figure 2. Setup for a single sided SiPM readout (left) and a double sided SiPM readout (right).

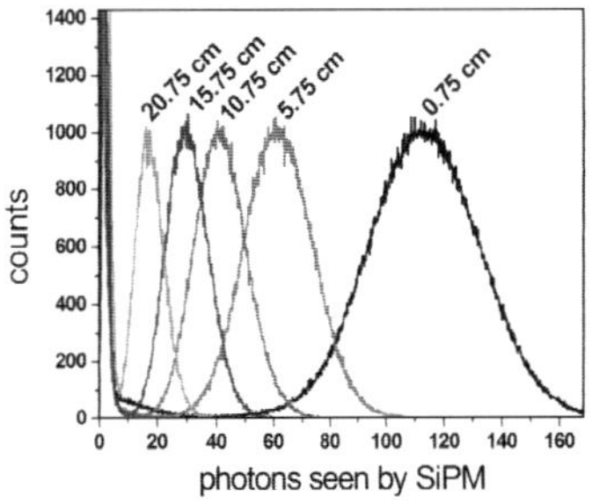

Figure 3. Response of a 23 cm long, ⊘ = 2 mm fiber to α particles from an ^{241}Am source

3.1. Response to alpha particles
For the characterization of the light yield and the attenuation coefficient, alpha particles from an collimated ^{241}Am source with a mean effective energy of 4.0 MeV were used. The main advantage of alpha particles compared to gamma sources is, that they deposit locally their complete energy even in very thin fibers. One obtains a narrowly peaked response, which can be fitted with a Gaussian distribution. Fig. 3 shows the response to an α-source for a ⊘ 2 mm fiber with increasing distance between source and SiPM. The highest signal amplitude is observed at closest distance to the SiPM. The attenuation of the light yield LY as a function of the distance d between source and PMT can be described by the following relation : $\mathrm{LY(d)} = \mathrm{LY}_0 \cdot \mathrm{e}^{-\mu \cdot d} = \mathrm{LY}_0 \cdot \mathrm{e}^{-\frac{d}{\lambda}}$ with LY_0 as initial light yield, μ as attenuation coefficient and λ as attenuation length, respectively.

4. Characterization and optimization of light yield and light attenuation of LuAG:Ce fibers
To optimize the light yield at reduced light attenuation several production runs have been performed.

4.1. Influence of the speed of pulling down and other growth parameters
In a first optimization run the influence of different growth parameters was investigated, like the speed of pulling down, the orientation of the seed and the sensitivity of the fiber quality to the distance to the seed. Table 2 shows some of the properties of the investigated fibers, while the corresponding attenuation curves are shown in fig. 4. The results do not indicate a significant influence of the seed orientation but favor a slow pulling speed, since it provides the crystal lattice more time to relax leading to less bubbles and cracks. In addition, fibers samples cut close to the seed show smaller attenuation coefficients.

Table 2. Properties of 5 investigated LuAG:Ce fibers.

fiber	μ cm^{-1}	orientation of seed	dist. to seed	speed of pulling down
# 1	0.69	< 100 >	11.0	high
# 2	0.77	< 100 >	22.5	high
# 3	0.13 - 0.4	< 100 >	2.5	avg.
# 4	0.46	< 111 >	22.5	low
# 5	0.68	< 100 >	22.5	low

Journal of Physics: Conference Series **587** (2015) 012067 doi:10.1088/1742-6596/587/1/012067

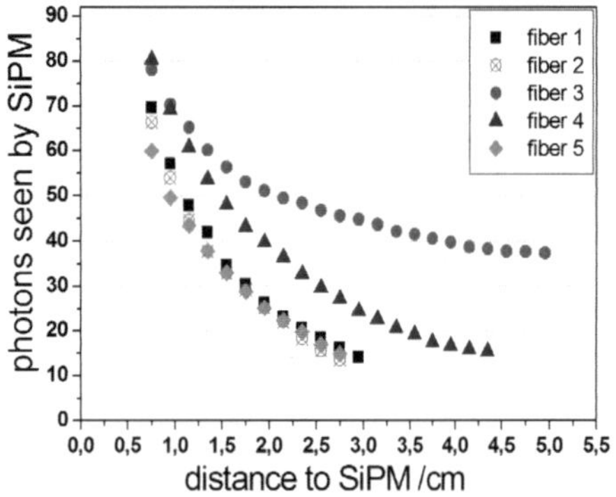

Figure 4. Attenuation curves for 5 5.5 cm long LuAG:Ce fiber samples manufactured under different growing conditions.

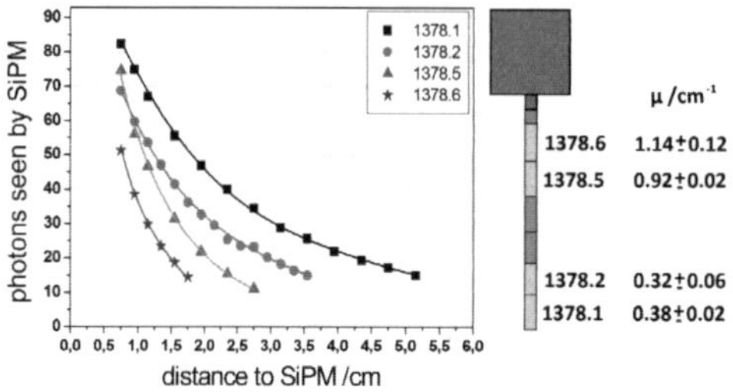

Figure 5. Attenuation curves for 4 10 cm long samples cut at different positions from an initially > 60 cm long single fiber. Sample 1378.1 and 1378.2 are cut 0.5 cm and 10.5 cm apart from the seed, while the samples 1378.5 and 1378.6 are cut 40.5 cm and 50.5 cm apart from the seed, respectively.

4.2. Influence of the position within the grown fiber

The μ-PD method allows the growth of fibers even longer than 1 m in length. To study the variation of the fiber quality over the full fiber length, a fiber with a diameter of 1 mm and a length > 60 cm was grown and cut into several 10 cm long samples, starting 0.5 cm apart from the seed. Fig. 5 shows the attenuation curves for 4 of these samples indicating a large variation. Attenuation coefficients of 0.38 cm^{-1} and 0.32 cm^{-1}, respectively, are measured for the first two samples. In contrast, the last two ones, cut further apart from the seed, show significantly larger values of 0.92 cm^{-1} or even 1.14 cm^{-1}. The significantly decreased fiber quality can be addressed to a variation of the Ce-concentration inside the fiber. In the beginning of the pulling process, less Ce might be implemented into the LuAG-matrix located close to the seed while Ce - Ions could become more abundant in the crucible at a later stage leading to a higher Ce-concentration causing more surface defects, cracks and bubbles [9], affecting the light attenuation. This hypothesis is supported by the attenuation corrected absolute light yield. For the first two samples a corrected light yield of 88 - 110 photons can be obtained in contrast to 121 and 149 photons for the second pair of fibers. This increase is expected for a higher Ce-concentration as described for bulk LuAG:Ce crystals [12]. The results are in agreement with a visual optical inspection (see Fig. 6). Sample 1378.1, which was cut close to the seed, appears in light color and shows almost no cracks. In contrast, sample 1378.6 has a darker color, indicating a higher Ce-concentration and showing many small cracks.

1378.6
1378.1

Figure 6. Picture of a fiber, cut close to the seed (1378.1) and 50.5 cm away from the seed (1378.6)

4.3. Influence of the Ce - concentration

To investigate directly the influence of the Ce-concentration, samples were grown with different Ce-concentrations. Fig. 7 shows the obtained attenuation curves for LuAG:Ce fibers with 1 mm and 2 mm diameter, respectively. The correlation of the light yield and the attenuation coefficient with the Ce-concentration is shown in fig. 8 for fibers with two different diameters. As

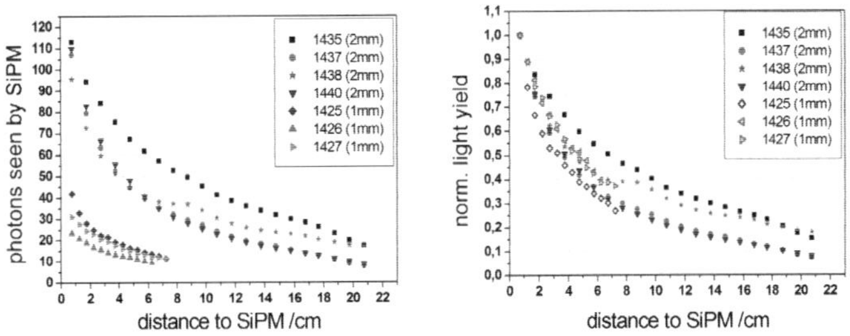

Figure 7. Attenuation curves for ⊘ 1 mm and 2 mm LuAG:Ce fibers containing different Ce-concentrations. The detected light yield is described by the absolute number of observed photons (left) or normalized to the value at closest distance to the SiPM (right).

expected, a lower Ce-concentration decreases the attenuation coefficient but to the disadvantage of a reduced light yield, as known for LuAG:Ce bulk crystals.

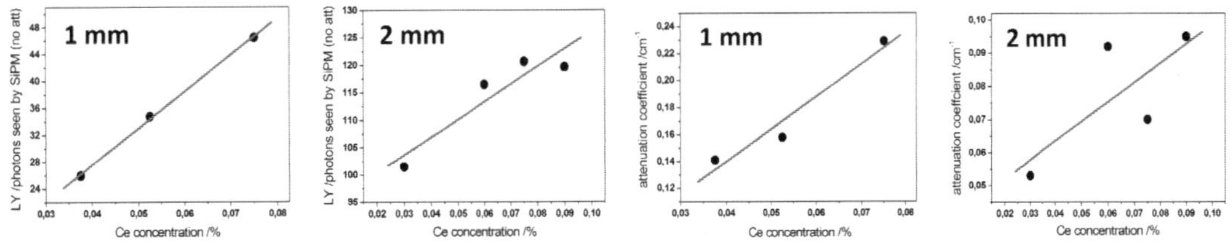

Figure 8. Influence of the Ce-concentration on the light yield and the attenuation coefficient for ⊘ 1 mm and 2 mm LuAG:Ce fibers

4.4. *Visual inspection of the fiber quality*

The optimization discussed in this chapter can also be confirmed in a visual inspection of the different fiber samples shown in fig. 9. It becomes clearly visible, that a lower pulling speed and a reduced Ce-concentration avoid significantly visible cracks and bubbles along the fiber causing primarily strong intrinsic attenuation.

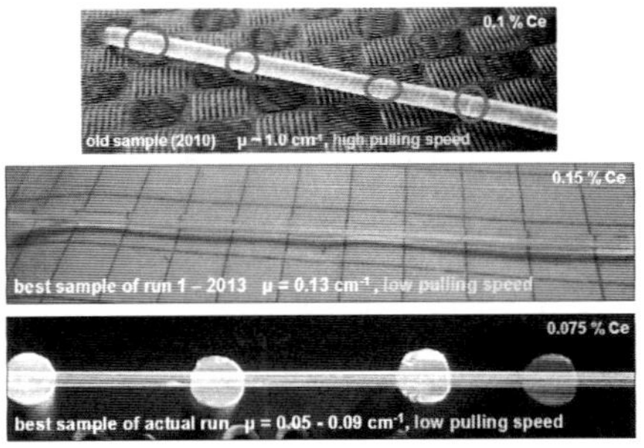

Figure 9. Pictures of LuAG:Ce fibers from different production runs. upper part: production run from 2010 with a high pulling speed, middle: production from 2013 with a low pulling speed, lower part: most recent production run combining a low pulling speed with a reduced Ce-concentration.

5. Characterization and optimization of light yield and light attenuation of LYSO:Ce fibers

So far only a few LYSO:Ce fiber samples with a diameter between 0.35 mm and 0.65 mm and a length of up to 10 cm have been produced and investigated. All fibers investigated so far have a

Y-content of 6 % and a Ce-concentration of around 0.05 %. While the first fibers had a quite high attenuation coefficient of up to 1.8 cm^{-1}, it was possible to reduce the attenuation coefficient down to $\mu = (0.68 \pm 0.02)\ cm^{-1}$ for the last production run. The light yield measured for the four available samples from this run is very homogeneous. Fig. 10 shows the light yield as a function of the source position for a typical LYSO:Ce fiber of the latest production run. Up

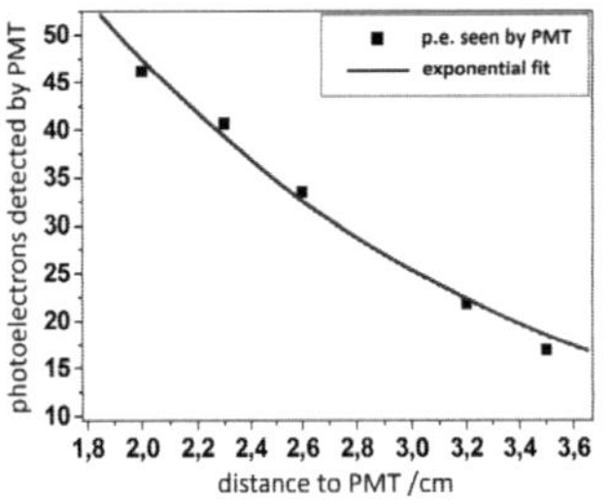

Figure 10. Position dependence of the light yield for a typical LYSO:Ce fiber.

to now LYSO:Ce fibers show a very promising performance but compared to the actual fiber quality of LuAG:Ce there is still a lot of space for improvements. Due to this, the optimization of LYSO:Ce fibers will start soon. For this process, a lot of the experiences made with LuAG:Ce can be used.

6. Coincidence readout of both fiber ends

For the developed high quality fibers made of LuAG:Ce, a readout at both fiber ends is possible, even for 23 cm long fibers. Fig 11 shows the recorded signal distribution at both SiPMs placing a ^{241}Am source either close to SiPM 1 or in the middle of a 23 cm long fiber. The correlation between the signals of both SiPMs is shown for both cases in addition. Even after a distance of 23 cm, a signal well separated from the background can be observed. Fig. 12 (left) shows the obtained attenuation curves for both 50 μm SiPMs and the behavior of the sum of both signals. In the latter case, one would expect a symmetric function for an homogeneous fiber. The asymmetry indicates an increased attenuation near SiPM 2. Presenting the logarithm of the ratio of the two amplitudes A1 and A2, respectively, one can directly determine the attenuation coefficient according to the relation $ln(A1/A2) = ln([LY_0 \cdot e^{-\mu \cdot x}]/[LY_0 \cdot e^{-\mu \cdot (L-x)}]) = -2\mu \cdot x + \mu \cdot L$, which predicts a linear function for a constant value of μ, depending on the distance x to SiPM 1. Fig. 12 (right) indicates instead three regions of different attenuation coefficients and confirms the increase of the attenuation coefficient close to SiPM 2. Nevertheless, knowing the attenuation characteristics of the fiber, the logarithm of the amplitude ratio can be used to reconstruct the point of interaction within the fiber. For the coincidence readout, a time resolution of $\sigma \sim 800$ ps was achieved for a typical light output of 120 photons seen by the 50 μm SiPM.

7. Response to low energetic gamma sources

An important aspect is the response of the fibers to low energetic gamma rays, in particular for medical applications. In case of the detection of the 511 keV γ-rays of a ^{22}Na source in coincidence with an external scintillation detector, the energy spectra show clearly visible indications of the full energy photo peak (see fig. 13 (left)). This supports the perspectives for the use in medical applications, like PET scanners. The overall response of 511 keV γ-rays differs significantly from the spectrum obtained using a ^{60}Co source. The right part of fig. 13 illustrates the measurement using the 662 keV γ-rays emitted from a ^{137}Cs source. For a distance of 1 cm between γ-source and SiPM, a light yield of 270 $\pm$ 25 ph/MeV has been determined with a 50 μm SiPM for $\oslash$ 2 mm fiber wrapped with teflon.

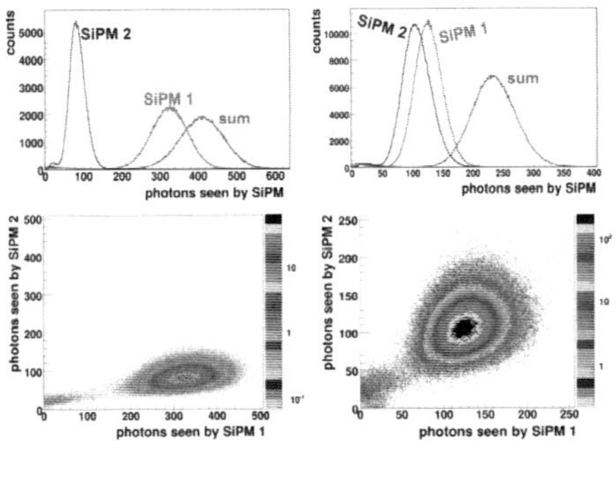

Figure 11. Respose of both SiPMs to a ^{241}Am source placed close to SiPM 1 (left) and in the center of a 23 cm long LuAG:Ce fiber (right), respectively. In addition, the correlation between the signals of both SiPMs is shown for both cases (lower row).

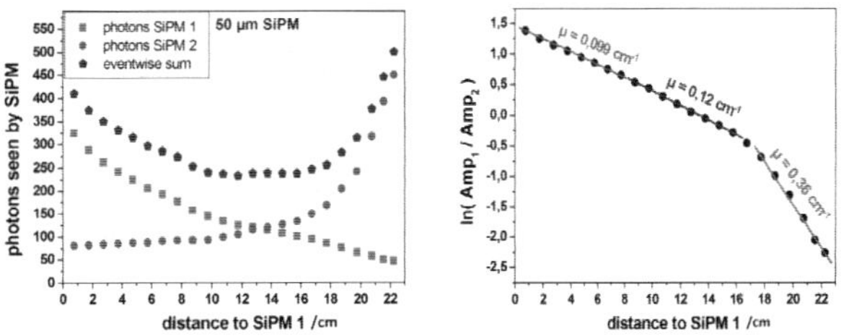

Figure 12. Attenuation curves individually for both SiPMs and the shape of the sum of both signals (left). Dependence of the logarithm of the ratio of the signal amplitudes of both SiPMs as a function of the source location relative to SiPM1 (right).

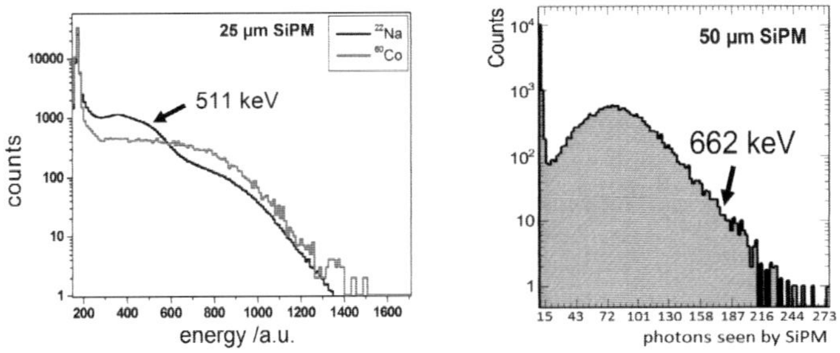

Figure 13. Response of a ⊘ 2 mm LuAG:Ce fiber to a ^{22}Na and a ^{60}Co source (left) and to a ^{137}Cs source (right).

8. Conclusion and Outlook

Over the last years, there was a significant improvement of the quality and homogeneity of LuAG:Ce fibers. Fig. 14 (left) shows the attenuation curves for different LuAG:Ce fiber productions from 2013 and 2014 in comparison to a typical curve for LYSO:Ce fibers and to rectangular LYSO:Ce rods cut from a large conventional crystal. The illustration shows, that with an attenuation coefficient of 0.05 cm^{-1} the quality of the best fibers from the most recent production run comes close to the reference value of 0.02 cm^{-1} obtained with the rectangular rods. In addition the right part of fig. 14 summarizes the development of the attenuation coefficient over the last years, which documents an improvement by more than a factor 20. Based on the experience with the growing parameters from the LuAG:Ce production, the optimization of LYSO:Ce fibers can be pushed forward.

Acknowledgement

The Project is supported by the European Community-Research Infrastructure Activity under FP07: Hadron Physics 3, WP22 and by the PICOSEC Marie Curie project of the European Communitys Seventh Framework Program.

Journal of Physics: Conference Series **587** (2015) 012067 doi:10.1088/1742-6596/587/1/012067

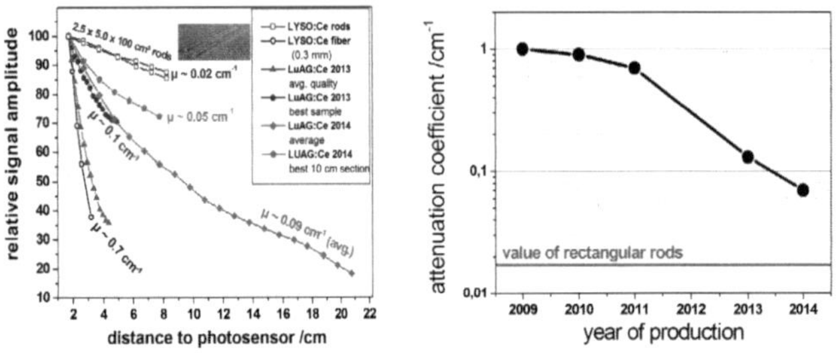

Figure 14. Comparison of the attenuation curves of different fiber production runs of rectangular LYSO:Ce rods cut from a large crystal and polished (left). Development of the fiber quality over the last years (right).

References

[1] P. Lecoq, New crystal technologies for novel calorimeter concepts, J. Amer. Ceram. Soc 160:1, 12016 (2009)

[2] K. Pauwels et al., Single crystalline LuAG fibers for homogeneous dual-readout calorimeters, Journal of Instrumentation 8, P09019 (2013)

[3] T. Meyer, Endo-TOFPET-US: A multimodal ultrasonic probe featureing time of flight PET in diagnostic and therapeutic endoscopy, Nucl. Instrum. Methods. Phys. Res. A 718, 121 (2013)

[4] W. Moses, Trends in PET imaging, Nucl. Instrum. Methods Phys. Res. A 471: 1-2, 209 (2001)

[5] K. Lebbou et al., Fiber single crystal growth from the melt for optical applications, J. Amer. Ceram. Soc 89:1, 75 (2006)

[6] T. Fukuda and V. Chani, Shaped Crystals - Growth by Micro-Pulling-Down Technique, Springer: Advances in Materials Research 8 (2007)

[7] Hamamatsu Photonics Deutschland GmbH: MPPC (multi-pixel photon counter) datasheet, *http : //www.hamamatsu.com/resources /pdf/ssd/s*12572 − 025*_etc_kapd*1043*e*03*.pdf* (2014)

[8] Stefan Diehl, Rainer W. Novotny, Nicolas Aubry, Sara Faraj and Gabriel Ferro, Development and Characterisation of Inorganic Scintillating Fibers Made of LuAG:Ce and LYSO:Ce, IEEE Transactions on Nuclear Science 61:1, 353 (2014)

[9] X. Xu et al., Ce-doped LuAG single-crystal fibers grown from the melt for high-energy physics, Acta Materialia 67, 232 (2014)

[10] C. Dujardin et al., LuAG:Ce fibers for high energy calorimetry, Journal of Applied Physics 108, 013510 (2010)

[11] M. Nikl et al., Development of LuAG-based scintillator crystal - A review, Progress in Crystal Growth and Characterisation of Materials 59, 47 (2013)

[12] A. G. Petrosyan et al., Bridgmann growth and site occupation in LuAG:Ce scintillator crystals, Journal of Crystal Growth 312, 3136 (2010)

Journal of Physics: Conference Series **587** (2015) 012068 doi:10.1088/1742-6596/587/1/012068

EndoTOFPET-US - A Miniaturised Calorimeter for Endoscopic Time-of-Flight Positron Emission Tomography

Milan Zvolský on behalf of the EndoTOFPET-US Collaboration
Deutsches Elektronen-Synchrotron DESY, Notkestr. 85, 22607 Hamburg, Germany
E-mail: milan.zvolsky@desy.de

Abstract. In the scope of the EndoTOFPET-US project, a novel multimodal device for Ultrasound (US) Endoscopy and Positron Emission Tomography (PET) is being developed. The project aims at detecting and quantifying morphologic and functional markers and developing new biomarkers for pancreas and prostate oncology. Exploiting the Time-of-Flight (TOF) information of the gamma rays allows for a more sensitive, more precise and lower radiation-dose imaging and intervention on small internal structures. The detection of the gamma rays is realised with the help of scintillator crystals with Silicon Photomultiplier (SiPM) read-out, aiming at a coincidence time resolution of 200 ps and a spatial resolution of $\approx$ 1 mm. For the endoscopic detector, digital SiPMs are utilised for the first time in an instrument planned for clinical applications. The functionality of the instrument as well as the challenges that accompany the high miniaturisation of the endoscopic detector and the asymmetric and variable geometry of the system, are presented. The demands on the system involve the fields of scintillating crystallography, ultra-fast photon detection, highly integrated electronics, system integration as well as image reconstruction.

The single detector components have been fully characterised and are performing up to specifications. Two dedicated ASIC chips have been developed for the project. The first PET images have been acquired with a test setup that consists solely of hardware and software developed within the collaboration and demonstrate that the data acquisition and reconstruction chain is operational. In this talk, the characterisation of the single components and the status of the detector integration and comissioning is presented.

1. Introduction

The EndoTOFPET-US detector is being developed for the evaluation of new biomarkers and for diagnostic interventional imaging [1, 2, 3]. It consists of a miniaturised PET detector head mounted on an ultrasound endoscope as well as an external detector plate being located next to the patient. The detection of the coincident gamma rays is carried out with scintillation crystals read out by silicon photomultipliers. The first endoscopic prototype is being developed for prostate exams whereas the external plate can be utilised for both prostatic and pancreatic exams. This report reviews the design of the detector system and its software and mainly focuses on the latest results of the project.

2. Internal Probe

2.1. Design

The endoscpopic detector comes in two different versions, one for the prostate exam and one for the pancreatic exam. The first prototype will be designed for the prostate case. In this case the endoscopic PET detector is mounted on the trans-rectal ultrasound (US) endoscope EUP-U533 from Hitachi. The PET head extension will be clamped onto the endoscope without alterating the endoscope itself. The PET head must not exceed a volume of $23 \times 23 \times 40$ mm^3 due to anatomical constraints. Due to the immediate proximity of the detector to the organ under study, a high granularity is crucial for an excellent spatial resolution. This is realised by the use of two matrices of 9×18 LYSO scintillator crystals from Proteus with a size of $0.71 \times 0.71 \times 15$ mm^3. They are coupled with optical glue to two boards of multi-channel digital SiPMs (MD-SiPMs). A front-end printed circuit board (PCB) provides a digital output via an interconnection to the data acquisition (DAQ) card. Besides, a water cooling pipe is embedded in the detector housing which is necessary to ensure a low dark count rate (DCR) despite the heat generation by the electronics. A sketch of the US endoscope and the PET extension is depicted in Fig. 1 (left), along with a drawing of the outer plate design.

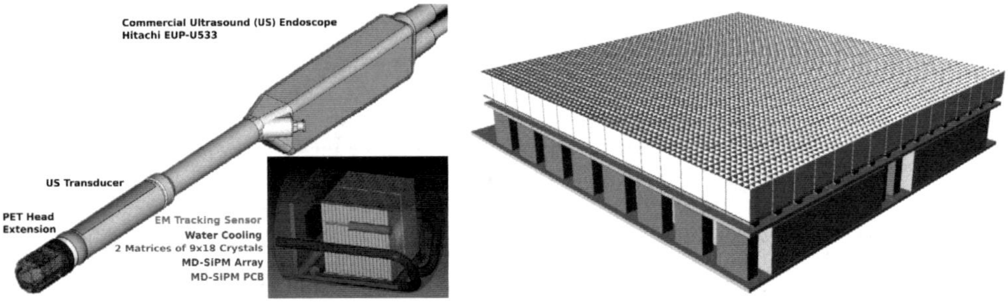

Figure 1: *(left)* Sketch of the design of the endoscopic ultrasound endoscope with the PET head extension. The inlet shows a zoom of the PET head, depicting the crystal matrices, MD-SiPMs, PCB and cooling pipes. *(right)* CAD drawing of the current design of the external PET detector plate.The green plate underneath the detector depicts the front-end boards. Above that, two aluminium plates (grey) house the cooling and mechanically fix the printed circuit boards (PCBs) on which the ASICS are mounted. The uppermost structure symbolises the crystal matrices and the SiPM arrays.

2.2. Multichannel digital Silicon Photomultipliers

The photodetectors of the internal probe are multi-channel digital Silicon photomultipliers (MD-SiPMs) which have been custom developed within the collaboration [4, 5]. The sensor has a size of 7.2×14.4 mm^2 and is composed of 9×18 MD-SiPMs or clusters, each of which contains of 16×26 single photon avalanche diode (SPAD) pixels with a size of 30×50 μm^2. The SPADs are equipped with a one-bit counter which provides the digital count of pixels fired. They are read out column-wise with multiple time-to-digital converters (TDCs) which enables independent photon time-of-arrival measurements. A reasonable trade-off between fill factor and number of time-over-threshold measurements leads to a number of 48 TDCs per column of clusters and a fill factor of 57%. Combining the information of the different timestamps statistically ensures that the lower bound of the theoretically achievable CTR is always reached and thus improves the CTR significantly [6, 7, 8]. Digital SiPMs typically suffer from high dark count rate (DCR) which, however, can be significantly reduced by cooling and by masking noisy pixels.

This behaviour is shown in Fig. 2. As can be seen, a DCR of 41 MHz without masking at a temperature of 20°C and 3 V excess bias can be reduced to 23 MHz with $\approx$ 10% masking. The photon detection efficiency (PDE) in this scenario is $\approx$ 12%. Fig. 2 (right) shows the benefits of additional cooling. The single pixel time resolution of the prototype sensor has been determined to be 120 ± 13 ps FWHM excluding the laser and clock jitter.

The 9×18 MD-SiPM chip has been fabricated and tested. It features in-situ statistical analysis of the gamma events, real-time noise rejection, a high-speed interface with the DAQ as well as multi-channel timestamping.

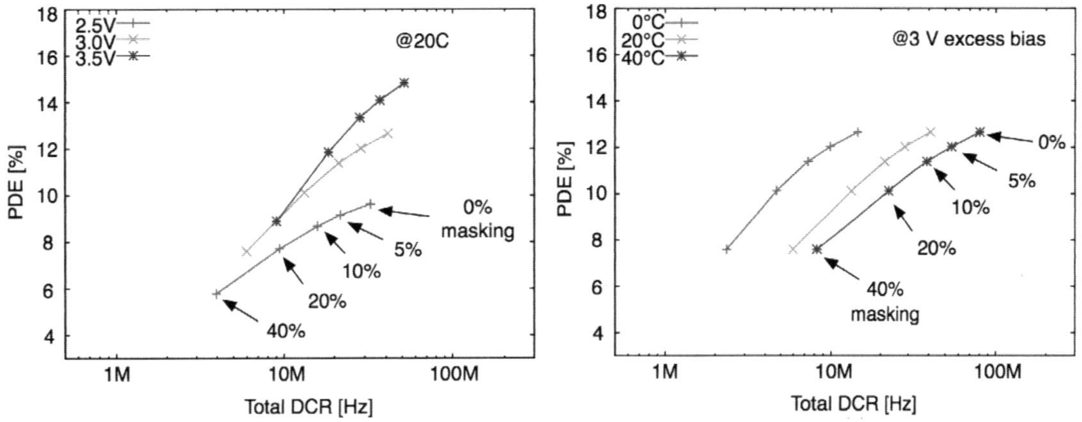

Figure 2: Photon detection efficiency (PDE) of the 4×4 MD-SiPM prototype as a function of the dark count rate (DCR) for different masking levels and for different excess bias voltages (left) and different temperatures (right).

3. External Plate

3.1. Design

The outer detector plate has an area of 23×23 cm^2 and consists of 4096 LYSO crystals from Crystal Photonics Inc. with a size of $3.5 \times 3.5 \times 15$ mm^3. The crystals are separated by a 100 μm thick reflector foil from Vikuiti 3M^{TM}. They are coupled to analogue xSiPMs which are discrete through-silicon via (TSV) MPPC arrays from Hamamatsu. Dedicated fast 64-channel application specific integrated circuits (ASICs) are mounted directly on the detector plate in order to optimise the timing properties. Two cooling plates are embedded in the detector housing which consist of aluminium plates with integrated water pipes. The design of the outer plate is sketched in Fig. 1 (right). The frontend system comprises two types of boards, the board housing the ASICs (FEB/A) and the board interfacing to the DAQ (FEB/D). The FEB/A which is housing two 64-channel ASICs has one connector for each MPPC matrix allowing for easy replacement of damaged components and take care of the digitization of MPPC signals. The connectors between the FEB/A daughter boards and the FEB/D motherboard provide connections for input voltages, MPPC bias voltages, clocks, control bus and output links. The current design foresees 2 FEB/D boards mounted on the bottom of 2 cooling plates.

3.2. ASICs

The aim is to develop a front-end TOF ASIC, being able to read out the time stamping and energy measurements. Special requirements are the time binning of 50 ps and a linear input bias range of 0.5 V as well as a low power consumption to minimise cooling. Two dedicated fast ASIC prototypes have been developed, StiC [9] and TOFPET ASIC [10]. In both cases, the ASICs provide a fine time measurement and a time-over-threshold (ToT) measurement for each input pulse, where the latter one determines the pulse energy. The ASIC inhibits a self-triggering mechanism based on a double-threshold system. For the time measurement, a low threshold is used, whereas a higher threshold accomodates for dark count rejection and ToT measurement. Tab. 1 summarises some key parameters of the two ASICs.

	STiC	TOFPET-ASIC
Jitter (at >5pC)	< 30 ps	< 25 ps
Input bias lin. range	0.7 V	0.5 V
TDC time bin width	50 ps	50 ps
Power consumption	19 mW/ch.	8 mW/ch
Output rate	160 MBit/s	160 MBit/s

Table 1: Key parameters of the two ASICs, StiC and TOFPET.

Fig. 3 shows the measured coincidence time resolution (CTR) for the STiC (left) and the TOFPET (right) ASIC. The measurements with the 16-channel STiC chip are performed using a ^{22}Na source and two $3.1 \times 3.1 \times 15$ mm LYSO crystals coupled to MPPCs. The MPPCs are connected to the STiC chip using differential readout. The energy resolution of the 511 keV peak is $\approx 12\%$. The measured CTR at optimal HV-settings is 220 ps FWHM. The final 64-channel STiC chip has been manufactured and tests are ongoing. The measurements with the 64-channel TOFPET ASIC also use a ^{22}Na source and two $3.1 \times 3.1 \times 15$ mm LYSO crystals coupled to MPPCs in coincidence. They are connected to the TOFPET chip using single-ended readout. Using a preliminary calibration, the CTR in this case was measured to be 270 ps FWHM, see Fig. 3 (right).

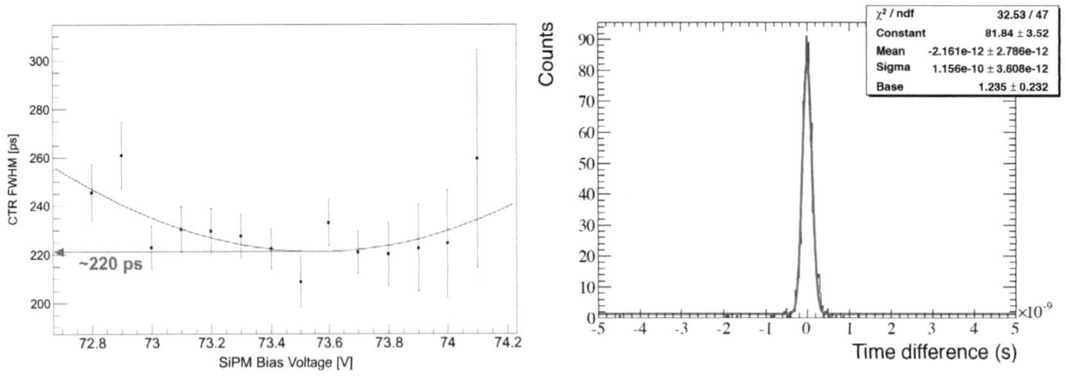

Figure 3: *Left:* Coincidence time resolution (CTR) for different SiPM bias voltages for the STiC chip. *Right:* Distribution of the time difference between both modules, i.e. the CTR for the TOFPET chip.

3.3. SiPMs

A mass characterisation of the MPPC arrays has been performed in order to assure the quality of the devices as well as to adjust the operation bias voltage of all individual detector channels to the same gain. For this each MPPC array was characterized under low intensity laser light for 30 different applied bias voltages. For each voltage value a dark run is recorded to extract DCR and crosstalk. The temperature dependence has been measured to be $\Delta V_{bd}/\Delta T = 70$ mV/K and the measurements have been corrected for that temperature dependency.

A typical single photo-electron spectrum is depicted in Fig. 4 (left). It is acquired by illuminating the MPPC channels with blue laser light, attenuated to low intensity to ensure single-photon detection. The MPPC is mounted on a motherboard incorporating linear amplifiers, high voltage filters and a temperature sensor. The output signal is amplified by a ten times voltage amplifier and read out by a VME-based charge-to-digital converter (QDC) with a resolution of 25 fC per QDC bin. The signal from the MPPC is integrated in a gate with effective length of 100 ns. The distance between the individual peaks of the single photo-electron spectrum is defined as the gain. It is plotted as a function of the applied bias voltage in Fig. 4 (right). From a linear fit, the breakdown voltage is extrapolated as the voltage at which the gain equals to zero. The knowledge of the breakdown voltage is of importance since all channels will operate at a fixed excess bias voltage to assure a homogenous detector response to incoming photons.

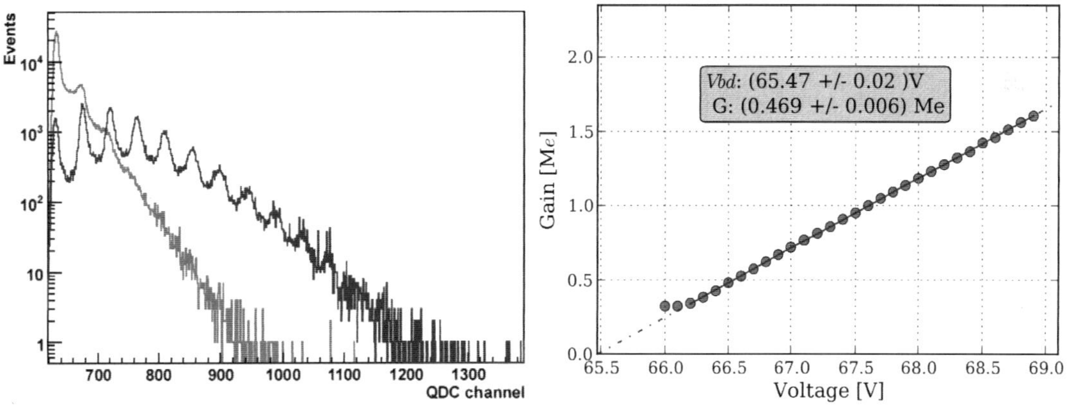

Figure 4: (*left*) Single photo-electron spectrum (blue) with a multiple Gaussian fit (red asterisks). The red histogram is a spectrum from a dark run from which DCR and cross talk can be extracted. (*right*) Gain in million electrons as a function of the applied bias voltage as extracted from the aingle photo-electron spectra. The blue line is a straight line fit on the data points whose errors are negligible on this scale. The breakdown voltage (Vbd) and gain (G) at 1 V overvoltage, obtained from the fit, are given in the inlet.

Fig. 5 (left) shows the difference between the maximum and minimum breakdown voltage per matrix for all 269 MPPC matrices. The mean difference is 140 mV while the maximum difference is smaller than 300 mV for almost every matrix. The deviation is thus below 0.5 V required by the ASICs. In Fig. 5 (right) the DCR for each MPPC channel is shown. The threshold for accepting or rejecting the MPPC array is 3 MHz and the devices above this limit (red entries) have been sent back to the producer.

3.4. Crystals

The key parameter of the the scintillation crystals is their light yield (LY) since it influences its time resolution. The LY of the external plate crystal matrices is measured with a photomultiplier tube (PMT) and a ^{137}Cs source. Several response spectra are shown in Fig. 6 (left). In order to account for the differences between a loosely connected PMT and a glued and wrapped SiPM, the resulting LY is corrected for the air gap and the missing glue and wrapping. The average LY amounts to 32320 photons/MeV $\pm 3.5\%$ as can be seen from Fig. 6 (right). The energy resolution is $\approx 13\%$ for the external plate matrices, and similar for the internal probe matrices.

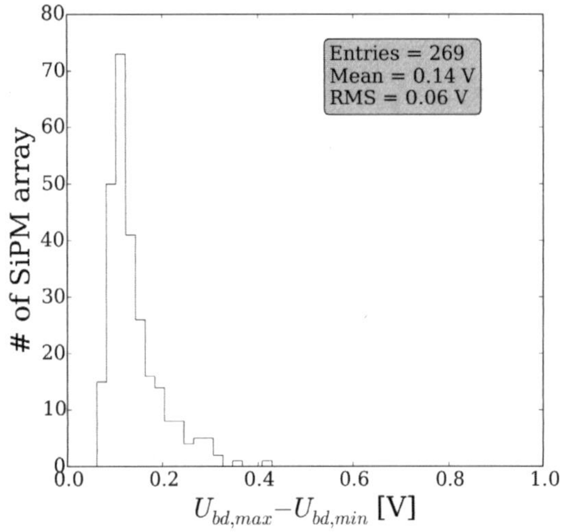

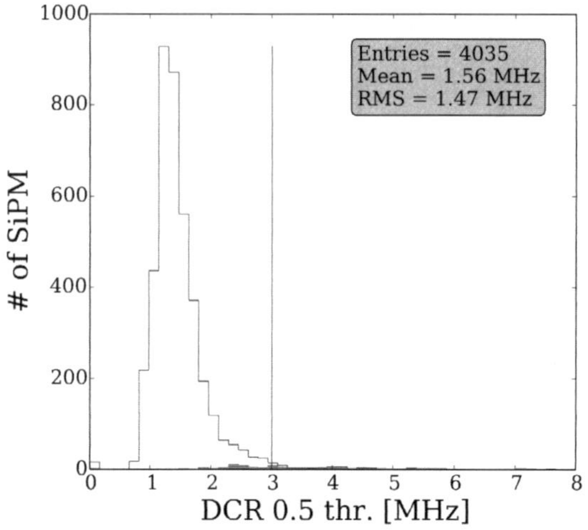

Figure 5: *Left:* Distribution of the maximum difference maximum and minimum breakdown voltage U_{bd} in one array. *Right:* Distribution of the DCR at 0.5 pixels fired at 2.5 V excess bias for all SiPM channels. The red line indicates the threshold of acceptable DCR. Entries marked in red correspond to arrays with at least 1 channel that excesses this limit.

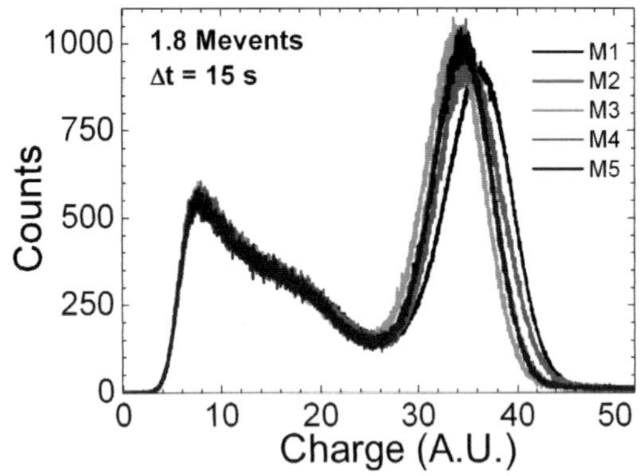

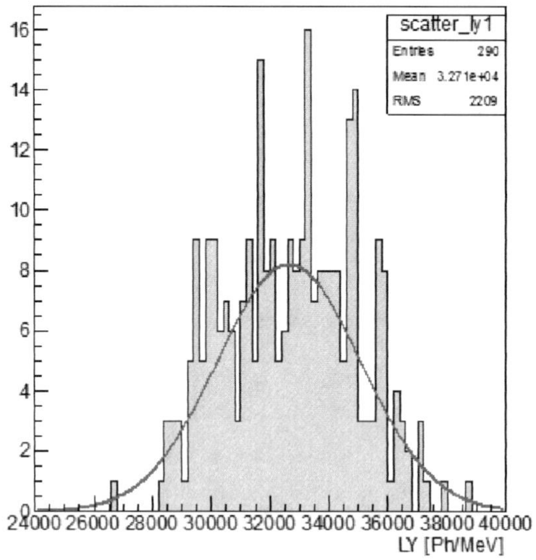

Figure 6: *Left:* Charge spectra of a ^{137}Cs source (in arbitrary units) for 5 different matrices. *Right:* Light yield (LY) for 290 crystal matrices.

The gluing of the crystals matrices to the MPPC arrays is performed with the help of a automatised alignment setup with additional visual inspection in order assure the alignment and absence of air bubbles. As an additional quality assurance test, current-voltage (IV) measurements are performed before and after the gluing. After the gluing, the crystals are wrapped with a reflextor foil and covered with a thin plastic cap.

3.5. Modules

For the modules (crystal matrices glued to MPPC arrays), the light output, i.e. the number of fired pixels is determined. The experimental setup is similar to the one described in Sec. 3.3 with a ^{22}Na source instead of a laser being used. Temperature monitoring and correction is applied and every channel is operated at the same gain ($G = 1.25 \cdot 10^6$). The mean energy resolution at 511 keV is $\approx$ 12.8%. The mean number of fired pixels at 511 keV is determined to be 1876 ± 118 as can be seen in Fig. 7 (left). The mean spread within one module is $\approx$ 15%.

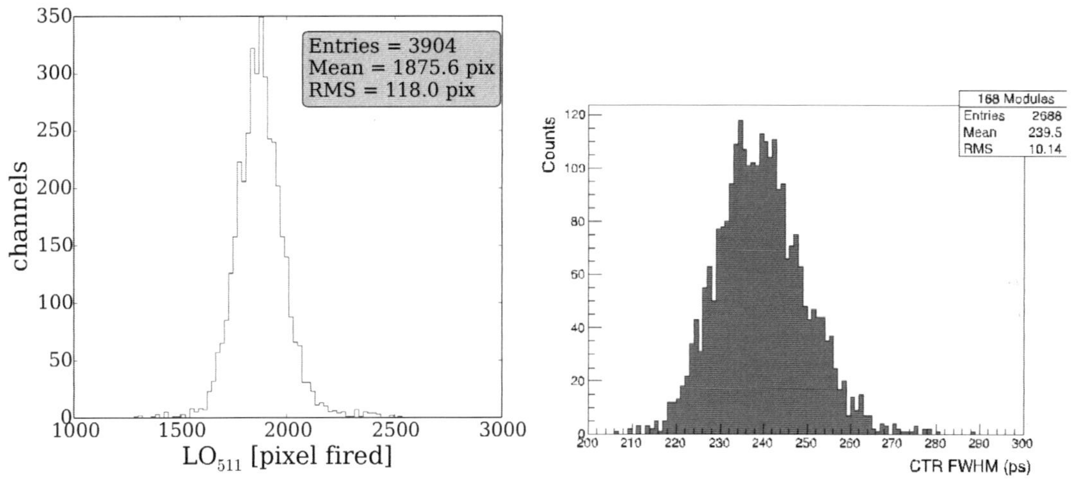

Figure 7: *Left:* Distribution of the light output for 511 keV photons in number of fired pixels for all module channels. *Right:* Distribution of the coincidence time resolution (CTR).

Besides, the coincidence time resolution (CTR) is measured for all modules. For this, the time resolution of each module is measured in coincidence with a reference module. It has been determined with the time-over-threshold method with an ultra-fast amplifier-discriminator chip (NINO)[11] and a time-to-digital converter. Fig. 7 (right) shows the distribution of the CTR for all channels. From this, a mean CTR of 239.5 ± 10 ps can be extracted.

4. DAQ & software environment

4.1. DAQ

The data acquisition (DAQ) backend is implemented in a PCI-e board interfacing directly to the DAQ computer. It includes a powerful Virtex FPGA suitable to implement the trigger and data acquisition firmware. Four high speed links to the external plate and up to two links to the endoscopic probe are available on board. Links for control and configuration of the frontend systems are also implemented, as well as central distribution of clock and sync signals. The firmware needed to move data from the external plate ASIC up to the data acquisition computer has been developed and fully tested. The DAQ firmware to communicate with the endoscopic probe system was developed and has been successfully tested. In order to control the communication between the multiple imaging and tracking devices, CAMPCom, a lightweight and portable communication framework for multimodal image-guided therapy [12, 13] has been developed within the project. The communication framework involves the full monitoring and quality control of all processes and enables the slow control. Moreover, a graphical user interface for the medical doctor including the visualisation of the PET/US image is available.

4.2. Tracking
The external plate will be held and thus tracked mechanically by a moveable and lockable arm. It will be additionally tracked by optical tracking system. The expected accuracy is ≈ 0.5 mm and $0.5°$. The transrectal endoscope will be tracked optically as well. Electromagnetic tracking is not able to provide the required tracking accuracy below 1 mm.

4.3. Image reconstruction
A dedicated image reconstruction based on the ML-EM algorithm has been developed within the project and accounts for the limited field of view and the freehand nature of the acquisition that results in an undefined and continuously changing volume of interest [14]. In order to solve the massively parallel problem of inverting huge matrices the computation is performed on powerful graphic processing units (GPUs). They enable a speedup by a factor of $\mathcal{O}(10)$ with respect to conventional central processing unit (CPU) computation, providing an on-line reconstructed image within the order of minutes.

5. Detector integration
As a first proof of principle, the functionality of the full data acquisition, readout and analysis chain is tested with only a few detector channels in coincidence. The 16-channel prototype chip STiC2 was used to read out eight channels of two detector modules. The two detector modules are rotated by 180 degrees in 30 distinct steps around two ^{22}Na sources with a distance of ≈ 1 cm. 50k events have been acquired per detector pose. Coincidence events are found by selecting only events from the photopeak and applying a coincidence time window of 2 ns. In Fig. 8 the measurement setup is shown on the left and the reconstructed image of the two-point source can be seen on the right hand side. The reconstructed spatial resolution of the image is $\approx 3-4$ mm which is expected since the image quality is driven by the low number of channels and the crystal size of 3.5 mm. It however gives a first proof of principle that the complete acquisition, readout and image reconstruction chain is operational.

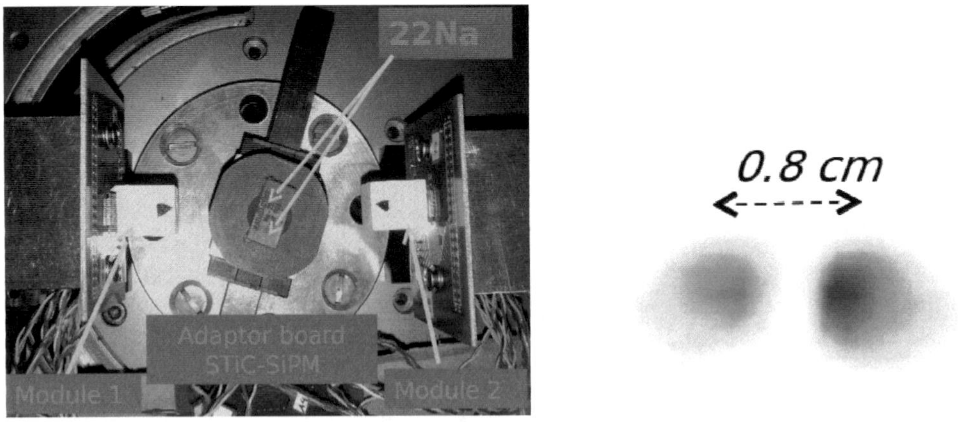

Figure 8: *Left:* Setup for the coincidence measurements *Right:* Reconstructed image of the two point sources.

6. Conclusion & Outlook
The EndoTOFPET-US project is a very challenging project due to the extreme miniaturisation of the endoscopic detector, the high demands on the timing characteristics of crystals, SiPMs and ASICs as well as the required reconstructed spatial image resolution of 1 mm. Quality assurance

measurements of the MPPCs and crystals have been performed and its key parameters such as breakdown voltage, gain, DCR and light yield have been extracted. All detector modules have been assembled and fully characterised. It was shown that the measured coincidence time resolution is close to design goal of 200 ps. Two dedicated ASICs have been developed within the project and are shown to perform up to specifications.

The final decision on the choice of the ASIC will be based on measurements to be made with the 64-channel chips fully connected to the detector modules. After the mechanical integration, integration tests of the full firmware reading simultaneously the external plate ASICs and the probe SPAD will be performed. Then, the tracking system will be integrated and the fully operating detector system, including DAQ, reconstruction, PET-US fusion etc will be validated. According to the schedule pre-clinical tests are supposed to start in autumn of 2014.

Acknowledgements

The author thanks the EndoTOFPET-US collaboration for the privilege of representing it at this conference. EndoTOFPETUS has received funding from the European Union 7th Framework Program (FP7/ 2007-2013) under Grant Agreement No. 256984 and is supported by a Marie Curie Early Initial Training Network Program (PITN-GA-2011-289355-PicoSEC-MCNet).

Bibliography

[1] M. Zvolsky. EndoTOFPET-US - A Miniaturised Calorimeter for Endoscopic Time-of-Flight Positron Emission Tomography. *Proceedings of the CHEF2013 Conference - Eds. J.-C. Brient, R. Salerno, and Y. Sirois*, pages 433–444, 2013.

[2] E. Garutti. EndoTOFPET-US - A novel multimodal tool for endoscopy and Positron Emission Tomography. In *Nuclear Science Symposium and Medical Imaging Conference (NSS/MIC), 2012 IEEE*, pages 2096–2101, 2012.

[3] B. Frisch. Combining endoscopic ultrasound with Time-Of-Flight PET: The EndoTOFPET-US Project. *Nuclear Instruments and Methods in Physics Research A*, 732:577–580, December 2013.

[4] S. Mandai, V. Jain, and E. Charbon. A fully-integrated $780 \times 800 \mu m^2$ multi-digital silicon photomultiplier with column-parallel time-to-digital converter. In *Proceedings of the ESSCIRC*, pages 89–92, 2012.

[5] S. Mandai and E. Charbon. Multi-channel digital SiPMs: Concept, analysis and implementation. In *Nuclear Science Symposium and Medical Imaging Conference (NSS/MIC), 2012 IEEE*, pages 1840–1844, 2012.

[6] M. Fishburn and E. Charbon. System Tradeoffs in Gamma-Ray Detection Utilizing SPAD Arrays and Scintillators. *IEEE Trans. Nucl. Sci. Vol. 57, No. 5*, 2010.

[7] S. Seifert, H.T. van Dam, and D.R. Schaart. The lower bound on the timing resolution of scintillation detectors. *Phys Med Biol. 57 1797-1814*, 2012.

[8] S. Mandai, E. Venialgo, and E. Charbon. Timing optimization utilizing order statistics and multichannel digital silicon photomultipliers. *Optics letters*, 39(3):552–4, February 2014.

[9] W. Shen, K. Briggl, W. Chen, P. Fischer, A. Gil, T. Harion, M. Ritzert, and H.-C. Schultz-Coulon. STiC - A mixed mode chip for SiPM ToF applications. In *Nuclear Science Symposium and Medical Imaging Conference (NSS/MIC), 2012 IEEE*, pages 877–881, 2012.

[10] M.D. Rolo, R. Bugalho, F. Goncalves, G. Mazza, A. Rivetti, J.C. Silva, R. Silva, and J. Varela. TOFPET ASIC for PET applications. *Journal of Instrumentation*, 8(02):C02050, 2013.

[11] F Anghinolfi, P Jarron, F Krummenacher, E Usenko, and M C S Williams. NINO: An Ultrafast Low-Power Front-End Amplifier Discriminator for the Time-of-Flight Detector in the ALICE Experiment. *IEEE Trans. Nucl. Sci.*, 51(5):1974–1978.

[12] CAMPCom: A Lightweight and Portable Communication Framework for Multimodal Image-Guided Therapy. `http://campar.in.tum.de/Main/CAMPCom`. Accessed: 2013-08-16.

[13] A. Schoch, B. Fuerst, F. Achilles, S. Demirci, and N. Navab. A lightweight and portable communication framework for multimodal image-guided therapy. *The Sixth International Workshop on Systems and Architectures for Computer Assisted Interventions (SACAI)*, September 2013.

[14] A. Cserkaszky. Reconstruction of freehand PET examinations. *Proceedings of the WMIC2013 Conference*, 2013.